Landscapes and Societies

Siena: Details of the fresco (1338–39) by Ambrogio Lorenzetti painted inside the Palazzo Pubblico (City Hall of Siena). The fresco shows the effects of 'good government' in town (*upper large part*) and in the countryside (*lower left*), and during 'bad government' in the countryside (*lower right*).

I. Peter Martini • Ward Chesworth
Editors

Landscapes and Societies

Selected Cases

Springer

Editors
I. Peter Martini
Ward Chesworth
University of Guelph
School of Environmental Sciences
Axelrod Building
N1G 2W1 Guelph, Ontario
Canada
pmartini@uoguelph.ca

wcheswor@uoguelph.ca

ISBN: 978-90-481-9412-4 e-ISBN: 978-90-481-9413-1
DOI: 10.1007/978-90-481-9413-1
Springer Dordrecht Heidelberg London New York

Library of Congress Control Number: 2010937238

Cover illustrations: Views of landscapes and traces of human activities through time. *Main upper frame*: downtown Florence with (a) on the left (*book spine*) the fortress-like Town Hall (Palazzo Vecchio) built by the architect Arnolfo di Cambio between 1299 and 1304, and (b) in the center, the cathedral (Duomo) built between 1296 and 1436, designed by Arnolfo di Cambio, with the famous dome (cupola) engineered by Filippo Brunellechi, and the bell tower built between 1334 and 1359 to the design of Giotto di Bondone. *Lower frames from left to right*: (1) The Wadi Afar cave occupied in early Holocene, Acacus Mountains, Western Lybia desert; (2) The Roman Dam at Harbaqa in central Syria; (3) Terraced upper river valley with village on adjoining hillside, AsukaTanada, Japan (the 'tanada' are watered from a diverted mountain stream); (4) Reclaimed land from the North Sea: artificial seaward shift of the coastline south of the Hague, The Netherlands

Cover design: deblik, Berlin

Printed on acid-free paper

Springer is part of Springer Science+Business Media (www.springer.com)

Dedicated with love to our children—the next generation – Anthony[†] *and Lisa Martini and Aaron, Delia, Iona and Amanda Chesworth.*

† Anthony died suddenly at 45 in 2008. He was responsible for bringing the editors together on this book.

Preface

Relationships between societies and environments have spawned an extensive literature going back at least to Plato. Some fundamental concepts crop up repeatedly, yet individual societies have their idiosyncrasies in adapting to or in modifying their environments to meet their needs. Collateral damage is a constant accompaniment to our modifications. History shows that a relationship between society and landscape that proved successful in one era, may fail at a later time. Because of factors such as climate, economics, political and strategic conditions, to say nothing of pestilence, war and conquest, adaptation is a continuing process. Human history is punctuated by the continual need to adjust to fire, flood, earthquake, and volcanic eruption, to name only the most obvious. In fact, at this late stage in our history we do not so much as adapt to changing circumstances, as constantly readapt.

As societies evolve and enlarge they may reach the Malthusian limit of the human carrying capacity of the local landscape. In the past when the human population was small and there was lots of room we commonly escaped the problem by moving on. If we stayed put, the options were to increase our ecological footprint to bring in the resources of a wider area by trade, theft or war, the most highly organized form of theft. Alternatively, we have found technological solutions to the problem, though on a finite planet with our numbers still growing, we are forever looking for newer and better technological fixes. Society is always a work in progress.

The simplest societies, at least in theory, have the greatest adaptability. Societies that become increasingly stratified and custom-bound have a harder time. The more evolved a society, the greater its impact on the Earth. History is unequivocal, complex societies fail—whether by exhausting resources (soils most critically), fouling their nests with wastes, or simply by failing to adapt to change in a sustainable way. Rise, decline, sometimes rejuvenation, and fall is the pattern. Erstwhile successful societies have continually disappeared in history, only to be replaced by equally ephemeral ones. While the second law of thermodynamics remains un-repealed, no culture or civilization will last forever.

Consequently, this book presents the work of various scholars who were asked to describe and analyze their respective examples in a multidisciplinary way. Such analyses of the relationships between landscapes and societies at different times and in different places is useful not only in its intrinsic value in contributing to the history of Homo sapiens, but potentially in informing our present situation and helping us to recognize opportunities and risks in planning for a more sustainable future than we appear to be facing at present.

The various chapters examine changes in selected landscapes and societies from early Holocene times to the Present. In **Part I** we favor the Mediterranean centered area because in a relatively small area a great variety of landscapes have interacted with all manner of societies from simple nomadic and agrarian ones to the great civilizations of Egypt, Greece and Rome. The interaction has been so great and so prolonged that an entire natural biome has become anthropic. **Part II** contains analyses of societies from other parts of the globe. From Asia there are reports from the source of eastern civilizations, China, as well as Japan, and Sri Lanka. There is also together with a novel analysis of the effect of disease, particularly malaria, on the expansion of the human footprint into the Pacific. Central America provides the example of one of the great civilizations of the world, the Maya, which developed independently of the founding civilizations of the Old World, and met its demise as a result of environmental change, internecine conflict and finally invasion from Europe. From North America the extremely variable, harsh Arctic landscape capable of sustaining only relatively small number of people, comes the example of highly resilient Inuit communities dealing with one of the world's most extreme landscapes in the Arctic, as well as an examination of the modern technological civilization of the western USA, where societies survive for the present by transferring water across mountain chains. All these cases reveal the mutability of human societies and their dynamic, reciprocal relationship with changing landscapes.

We are grateful for the enthusiasm, hard work and scholarship of our collaborators, and for the patient and graceful way they dealt with reviewers and editors. A volume like this would not have been possible without the help of many people—colleagues, students, technicians, family members—and supporting organizations, and we sincerely thank them all. In particular, we would like to thank the scientific reviewers who helped considerably in focusing the message of the various chapters, Mario Panizza who helped during the early stages in the selection of a few contributors, and the editorial staff of Springer and the Publishing Editors in charge of this volume, in particular Petra D van Steenbergen and Cynthia de Jonge among them, for support during the preparation of the book and in seeing it successfully published. We also thank all publishers who allowed the reproduction of various figures as noted in the appropriate captions.

I.P. Martini and W. Chesworth

Contents

List of Contributors

Barnes Gina L. *Department of Art & Archaeology at SOAS (School of Oriental & African Studies), University of London, Thornhaugh Street, London WC1H 0XG, UK – gb11@soas.ac.uk*

Beach Timothy *Center for the Environment School of Foreign Service, Georgetown University, 37th and O Streets, NW Washington, DC 20057 USA – beacht@georgetown.edu*

Bellini Cristina *Dipartimento di Biologia Vegetale, Università di Firenze, Via G. La Pira 4, 50121, Firenze, Italy – cri_mini@yahoo.com*

Benvenuti Marco *Dipartimento di Scienze della Terra, Università di Firenze, Via G. La Pira 4, 50121 Florence, Italy – oredep@unifi.it*

Brookfield Michael *Institute of Earth Sciences, Academia Sinica, Nankang, Taipei 11529, Taiwan – mailing address: 45 Dean Avenue, Guelph, On N1G 1K6, Canada – mbrookfi@hotmail.com*

Brooks Nicholas *Tyndall Centre for Climate Change Research, School of Environmental Sciences, University of East Anglia, Norwich NR4 7TJ, UK – nick.brooks@uea.ac.uk*

Censini Gianfranco *Georisorse Italia, Via E. Fermi, 8, 53048 Sinalunga, Siena, Italy –g.censini@georisorse.it*

Chester David K. *Department of Geography, University of Liverpool, Liverpool, L69 3BX, UK – jg54@liverpool.ac.uk*

Chesworth Ward *School of Environmental Sciences, University of Guelph, Guelph, ON, N1G 2W1 Canada – wcheswor@uoguelph.ca*

Cioacă Adrian *Faculty of Geography and Geography of Tourism, Universitatea "Spiru Haret", Bucharest, Romania – acioaca@gmail.com*

Costantini Armando *Dipartimento di Scienze della Terra, Università di Siena, Via Laterina 8, 53100 Siena, Italy – acostantini@unisi.it*

Cremaschi Mauro *Dipartimento di Scienze della Terra "A. Desio", Università degli Studi di Milano, Via Mangiagalli 34, I-20133 Milano, Italy – mauro.cremaschi@unimi.it*

Depalmas Anna *Dipartimento di Scienze Umanistiche e dell'Antichità, Università di Sassari, Piazza Conte di Moriana 8, 07100 Sassari, Italy – depalmas@uniss.it*

Dharmasena P.B. *Field Crop Research and Development Institute, Mahailluppallama, Sri Lanka – dharmasenapb@ymail.com*

Dinu Mihaela S. *Romanian American University, Bucharest, Romania – mihaeladinu2003@yahoo.com*

Duncan Angus M. *Research Driectorate, University of Bedforshire, Park Square, Luton, LU 1 3 JU, UK – angus.duncan@beds.ac.uk*

Dunning Nicholas P. *Department of Geography, University of Cincinnati 401 Braunstein Hall, Cincinnati, OH 45221-0131 USA – nicholas.dunning@uc.edu*

el-Rishi Hwedi *Benzaid Department of Geography, University of Garyounis, Benghazi, Libya – hamerrishi@yahoo.com*

Fouache Eric *Département de Géographie, Paris Ouest Nanterre la Défense (Paris 10), Paris – mailing address: 38 rue Louis Ricard, 76000 Rouen, France – eric.G.fouache@wanadoo.fr*

Goudie Andrew S. *Mater, St Cross College, St Giles, Oxford, OX1 3LZ Great Britain – andrew.goudie@stx.ox.ac.uk*

Hunt Chris *School of Geography, Archaeology & Palaeoecology, Queen's University Belfast, Belfast BT7 1NN, UK – c.hunt@qub.ac.uk*

James Peter A. *Department of Geography, University of Liverpool,Liverpool, L69 3BX, UK – p.a.james@liv.ac.uk*

Jungerius Pieter D. *Stichting Geomorfologie en Landschap, Oude Bennekomseweg 31, 6717 LM Ede, The Netherlands – p.d.jungerius@uva.nl*

Junjie Xu *College of Environmental and Urban Sciences, Peking University, Beijing, 100871, China – tonyjjxu@163.com*

Li Minglin *College of Environmental and Urban Sciences, Peking University, Beijing, 100871, China – lml0906@163.com*

Mariotti-Lippi Marta *Dipartimento di Biologia Vegetale, Università di Firenze, Via G. La Pira 4, 50121, Firenze, Italy – mariotti@unifi.it*

Marriner Nick *Aix Marseille University, CEREGE Europôle méditerranéen de l'ARBOIS, F 13555 Aix-en-Provence, France – marriner@cerege.fr*

Martini I. Peter *School of Environmental Sciences, University of Guelph, Guelph, ON, N1G 2W1 Canada – pmartini@uoguelph.ca*

Melis Rita T. *Dipartimento di Scienze della Terra, Università di Cagliari, Via Trentino 51, 09127 Cagliary,Italy – rtmelis@unica.it*

Mo Duowen *College of Environmental and Urban Sciences, Peking University, Beijing, 100871, China – dmo@urban.pku.edu.cn*

Morhange Christophe *Aix Marseille University, CEREGE Europôle méditerranéen de l'ARBOIS, F 13555 Aix-en-Provence, France– morhange@cerege.fr*

Pallecchi Pasquino *MiBAC, Soprintendenza per i Beni Archeologici della Toscana, Largo del Boschetto, 3, 50143 Firenze, Italy – pasquino.pallecchi@beniculturali.it*

Park Robert W. *Department of Anthropology, University of Waterloo, 200 University Ave. West, Waterloo ON, N2L 3G1 Canada – rwpark@uwaterloo.ca*

Parker Adrian G. *Department of Anthropology and Geography, School of Social Sciences and Law, Oxford Brookes University, Headington, Oxford, UK – agparker@brookes.ac.uk*

Pavlopoulos Kosmas *Faculty of Geography, Harokopio University, 70 El. Venizelou str., 17 671 Athens, Greece – kpavlop@hua.gr*

Sagri Mario *Dipartimento di Scienze della Terra, Università di Firenze, Via G. La Pira 4, 50121 Firenze, Italy – sedim@unifi.it*

Sallares Robert *UMIST, 184 Albert Avenue, Prestwich, Manchester M25 0HF, UK – RSallares@aol.com*

Sarti Giovanni *Dipartimento di Scienze della Terra, Università di Pisa, Via S. Maria 53, 56126 Pisa, Italy – sarti@dst.unipi.it*

Thordarson Thorvaldur *Earth and Planetary Sciences, School of GeoSciences, University of Edinburgh, Grant Institute, The Kings Buildings, West Mains Road, EH9 3JW, Edinburgh, UK – Thor.Thordarson@ed.ac.uk*

Wilkinson Tony J. *Dept. of Archaeology, Durham University, South Road, Durham, DH1 3LE, UK – t.j.wilkinson@durham.ac.uk*

Wohl Ellen *Department of Geosciences, Colorado State University, Fort Collins, Colorado, 80523-1482 USA – ellenw@warnercnr.colostate.edu*

Zerboni Andrea *Dipartimento di Scienze della Terra "A. Desio", Università degli Studi di Milano, Via Mangiagalli 34, I-20133 Milano, Italy – Andrea.Zerboni@unimi.it*

Zhao Zhijun *Institute of Archaeology, Chinese Academy of Social Sciences, 27 Wangfujing Street, Beijing, 100710, China – zzzhangy@public3.bta.net.cn*

List of Reviewers

Adams Russell *McMaster University, Hamilton, ON, Canada – radams@mcmaster.ca*

Beach Timothy *Georgetown University, Washington, DC, USA – BeachT@Georgetown.edu*

Bintliff John *Leiden University, Leiden, The Netherlands – J.L.Bintliff@arch.leideneuniv.nl*

Bloomfield Gerry *University of Guelph, Guelph, Canada – gbloomfi@uoguelph.ca*

Brenner Mark *Department of Geological Sciences, University of Florida, Gainesville, Florida, USA – brenner@ufl.edu*

Brookfield Michael *Institute of Earth Sciences, Academia Sinica, Nankang, Taipei 11529, Taiwan – mbrookfi@hotmail.com*

Brown Robert *University of Guelph, Guelph, Canada – rbrown@uoguelph.ca*

Castaldini Doriano *Università degli Studi di Modena e Reggio Emilia, Modena, Italy – oriano.castaldini@unimore.it*

Chesworth, Ward *School of Environmental Sciences, University of Guelph, Guelph, ON, N1G 2W1 Canada – wcheswor@uoguelph.ca*

Clarke Joanne *University of East Anglia, Norwich, UK – Joanne.Clarke@uea.ac.uk*

Cremaschi Mauro *Università Statale di Milano, Milano, Italy – mauro.cremaschi@unimi.it*

Dahlin Bruce H. *Shepherdstown University, Shepherdstown, WV, USA – dahlin.bruceh@gmail.com*

De Dapper Morgan *Ghent University, Gent, Belgium – Morgan.DeDapper@UGent.be*

Demarest Arthur A. *Vanderbilt University, Nashville, TN, USA – Arthur.A.Demarest@vanderbilt.edu*

Dibben Christopher *University of St. Andrews, St. Andrews, Scotland – cjld@st-andrews.ac.uk*

Dodson John *Australian Nuclear Science and Technology Organization, Sydney, Australia – jdd@ansto.gov.au*

Ellis Christ *Department of Anthropology, University of Western Ontario, Canada – cjellis@uwo.ca*

Groenevelt Peter *University of Guelph, Guelph, Canada – pgroenev@lrs.uoguelph.ca*

Guðrún Larsen *University of Iceland, Reykjavík, Iceland – glare@raunvis.hi.is*

Kilburn Christopher *University College London, London, UK – c.kilburn@ucl.ac.uk*

Li Xiaoqiang *State Key Lab of Loess & Quaternary Geology, Xi'an, China – lxq@loess.llqg.ac.cn*

Madduma Bandara C.M. *University of Peradanaya, Peradanaya, Sri Lanka – madduband@yahoo.com*

Martinez Cortizas Antonio *Santiago University, Santiago de Compostela, Spain – antonio.martinez.cortizas@usc.es*

Martini I. Peter *University of Guelph, Guelph, Canada – pmartini@uoguelph.ca*

Montgomery David *Earth and Space Sciences, University of Washington, Washington DC, USA – dave@ess.washington.edu*

Panizza Mario *Dipartimento di Scienze della Terra, Università di Modena e Reggio Emilia, Modena, Italy – pit@unimore.it*

Pearson Richard *Sainsbury Institute for the Study of Japanese Arts and Culture, Norwich, UK – pearsonrj@shaw.ca*

Pirazzoli Paolo A. *CNRS (Centre national de la recherche scientifique), Paris, France – pirazzol@cnrs-bellevue.fr*

Sarti Giovanni *Università di Pisa, Pisa, Italy – sarti@dst.unipi.it*

Usai-Salvatori D. *Istituto Italiano per l'Africa e l'Oriente (Is.I.A.O.), Rome, Italy – usai.salvatori@alice.it*

van der Meulen Michiel J. *Geological Survey of the Netherlands, Utrecht, The Netherlands – michiel.vandermeulen@tno.nl*

Wilkinson Tony J. *Durham University, UK – t.j.wilkinson@durham.ac.uk*

Part I
Introduction

A synopsis of the book is presented in terms of the basic concepts emphasized within its 25 chapters. A variety of case histories are dealt with covering various times and places. In addition, important terms used throughout the book, landscape, culture, civilization and society amongst them, are examined. Finally, an analysis is made of the extraordinary effect agriculture has had in transforming the surface of the Earth into an increasingly anthropic construct.

References

Fagan B (2009) Floods, famines and emperors. Basic Books, New York, 352 pp

Powell JW (1875) Exploration of the Colorado River of the West and its tributary. Government Printing Office, Washington [Reprinted several times with the title of "Exploration of the Colorado River and its Canyons", latest in 2003 by Penguins Books, London, 409 pp]

Chapter 1
Summaries of the Contributions and a Few Considerations

I. Peter Martini and Ward Chesworth

Cultural, sociological and political factors determine the nature of human societies, but the environment clearly played a dominant role in our early histories and continues to influence society. The objective of this book is to examine this important factor in our history, not only from the standpoint of how the environment influences humanity, but also from the perspective of how we affect the environment. A case history approach is taken with revealing examples chosen to cover prehistoric to recent times. Particular attention is paid to the ways in which societies modified their territories in response to changing conditions—matter of some topicality now.

Meaning of the terms landscape and society are considered in more detail in the next chapter. For present purposes landscape will be taken to mean a geological/geomorphologic/environmental entity within the terrestrial biosphere possessing attributes dependent on climate, hydrology, soils, organisms and historical development. Society is considered to represent a group of people living together within the landscape(s) of a region, and sharing customs, language, laws or institutions. All societies have their complexities, but the ones considered the most complex in terms of institutions and material development are the ones we call civilizations, a word used in this book as a non-judgmental, neutral designation.

Much has already been written on these topics, and the subject is clearly of intrinsic interest, but potentially there is a clear practical value in such studies. It may be that in the give and take between humans and the biosphere in the past, there are lessons to be learned that could serve us well during the current period of global change, for historical purposes as well as for properly planning out present activities and our future. Clearly from ancient times much has changed, particularly with increasingly powerful technologies enabling modifications to the landscape that dwarf most anthropic changes of former times. However, although our sphere of influence has progressed from the village and its immediate environs to McLuhan's Global Village (McLuhan 1962), there remain some basic principles of society/landscape interaction that are brought out in the chapters of this book.

The choice of case-histories is potentially very large. Our selection has been guided by the need to illustrate as concisely as possible, human populations in a wide variety of landscapes over the time period from Neolithic to the present when *Homo sapiens* became the dominant large mammal in the biosphere. We have favored the Mediterranean area because in a relatively small region these requirements of time and space are well satisfied. In addition, much geological, archaeological, and historical information is available.

1.1 The Chapters: Brief Summaries and Considerations[1]

Each chapter analyzes how certain landscapes may have improved or damaged the prospects of a given society at different times, and, conversely, how that

I. P. Martini (✉)
School of Environmental Sciences, University of Guelph
Guelph, ON N1G 2W1, Canada
e-mail: pmartini@uoguelph.ca

[1] Edited summaries submitted by authors are in regular fonts, editors' comments are in *Italics*. Any inaccuracy may be due to editing.

I. P. Martini, W. Chesworth (eds.), *Landscapes and Societies*,
DOI 10.1007/978-90-481-9413-1_1, © Springer Science+Business Media B.V. 2010

society coped with its landscape and tried to modify it, well or badly, to the society's advantage. The approach is multidisciplinary though each chapter has its own bias and approach depending on the interests of the authors. This results in a variety of points of view regarding the complex two-way interaction between landscape and human beings, to our ultimate benefit in understanding the issue.

In *Chap. 2 (A Semantic Introduction)* Chesworth provides a definition of commonly used terms: landscape society, culture, civilization.

In *Chap. 3 (Womb, Belly and Landscape in the Anthropocene)* Chesworth examines some major effects humans had on the Earth's surface during most of the Holocene. Slightly more than 10,000 years ago agriculture developed, as the new means of food production, which currently dominates the human economy, and indeed, the ecology of the Earth. We have become the dominant large bodied species in the biosphere and have gradually risen to prominence in all biomes. We have taken over the temperate grasslands and much of the temperate forests and converted them to arable farmland. A pristine Mediterranean biome no longer exists in its type locality. Approximately 60% of the world's soils have now been modified by human use, half of that amount, more or less completely. This anthropic makeover of the landscape, justifies the recognition of this latest geological age under the rubric of the Anthropocene (Crutzen 2002). The major question of our time is whether this makeover is so destructive of the biosphere, that the civilization that arose from the practice of farming is not sustainable by farming over the long term.

Human modification of the landscape in a substantial way began when Homo erectus mastered fire. Homo sapiens inherited that mastery and employed it as part of its toolbox in clearing the landscape in the revolutionary new food production system called agriculture. In exploiting soil, water and genetic resources this Neolithic Revolution began a wholesale transformation of the landscapes of the Earth, that provides justification for designating all time since as a distinctive geological era, the Anthropocene. Agricultural sophistication and population numbers increased through time in a punctuated fashion consisting of rapid advances and temporary sharp retreats. By increasing the area of land under cultivation, we were always able, in a general way, to provide food for our growing population. Later, when only marginal land was left to exploit, we avoided collapse by increasing the productivity of the land. Now, as famine beckons again, we hope that the radical technique of genetic engineering will save us. The problem is that population always has the power to outgrow the food supply, as the Reverend Thomas Malthus said over 200 years ago—an insight that has never been falsified.

In *Chap. 4 (Human Responses to Climatically-Driven Landscape Change and Resource Scarcity: Learning from the Past and Planning for the Future)* Brooks examines a number of periods of century-scale rapid climate change (RCC) that have been identified throughout the Holocene period (the past 10,000 years). The best known of these occurred around 8200 and 4200 years before present (BP). However, less well-known, but qualitatively similar events occurred around 5900 and 5200 BP. This chapter reviews the evidence for these events and for large changes in climate and regional environments in the intervening period. The author concludes that the 5900 and 5200 BP events bracketed a period of profound change that represents the last widespread reorganization of the global climate system, and that they played a key role in the termination of the so-called Holocene Climatic Optimum. The chapter goes on to review the evidence for human responses to climatic and environmental change in northern Africa and western Asia during the 6th millennium BP, a crucial period in the development of human civilization. Finally, the chapter discusses what we can learn about human adaptation to RCC by examining the 6th millennium BP, and draws some lessons for the twenty-first century.

The development and collapse of societies are issues too complex for a single explanation. Historical and archaeological information, however, indicate that the collapse of past complex societies is the rule. Although there is no reason to believe that present societies will prove to be exceptions to the rule, past experience may help us to delay the inevitable, or indeed help in any transition to what comes next. Some of the reasons for decline and collapse are: natural disasters (including climatic change), depletion of resources, inability to respond rapidly to environmental or historical changes, including adaptation to a new resource base, influence of neighboring societies and intruders, internal class conflicts, mismanagement of critical issues, social disfunction, religious factors, and economic issues (Karling 2007). Climate change has been the dominant environmental factor to affect humanity during the Holocene, profoundly affecting the development of cultures and civilizations, and even bringing about their demise. The conventional view is that climatic stability and a benign environment are necessary for the development of stable societies, and that the opposite attributes lead to collapse (Redman 1999; Diamond 2005). Brooks' contrarian view is that sudden variations in climate within the overall Holocene trend (he emphasizes the 5200 BP cold, arid episode) led to a concentration of populations from which complex stratified societies developed.

In *Chap. 5 (Human Communities in a Drying Landscape: Holocene Climate Change and Cultural Response in the Central Sahara)* Cremaschi and Zerboni analyze the effect of mid-Holocene droughts on the landscape and human communities in arid lands of Libya and central Sahara. They consider three different adjoining physiographic units that belong to the same geographic and hydrologic ecosystems: the Tadrart Acacus massif, the dune field of Uan Kasa, and the fluvial valley of the Wadi Tanezzuft. The differing reactions of each physiographic unit to climate change is analyzed from a geoarchaeological perspective. The human groups that settled in the region responded with different adaptations to the drying environment, thus giving rise to specific social dynamics. The termination of the African Humid Period could be considered as the turning point for the environments and settlements in the central Sahara. The onset of aridity corresponds to the abandonment of the Uan Kasa region and to a strong reduction in the occupancy of caves and rock-shelters in the Acacus range. On the contrary, along the Wadi Tanezzuft the persistence of surface water reservoirs determined a change in the stream pattern, and an alluvial plain formed. Furthermore, a large oasis came into existence between ca. 4000 and 2000 year BP, to be exploited by Late Pastoral-Neolithic communities, and later by Garamantians. This marked the introduction of soil management and agricultural practices. The Tanezzuft oasis suffered a drastic reduction in size during the first centuries AD, at the time of the abandonment of the Garamantian settlements.

Cremaschi and Zerboni describe and compare the different responses of their three chosen areas and inhabitants to the overall Holocene desertification of the north and central Sahara. In this case availability of water was the dominant factor, but they conclude that reaction of local physical environments and societies differ independently of whether climate change, overgrazing, or other human activities are considered.

In *Chap. 6 (The Desertification of the Egyptian Sahara During the Holocene (the last 10,000 years) and Its Influence on the Rise of Egyptian Civilization)* Brookfield examines the overall continent-wide climatic changes of North Africa and the adaptation and evolution of early to middle Holocene societies culminating in the Pharaonic civilization along the Nile River. With the last retreat of the glaciers of the Northern Hemisphere, wetter climates, comparable to those of the Sudanian savannah, rapidly advanced northwards reaching the northern Sahara (over 15°N of latitude) by 10,000 year BP. There, between about 10,000 and 7700 year BP, human societies developed from hunting and gathering to multi-resource pastoralists. From about 7700 BP until 7000 BP, rapid climatic deterioration caused the progressive restriction of these pastoral societies from northern to southern Egypt (a distance of over 1000 km) and by 5000 BP such societies were practically limited to the northern Sudan. During this climatic deterioration, from about 7000 BP to about 5000 BP, societies rapidly developed, in refuges like the Nile Valley, from multiresource pastoralism to complex Pharaonic societies.

Brookfield provides a generalized picture of what occurred to the Holocene populations of North Africa and how they adapted and were moulded by the numerous climatic changes that occurred, in particular the several droughts. The migration of many people into refugia like the fertile Nile Valley led to the development of the highly stratified Pharaonic civilization. Whereas it is possible to provide a general history of these transitions, as the societies become more complex it is more difficult to relate cultural changes to environmental changes accurately. This must await more precise numerical dating methods that will allow ordering of events on a centennial scale.

In *Chap. 7 (Paleoenvironments and Prehistory in the Holocene of SE Arabia)* Goudie and Parker present details of geomorphologic features of the Rub al Khali of SE Arabia. Included are dunes, lakes, and fans. Their careful numerical dating allows a reconstruction of the environmental changes that occurred in the late Quaternary. These changes can be linked to various societal changes that occurred in response to wet and dry phases. Important events included Upper Pleistocene aridity, lower Holocene wetness, an abrupt, short-lived drying at ca 8.2 ka, another drying at 6.0 ka, and a pronounced dry event from 4000 ka. Populations were able to respond to these changes by emigration, returning when conditions ameliorated.

Goudie and Parker consider the major drying events that have affected northern Africa during the Holocene (8200; 5200 and 4200 cal year BP) and try to determine their influence on the societies of southeast Arabia. The populations of some societies permanently diminished while others responded by repeatedly leaving and re-colonizing an area as environmental conditions worsened or ameliorated respectively. This implied that localities existed in the region that were less affected by the climatic change and were sufficiently fertile to accommodate the influx of displaced people, peacefully or otherwise. The same kind of thing occurs nowadays in various parts of the world, as environmental, and particularly economic, refugees make for the richer countries. Food-aid to poorer countries helps populations to stay in place, but in the long run, it is incapable of helping them to live sustainably within their local landscape.

In *Chap. 8 (Human Paleoecology in the Ancient Metal-Smelting and Farming Complex in the Wadi Faynan, SW Jordan, at the Desert Margin in the Middle East)* Hunt and el-Rishi report on the remote, arduous Wadi Faynan, between the Edom Mountains and the Wadi Araba in southern Jordan. It was a major center for copper production from the Chalcolithic to the Byzantine period. A field system dating from the Classical period shows that some kind of agricultural activity accompanied the industrial operations. The authors review the human paleoecology of the Faynan complex and provide new georchaeological evidence of the environment in which the metal producers worked and what they ate. They conclude that during the Classical period extensive arable farming was unlikely—ploughsoils and the pollen of cultivated species are very rare, while mollusc assemblages are inconsistent with crops. The Faynan system required large numbers of draught animals to supply the enormous quantities of fuel required for smelting, to transport the ore from mines to smelting sites, to export finished copper and to import foodstuffs. Most of the latter could not have been grown locally because of lack of water, as well as low yields caused by the ubiquitous metal pollution. Consequently, it is suggested that agricultural activity was restricted to supplying some forage for the large population of pack animals that the mining operations required, and for feeding penned, imported livestock. Most food must have been imported, and since it was not locally grown, would not have bioaccumulated heavy metals. Benefit to the health of the metal workers was a collateral result of this dependence on relatively unpolluted, imported food. Inhalation and ingestion of airborne particles and contaminated water remained however, as hazards to the working population (Pyatt et al. 2005).

Since antiquity, communities have been established in locations inherently incapable of supporting the population. Mining communities are a case in point, and the Wadi Faynan mining community of SE Jordan is a particularly good example. Established in a harsh desert area (utilizing slaves as well) virtually all the necessary food and supplies for humans and much for the working animals was imported in exchange for ore. Nowadays, with the help of modern technology, willing populations establish a civilized existence by bringing in water and greening the desert, as modern Israel shows. Although admirably successful at least in the short term, these experiments remain precarious, being unsustainable by the local natural landscape.

In *Chap. 9 (Empire and Environment in the Northern Fertile Crescent)* Wilkinson covers the first millennia BC and AD in the Near East, and relates the cultural landscapes of the later territorial empires of the Near East to the local, including the atmospheric environment. During this time, landscape changed in ways that were new to the area. Massive canal systems were constructed on a monumental scale by the Sasanians for example, though more representative are smaller scale features such as extensive irrigation systems of dispersed farms, villages and villas. The latter commonly resulted in the colonization of new lands between pre-existing "tell" settlements or within the uplands. These dispersed settlements commonly resulted in terrain that had previously been fairly stable, becoming eroded and thereby contributing increasing amounts of sediment to the lowlands as valley fills. Landscape degradation was not simply the result of local phenomena, population pressure or agriculture for example. The needs of the empire had to be served, which meant that goods and services not obtainable locally, had to come from distant locations. These distant locations suffered environmental degradation for the greater good of other parts of the empire. The extension of settlement into marginal lands was frequently associated with the construction of extensive canal systems in areas that had not been irrigated before. These allowed environmental limitations to be over-ridden, but also imposed their own proxy environmental record by withdrawing significant amounts of water from streams. By creating the preconditions for soil erosion, the spread of human settlement in the landscape exacerbated the effects of rainstorms and deluges when they occurred.

Population pressure and large structured societies such as the Empires of the Fertile Crescent imposed maximum exploitation on fertile lands and occupational expansion onto marginal landscapes. This required modifications to the landscape and redistribution of waters for irrigation. The consequence was a degradation of the environmental conditions not just for local sustenance but also for the demands of burgeoning distant populations of the empire. This environmental degradation at a distance is now the normal global pattern as our demands span the globe and endanger the sustainability of producing societies located far from the societies consuming the fruits of their production.

In *Chap. 10 (The Interplay Between Environment and People from Neolithic to Classical Times in Greece and Albania)* Fouache and Pavlopoulos examine the Neolithization of Greece and Albania, during the Holocene climate optimum. Palinological investigations indicate agricultural clearances as early as that period through it is not until the Bronze Age that arable agriculture and animal husbandry notably shaped the landscape. Natural disasters such as volcanic eruptions and earthquakes have actively contributed to the geomorphologic evolution here for millions of years, and continue to do so. What is new is that the opening of the landscapes by human beings led to the development of erosion in often spectacular yet discontinuous ways as a result of original combinations of climatic and/or anthropic forces. The most spectacular consequences were the alluviation of valley bottoms and the progradation of deltas and coastal plains.

Anthropic deforestation particularly since mid-Holocene, as well as other activities such as grazing, have been important factors in helping to modify pristine landscapes, augmenting the rate of natural processes in doing so. Highly anthropicized nations such as Greece clearly show the results: hydraulic regimes (runoff in particular) have been changed, intensity and rates of erosion from slopes have been increased, and sediments redistributed. The results consist, among others, of modification in soil characteristics with development of catenas at different scales from local mounds to mountainous areas characterized by eroded hilltops and thickened mid-slope and slope-foot soils. At a larger scale, new landscapes are produced such as silted valley floors, and extended coastal and deltaic areas that locally prograde seaward.

In *Chap. 11 (The Nuragic People: Their Settlements, Economic Activities and Use of the Land, Sardinia, Italy)* Depalmas and Melis examine some aspects of the little known Nuragic population of Bronze Age Sardinia. The people were primarily sheep-herders and farmers, though toward the end of their culture, perhaps under Phoenician and other influences, they adapted to fishing along the coast. They adapted to various landscapes, first the fertile slopes of lava plateaus near springs, later the tops of the plateaus, as well as higher lands, and along the few narrow floodplains of the island. They seem to have mostly avoided the larger lowlands of the Campidano tectonic basin perhaps because it was partly covered by wetlands, so that it was prone to malaria: a rampant scourge during subsequent Roman times (Burke and Worcester 1996; Hays 2005). They take their name from the high stone towers called 'nuraghi', or 'nuraghe', which they constructed. Two main types were built, one resembling a high platform with a corridor as entrance, and a second the 'tholos' type, with a truncated conical, internally domed tower. More ancient smaller settlements had a single nuraghe, larger more recent ones had two or more. A settlement of huts surrounds some. About 9000 towers were constructed from Middle to Late Bronze Age. Analysis of the relationship between the monuments and the settled areas indicates that the presence of nuraghe was mainly determined by the environmental resources and the possible economic exploitation of the area, rather than any defensive consideration. The Nuragic people used a wide range of bronze and stone tools as well as animal power, and caused a significant degradation of the environment, primarily in the final stages of the Nuragic period.

Relatively stable, mid-latitude climatic conditions of isolated regions, such as the island of Sardinia, have fostered the development of stable ancient cultures, such as the Nuragic. During their existence, they did not change much because they did not need to change. As the population grew, the Nuraghi people slowly expanded to occupy adjacent landscapes—mostly highlands and high plains, and avoiding some the largest lowlands containing wetlands locally infested by malaria. Such cultures, lacking internal reasons for change, are more prone to modifications and amalgamations from outside—trade, immigration and conquest, for example, however some inherent original traits may persist for millennia.

In *Chap. 12 (Floods, Mudflows, Landslides: Adaptation of Etruscan–Roman Communities to Hydrogeological Hazards in the Arno River Catchment (Tuscany, Central Italy))* Benvenuti et al. analyze the different histories of three Etruscan-Roman sites, Pisa, Gonfienti, and Lago degli Idoli, located respectively in the vicinity of the estuary, in mid-course, and in the headwaters of the drainage basin of the Arno River, central Italy. Pisa had fluvial harbors that were occasionally affected by catastrophic floods, but both the Etruscans and the Romans continued to repair their facilities to avoid losing those economically important, irreplaceable sites. Gonfienti was mainly a commercial and agricultural settlement that was repeatedly damaged by floods and mudflows and was eventually abandoned during Etruscan times. Finally the Lago degli Idoli was a forested spring-site affected by local landslides, which none the less remained an essentially pristine locality. It was sacred to the Etruscans, but there is no archaeological evidence to indicate that the more pragmatic Romans held it in religious esteem.

Localization of settlements of ancient cultures and civilizations as they developed were carefully chosen to suit their needs, given the technology available. Natural processes and/or human activities induced damages. Wherever possible, changes to the landscape and environment, inconvenient to the human inhabitants, were remedied. Where this became too difficult, flexible populations abandoned the sites and migrated. More structured societies avoided abandonment by repairing and rebuilding, provided that expenditure of effort and resources were justified by the returns from a region, including considerations of viability or defense. Religion offered less practical reasons for the occupation or abandonment of a site. The sacred spring at Lago degli Idoli is a good example—revered by the Etruscans, was given no religious status and not visited by the Romans.

In *Chap. 13 (Landscape Influences on the Development of the Medieval–Early Renaissance City-States of Pisa, Florence, and Siena, Italy)* Martini et al. recognize that growth and decline of any advanced society is a complex matter. Political and sociological factors generally dominate in all but the earliest stages, though environmental conditions remain influential. The environmental factor is examined here in the context of the development of the neighboring city–states of Pisa, Siena and Florence during medieval to early-Renaissance times. Pisa developed first because its harbor was used by the early crusaders. During medieval times, Siena was next, favored by its hilly position which lay on a major pilgrim route to Rome. Florence flourished last, favored by its position along the Arno River. Pisa was handicapped by the siltation of its harbors and partly because of the extensive wetlands surrounding the city. The difficulties of Siena related to the limited space available for expansion within a walled hill-top town. In addition, there was a debilitating absence of water. Both reasons stunted any potential growth and curtailed the city's industrial capabilities. Florence suffered from river floods occasionally, but having access to abundant hydraulic power could run mills for its industries. It finally prevailed over its two neighbors.

Socio-economic and political factors and technology generally determine the fate of advanced societies, but some are also strongly influenced by the landscape they develop in. This is illustrated, for example, by the medieval-early Renaissance city-states considered here. Pisa and Florence are located in the same drainage basin, with Siena on a nearby hill. Landscape characteristics favored a particular society over its neighbors for a certain time, but became a liability later.

In *Chap. 14 (Paleo-Hazards in the Coastal Mediterranean: A Geoarchaeological Approach)* Morhange and Marriner examine the roles of human impacts and natural hazards in shaping the evolution of Mediterranean coastlines during the Holocene; where, when and how societies transform the coastal zone; at what scales and rhythms the changes take place; and what coastal sedimentary archives tell us about human–environment interactions. During the past 20 years, geoarchaeological research in the Mediterranean has attempted to understand the interplay between culture and nature, and more particularly how environmental potential and processes (sea-level changes and sediment supply for example) have played a role in Holocene human occupation of the coastal zone. This approach has drawn on the multidisciplinary study of sedimentary archives to attempt to differentiate between anthropogenic and natural forcings. The authors argue that the latter has played an increasingly secondary role with time. Archaeologists today are increasingly aware of the importance of the environment in understanding the natural frameworks in which ancient societies lived, and multidisciplinary dialog has become a central pillar of most large-scale Mediterranean excavations. Three important spatial scales of analysis have emerged, local, regional, and Mediterranean.

Climatic changes and human activities bring about erosion and redistribution of sediment. One effect is the siltation of harbors, with the resulting sedimentary deposits containing artefacts that illustrate habits and technologies of the various societies that made and used them. One observation to note (here and equally with Chap. 8) is that pollution is not a modern invention—witness the toxic lead concentrations in the pre-Hellenistic site of Alexandria, Egypt (Véron et al. 2006).

In *Chap. 15 (Mount Etna, Sicily: Landscape Evolution and Hazard Responses in the Pre-Industrial Era)* Chester et al. report on the Etna area of Sicily. In terms of the relationships between people, land and the creation of distinctive landscapes Etna is fascinating because, in spite of the ever present threat of volcanic eruptions and earthquakes, the flanks of the volcano attracted settlers in large numbers throughout the pre-industrial era. In this chapter the authors argue that this contrast with the barren and increasingly highly eroded lands of the Sicilian interior, was present in pre-historic times, but became more marked in the 2000 years which stretched from the Classical era to the late nineteenth century. Indeed, elements of contrast may still be seen in the landscape today. The reasons for Etna's attraction not only relate to the agricultural and irrigation potential of the region, but also to a range of social and economic factors, not least the ability of people successfully to develop indigenous and distinctive means of coping with the ever present threat posed by earthquakes and lava incursions.

The choice of a society to colonize a dangerous volcanic region is largely a pragmatic one. The benefits—good agricultural terrain with fertile soils, geothermal resources in places—outweigh the risk of eruptions and earthquakes. In areas of explosive volcanism (Vesuvius for example), tragedies that recur over a long time, and loss of entire villages or displacement of societies once every several centuries, are apparently acceptable risks. Etna is a quieter volcano, and although frequent eruptions and earthquakes occur that are damaging to local features, they seldom lead to a high mortality. People have adapted to them, and in some places have even succeeded in diverting the path of lava flows from important sites.

In *Chap. 16 (Romanian Carpathian Landscapes and Cultures)* Cioacă and Dinu examine human–landscape interactions in the mountainous areas of the Romanian Carpathians through geomorphic and geoarcheaologic analyses. Bottom valleys of intermontane basins, mountain slopes and high planes were progressively populated by Neolithic societies as population pressure imposed, and climatic conditions allowed, the higher terrains to be inhabited. As new lands were utilized, husbandry evolved technically to adapt to the characteristic of the various terrains. Intermontane basins and valleys led initially to isolation of their populations that therefore acquired distinctive characteristics. Isolation ended as communication increased between valleys, the peri-Carpathian planes, and the rest of Europe. Eventually Romania was conquered and colonized by Romans. The amalgamation of the ancient Dacian culture and new customs introduced from Rome produced the Daco-Roman civilization that was the precursor of current Romanian civilization. Although major changes in landscape and human habits have occurred through the ages, great parts of the Carpathian forests and the "wood culture" that developed in them persist today.

Cioacă and Dinu provide an insight into the adaptation of ancient societies to mountainous areas. These remain partly isolated in intermontane basins and valleys, and partly expand progressively from the more fertile and accessible valley bottoms onto the surrounding highlands and adjacent lands. As communication and commerce expanded to wider areas, continent-wide influence and eventually conquest and amalgamation with other societies led to the development of the modern civilization that nonetheless partly preserves some of the original attributes imparted by the local landscape.

In *Chap. 17 (Sea-Level Rise and the Response of the Dutch People: Adaptive Strategies Based on Geomorphologic Principles Give Sustainable Solutions)* Jungerius examines the complex and schizophrenic relationship of the inhabitants of the Netherlands with the sea. On the one hand, half of the land is a creation of the sea as ally. On the other, the sea is the enemy, with loss of land due to sea-level rise. For the last millennium, the inhabitants of the Netherlands have had to adapt their strategy of survival against the threat several times. The first strategy was learning to cooperate and coordinate defense against this common foe. The next and simplest strategy was preventing the floods from inundating the land by building dikes around polders. Pumping excess water out of the polders with wind mills and later mechanical pumps was the following step. Then it became necessary to strengthen the coastal dunes, the original, natural defense line. This was effectuated by supplying sand. For the fifth strategy the roles were reversed: the sea was forced to retreat and the bottom of the North Sea was converted to new land. Strategies based on geomorphologic principles used by nature to build the coast have invariably been shown throughout time to give the most sustainable results.

Jungerius provides insight into the ongoing, millennial battle of the inhabitants of the Netherland against the North Sea. This is not the only place in the world where land reclamations from the sea have been made, another example being the very extensive reclamations of east China along the Yellow Sea coast since ancient times. It stands apart from others however, because of the sophisticated technology that has been applied locally and the careful management of the waters and of the ripening of the reclaimed land for agricultural purposes. The strategy has not necessarily been to build only immobile, impermeable barriers to completely seal off marine and river flood waters from every locality at all times. Technogically more sophisticated barriers that open under normal sea conditions have also been built in places. Passage of tidal water is thereby allowed, and ships can pass between sea and canal. During major sea storms the barriers are closed. Furthermore certain areas have been set aside to be flooded during storms or river floods to alleviate pressure against the existing sea-dams. Of interest also is the defensive use of water during wartime. Inundation of vast areas can be effected to impede or retard the advance of enemy armies. Amsterdam, for example, was surrounded by such floods which acted like a vast moat during the Second World War (WWII). The water was kept shallow—about 10 cm depth—so that boats could not be used by the enemy. Similarly large areas of Eastern China were inundated by cutting the levees of the Yellow River to retard the advance of Japanese armies just before, and during WWII, as well as in ancient times against earlier enemies.

In *Chap. 18 (Perception of Volcanic Eruptions in Iceland)* Thordarson examines the perception the inhabitants of Iceland have of volcanism. Iceland is one of the most volcanically active regions on Earth, and its inhabitants have been exposed to volcanic eruptions and their consequences since the time of settlement about 1140 years ago. It is not surprising therefore, to discover that the nature and consequences of volcanic eruptions are imprinted on the cultural landscape of the nation. This evidence along with excellent preservation of historical records provides an ideal platform for evaluating social aspects of volcanism. A first pass assessment of these records indicates that throughout the country's history, an eminently practical, down-to-earth and matter-of-fact view towards volcanic eruptions has prevailed among the Icelanders.

Iceland is a land of fire (volcanoes), tremors (earthquakes), and ice (glaciers and outburst floods). Its environment is harsh but ameliorated by the warm North Atlantic Current. Volcanic activity provides thermoelectric power, and warm water for heating houses. Soils are fertile and are cropped to hay as fodder principally for sheep and horses, and some barley in the south. The sea around the island is a well managed, rich fishery. Occasional explosive volcanic events and climatic changes have created havoc and famine leading to major emigrations from the island. On the whole however, the people have adapted well and communally, to the often harsh life on this rugged landscape. The harshness comes through in the flourishing literature of the island, from the sagas of the Norse colonists, to writers of modern times such as Halldór Laxness, who won the Nobel Prize for Literature in 1955. Illustrations are another source of knowledge about the Icelanders perception of their lives, the problems they face being represented as monsters roaming the land.

In *Chap. 19 (Holocene Environmental Changes and the Evolution of the Neolithic Cultures in China)* Mo et al. report on the Neolithic cultures of east and north China. The Neolithic cultures first appeared about 10,000–8000 years BP and flourished preferentially in the areas of the Yangtze, Yellow, and Western Liao rivers because of mild climate, copious precipitations, and fertile soils. Low-lying floodplains were most suitable to the development of rice culture, while fertile loessial soils on higher ground became the locus of dryland agriculture based on foxtail and broom millet. Initially the various cultures of the different regions continued to evolve and flourish almost simultaneously. From 6000 to 5000 year BP many new cultural changes occurred as production techniques.

Most significantly, the society became stratified. From 5000 year BP onward, the evolution of the Neolithic cultures of the various regions diverged due to local response to strong climatic and other environmental changes. The authors catalogue resulting crop failures in certain areas and increasing flooding in others, as well as internal and external social conflicts that led to cultural regression and decrease in populations in most areas. Environmental conditions, particularly soil fertility, ultimately led to the prevalence of the culture of the Zhongyuan region of the middle Yellow River area. This culture gradually influenced the culture and politics of the other regions, and became the center of Chinese civilization.

Mo et al. synthesize information about the complex, numerous societies that evolved and waned during Neolithic times on the various landscapes of eastern China. Temperature, precipitations and soil fertility greatly affected the progress of humankind in the region. As today, so in the past, a rice-culture was prevalent in the warmer, wetter lowlands of the southeast affected by monsoonal precipitations, while wheat culture prevailed in the northeast due to the wide expanses of very fertile loessial soils in lowlands and hills. It was the fertile valley of the middle-lower Yellow River in the north that sustained the consistently evolving cultures that became the cradle of modern Chinese civilization. The dark side is that the easily eroded, re-worked loessial soils of the valley, were frequently ravaged by floods, and large populations were continually threatened, then as now.

In *Chap. 20 (Landscape and Subsistence in Japanese History)* Barnes reviews the evolution of Japanese civilization in terms of adaptations to life in a tectonically active island arc, dominated by volcanoes and mountains and with what little level ground there is, mostly confined to the coast. The early hunter-gatherers, focussing on inland mountains and on coastlines, contrast with the imported wet-rice technology focussed on the lowland plains. Why different ethnic groups did not develop to exploit these radically different environments is a topic explored in depth. The bifurcation of the historic agricultural system into wet-rice cultivation and dry-field production is another important topic. The modern situation differs from the historical in that food importation has had devastating consequences for local farming. The future of food production in Japan relies not only in reversing the decline of indigenous production but also on the retreat to higher ground as sea-level rise takes its toll on the bottomlands—now the locus not only of the major paddy areas but also of urban and industrial developments.

Barnes offers a brief, clear review of the evolution of the Japanese civilization in a tectonically active region consequently subjected to numerous earthquakes and landslides. It is also vexed by powerful sea storms and the occasional tsunami along the coasts. The country is for the most part mountainous with only a third as coastal plains. These are part removed from agricultural production by expanding urban areas and their associated infrastructures such as routes of communication, power lines and so on. From early times Japanese populations adapted well to the complex landscape and flourished, managing the production of rice. Japanese society, like many others evolved on islands, remained isolated and self sufficient for a long time and acquired characteristic and lasting traits and customs. Once opened to the world it rapidly developed into a modern, successful, technological society. As such, it has lost self sufficiency with regard to food and mineral resources and must rely on imports in exchange for manufactured goods.

In *Chap. 21 (Evolution of Hydraulic Societies in the Ancient Anuradhapura Kingdom of Sri Lanka)* Dharmasena analyzes the development of a sophisticated hydraulic civilization in the dry lowland of the northern part of Sri Lanka. The first of many invasions from mainland India in 3000 BC introduced the religious beliefs that governed the sacred town of Anuradhapura in the north, where the hydraulic civilization was centered. Early societies of northern Sri Lanka developed agricultural practices near rivers. They were dependent on unreliable monsoon rains and vexed by recurring droughts and floods. Irrigation canals were soon built and small scale systems of irrigation were in operation possibly by 500 years BC. Subsequently the Sinhalese constructed a complex system of canals and water storage reservoirs locally called 'tanks', after the Portuguese 'tanque', to provide a technically sophisticated irrigation system to regulate water distribution. Completed by the first century AD, several large-scale irrigation works existed fed both by harvesting the runoff of catchment basins and where possible by diverting water from streams. The organization of small tanks into a cascading sequence within micro-catchments allowed great efficiencies in water use. For instance, drainage from the paddy fields in the upper part of the cascade flowed into a downstream tank for reuse in the paddy fields below.

The original ancient hydraulic civilization of the dry zone disappeared after the twelfth century AD. Breakdown of the efficient irrigation management system, climatic change, malaria, depletion of soil fertility, foreign invasions, and famine are some of the reasons cited. However, the tank-cascade irrigation systems based on natural drainage basins is still considered a valid, lasting technology that, suitably modified for modern needs is being recommended for restoration.

Sri Lanka is a country with two very distinct geological/geomorphologic and climatic regions: a hilly, wet southern part and flat, dry northern lowlands. The expansion of the population throughout the northern dry lowlands was made possible by the implementation of an early, sophisticated irrigation system consisting of interconnected small reservoirs and good management and conservation practices of the water resource. This irrigation system has lasted for millennia, as have others such as the qanat system of Mesopotamia satisfying the need of a vital resource in hot, dry countries.

In *Chap. 22 (Disease in History: The Case of the Austronesian Expansion in the Pacific)* Sallares focuses on the role of pathogens in history, taking as a specific example the possible effects of malaria on the prehistoric migrations of speakers of Austronesian languages from Taiwan across the Pacific and Indian Oceans. Both the view that malaria assisted the Austronesian migrations and the view that it hindered them have been expressed in previous literature. In recent times many of the western Pacific islands, where malaria is present, have been less densely populated and have suffered lower degrees of environmental degradation as a result of human activity than the islands of Polynesia in the eastern Pacific where malaria has never been present. This chapter investigates the role of malaria in the Austronesian migrations and its relevance to landscape history in this part of the world. Human genetic mutations believed to be associated with resistance to malaria, provide the scientific evidence on which the argument is based. The conclusion is that malaria probably hindered rather than helped the Austronesian migrations.

Landscape is a contributing factor in the evolution and demise of cultures and migrations of populations. Elements of the landscape such as its productivity of natural, cultivated or raised macro-organisms are usually considered to be paramount when dealing with the well-being of human populations. The role of microorganisms, though less understood until recently, has been massive in determining the fortunes of many societies. Suffice to mention only the effect of the Black Death in fourteenth century Europe, in addition to the long history of malaria, as emphasized by Sallares, a scourge that persists in places around the world. Malaria is closely tied to elements of the landscape such as wetlands and some tropical forests, and had important effects in shaping some societies. Italy provides a number of telling examples of the deaths from malaria of war-leaders, of powerful political people such as popes and cardinals, as well as famous artists such as Dante Alighieri (Sallares 2002). A more concrete effect is seen where communication routes have been planned to avoid infected areas. Malaria has clearly affected the migration and peopling of the Western Pacific islands. Deforestation of some islands for agricultural purposes changed the landscape so that wetlands formed in places. These became breeding grounds of the mosquito hosts of the malarial microorganism. This led to health problems and loss of population due to death and/or migration. In a seemingly ironic, but common fate, islands and other localities in the world have undergone more anthropic change in the absence of malaria, than in its presence.

In *Chap. 23 (Farms and Forests: Spatial and Temporal Perspectives on Ancient Maya Landscapes)* Dunning and Beach analyze some aspects of the Maya civilization. The Maya Lowlands, once thought of as a relatively homogeneous region, is actually a mosaic of habitats to which the ancient Maya adapted many and varied systems of water management and agriculture. The overall trajectory of ancient Maya civilization is probably best represented by a metastable equilibrium model in which periods of relative environmental and population stability were punctuated by episodes of sudden change. Older models correlated peaks in population with maximal environmental degradation in the Late and Terminal Classic periods (600–900 AD). Several recent cases, however, indicate that the greatest impact was caused by relatively small populations. Conversely large, densely settled populations sometimes stabilized the landscape through careful conservation measures. Episodic periods of severe drought played the greatest havoc with the ability of the Maya to adopt successful long-term strategies to meet their food and water needs. Interior regions of the

Maya Lowlands with less secure water sources were the most likely to experience extreme cycles of population growth and decline. In the long term, coastal and low-lying areas with perennial wetlands generally experienced less severe perturbations, but even here the Maya had to adapt to rising sea and groundwater levels and widespread landscape burial.

The Maya civilization has been long studied, but new discoveries continue to be made and fresh insight achieved. This complex, structured society became progressively more populous and less resilient and mobile in its relationship with a changing environment. Changing climatic conditions, and spells of drought, combined with intense deforestation that allowed increased soil erosion, produced a progressively deteriorating environment. As a sophisticated society the Maya tried to combat the inevitable by innovative agricultural practices, building reservoirs to supply water during periods of drought, and maintaining them free of silt. As environmental conditions deteriorated, conflicts between communities increased and military predation occurred. Some sites could persist for a time, while others had to be abandoned and the population migrated. Some areas may have been overwhelmed to the point of collapse by an unsustainable influx of refugees. If history teaches lessons, this is one of them, and the problem of environmental refugees is becoming a serious one again in the modern world.

In *Chap. 24 (Water Follows the People: Analysis of Water use in the Western Great Plains and Rocky Mountains of Colorado, USA)* Wohl reviews the challenges presented by the semiarid climate of the State of Colorado in western USA, to development of contemporary sustainable societies. The first settlers of European descent who reached this region in the mid-nineteenth century immediately began to engineer water supplies, rather than adapting their lifestyles to aridity as had wildlife and indigenous peoples. Extensive manipulation of surface and ground water supplies has facilitated rapid growth of irrigated agriculture and locally dense human populations, but has also produced unintended effects including water pollution and endangerment of native plant and animal species. Water shortages resulting from persistent drought and increased population and water consumption, are forcing contemporary societies in the region to use water more efficiently, to allocate water differently among competing human users, and to maintain sufficient surface and ground water supplies to support endangered species and river ecosystems. The South Platte River basin of Colorado serves as the primary case study to explore these regional issues.

Wohl focuses her analysis on the water management of modern societies in the semi-arid USA, specifically in the Rocky Mountains and foothills of Colorado. Her title ironically reflects the misleading nineteenth century slogan 'water follows the plough', concocted to entice settlers to homestead in the arid west. The low precipitation along the orographically unfavorable eastern mountain flank and nearby plains was not sufficient to support an expanding agriculture modelled on the practices of the well watered east. Modern technology has allowed the tapping of water from the humid western side of the mountain chain and its transmountain transport to the eastern side. This is one more example of humans creating, nowadays mostly for the purpose of short term financial gain, an artificial condition not sustainable by the local landscape, instead of trying alternatives, dryland techniques of water agriculture amongst them.

In *Chap. 25 (Frozen Coasts and the Development of Inuit Culture in the North American Arctic)* Park examines the effects on past and present inhabitants of the extreme seasonal changes in landscape in the North American Arctic due to sea-ice formation and breakup. Archaeological research has demonstrated that over time the people of this daunting region developed a complex adaptation to the frozen-coast environment, which includes the sea ice. Early aspects of this adaptation included the hunting of small sea mammals in open water from the shore or from the floe edge, followed by the hunting of larger sea mammals in open water from boats. The later and more complex aspects of the adaptation included the hunting of ringed seals at their breathing holes and living for some or all of the winter far out on the sea ice. The successive stages of this adaptation to the frozen-coast landscape can be identified in the archaeological record of the Arctic through the sequential appearance of technological artefacts such as harpoons, drag floats, snow knives and lamps.

Some landscapes cannot be drastically changed by its inhabitants: adaptation of the human lifestyle is the only answer. This is the case for Arctic regions where ice is a major component of the landscape. Ice-distribution, sunlight, and temperature drastically change seasonally, and thereby totally modify the landscape and its natural resources. Hunting and gathering, and in modern times, import from the south, are the only means of sustenance in North America. In Europe and Asia, husbandry of reindeer is also practiced. Park examines the resilience of the ancient northern people of North America and their delicate adaptation to this variable, frozen landscape. Adaptation was possible because the people could rapidly change their habits and locations according to the seasons and to longer time periods. European tools and habits were eventually introduced, and the forced agglomeration of people into unmovable villages occurred. Old time famine could now be avoided, but severe, mainly psychological health problem, have become endemic. In addition, global atmospheric circulation has ensured that the overall environment of these remote communities is being subtly polluted from distant sources in the industrialized south. As with other indigenous northern populations of the continent, a new Inuit culture is evolving that cannot be sustained by the local landscape. This makes for a difficult transitional period.

1.2 Commentary

The basic problem for any species in the biosphere is ecological: how to derive sufficient sustenance from its environment that it has a good chance to survive and pass on its genes (Hardin 1993). Humankind is no exception to this fundamental requirement. Where we are exceptional is in succeeding so spectacularly. Big, fierce animals need a large territory to support them and are consequently rare in nature (Colinvaux 1978). We have developed a foraging strategy of so successful a kind, that although our Paleolithic ancestors passed through the bottleneck of near-extinction (Wilson 2006) we have become the one big, fierce animal that is not rare. On the contrary, we are so good at the ecological practice of competitive exclusion (Rees 2004) that we are now as a society (or pack) the most successful predator in all biomes, terrestrial and marine. Competitive exclusion for *Homo sapiens* means that we take over all major landscapes as human habitat, from lowlands (Jungerius[2]) to highlands (Cioacă and Dinu), from coasts (Morhange and Marriner) to continental interiors (Wohl), from frigid (Park) to tropical (Dunning and Beach), from tectonically quiet (Depalmas and Melis; Dharmasena) to tectonically active (Barnes; Chester; Fouache and Pavlopoulos; Thordarson), from sparsely populated (Goudie and Parker) to highly urbanized (Barnes; Jungerius; Martini et al.) and from dry (Brookfield; Brooks; Cremaschi and Zerboni; Hunt and el-Rishi; Wilkinson) to wet (Barnes). Even active volcanoes are no absolute barrier to human settlement (Chester et al.; Thordarson).

Elevation and relief have often presented problems to human settlement and the general pattern of landscape adaptation has been to populate flat coastal and near coastal areas first, as well as deltas and riparian zones (Benvenuti et al.; Cioacă an Dinu; Mo et al.). Desert regions have also presented difficulties, especially where an acceptance of the natural state of a landscape (Brooks; Cremaschi and Zerboni) is replaced by a determination to use advanced technological resources to fix the problem (Wilkinson). Water has always been problematic in areas of karst (Dunning and Beach; Fouache and Pavlopoulos), where drought is produced by excessive drainage through the highly porous limestone landscapes.

Once we lived by hunting, gathering and scavenging, then some 10,000 years ago, our foraging strategy changed in a revolutionary way. We took charge of food production in the technology called agriculture. All our subsequent history is a kind of footnote to that Neolithic Revolution that took place first in SW Asia, North Africa and China (Brooks; Mo et al.; Wilkinson), slightly later in the Americas (Dunning and Beach), and later still as the technique diffused out from the heartlands—for example 7000 years ago in the western Mediterranean (Chesworth Chap. 3), about 5000 years ago in Sri Lanka (Dharmasena). The food production system drives all human economic activity (Smith 1776/1990), and as such is the basis of our various socio-political systems. Once farming started, technologies developed that allowed humanity to avoid (no doubt temporarily) the usual control on the size of a population—that is, the negative feedback that Darwin (1859/2010) called natural selection and Spencer (1864–67/2002) the survival of the fittest. All of our subsequent history, and our current predicament, is colored by that fact. The first population explosion began with the Agricultural or Neolithic Revolution and subsequent spurts in population growth have followed each expansion of farming onto new landscapes, and each advancement in agricultural technique. Our agro-footprint has been extended onto virtually all arable

[2] Undated references are to chapters of this book.

landscapes of the temperate zone (Barnes; Cioacă and Dinu; Mo et al.; Wohl) into warmer climatic regimes (Dharmasena; Dunning and Beach) and even into the sub-Arctic (Thordarson). Some landscapes still retain an attachment to hunter-gatherer economies (Barnes; Park), and Romania (Cioacă and Dinu) is interesting in having an area of unique woodland culture where agriculture is practiced in harmony with a kind of forest-craft which pre-dated it. One way or another, humanity has adapted and developed appropriate foraging strategies for all climatic zones.

By the time the Agricultural Revolution had reached the Atlantic coast of Europe, the first civilizations and cities had appeared in Mesopotamia and Egypt (Brookfield; Brooks; Chesworth Chap. 3), the first scripts had been invented, and new technologies involving new materials had come into being. Methods of cultivation and water-management had advanced and were changing landscapes in widening circles of influence spreading out from the original centers of civilization. Cities were a new feature of the landscape, and in spite of superficial differences, are recognizably the same the world over, with monumental architecture, elaborate communication systems, a division of labor, a stratified social system with a hierarchy that commonly hardens into a caste or class system.

The stratified nature of urban civilization created new needs amongst the populace, and made new demands on the landscape. Subdivision of labor was formalized and specialized artisans of various types plied their trades and sought for, and worked with earth materials from clay to base and noble metals, as well as gemstones, to make pots, implements, jewellery and weapons. Mining communities grew up to serve their needs, and an industrial element was added to the growing human footprint on the landscape. Hunt and el-Rishi (this volume) describe an early example from Jordan—a prehistoric foretaste of the Industrial Revolution that began to transform human society centuries later. Excavations and mine sites became a common feature of Middle Eastern landscapes, where the rift that runs from the borders of Turkey, down the Jordan valley, the Dead and Red seas and into Africa. It is the site of numerous red-beds copper deposits that have been worked since Chalcolithic and Bronze Age times.

One of the clear trends since the beginnings of civilization has been the increase in human technical abilities, associated with and dependent on the development of new energy sources. Energy is the key to modifying a landscape—the heat of the sun in nature, often adapted by humans in the form of water and wind power, and supplemented by human and later animal muscle. Modification of landscapes by cultivators started with scratching and digging sticks and progressed through the simple wooden plough to the metal mold-board plough pulled by oxen, and much later, the horse. By the Iron Age even the heaviest soils could be tackled, and the temperate clay plains of the world came under agricultural attack for the first time.

With the exploitation of fossil fuels on an industrial scale, starting with coal in the eighteenth century, anthropic landscape-changes on a massive scale began to appear for the first time in Europe since the great earthworks of the Neolithic. The Netherlands provide a case in point. In their constant battle with rising sea levels, the Netherlanders soon saw the advantage of adopting the Newcomen and Watt stationary steam engines to increase their power to drain territory and change seascape to landscape (Jungerius). Interestingly, after centuries of experience the engineers in the Netherlands have concluded that the best defenses were provided by the natural dunes and that the best strategy here is to work with nature and fortify the dunes. This is an attitude that brings to mind the fatalistic approach of North African societies, who in a sense, worked with nature (if in a passive way) when they adjusted their lives to aridification (Brooks; Cremaschi and Zerboni; Goudie and Parker).

The uncontrolled growth of human numbers has created the need to commandeer ever more territory to grow food. The point of saturation has now been reached in that all the best landscapes for arable agriculture have now been taken over according to the Millennium Ecosystem Assessment (2005) and modified by the farmer. The Mediterranean region was the first natural biome to become a more or less completely anthropic one (Sect. B2, this volume), a fate that has since overtaken the Mediterranean biome in California, Chile, Australia and South Africa and other areas including those reported in this book. It is also pushing agriculture into the marginal and inappropriate landscapes of arid regions, such as great parts of the western United States (Wohl). Water has always been a problem for farms and urban settlements in this region, and for that reason, river diversions and reservoir construction have been common since the nineteenth century. Now, engineering technology is capable and powerful enough to bring water from the far side of the Rocky Mountains (Wohl). But at what stage do we decide with the North African herdsmen and nomads that the landscapes of the Desert biome are inappropriate places for high human and animal populations?

Biome-modification is practiced by all organisms but we do it on a larger scale and more thoroughly than any others. In this, we appear to have overshot the ability to support ourselves sustainably by about 25%. The consequences are:

(a) Habitat destruction particularly by deforestation of the landscape, and by competitive exclusion generally. Coastal plains were particularly vulnerable to biome modification (Barnes; Mo et al.; Morhange and Marriner). The temperate grassland and much of the temperate forest biomes have also been appropriated by *Homo sapiens* (Chesworth Chap. 3). A complete utilization of the Earth's photosynthetic capacity is threatened, with dire consequences for biodiversity.
(b) Over-exploitation of abiotic resources such as soil, leading to erosion (Fouache and Pavlopoulos; Wilkinson), acidification (Chesworth Chap. 3), salinization (Brook; Wohl), compaction, drainage impedance and soil fertility losses (Mo et al.). Problems of water supply become increasingly prevalent most obviously in arid areas, and water-management becomes necessary, starting with the irrigated, canalized landscapes of the Neolithic societies in the valleys of the Tigris and the Euphrates and in North Africa (Bookfield; Brooks), but less obviously in ostensibly humid areas such as the East Anglian region of the UK (Chesworth Chap. 3) or the Siena region of Italy (Martini et al.). Since the Industrial Revolution, over-exploitation of minerals and fossil fuels has become a major problem.
(c) Over-exploitation, overkill and even mega-kill by hunting, fishing, and collateral damage of the biotic resource. Die off of indigenous animal communities in island habitats of the Mediterranean, as well as elsewhere, seem always to coincide with settlements of such landscapes by humans.
(d) Effects of introduced species on native species (the agricultural tool-kit expanding out of the Levant eventually into the Americas, as well as adventitious introductions). This is particularly critical in terms of introduced microbes, with the famous example of the Spanish conquistadors, who, though few in number were able, with the considerable help of their European microbial allies, to conquer the Aztecs and the Incas. The introduction of malaria in the Pacific (Sollares), of smallpox into North America in the mid 1600s, and the encouragement of schistosomiasis by irrigation works in Africa, Egypt are striking examples (Kloos 2002).

The geologists dictum 'the present is the key to the past' no longer holds in at least one important respect. Until the present, no large organism has become a dominating geological force at the surface of the Earth. Now, for better or worse, *Homo sapiens* has achieved that distinction and we have entered a new division of geological time: the Anthropocene. Every chapter of this book is testament to the new era, and provides multiple examples of the modifications that a thinking, self-aware and ingenious primate has made.

References

Burke PF, Worcester MA (1996) Malaria in the Greco-Roman World. A historical and epidemiological survey. In: Temporini H, Haase W. (eds) Medizin und Biologie [Forts], Aufstieg und Niedergang der römischen Welt: Geschichte und Kultur Roms im Spiegel der neueren Forschung. 2, Principat, Bd 37, Philosophie, Wissenschaften, Technik, Teilbd 3, pp 2252–2281

Colinvaux PA (1978) Why big fierce animals are rare: an ecologist's perspective. Princeton University Press, Princeton, NJ, 256 pp

Crutzen PJ (2002). The "anthropocene". Journal de Physique IV 12:1–5

Darwin C (1859/2010 3) On the origin of species (Reprinted many times including in 2010 by New York University Press, New York, 360 pp; and available in several Internet sites including http://www.literature.org/authors/darwin-charles/the-origin-of-species/

Diamond J (2005) Collapse – How societies choose to fail of succeed. Viking, New York, 575 pp

Hardin G (1993) Living within limits: Ecology, economics and population taboos. Oxford University Press, Oxford, 339 pp

Hays JN (2005) Epidemics and pandemics: their impacts on human history ABC-CLIO, Santa Barbara, California, 499 pp

Karling HM (ed) (2007) Global climate change revisited. Nova Science Publishers, Inc., New York, 163 pp

Kloos H, David R (2002) The paleoepidemiology of schistozomiasis in ancient Egypt. Human Ecology Review 9:13–25

McLuhan HM (1962) The Gutenberg Galaxy: The making of a typographic man. University of Toronto Press, Toronto, 294 pp

Millennium Ecosystem Assessment (2005) Ecosystems and human well-being: Synthesis. Island Press, Washington, DC, 137 pp

Pyatt FB, Pyatt AJ, Walker C, et al. (2005) The heavy metal content of skeletons from an ancient metalliferous polluted area in southern Jordan with particular reference to bioaccumulation and human health. Ecotoxicology and Environmental Safety 60:295–300

Redman CL (1999) Human impact on ancient environments. The University of Arizona Press, Tucson, 239 pp

Rees W (2004) Waking the sleepwalkers: a human ecological perspective on prospects for achieving sustainability. In: Chesworth W, Moss MR, Thomas V (eds) The human ecological footprint. University of Guelph, Guelph, Canada, pp 1–34

Sallares R (2002) Malaria and Rome: a history of malaria in ancient Italy. Oxford University Press, Oxford, 341 pp

Smith A (1776/1990) An inquiry into the nature and causes of the wealth of nations. Encyclopædia Britannica, Inc., Chicago 515 pp [Also available on Internet at http://www.adamsmith.org/smith/won-index.htm (accessed December 2009)]

Spencer H (1864-67/2002) The principles of biology. William and Norgate, London, 2 v. Reprinted 2002 by Ross and Perry, Washington, DC.

Tainter JA (1988) The Collapse of Complex Societies. Cambridge University Press, Cambridge, 250 pp

Véron A, Goiran JP, Morhange C, et al. (2006) Pollutant lead reveals the pre-Hellenistic occupation and ancient growth of Alexandria, Egypt. Geophyscal Research Letters, 33, L06409, doi:10.1029/2006GL025824.

Wilson EO (2006) The creation: An appeal to save life on earth. W.W. Norton & Company, Inc., New York, 175 pp

Chapter 2
A Semantic Introduction

Ward Chesworth

The objective here is to take a brief look at semantic questions surrounding several words that are frequently used in this book, and which relate directly to the two nouns in the title: "Landscapes and Societies". This includes the obviously similar terms land, culture and civilization, but also three words concerned with the relationship between the two—determinism, adaptation and sustainability.

2.1 Landscape and Land

The Merriam-Webster definition of landscape is "the landforms of a region in the aggregate", which is almost the same as the definition of topography in the Oxford English Dictionary (OED): "the features of a region or locality collectively". For the OED landscape is: "a tract of land with its distinguishing characteristics and features, esp. considered as a product of modifying or shaping processes and agents (usually natural)." That is a better definition, and since landscape is there defined as "a tract of land", I will take landscape and land to be synonyms. However, it's a bare-bones definition, and the details need to be fleshed out. An examination of the phrase in parenthesis is also necessary.

The ideas of Aldo Leopold on the topic of land, his "land organism", are germane to the discussion. According to Leopold (1993, p. 46) the complexity of the land organism is "the outstanding scientific discovery of the twentieth century". He wrote that "the individual [human being] is a member of a community of interdependent parts [which includes] soils, waters, plants, and animals, or collectively: the land." The concept owes much to earlier work by geographers. Carl Sauer for example, wrote of the "phenomenology of landscape" and emphasized that "the works of man [were] an integral expression" of it (Sauer 1925, p. 21).

Notice Leopold's introduction of the word community into the debate. As an ecologist Leopold was well aware of our Darwinian drive to compete within the land organism, but in the interests of conservation he was keen to develop a "land ethic" that would encourage human beings to cooperate as well as compete. He hoped that seeing the land organism as a community, with the strong need for cooperation that holds a community together, would provide a firm foundation for the ethic he desired.

Even without Leopold's ethical agenda, it is still necessary to see landscape as a 'community' or, to use a less anthropomorphic phrase, a complex of interrelationships. Only with this perspective is it possible to appreciate how a landscape works, and how interdependent its many parts are. It also encourages us to be humble—the complexity of landscape is so great that we do not yet seem to have a sufficient grasp of it to live sustainably on it.

A final matter to clear up is the phrase "usually natural" in the OED definition. Although as good Darwinians, we insist that human beings are an integral part of the natural world, we normally distinguish our activities and effects as being artificial rather than natural, a convention I will follow.

The fact is that little if any of the Earth's surface is free from human influence so that pristine nature if it still exists, is rare. Our mark is seen even in the inhospitable environments of the high mountains and high

W. Chesworth (✉)
School of Environmental Sciences, University of Guelph
Guelph, ON N1G 2W1, Canada
e-mail: wcheswor@uoguelph.ca

I. P. Martini, W. Chesworth (eds.), *Landscapes and Societies*,
DOI 10.1007/978-90-481-9413-1_2, © Springer Science+Business Media B.V. 2010

latitudes. Indeed it has been said that in the Holocene, *Homo sapiens* became a geological force on the planetary surface comparable to those such as volcanism, tectonism, glaciation and weathering that have dominated the landscape surface since Precambrian times (Chesworth 1996). We have begun a new geological epoch, the Anthropocene, says Crutzen (2002).

The Millennium Ecosystem Assessment Project (2005, p. 18) in its definition of landscape clearly recognizes the human component: a landscape is "an area of land that contains a mosaic of ecosystems, including human-dominated ecosystems. The term cultural landscape is often used when referring to landscapes containing significant human populations or in which there has been significant human influence on the land".

In light of such considerations, I will take land or landscape to be part of the macroscopic three dimensional continuum of landforms that make up the Earth's land surface. It is composed of a chaotic complex of rocks, minerals, soils, and amorphous solids, surface water, ground water, organisms and products of their decay, all of which interact and change the landscape, with much feedback, in response to the forces of weathering, tectonism and human activity. Landscapes are transient features, maintained in a state of disequilibrium on scales ranging from micro to macroscopic, by external forces powered by the energy of the sun, and internal ones driven by the decay of unstable nuclides within the Earth (Chesworth 1991). The movement of water on and in a landscape is the principal natural means by which a gravitational equilibrium may be approached, and the erosion of the loose skin of the land surface, the soil, is the chief way that landscapes are worn down towards a base level. Since the practice of agriculture requires the manipulation of both soil and water, it was inevitable that farmers would become a geomorphic force as their enterprise prospered and became more intensive. Human activities of all kinds now mark virtually all landscapes with an indelible footprint.

2.2 Landscape Quality

Until about the eighteenth century, the quality of a landscape was judged solely by how well it served the purposes of humanity. Landscapes with deep, loamy soils were considered fair and fruitful, while areas with poor drainage were foul and waste. From the Age of Enlightenment onwards, landscapes were also considered to have an intrinsic quality of their own, regardless of their usefulness to us.

To the Romantics, beginning with Rousseau, and to the landscape gardeners of the English school, this quality was aesthetic in nature. To the scientists of the late nineteenth century, and increasingly thereafter, it was the role in the biosphere that defined the quality of the landscape and its components. The ecologist for example judges the landscape in terms of the different types of habitat that it provides, and that ensures a particular level of biodiversity. However, farmers, no matter what their artistic or aesthetic propensities, are practical people with a utilitarian perspective. Their assessment of land is based above all else on the age-old one of its ability to provide food and fibre to the human population. Opinion in society at large is also dominated by practical requirements—disposal of wastes, provision of foundations for roads and buildings, maintenance of a built environment conducive to human health, and so on (Thomas 1996).

Mention of human health brings to mind again Aldo Leopold and his concept of the "health" of land, another way of looking at landscape quality. Writing in the 1930s, he said that "the land consists of soil, water, plants, and animals, but health is more than a sufficiency of these components. It is a state of vigorous self-renewal in each of them, and in all collectively. Such collective functioning of independent parts for the maintenance of the whole is characteristic of an organism. In this sense land is an organism, and conservation deals with its functional integrity or health" (Leopold 1993). This is a clear anticipation of the idea of Gaia introduced by Lovelock (1979/2000) with notable support from Lynn Margulis (Margulis and Sagan 1997).

At its simplest the Gaia Hypothesis considers that the Earth acts as a kind of superorganism, maintaining itself as a healthy abode for life. The state of the planetary surface is constantly adjusted by biological feedback mechanisms whenever inhospitable influences make themselves felt. Rather than the Gaia Hypothesis it might be more accurate to speak of the Gaia Syndrome, since as Kirchner (1989) points out, the Gaia idea incorporates many hypotheses, which he recognizes severally as Influential, Co-evolutionary, Homeostatic, Teleological and Optimizing Gaia (Kirchner 1991). His basic criticisms are (a) that where Lovelock's idea is right, for example the feedback mechanisms invoked, it is not original, and (b) where it is original, for example in claiming that Gaia has the teleological objective of maintaining homeostasis in the interests of life, it is wrong. Lovelock in fact, dropped the claim of teleology (that is that Gaia is specifically constituted to achieve the end-point of a planet in homeostasis) in his later publications.

The most notable discussion of Lovelock's ideas in terms of a consideration of landscape is by van Breemen (1993a). For the sake of argument he divides soil properties into "favorable" (or pro-Gaia) and "unfavorable" (or anti-Gaia). In his usage a favorable property as one that "helps to increase the net primary production on a definable part of the landscape with a more or less uniform vegetation, of a size in the order of $10–10^3\,m^2$". Favorable properties are found for example in a soil (such as a loamy textured Luvisol) with a high inherent fertility, and a structure that includes a heterogeneous system of interconnected pores. A structure of this kind will simultaneously provide good anchoring for roots, good water-holding capacity and good aeration. Unfavorable properties occur in soils with little rooting-depth (Regosols and Leptosols), a texture conducive to excessive drainage and droughtiness (Arenosols), heavy soils subject to waterlogging (Vertisols), and soils developed in extremely cold (Cryosols) or dry (Solonchaks, Solonetz) environments.

There is no conclusive evidence that natural soil-forming processes are Gaia-directed in any way to lead towards favorable properties and thus to a soil particularly comfortable and hospitable towards life on Earth. The fact is that the land surface is constantly modified by weathering, soil-forming and soilwasting processes to produce a kind of dump of natural wastes that is in a state of continuous recycling. Life on Earth has found this to be collaterally useful, and has evolved into a "best fit" to the properties of the resulting soils. Volk (2002) convincingly develops this Darwinian explanation for the comfortable look that life has in a landscape, and as van Breemen (1993b) says, natural selection explains everything without recourse to teleology.

Consequently, the state of a landscape, whether it is considered its quality, to use a neutral term, or its health, to use a metaphorically loaded one, is perfectly well explained without the invocation of a mystical Gaia. The co-evolution of biotic and abiotic components of the Earth's surface, under the influence of natural selection, is a completely adequate explanation.

2.3 Society, Culture and Civilization

These are amongst the most difficult words in English to pin down. The difficulty arises not only because there is considerable overlap between them, but because each has an abstract meaning in addition to a concrete one. Society is the obvious place to start since it subsumes the other two.

Society is clear in two main senses says Raymond Williams (1983, p. 291) "as our most general term for the body of institutions and relationships within which a relatively large group of people live; and as our most abstract term for the condition in which such institutions and relationships are formed."

The word society has a long pedigree coming via Norman French from the Latin 'societas' with the connotation of community, companionship or fellowship. By the sixteenth century it had acquired the meaning of a group of people living together in a country or region, and sharing customs, laws or institutions. This is where there is considerable overlap with culture, again from a Latin root, 'cultura', meaning cultivation or tending of land, though including also the cultivation of the spirit (Williams 1983). In several western European languages it developed to connote the cultivation of the intellect, and the arts in particular. By the nineteenth century the word had acquired the meaning of a society in a particular place and time, characterized by a common language, distinctive ideas, customs, social behaviour, artefacts, and general way of life.

As culture is nested within society, so civilization is nested within culture. The basic characteristic of civilization is a society at a particular stage of complexity. For example the OED says that civilization is "a developed or advanced state of human society". One implication is that at a certain level of complexity a society grades into a civilization. The direction that complexity takes is most readily defined in concrete terms. For example Richard Wright defines the term as "a special kind of culture: large, complex societies based on the domestication of plants, animals, and human beings … typically [with] towns, cities, governments, social classes, and specialized professions" (Wright 2004, p. 33).

All three may be used as singular nouns to mean society, culture or civilization in general. From the late eighteenth to early nineteenth centuries they also acquired plural meanings so that it became possible to refer to Greek, Roman, Chinese or any number of societies, cultures or civilizations—tacit recognition that human communities differed from each other on a regional basis. This raises a question central to the objectives of this book: to what degree is human society conditioned by landscape. Or, another way of asking the question: to what degree is it possible to state that human society is deterministic in a geological, geographical or geomorphologic sense.

2.4 Determinism

To many historians and pre-historians determinism is a dirty word. A typical criticism is embedded in Northrop Frye's (1957) wise-crack that determinism is a fallacy in which "a scholar with a special interest in geography or economics expresses that interest by the rhetorical device of putting his favorite subject into a causal relationship with whatever interests him less". Jared Diamond's "Guns Germs and Steel", in which the broad patterns of human history and prehistory are determined by "biogeography, crop cytogenetics, microbial evolution, animal behaviour, and other fields remote from historians' training" (Diamond 1997), has recently reignited discussion of the topic. The book was received well by the general public and the author received a Pulitzer Prize for it but many scholars were highly critical of what they considered to be its deterministic approach. Judkins et al. (2008) is a typical critique and references earlier ones. William H. McNeill (1997) considers Diamond's book "a clever caricature rather than a serious effort to understand what happened across the centuries and millennia of world history", a "sort of geographical reductionism" that simplifies "the tangled web of recorded history to four natural processes". Three of the four are specifically landscape or physiographic attributes in the strict sense, while one (the first) is concerned with the Leopoldian extension of land to include the biospheric aspects:

a. availability of domesticable plant and animal species, since food production and the agricultural surplus is necessary for the support of non-farming specialists and large populations that might give a military advantage;
b. mountains, deserts and day lengths, varying with latitude and "affecting rates of diffusion and migration, which differed greatly among continents";
c. distances across open water, "influencing diffusion between continents";
d. continental differences in area or total population size.

Distilled to its essence McNeill's objection to Diamond's thesis is that it tends to rule out, or greatly diminish the role of what he calls 'cultural autonomy', that is, the "personal and collective behaviour shaped by shared meanings [that] distinguishes us from other species. It is the hallmark of humanity." McNeill says that Diamond ignores freedom of choice in favor of "the tyranny of natural environments".

McNeill is willing to grant that Diamond's type of determinism is indeed applicable to the early phases of the history of human society "when technical skills and organizational coordination were still undeveloped" and we were "closely constrained by the local availability of food". Now, however, "the vast differences in the wealth and power that different human societies have at their command today reflect what long chains of ancestors did, and did not, do by way of accepting and rejecting new ways of thought and action, most of which were in no way dictated by, or directly dependent on, environmental factors."

2.5 Adaptation

There are two meanings to adaptation that are applicable to the subject matter of this book: a general meaning and a specialist, biological meaning.

The more general one of the two is exemplified by a definition taken from the OED: the process of modifying a thing so as to suit new conditions. The "thing" here is the Earth's land surface, and the "new conditions" are those that result from human activities, requirements and desires. Interpreted this way, it is the reverse of determinism—the opposite direction of the road from landscape to humanity implied by the latter word.

The specialist meaning in modern biology comes from the Darwinian theory of evolution. Thus in ecology adaptation is the way that an organism, including *Homo sapiens*, adjusts to its environment in order to improve its chances of survival, The adjustments may include behavioural, physiological or structural changes to the organism, either singly or together.

In the case of human adaptation, physiological and structural change does not appear to have been particularly active since at least the appearance of the Cro-Magnons. Consequently any adaptations since the beginning of agriculture, and the origin of urban and industrial civilization, have been essentially behavioural. We have adjusted our behaviour to fit us for life on virtually every type of landscape, in all terrestrial biomes.

A specific form of adaptation (sometimes referred to as exadaptation) amongst human beings is the development of tools and technologies that act like prosthetic devices in enabling us to deal with our environment in ways that our physiology would otherwise not allow (Catton 1980). This is manifestly obvious in the tools we have devised to modify landscape to our needs, from the digging stick of the early farmer to the massive excavator of the modern civil engineer.

2.6 Sustainability

Sustainability is the property of being able to continue to support the existence of an entity such as society (Brown 1981). The basic concept is simple: a sustainable system is one that lasts.

Simple though the concept is, problems begin when we try to define how long a system must last in order for it to be labelled sustainable. Forever is not an option on a finite planet governed by the laws of thermodynamics, so we must set a pragmatic limit. The fact is that no human society has persisted in unbroken succession for longer than about a thousand years, so current attempts to devise systems of managing the terrestrial landscape in a way sustainable to the interests of *Homo sapiens* are probably doomed to failure. However, it is possible to speak of a society as living sustainability within its environment for a specified length of time—meaning that resources and wastes were managed adequately enough to allow the society to persist for that period. The point is that the term sustainability only makes sense in the real world when a time limit is imposed.

Our current power to modify the environment has now become problematic and the question of the sustainability of human society is of growing concern. We have more or less taken over a third of "human friendly" landscapes, and have notably modified another third. The temperate grasslands and the Mediterranean biome have been completely wrested from their original inhabitants, and the temperate forests are moving along the same path of human makeover. In commandeering the habitats of other species we threaten the integrity of the biosphere, our life-support system, and many voices in the ecological community consider that industrial civilization in particular, amongst human societies, is unsustainable (Rees 2008). Indeed, it has been said that the human species has been adapted to the needs of short term interests and is itself inherently unsustainable.

References

Brown L (1981) Building a sustainable society. Norton, New York, 433 pp

Catton W (1980) Overshoot. University of Illinois Press, Urbana, 298 pp

Chesworth W (1991) Geochemistry of micronutrients. Chapter 1 in "Micronutrients in Agriculture" (2nd ed.) Soil Science Society of America, Madison, Wisconsin, pp 1–30

Chesworth W (1996) Agriculture as a Holocene process and its bearing on sustainability. In: Perez Alberti A, Martini IP, Chesworth W, Martinez Cortizas A (eds) Dinamica y evolucion de medios Cuarternarios. Xunta de Galicia, Santiago de Compostela, pp. 31–36

Crutzen PJ (2002) The anthropocene. Journal de Physique IV 12:1–5

Diamond J (1997) Guns, Germs and Steel. W.W. Norton, New York, 480 pp

Frye N (1957) Anatomy of criticism; four essays. Princeton University Press, Princeton, NJ, 383 pp

Judkins G, Smith M, Keys E (2008) Determinism within human–environment research and the rediscovery of environmental causation. The Geographical Journal 174:17–29

Kirchner JW (1989) The Gaia Hypothesis: can it be tested? Reviews of Geophysics, 27:223–235

Kirchner JW (1991) The Gaia hypotheses: are they testable? Are they useful? In: Schneider SH, Boston PJ (eds) Scientists on Gaia. MIT Press, Boston, pp 38–46

Leopold LB (1993) Round River: From the journals of Aldo Leopold. Oxford University Press, New York, 173 pp

Lovelock J (1979/2000) Gaia, A new look at life on Earth. Oxford. Oxford University Press, New York, 157 pp

Margulis L, Sagan D (1997) Slanted Truths: Essays on Gaia, symbiosis, and evolution. Copernicus, New York, 368 pp

McNeill WH (1997) History upside down. New York Review of Books 44:305–311

Millennium Ecosystem Assessment Project (2005) Ecosystems and human well-being: Synthesis. Island Press, Washington, DC, 137 pp
Rees WE (2008) Human nature, eco-footprints and environmental injustice. Local Environment 13:685–701
Sauer K (1925) The morphology of landscape. University of California Publications in Geography 2:19–54
Thomas K (1996) Man and the natural world: Changing attitudes in England, 1500–1800. Oxford University Press, New York, 425 pp
van Breemen N (1993a) Soils as biotic constructs favoring net primary productivity. Geoderma, 57:183–211
van Breemen N (1993b) Author's epilogue. Geoderma, 57:229–230
Volk T (2002) Toward a future for Gaia theory. Climate Change 52:423–430
Williams R (1983) Keywords: a vocabulary of culture and society. Fontana Paperbacks, London, 341 pp
Wright R (2004) A short history of progress. Anansi Press, Toronto, 211 pp

Chapter 3
Womb, Belly and Landscape in the Anthropocene

Ward Chesworth

"the encounter between womb and belly and earth and water"
Marvin Harris's (1980, p 40) definition of agriculture

The poet William Cowper, enraptured by the beauty of the English countryside, wrote "God made the country and man made the town".

He was wrong. Driven by the urgings of womb and belly we made them both. Agriculture, involving the increasingly large scale manipulation (husbandry) of the soil, is what brought it about, and it began with the Neolithic Revolution about 10,000–12,000 years ago. Since then, all the usual geological processes that shape the land surface "have been joined by a new and immensely significant one, never before seen on the planet, and one without which civilization would not exist: agriculture" (Chesworth 2002, p. 5). It was not the first technology available to humans for modifying the planetary surface. That was fire, first controlled by *Homo erectus* (Stearns 2001), possibly as early as 1.8 million years ago, though evidence for this early date (from Kenya) is controversial. Australia appears to have been transformed by burning soon after Homo sapiens arrived there some 40,000 years ago, and long before agriculture was invented (Head 2000). Similarly pre-agricultural peoples cleared forests in Western Europe by burning, and as a consequence created the acid, ericaceous heathlands referred to in the north of England as moors, maybe as early as 20,000 years ago (Mabey 1980). However, the human activity that ultimately led to the wholesale transformation of the land was undoubtedly agriculture, a process that will be examined in this chapter.

W. Chesworth (✉)
School of Environmental Sciences, University of Guelph
Guelph, ON N1G 2W1, Canada
e-mail: wcheswor@uoguelph.ca

I start with the historical context, then proceed to consider the three important ways in which farming has changed landscapes around the world. This leads to a discussion of our ecological footprint, how and why it grew so large, and whether it represents a sustainable state. At the end of the chapter, and with apologies to Shakespeare, I conclude that we bestride the landscape like an ecological colossus.

3.1 Historical Background

From the outset, the development of human society had a reciprocal relationship with the land. In the beginning, the nature of the land and its environmental ambience, dictated where farming developed and how it expanded. At the same time, farmers were modifying the landscape to suit their needs, though perhaps at the start, not particularly systematically. Gradually, humanity developed technologies and sources of power that have allowed us to an increasing degree to mould the land to our dictates. Figure 3.1 provides the historical and prehistorical context for this section of the chapter.

3.1.1 *The Neolithic Revolution*

The Neolithic Revolution brought about the greatest material transformation in human history. In spite of the adjective Neolithic (literally "new stone age"), "it is not stone working methods but rather the technology of food production that distinguishes this period

I. P. Martini, W. Chesworth (eds.), *Landscapes and Societies*,
DOI 10.1007/978-90-481-9413-1_3, © Springer Science+Business Media B.V. 2010

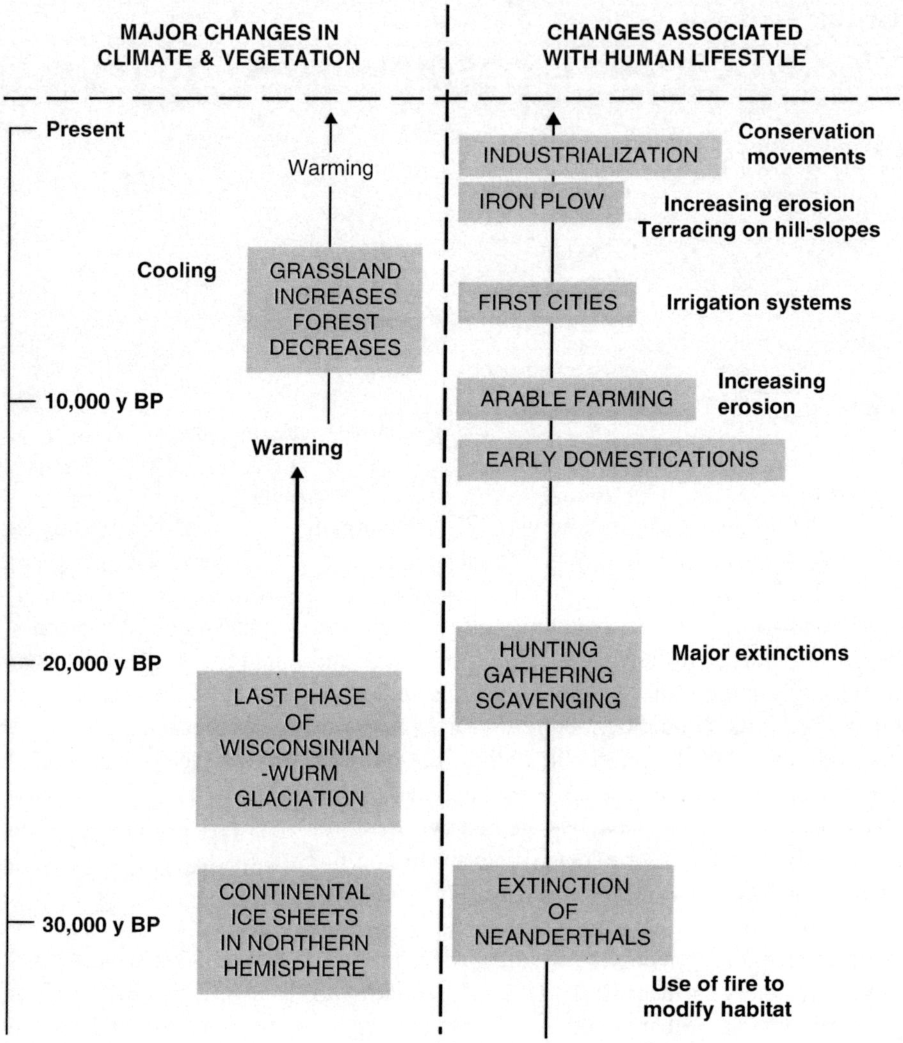

Fig. 3.1 The context of human-landscape interactions over the last 30,000 years. Control of fire goes back to *Homo erectus* and so has always been available to *Homo sapiens* to modify landscapes. (After Chesworth 2008)

from previous phases of human history" (Harris 1971, p. 174). Ultimately, by producing a surplus of food in excess of the needs of the farming families themselves, it had one momentous consequence: civilization. In effect, civilization and the written history of the human species are footnotes to the Neolithic Revolution (Chesworth 2008). Civilizations are "a special kind of culture: large, complex societies based on the domestication of plants, animals, and human beings … typically [with] towns, cities, governments, social classes, and specialized professions (Wright 2004, p. 33)." They bring with them "metallurgy, writing, the city and the state" according to Hobsbawm (1996, p. 15), and have "changed *Homo sapiens* from a rare to an abundant species" (Harris 1971, p. 53). In making these changes possible, agriculture opened "a radically new phase of human history" and is "perhaps the most basic of all human revolutions" (McNeill 1991, p. 40).

East (1924) states that soil, "common brown earth, is the limiting factor which shall determine not only the number of people the world can contain but also the comfort and therefore the final trend of their civilization." Aldo Leopold agrees, though significantly he broadens the emphasis from soil to land. In a talk that he gave to a meeting of the American Association for the Advancement of Science in New Mexico in 1933, he said "the reaction of land to occupancy determines the nature and duration of civilization" (Leopold 1933, p. 40).

Two points can be emphasized at this juncture. First, civilization owes its origin to the exploitation of the resources and processes of the landscape. Second, in cultivating the landscape, farmers may provoke changes that bring about the collapse of civilization. In other words, landscape is the alpha and omega of civilization.

Ideally, the landscape requirements of agriculture are a soil of reasonable depth, half a meter say, with a high inherent fertility, and the capacity to maintain, store, and deliver water and soil nutrients to plants. Storage and delivery requires the presence in the soil of components capable of holding nutrients in readily available form (that is, a form easily mobilized into the aqueous phase). Organic matter, and the right type of clay mineral play this role, and weathering and the erosional cycle are the natural maintenance systems of soil fertility that the early farmers depended upon. Strip away all the elaborations of modern, industrialized agriculture, and these are the basics revealed underneath.

A characteristic locality chosen by the earliest farmers was essentially a landscape at base level, a situation that would tend to minimize erosional losses of soil. A lake, wetland or the lower reaches of a river, were typical, having as Sherratt (1980) points out, damp, silty, and (commonly) easily worked soils. At this stage the footprint on the landscape was punctual and miniscule, being confined to localized sites in highland regions such as the belt stretching from southern Turkey, through northern Syria and Iraq to the Zagros Mountains that mark the western highland rim of the "Fertile Crescent" (Jarrige and Meadow 1980). The Neolithic village of Çatalhöyük in southeastern Turkey is typical (Hodder 2007). Not only does the soil tend to be young in these earliest agricultural regions (since mountain slopes are prone to erosion) and fertile (because of the continuous exposure of fresh parent material in the erosional cycle), but orographically controlled rainfall is adequate to sustain a crop.

After considerable trial and error in the highlands, farming was introduced into the valleys of the Tigris and Euphrates, where extensive areas of soil with a high inherent fertility (river alluvium derived from the erosional material washed down from the mountains) occur. The rivers themselves delivered water more reliably than the serendipity of rainfall, and the resulting surplus of food became large enough to sustain a non-farming population, which allowed the development of those cultures that we call civilized.

The earliest civilizations in the old world, in the valleys of the Tigris and Euphrates, the Nile, Indus and Huang Ho, are all based on large river systems, draining an area where freshly exposed and lightly weathered geological materials provide nutrient-rich sediment for distribution lower down the valley. There they weather predominantly to fertile Fluvisols. Because of the association with rivers, they have been referred to as hydraulic civilizations, the earliest in the world being Sumeria, at the head of the Persian Gulf in what is now Iraq. As with all civilizations since, it was based on the farmers' surplus—allowing the support of elites, a division of labor, and a stratification of society, that led to a rise of urban and administrative centers (cities in other words) as well as trade with neighboring peoples for raw materials that Sumer lacked (Crawford 1991).

Sumeria is considered to be the site of the fabled Garden of Eden with the name Eden believed to be cognate with the Sumerian word Edin. This referred to the wild uncultivated grassland of southern Mesopotamia. Taking over grassland for the purpose of farming, is a feature that runs throughout human history. By now, virtually all the major temperate grasslands of the world such as the prairies of North America, the steppes of Eurasia, and the pampas of Argentina have been commandeered for agricultural purposes.

Irrigated landscapes are a characteristic of the hydraulic civilizations. Irrigation allowed cultivation to extend into the arid lands that bordered the river valleys. Consequently the agricultural footprint became much more extensive and less punctual than in the highland sites, and collaterally, soil erosion and irrigation-induced salinization became a significant feature of the human footprint on the landscape.

The subsequent spread of civilization away from the great river valleys of the old world carried it into areas such as southern and western Europe, where atmospheric precipitation was reliable and irrigation unnecessary. The farmers migrating from the Levant brought their techniques, crops and domesticated animals into Europe at the rate of about 1 km/year (Sokal et al. 1991). The diffusion was made easy by the absence of north-south barriers to latitudinal migration, while moving crops in an essentially east-west direction minimized the amount of climatic variation to which the crop varieties needed to be adapted (Diamond 1997).

3.1.2 Expansion into Europe

Two routes brought agriculture to the Atlantic coast—one via the Mediterranean that exploited humid coastal

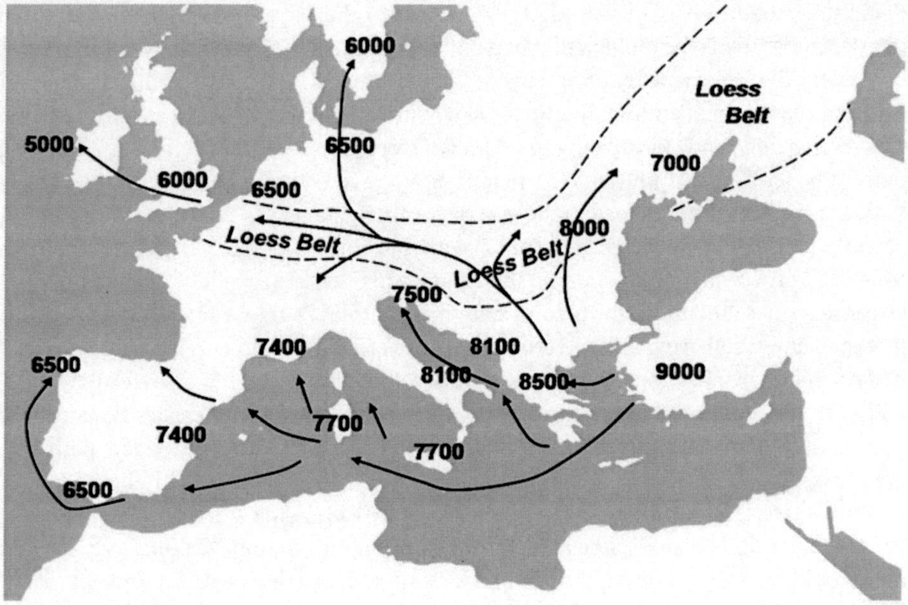

Fig. 3.2 The Neolithic spread of agriculture into Europe showing two major routes to the Atlantic Coast from the Levant—east to west across the Mediterranean, and a more northerly route along the central European loess belt. (After Renfrew 1994)

localities, commonly near river mouths, with damp, silty soils, the other via the soils of the alluvial plain of the Danube valley and the loessial deposits of central Europe (Fig. 3.2). These were light enough to be worked by hoe, ard or wooden plough, while the heavier, clay-rich soils of the river valleys remained off-limits until the invention of the iron plough harnessed to draft animals. This technology was introduced between 2000 and 2500 years ago, and remained at the cutting edge of cultivation until the eighteenth century. Given an appropriate climate, it opened up new landscapes for exploitation, and given a reasonable climate, the main restriction on a soil became drainage and hydrodynamics, rather than ease of husbandry.

By about 7000 years ago, the distribution of the Neolithic farming population around the Mediterranean must have had the aspect of the much later scatter of Greek colonies and trading posts that Plato described as resembling frogs sitting round a pond. Pioneer cultivators settled in those locations easily accessible by boat, such as river mouths and adjacent valleys where warm, wet and easily worked soils were to be found. Here, the earliest sites of arable farming were all close to the coast, at elevations below the present 200 m mark, though it is important to note that there has been a sea level rise since the early Neolithic and some sites may now be on the seabed. In Spain, the Meseta, would have been left to non-farming populations as being too cold for the unadapted package of crops brought from the Levant. Evidence from pollen and lithogenic dust in the ombrotrophic Pena da Cadela bog in the Xistral Mountains near Lugo, Spain, indicates agricultural activity in the northwestern region of the peninsula by about 6500 years ago (Martinez Cortizas et al. 2005). In other words, farming appeared on this part of the Atlantic coast some 1000 years later than on the Mediterranean coast of Spain.

Further north, agriculture reached the Atlantic slightly later as the history of farming in the British Isles will show.

3.1.3 Synoptic History of British Agriculture

For this part of the narrative I will develop the template provided by Evans (1992). Britain as a whole was typical of Atlantic Europe in being largely wooded in its pre-Neolithic state, though there is evidence of the Paleolithic use of fire to clear the land, possibly as a means of controlling the movements of wild animals hunted for food. The first farmers entered the island from across the North Sea between 4000 and 5000 years ago. Early sites, for example along water courses on the Cretaceous chalklands of southern England, and on silty soils around the wetlands of Holderness in Yorkshire, conform to usual requirements: a high water table with damp, light soils, workable by rudimentary equipment. The chalklands of the south appear to have been cleared the earliest, and the anthropogenic erosion that began then has since accounted for the removal of a cumulative 15–25 cm of soil, washed from the higher ground into what are now dry valleys. In Holderness the early farmers colonized sites north of the Humber River, though elsewhere, in what are now the North Yorkshire moors, clearance of trees by fire produced an acid, heath-like landscape, unsuitable for arable agriculture, on which podzolic soils, commonly gleyed, evolved out of the original forest soils (luvisolic or umbrisolic).

Under Roman occupation, population in England was about 4,000,000. Agricultural production intensified with autumn planting of wheat, with the result that the soil was unprotected during the stormy winter months. More benign Roman introductions were manuring and marling, the latter, like the former, serving to improve the fertility of soil, while also correcting excessive drainage and acidity in coarse grained soils. A lasting effect of marling, a practice that continued into the twentieth century, can be seen in Cheshire in the Midland Plain. Water filled marl-pits litter the landscape like post-glacial kettles. They are distinguishable from the latter by having a steep bank where the marl (calcareous clay) was excavated, opposite a gently sloping bank down which a cart could be backed and loaded. A more occult change is also found where sandy soils (in the region of Delamere Forest for example) were cleared and marled for centuries. Illuviation of the fine grained component of the marl has produced a clay-rich B horizon, locally known as the foxes bench. On a profile scale this has an important influence on near-surface water movement, in places generating pseudo-gley, a subtle example of the way in which human activities may alter local hydrology.

When the Romans withdrew, agricultural productivity diminished considerably as indicated by the fact that erosion was greatly diminished. Population dropped to about a half, then gradually grew until the Black Death in the mid 1300s. By this time there was no extensive forest left in southern Britain. Fallowing had now been introduced to deal with problems of soil fertility. This practice leaves fields vulnerable to erosion until a cover of vegetation has been established. After the plague, population was down to about 2–2.5 million, again with a consequent decrease in erosion. From 1500, population began to build again and after 1750, with the prosperity brought about by the agricultural and industrial revolutions, it soared. The introduction of the turnip and the four-fold rotation from Holland, as well as improvements in ploughing and drilling (Timmer 1969) were the principal innovations, though enclosure of the land, begun much earlier, was the main protection against the erosion caused by cultivation. At this time, agriculture in England may have come closer to being sustainable that it has been before or since. Jaquetta Hawkes (1953, p. 162) compared twentieth century unfavorably with eighteenth century agriculture in this regard: "The struggle of 200 generations of cultivators had its culmination in the high farming of the eighteenth and early nineteenth centuries. Now those thousands of years of wooing fertility under the sun and rain were to be half forgotten in a third way of living which resembles the first, that of the hunters in its predatory dependence on the natural resources of the country."

In the late eighteenth century, parish records reveal another significant drop in population with the usual knock-on effect on the landscape, of decreasing the amount of erosion. Witham and Oppenheimer (2004) have assembled documentary and proxy-climatic data that implicates the volcanic haze of acid, toxic aerosols from the Laki Crater eruption in Iceland in 1783–1784. They claim that this fumigated Britain (and much of Western Europe) during the notably hot 1783 English summer and the subsequent severe winter.

An alteration to the landscape that had begun much earlier, reached its culmination in the eighteenth and early nineteenth centuries—the practice of enclosure; that is, of enclosing farmland by means of hedgerows. Ultimately, the open field agricultural landscape which had been typical since Anglo Saxon times became the patchwork quilt of smaller fields common until the twentieth century. Enclosure caused great social upheaval amongst the rural population since it excluded many of the poor from land that had once been held in common. Physically, its effect was to cut back on the modification of the landscape by wind erosion.

In the eighteenth and nineteenth centuries improvements in communication by canal and later by rail, gave the farmer access to distant markets, and farming, which had become most intense in the vicinity of cities, then intensified in areas outside the urban catchments. In addition, the steam engine vastly increased the farmer's ability to remodel the landscape and to work heavy, clay-rich soils. The Fowler method used a steam engine to haul ploughs by cable across a field, and could break up about six acres of ley, compared to horse drawn ploughing which could only manage one (Fussell 1976). Water-logging remained a problem for clay-rich soils however, until the development of tile-drains in the second half of the nineteenth century (Fussell 1976), a technology which invariably leads to an increase of clay-movement, and the turbidity of watercourses that collect the drainage (Stone and Krishnappan 1997).

In the twentieth century increased erosion was encouraged by socio-political decisions. In Britain in the 1920s for example, the Ministry of Agriculture encouraged the farmers of East Anglia to grow sugar beet, a crop that provides poor protection to the soil. Both the first and second world wars forced the Government to encourage increased food-production in the interests of self-sufficiency. The result was to push arable farming onto marginal soil (for example onto the hilly flanks of the Pennines) with a consequent increase in the erosion of agriculturally marginal soils. More recently, the agricultural policies of the European Union have affected the British landscape, not only in determining what crops might be grown, but also in whether farmland would be kept in production at all. Now, the new pressure on the landscape is coming from the decision of the EU to produce 10% of the gasoline and diesel needs of the Community by 2020, in the form of biofuels (Schnepf 2006).

So-called market forces created a greater problem of erosion after about 1960, when a more intensive, monocropping form of industrial farming became common. This practice is made possible by the application of artificial fertilizers, essentially an invention of the nineteenth and early twentieth centuries, which slowly replaced older techniques of manuring and rotation of crops. Hedgerows were now removed to facilitate the use of combine harvesters and other heavy machinery. It is more than a little ironic that the removal of hedgerows should cause problems, in view of the fact that their introduction during the enclosure movements of former times was itself attended by considerable social upheaval. At the present day, water erosion, resulting from the planting of autumn cereals, has replaced wind erosion as the principal landscape problem in the United Kingdom.

A future development in British agriculture that is likely to have its effect on the landscape is an increased use of irrigation. Irrigation became common in some areas after Word Ward 2. Currently 60% of irrigated farms and market gardens are in eastern England, which has always been the drier half of the country. An even drier climate has been projected for this region in the future, and a greater need to irrigate land is therefore foreseen (Knox et al. 2000).

3.2 Landscape Change and the Soil: Anthropic Change in the Landscape

On the evidence of the foregoing section it is clear that the conversion of landscape from its pristine to its current, human-dominated state involves many processes. Farmers clear, till, fertilize, irrigate and harvest, thereby changing and remolding the land to a degree that is irreversible at anything less than the geological long term. Land-cover conversion, land degradation and land-use intensification are the main processes of change (Lambin 1997) and our impact has been amplified throughout history by technological progress, associated especially with the exploitation of fossil fuels in the last 200 years. These three processes are dealt with below.

The decisions and policies of ruling elites, as well as natural disasters, have also added a serendipitous element to the changes that have taken place. Above all, the needs of the human population, growing exponentially until about a decade ago, has pushed agricultural changes onto at least two thirds of the farmable landscapes of the world. Our contribution to change at the planetary surface has become significant enough that the last 10,000 years such that Crutzen's (2002) term Anthropocene may be extended to cover that period.

With the urban population now about 50% of the total number of human beings on the planet, it may seem that equal weight should be given to urbanization as a landscape modifying factor. But the amount of the landscape that is urban is less than 2% of the total, while only a tenth of that can be considered densely built-up with a strong direct impact (Meyer and Turner 1992). Even so, the ecological footprint of a large city can amount to hundreds of times the area of the city itself, so it's ultimate effect on the landscape as a whole is very great. However, most of that footprint is accounted for by agriculture and associated distribution systems that are necessary to feed the urban dwellers.

3.2.1 Conversion of the Land Cover

On entering new territory, farmers do what they have always done. They remove existing vegetation as the first stage of preparing the land for a crop. Removing the original vegetation from a landscape is a direct attack on the integrity of the natural biome, which may have persisted before human intervention for thousands of years. A relatively stable biome (state of biostasis: Erhart 1964) is replaced by an ephemeral state with vegetation changing annually. Currently, the most active locus of change is in the tropical forests of the world, where FAO (2005) estimated a conversion rate of 15.5 million ha/year for the decade 1981–1990.

The most complete conversions have taken place in three major areas of the temperate zone: the grassland and forest biomes, and the much smaller Mediterranean biome. As stated earlier, conversion of grassland goes back to the first large scale agriculture in the world's oldest civilization, Sumeria, and has since become a major theme in human history. Grassland is the natural vegetation of the drier parts of the loess deposits of the planet, with forest occupying the wetter zones. The grassland plains of the northern hemisphere are now crop and range landscapes, while the forest in Europe and eastern North America has been largely converted to arable land. The Mediterranean biome in its type locality of southern Europe, the Levant and North Africa, has been transformed over 8000 or 9000 years into a completely anthropic landscape, and areas of the biome in California, Chile and southwestern Australia have proceeded substantially along that route within the last 200 years.

Table 3.1 An estimate of the degree to which land has been converted from its original state. (Data from FAO 2005)

	Undisturbed (%)	Disturbed (%)	Human dominated (%)
Europe	15.6	19.6	64.9
Asia	42.2	29.1	28.7
Africa	48.9	35.8	15.4
North America	56.3	18.8	24.9
Latin America & Caribbean	62.5	22.5	15.1
Australasia	62.3	25.8	12.0
Antarctica	100.0	0.0	0.0
World	51.9	24.2	23.9
World minus area of Rock, Ice & Barrens	27.0	37.0	36.0

The Mississippi watershed is a good example of the changes that have taken place in the grassland biome over the last 150 years or so. It has become the breadbasket of the United States and even of the world via food aid programs. As with other major temperate zone grasslands, the inherent fertility of the soils comes predominantly from loessial parent materials, commonly reworked by fluvial processes as in northern China, and along the Danube in Europe. In the eastern part of the Mississippi watershed, where rainfall is generally above 500 mm/year, the natural vegetation was the tall grass prairie. This has been almost completely converted into fields of wheat, corn and soya. Further west, as far as the hundredth meridian, agriculture expanded rapidly during the latter half of the nineteenth century once the railroad was available to bring produce to market. Beyond the hundredth meridian, in arid land that John Wesley Powell recognized as being unsuitable for the agricultural technologies developed in the eastern states and Europe (deBuys 2004) farming has none the less expanded as a consequence of irrigation technologies. Even the desert biome has become a farmed landscape, thanks particularly to exploitation of the Ogallala and other aquifers.

The tropical grassland, or savannah, has escaped the wholesale conversion to human use suffered by the temperate grassland, but it has not been totally immune. As rangeland, it had been used in the Sahel zone of Africa from time immemorial by nomadic herdsmen following seasonal changes in the rain belt, more or less sustainably. Indeed, part of the savannah (for example in the 'cerrado' of Brazil) may itself be anthropogenic, an artefact of the removal of forest by burning (Mistry and Berardi 2006).

Conversion of arid and semi-arid landscapes to farmland has also been a constant theme in agriculture almost from the start, beginning with the great hydraulic civilizations in the old world. In modern agriculture it has made California the horticultural capital of North America, and has turned the arid region south of the Aral Sea into cotton plantations. I will return to this subject under the next heading.

3.2.2 Land Degradation

The term land degradation applies to landscapes that have already been converted to human use. The implication is that the productivity of the agro-ecosystem has been diminished by human activities. Soil erosion is the main process of degradation, with other significant processes being acidification, salinization, organic matter loss, nutrient decline, compaction, and waterlogging. Agricultural soils are particular prone to degradation during severe weather events as the Iowa floods of 2008 emphasized.

Tilling the soil, breaks up its natural coherence so that it becomes more easily moved by wind and water. A pulse of eroded soil can be detected wherever the beginnings of agriculture in a region can be reliably dated. In Central America for example, soil erosion contributed to lake sediments in Mexico from the earliest agriculture there (Butzer 1993; O'Hara et al. 1993). Environmental degradation did not begin with the Spanish therefore, in spite of earlier claims to the contrary. Even so, this is not to claim that the beginnings of agriculture in Spain were any different. As Martinez Cortizas et al. (2005) show, the first farmers in northwest Spain left their mark in eroded soil added as windblown sediment to an ombrotrophic bog some 5000 years ago. Over millennia the changes to the landscape may become immense. In the last 5000 years the coastline of the Persian Gulf for example has advanced seaward by some 200–300 km. Here, the farmer has aided the natural erosional cycle by deforestation in the Taurus Mountains as well as by disturbing the soils of the Tigris and Euphrates valleys with an annual cultivation.

In 'Dirt: The erosion of civilizations' David R. Montgomery has a useful discussion of land degradation in ancient Greece (Montgomery 2007). Plato (427–347 BC) believed that the hills around Athens had been deforested, and the landscape became only a skeleton of its former self. Both Plato and Aristotle (384–322 BC) believed that farming during the age of Homer (the Bronze Age in modern terminology) was to blame. Modern dating essentially confirms this. In the Argolid for example, a first pulse of erosion marks the introduction of farming between about 6500 and 5500 years ago. Between 4300 and 3600 years ago, the plough was introduced and population growth pushed agriculture onto steeper slopes, producing another pulse of erosion. The need to feed a growing population is a common driver of landscape change, and when the Romans came with their improved agriculture and consequent population growth, another major episode of soil erosion occurs. In total, since the beginnings of cultivation as much as a meter of soil may have been stripped from lowland parts of the Argolid landscape (Montgomery 2007). By comparison, the rate of soil formation in the humid temperate zone is generally between 2 and 3 cm per 1000 years, say between 15 and 20 cm since farming began in the Argolid. In other words the Greeks have been losing soil there two orders of magnitude faster than it forms.

Land degradation is most severe in semi-arid regions, where wind erosion and irrigation-induced salinization again, are the common problems (Lambin 1997). Together with loss of biodiversity, the overall effect has been described as desertification (Binns 1990). It is estimated that 69.5% of the world drylands are affected by various forms of land degradation (Dregne et al. 1991). In recent history, the classic example was the Dustbowl of the American Midwest in the 1930s. There, silty soils in Oklahoma, Texas, and Kansas, were exposed to wind erosion in an area that suffered periodic drought. The problem was exacerbated by the increased power of ploughing technology—steam traction was introduced to the region in the early twentieth century.

The US Department of Agriculture (USDA 2008) classifies land in terms of an erodibility index: the ratio of inherent erodibility to the soil loss tolerance. Inherent erodibility for a given soil is the rate of erosion (tons per acre per year) that would occur on land that was continuously clean tilled throughout the year. The soil loss tolerance, or T value, is an estimate of the rate of soil erosion that can occur on a given soil without significant long-term productivity loss. Thus, the erodibility index captures both the propensity of a soil to erode and the potential for damage from erosion. Land can be highly erodible based on potential for waterborne erosion, wind erosion, or both. A little more than 100 million acres of US cropland are highly erodible, that is about 25% of all cropland. Highly erodible land is a special category with an erodibility index of eight or larger.

Many techniques have been employed to deal with the problem of enhanced soil erosion due to farming. The most successful is to re-establish a permanent land cover. Reforestation is the common technique and is undoubtedly successful though aesthetically the new plantations tend to look regimented and to lack diversity. The Azores provide a different solution to the problem. They were uninhabited until the fifteenth century when they were discovered by Portuguese mariners a few decades before the first voyage of Columbus. The islands were lush and wooded, covered in deep, fertile soils developed on volcanic ash, and the Portuguese settlers did what farmers do on entering virgin territory, they chopped down trees to make way for crops. By the early twentieth century they had essentially removed all trees from the major islands and their crops had become a significant source of food for the mainland. Erosion by streams and wind was gradually carrying the soil into the sea, and the decision was made to switch from crops to animals and arable land was converted to pasture. Grass is as good as trees for anchoring soils, so this was a successful conservation measure. However, as usual there were unintended consequences. The wastes produced by a large population of cattle has became so great that the eutrophication of surface waters and the nitrate contamination of ground waters has become a problem.

In fact, the addition of nitrogen into the natural nutrient cycle is a problem worldwide. The numbers are truly astonishing. 140 million tonnes/year of N is generated in the soil naturally. Human activities add 210 million tones each year—50% more than the natural processes. In the worst cases we can follow the impact all the way to the ocean. The consequences are clearly seen at the mouth of the Mississippi where a 500 km dead zone has spread along the shore of the Gulf of Mexico. In other words agriculture is now affecting seascapes as well as landscapes.

Since tillage is a major contributor to soil erosion, an alternative strategy in soil conservation is to farm without tilling the land. Unfortunately it requires an "agent-orange" approach to agriculture with the farmer spreading pesticides and other chemicals on the landscape in order for the crop to produce high yields. Crop varieties now need to be selectively bred to withstand the biocides designed to eliminate competition—a new twist on the survival of the fittest: the survival of the genetically engineered. Again, the technique has malign consequences with chemical contamination being the most obvious. When she warned about our excessive reliance on agricultural chemicals in 'Silent Spring', Rachel Carson was a voice crying in the wilderness (Carson 1962). Even today, her warnings are largely discounted by agribusiness.

The second common problem referred to in the exploitation of arid zone soils, irrigation-induced salinization was probably first encountered by the early farmers in Mesopotamia (Chew 2007). The problem is that in these regions evapotranspiration naturally exceeds water-input, so that evaporation causes the deposition of salts in the upper part of the solum. More water may then be added to flush the salts down to the groundwater. Essentially the same type of system invented in Mesopotamia whereby a system of distributary canals carries water into dry regions away from the riparian zone is employed in the south-western United States, along the Rio Grande in New Mexico for example, and salinization is still a problem.

Compared with the classical regions of the first civilizations irrigation is on a massive scale in modern farming. A major difference is that in addition to surface waters, ground water is also exploited, much of it being fossil water and non-renewable. Consequently the agricultural landscape of the dryer regions of the USA, China, Pakistan, India, North Africa and Australia can only be considered as transient phenomena in human history. Induced salinization is endemic to all these regions. Furthermore, drainage of saline groundwaters from irrigated areas may cause problems to landscapes further dowstream. In California for example, a major drain led to the Kesterton Dam, an artificial lake, which is now a toxic wasteland.

Climate change is bringing about a further complication. Rainfall is decreasing in desert areas worldwide: in the last 25 years there has been a 16% drop in the Dashti Kbir desert of Iran, 12% in the Kalahari, and 8% in the Atacama (Ezcurra 2006). In addition, much of the irrigation water used in the Americas and central Asia comes from rivers fed by mountain glaciers and snow-packs. Accumulating evidence indicates that this source is also shrinking on account of a decline in atmospheric precipitations.

The conclusion must be that irrigated farmland is a particularly unsustainable type of anthropogenic landscape, and the danger is that since we have become increasingly dependent on food from irrigated land, future policy makers will look at ways of increasing the ecological footprint of the dry-land farmer by bringing water from ever more distant sources. In the case of agriculture in the American Southwest, this could threaten the integrity of the Great Lakes by water diversions out of the catchment area. Eco-disaster by water diversion, as for example in the case of the Aral Sea, is a cautionary lesson that needs to be emphasized.

3.2.3 Intensification of Land Use

Intensification of land use is normally associated with a need to grow more food, a need that may be driven by such factors as population growth, market demand, the drive to become self-sufficient as in time of war, or by political decisions in support of strategies of advantage to a particular social group or groups (Blaikie and Brookfield 1987). Intensification may be brought about by many mechanisms amongst which are introduction of new crop varieties, fertilization, irrigation, drainage and a general improvement in husbandry techniques (Netting et al. 1993). All of these were involved in the intensification of agriculture known as the 'green revolution', introduced into the developing world in the 1970s.

Often, intensification takes place in regions of material poverty, where population pressure has pushed people onto marginal or unsafe land. Low lying estuaries are a case in point, with Bangladesh being a prime example. Sometimes the motive is profit, rather than subsistence. The insatiable desire of the affluent world for sea-food for example has made many farmers in Southeast Asia rich. Mostly it involves the conversion of the mangrove habitat into shrimp farms. In the southern Thailand between 1975 and 1993, the mangrove habitat decreased by about a half (from 312,700 to 168,683 ha) as a result (Barbier and Sathirathai 2004). The long term effect is to expose sulphides in the subsoil of the mangrove regions, to weathering and the formation of thionic soils. These have soil waters with a pH of 3 or less, which mobilizes aluminum and produces an essentially sterile coastal landscape. One deadly consequence of this is that protection of coastal communities from storm surges is largely removed—witness the carnage of the 2004 tsunami in the Indian Ocean.

3.3 Discussion: The Human Ecological Footprint on the Land

Figure 3.3 is a snapshot of the effect of humanity on soil at present. It has been called an agrobleme (Jackson 2004), literally an agricultural scar, or since it is our total ecological footprint on the land, including the cities and the complex infrastructure of modern society, anthrobleme or human scar (Chesworth 2008). The term ecological footprint was invented by William Rees and his co-development of footprint analysis is described in Wackernagel and Rees (1996). It constitutes the area of the biosphere, including both terrestrial and aquatic ecosystems, required to produce the resources that a population consumes and to assimilate the wastes, or at least a sub-set of the wastes, that a population produces. In the form of maps such as Fig. 3.3, it summarizes the direct connection between human beings and the landscape—Marvin Harris (1971) encounter of the landscape with our reproductive prowess and our need for sustenance ("womb and belly" in his words). The size of our footprint at this time is a consequence of the growth of the human population. This is a response—with feedback—to the success of agriculture in providing a growing food supply. In turn, this was made possible by farming populations colonizing new soils and increasing their productivity—our main strategies for keeping ahead of Malthusian disaster since the Neolithic, and especially since the Industrial Revolution.

Before cities developed, the footprint was not much bigger than the physical area needed for the farm itself. As it is in peasant farming at the present time, that minimal area would have been augmented by an area of non-cultivated land used for forage and bedding for

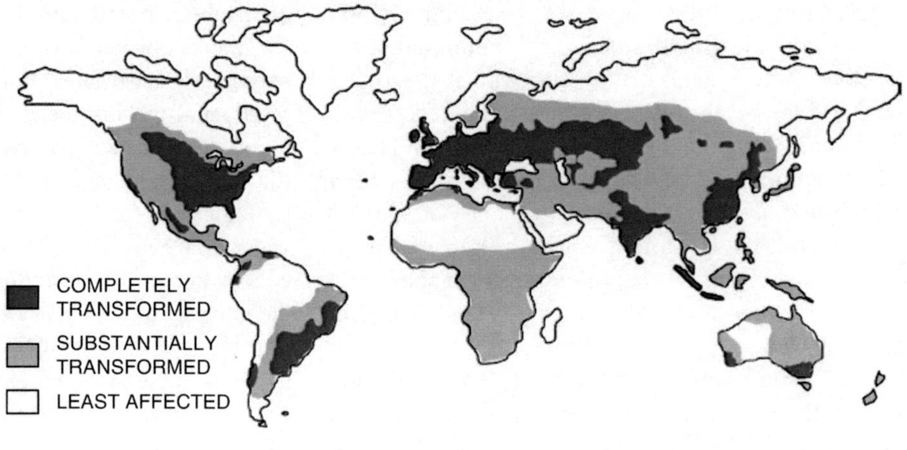

Fig. 3.3 The human ecological footprint or scar (anthrobleme) on the landscapes of the Earth. It is largely a product of the last 10,000 years. (After Chesworth 2008)

farm animals, for collection of fuel and wild foods and for the hunting and trapping of bush meat. Judging by the practice of peasant farmers in the Himalayas (personal observation during field work in Nepal from 1986 to 1989), and by small-scale farmers in northern Spain (Felipe Macias, personal communication), at least 3 ha is used to support 1 ha of farmed land. Borgstrom (1972) calls the extra area of land needed to support the area actually farmed, 'ghost acreage'. Estimating ghost acreage for a Neolithic farm can be no better than a back-of-the-envelope calculation at present, but we might start with the smallest area a subsistence farmer can survive on at the present day. Rees (2004, p. 90) notes that "the area of world-average cropland used to produce the diets of today's high-income consumers can be as high as 1.5 ha (3.7 acres) per capita, typically four to eight times the cropland required by the poorest of the world's poor." Consequently, as a generous estimate, I will take half a hectare as the area of a Neolithic, subsistence farm. If now we take the farming population to have reached about 5,000,000 at some time in the early Neolithic, say about 7000 or 8000 years ago (Cohen 1995), a minimum cropped area of 2.5 million ha, backed up by a possible 7.5 million "ghost" hectares would be implied. This amounts to 10^7 ha as a very approximate figure for the agricultural footprint at that period. In the world today, about 1.5 billion ha of land are under cultivation (WRI 2008), which means that the Neolithic agricultural footprint could have been no bigger than 0.01% of its present value.

The current size of the anthrobleme has involved a virtual takeover of two biophysical units of the biosphere, a major one—the temperate grassland biome—and, in terms of area, a minor one—the Mediterranean biome. In addition, the temperate forest biome has also been largely commandeered for our use (Chesworth 2008). Essentially, we are capable of making all terrestrial biomes the habitat of *Homo sapiens*. As H. G. Wells (1939, p. 16) remarks, we are a species "who will leave nothing undisturbed from the ocean bottom to the stratosphere".

3.3.1 Sustainability

Jackson (2004, p. 89) says "sustained life is a property of an ecological system rather than a single organism or species". It is worth adding that agricultural productivity, however manipulated by humanity, is a product of the biosphere. Jackson also says that the base of the ecological system is soil, which "for all practical purposes a non-renewable resource, is in serious decline". The implication must be that farming as currently practiced is an unsustainable activity, and that civilization which is totally dependent on agriculture is an unsustainable form of human society.

Since the Neolithic, agriculture has become an increasingly powerful forcing factor on processes at the earth's surface. It attacks the vulnerable skin of the landscape and routinely increases physical and chemical change by one or two orders of magnitude over natural values (and by even greater amounts in extreme cases). The farmer has always depended on renewable resources (sunlight, water, and seed for example), but also on the natural capital of the land, which is non-renewable on the scale of human generations. We have continually drawn down this natural capital since the Neolithic (Jackson 2004) by deforestation, soil erosion, dust storms, salinization, and desertification. These are only the more dramatic examples of change produced by our agricultural methodology, which "has had serious flaws in it from the start" (Martin 1975, p. 180). To begin with, the damage to the biosphere was sustainable because the agricultural population was small—a few million at most, 10,000 years ago. But, this powerful new method of food production set the human population on a growth curve that maintained an exponential rate of increase until the late twentieth century. We now overshoot any chance of long term sustainability by about 20% (Rees 2004).

As hunters and gatherers, our numbers were kept in check by the Darwinian competition that defines the biosphere. The domestication of crops and animals seemed to allow us to escape that fate and our inventiveness with regard to managing the fertility of soil, commanding inorganic sources of energy, and developing efficient transport systems, have kept us growing until now. This modern period of exuberant growth (Catton 1982) is what has led the present generation of neo-classical economists to believe that there are no limits to the human economy and it will grow for as far into the future as we can see (Simon 1998).

However, there are limits to the growth of the human economy Fig. 3.4, so we need to consider the likely outcome when a population approaches them, and Malthus is the reliable guide in spite of repeated claims to the contrary. To reiterate, he believed that

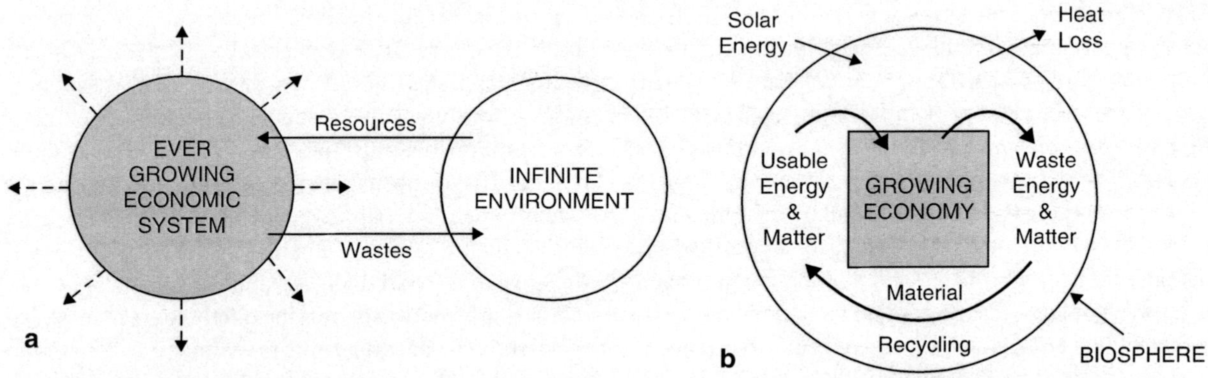

Fig. 3.4 Relationship between ecology and economy. **a** The "no limits" view of the classical and neo-classical economist. The environment is considered an infinite resource and waste receptacle—an "externality" that may be ignored in economic calculations. Consequently it is a model that contains no information about ecosystem structure. **b** The ecological model shows the economy nested within a finite biosphere. The economy can only grow at the expense of the biosphere. (After Rees 2004)

since population has the power to grow exponentially, while food supply tends to increase only arithmetically. Without controls the potential exists for society to be reduced to a level of subsistence—"misery" in Malthus' usage. In the limit, this could lead to collapse. Examples can be found over and over again in human history (Diamond 2005), and a cautionary example used by many scholars of the subject is provided by the history of Easter Island.

The original inhabitants of Easter Island probably got there by chance as part of the diaspora of Melanesian navigators who populated the South Pacific in the first few centuries of the Common Era. Because of the isolated nature of the place, the settlers were essentially marooned once they got there. That was fine to begin with because the island was well-forested, the soils were derived from volcanic ash, and were fertile. In addition there were lots of fish in the sea, and the Melanesian navigators had the requisite skills and equipment to hunt them. The new colonists cleared land for agriculture, built sturdy boats, built up a complex society, and prospered for a couple of hundred years. Eventually the population outgrew the resources of the island itself, and having chopped down the last tree, could no longer build their fishing boats. As a consequence a society that never recognized the limits of its existence, collapsed into barbarism and cannibalism (Tainter 1988).

The fact of limits should have been obvious from the great circumnavigations of the European Enlightenment. The earth is finite, and over the last 250 years modern science has revealed two important sets of constraint to emphasize the fact. First there are the conservation principles of matter and energy that tell us that we do not have access to infinite amounts of either. Second we are animals like other animals, shaped by a struggle for survival in a biosphere of finite resources. This struggle kept the size of our population in check until agriculture, especially in its fossil-fueled, modern form, gave us the ability to compete so successfully with other species that we could exclude them from their own evolutionary niches and take over. These constitute what Hardin (1995) refers to as the "default conditions" of human life.

By following a path of development that fails to recognize such important constraints we have appropriated, more or less completely, whole biomes to produce our food. In the process our footprint on the landscape has grown from a globally imperceptible one in the Paleolithic and early Neolithic, to the situation shown in Fig 3.3. Under the influence of agriculture, forest and steppe have given way to manicured farmland, rangeland, waste and desert, as population reaches 7 billion. Kennedy (1993) for one believes that Malthus prediction of 200 years ago is now becoming a reality on a global scale.

Clearly this is an unsustainable way to live on a planet, yet two areas seem to have managed the trick: Egypt and northern China. Both have farmed specific regions of the Earth's surface continuously for more than 5000 years, and we might conclude that if they can do it, we all can. However it was not accomplished by some great feat of human ingenuity, it was all on account of the contingencies of local geology. Egyptian agriculture has been sustained by the Nile providing a reliable source of water and a predictable annual flood to add the sediment that keeps the soils fertile. In essence, the fertilizer comes from the distant, nutrient-rich volcanic rocks, weathering and eroding in the Ethiopian Highlands. The early center of Chinese civilization is similarly supported by the waters of the Huang Ho and the fertile sediment of reworked loess, which it spreads over its flood plain, and which originates as wind-borne dust from the Gobi Desert.

In all but very few instances, the natural fertility exploited by early farmers was derived from geological source materials fresh enough to contain a reasonable store of nutrients that could be released by weathering. In addition to the two examples already given, societies in Mesopotamia, the Indus and Ganges valleys, and in the Andes amongst others, depended on soils developed on sediments derived from young fold-mountains. That only the two instances mentioned in the preceding paragraph have persisted as major agricultural producers may have many causes—some river systems are difficult to manage, some are affected by neo-tectonic activity, some by climatic changes—but there seems little doubt that all the early civilizations grew out of, and were supported by, the local geological environment.

Unfortunately there are not enough of these geologically and geomorphologically benign environments to go around and our attempts to engineer less benign environments (witness the disaster of the Aral Sea) have not proven to be sustainable. However, engineers are can-do people and will tell you that given enough energy any problem can be solved. This brings us up against limits again: the pool of energy available to do work is limited, and in keeping with the second law of thermodynamics, will continually diminish. In building our modern industrial civilization we have consumed most of the low-hanging, low-entropy fruit, especially oil and gas, the twin prime movers of human society since the Second World War. Indeed, regarding this last point, Bartlett (2004) has produced what amounts to a minimalist definition of the landscape. He says that land is what we use "to convert oil into food".

Hoyle (1964, p. 99) makes the point in this way: and there may not be enough left to allow a second kick at the can if the civilization we have achieved collapses. "It has often been said that, if the human species fails to make a go of it here on Earth, some other species will take over the running. … this is not correct. We have, or soon will have, exhausted the necessary physical prerequisites so far as this planet is concerned. With coal gone, oil gone, high-grade metallic ores gone, no species however competent can make the long climb from primitive conditions to high-level technology. This is a one-shot affair. If we fail, this planetary system fails so far as intelligence is concerned."

3.4 Conclusions

"Human social life is a response to the practical problems of earthly existence" says Marvin Harris (1980, p. 22), and our agricultural use of landscape is a response to the fundamental problem faced by any species, of finding enough to eat. Unfortunately, agriculture is strategically situated to attack the biosphere from within and since Neolithic times we have grown to the stage where we are drawing down biospheric and other resources by a margin of about 20% beyond the level that might ensure sustainability. Our footprint on the landscape is colossal and clearly involves the wholesale conversion of the habitats of other species, into our own. Inevitably the biodiversity of the earth is diminished, perhaps to the degree that we may provoking the sixth great extinction in geological history (Lewin and Leakey 1996; Larsen 2004).

Only by adopting the no-limits ideas of economists such as Julian Simon can we ever come to believe with 'The skeptical environmentalist' that "Food will get cheaper and ever more people will be able to consume more and better food" (Lomborg 2001, p. 109). There is a limit to the amount of land we can consume in pursuit of this rosy future and if we haven't reached it yet, we appear to be close. "The constraints of the biosphere are fixed. The bottleneck through which we are passing is real. It should be obvious to anyone not in a euphoric delirium that whatever humanity does or does not do, earth's capacity to support our species is approaching the limit" (Wilson 2002, p. 18).

Lomborg's "euphoric delirium" is based on a short term perspective of course—a perennial problem in the history of farming. Tudge (1985) observes that agriculture has never been seen as simply a means of feeding people. It has always been used to serve the immediate needs of generating profit, propping up ancient systems of fealty, or quantifying the importance of some chief in heads of cattle.

Long term sustainability requires a different cast of mind. We need to face the fact that the only response to our unsustainable use of planetary capital, that seems at all likely at this juncture, is collapse. Consider how we might approach the problem of engineering a soft-landing into sustainability. We could begin by basing our economics on ideas consistent with ecology as Daly (2005) suggests. We could reduce or in some cases eliminate our consumption of non-renewable resources. For example we might forego fossil fuels and use contemporary sunlight as our energy source, either directly or in its secondary manifestations as wind, water or tidal power. We could perennialize our grain-crops and convert them into nitrogen-fixing plants, thereby cutting down on soil erosion as well as minimizing the need for fertilizer (Jackson 2004). But no matter what combination of strategies we put together we find that a solution remains impossible because we have not solved the fundamental problem of the size of our ecological footprint caused by a growing population (Ehrlich and Ehrlich 2008).

There are two major reasons that make the population problem intractable. The first is the fact that for millions of years we have been selected to propagate. It's what we do as organisms, part of our hard-wiring. Over-riding this drive can perhaps be accomplished but history suggests that short of war, pestilence, a perpetual state of totalitarianism (*vide* the one-child-per-family program in China) or natural disaster, planned population control is essentially impossible. The second difficulty is the prevailing attitude encouraged by major religions. Believers have been taught by popes and ayatollahs to consider birth-control sinful, and a bar to a blissful afterlife in paradise.

The fact remains, a major adjustment to human society will take place. Viewed dispassionately, it appears more likely to be in the nature of a collapse than otherwise. No doubt we will continue to hope for the best—another one of our human traits—and dream of, and strain every sinew to achieve, a sustainable future for humanity. Successful or not, "sustainability is the greatest collective exercise the human race will ever have to undertake" says William Rees (2007).

References

Barbier E, Sathirathai S (eds) (2004) Shrimp farming and mangrove loss in Thailand. Edward Elgar Publishing, Cheltenham, UK, 288 pp

Bartlett A (2004). The essential exponential for the future of our planet. University of Colorado, Boulder, Colorado, 302 pp

Binns T (1990) Is desertification a myth? Geography 75:106–113

Blaikie P, Brookfield HC (1987) Land degradation and society. Methuen, London, 296 pp

Borgstrom, G (1972) The hungry planet; the modern world at the edge of famine. 2nd ed. Macmillan, New York, 552 pp

Butzer KW (1993) No Eden in the New World (evidence of environmental degradation due to prehispanic agricultural practices in Mexico). Nature 362, no. 6415:15–18

Carson R (1962) Silent Spring. Houghton Mifflin, Boston, 368 pp

Catton WR (1982) Overshoot: The Ecological Basis of Revolutionary Change. University of Illinois Press, Urbana, 298 pp

Chesworth W (2002) Sustainability and the end of history. Geotimes 47: 5 and 52

Chesworth W (2008) Biomes and their soils, pp 61-68; Neolithic Revolution pp 488-489; Weathering systems in Soil Science pp 825-829. In: Chesworth W(ed) Encyclopedia of soil science. Springer, Dordrecht, 902 pp

Chew SC (2007) The recurring Dark Ages: ecological stress, climate changes, and system transformation. AltaMira Press, Lanham, MD, 295 pp

Cohen JE (1995) How many people can the earth support? Norton, New York, 532 pp

Crawford HEW (1991) Sumer and the Sumerians. Cambridge University Press, Cambridge, 182 pp

Crutzen PJ (2002) The anthropocene. Journal de Physique IV 12:1–5

Daly HE (2005) Beyond growth: the economics of sustainable development. Beacon Press Boston, 253 pp

deBuys W(ed) (2004) Seeing things whole: The essential John Wesley Powell. Island Press, Washington DC, 402 pp

Diamond J (1997) Guns, germs and steel. W.W. Norton, NewYork, 480 pp

Diamond J (2005) Collapse: How societies choose to fail or succeed. Viking Press, New York, 591 pp

Dregne H, Kassas M, Rozanov B (1991) A new assessment of the world status of desertification. Desertification Control Bulletin 20:6–19

East WG (1924) The geography behind history. Nelson, London, 203 pp

Ehrlich PR, Ehrlich AH (2008) The dominant animal: human evolution and the environment. Island Press, Washington, DC, 428 pp

Erhart H (1964) Consequences logiques de la bio/rhexistasie sur le chimisme des mers anciennes et sur les sediments qu'elles ont engendres. Comptes Rendus de l'Académie des Sciences 259:2669–2671

Evans R (1992) Erosion in England and Wales – the present is the key to the past. In: Bell M, Boardman J (eds) Past and present soil erosion. Oxbow Books, Oxford, pp 53–66

Ezcurra E (2006) Global deserts outlook. Nairobi, Kenya. United Nations Environment Programmes (UNEP), Nairobi, 150 pp

FAO (2005) State of the World's forests. Food and Agriculture Organization of the United Nations, Rome, 153 pp

Fussell GE (1976) Farms, farmers and society: Systems of food production and population numbers. Coronado Press, Lawrence, Kansas, 332 pp

Hardin G (1995) Living within limits: Ecology, economics and population taboos. Oxford University Press, Oxford, 339 pp

Harris M (1971) Culture, man, and nature – An introduction to general anthropology. Crowell, New York, 660 pp
Harris M (1980) Cultural materialism: the struggle for a science of culture. Vintage Books, New York, 381 pp
Hawkes J (1953) A land. Cresset Press, London, 247 pp
Head L (2000) Second nature: the history and implications of Australia as aboriginal landscape. Syracuse University Press, Syracuse, NewYork, 272 pp
Hodder, I. 2007 Çatalhöyük in the context of the Middle Eastern Neolithic. Annual Review of Anthropology 36:105–120
Hobsbawm EJ (1996) The age of revolution 1789–1848. Vintage Books, New York, 356 pp
Hoyle F (1964) Encounter with the future. Trident Press, New York, 108 pp
Jarrige JF, Meadow RH (1980) The antecedents of civilization in the Indus valley. Scientific American 243:122–133
Jackson W. (2004) Agriculture: the primary environmental challenge of the 21st century. In: Chesworth W, Moss MR, Thomas VG (eds) The human ecological footprint. Faculty of Environmental Sciences, University of Guelph, Guelph, ON, pp 85–99
Kennedy P (1993) Preparing for the twenty-first century. Random House, New York, 428 pp
Knox JW, Morris J, Weatherhead EK, Turner APF (2000) Mapping the financial benefits of sprinkler irrigation and potential financial impact of restrictions on abstraction: A case-study in Anglian Region, Journal of Environmental Management 58:45–59
Lambin EF (1997) Modelling and monitoring land-cover change processes in tropical regions. Progress in Physical Geography 21:375–393
Larsen J (2004) The sixth great extinction: a status report. Earth Policy Institute, Washington, DC, Rept. 2004.4., 4 pp
Leopold A (1933) The conservation ethic. Journal of Forestry, 31:634–643
Lewin R, Leakey R (1996) The sixth extinction: Patterns of life and the future of humankind. Anchor Press, Norwell MA, 288 pp
Lomborg B (2001) The skeptical environmentalist. Cambridge University Press, Cambridge, 515 pp
Mabey R (1980) The common ground: A place for nature in Britain's future? Hutchinson, in Association With the Nature Conservancy Council, London, 280 pp
Martin A (1975) The last generation: The end of survival. Fontana, Glasgow, 188 pp
Martinez Cortizas A, Mighall T, Pontevedra Pombal X (2005) Linking changes in atmospheric dust deposition, vegetation change and human activities in northwest Spain during the last 5300 years. The Holocene 15:698–706
McNeill WH (1991) The Rise of the West: A history of the human community: With a retrospective essay. University of Chicago Press, Chicago, 828 pp
Meyer WB, Turner BL (1992) Human population growth and global land-use/cover change. Annual Review of Ecology and Systematics 23:39–61
Mistry J, Berardi A (2006) Savannas and dry forests: Linking people with nature. Ashgate Publishing, Burlington VT, 274 pp
Montgomery DR (2007) Dirt: the erosion of civilizations. University of California Press, Berkeley and Los Angeles, 280 pp
Netting RM, Stone GD, Stone MP (1993) Agricultural expansion, intensification, and market participation among the Kofyar, Jos Plateau, Nigeria. In: Turner II BL, Hyden G, Kates RW (eds) Population growth and agricultural change in Africa, University Press of Florida, Gainesville, pp 206–249
O'Hara SL, Street-Perrott FA, Burt TP (1993) Accelerated soil erosion around a Mexican highland lake caused by prehispanic agriculture. Nature 362:48–51
Rees WE (2003) A blot on the land. Nature 421:898
Rees WE (2004) the eco-footprint of agriculture: a far-from-(thermodynamic)-equilibrium interpretation. In: NABC Report 16: Agricultural Biotechnology: Finding Common International Goals. National Agricultural Biotechnology Council, Cornel University, pp 87–109
Rees WE (2007) Human eco-footprints: straying off the sustainability trail. The Kenneth R. Farrell Distinguished Public Policy Lecture, delivered at the University of Guelph, May 6, 2007.
Renfrew C (1994) World linguistic diversity. Scientific American, 270 (1):116–122
Schnepf R (2006) European union biofuels policy and agriculture: An overview. CRS Report for Congress. Congressional Research Service. http://italy.usembassy.gov/pdf/other/RS22404.pdf (Accessed November 19, 2008).
Sherratt A (1980) Water, soil and seasonality in early cereal cultivation. World Archaeology 11:313–330
Simon JL (1998) The ultimate resource 2. Princeton University Press, Princeton, NJ, 78 pp
Sokal RR, Oden NL, Wilson C (1991) Genetic evidence for the spread of agriculture in Europe by demic diffusion. Nature 351:143–145
Stearns PN (2001) The encyclopedia of world history: Ancient, medieval, and modern. Houghton Mifflin Company, Boston, 1243 pp
Stone M, Krishnappan BG (1997) Transport characteristics of tile-drain sediments from an agricultural watershed. Water, Air, & Soil Pollution 99:89–103
Tainter JA (1988) The collapse of complex societies. Cambridge University Press, Cambridge, U.K., 250 pp
Timmer CP (1969) The turnip, the new husbandry, and the English Agricultural Revolution. The Quarterly Journal of Economics 83:375–395
Tudge C (1985) Editorial. New Scientist 111:42
USDA (US Department of Agriculture) (2008) Soil erodibility index. Electronic Directives System: http://www.info.usda.gov/M 180 511 A 01.htm (Accessed November 19, 2008)
Wackernagel MRW and Rees WE (1996) Our ecological footprint: Reducing human impact on the Earth. New Society Publishers, Gabriola Island, BC, 160 pp
Wells HG (1939) The fate of *Homo sapiens*; An unemotional statement of the things that are happening to him now, and of the immediate possibilities confronting him. Secker and Warburg, London, 329 pp
Wilson EO (2002) The future of life. Knopf, New York, 254 pp
Witham CS, Oppenheimer C (2004) Mortality in England during the 1783–1784 Laki Craters eruption. Bulletin Volcanologique, 67:15–26
WRI (World Resources Institute) (2008) Agricultural Inputs. http://earthtrends.wri.org/pdf_library/data_tables/agr2_2003.pdf Accessed November 19 2008.
Wright, Ronald (2004) A short history of progress. Anansi Press, Toronto, 211 pp

Part II
The Mediterranean and European World–Arid Mediterranean Lands

Varied, well studied and thoroughly documented landscapes and societies occur in the Mediterranean centered area and in Europe generally. This is where the civilization born in Mesopotamia and Egypt, grew into the civilization we think of as western. A first major contrast is observed between societies that have developed in the southern (North Africa and Near East) and the northern areas (Europe). Both experienced strong climatic changes during the Holocene, but southern ones, being closer to the critical temperate-tropical boundary, have been greatly affected by desertification, whereas northern ones were most affected by human makeover through intense agricultural practices, deforestation in particular.

Desertification has overwhelmed most other influences in North African and Near East lands. Nevertheless, during the transition from the lower Holocene wet, savannah conditions to the final vast desert biome, local differences occurred as societies adapted to their landscapes. In large wadis, where rivers had once flowed, oases commonly persisted, for example, allowing some societies to survive in downsized form. People also migrated toward major river valleys such as the Nile, Tigris and Euphrates where farming techniques developed in earlier localities evolved into an irrigated agriculture capable of sustaining large populations. From that came the first population 'explosions' and the development of structured societies and the first civilizations. Relatively large populations, in part enslaved, during Roman times, also persisted in harsh desert areas where mineral resources were mined.

The temperate lands of south and mid Europe sustained a proliferation of cultures several leading to major civilizations such as the Greek, Etruscan, and Roman ones. Essentially every landscape was utilized baring high mountain peaks and, to a lesser degree, malarial lowlands. Floodplains, hill country, intermontane basins and even volcanic areas were preferred settlement localities. The variety of landscapes settled, led to a great differentiation of societies and cultures, developmental differences being encountered even within the drainage basin of a single river. The intense exploitation of a terrain and its by growing populations led to a significant anthropic modification of the landscape. This triggered intense erosion, siltation of bottom lands, progradation of coasts and deltas, and siltation of harbors. Through custom or serendipity, if not management, some original landscape characteristics, forests included, have been locally preserved, as in Romania. Conversely landscapes that never existed before were created by human activity, the reclamation of land from the sea in the Netherlands being an outstanding example.

Chapter 4
Human Responses to Climatically-driven Landscape Change and Resource Scarcity: Learning from the Past and Planning for the Future

Nicholas Brooks

4.1 Introduction

The study of past climatic and environmental changes and human responses to such changes is increasingly relevant today, as societies across the world begin to confront anthropogenic climate change resulting principally from the burning of fossil fuels and the resulting emission of greenhouse gases (Raupach et al. 2007; Somerville et al. 2007). While there is widespread agreement among scientists and policy makers that efforts should be made to prevent global mean surface temperature rising by more than 2°C above late pre-industrial values, current policy regimes risk committing the world to a global warming of 4°C or more by 2100 (Anderson and Bows 2008). Although the precise consequences of a warming above 2°C are uncertain, such a warming is likely to be associated with systematic climatic reorganization and the transformation of landscapes and biogeochemical systems at scales ranging from the global to the local (IPCC 2007).

Evidence from the past indicates that climate can change rapidly at global and regional scales, even in the absence of obvious external driving mechanisms (Alley 2003). In this context "rapid" climate change is defined as involving changes in climate and climate-sensitive natural systems of sufficient rate and magnitude to be represented by large anomalies in proxy records of climate spanning periods of decades to centuries, noticeable over timescales of the order of a human lifetime, and potentially problematic for the functioning of extant ecological and social systems. Examples of past rapid climate change (hereafter abbreviated to RCC) include cooling, warming, changes in atmospheric and oceanic circulation, increases and decreases in sea level, changes in rainfall patterns, the collapse of ecological systems, coherent advances and recessions of glaciers and ice sheets over large areas, and shifts from arid to humid conditions and vice versa (Alley et al. 2003; Brooks 2006; Magny et al. 2006; Jansen et al. 2007). Numerous studies suggest that comparable changes are likely or plausible throughout the twenty-first century and beyond, including large and rapid increases in regional mean and maximum temperatures, shifts in rainfall patterns, changes in the behaviour of El Niño and monsoon systems, increases in global mean sea level of the order of 1 m per century or more, extreme climatic and environmental desiccation, the loss of water resources associated with the melting of snow and ice at high altitudes, changes in river flow, ecological collapse and species extinction, the loss of natural resources (including forests, coral reefs, fisheries, and pastures), and changes in the distribution and availability of agricultural land (Thomas et al. 2005; Warren 2005; IPCC 2007; Rahmstorf 2007).

As a result of anthropogenic climate change, human societies across the globe will have to adapt to potentially profound changes in their environments, and in the distribution and availability of key natural resources, particularly water and productive land. While there is a burgeoning literature on adaptation to climate change, most studies focus on recent, current or near-future adaptation to small, incremental changes in climate, for example as manifest through

N. Brooks (✉)
Tyndall Centre for Climate Change Research
School of Environmental Sciences, University of East Anglia
Norwich, NR4 7TJ, UK
e-mail: nick.brooks@uea.ac.uk

See Plate 1 in the Color Plate Section; also available at: extras.springer.com

I. P. Martini, W. Chesworth (eds.), *Landscapes and Societies*,
DOI 10.1007/978-90-481-9413-1_4, © Springer Science+Business Media B.V. 2010

incremental increases in risk associated with climatic extremes (Brooks et al. in press). Little attention is paid to how societies might respond or adapt to large and/or rapid changes in climatic or environmental conditions that might fundamentally alter the nature of the environmental constraints and opportunities that mediate human activities and economic development (Brooks et al. in press).

While there are no precise past analogues for human impacts and adaptation associated with a global warming of more than 2°C above late pre-industrial values, broad analogues for the kinds of changes in climate and environment that might be experienced at regional scales over the course of the twenty-first century and beyond may be found in the very recent geological past. Episodes of RCC have occurred throughout the Holocene period (representing approximately the past 10,000 years), and these episodes are becoming increasingly well understood as a result of extensive paleoenvironmental research (Mayewski et al. 2004). Furthermore, the impacts of these changes on human societies can be interrogated through the combined interpretation of paleoenvironmental and archaeological records (Brooks 2006).

This chapter seeks to identify how human societies responded to RCC in the middle Holocene period, with a specific focus on the 6th millennium BP (all dates are given in calendar years before present, abbreviated to BP, with dates originally provided as uncalibrated radiocarbon years converted to calendar years BP using the Intcal calibration of Reimer et al. (2004)). While much has been published about episodes of RCC around 8200 and 4200 BP (Cullen et al. 2000; Alley and Ágústsdóttir 2005; Rosen 2007), the 6th millennium BP has been relatively neglected. However, it is during the 6th millennium BP that the Earth's climate last appears to have undergone a widespread and systematic reorganization, as discussed in detail below. The premise of this chapter is that, as we enter a period of global climatic reorganization driven by human modification of the atmosphere, we might learn some valuable lessons about how societies might (and might not) adapt to RCC by examining human responses to RCC during this last period of global climatic transition. While the mechanisms driving global climate change during these two periods are very different (Brooks 2006; Raupach et al. 2007), and while such a comparison cannot provide precise analogues, some very general lessons might be drawn by comparing human responses to RCC in specific regions and contexts separated widely in time and space. The 6th millennium BP was a time not only of profound climatic and environmental change, but was a formative period in human history, during which the earliest large, urban, state-level societies ("civilizations") emerged (Brooks 2006).

The focus of this chapter is on the northern hemisphere subtropics and adjacent regions (abbreviated to NHST), and on northern Africa and Western Asia in particular. These two regions have been selected for a number of reasons. First, they furnish us with clear and abundant evidence of profound environmental changes that can be linked with episodes of RCC at the beginning and end of the 6th millennium BP. Second, extensive archaeological research has provided us with ample evidence of cultural changes in these regions which coincide with periods of regional and global RCC. Third, the societies that developed in these regions during the 6th millennium BP collectively share some traits with modern societies, including high population densities in large urban centers, a high degree of social and economic differentiation, formal systems of government, complex networks of trade, intensive agricultural production and, outside of urban areas, a high dependence on small scale agriculture and pastoralism.

Evidence for RCC in the NHST, where the 6th millennium BP was a time of widespread climatic and environmental desiccation (Brooks 2006) is placed in the context of wider, global climatic change as apparent in records from a variety of other regions at both low and high latitudes. Human responses to these changes are inferred from the archaeological literature, focusing as far as possible on evidence from specific locations where key changes in livelihoods, mobility, settlement patterns and other aspects of human behaviour either can be linked directly to, or are coincident and compatible with, changes in climatic and environmental conditions evident in local or regional records.

The analysis that follows begins with a synthesis of paleoclimatic information from a wide range of sources, the aim of which is to provide a broad chronological and phenomenological framework of Middle Holocene climatic and environmental change. This synthesis is followed by a discussion of archaeological data from northern Africa and western Asia, focusing on changes evident in the archaeological record that are suggestive of adaptation to RCC and its impacts on landscapes. This discussion focuses on three key periods: the transition from the 7th to the 6th millennium BP, the early-middle 6th millennium BP, and the end of the 6th millennium BP. Finally, a variety of apparent recurring responses to RCC and climatic desiccation are identified. These are used to draw some broad lessons about adaptation to climatic and environmental change that are relevant in the context of contemporary anthropogenic climate change.

4.2 Linked Climatic and Environmental Change in the Middle Holocene

4.2.1 Rapid Climate Change Events and the Approach to the 6th Millennium BP

The climatic amelioration that followed the Last Glacial Maximum (LGM) some 21,000 years ago (Jansen et al. 2007) was driven largely by increases in summer solar heating (insolation) of the northern hemisphere middle and high latitudes in the boreal summer, due to cyclical changes in the Earth's orbital dynamics (Fleitmann et al. 2003; Tuenter et al. 2003). This increase in summer insolation led to the decay of the northern hemisphere ice sheets, and reached a maximum around 10,000 years ago (de Menocal et al. 2000; Guo et al. 2000). Stronger summer solar heating of low to middle latitude land masses also resulted in more vigorous monsoon systems, and by about 10,000 BP northern hemisphere monsoon rains extended many hundreds of kilometers further north than during the LGM, transforming areas that had been hyperarid desert into well-watered savannah, scrub and open woodland (Hoelzmann et al. 1998; Gasse 2000; de Menocal et al. 2000). Rainfall throughout the currently arid NHST zone was much greater than today, and regions such as the Sahara supported abundant flora and fauna, which in turn sustained significant human populations (Hoelzmann et al. 1998; Wei and Gasse 1999; Linstädter and Kröpelin 2004; Brooks et al. 2005). While summer insolation was declining by around 9000 BP, the warm, humid Holocene Climatic Optimum lasted for some five millennia throughout most of the NHST (de Menocal et al. 2000).

Superimposed on these long-term trends in insolation were short-lived episodes of climatic disruption manifest over regional and hemispheric scales, with broadly synchronous changes evident in records from different geographic regions. These episodes recurred approximately every 1000–2000 years, typically lasting for decades to several centuries, and were associated with cooling at middle and high northern latitudes and enhanced aridity in the NHST (Bond et al. 1997; Gasse 2000). The mechanisms behind these events are not well understood, although it has been proposed that they are driven by perturbations to the North Atlantic circulation, and also that they may be associated with solar variability (Neff et al. 2001; Mayewski et al. 2004). The extensively studied event around 8200 BP has also been linked with the outburst of glacial Lake Agassiz, formed by the melting of the Laurentide Ice Sheet, via the Hudson Bay (Alley and Ágústsdóttir 2005). Bond et al. (1997) identify eight cold events from an analysis of the distribution and age of ice rafted debris in the North Atlantic, at 11,100; 10,300; 9400; 8100; 5900, 4200; 2800; and 1400 BP.

A number of these Bond events coincided with arid episodes in the NHST, during which monsoons weakened and rainfall declined. For example, Wang et al. (2005) examine changes in the isotopic composition of rainwater over the past 9000 years by analysing variations in $\delta^{18}O$ in a stalagmite from Dongge Cave in southern China (~25°N; 108°E; Fig. 4.1), and identify weak monsoon events centered at 8300; 7200; 6300; 5500; 4400; 2700; 1600; and 500 BP. Within the margins of error in these data, and considering their apparent duration, the weak monsoon events at 8300; 4400; 2700; and 1600 BP may be viewed as broadly coincident with Bond events. Cooling and aridity signals in widely separated geographical regions are particularly well established around 8200 and 4200 BP (Cullen et al. 2000; Gasse 2000; Marchant and Hooghiemstra 2004; Alley and Ágústsdóttir 2005; Parker et al. 2006). In the record of Wang et al. (2005) weak monsoon events around these times span the periods from 8400–8100 and 4400–3900 BP. Other regional records suggest periods of enhanced aridity broadly coincident with other Bond events, although these are generally less well established and less spatially coherent (di Lernia 2002; Brooks 2006). In addition, data from across the globe increasingly indicate an episode of cooling and aridity around 5200 BP, as discussed below.

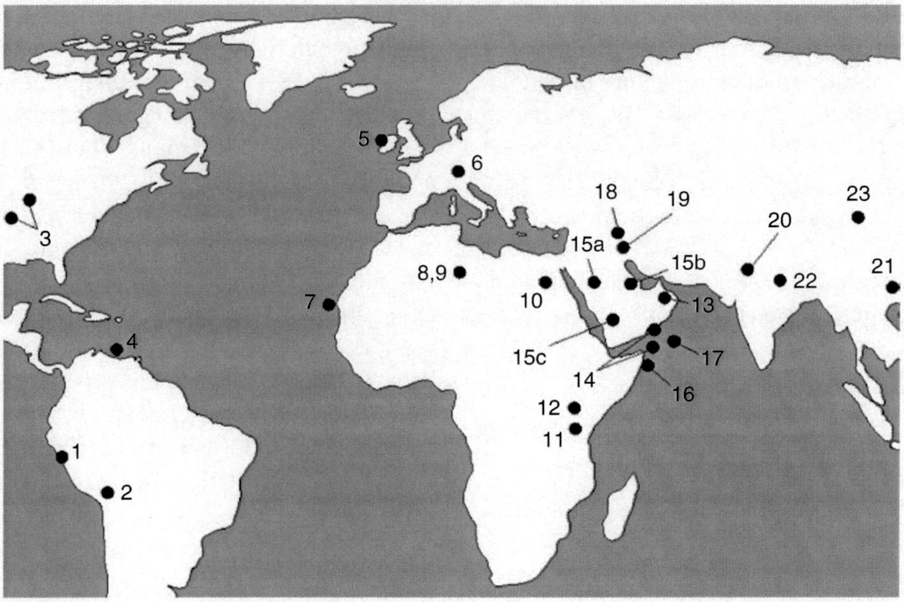

Fig. 4.1 Locations of sites mentioned in the text yielding evidence of rapid climatic or environmental change during the 6th millennium BP, some of which are also referred to in the discussion of earlier RCC. Geographical locations are given below. Where no specific geographical name is given (sediment core locations or multiple sites) a source is cited. *1* northern coastal Peru; *2* Lake Titicaca; *3* Upper Midwestern USA sites described by Baker et al. (2001a); *4* Cariaco Basin; *5* County Mayo; *6* Lake Constance; *7* site 658C described by de Menocal et al. (2000); *8* Acacus Mountains; *9* Erg Uan Kasa and Edeyen Murzuq; *10* Farafra Oasis; *11* Mount Kilimanjaro; *12* Mount Kenya; *13* Hoti Cave; *14* Arabian Peninsula sites described by Fleitmann et al. (2007); *15* Arabian Peninsular sites described by Lezine et al. (1998); *16* Marine sediment core site described by Jung et al. (2004); *17* site 74KL described by Sirocko et al. (1993); *18* Lake Zeribar; *19* Lake Mirabad; *20* Lake Lunkaransar; *21* Dongge Cave; *22* Region east of Ganga Plain; *23* Tengger Desert

Whereas warm, humid conditions persisted for several millennia in the NHST after the 8200 BP event, this episode appears to have heralded the onset of a more variable climate at low and middle northern latitudes. Based on an analysis of Sr and Nd ratios in dust deposited in marine sediments off Somalia (Site 905P, Fig. 4.1), Jung et al. (2004) suggest that the period around 8500 BP, possibly encompassing the 8200 event, marked an aridification step in the region around the Arabian Sea. Jung et al. (2004) interpret subsequent increased climatic variability as representing an intermediate step between the very humid conditions of the very early Holocene and the arid climate pertaining today, during which weaker orbital forcing of the monsoon could not sustain a fully wet climatic regime. This interpretation is consistent with paleoenvironmental records from the Saharan region, which indicate a number of dry episodes, and a general trend towards drier conditions and a more variable climate, after about 8000 BP (de Menocal et al. 2000; Fig. 4.1). di Lernia (2002) provides detailed evidence from the Libyan Fezzan (Fig. 4.1), and summarizes evidence from other localities in North Africa, for an arid episode between about 7300 and 6900 BP. In Egypt, rainwater-fed playas declined after about 7000 BP, with progressive desertification apparent in records from southern Egypt (Nicoll 2004).

In many parts of the world the end of the 7th millennium BP heralded the beginning of a long period of climatic disruption focused on the 6th millennium BP (Fig. 4.1). Linstädter and Kröpelin (2004) find evidence for the collapse of the summer monsoon in the Gilf Kebir region of southwestern Egypt around 6300 BP (coincident with the weak monsoon event identified by Wang et al. (2005) in speleothem records from Dongge Cave in southern China), while Wendorf and Schild (1998) place the onset of hyper-aridity in the eastern Sahara at around 6200 BP. Fluvio-lacustrine activity ceased on the Abu Tartar plateau in the Western Desert of Egypt between Dakhla and Kharga oases around this time (Bubenzer et al. 2007). In the Thar Desert of northwestern India, water levels in Lake Lunkaransar fell abruptly around 6400 BP (Enzel et al. 1999). Thompson et al. (2006) place the beginning of the Neoglaciation—a period of substantial glacier advance evident across the globe—around 6400 BP.

4.2.2 NHST Aridification and Widespread Climatic Reorganization in the 6th Millennium BP

Trends towards aridity in the NHST appear to have accelerated around 6000 BP, with declines in lake levels and vegetation changes indicating a shift to drier conditions around the time of the 5900 BP Bond event in northern Africa, North America, and various locations in Asia (Gasse and van Campo 1994; Damnati 2000; Guo et al. 2000; Baker et al. 2001a; Brooks 2006). For example, climatic deterioration is evident between 6380 and 5950 BP in records from the Tengger Desert of northwestern China (Zhang et al. 2000) (Fig. 4.1). At Farafra Oasis adjacent to the Nile Valley a major phase of playa formation ceased around 6000 BP (Hassan et al. 2001; Fig. 4.1), while lake levels in many other Saharan regions began to decline around 6000 BP (Damnati 2000). Mayewski et al. (2004, pp. 248–250) identify the 6th millennium BP as a period of extensive "rapid climate change" involving the "co-occurrence of high-latitude cooling and low-latitude aridity".

There is currently some debate as to whether the desiccation of the NHST that commenced in earnest around 6000 BP was characterized by gradual or abrupt change. Climatic variations inferred from speleothem records recovered from caves in Oman and Yemen (Fig. 4.1) indicate a gradual retreat of the summer ITCZ and weakening of the Indian summer monsoon in response to slowly decreasing insolation (Fleitmann et al. 2007). From these results, Fleitmann et al. (2007) conclude that abrupt changes in monsoon climate were short lived and superimposed on a much more gradual long-term trend of decreasing monsoon precipitation. Kuper and Kröpelin (2006, p. 807) similarly argue that desiccation in the eastern Sahara was a gradual process, and that apparent "abrupt drying events elsewhere in the Sahara may be explained by fading rainfall at a specific latitudinal position at a certain moment, or by dropping local groundwater." It might therefore be concluded that apparent abrupt changes were simply manifestations of highly localised environmental responses to gradual changes in regional climatic regimes. For example, a permanent retreat of the summer monsoon south of about 23°N resulting from a gradual, long-term southward migration of the monsoon zone might explain the termination around 6100–6000 BP of a speleothem record from Hoti Cave in northern Oman (Fig. 4.1; Neff et al. 2001).

In contrast, other authors interpret paleoenvironmental records as indicating stepwise climatic desiccation at regional scales. For example, Jung et al. (2004) describe an aridification step around 6000 BP in the terrestrial regions adjacent to the Arabian Sea, indicated by a sharp increase in the ratio of ^{87}Sr to ^{86}Sr in clay sediments from an Arabian Sea sediment core (Fig. 4.1). de Menocal et al. (2000) argue for an abrupt termination of humid conditions in northern Africa around 5500 BP, based on sediment core data from Site 658C in the Eastern Tropical Atlantic off Cap Blanc, Mauritania, which indicate a sharp increase in aeolian dust deposition off the coast of Mauritania after about 5700 BP (Fig. 4.2a), combined with results from climate simulations.

The record of de Menocal et al. (2000) of dust fluxes to the eastern tropical Atlantic provides strong evidence for rapid, climatically-driven landscape change at the regional scale. After about 8000 BP, the record indicates a gradual, long-term increase in dust flux until the early 6th millennium BP, on which are superimposed significant, century-scale variations (Fig. 4.2a). This long-term increase in dust flux suggests a similarly gradual process of desiccation driven by a gradually weakening and/or retreating monsoon in the Saharan region, whereas the century-scale variations indicate the onset of a less stable climatic regime after 8000 BP. Increases in dust deposition could result from changes in atmospheric mobilization and transport mechanisms, and in atmospheric circulation and transport pathways (Brooks 2000). However, the most likely explanation for high dust fluxes is increased aridity and reduced vegetation cover in the Sahara, resulting in a greater abundance of exposed erodible sediments—particularly given the widespread evidence of drier conditions after 8000 BP (Nicoll 2004; Cremaschi 2002; di Lernia 2002). The very sharp increase in dust deposition at around 5700–5600 BP (Fig. 4.2a) suggests a large and rapid increase in the availability of erodible material. This is unlikely to be the result of a gradual process of desiccation driven

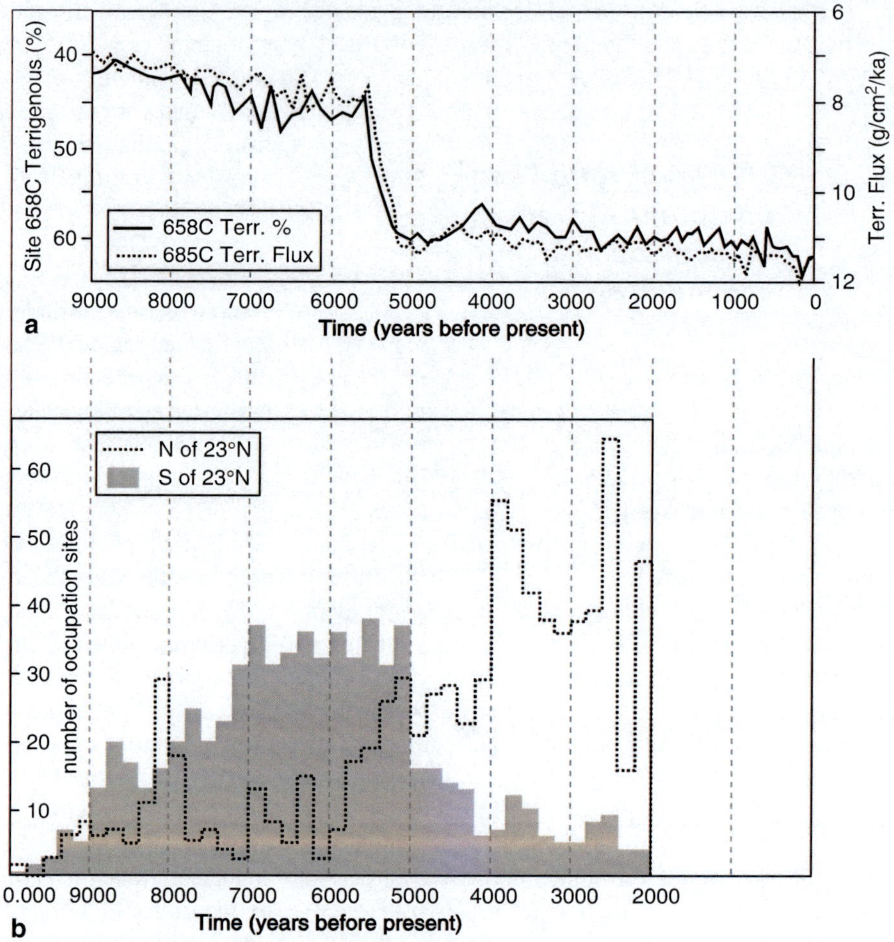

Fig. 4.2 **a** Terrigenous dust flux at Ocean Drilling Program Site 658C off Cap Blanc, Mauritania, West Africa (From Vernet and Faure (2001) with permission from authors & after de Menocal et al. 2000). **b** Changes in number of radiometrically dated occupation sites over time in the Sahara north of 23°N and south of 34°N (*solid line/shaded area*), and in the Sahara and Sahel south of 23°N and north of 13°N (*dashed line/clear area*), based on 1040 dates. (From Vernet and Faure 2001)

by a leisurely retreat of the African Monsoon, and is more likely to have been associated with a widespread and rapid (decadal to century scale) collapse of vegetation cover and soil moisture. Cremaschi (2002) and di Lernia (2002) provide evidence for a dry interval in the Acacus Mountains (Libyan Fezzan; Fig. 4.1) and adjacent regions between 5850 and 5600 BP, with a transition to arid climate vegetation after 5800 BP in the adjacent sand seas (Erg Uan Kasa and Edeyen of Murzuq), heralding the onset of permanently arid conditions which were well established by around 5000 BP. The region-wide nature of the inferred desiccation, extending outside of Africa and into Western Asia, is indicated by comparable records of increased dust fluxes into the Arabian Sea over the same period (Jung et al. 2004). Similar evidence is presented in earlier work by Sirocko et al. (1993), who identify a general increase in dust flux to the Arabian Sea after 5500 BP (Fig. 4.1), suggesting widespread desertification in the adjacent terrestrial regions at a similar time to the apparent collapse of Saharan vegetation.

Interpreted alongside the much more gradual changes in monsoon rainfall evident in the records from southern Arabia (Fleitmann et al. 2007), the inferred rapid collapse of the monsoon in the Sahara suggests that the hydrological cycle in the latter region may have been sustained by feedback mechanisms such as moisture recycling by vegetation. Such mechanisms may have sustained the monsoon in a weak insolation regime in the absence of external perturbations. However, disruption during a transient cold, arid episode might have resulted in the relatively rapid collapse of sensitive and fragile coupled monsoon-vegetation systems, with the weak insolation regime insufficient to drive a full recovery of this system (Brooks 2004). Monsoon collapse in the Sahara might therefore have been triggered by the 5900 BP Bond event, although it is unclear whether the apparent lag of some two centuries between this event and the rapid increase in dust deposition can be accounted for in terms of margins of dating error and lagged responses associated with the time required for vegetation dieback to translate into an increase in the availability of erodible material. A lagged response is perhaps plausible given the timescales associated with the transition from low to high dust fluxes—whereas the onset of this transition was rapid, dust fluxes continued to increase for some 4–5 centuries before stabilising at the end of the 6th millennium BP.

Drying in Western Asia at the time of the Saharan desiccation is also identified by Lezine et al. (1998), who present and review evidence for lacustrine activity in central, southern and eastern Arabia and find the longest-lived lacustrine environments drying in the late 7th or the 6th millennium BP. Within the wider NHST region, but outside of the monsoon zone, Stevens et al. (2006) interpret a multi-proxy record from Lake Mirabad in the Zagros Mountains of Iran as indicating a shift in rainfall regimes around 5800 BP, followed by a 600 year drought from 5700 to 5100 BP, reflecting similar results from Lake Zeribar some 300 km to the northwest (Fig. 4.1).

Regional desiccation in South Asia at this time is indicated by the drying of Lake Lunkaransar and the disappearance of vegetation in the surrounding dunes in the Thar Desert by around 5500 BP (Enzel et al. 1999), loess sedimentation on the Ghaggar-Hakra plain (Schuldenrein et al. 2004), and the cessation of fluvial activity east of the Ganga plain (Srivastava et al. 2003) (Fig. 4.1).

Evidence for widespread regional changes in climate in the 6th millennium BP is not restricted to the Afro-Asiatic arid belt. Ice cores from Mount Kilimanjaro and a lake record from Mount Kenya (Fig. 4.1) reveal significant $\delta^{18}O$ depletions between 6500 and 5200 BP, indicating drying and cooling and/or anomalously heavy snowfall (Barker et al. 2001c; Thompson et al. 2002). Haug et al. (2001) find evidence in sediment records in the Cariaco Basin off the Venezuelan coast for a reduction in river discharge commencing around 5800 BP and culminating around 5200 BP, indicating a long-term weakening of the summer monsoon regime and drying in northern South America (Fig. 4.1), revealing much more widespread changes throughout the Intertropical Convergence Zone (ITCZ). Prairie replaced forest in the Upper Midwest of the USA in the early 6th millennium (Fig. 4.1), suggesting changes in large-scale atmospheric circulation and the replacement of an essentially monsoonal circulation with a zonal circulation more typical of the mid latitudes (Baker et al. 2001a).

Evidence for more systemic changes in global climate in the early 6th millennium BP also comes from northern coastal Peru, where a variety of data indicate what Sandweiss et al. (2007) refer to as a major change in the tropical Pacific climate at about 5800 BP. In particular, El Niño events appear to have become more frequent after 5800 BP, following a period of several millennia during which El Niño was weak or absent, although a periodicity similar to that of today was not established until around 3000 BP (Sandweiss et al. 2007). The shift to a more regular El Niño was associated with increased coastal upwelling and a fall in sea surface temperatures of around 3°C off the coast of Peru (Reitz and Sandweiss 2001; Andrus et al. 2004), which would have increased atmospheric stability and is likely to have suppressed rainfall (Brooks 2006). During the 6th millennium BP, Lake Titicaca (Fig. 4.1) fell to its lowest level in 25,000 years (Baker et al. 2001b), providing further evidence of RCC in the Americas.

In Europe, there were three successive increases in the level of Lake Constance between 5600 and 5300 BP (Fig. 4.1), during a period of climatic transition coinciding with the so called Neoglaciation, when glaciers advanced across the continent (Magny et al. 2006). Thompson et al. (2006) conclude that glacial advance in western North America, Europe, the Himalayas, New Zealand, and the Andes occurred throughout the 6th millennium BP. Within this period, Casseldine et al. (2005) also find evidence for an especially dry period in County Mayo, Ireland (~54°N, 10°W) from 5800 to 5200 BP (Fig. 4.1).

Evidence that the apparent widespread climatic reorganization in the 6th millennium BP was associated with changes in North Atlantic circulation is presented by McManus et al. (2004), who employ ratios of ^{231}Pa to ^{230}Th in sediments from a core from the subtropical North Atlantic Ocean as a kinematic proxy for the meridional overturning circulation. $^{231}Pa/^{230}Th$ ratios fall sharply in the early 6th millennium BP, around the time of the sharp increase in dust deposition off the West African coast, and reach a minimum some

1000 years later, when they were lower than at any time since the end of the Younger Dryas, when rapid cooling at high latitudes was associated with aridity in the NHST (Gasse 2000).

4.2.3 The 5200 BP Cold, Arid Crisis and the end of the Holocene Climatic Optimum

In recent years evidence has mounted for an abrupt cold, arid event around 5200 BP, with indications of such an event from both high and low latitude sites across the globe (Fig. 4.3). Although this event was not identified by Bond et al. (1997), it exhibits the characteristics of other Bond events, namely decadal to century-scale cooling coupled with enhanced aridity. While evidence for cooling and aridity around 5200 BP is generally particularly pronounced at middle to high latitudes and the NHST respectively, cooling is evident at lower latitudes and also in the southern hemisphere, and drying at higher northern latitudes. For example, $\delta^{18}O$ records from ice cores from Mount Kilimanjaro and speleothems in Soreq Cave in Israel (Fig. 4.3) indicate sharp, well defined and relatively short lived temperature and rainfall minima respectively, centered on 5200 BP and exceeding in magnitude any other such fluctuations over the past 11,000 years (Bar-Matthews et al. 1999; Thompson et al. 2002). The centuries immediately following 5200 BP coincided with the termination of humid conditions throughout the Sahara (Gasse 2000, 2002; Lancaster et al. 2002; Brooks et al. 2005), whereas the final alluvial deposits at Nekhen in the Nile Valley date to around 5200 BP (Fig. 4.3; Hoffman et al. 1986). Elsewhere in the Afro-Asiatic arid belt, Zhang (2000) identifies an abrupt temperature drop in the Tengger Desert of northwestern China between 5340 and 5290 BP that signalled the end of warm, humid conditions.

Evidence of rapid climate change in Europe is found in the Tyrolean Alps (Fig. 4.3) at 5300–5050 BP, when a persistent snow cover on previously deglaciated areas, associated with a sudden and persistent lowering of the equilibrium-line altitude resulted in the burial and subsequent preservation of the "Iceman", a corpse revealed by the ablation of a small glacier in summer 1991 (Baroni and Orombelli 1996). This event is approximately coincident with an abrupt rise in the level of Lake Constance reconstructed by Magny et al. (2006) from sediment and pollen analysis of a sequence from Arbon-Bleiche (Fig. 4.3), which was associated with the abandonment of a Neolithic lake shore village.

Casseldine et al. (2005) identify a very short-lived, extreme depositional event in County Mayo, Ireland occurring between 5300 and 5050 BP which they suggest was associated with a severe storm event or a series of such events. This event occurred after a short period characterized by extremely dry conditions and is unique within the Holocene record at this site. Casseldine et al. (2005) also review other paleoenvironmental records from Ireland, which include evidence for hiatuses in lake sedimentation around 5250 BP at Lough Gur in County Limerick (Ahlberg et al. 2001),

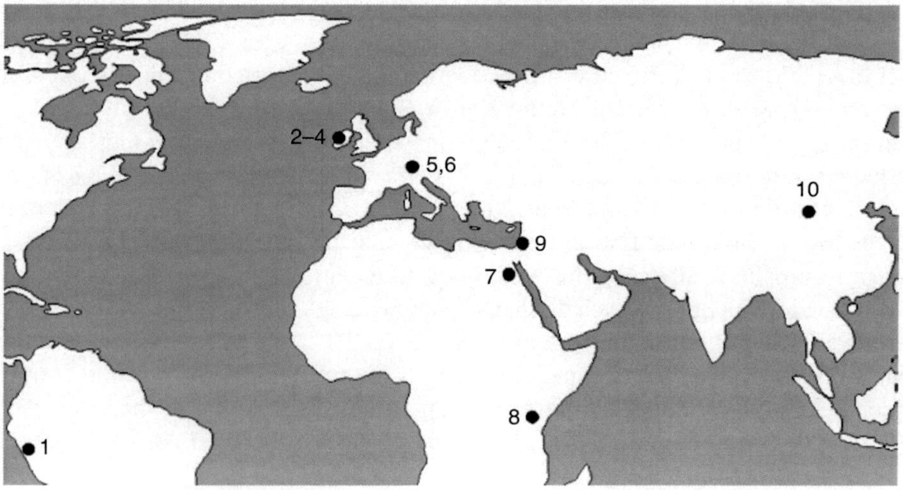

Fig. 4.3 Locations of sites mentioned in the text yielding evidence of rapid climatic or environmental change around 5200 BP. *1* Huascarán; *2* Achill Island; *3* Lough Gur; *4* Inis Mor; *5* Tyrolean Alps; *6* Lake Constance; *7* Nekhen; *8* Kilimanjaro; *9* Soreq Cave; *10* Tengger Desert

and Inis Mor on the Aran Islands (Fig. 4.3; Molloy and O'Connell 2004).

Ice cores from Huascarán in the northcentral Andes of Peru (Fig. 4.3), complemented by pollen evidence indicating a lowering of the tree line, identify the onset of a period of long-term cooling around 5200–5000 BP after several millennia of relative warmth (Thompson et al. 1995), indicating that the 5200 RCC was not restricted to the northern hemisphere.

4.2.4 Summary

The transition to a full postglacial climate characterized by aridity at low latitudes and historical patterns of glaciation at higher latitudes was a very long process, commencing with the collapse of the remnant ice sheets around 8200 BP and culminating well into the late Holocene in some low-latitude regions. However, it appears that the 6th millennium BP represents a key transitional period during which the global climate underwent a major reorganization. This transition appears to have been bracketed by RCC events at around 5900 and 5200 BP, with evidence from across the globe indicating climatic and environmental transitions from 5800 to 5200 BP. The 5900 RCC is evident in the marine records published by Bond et al. (1997), and follows a period of several centuries in which evidence of a shift towards aridity is widespread throughought the NHST. The 5200 RCC has come to light from a number of terrestrial records, which indicate cooling and aridity signalled by glacial advance in Europe and the final collapse of rainfall throughout large swathes of the NHST. It is proposed that these two transient episodes of RCC played a key role in the transition to long-term aridity in the NHST, with the 5900 BP event helping to trigger widespread desertification that began in the early 6th millennium BP and continued into the middle of the millennium, and the 5200 BP event representing the effective termination of the Holocene Climatic Optimum. While truly modern climatic and environmental conditions took a further two millennia or more to evolve in some NHST locations (Brooks et al. 2005; Brooks 2006), the end of the 6th millennium BP saw the establishment of a global climate that was broadly similar to that pertaining throughout the historical period of the past five millennia.

4.3 Human Responses to Climatic and Environmental Change in the Middle Holocene

4.3.1 Linking Environmental and Cultural Change

The 6th millennium BP was a formative period in the development of human societies. The earliest large, complex, urban, state-level societies emerged in Mesopotamia and Egypt in the middle and late 6th millennium BP respectively, and the foundations of complex societies were laid in South Asia, northern China, and northern coastal Peru during the same period (Brooks 2006). Brooks (2006) has argued that climate change and associated changes in the availability and distribution of resources played a key role in these developments. While RCC at 8200 and 4200 BP has received a great deal of attention in the paleoenvironmental literature (Marchant and Hooghiemstra 2004; Alley and Ágústsdóttir 2005; Rohling and Pälike 2005; Wang et al. 2005) and increasingly in the archaeological literature (Weiss et al. 1993; Cullen et al. 2000; Wenxian and Tungshen 2004; Clare et al. 2008), these events may ultimately have been less important for the long-term development of human societies than lesser known episodes of RCC in the 6th millennium BP, whose impacts on landscapes and resources were more sustained.

The following discussion seeks to infer specific human responses to RCC by considering archaeological data within their paleoenvironmental contexts. The discussion focuses on the Sahara, the Nile Valley, and Mesopotamia, which provide us with evidence of profound changes in environmental conditions that can be linked with specific cultural changes at particular localities, and/or evidence of cultural transitions that coincide with episodes of regional-scale RCC and which are consistent with their expected impacts. Each of these three regions is discussed in turn, with particular attention given to (a) the widely recorded shift towards aridity throughout the NHST in the late 7th millennium BP and around the time of the 5900 BP Bond event, (b) the period from around 5800–5500 BP, when marine and terrestrial records indicate rapid landscape change associated with severe desertification in Afro-Asiatic arid belt, and (c) the late 6th millennium BP, with particular attention to climatic deterioration around 5200 BP.

4.3.2 The Sahara: Climate, Livelihoods, Mobility and Social Organization

With the exception of the Nile, the Sahara contains no permanent rivers, and outside oasis and highland areas productive landscapes supporting human populations were transformed to hyperarid desert with the collapse of both summer monsoonal and winter rainfall as the Holocene Climatic Optimum drew to a close (Brooks et al. 2005; Brooks 2006). Consequently, the Sahara has been described as a "theoretically ideal 'training ground' for the analysis of social responses to major environmental alterations" (di Lernia 2006, p. 51) and "a unique natural laboratory for the reconstruction of the links between changing climate and environments, and human occupation and adaptation" (Kuper and Kröpelin 2006, p. 803). It is no coincidence that the Sahara and the Nile Valley provide us with some of the clearest evidence of human responses to severe, climatically-driven, landscape change.

There is considerable evidence linking cultural and environmental changes in the Sahara prior to the 6th millennium BP. Changes in the number and distribution of occupation sites over time indicate an expansion of human occupation in the Sahara north of 23°N from the onset of humid conditions around 10,000 BP until about 7000 BP, with a temporary southward shift in occupation around the time of the 8200 RCC (Fig. 4.2b; Vernet and Faure 2001). The cessation of this expansion around 7000 BP coincided with the arid episode from 7300 to 6900 BP evident in records from across northern Africa (di Lernia 2002). In the Fezzan region of the Libyan central Sahara (Fig. 4.1), which appears to have acted as refuge during dry periods throughout the middle and late Holocene (Brooks et al. 2005; Brooks 2006), there is abundant evidence of adaptation to aridity in the form of in-migration, the emergence of transhumance, and the development of vertically differentiated pastoralism, with sheep and goats in highland areas and cattle in lowland areas where activity was focused around the fringes of lakes (di Lernia 2002).

Cremaschi (2002) and di Lernia (2002) present a range of evidence indicating a severe and abrupt transition to aridity in the Fezzan between 5850 and 5600 BP, coinciding with the onset of wider Saharan aridity as inferred from the dust record of de Menocal et al. (2000) as described above. In the Fezzan this transition marked the end of the Holocene humid episode; while conditions remained more humid than at present in some oasis localities until the late 4th or 3rd millennium BP (di Lernia et al. 2002; Drake et al. 2004), the majority of lowland areas desiccated as lakes dried and present day arid climate vegetation was established in the middle to late 6th millennium BP (Cremaschi 2002; di Lernia 2002).

di Lernia (2002) identifies two types of response to this humid-arid transition in the Fezzan. The first involved increased mobility in the form of year-round, large-scale movement of pastoralists, evident from sites in the Acacus Mountains and the remains of transient camps in adjacent regions, including on dry lake beds. Cattle disappeared from most areas as the emphasis shifted to sheep and goats (whose presence is demonstrated by dung layers in the Acacus sites), which are more tolerant of aridity. Exotic raw materials, stone tools and ceramics indicate that these wide ranging groups maintained contacts with regions many hundreds of kilometers distant.

The second response involved the abandonment of areas subject to desertification by certain pastoral groups, which shifted their focus to oasis areas such as the Wadi Tanezrouft and the Wadi al-Ajal (di Lernia 2002; Mattingly et al. 2003). In these areas the density of occupation/settlement sites increased as populations became more sedentary. Cattle herding persisted in some localities, although the shift in emphasis to sheep and goats is also apparent in these areas. An increase in plant-related tools suggests more intensive exploitation of the landscape and a possible move towards cultivation (di Lernia 2002). The ultimate outcome of this process of in-migration and innovation was the emergence of the Garamantian Tribal Confederation in the 3rd millennium BP, with a transition to irrigated agriculture, urban living and organized political life occurring only after the final disappearance of surface water, and made possible by the peculiar topographic and geographic characteristics of certain scarp-foot oasis areas in this region (Mattingly et al. 2003; Brooks et al. 2005; Brooks 2006).

In the data presented by Vernet and Faure (2001) there is no significant change in the number of dated occupation sites in the Sahara north of 23°N throughout the 6th millennium BP (Fig. 4.2b). However, a steady increase in the number of such sites in the southern Sahara and Sahel, south of 23°N, after 6000 BP suggests an expansion of population in this region as conditions became drier. This might be the result of drier conditions facilitating the southward expansion of pastoralism as humid climate barriers such as disease were removed (Smith 1984). Cattle appear in the archaeological record in the early 6th millennium BP in Mali and Mauritania, south of 23°N, lending weight to this interpretation (Jousse 2004).

The transition from the 6th to the 5th millennium BP is marked by a dramatic and permanent decline in the number of dated occupation sites north of 23°N, and by a less pronounced and temporary decline south of 23°N (Fig. 4.2b). Within the margins of dating error, and given the temporal resolution of the occupation site data, these apparent demographic changes may be said to be broadly coincident with the 5200 BP RCC. The decline in occupation sites both north and south of 23°N suggests that the abandonment of desiccating areas was associated with population decline and/or the concentration of human populations in favorable areas—environmental refugia—where water, pasture and wild food resources were still available, echoing the situation in the Fezzan earlier in the 6th millennium as described by di Lernia (2002).

A further, dramatic increase in sites south of 23°N is evident around 4000 BP, suggesting further expansion of human populations in the southern regions following the 4200 BP RCC. It is likely that "pull" factors operated to encourage certain groups to move into new areas, for example the retreat of lake shorelines and the resulting availability of new land. For example, "... the first settlers of the Chad Basin in Nigeria entered a 'virgin' area which had previously been flooded" by Megalake Chad in the 4th millennium BP (Breunig and Neumann 2002, p. 150). However, this expansion follows a second decline in the number of sites north of 23°N around 4200 BP, suggesting that the southward migration of groups fleeing aridity in more northerly regions also contributed to demographic changes at this time. Such southward migrations are "archaeologically traceable from Mauritania to Niger" according to Breunig and Neumann (2002, pp. 149–150), who also associate them with environmental deterioration.

Climatic and environmental changes in the Middle Holocene Sahara, and in the 6th millennium BP in particular, also appear to have been associated with wider changes in social organization. Such changes have been inferred from the Saharan funerary record, particularly from the wide variety of monumental stone funerary structures which today are found in great abundance throughout most of the Sahara (Gauthier 2009). Monumental funerary architecture appears to have spread throughout the Sahara in association with cattle pastoralism, and the earliest known monumental burials, dating to the mid to late 8th millennium BP, contain only faunal (principally cattle) remains (Sivili 2002; di Lernia 2006). Funerary monuments were being used for both human and animal internments by the early to mid 7th millennium BP, placing the earliest monumental human burials in the centuries following the 7300–6900 BP arid event in northern Africa (di Lernia 2002). Excavated monuments dating from the middle of the 6th millennium BP onwards (coincident with or rapidly following the abrupt transition to aridity), are associated with human burials (Sivili 2002).

In a drying environment, monumental funerary structures may have served to assert rights over land at a time of heightened climatic insecurity, as increasingly mobile pastoralists moved to new areas in search of diminishing resources. di Lernia (2006) explicitly links the apparently rapid spread of pastoralism throughout the central Sahara in the 7th millennium BP with periods of aridity and a "leap-frogging" of pastoral groups from one location to another in search of resources. The transition from animal to human burials around this time has been interpreted as an expression of incipient social stratification (di Lernia et al. 2002), a phenomenon which has been linked with increased social organization or complexity in response to climatic desiccation and resource scarcity throughout the middle Holocene NHST (Brooks 2006). In the Acacus region, monumental human burials emerged after the shift towards permanent aridity around 5700 BP, a time that di Lernia (2002) speculates was characterized not only by increasing social stratification but also by conflict, the latter suggested in rock paintings and engravings. Within this context it may be significant that the earliest monumental human burials in the Sahara were of single adult males (di Lernia et al. 2002).

While the evidence for environmentally mediated conflict may be equivocal, it seems beyond doubt that in the middle Holocene Sahara, and during the 6th millennium in particular, severe climatic desiccation was a dominant factor driving migration, livelihood innovation, and changes in social organization. Referring to the last of these, di Lernia et al. (2002, p. 296) conclude that "The stone tumuli nowadays punctuating the Saharan landscape may represent the solitary funerary testimony of ... dramatic climatic change."

4.3.3 Egypt: Confinement and Social Stratification

As the only permanent river in the Sahara, the Nile was a natural focus for populations adapting to increasing regional aridity in the middle Holocene, and there is considerable evidence that the emergence of Egyptian Dynastic civilization owed much to developments in the adjacent desert. As in the wider Saharan region, livelihood innovation, population movements and changes in social organization are evident during this period, with the 6th millennium BP representing the formative period of Egyptian Dynastic Culture (Midant-Reynes 2000; Wengrow 2001).

The beginnings of the Egyptian Predynastic period may be dated to the late 7th millennium BP and the emergence of the Badarian culture in the Upper (southern) Egyptian Nile Valley, a culture whose origins, while multiple and complex, were probably most heavily influenced by events in the Western Desert (Hendrickx and Vermeersch 2000). It is during the Badarian that evidence of social hierarchies first appears in Upper Egypt, in the form of variations in the size of burials and the wealth associated with them (Wilkinson 1999; Midant-Reynes 2000). The final centuries of the 7th millennium BP also yield the earliest funerary evidence of status differentiation in Lower (northern) Egypt (Wilkinson 1999). Echoing the emergence of social stratification in the wider Saharan region (di Lernia et al. 2002), these developments took place at a time of climatic and environmental desiccation in the areas adjacent to the Nile (Wendorf and Schild 1998; Linstädter and Kröpelin 2004; Nicoll 2004; Bubenzer et al. 2007). As in the central Sahara, the emergence of the earliest social hierarchies in Egypt thus took place at a time of environmental deterioration, as human populations were facing a contraction in their geographical ranges and livelihood options.

After 6000 BP, when a phase of playa formation ceased at Farafra Oasis (Hassan et al. 2001), aridity accelerated in the areas adjacent to the Nile (Fig. 4.4), as in the wider Sahara region. Sahelian and Sudanian vegetation was replaced with *Acacia* and scrub in the vicinity of Oyo, grasslands in southern Egypt diminished, the Wadi Melik ran dry, and input to the Nile from Wadi Howar was reduced (Nicoll 2004). Lakes in the Fayum waned after 5700 BP, the Selima area dried around 5600 BP, and most rainwater-fed playas were significantly desiccated by 5500 BP, with drying of playas at Bir Kiseiba around 5400 BP (Nicoll 2004). Hand-dug wells dating to the 6th millennium BP at playa sites throughout the eastern Sahara attest to diminishing surface water, and there is evidence that many locations in southern Egypt were abandoned after 6000 BP (Nicoll 2004). Malville et al. (1998) identify the onset of hyperaridity at Nabta Playa in southwestern Egypt around 5500 BP, and speculate about the impact on the societies of the Nile Valley of the arrival of nomadic groups from the adjacent desert areas. The strong influence of pastoralists from the desert regions adjacent to the Nile on the development of what was to become Egyptian Dynastic culture may be traced through the evolution of iconography and ritual objects originating in the mobile pastoral cultures of the Predynastic (Wengrow 2001).

Midant-Reynes (2000, p. 232) describes a "gradual movement of human settlements from the deserts towards the [Nile] river valley" throughout the 6th millennium BP as the adjacent regions desiccated. Within this context, the first permanent occupation at the key Nile Valley settlement of Hierakonpolis (Fig. 4.4) took place around 6000 BP, followed by a dramatic increase in the town's population between about 5800 and 5400 BP, as Hierakonpolis became a dominant regional political center (Hoffman et al. 1986).

The area of settlement at Hierakonpolis contracted between the Naqada I or Amratian period (ca. 5800–5500 BP) and the Naqada II or Gerzean period (ca. 5500–5200 BP), with settlement in the latter period more tightly focused on the banks of the Nile (Midant-Reynes 2000). Midant-Reynes (2000, p. 232) describes the population at Hierakonpolis as being "gradually squeezed into the confines of the alluvial plain, abandoning the desertified wadis" in the final stages of the Predynastic, towards the end of the 6th millennium BP. As the surrounding areas dried, it appears that cattle herders were adapting to a sedentary lifestyle at Hierakonpolis during the Naqada II period from about 5500–5200 BP. Hoffman (1982) describes enclosures with stalls and troughs associated with a large basin containing the remains of barley, and close to a small settlement interpreted as a herding station, evidence summarized by Wengrow (2001, p. 97) as suggesting "an attempt to combine the maintenance of herds with increased sedentism, probably through an increase in artificial feeding with cultivated grain." This is significant as it indicates that previously mobile pastoralists were adapting to a sedentary lifestyle in a riverine area

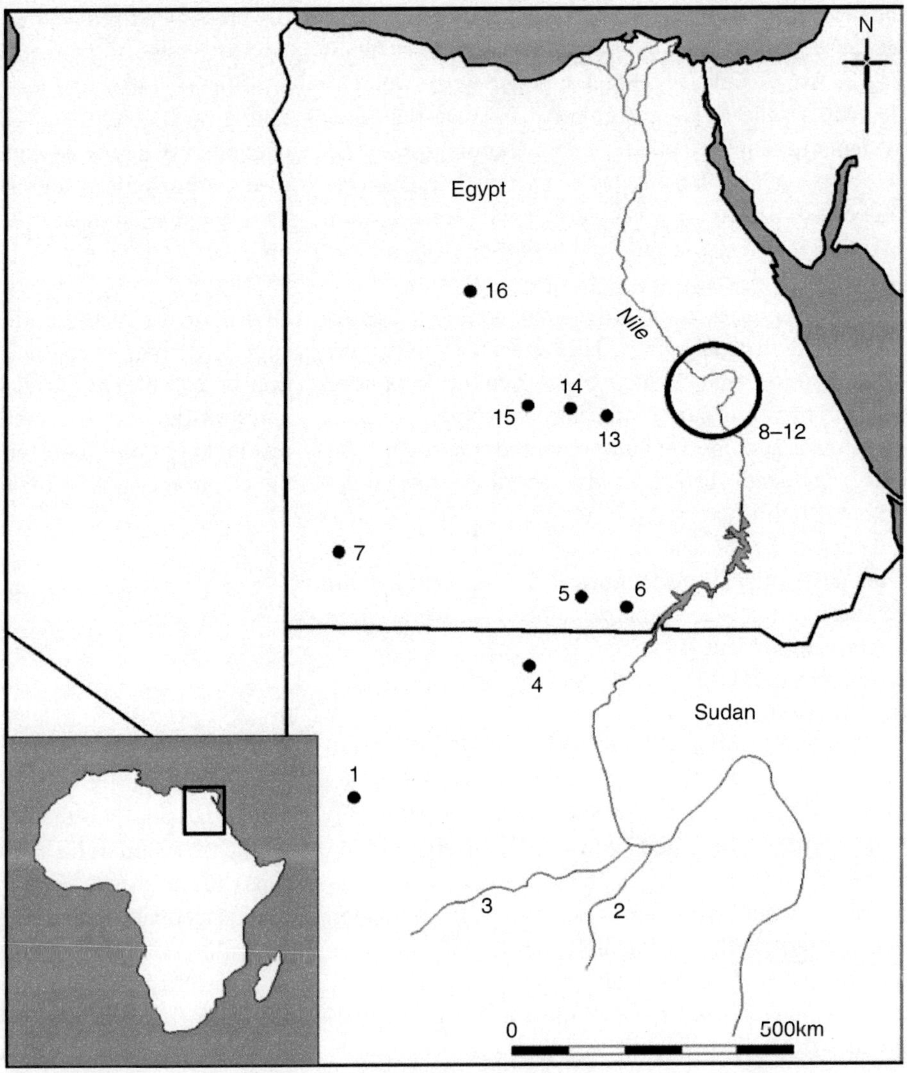

Fig. 4.4 Locations of sites in the eastern Sahara and Nile Valley mentioned in the text. *1* Oyo; *2* Wadi Melik; *3* Wadi Howar; *4* Selima; *5* Bir Kiseiba; *6* Nabta Playa; *7* Gilf Kebir; *8* Hierakonpolis; *9* Abydos; *10* Nekhen; *11* Mahasna; *12* Armant; *13* Kharga Oasis; *14* Abu Tartar Plateau; *15* Dakhla Oasis; *16* Farafra Oasis

with permanent surface water, at a time when adjacent areas were becoming hyper-arid, echoing the confinement of cattle herding to oasis areas in the Fezzan (di Lernia 2002). At other sites in Upper Egypt human activities also moved away from the marginal areas at the fringe of the desert towards the Nile flood plain (Midant-Reynes 2000).

This period of population growth, adaptation to sedentary lifestyles, and early urban development in the Upper Egyptian Nile Valley coincided with further moves towards the formal hierarchies that would characterize the Dynastic period. Wilkinson (1999, pp. 29–30) describes evidence of inherited status in Naqada I child burials at Mahasna and Armant (Fig. 4.4), indicating that the heredity principle operated "to the benefit of a restricted elite" at this time. Wilkinson (1999, p. 31) argues that "an ideology of power was being formulated by the ruling lineages of Upper Egypt" towards the end of the Naqada I period (ca. 3800–3500 BP), based on the first appearance of motifs associated with royal power in tombs at Hierakonpolis. Hierakonpolis also boasts the earliest funerary murals, in the so-called Painted Tomb, which is likely to date from the Naqada II period, and which depict boats, hunting, and images clearly representing violence and authority that are reminiscent of later iconography from the very late Predynastic and early Dynastic periods (Midant-Reynes 2000).

The unification of Egypt in the Late Predynastic Naqada III period (5200–5000) BP took place within a century or two of a period of further regional environmental deterioration in the eastern Sahara that coincided with the more widespread 5200 BP RCC event. Localised evidence of environmental change is apparent in the vicinity of Hierakonpolis, where alluvial deposits of the Nekhen Formation indicate that wadi systems became effectively inactive around 5200 BP (Hoffman et al. 1986). Away from the Nile, desiccation of the Gilf Kebir was complete with the cessation of the winter rains around 5300 BP (Linstädter and Kröpelin 2004).

These centuries saw a further concentration of settlement along the banks of the Nile and a shift in emphasis from pastoralism to intensive agriculture supported by irrigation. It is during this period that the first kings of Upper Egypt are formally recorded, the royal cemetery was established at Abydos, and the earliest monumental public architecture appears at Nekhen near Hierakonpolis (Midant-Reynes 2000). Given the shift from pastoralism to agriculture, it is significant that the ceremonial King Scorpion mace head recovered from Hierakonpolis and associated with one of the earliest rulers of a unified Egypt, depicts the king opening an irrigation channel.

The trends of increased social stratification and organization in the Nile Valley that ultimately resulted in the unification of Upper and Lower Egypt appear to have been associated with increased competition and conflict. The iconography of the Late Predynastic period includes numerous depictions of crenellated walls, interpreted by Midant-Reynes (2000, p. 203) as indicating "an ideological landscape that was evidently marked by a tendency towards violence." Some settlements of this period, such as Elkab on the opposite bank of the Nile to Hierakonpolis, were surrounded by enclosure walls, and the town of Nekhen was heavily fortified. The famous Narmer Palette portrays the unification of Egypt as a violent act, although the extent to which this represents the reality of unification has been questioned (Midant-Reynes 2000).

While there is still much debate concerning the details of Egyptian unification, it was almost certainly associated with an expansion of the Upper Egyptian Naqada culture at a time of RCC and deteriorating regional environmental conditions, when "more and more non-productive members of a growing population were grouped together in the agricultural regions of the flood plain, exerting demographic pressure which would provide a decisive impetus for the Naqada expansion" (Midant-Reynes 2000, p. 237). This in turn appears to have represented the culmination of trends towards increased settlement in the Nile Valley as adjacent areas became unproductive, and greater social stratification that was most likely associated with certain groups exploiting geographical, environmental, economic or social advantages to exert greater control over diminishing resources. Egypt appears to provide us with an exemplary case study of how the confinement of a growing population within an increasingly restricted geographical area due to climatically-driven environmental deterioration and resource scarcity can stimulate competition and increased social organization or complexity, resulting in social differentiation and stratification (Brooks 2006). In the Nile Valley this resulted not only in profound cultural change, but also the creation of one of the world's earliest states.

4.3.4 Mesopotamia: Shifting Settlement Patterns and Urbanization

In Mesopotamia the late 7th and early 6th millennia BP were associated with the transition from the Ubaid to the Uruk culture. In northern Mesopotamia (encompassing northern Iraq, northeastern Syria, southeastern Anatolia and the northern Zagros Mountains) the terminal Ubaid spanned several centuries. Ubaid occupation ceased around 6300 BP in the climatically marginal middle Khabur River region of northeastern Syria (Fig. 4.5), after some five centuries of occupation, and widespread abandonment of sites occurred around 5900 BP in the wider Khabur region and at the site of Hamman et-Turkman (Akkermans and Schwartz 2003). As certain sites were being abandoned at the end of the Ubaid, population agglomeration was occurring at other locations. For example, Ur et al. (2007) present evidence that the settlement of immigrant groups in distinct clusters was the driving force behind the beginnings of an early phase of urbanism at Tell Brak (Fig. 4.5) in northeastern Syria between 6200 and 5900 BP.

According to Akkermans and Schwartz (2003, p. 159) Ubaid groups in northern Mesopotamia eschewed irrigation agriculture, and instead "clearly favored the fertile, well-watered lands in regions suited to dry farming", with Ubaid sites in northeastern Syria "primarily found along the Euphrates and its tributaries". In this context it is significant that the abandonment of many of these sites during the termi-

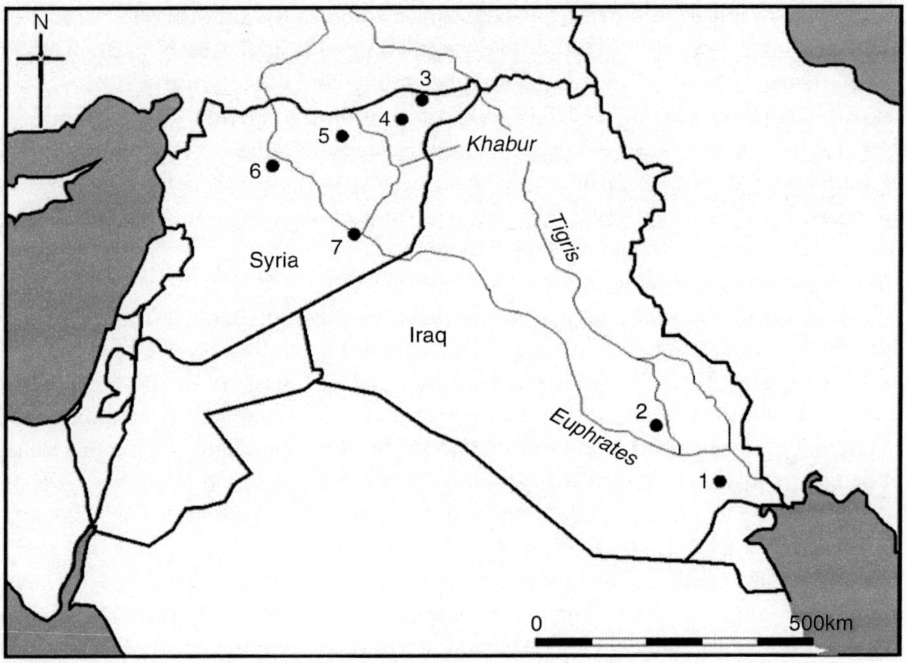

Fig. 4.5 Locations of sites in Mesopotamia mentioned in the text. *1* Uruk-Warka; *2* Abu Salabikh; *3* Tell Leilan; *4* Tell Brak; *5* Hamman et Turkman; *6* Habuba Kabira; *7* Qraya

nal Ubaid coincided with a greater emphasis on animal husbandry and pastoralism (alongside the hunting of wild game), with intensified use of desert steppe environments beyond the river valleys (Akkermans and Schwartz 2003). Given the environmental context of RCC and regional aridity during this period (Jung et al. 2004; Stevens et al. 2006), these factors strongly suggest that people were adapting to drier conditions by changing their livelihoods, settlement patterns and use of the landscape, an interpretation also supported by the fact that the large urban centers that were emerging in this period, such as Tell Brak, were situated away from riverine environments where arid conditions would have had an adverse impact on unirrigated agriculture (Akkermans and Schwartz 2003). During the long Ubaid period there was little social differentiation, no permanent leadership or centralised authority, little variation in settlement size or economic activity within and between settlements, and no overt displays of wealth or status (Akkermans and Schwartz 2003). The onset of the Uruk period marked a shift to more unequal societies, with a greater emphasis on the individual, and explicit displays of social differentiation, wealth and political power through phenomena such as prestige goods and monumental public architecture.

As in the Nile Valley, the trend towards more differentiated and hierarchical societies continued throughout the 6th millennium BP in Mesopotamia, particularly in the southern alluvial regions (southeastern Iraq and the adjacent Susiana Plain on southwestern Iran). Here, trends towards population agglomeration, urbanization and inequality culminated in the development of the city of Uruk-Warka (the ancient city of Uruk being situated near the modern settlement of Warka), the largest urban center in the region (Fig. 4.5; Pollock 2001). Clear social and political hierarchies are apparent from Uruk inscriptions and iconography, associated with a strong political authority, a clear division of labor and a complex administrative system (Wright 2001). Pollock (2001) describes a repressive and violent political environment in the region around Uruk-Warka, coupled with evidence of demographic volatility and unstable settlement patterns, and contrasts this with apparently more stable and less repressive conditions of adjacent areas in southern Mesopotamia.

The development of Uruk-Warka and other urban centers was broadly coincident with the acceleration of aridity in Western Asia and North Africa between about 5800 and 5500 BP, as indicated by marine sediment records (Sirocko et al. 1993; de Menocal et al. 2000; Jung et al. 2004). This period of apparently rapid regional environmental deterioration coincided with the abandonment of villages in the Susiana Plain and between Abu Salabikh and Tell al-Hayyad (Fig. 4.5), with the latter area given over to pasture and foraging as river courses changed significantly (Algaze 2008).

Algaze (2008) interprets these episodes of abandonment as the creation of buffer zones between competing settlement centers. However, given the timing of these phenomena and the wider context of environmental change, we might ask whether such changes in settlement patterns were in whole or in part associated with migration to favorable localities in areas that were otherwise becoming less habitable due to environmental deterioration. Of course environmental and political drivers may well have interacted, with environmental deterioration resulting in the control of dwindling resources by well placed elites, and/or increasing conflict and repression as a less productive environment undermined food security and the ability of people to provide tribute to elite groups. Kennet and Kennet (2006) argue that increasing aridity reduced the productive potential of land and encouraged population agglomeration in favorable areas, and also increased competition for resources, stimulating urban development and the emergence of coherent political entities. They also propose that a stabilization and subsequent fall in sea levels from around 6000 BP, after several millennia of sea-level rise, exposed new coastal land in southern Mesopotamia which resulted in more extensive flood plains and a higher water table, permitting irrigated agriculture which in turn supported population growth through the production of food surpluses. It is possible that changes in sea level and aridity interacted; expansion of settlements and associated population growth encouraged by falling sea levels and the availability of newly productive areas might have increased the population exposed to subsequent arid crises, stimulating competition and population movement.

A key development in mid-6th millennium BP Mesoptamia was the Uruk expansion, indicated by the establishment in northern Mesopotamia of settlements characterized by southern Mesopotamian Uruk material culture (Akkermans and Schwartz 2003). While the Uruk expansion has been interpreted as an imperial expansion by a polity centered on Uruk-Warka seeking to establish control of key trading points on its periphery (Algaze 1993), it is increasingly argued that cities and states in mid-6th millennium BP Mesopotamia emerged in a "culturally homogenous but politically fractious" environment (Algaze 2008, p. 114). More recently, it has been suggested that this expansion may have been driven by the search for new cultivable lands by impoverished farmers from southern Mesopotamia in the early-middle 6th millennium BP (Akkermans and Schwartz 2003).

Uruk migrants may have been displaced from southern Mesopotamia by increasingly powerful elites (Ackermans and Schwartz 2003), and Pollock (2001, p. 220) goes so far as to argue that the Uruk colonists were "essentially refugees" fleeing "heavy demands for their labor and products" in a society that "was not, for many people, a pleasant place to live". Johnson (1988–1989) suggests that the Uruk expansion may have been driven by demographic crises and unrest in southern Mesopotamia, associated with competition and conflict that caused displacement and migration, although this interpretation remains controversial (Akkermans and Schwartz 2003). While the nature of the Uruk expansion, and of relations between southern colonists and indigenous northern populations is still hotly contested, recent evidence is suggestive of violent conflict at key sites in northern Mesopotamia where indigenous material culture was replaced by, or combined with, Uruk-style artefacts, around the time of the Uruk expansion (Lawler 2007). While the nature of the Uruk expansion remains enigmatic, and while firm evidence linking cultural changes with environmental change during this period in Mesopotamia is lacking, the coincidence of demographic change and likely social disruption with widespread regional desiccation demands further investigation. A further notable coincidence in Mesopotamian history is the collapse of the Uruk culture at the time of the 5200 BP RCC. According to Wright (2001, p. 146) "From about 3350 BC until 3100 BC, the predominant dynamic [in Greater Mesopotamia] is one of differential growth, accelerating regional conflict, the emergence of very large polities, and their collapse." Uruk-related material culture disappeared throughout most of northern Mesopotamia around this time, and Uruk colonies in the north, such as Habuba Kabira (founded in the mid 6th millennium BP and occupied for several centuries) and Qraya (Fig. 4.5), were abandoned (Akkermans and Schwartz 2003). This period saw the "abandonment of many smaller settlements and formerly inhabited areas" throughout greater Mesopotamia, and a further increase in the importance of mobile pastoralism and transhumance (Wright 2001, p. 146; Hole 1999). The emergence of hierarchical settlement clusters for the first time in some riverine areas (Wright 2001) suggests population agglomeration in favorable locations on a scale that required significant social organization, echoing the situation in the Nile Valley at this time.

Despite the regional collapse of the Uruk culture, "the region around Uruk-Warka played host to a sudden tenfold increase in settlement density at about 3200 BC" (Matthews 2003, p. 110, citing Nissen 1988, pp. 66–67). Nissen (1988), deploying similar arguments to those used more recently by Kennet and Kennet (2006), has proposed that this increase in population was due to climatic desiccation, which lowered water tables and prevented constant inundation, resulting in the availability of new, fertile land in the Uruk-Warka region. Such a phenomenon may well have acted as a pull factor, enticing people to settle in this region, and this may have been complemented by the push factor of reduced productivity elsewhere. While the relative roles of pull and push factors cannot easily be assessed, it is worth recalling Pollock's (2001) description of a repressive and violent political economy in the Uruk-Warka region, and noting that the iconography of this period is representative of coercive authority and political violence, with texts indicating controlled labor (Wright 2001). If late 6th millennium BP in the Uruk-Warka region was as unpleasant as has been suggested, we might infer that the push factor of environmental deterioration in neighboring areas probably played the dominant role in encouraging people to congregate in a geographically favorable, but socially and politically unappealing location.

Links between climatically driven environmental change and cultural change are less well established for Mesopotamia than for the Sahara and Egypt, and will remain so until the current lack of published, high resolution paleoclimatic records from within the Mesopotamian region is addressed. Nonetheless, the coincidence of periods of widespread societal disruption and cultural transition with episodes of RCC, and the compatibility of changes in settlement patterns and livelihoods with expected responses to environmental deterioration, is striking. Wright (2001) identifies key periods of social disruption and cultural transition centered on the 5900 BP and 5200 BP RCC episodes, from 6100–5750 BP (4150–3800 BC) and 5300–5050 BP (3350–3100 BC). These periods, and the 6th millennium as a whole in Mesopotamia, deserve much more attention in studies of linked environmental and cultural change.

4.4 Synthesis: Recurring Responses to RCC and Climatic Desiccation

The strength of the evidence linking cultural change with climatic and environmental change varies across the regions discussed above. Evidence from the Sahara is particularly compelling, as a result of (a) the explicit consideration of the links between climatic and cultural change in archaeological studies, (b) consistent evidence of such links apparent at different spatial scales, and (c) the demonstrable severity of landscape change in the Middle Holocene in this region. In Egypt the likely role of desiccation in the areas adjacent to the Nile has long been recognized as an important factor in the development of Dynastic civilization (Hoffman et al. 1986; Midant-Reynes 2000; Wilkinson 2003), and new evidence continues to strengthen this link. Crucially, this evidence explicitly links demographic and cultural changes with evolving environmental constraints in a drying landscape, through archaeological and paleoenvironmental data from specific locations in the Nile Valley and the surrounding desert. In contrast, evidence of environmental and landscape change that can be linked with archaeological data in Mesopotamia is scarce, making the links between environmental and cultural change in this region less secure. Nonetheless, many of the trends and transitions apparent in the Mesopotamian archaeological record are consistent with responses to increased aridity, and are coincident with periods of RCC and wider regional desiccation. In addition, they represent broadly similar cultural changes to those apparent in the other two regions examined here, as discussed below. In all the regions examined, episodes of profound demographic and cultural change, and transitions from one archaeological period to another, are coincident with RCC and associated environmental transformation and deterioration, a pattern that is repeated throughout the NHST (Brooks 2006).

The intention here is not to propose a deterministic model of environmentally-driven cultural change, but rather to identify a set of common responses that, while not universal, nonetheless recur across space and time. While such common responses are strongly suggested by archaeological and paleoenvironmental records, it is clear that societies in different geographical regions did not all respond in the same way to any given RCC event. Furthermore, different response or adaptation trajectories were followed during different periods in any given region. The nature of responses to RCC at any given time and place was mediated by local environmental and cultural circumstances, and the associated opportunities for, and constraints on, adaptation.

The responses to RCC that were possible at any given time and location depended on the magnitude of the associated changes in the local environment, and the impact these changes had on key resources. For example, drier conditions in the Sahara throughout the 7th millennium BP appear to have encouraged the spread of mobile pastoralism. A greater reliance on mobile pastoralism is also seen in parts of Mesopotamia during periods of regional-scale aridity. However, drier conditions in Egypt at the end of the 6th millennium BP were associated with a reduction in mobility and a shift in emphasis from pastoralism to agriculture. As in much of the Sahara, environmental change in the regions adjacent to the Nile at the end of the 6th millennium BP was so severe that mobile pastoralism ceased to be viable.

The geographic differentiation of responses might be also be illustrated by the beginning of the drift of population towards the Nile in the very early 6th millennium, and the abandonment of riverine environments in the Middle Khabur region in northeastern Syria in favor of the steppe at the same time. These apparently opposing trends do have in common the apparent abandonment of areas where desiccation placed additional constraints on livelihood options, and associated population agglomeration in areas seen as more favorable. We might also point to contemporaneous adaptation strategies in the Fezzan based on the sedentarization of cattle herding on the one hand, and on enhanced mobility based on sheep and goat herding on the other (di Lernia 2002). Whether people were adapting to aridity by moving from river to steppe or desert to river, or through increased mobility or greater sedentism, this adaptation involved fundamental changes in livelihoods and the ways in which landscapes were exploited.

The mediation of adaptation by local environmental factors should also be highlighted. For example, irrigated agriculture could be adopted only where fertile land existed alongside readily exploitable water resource, as was the case in river valleys and in the peculiar context of the scarp foot oases of the Libyan Fezzan (Mattingly et al. 2003).

In addition to environmental factors, adaptation was probably mediated by pre-existing practices and cultural preferences, as suggested by the fact that the late Ubaid cultures in the Middle Khabur region of northern Mesopotamia did not adapt their agriculture to drier conditions by developing irrigation systems, but instead shifted their focus to the adjacent steppe. This may have resulted from a lack of expertise in irrigation, cultural preferences for herding rather than intensive agriculture, local environmental factors that mitigated against irrigation, a perception that steppe pastoralism was a more obvious and easier choice, or a combination of these and other factors.

One very robust outcome of the cultural changes that accompanied widespread desiccation in the NHST was increased social stratification, which reached its ultimate expression in the extremely hierarchical society of early Dynastic Egypt and in the coercive early states of southern Mesopotamia. The earliest evidence of social hierarchy dates to the transition from the 7th to the 6th millennium BP not only in Egypt and Mesopotamia, but also in China (Liu 2004), and is contemporaneous with the spread of social stratification throughout the Sahara evident in the form of monumental funerary architecture associated with human internments. The move towards more unequal societies is most readily understood as a manifestation of increased social organization or complexity. Brooks (2006) argues that increased social complexity in middle Holocene northern Africa, western Asia, South Asia, north-central China and northern coastal Peru resulted in large part from ad hoc adaptation to resource scarcity. As the environments of these regions became more arid, variable and marginal, greater social organization would have been required to ensure food security, and control over diminishing resources would have provided mechanisms for the emergence of elites. As resources became scarcer there would have been more opportunities for groups occupying favorable geographic, environmental, economic or social positions to exercise influence over access to resources, and to demand some form of tribute, whether economic or in terms of enhanced social status, from those seeking such access.

If increasing social complexity, and by implication the development of civilization itself (Brooks 2006), represented a type of adaptation to climatically-induced environmental deterioration and resource scarcity, it was an adaptation associated with significant costs to large sections of society, principally a loss of individual autonomy and "economic and social subjugation" (Kennet and Kennet 2007, p. 252). The shift from the relatively egalitarian societies of the 7th millennium BP to more complex social systems in which minority elites exerted power and influence over the majority of the population is summed up by Algaze (2001 p. 212), who states that "Early Near Eastern villagers domesticated plants and animals. Uruk urban institutions, in turn, domesticated humans."

4.5 Lessons from the 6th Millennium BP for the Twenty-First Century

The 6th millennium BP provides us with an opportunity to examine actual societal responses to RCC, and to ground the intensifying international adaptation debate (Giddens 2009) in empirical evidence. Despite the vast changes that have affected human societies over the past 6000 years, a large proportion of the world's population still depends on unirrigated subsistence or small-scale agriculture, and pastoralism remains an important pillar of many livelihoods. Around half the world's population currently lives in urban centers (UNFPA 2007) – the culmination of processes of urbanization that began in the 6th millennium BP. Both urban and rural populations, and the social and economic systems that bind them together, will need to adapt to climate change over the coming decades and centuries.

While there is a burgeoning industry emerging around issues of adaptation funding and policy, the bureaucratization of adaptation by governments and development agencies will not wholly replace adaptation based on the independent actions of individuals and groups acting in response to environmental and other pressures, and within the context of existing opportunities and constraints. Autonomous adaptation will continue to occur on an ad hoc basis, albeit within contexts influenced to a greater or lesser extent by policy. Where adaptation is policy driven, these policies will need to consider the implications of potential RCC. As discussed at the beginning of this chapter, the twenty-first century is likely to bear all the hallmarks of a period of rapid climate change. Qualitatively similar processes of desiccation and environmental deterioration to those operating in the 6th millennium NHST are indicated by climate projections for the Maghreb, the greater Kalahari region, Central Asia, and parts of North America, while other regions will be affected by loss of land and natural resources resulting from processes such as sea level rise and ocean acidification (Thomas et al. 2005; Christensen et al. 2007; Nicholls et al. 2007). Current adaptation plans, programmes and policies do not address such challenges (Brooks et al. in press), and policymakers and development practitioners are ill equipped to consider the implications of RCC. The adaptation evidence base therefore needs to be extended to encompass case studies of RCC, making the 6th millennium BP particularly instructive. In northern Africa, western Asia and elsewhere in the NHST, RCC associated with environmental desiccation and resource scarcity in the 6th millennium BP appears to have been associated with the following cultural processes or responses:

a. Changes in the nature of livelihoods and the way in which landscapes were exploited, for example transitions from pastoralism to agriculture and vice versa, and changes in the types of pastoralism practiced.
b. Migration out of areas where resource availability fell below critical thresholds, illustrated by the abandonment of much of the Sahara and riverine settlements in the middle Khabur region of northern Mesopotamia.
c. Population agglomeration in favorable locations where resources remained available (environmental refugia) in otherwise less productive landscapes, such as the Acacus region in the central Sahara, the Nile Valley, and settlements such as Tell Brak and Uruk-Warka in Mesopotamia. A concentration of population in riverine areas in otherwise desiccating environments during the 6th and 5th millennia BP is indicated by archaeological records from across the NHST and also northern coastal Peru (Brooks 2006).
d. Increased territoriality associated with competition and conflict, suggested by evidence of violence and the fortification of settlements, such as in Upper Egypt and parts of Mesopotamia.
e. Changes in social organization driven by the need to manage limited resources and high population densities in restricted geographical areas, and widely manifested in terms of greater social stratification and inequality, evident across the NHST (Brooks 2006).

f. Cultural fragmentation and collapse, for example during the Ubaid-Uruk transition and at the end of the Uruk period in Mesopotamia. Collapse of centralised authority and societal disruption is also well established at the time of the 4200 RCC in Egypt (Hassan 1997), Mesopotamia (Weiss et al. 1993; Cullen et al. 2000) and China (Wenxiang and Tungshen 2004).

Given the potential severity of anthropogenic climate change and its environmental consequences in many parts of the world, societies need to anticipate and manage these types of responses, which appear to be robust across time as well as space, and which have been echoed in more recent times. For example, during the famine triggered by severe and protracted drought in the 1970s in the Sahel (Brooks 2004), livelihoods collapsed and people from rural areas migrated to urban centers, echoing processes of abandonment and population agglomeration apparent in the distant past. Some smallholders who lost their livelihoods became laborers on larger farms, resulting in greater economic dependency that qualitatively reflects the social stratification of the 6th millennium BP (Heyd and Brooks 2009; Brooks et al. in press). It has also been argued that the social impacts of drought on marginalised populations exacerbated existing political grievances and played a role in precipitating conflict between these groups and central governments (Brooks et al. 2005).

Whereas climate change is often (and generally correctly) viewed as one factor among many influencing the development of human societies over a variety of timescales, evidence from the middle Holocene suggests that where climate change is rapid and severe, with large adverse impacts on resources, it can dominate over other drivers of cultural change. An example of such an instance is the extreme desiccation of the Sahara, which meant that the only option for people in many areas affected by climatically-driven landscape change was to move. Examples such as this remind us that there are limits to the potential for adaptation, and that adaptation is not simply a question of identifying, elaborating and implementing the right measures. Where adaptation is possible *in situ*, the nature of adaptation will be influenced by local geographic, environmental and cultural factors as indicated above. This is an important consideration for the identification of adaptation measures, which needs to go beyond top-down, technocratic approaches and pay close attention to local contexts.

The evidence from the 6th millennium BP indicates that past transient RCC and associated social and cultural transitions have tended to occur over timescales of centuries, reinforcing the lesson that adaptation is a continuous process and not a one-off action. Furthermore, adaptation to large environmental changes can result in fundamental changes to the structure and organization of human societies. Increased social inequality, generally associated with a rise in authoritarianism and a loss of individual autonomy, as well as conflict in some instances, appears to be a very robust consequence of adaptation to resource scarcity. This is an important consideration for development agencies whose remit includes the promotion of human rights and poverty reduction as well as adaptation, and also emphasises that adaptation can be associated with costs to societies and individuals that are not evenly distributed. These lessons about the potential for adaptation to disrupt and transform societies are at odds with current assumption about adaptation as a means of broadly preserving existing social and economic systems and promoting development and human welfare, and need to be integrated into adaptation planning (Brooks et al. in press).

4.6 Conclusions

The 6th millennium BP represents the last phase of major global climatic reorganization, and provides us with striking examples of cultural changes likely to have been associated with responses to RCC from across the globe. Episodes of RCC around 5900 and 5200 BP appear to have played a key role in the termination of the Holocene Climatic Optimum, but have been relatively neglected compared with qualitatively similar events at 8200 and 4200 BP. These less well known episodes coincided with profound cultural transitions, and may be more important than the 8200 and 4200 BP events in influencing the long-term development of human societies.

The extent to which specific cultural changes can be associated with particular episodes of RCC through existing archaeological and paleoenvironmental records varies greatly, and the links between climatic and cultural change in the 6th millennium BP will continue to be hotly debated. Nonetheless, even where local, high resolution records that can link cultural changes with specific changes in the environment are lacking, the archaeological evidence often reveals processes that are consistent with adaptation to RCC and its impacts. These processes include changes in livelihoods; abandonment of less productive areas; population agglomeration in environmental refugia; increases in mobility and sedentism depending on local preferences, opportunities and constraints; and increased social organization and stratification. The last of these illustrates that adaptation may be associated with costs, while the widespread evidence of social disruption during periods of RCC tells us that adaptation is far from being a smooth process.

While the mechanisms driving global climate change in the 6th millennium BP were very different to those operating today, an examination of human-environment interactions during this period, as inferred from archaeological records interpreted within their paleoenvironmental contexts, gives us some valuable insights into how human societies respond to rapid changes in climate, landscapes and resource availability. These insights are urgently required as the Earth enters a period likely to be characterized by widespread climatic reorganization and large-scale changes in resource distribution after five millennia during which the configuration of the global climate has been relatively stable. In particular, the evidence base that informs adaptation policy needs to be extended to include examples of human responses to rapid and severe climate change if it is to be representative of the kinds of changes likely to occur over the coming decades and centuries. Archaeology and paleoenvironmental studies therefore have a vital role to play in framing and informing climate change policy.

References

Ahlberg K, Almgren E, Wright HE, Ito E (2001) Holocene stable-isotope stratigraphy at Lough Gur, County Limerick, Western Ireland. The Holocene 11:367–372

Akkermans PMMG, Schwartz GM (2003) The Archaeology of Syria: From complex hunter-gatherers to early urban societies (ca 16,00-300 BC). Cambridge University Press, Cambridge, UK, 467 pp

Algaze G (1993) The Uruk world system: The dynamics of expansion of Early Mesopotamian Civilization. University of Chicago Press, Chicago, 169 pp

Algaze G (2001) Initial social complexity in southwestern Asia. The Mesopotamian advantage. Current Anthropology 42:199–233

Algaze G (2008) Ancient Mesopotamia at the dawn of civilization: The evolution of an urban landscape. University of Chicago Press, Chicago, 230 pp

Alley R, Ágústsdóttir AM (2005) The 8k event: cause and consequences of a major Holocene abrupt climate change. Quaternary Science Reviews 24:1123–1149

Alley RB, Marotzke J, Nordhaus WD, et al. (2003) Abrupt climate change. Science 299:2005–2010

Anderson K, Bows A (2008) Reframing the climate change challenge in light of post-2000 emission trends. Philosophical Transactions of the Royal Society A. doi:10.1098/rsta.2008.0138

Andrus CFT, Crowe DE, Sandweiss DH, et al. (2004) Otolith d18O record of Mid-Holocene sea surface temperatures in Peru. Science 22:1508–1511

Baker RG, Bettis EA, Denniston RF (2001a) Plant remains, alluvial chronology, and cave speleothem isotopes indicate abrupt Holocene climatic change at 6 ka in Midwestern USA. Global and Planetary Change 28:285–291

Baker PA, Seltzer GO, Fritz SC, et al. (2001b) The history of South American tropical precipitation for the Past 25,000 years. Science 291:640–643

Barker PA, Street-Perrott FA, Leng MJ, et al. (2001c) A 14,000-Year Oxygen Isotope Record from Diatom Silica in Two Alpine Lakes on Mt. Kenya. Science 292:2307–2310

Bar-Matthews M, Ayalon A, Kaufman A, Wasserburg GJ (1999) The Eastern Mediterranean paleoclimate as a reflection of regional events: Soreq cave, Israel. Earth and Planetary Science Letters 166:85–95

Baroni C, Orombelli G (1996) The Alpine "Iceman" and Holocene climate change. Quaternary Research 46:78–83

Bond G, Showers W, Cheseby M, et al. (1997) A pervasive millennial-scale cycle in North Atlantic Holocene and glacial cycles. Science 278:1257–1266

Breunig P, Neumann K (2002) From hunters and gatherers to food producers: new archaeological and archaeobotanical evidence from the West African Sahel. In: Hassan FA (ed) Droughts, food and culture: Ecological change and food security in Africa's Later Prehistory. Kluwer, New York, pp 123–155

Brooks N (2000) Dust-climate interactions in the Sahel-Sahara zone of northern Africa, with particular reference to late twentieth century Sahelian drought. PhD Thesis, Climatic Research Unit, University of East Anglia, Norwich, 331 pp

Brooks N (2004) Drought in the African Sahel: long-term perspectives and future prospects. Tyndall Centre Working Paper No. 61, University of East Anglia, Norwich, UK

Brooks N (2006) Cultural responses to aridity in the Middle Holocene and increased social complexity. Quaternary International 151:29–49

Brooks N, Brown K, Grist N (in press) Development futures in the context of climate change: Challenging the present and learning from the past. Development Policy Review 27: 741–765

Brooks N, Chiapello I, di Lernia S, et al. (2005) The Climate-Environment-Society Nexus in the Sahara from Prehistoric Times to the Present Day. The Journal of North African Studies 10:253–292

Bubenzer O, Besler H, Hilgers A (2007) Filling the gap: OSL data expanding 14C chronologies of Late Quaternary environmental change in the Libyan Desert. Quaternary International 175:41–52

Casseldine C, Thompson G, Langdon C, Hendon D (2005) Evidence for an extreme climatic event on Achill Island, Co. Mayo, Ireland around 5200–5100 cal. yr BP. Journal of Quaternary Science 20:169–178

Christensen JH, Hewitson B, Busuioc A, et al. (2007) Regional climate projections. In: Solomon S, Hewitson B, Busuioc A et al. (eds) Climate change 2007: The physical science basis. Contribution of Working Group I to the Fourth Assessment Report of the Intergovernmental Panel on Climate Change. Cambridge University Press, Cambridge, pp 847–940

Clare L, Rohling EJ, Weninger B (2008) Warfare in Late Neolithic/Early Chalcolithic Pisidia, southwestern Turkey. Climate induced social unrest in the late 7th millennium cal BC. Documenta Praehistorica

Cremaschi M (2002) Late Pleistocene and Holocene climatic changes in the central Sahara. The case study of the southwestern Fezzan, Libya. In: Hassan, FA (ed) Droughts, food and culture: Ecological change and food security in Africa's Later Prehistory. Kluwer, New York, pp 65–81

Cullen HM, deMenocal PB, Hemming S, Hemming G (2000) Climate change and the collapse of the Akkadian empire: Evidence from the deep sea. Geology 28:379–382

Damnati B (2000) Holocene lake records in the Northern Hemisphere of Africa. Journal of African Earth Sciences 31:253–262

de Menocal P, Ortiz J, Guilderson T, et al. (2000) Abrupt onset and termination of the African humid period: Rapid climate responses to gradual insolation forcing. Quaternary Science Reviews 19:347–361

di Lernia S (2002) Dry climatic events and cultural trajectories: adjusting Middle Holocene pastoral economy of the Libyan Sahara. In: Hassan FA (ed) Droughts, food and culture: Ecological change and food security in Africa's Later Prehistory. Kluwer, New York, pp 225–250

di Lernia S (2006) Building monuments, creating identity: Cattle cult as a social response to rapid environmental changes in the Holocene Sahara. Quaternary International 151:50–62

di Lernia S, Manzi G, Merighi F (2002) Cultural variability and human trajectories in later prehistory of the Wadi Tenezzuft. In: di Lernia, S, Manzi, G. (eds) Sand, stones and bones: The archaeology of death in the Wadi Tannezzuft Valley (5000–2000 BP). Centro Interuniversitario di Ricerca per le Civilta` e l'Ambiente del Sahara Antico e Delle Zone Aride, Universita` Degli Studi di Roma and Department of Antiquities, Libya, pp 281–302

Drake N, Wilson A, Pelling R, et al. (2004) Water table decline, springline desiccation, and the early development of irrigated agriculture in the Wadi al-Ajal, Libyan Fezzan. Libyan Studies 35:95–112

Enzel Y, Ely LL, Mishra S, et al. (1999) High-resolution Holocene environmental changes in the Thar Desert, northwestern India. Science 284:125–128

Fleitmann D, Burns SJ, Mudelsee M, et al. (2003) Holocene forcing of the Indian Monsoon recorded in a stalagmite from southern Oman. Science 300:1737–1739

Fleitmann D, Burns SJ, Mangini A, et al. (2007) Holocene ITCZ and Indian monsoon dynamics recorded in stalagmites from Oman and Yemen (Socotra). Quaternary Science Reviews 26:170–188

Gasse F (2000) Hydrological changes in the African tropics since the Last Glacial Maximum. Quaternary Science Reviews 19:189–211

Gasse F (2002) Diatom-inferred salinity and carbonate oxygen isotopes in Holocene waterbodies of the western Sahara and Sahel (Africa). Quaternary Science Reviews 21:737–67

Gasse F, Van Campo E (1994) Abrupt post-glacial climate events in West Asia and North Africa monsoon domains. Earth and Planetary Science Letters 126:435–456

Gauthier Y (2009) Orientation and distribution of various dry stone monuments of the Sahara. In: Rubiño-Martín JA, Belmonte JA, Prada F, Alberdi A. (eds) Cosmology Across Cultures, ASP Conference Series 409:317–330

Giddens A (2009) The Politics of Climate Change. Polity Press, Cambridge, UK, 256 pp

Guo Z, Petit-Maire N, Kröpelin S (2000) Holocene non-orbital climatic events in present-day arid areas of northern Africa and China. Global and Planetary Change 26:97–103

Hassan F (1997) Nile floods and political disorder in early Egypt. In: Dalfes HN, Kukla G, Weiss H. (eds) Third Millennium BC Climate Change and Old World Collapse, NATO ASI Series I 49. Springer, Berlin, Heidelberg, pp 711–723

Hassan FA, Barich B, Mahmoud M, Hemdan MA (2001) Holocene playa deposits of Farafra Oasis, Egypt, and their palaeoclimatic and geoarchaeological significance. Geoarchaeology 16:29–46

Haug GH, Hughen KA, Sigman DM, et al. (2001) Southward migration of the Intertropical Convergence Zone through the Holocene. Science 293:1304–1308

Hendrickx S, Vermeersch P (2000) Prehistory: From the Palaeolithic to the Badarian culture (*c.*700,000 to 4000 BC) In: Shaw I. (ed) The Oxford history of ancient Egypt. Oxford University Press, Oxford, 525 pp

Heyd T, Brooks N (2009) Exploring cultural dimensions to climate change. In: Adger WN, Lorenzoni I, O'Brien K. (eds) Adapting to climate change: Thresholds, values and governance. Cambridge University Press, Cambridge, pp 269–282

Hoelzmann P, Jolly D, Harrison SP, et al. (1998) Mid-Holocene land-surface conditions in northern Africa and the Arabian peninsula : A data set for the analysis of biogeophysical feedbacks in the climate system. Global Biogeochemical Cycles 12:35–51

Hoffman MA, (1982) The predynastic of Hierakonpolis: An interim report. Egyptian Studies Association, Publication No. 1, Cairo University Herbarium, Faculty of Science, Giza, Egypt, and the Department of Sociology and Anthropology, Wetern Illinois University, Macomb, I154 pp

Hoffman MA, Hamroush HA, Allen RO (1986) A model of urban development for the Hierakonpolis region from predynastic through Old Kingdon times. Journal of the American Research Center in Egypt 23:175–187

Hole F (1999) Economic implications of possible storage structures at Tell Ziyadeh, NE Syria. Journal of Field Archaeology 26:267–283

IPCC (2007) Impacts, vulnerability and adaptation. Cambridge University Press, Cambridge and New York, 976 pp

Jansen E, Overpeck J, Briffa KR, et al. (2007) Palaeoclimate. In: Solomon S, Qin D, Manning M. (eds) Climate change 2007: The physical science basis. Contribution of Working Group I to the Fourth Assessment Report of the Intergovernmental Panel on Climate Change. Cambridge University Press, Cambridge, United Kingdom and New York, NY, USA, 976 pp

Johnson G (1988-1989) Late Uruk in Greater Mesopotamia: expansion or collapse? Origini 14:595–613

Jousse H (2004) A new contribution to the history of pastoralism in West Africa. Journal of African Archaeology 2:187–201

Jung SJA, Davies GR, Ganssen GM, et al. (2004) Stepwise Holocene aridification in NE Africa deduced from dust-borne radiogenic isotopes. Earth and Planetary Science Letters 221:27–37

Kennet DJ, Kennet JP (2006) Early State Formation in Southern Mesopotamia: Sea Levels, Shorelines, and Climate Change? Journal of Island and Coastal Archaeology 1:67–99

Kennet DJ, Kennet JP (2007) Influence of Holocene marine transgression and climate change on cultural evolution in Southern Mesopotamia. In: Anderson D, Maasch KA, Sandweiss DH. (eds) Climate change and cultural dynamics: A global perspective on Mid-Holocene transitions. Academic Press, London, pp 229–264

Kuper R, Kröpelin S (2006) Climate-controlled Holocene occupation in the Sahara: Motor of Africa's evolution. Science 313:803–807

Lancaster NG, Kocurek A, Singvi V, et al. (2002) Late Pleistocene and Holocene dune activity and wind regimes in the western Sahara Desert of Mauritania. Geology 30:991–94

Lawler A (2007) Murder in Mesopotamia? Science 317: 1164–1165

Lezine AM, Salliège JF, Robert F, et al. (1998) Holocene lakes from Ramlat as-Sab'atayn (Yemen) illustrate the impact of monsoon activity in southern Arabia. Quaternary Research 50:290–299

Linstädter J, Kröpelin S (2004) Wadi Bakht revisited: Holocene climate change and prehistoric occupation in the Gilf Kebir region of the Eastern Sahara, SW Egypt. Geoarchaeology 19:735–777

Liu L (2004) The Chinese Neolithic: Trajectories to early states. Cambridge University Press, Cambridge, 328 pp

Magny M, Leuzinger U, Bortenschlager S, et al. (2006) Tripartite climate reversal in Central Europe 5600–5300 years ago. Quaternary Research 65:3–19

Malville JM, Wendorf F, Mazar AA, et al. (1998) Megaliths and Neolithic astronomy in southern Egypt. Nature 392:488–491

Marchant RA, Hooghiemstra H (2004) Rapid environmental change in African and South American tropics around 4000 years before present: a review. Earth Science Reviews 66: 217–260

Matthews R (2003) The Archaeology of Mesopotamia: Theories and approaches. Routledge, London, 253 pp

Mattingly DJ, Reynolds T, Dore J (2003) Synthesis of human activities in Fazzan. In: Mattingly DJ. (ed) The archaeology of Fazzan Volume 1, Synthesis. Department of Antiquities, Tripoli and Society for Libyan Studies, London, pp 327–373

Mayewski PA, Rohling EE, Stager JC, et al. (2004) Holocene climate variability. Quaternary Research 62:243–255

McManus JF, Francois R, Gherardi J-M, et al. (2004), Collapse and rapid resumption of Atlantic meridional circulation linked to deglacial climate changes. Nature 428:834–837

Midant-Reynes, B (2000) The prehistory of Egypt: From the first Egyptians to the first Pharoahs. Blackwell Publishers Ltd, Oxford, 344 pp

Molloy K, O'Connell M (2004) Holocene vegetation and land-use dynamics in the karstic environment of Inis Oírr, Aran Islands, western Ireland: pollen analytical evidence evaluated in the light of the archaeological record. Quaternary International 113:41–64

Neff U, Burns SJ, Mangini A, et al. (2001) Strong coherence between solar variability and the monsoon in Oman between 9 and 6 kyr ago. Nature 411:290–293

Nicholls RJ, Wong PP, Burkett VR, Codignotto JO (2007) Coastal systems and low-lying areas. In: Parry ML, Canziani OF, Palutikof JP, et al. (eds) Climate change 2007: Impacts, adaptation and vulnerability. Contribution of Working Group II to the Fourth Assessment Report of the Intergovernmental Panel on Climate Change. Cambridge University Press, Cambridge, UK, pp 315–356

Nicoll K (2004) Recent environmental change and prehistoric human activity in Egypt and Northern Sudan. Quaternary Science Reviews 23:561–580

Nissen HJ (1988) The early history of the ancient Near East 9000–2000 BC. Chicago University Press, 224 pp

Parker AG, Goudie AS, Stokes S, et al. (2006) A record of Holocene climate change from lake geochemical analyses in southeastern Arabia. Quaternary Research 66:465–476

Pollock S (2001) The Uruk period in southern Mesopotamia. In: Rothman MS. (ed) Uruk Mesopotamia and its neighbors: Cross-cultural interactions in the era of state formation. SAR Press, Santa Fe, pp 181–231

Rahmstorf S (2007) A Semi-Empirical Approach to Projecting Future Sea-Level Rise. Science 315:368–370

Raupach MR, Marland G, Ciais P, et al. (2007) Global and regional drivers of accelerating CO_2 emissions. Proceedings of the National Academy of Sciences USA 104:288–310

Reimer PJ, Baillie MGL, Bard E, et al. (2004) IntCal04.14c. Radiocarbon 46:1029–1058

Reitz EJ, Sandweis DH (2001) Environmental change at Ostra Base Camp, a Peruvian pre-ceramic site. Journal of Archaeological Science 28:1085–1100

Rohling EJ, Pälike H (2005) Centennial-scale climate cooling with a sudden cold event around 8200 years ago. Nature; 434:975–979

Rosen AM (2007) Civilizing climate. Altamira Press, Lanham, UK, 209 pp

Sandweiss DH, Maasch KA, Andrus CFT, et al. (2007) Mid-Holocene climate and culture change in coastal Peru. In: Anderson DG, Maasch KA, Sandweiss DH. (eds) Climate change and cultural dynamics: A global perspective on Mid-Holocene transitions. Academic Press, Elsevier, London, pp 25–50

Schuldenrein J, Wright RP, Mughal MR, et al. (2004) Landscapes, soils, and mound histories of the Upper Indus Valley, Pakistan: new insights on the Holocene environments near ancient Harappa. Journal of Archaeological Science 31:777–797

Sirocko F, Sarnthein M, Erienkeuser H (1993) Century-scale events in monsoonal climate over the past 24,000 years. Nature 364:322–324

Sivili S (2002) A historical background: Mortuary archaeology in the Sahara between colonialism and modern research.

In: di Lernia S, Manzi G (eds) Sand, stones and bones: The archaeology of death in the Wadi Tannezzuft Valley (5000–2000 BP). Rome: Centro Interuniversitario di Ricerca per le Civiltà e l'Ambiente del Sahara Antico e delle Zone Aride, Università degli Studi di Roma and Department of Antiquities, Libya, 356 pp

Smith AB (1984) Environmental limitations of prehistoric pastoralism in Africa. The African Archaeological Review 2:99–111

Somerville R, Le Treut H, Cubasch U, et al. (2007) Historical overview of climate change. In: Solomon S, Qin D, Manning M, et al. (eds) Climate change 2007: The physical science basis. Contribution of Working Group I to the Fourth Assessment Report of the Intergovernmental Panel on Climate Change. Cambridge University Press, Cambridge, pp 93–127

Srivastava P, Singh IB, Sharma S, et al. (2003) Late Pleistocene-Holocene hydrologic changes in the interfluve areas of the central Ganga Plain, India. Geomorphology 54:279–292.

Stevens LR, Ito E, Schwalb A, et al. (2006) Timing of atmospheric precipitation in the Zagros Mountains inferred from a multi-proxy record from Lake Mirabad, Iran. Quaternary Research 66:494–500

Thomas DSG, Knight M, Wiggs GFS (2005) Remobilization of Southern African Desert dune systems by twenty-first century global warming. Nature 435:1218–1221

Thompson LG, Mosley-Thompson E, Davis ME, et al. (1995) Late glacial stage and Holocene tropical ice core records from Huascarán. Peru Science 269:46–50

Thompson LG, Mosley-Thompson E, Davis ME, et al. (2002) Kilimanjaro ice core records: Evidence of Holocene climate change in tropical Africa. Science 298:589–593

Thompson DP, van Hollen D, Osborn G, Ryane C (2006) Did neoglaciation begin as early as 6400 cal years ago? Geological Society of America Abstracts with Programs 38:236

Tuenter E, Weber SL, Hilgen FJ, et al. (2003) The response of the African summer monsoon to remote and local forcing due to precession and obliquity. Global and Planetary Change 36:219–235

UNFPA (2007) State of World Population 2007: Unleashing the Potential of Urban Growth. UNFPA, New York, 99 pp

Ur JA, Karsgaard P, Oates J (2007) Early urban development in the Near East. Science 317:1188

Vernet R, Faure H (2001) Isotopic chronology of the Sahara and the Sahel during the late Pleistocene and the early and Mid-Holocene (15,000–6000 BP). Quaternary International 68–71:385–387

Wang Y, Cheng H, Edwards RL, et al. (2005) The Holocene Asian Monsoon: Links to solar changes and North Atlantic climate. Science 308:854–857

Warren R (2005) Impacts of global climate change at different annual mean global temperature increases. In Schellnhuber HJ, Cramer W, Nakicenovic N, et al. (eds) Avoiding dangerous climate change. Cambridge University Press, Cambridge, pp 93–131

Wei K, Gasse F (1999) Oxygen isotopes in lacustrine carbonates of West China revisited: implications for post glacial changes in summer monsoon circulation. Quaternary Science Reviews 18:1315–1334

Weiss H, Courty M-A, Wetterstrom W, et al. (1993) The genesis and collapse of third millennium North Mesopotamian civilization. Science 261:995–1003

Wendorf F, Schild R (1998) Nabta Playa and its role in northeastern African prehistory. Journal of Anthropological Archaeology 17:97–123

Wengrow D (2001) Rethinking "cattle cults" in early Egypt: towards a prehistoric perspective on the Narmer Palette. Cambridge Archaeolological Journal 11:91–104

Wenxiang W, Tungshen L (2004) Possible role of the "Holocene Event 3" on the collapse of Neolithic cultures around the Central Plain of China. Quaternary International 117:153–166

Wilkinson T (1999) Early dynastic Egypt. Routledge, London, 440 pp

Wilkinson T (2003) Genesis of the Pharaohs. Thames and Hudson Ltd, London, 208 pp

Wright HT (2001) Cultural action in the Uruk world. In: MS Rothman (ed) Uruk Mesopotamia and its neighbors: Cross-cultural interactions in the era of state formation. SAR Press, Santa Fe, pp 123–147

Zhang HC, Ma YZ, Wünnemann B, Pachur H-J (2000) A Holocene climatic record from arid northwest China. Palaeogeography, Palaeoclimatology, Palaeoecology 162:389–401

Chapter 5
Human Communities in a Drying Landscape: Holocene Climate Change and Cultural Response in the Central Sahara

Mauro Cremaschi and Andrea Zerboni

5.1 Introduction

Dagge and Hamad (2008) report that Syria is in the midst of an environmental emergency occasioned by a drought that has affected the region since 2006. Massive crop failure with harvests down by a half has produced conditions where famine and even societal collapse might be expected. Fortunately, the Syrian government has been able to ameliorate the problem with imports of food and potable water. In the distant past disasters could not have been avoided in this way, and famine would have resulted in the disaggregation of the community or at least would have forced a drastic change in survival strategies. Episodes similar to those now affecting Syria have caused societal crises throughout prehistory and history, with climate change and human abuse of landscape and natural resources playing crucial roles (Diamond 2005). Of course, local differences need to be understood and treasured. The reaction of the physical environment (the landscape) to stress is not uniform across the globe, regardless of whether we consider climate change, the impact of grazing, or any other human activity.

The present day South Mediterranean arid environments (the subtropical arid belt) offer the opportunity to investigate the human–landscape relationships for the last 10 millennia. During the Holocene, the region experienced a major change in landscape: the shift from desert to savannah and back again. The environmental effects of this cycle are comparable to the ecological change which affected the mid-latitudes at the time of the waning of the Pleistocene glaciers. For instance, in the Saharan region environmental conditions favorable for animal and plant life promoted a high human concentration since the beginning of the Holocene. Consequently, the later aridification affected a densely populated landscape.

The case study analyzed in this chapter deals with the effect of drought on the landscape and human communities in the central Sahara. We report and discuss the environmental changes that occurred in SW Fezzan (Libya; Fig. 5.1), by considering three different physiographic units: the Tadrart Acacus massif and the two adjoining regions lying at the opposite sides of the mountain range, the Erg Uan Kasa, and the fluvial valley of the Wadi Tanezzuft (Fig. 5.2). Furthermore, we will analyze how in the same geographic and hydrologic ecosystems, different geomorphologic units react to climate change. In addition and subsequent to environmental modifications, the human dwellers gave rise to different and complex social dynamics and adaptive strategies.

5.2 Paleoclimate of Central Sahara

From the early to the middle Holocene, the SW Fezzan area enjoyed, as did the entire Saharan region, a period of high rainfall (Cremaschi 1998, 2002; deMenocal et al. 2000; Gasse 2000; Hoelzmann et al. 2004; Mayewski et al. 2004; Wendorf et al. 2007). The latter was driven by the expansion of the summer monsoon from the Gulf of Guinea and the migration of the ITCZ (Intertropical Convergence Zone) to northern positions

M. Cremaschi (✉)
Dipartimento di Scienze della Terra "A. Desio",
Università degli Studi di Milano, Via Mangiagalli 34,
I-20133 Milano, Italy
e-mail: mauro.cremaschi@unimi.it

See Plates 2, 3, 4 in the Color Plate Section; also available at: extras.springer.com

I. P. Martini, W. Chesworth (eds.), *Landscapes and Societies*,
DOI 10.1007/978-90-481-9413-1_5, © Springer Science+Business Media B.V. 2010

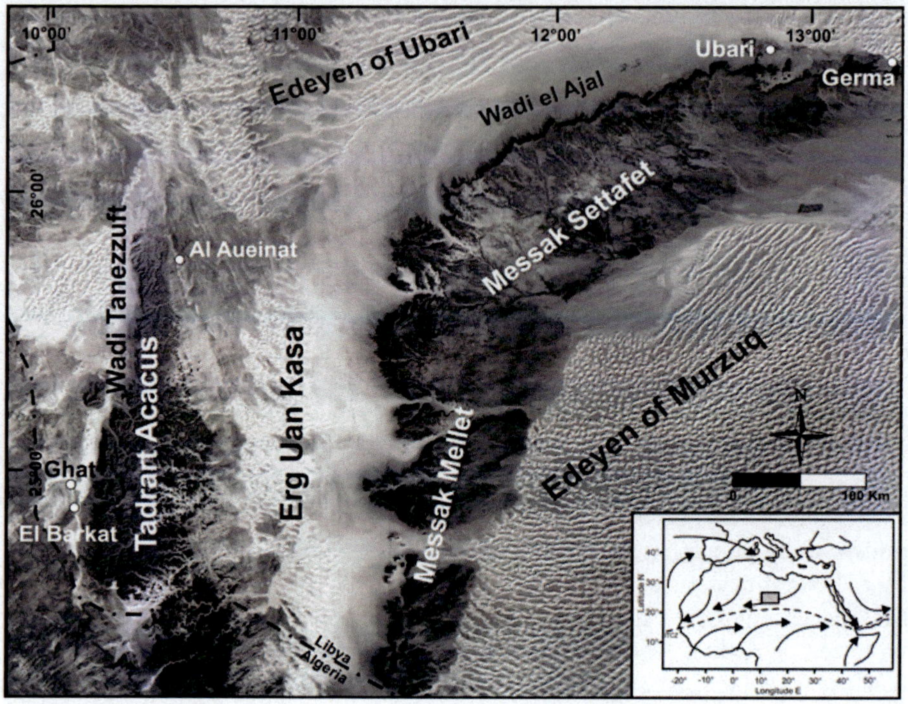

Fig. 5.1 Landsat 7 satellite imagery of the SW Fezzan, indicating the localities mentioned in the text. In the insert the position of the area in a regional context is shown; the present position of the ITCZ (Intertropical Convergence Zone) and the Northern Hemisphere summer atmospheric circulation pattern are also reported (main winds are indicated as *arrows*)

(deMenocal et al. 2000; Gasse 2000). The main effect of this climate change was the recharge of the local aquifers (Cremaschi 1998; Zerboni 2006) and the activation of springs, rivers, and lakes, and the growth of luxuriant vegetation (Mercuri 2008), determining conditions suitable for animal life. Subsequently, the entire area was settled by Epipaleolithic and Mesolithic groups (in this region defined as Early and Late Acacus respectively) and later by Pastoral-Neolithic communities (Cremaschi and di Lernia 1998; di Lernia 2002).

The so-called African Humid Period was interrupted by several transitory dry spells (Mayewski et al. 2004; Kutzback and Liu 2007; Cremaschi et al. 2010) and the termination of the wet phase is dated at ca. 5000 year BP (Cremaschi 2002; Mayewski et al. 2004; Zerboni 2006), due to the abrupt reduction in the intensity of the African monsoon (deMenocal et al. 2000). This event led to the present desert conditions following different modalities as the varying physiographic features responded differently to aridification (Gasse 2000; Kröpelin et al. 2008). For much the same reason, the human communities living in the region reacted in different ways to face the constraint of a landscape advancing toward aridity (di Lernia 2002; Brooks et al. 2005; Brooks 2006; Kuper and Kröpelin 2006).

The present climate of the SW Fezzan is hyperarid. The mean annual temperature is 30 °C and the mean annual rainfall is between 0 and 20 mm, which is mostly distributed in spring and summer (Fantoli 1937; Walther and Lieth 1960). Occasional rainstorms are recorded also in the winter season (Fantoli 1937).

5.3 Geological and Geomorphologic Background

The Tadrart Acacus and the adjoining erg Uan Kasa and Wadi Tanezzuft are located in SW Libya (Fig. 5.1), in the Fezzan region, well inside the hyperarid belt of the Sahara desert, between 26° and 24° latitude N. Geologically, this region belongs to the western fringe of the wide geosyncline of the Murzuq Basin (Goudarzi 1970; Kalefa El-Gahali 2005), whose base, mainly composed of Paleozoic sandstone and marls, lies upon the intrusive formation of the Tassili massif, located in Algeria. The Paleozoic formations are covered by the Mesozoic sandstone and shale of the Messak Settafet ridge and Murzuq Basin (Kalefa El-Gahali 2005). The main geologic structural pattern consists of a monocline characterized by an E–NE tilted cuesta in

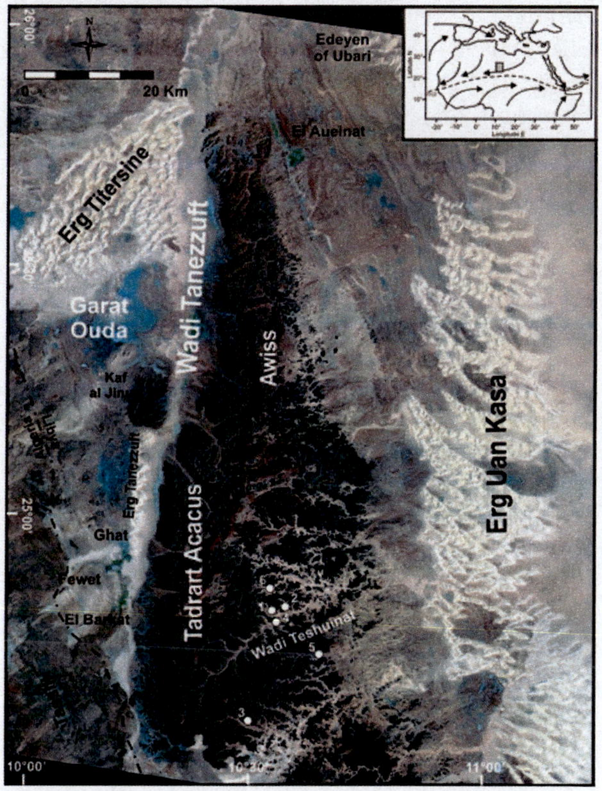

Fig. 5.2 Landsat 7 satellite imagery of the investigated area, indicating the localities mentioned in the text; *dots* represent the main archaeological sites in the Tadrart Acacus area: *1* Uan Afuda; *2* Uan Tabu; *3* Wadi Afar; *4* Uan Muhuggiag; *5* Uan Telocat; *6* Wadi Sennadar

the Tassili and Tadrart Acacus areas (Goudarzi 1970; Cremaschi 1998), and flat landscape in the Erg Uan Kasa. The Wadi Tanezzuft valley is oriented according the direction of the rock strata and is bounded by the Tassili mountain to the west and the cliff of the Acacus massif to the east. The wadi is underlain by the almost impermeable Tanezzuft Formation composed of shales and thin-bedded sandstones. On the western side of the valley, along the geological contact between the highly permeable Tassili sandstone and the Tanezzuft Formation suitable conditions exist for development of numerous springs that made this area rich in water and palm groves (Desio 1937).

The Tadrart Acacus massif covers an area of 4800 km^2 with a maximum elevation of 1100 m asl in the western part and is composed of Paleozoic sedimentary rocks (Goudarzi 1970; Kalefa El-Gahali 2005). It is dissected by a fossil drainage network whose NE–SW, E–W, and N–S trending pattern is controlled by the tectonic structure (Galeĉiĉ 1984; Cremaschi 1998). A scarp delimits the massif towards the west, whereas to the east it grades through a pediment to the dunefield of the Erg Uan Kasa. The mountains consist of lower-middle Silurian shales (the Tanezzuft Formation), outcropping at the western fringe along the Tanezzuft valley, apparently conformably overlying upper Silurian and lower Devonian sandstones (Acacus and Tadrart formations). These formations display different hydrological behaviour: the Tanezzuft Fm. has a very low hydraulic conductivity and transmissivity and acts as an aquiclude, whereas the Acacus and Tadrart formations are highly permeable also being highly fractured. Numerous fossil springs identifiable by the presence of calcareous tufa had developed in the past along the contact between these formations. At present few springs are still active in the inner part of the massif, feeding small pools locally called 'gheltas'.

The Erg Uan Kasa is a 200 km long, 20–35 km wide, N–S oriented sand sea; it is located between the eastern fringe of the Tadrart Acacus massif and the western escarpment of the Messak Settafet plateau. It consists of parallel, NW–SE trending alignments of complex linear sand dunes, more than 100 m in height, and separated by flat, wide interdune corridors (Cremaschi 1998). Its substrate consists of Devonian and Carboniferous shales (Wadi Ubarracat Fm.) and Carboniferous-Permian thin bedded limestone (Wadi Tesalatin Fm.) that deeps about 10° toward the northeast (Kalefa El-Gahali 2005).

The Wadi Tanezzuft is a 200 km long, N–S oriented valley with an ephemeral stream (Fantoli 1937). The catchment basin is delimited to the west by the Tassili and to the east by the Tadrart Acacus massifs. From a geomorphologic point of view, the Wadi Tanezzuft flows along the geological homocline constituted by the Tassili sandstones to the west, the Acacus sandstone to the east, and the Tanezzuft shale in the center. The shape of the valley is asymmetrical being steeper to the east and for that reason the widest part of the catchment basin lies in the Tassili region. The source of the river is located in the Takarkori area (southern Tadrart Acacus), and its northern reach has been recently identified in a wide endorheic depression at the western fringe of the Edeyen of Ubari (Perego et al. 2007). The main course of the wadi is surrounded by lowlands with inselberg/pediment type relief, playas, and sand seas.

The Tadrart Acacus massif, the Erg Uan Kasa, and the Wadi Tanezzuft are different neighboring physiographic units interconnected by the fact that the former constitutes the hydrographical catchment of the latter.

5.4 The Wet Holocene: Landscapes and Strategies

The onset of wet conditions at the beginning of the Holocene led to the systematic occupation of the central Sahara, with variations of human settlement patterns and adaptive strategies depending on the characteristics of each region. During subsequent millennia the strategies were modified, with climate as one of the driving factors.

5.4.1 Inside the Mountain Range of the Tadrart Acacus

5.4.1.1 Configuration and Paleoclimate of the Tadrart Acacus at the Beginning of the Holocene

The morphology of the Tadrart Acacus massif is rather similar to that described for the Navajo Sandstone in the Arizona and Utah deserts: the scarp of the valley shows slick-rock slopes and slab walls. The valleys and canyons dissecting the mountains are the relicts of a landscape that has been shaped mostly under a Tertiary equatorial climate (Busche and Erbe 1987). The present morphology of the mountain is the result of several processes the most important having been fluvial and pseudo-karst actions on siliceous rocks (Busche and Hagedorn 1980).

The original width of the valleys has been enlarged by backwasting of scarps and by progressive undersapping processes (Busche and Hagedorn 1980; Busche and Erbe 1987). Typical features developed such as bluffs, rock arches, and stacks. Wide and deep rockshelters were also formed in massive sandstone units by undersapping along bedding planes at the contact between the sandstones beds and the shales; alcove-shaped shelters occur at the head of the wadis. Today the formation of the rockshelters is inactive. In fact, the parts which can be reached by rain water or surface runoff display black desert varnish, but the accumulation of unvarnished blocks in front of them indicates that some roof collapses have recently occurred. Furthermore, the retreating of the valley slopes exposes relict tunnels of various caves that developed during the Tertiary. The caves were an effect of the silica karst processes acting at that time in response to the equatorial climate existing in the area. Therefore the walls flanking the wadis and the steep slopes of the canyon are pinpointed by a variety of rockshelters and caves that were systematically occupied by the communities living in the Tadrart Acacus since the Late Pleistocene. Moreover, the caves preserved the most reliable anthropogenic characteristics and natural proxies as archives to reconstruct the environmental changes of the region.

The most reliable proxy indicating the arrival of the monsoon pattern of rainfall in the Tadrart Acacus during the Holocene consists of calcareous tufa discovered, and recently re-dated, near former springs and pseudo-karst cavities (Carrara et al. 1998; Zerboni 2006; Cremaschi et al. 2010). Generally speaking, the formation of calcareous tufa requires a continuous soil cover, high metabolic soil activity induced by plants, and high, constant water supply; furthermore, in the present case, a saturation of the Tadrart Acacus aquifer up to the upper reaches of the mountain system would be required, which could only be sustained by high rainfall. U/Th dating indicates that conditions suitable for tufa formation began at ca. 9600 year BP and were mostly interrupted at ca. 8200 year BP (Cremaschi et al. 2010), coinciding with the well known, cold-dry event of North Atlantic origin (Fig. 5.3; Alley et al. 1997). Tufa did not form later, indicating that during

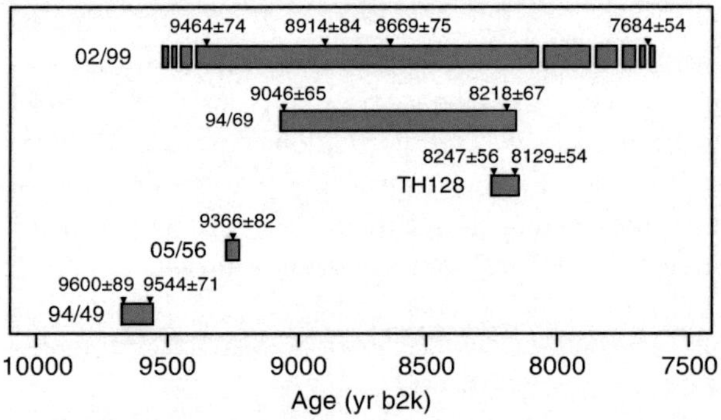

Fig. 5.3 U-series dating indicates that spring tufa sedimentation in the Tadrart Acacus took place between ca. 9600 and 8200 year BP, as consequence of the northward shift of the SW African Monsoon (*Bars* indicate the duration of the tufa deposition in each locality; obtained dates are also reported). (After Cremaschi et al. 2010)

the middle Holocene the water supply and the intensity of the monsoons did not reach again the intensity they had during the very early Holocene.

5.4.1.2 Living in Caves and Rockshelters

The most important stratigraphic sequences recently explored in the caves from the Tadrart Acacus massif (Cremaschi and di Lernia 1998) permit the delineation of the past environmental conditions of the area, their modifications, and the consequences on human communities. The evidence derives from the analysis of the natural and anthropogenic deposits of the Uan Afuda Cave (di Lernia 1999a), and the rockshelters of Uan Tabu (Garcea 2001), Uan Muhuggiag (Pasa and Pasa Durante 1962; Cremaschi and di Lernia 1998), Uan Telocat, Wadi Sennadar (Cremaschi and di Lernia 1998). The importance of these sites both for the archaeology of the central Sahara and for the environmental history of the Holocene has been long known, since pioneering multidisciplinary researches were introduced in the late 1950s (Pasa and Pasa Durante 1962), and a geoarchaeological approach was established at the beginning of the 1990s (Cremaschi and di Lernia 1998). Many radiocarbon dates are available (Fig. 5.4).

The basal parts of the most complete stratigraphic sequences show evidence for a dry Upper Pleistocene phase: at that time the so-called Ogollan desert was at its apogee (Rognon 1989). Reddish aeolian sand has been described at the base of the sections excavated in the Uan Afuda cave (Fig. 5.5) and in the Uan Tabu rockshelter. These units are correlated with the remains of fossil dunes deposited in the area, and have been used as evidence of a general desert expansion during the Late Pleistocene. The discovery at the base of the rubified dune in the Uan Afuda cave of Middle Paleolithic artefacts attributed to the Aterian cultural phase (di Lernia 1999a), and the TL and OSL dating of sand (90,000–69,000 year BP; Martini et al. 1998) allow the correlation of the last Pleistocene expansion of the desert with the MIS 4 isotopic stage.

The onset of wet conditions at the beginning of the Holocene is marked in the Uan Afuda sequence by the weathering of the dune (Fig. 5.5) through rubification and clay translocation; both processes require high water availability and seasonality in precipitation. The organic unit dated at ca. 9800 year BP that overlaps the weathered dune represents a limit *ante quem* for the development of pedogenesis (Cremaschi 1998; Cremaschi and Trombino 1999). This date accords with the recharge of the water system as indicated by the deposition of calcareous tufa.

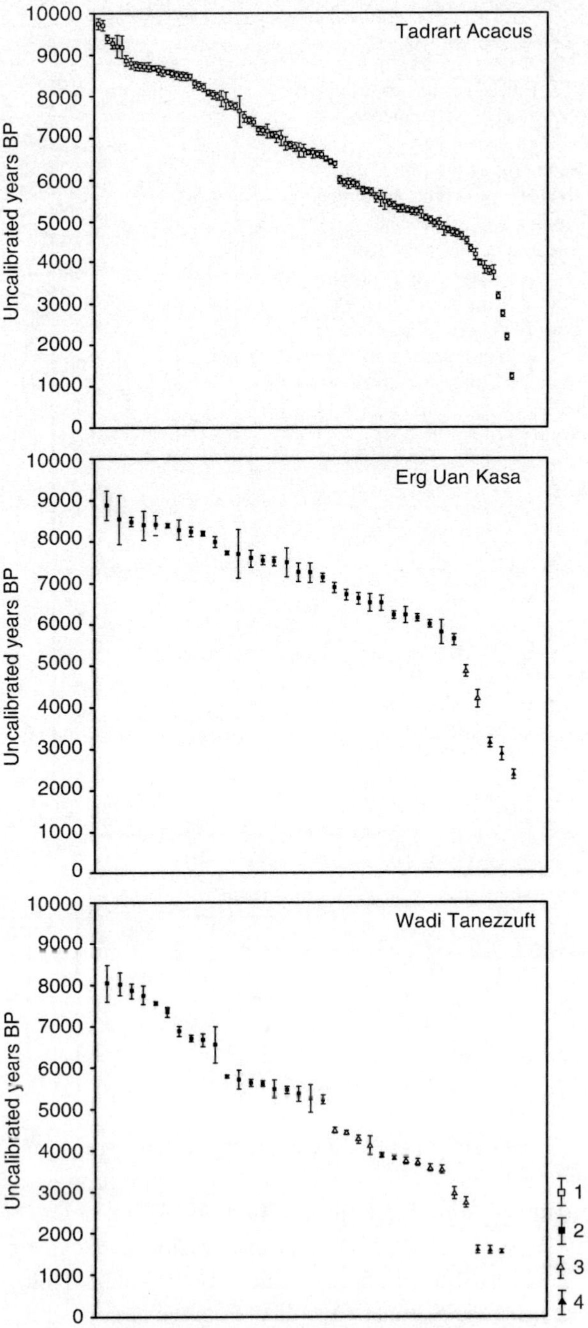

Fig. 5.4 Uncalibrated conventional and AMS ^{14}C dates for the Tadrart Acacus, Erg Uan Kasa, and Wadi Tanezzuft. (*1* anthropogenic deposit; *2* lake deposits; *3* fluvial deposits; *4* other: sebkha deposits, organic sediments, sub-fossil wood)

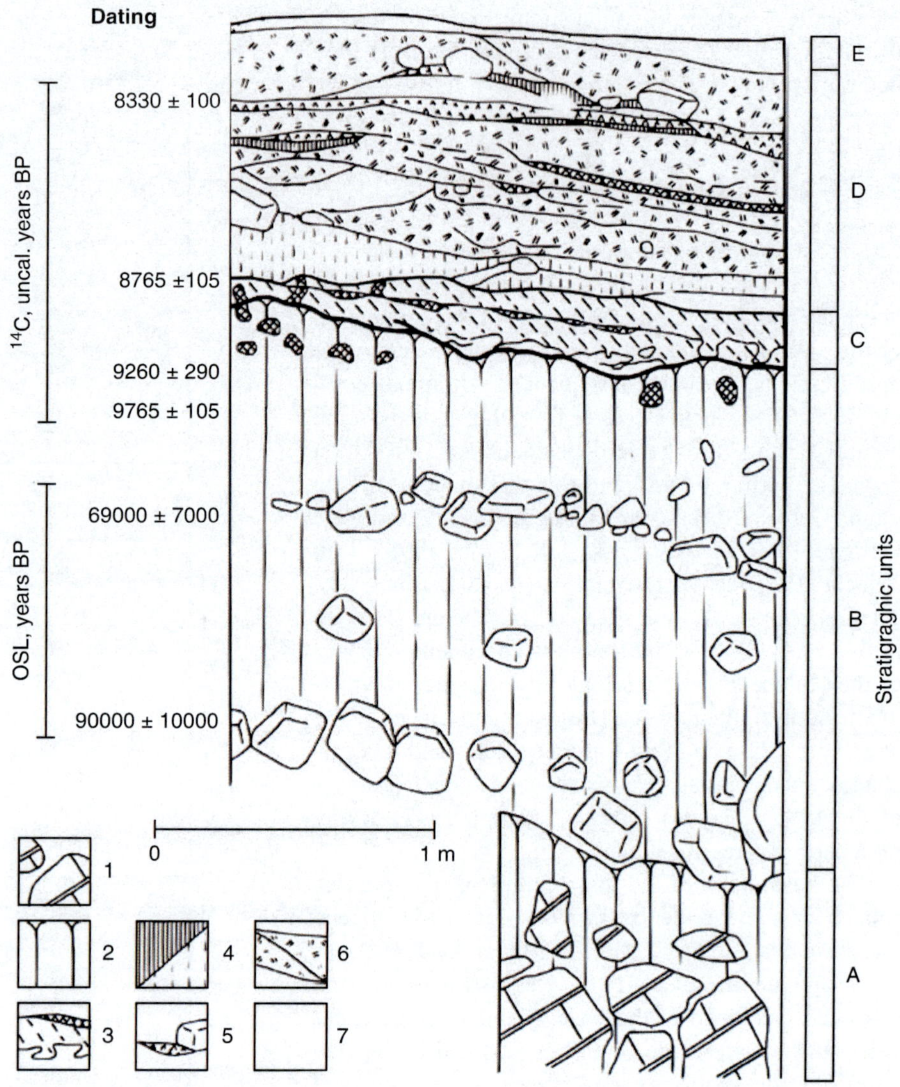

Fig. 5.5 The stratigraphic sequence of the Uan Afuda Cave (After Cremaschi and di Lernia 1998; di Lernia 1999a). (Units: *A* paleosol; *B* Pleistocene aeolian sand, weathered at the top; *C* colluvial sand; *D* organic matter, hearths, and loose sand; *E* modern aeolian sand. Key: *1* collapsed blocks; *2* aeolian sand; *3* colluvial sand and gypsum concretions; *4* loose sand with charcoal, humified organics; *5* hearth with stones and ash lenses; *6* undecomposed vegetal remains; *7* loose sand)

The beginning of a human presence in the study area dating to the Holocene is marked by anthropogenic sedimentation above the rubified dune. The deposits in question have survived wind erosion and can be regarded as lithostratigraphic units typical of the Holocene. Deposits consist of plant fragments, coprolites, finely subdivided organic matter in different stages of humification, ash, charcoal, chemical precipitations, and aeolian sand (Cremaschi 1998). Artificial accumulation of fodder, mainly grass and vegetal fragments, and its transformation through trampling, fire, and addition of organic material related to the penning practice at the site and to human dwelling, are the main processes responsible for the sedimentation inside the rockshelters.

The Holocene basal layers of the deposits in Uan Afuda Cave are of colluvial origin (gray brown laminated sand, including gypsum nodules). Toward their top the organics are progressively better preserved (massive dark brown organic sand, including lenses of preserved plant remains with gypsum concretions), indicating increasing aridity. An interruption in sedimentation is recorded in caves by an erosional surface dated between ca. 8000 and 7500 year BP, thanks to a humified dung layer. The pollen diagrams obtained from the fill of rockshelters (Mercuri 2008) suggest the same scenario of higher water availability. Reconstruction of the landscape between 9800 and 8000 year BP from these data suggests a patchwork of savannah and wooded grassland. The diversified flora indicated includes plants requiring permanent freshwater sources (*Typha, Potamogeton, Lemna, Scirpus*), thus confirming the occurrence of spring pools sustained by constant water supply by precipitation.

Evidence of an Epipaleolithic human presence in the mountain range is not as strong as in the lowlands surrounding the Tadrart Acacus (see following sections and Fig. 5.6). It is dated to the 10th millennium BP

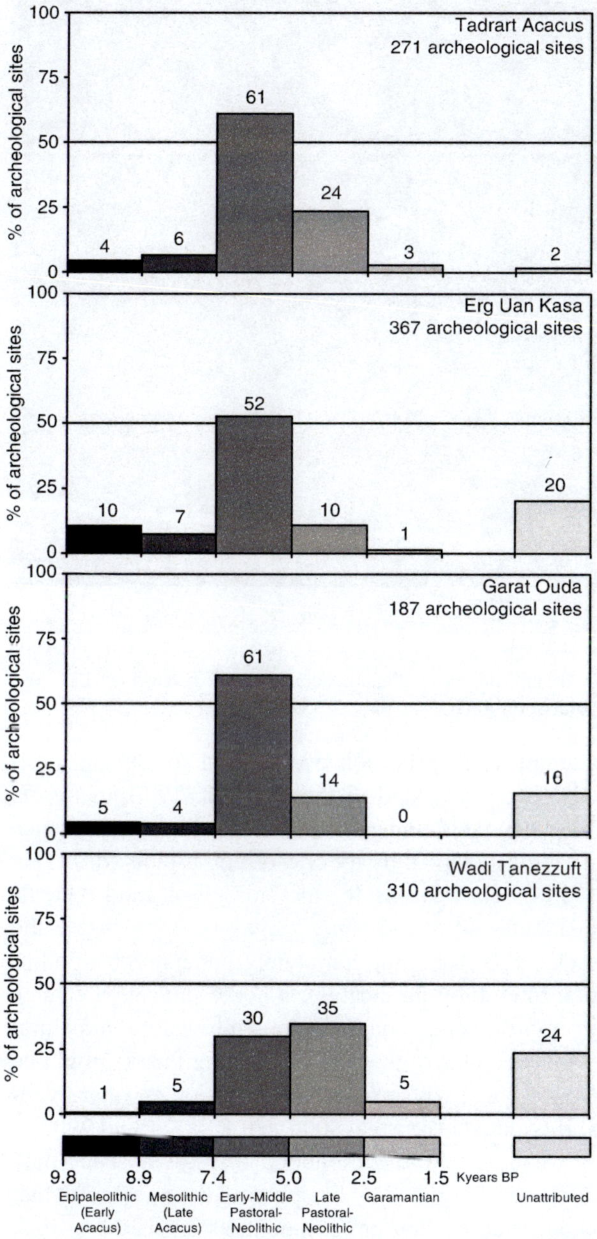

Fig. 5.6 Chronological distribution of the archaeological sites in the investigated areas (Tadrart Acacus, Erg Uan Kasa, Garat Ouda playa, and Wadi Tanezzuft)

(Cremaschi and di Lernia 1998). Stone structures, shallow hearths, lithics (including backed points), and less commonly grinding implements are found at the base of the sequence of Uan Afuda (di Lernia 1999c). No seeds of wild cereals were found in the layers belonging to this phase. The few faunal remains are mostly of *Ammotragus lervia* (the Barbary sheep, a wild capriovid) indicating selective hunting. Even if rare and poor in findings, Epipaleolithic sites inside the massif may be interpreted as base camps for nomadic hunter groups.

In the Uan Afuda and Uan Tabu sites the layers consisting of poorly decomposed organic matter enclose archaeological materials dating back to the Mesolithic phase. This period is marked by the introduction of pottery, displaying dotted wavy line and packed zigzag decorations; at Uan Afuda and Uan Tabu pottery production is documented since the first half of the 9th millennium BP (Cremaschi and di Lernia 1998; di Lernia 1999c; Garcea 2001). Mesolithic sites are also rich in grinding equipment and lithics, including backed tools and geometrics (lunates and triangles). The main difference from the previous phase is seen in a change in economic strategies from an economy of specialized hunting of *Ammotragus* to hunting diversified to include small and large mammals, fish, and birds (Cremaschi and di Lernia 1998). The most important event however, was the introduction of the exploitation of wild cereals (di Lernia 1999; Cremaschi and di Lernia 1999), testified in caves and rockshelters by large concentrations of straw and wild cereals seeds, either complete or at different stages of processing (Mercuri 2008). Recently, we found a good evidence of the use of wild plants in the Wadi Afar Cave, where entire layers of the cave fill are made of straw, grass fragments, and cereal seeds displaying different stages of processing. Moreover potholes served as silos were found; some are empty because of recent erosion, but a few, closed by large, flat stones, preserve a large amount of wild cereals (Fig. 5.7). In one pothole, seeds were found in a basket that was radiocarbon dated to ca. 8400 year BP.

Soil micromorphology of the Mesolithic anthropogenic sediments provides revealing information (Cremaschi and Trombino 1999, 2001). The stratigraphic units are mainly composed of a large quantity of poorly degraded, vegetal fragments (mainly grass), lenses of ash and charcoal from hearths, and small concentrations of sand with a lesser content of organics. The large quantity of straw (and phytolithes) and a general sub-horizontal lamination in thin section

Fig. 5.7 The Wadi Afar Cave: a pothole served as silo, still preserving a large quantity of wild cereals

(Fig. 5.8) indicate both anthropic accumulation of fodder and a continuous trampling; furthermore, the presence of almost entire caprovids coprolite (*Ammotragus*) and accumulation of fecal spherulites indicate the presence of these animals inside the rockshelters.

Considering the evidence from the wadis Afar and Uan Afuda, the intentional accumulation of straw and the presence of seeds remains in the deposits and inside the potholes (Fig. 5.6), can be interpreted to indicate that the rockshelter was a place used for processing and storage of wild cereals. Moreover, at Uan Afuda the occurrence of the pedofeatures related to trampling, the massive presence of coprolites and dung in the internal part of the cave, and the abundant presence of straw and other vegetal remains interpreted as fodder, may indicate forced enclosure of Barbary sheep and forms of taming (Cremaschi and di Lernia 1998; di Lernia 1999a, c).

In the Acacus area the preservation of natural resources (wild cereals and *Ammotragus*) is recorded between 8500 and 8000 year BP (the last part of the Mesolithic presence in the area), and it constitutes an

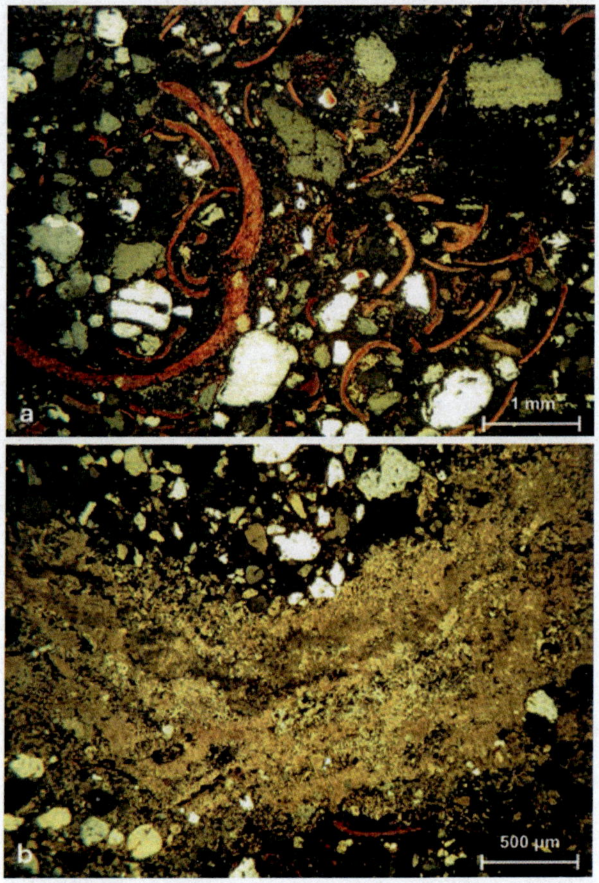

Fig. 5.8 Microphotographs of the sedimentary sequence of the Uan Afuda Cave. **a** Undecomposed straw (cross polarized light). **b** Crystallitic fabric of calcite, corresponding to ash lenses (cross polarized light)

attempt at conservation of the food for special periods (Cremaschi and di Lernia 1998, 1999; di Lernia 1999a, b). This evidences the complexity of subsistence strategies adopted by the Mesolithic groups (di Lernia 1999a). The reasons for the storage of food (cereals in natural silos and *Ammotragus* penning inside the rockshelters) are not completely understood. In addition to cultural implication, it could indicate a change in environmental conditions and a reduction in natural resources. One of the possible driving factors might be the dry phase with progressive reduction in water-availability dated here at ca. 8000 year BP (Zerboni 2006).

Changes in the exploitation of the landscape may also be investigated by deciphering the perception that people had of their environment, as displayed in rock art. The earliest expressions of the central Saharan rock art are mainly of mythological representations, including fantasy animals, but in some cases it is possible to identify realistic figures, which may be compared with the geoarchaeological evidence for ecological significance (Cremaschi et al. 2008). Large animals typical of the savannah environment (Large Wild Fauna style) are the most representative figures of the beginning of the Holocene. They indicate the interest in wild game of the Epipaleolithic and Mesolithic hunters. *Ammotragus* is depicted in the so-called Round Heads style. The 8000 year BP climatic deterioration appears to be supported by the representation of *Orix dammah*, a large antelope adapted to arid environments, in several rockshelters of the Tadrart Acacus. The early Holocene dry event could be expected to have facilitated the migration of the *Orix* inside the massif from the sandy lowlands, where they were confined earlier (Cremaschi et al. 2008).

Following the early Holocene dry spell, recharge of the water resources and the restoration of environmental conditions more suitable for life date from 7500 to 5000 year BP (Cremaschi and di Lernia 1998, 1999). The human population living in the area dramatically changed its adaptive strategy. The archaeological sequence of the Uan Muhuggiag rockshelter (Fig. 5.9) was generated through the occupation of Pastoral-Neolithic groups and is sufficiently well preserved to serve as a reference stratigraphic section (Pasa and Pasa Durante 1962; Cremaschi and di Lernia 1998). The older units of the deposit consist of gray sand that has suffered strong humification of the organic fraction with movement of solutes that indicates water percolation. Pollen content documents a savannah plant cover (Mercuri 2008). In the upper part of the sequence hearths are common, and organic matter of unrecompensed plant and grass fragments is progressively better preserved toward the top. A trend towards aridity is indicated by a peak in the concentration of *Panicum* pollen.

The Early Pastoral-Neolithic occupation of the Tadrart Acacus is mainly documented in the inner part of the massif. This probably indicates a sort of continuity with the settlement pattern adopted by the Mesolithic groups, as well as the existence of peculiar environmental conditions such as higher water availability. Food production has been confirmed for this phase (Cremaschi and di Lernia 1999), but processing of wild cereals and hunting continued to be practiced (Mercuri 2008). In the site of Uan Muhuggiag, as in other sites along the Wadi Teshuinat, archaeological deposits dating to the Middle Pastoral-Neolithic phase consist of alternating lenses of hearths, undecomposed vegetal remains, and coprolites. A seasonal occupation of the rockshelters by shepherds moving with their ovicaprines flocks is indicated (Cremaschi and di Lernia 1998). Micromorphology can help in better understanding the configuration of the settlements. Coprolites, spherulites, wood charcoals, bones, plant fragments, and phytolithes occur at different frequencies in each level, which may be interpreted as evi-

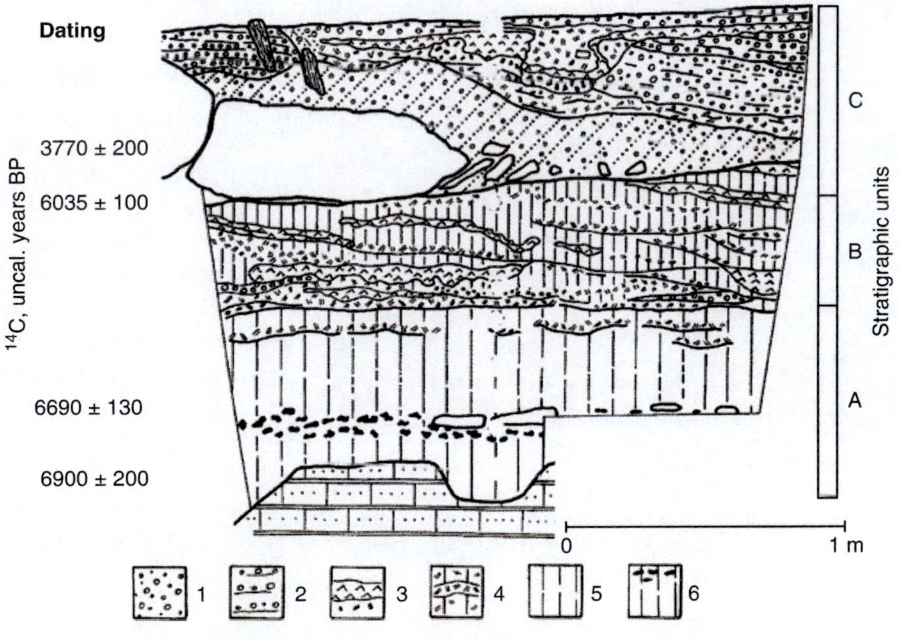

Fig. 5.9 The stratigraphic sequence of the Uan Muhuggiag rockshelter (After Cremaschi and di Lernia 1998). (Units: *A* humified sand; *B* straw and ash lenses; *C* organic sand. Key: *1* loose sand and coprolites; *2* dung; *3* ash lenses; *4* organic deposits, undecomposed vegetal remains; *5* humified organic deposits; *6* humified organic deposits with gypsum and carbonate concretions)

dence for differing uses of the rockshelter (Fig. 5.10). Charcoal, bones, and hearths indicate human dwelling inside caves, the occurrence of coprolites and their strong increase at the top, where charcoal and bone fragments decrease and disappear, shows that although penning is present from the base, it becomes dominant in the last phase of occupation when the rockshelters became stables. Changes in economic strategies show

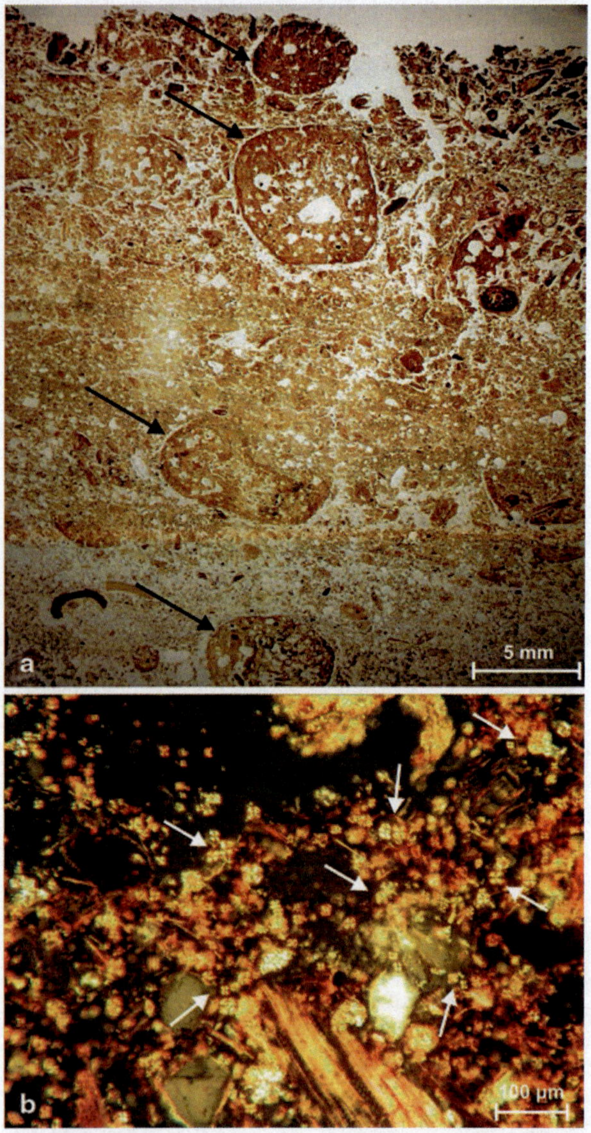

Fig. 5.10 Microphotographs of the sedimentary sequence of the Uan Muhuggiag rockshelter. **a** Undecomposed ovicaprines coprolites (indicated by *arrows*); the groundmass displays evidence of trampling (plane polarized light). **b** Arrows indicate clusters of spherulites (cross polarized light)

a parallel trend with the climatic variations inferred from the anthropogenic deposits. All this underlines the increase of aridity at the transition from the 6th to the 5th millennium BP (Cremaschi and di Lernia 1998).

In the Middle Pastoral-Neolithic phase goats were not the only component of flocks. Cattle were also present—large herds are represented on the walls of many caves used as stables by Pastoral-Neolithic shepherds. In addition to rock art representations, the landscape of the Tadrart Acacus still preserves signs of its pastoral exploitation, such as tracks on the rocky slopes of the massif left by the continuous trampling of herds during their migrations. The tracks are ancient and a Middle Holocene age is confirmed by a dark rock varnish that was deposited in the central Sahara starting around 5000 year BP (Zerboni 2008).

5.4.2 The Erg Uan Kasa

During the wet Holocene, the main effect of the enhanced water supply to the central Sahara was the rise of the water table and the formation of lakes and ponds. This has been particularly noticed in the lowlands at the eastern fringe of the Tadrart Acacus, occupied by the dunes of the Erg Uan Kasa. This region represents the discharge area of the main wadis dissecting the massif.

5.4.2.1 Geomorphology and Sediments

In the area of the Erg Uan Kasa, Holocene lake deposits and hydromorphic soils were found associated with archaeological sites. The sedimentary record here is heavily affected by wind erosion and represents only a small part of the former lakes (Cremaschi 2002).

A thick hydromorphic horizon, consisting of bleached and mottled sand and friable weathered sandstone (saprolite), systematically occurs at the base of the dunes and in the bedrock in the whole area covered by the erg. This phenomenon is the result of water-logging below the water table for a long period. Enhanced rainfall would explain the high water table. Because of the higher reflectance of the bleached sand, the hydromorphic horizon is visible on Landsat satellite imagery. The prevalence and extent of water-saturation would have been due to the ability conferred on the dunes by a high porosity, which made them potentially large reservoirs.

The availability of water promoted weathering along the dune slopes and formation of inceptisols. At the base of the dunes, springs gave rise, in suitable topographic and geomorphologic conditions, to small and shallow lakes or ponds (Cremaschi 1998, 2002). Lacustrine deposits are located in the interdune corridors, mainly at the fringes of the dunes, where they are protected from wind erosion. The sedimentary facies identified in the field (Fig. 5.11) consist of bioturbated organic sand with vegetal remains indicating shallow water; dark organic silt representing the shore facies; and biochemical calcareous silt, including a rich molluscan fauna composed of few species. Characteristic species are *Lymnaea natalensis, Valvata nilotica, Biomphalaria pfeifferi*, and *Afrogyrus oasiensis*, typical of permanent fresh or hypohaline waters (Girod 2005), and indicative of full lacustrine sedimentation. In the Erg Uan Kasa the thickness of the lacustrine sediments is generally low in comparison with the Holocene sequence found in the Edeyen of Murzuq where they consist of carbonatic mud up to 2 m thick (Zerboni 2006). Radiocarbon dates obtained from organic layers at the base of the sequence and from littoral facies indicate that in the erg the water table was rising from ca. 8900 year BP (almost coincident with the beginning of spring activity inside the Tadrart Acacus) (Fig. 5.4). The Uan Kasa lakes and those of the Edeyen of Murzuq (150 km east from the Erg Uan Kasa) reached their highest stand during the 8th and 7th millennia BP (Cremaschi 1998, 2002; Zerboni 2006). However, evidence of the negative fluctuation recorded in the Edeyen of Murzuq in coincidence with the 8200 year BP event is lacking in the Uan Kasa area (Cremaschi 1998; Zerboni 2006).

Fig. 5.11 Erg Uan Kasa. **a** Organic sand deposit at the base of the dune, corresponding to shore facies. **b** Discontinuously laminated, organic sand at the center of the basin, covered by calcareous silt, containing mollusc shells

The typical configuration of a lake deposit in the Erg Uan Kasa (Cremaschi 1998) consists of a basin up to 5 km long and 2 km wide, with sedimentary sequences on weathered bedrock ranging from 1 to 2 m thick. Marginal facies consist of organic and bioturbated silty sand, grading toward the center of the basin, into organic sand discontinuously laminated and superposed on a thin, hydromorphic, massive sand. At the center of the basin these deposits are covered by discontinuously laminated, calcareous silt containing molluscs (Fig. 5.11). The calcareous sediments are concentrated in the middle of the basin and they may indicate a phase of strong biological activity in the lake connected to a high water level. The organic deposits at the upper part of the sequence are interpreted as progression of the marginal facies toward the center of the basin as the lake level progressively dropped. The sequences are sometimes capped by a thick gypsum carbonate crust that marks the transformation of the lakes into sabkas.

5.4.2.2 Settlements in the Erg Uan Kasa

The archaeological sites are generally located close to former standing water bodies. More than 350 sites

(Fig. 5.6), ranging from the Early Acacus up to the Late Pastoral-Neolithic phase, have been identified mostly at the base of the dunes in correspondence to paleolake shores, marsh deposits, and paleosols. The environmental constraint for this was the hydrological behaviour of sand dunes under conditions of water saturation. This resulted in the formation of ecological niches at the fringe of the dunes rich in vegetation, suitable for animal life, and therefore attractive to prehistoric communities.

The Epipaleolithic sites consist of clusters of 'débitage', micro-blade cores, and formal tools such as backed tools, points, lunates, and particularly pedunculated Ounan points (Fig. 5.13; Cremaschi and di Lernia 1998). No primary archaeological structures or faunal remains are associated with these sites. Nevertheless, the occurrence of some specialized tools (hooks) may suggest fishing activities. The Epipaleolithic sites are not actually located along the shores of former lakes that are along the dune fringes, but rather they are buried within the organic deposits at the base of the lacustrine sequences. They are therefore related to the very beginning of lacustrine sedimentation in the area. No direct radiometric dates are available for the Epipaleolithic sites, but the ^{14}C age obtained from the organic sediments covering them should be considered as a limit *ante quem* for the human frequentation. These dates range from ca. 8700 to 8500 year BP, placing them at the very beginning of the wet phase in the area. During this phase the raw material employed in the lithic industry is largely represented by local lithotypes, such as very fine quartzarenite commonly outcropping in the Messak Settafet and Tadrart Acacus regions, and silcrete that outcrops inside the erg. The assemblage also includes Jurassic to Cretaceous flint obtained from remote areas (likely from northern regions). This indicates a high mobility of the hunter-gatherer groups promoted by the extension of a green corridor to the entire central Sahara (Hoelzmann et al. 2004).

Mesolithic sites are rare in the erg (Fig. 5.6) and can be distinguished in the field from Epipaleolithic ones because of the configuration and occurrence of specific archaeological materials (Cremaschi and di Lernia 1996). They are associated with microlithic equipment and pottery fragments decorated with a dotted wavy line, found together with large grinding equipment and clusters of poorly preserved fireplaces (Cremaschi and di Lernia 1996). These features suggest, for the 8th millennium BP, a change in exploitation of resources toward the introduction of wild plant processing, as also documented in the Tadrart Acacus (di Lernia 1999a; Mercuri 2008).

The Early and Middle Pastoral-Neolithic sites date back to the 7th and 6th millennia BP and represent 50% of the archaeological evidences identified in the area (Fig. 5.6). Their large number indicates the apogee of the Pastoral-Neolithic cattle herders and their good adaptation to the wet environment. Generally speaking, the sites are regularly related to the geomorphologic features of a high standing water level and the objects they are composed of (such as lithics, pottery, bone fragments, stones) are systematically scattered for hundreds of meters (sometimes up to few kilometers) in continuity along the former shores of the interdune lakes and on the slope above (Fig. 5.12).

Fig. 5.12 Erg Uan Kasa. **a** Configuration of a Pastoral-Neolithic archaeological site laying on the dune slope; the surface is dotted by artifacts. **b** Faunal remains and a complete Middle Pastoral-Neolithic pot residual of differential wind erosion

Fig. 5.13 Erg Uan Kasa. **a** Epipaleolithic stone tools. **b** Some examples of Middle Pastoral-Neolithic pottery

The spatial configuration of the archaeological features is well preserved and consists of clusters of several tens to hundreds of fireplaces, pits containing cattle bones, entire vessels, and grinding stones. Entire pots and grinding equipment were found semi-buried among fireplaces and probably represent hiding places (Fig. 5.12). By contrast, pits filled with animal bones are much more commonly located outside the perimeter of the sites.

The large size of the sites is interpreted to indicate a dense population, probably with recurrent occupation of the same places along the lake shores (Cremaschi and di Lernia 1998; di Lernia 1999b; Biagetti and di Lernia 2003). Support for this view is also provided by the permanent siting of heavy duty tools and large vessels, difficult to transport, and ready for reuse during subsequent visits. The sites of the lakes are therefore interpreted a terminus in the transhumance route connecting the lakes in the Erg Uan Kasa to the mountain ranges of Tadrart Acacus (Cremaschi and di Lernia 1998; di Lernia 1999a, b; Biagetti and di Lernia 2003). This strict connection is further demonstrated by the provenance of the raw material employed for lithics (quarzarenite) and grinding equipment (quartzarenite and sandstone), that derive from the Acacus area. However occurrence of grinding equipment obtained from granite and micaschist also indicates that transhumance may have extended at least to the western slopes of the Algerian Tassili (250 km to the west) or/and to the westernmost outcrops of the Tibesti granite (350 km to the east).

5.4.3 *The Wadi Tanezzuft and the Garat Ouda Paleolake*

5.4.3.1 Physiography

During the wet Holocene the Wadi Tanezzuft was a large river, fed by several influents from the Tassili and from the upper Tadrart Acacus. At its maximum extension, it was about 200 km long and ended in a large interdune lake close to the Edeyen of Ubari at about 60 km north of the Tadrart Acacus massif. In this region three main physiographic features still preserve paleoenvironmental records of the alluvial plain of the Tanezzuft valley (early to late Holocene), the interdune basins of the ErgTanezzuft (early Holocene), and the playas fed by lateral branches of the main river (early and middle Holocene).

The fluvial sedimentation inside the main valley was already active by the 7th millennium BP, but the deposits of this age are deeply buried in the middle of the wadi and outcrop only locally at its margins. In the early to middle Holocene (ca. 8000–6000 year BP) the discharge of the Wadi Tanezzuft was large enough to carry coarse sediments deposited as longitudinal gravel bars (Fig. 5.14) in a braided river system, turning downstream into large meanders (Perego et al. 2007), which architecture appears well preserved in satellite images. After 5000 year BP, as an effect of the reduced discharge, the grain size of the sediment transported also changed, as the gravel load was replaced by sand and mud, and an alluvial plain deposited along the main branches of the river. The overbank deposits belonging to the last aggradation phase are constituted by upward fining cycles of cross laminated sand, sandy silt, and sandy clay, often weathered into inceptisols or entisols. The deposition of silty sediments is dated between 4000 and 3000 year BP by the presence of several archaeological sites with fireplaces entombed within the alluvial deposits (Cremaschi and di Lernia 2001).

In the middle part of the wadi, a few kilometers north from the village of Ghat, the course of the river bordered the Erg Tanezzuft. Although the erg is not very wide, it is similar to the larger Erg Uan Kasa. It is composed of linear and star dunes separated by interdune corridors. Lacustrine deposits occur along the corridors at the base of the dune slopes, which consist of hydromorphic horizons and organic sand deposited during the early and middle Holocene.

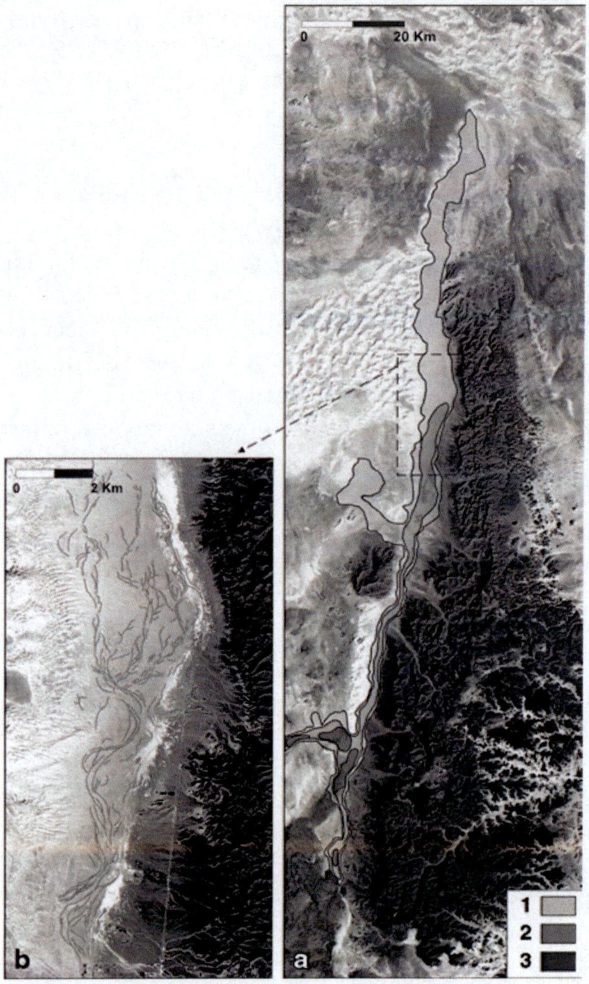

Fig. 5.14 The Wadi Tanezzuft as from Landsat 7 satellite imagery **a** On the left **b** a detail of the meander bars dating to the early-middle Holocene. The progressive contraction of the oasis is indicated in **a**. (Key: *1* extent of the late 6th millennium BP oasis; *2* extent of the 4th–3rd millennium BP oasis; *3* extent of the 2nd millennium BP oasis)

During the early Holocene the water supply from the Wadi Tanezzuft was very large and several minor channels branched out from the main stream into the lowlands and depressions of the floodplain (Cremaschi 2001; Cremaschi et al. 2005). A similar phenomenon is reported in northern Egypt where at the beginning of the Holocene the Nile bed silted up high enough to let the river overflow into the Fayum depression, feeding the ancient lake Moeris (Hassan 1986).

The Garat Ouda playa is the widest basin of the area. It hosted a lake some 80 km^2 wide in a depression dammed to the north and to the west by the south-eastern fringes of the Erg Titersine, and to the east and the south by Inselberg-type reliefs and flat-irons. Close to the dune, shore deposits are still preserved and consist of black organic sand some tens of centimeters thick intercalated with bleached sand. Fish bones and vegetal remains have been found in the organic horizons. On Landsat satellite images, the bright reflectance of the area of Garat Ouda is produced by light gray to white silt that fill the depression. The flat area covered by the silt corresponds to the alluvial plain formed during the early to mid-Holocene by the lateral branch of the Wadi Tanezzuft. In the southern part of the basin there is a system of meandering paleochannels that become the distributaries of a delta toward the middle of the alluvial plain. These channels are not evident on Landsat images but their shape is very clear in Ikonos images (Fig. 5.15). Furthermore, channel deposits (consisting of pink silty sand) are evident in the field, as is the fact that differential aeolian erosion has lowered the surface of the alluvial plain by more than 50 cm. On satellite images the architecture of the middle Holocene lower fluvial reaches and delta appear well preserved, and through the superimposition of the channels, a progressive migration of the main course towards the east can be discerned (Cremaschi et al. 2005).

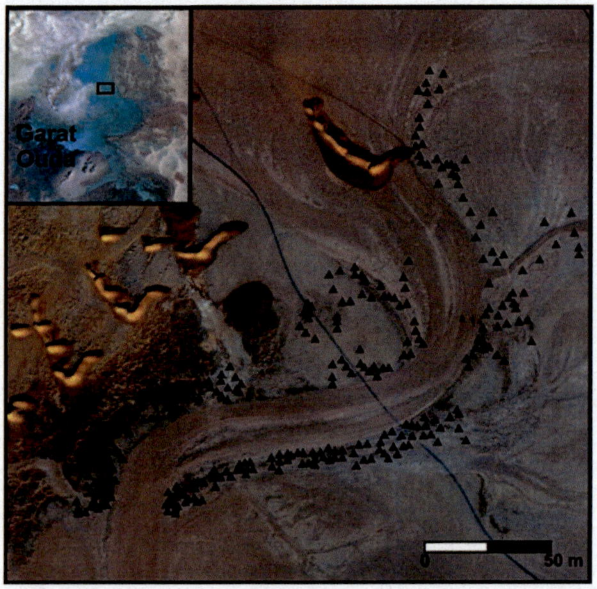

Fig. 5.15 A detail of the meandering paleochannel of the terminal reach of the left branch of the Wadi Tanezzuft in the area of Garat Ouda (Ikonos satellite imagery): the black triangles represent single middle Pastoral-Neolithic fireplaces, dotting the perimeter of the paleochannel. Insert: the location of the paleochannel in the Garat Ouda area

5.4.3.2 Settlements

A few Epipaleolithic and Mesolithic sites have been discovered at the margin of the early Holocene Tanezzuft river, mostly in the northern part of the valley. They consist of clusters of lithics eventually associated with fireplaces, large grinding equipment and pottery with dotted wavy line decoration. Commonly, Mesolithic sites are buried in the silt of the alluvial plain, indicating that the settlements were located close to the former bank of the Wadi Tanezzuft. The Early and Middle Pastoral-Neolithic sites occur in large number along the main wadi course and inside the small ergs surrounding it. The sites display similar features to contemporary sites in the Erg Uan Kasa (Cremaschi and di Lernia 2001). The Pastoral-Neolithic sites were mostly located in the vicinity of river banks or close to former lakes, especially in the central and northern part of the valley, whereas evidence dating to these phases are almost lacking south of the Erg Tanezzuft.

The distribution of sites in the Erg Tanezzuft is similar to that described for the Erg Uan Kasa and is related to the geological evidence of former standing water. Sites date almost to the Early and Middle Pastoral-Neolithic periods. At a site (97/180) located in an interdune corridor of the Erg Tanezzuft there is a catenary sequence of soils and sediments at the base of the dune slope, which may be regarded as representative for settlement in this area (Cremaschi and di Lernia 2001). On the floor of the corridor, lacustrine sediments occur consisting of black sand with mollusc shells. Upwards, along the slope, these sediments grade into hydromorphic soil with root casts, superposed on hydromorphic sand. Above these, there is red weathered sand, covered in turn by the mobile part of the dune. Around the fringe of the inner part of the corridor, which roughly corresponds to the shore of a former lake, there is a middle Pastoral-Neolithic semi-permanent camp. Radiocarbon dating indicates that the site was contemporaneous with the lacustrine sedimentation.

The archaeological record is particularly rich and well preserved in the area of the Garat Ouda playa, and indicates adaptation to a fluvial-lacustrine transitional environment. The Epipaleolithic sites are located along the former shore of the lake in the northern part of the area. The sites of the Pastoral-Neolithic periods are dominant and consist of more than 2500 fireplaces and grinding equipment distributed as a continuous belt along the lower fluvial reaches near the deltaic areas (Cremaschi et al. 2005). The main concentrations of archaeological features form small mounds up to 1 m high. This is a post-depositional effect of wind erosion, which lowered the area surrounding the sites but not the sites themselves protected by the concentration of stones and artefacts. The strong relationship between the former channels and the distribution of sites is clearly visible on Ikonos satellite imagery of a meander of the lowermost fluvial reaches feeding a delta located in the center of the Garat Ouda area (Fig. 5.15). Fireplaces systematically dot the two banks of the channel and some occur nearby (up to 30 m) on the surrounding alluvial plain. Several pits containing charred faunal remains have also been observed. Faunal remains are of fish, crocodile, and hippopotamus, and of cattle (as in the Erg Uan Kasa) better preserved along the paleochannel. Bones belonging to large animals are mainly found scattered throughout the archaeological sites, and fish bones are concentrated in small mounds that are the remains of garbage pits or fireplaces partially removed by wind erosion. This suggests a specific exploitation of the area of Garat Ouda through hunting and especially fishing integrating the Pastoral-Neolithic practice of herding.

5.5 Drought at 5000 Years BP

The main climate change toward aridity in the Sahara (termination of the African Humid Period) is dated at ca. 5000 year BP (deMenocal et al. 2000; Kröpelin et al. 2008). Dry conditions did not affect the whole region simultaneously, and effects on the landscape differed depending on geographical position and geomorphologic conditions. In SW Fezzan the desiccation of the lakes, together with other environmental evidence (such as rivers activity, pollen data, dendroclimatological studies), indicates that this occurred during the middle Holocene, and correlates with the weakening of the monsoon rainfall and its retreat to southern regions (Cremaschi 1998, 2002; Zerboni 2006, 2008). This dry event is well represented in the dendroclimatic record of the *Cupressus dupreziana* by two strongly negative spells in the tree-ring sequence, which dates it at ca. 5040–4850 cal year BP (Cremaschi et al. 2006). The influence of the termination of the monsoon rainfall on the Tadrart Acacus massif, the Erg Uan Kasa, and the Wadi Tanezzuft is described in the following sections.

5.5.1 Inside the Acacus

The transition to different environmental conditions at ca. 5000 year BP is evident in many caves of the Tadrart Acacus, and is marked by erosion and a change in the sedimentation pattern. Depositional trends that indicate aridity consist of erosional surfaces at the top of the Middle Pastoral-Neolithic layers, deposition of lenses of aeolian sand inside the caves capped by extensive dung layers, and collapse of the roofs of many rockshelters. Micromorphologically the dung layers are characterized by a platy discontinuous microstructure with few voids, including deformed coprolites and elongated plant fragments; this is interpreted as due to trampling and is indicative of a stable deposit (Fig. 5.10). In some cases sand lenses alternate with the organic strata, indicating a formation in a dry environment with marked seasonality. Dung deposits are dated from 5200 to 3770 year BP. In the sequence of Uan Muhuggiag, for instance, a discontinuity related to wind erosion is covered by a layer of dung consisting of unaltered plant and straw remains cemented by animal excrement. The fact that this material did not suffer significant bacterial degradation is attributed to a semi-arid, dry environment (Cremaschi 1998). Pollen analyzes from the Uan Muhuggiag site confirm the general paleoenvironmental trend of the sedimentary sequence with Pocaceae, Capparaceae, *Acacia*, and *Artemisia* at the top indicating dry steppe (Mercuri 2008). The same sedimentary pattern has been described in several other rockshelters, such as in the case of the Uan Telocat sequence, where a thick layer of preserved leaves alternate with sandy layers.

The Late Pastoral-Neolithic sites increased in number from the end of the 6th millennium BP up to the beginning of the 4th millennium BP (Fig. 5.6). This increase is both related to the archaeological visibility (the mountains archaeological sites are better preserved) and to a different settlement pattern that implied a seasonal occupation of rockshelters, used for penning activity. The main characteristic of this phase are the diversified use of rockshelters, the introduction of undecorated pottery, and the exclusive use of ovicaprines (Cremaschi and di Lernia 1998).

After the Late Pastoral-Neolithic phase and during the Garamantian period (5th century BC to 5th century AD), the caves and rockshelters of the Tadrart Acacus are still attended, although the sites are very limited in number. The exploitation of the mountain is mainly based on herding of flocks, and the sites in the massif were connected to the main settlements of the time (Liverani 2005), located in the oasis and corresponding to the main centers of the network of the Garamantian kingdom (a tribal people in the Fezzan, descended from Berbers and Saharan pastoralists).

Considering the rock art as an indication of the perception of landscape, a main discontinuity with the Pastoral-Neolithic phase is evident. During the wet early and middle Holocene, the representations were strictly related to the landscape and its exploitation (paintings and engravings of animals close to former water points, hunting scene, shepherd with flocks, and so on), whereas in this and later periods rock art became a tool to mark the territory (Cremaschi et al. 2008). The late Holocene dwellers of the massif left a few representations of humans and animals to indicate the seasonal availability of water, to signpost the main paths and passes crossing the mountain, and to mark their presence in specific areas. The main routes were marked by representation of four-horse chariots. Furthermore, they decorated with inscriptions in Tifinagh (a Berber alphabetic script) the most relevant localities leaving us a complex system of landmarks not yet completely understood.

The exploitation of the Tadrart Acacus valleys is at present limited to a few Tuareg groups (the so-called kel Tadrart), whose life is regulated by the availability of natural resources (water and grass); till the late 1970s they used to move on the massif, looking for water, and building their base camps close to active 'gheltas'. Today, notwithstanding the harsh climatic conditions, their style of life is becoming more sedentary, as the excavation of deep wells at the month of the main wadis of the massif guarantees the availability of water for man and goats throughout the year.

5.5.2 Drying and Abandonment of the Erg Uan Kasa

Lacustrine sedimentation is no longer recorded in the Erg Uan Kasa after ca. 5500–5200 year BP (Fig. 5.4) indicating that shortly after this period the lakes dried out (Cremaschi 1998, 2002; Zerboni 2006). Interruption of the monsoon precipitation radically changed the environmental conditions of the dune corridors in the Erg Uan Kasa. Water availability rapidly decreased and the water stored in sand was drawn to the topographic surface by capillary uprise. Then it evaporated and salts precipitated. Most of the lacustrine basins, located in the northern and central parts of the erg were sealed by gypsum and alkali crusts. The effects of incoming drought were slightly different in the southern part of the Erg Uan Kasa where the interdune corridors correspond to the downstream direction of the main valleys cutting the Tadrart Acacus massif. In this area the run off of water from the massif lasted for several centuries and the result was that the early to mid-Holocene lacustrine deposits were buried by fluvial silty sediments, as an effect of enhanced slope degradation in the mountain range (Fig. 5.16). A subsequent single, very short, wet episode in the southern part of the erg dates at ca. 2400 year BP (Cremaschi 1998). It is probably related to a limited resumption of the monsoon. However, the dominant process soon became wind erosion and this removed most of the deposits and soils formed during the wet Holocene (Cremaschi 2002, 2003).

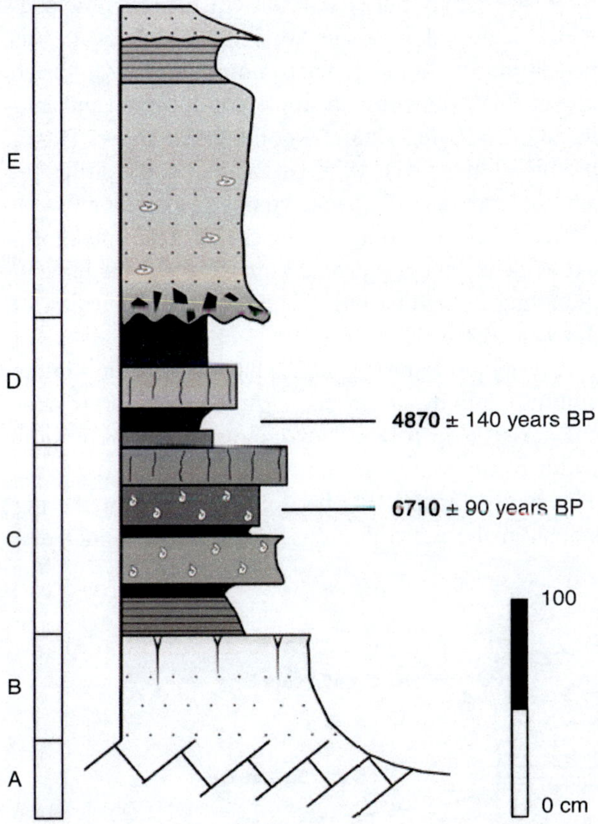

Fig. 5.16 Stratigraphic sequence from a southern interdune corridor of the Erg Uan Kasa (site 03/508); it is representative for the transition from a lacustrine to fluvial sedimentary environment. Uncalibrated radiocarbon dates are reported (*A* bedrock (sandstone); *B* hydromorphic horizon; *C* organic deposit and carbonatic mud containing molluscs (shallow lake facies); *D* silt intercalated by organic mud (alluvial plain); *E* silty to sandy deposits (channel bars))

The Pastoral-Neolithic occupation which followed the desiccation of the Erg Uan Kasa was dramatically reduced. The transition from Middle to Late Pastoral-Neolithic phases recorded not only a decrease in site concentration, but also a major change in the configuration of the settlements. The Middle Pastoral-Neolithic sites were large and included organized settlement structures and facilities, whereas those of the later phase appeared to be quite small, almost lacking any complex archaeological feature, except for a few scattered fireplaces, small concentrations of tethering stones, and scarce artefacts (Cremaschi and di Lernia 1998). The Late Pastoral-Neolithic groups left the margin of the dunes where the shores of the desiccated lakes were located, and moved towards the center of the interdune corridors. The sites of this phase lie on erosional surfaces. In any case, clusters of fireplaces could indicate that the erg was never completely abandoned even under very dry conditions and may have been marginally exploited by nomads for hunting and pastoralism (Kuper and Kröpelin 2006). Later, it was crossed by the early historical caravan routes, as testified by the Garamantian fortified sites at the western margin of the erg.

5.5.3 Origin and Decline of the Tanezzuft Oasis

During the middle Holocene, after the surrounding areas had already dried out, the activity of the Wadi Tanezzuft persisted for millennia (Fig. 5.4), although the river became shorter and endorheic, no longer reaching the terminal lake located close the sand sea of Ubari. Its fringe was located at the northern end of the Tadrart Acacus massif.

As an effect of the reduced discharge, the grain size of the sediment also changed. Gravel was replaced by sand and mud, and an alluvial plain developed along the main branches of the river over gravel bars and burying archaeological sites. The deposits belonging to the alluvial plain have been radiocarbon dated from ca. 4200 to 2900 year BP (Cremaschi 2001). Sandy-loamy sediments are interlayered with thin soils with hydromorphic features and crossed by root casts. Both are evidence of a greater groundwater availability and plant cover on the alluvial plain.

The subsequent increase in dryness reduced the flow along the Wadi Tanezzuft such that water supply to the Garat Ouda delta-lake system ended between 5200 and 4800 year BP and the lake dries up (Cremaschi 2001; Cremaschi et al. 2005). Abruptness of the change allowed for good preservation of geomorphologic features, also because no reactivation occurred afterwards.

During the mid-late Holocene drying phase, the Wadi Tanezzuft became an oasis (Cremaschi 2001, 2003). In this context the term oasis should be understood as an isolated physiographic unit within the mosaic of the desert physiographic features of the Saharan-Arabic arid belt. It consists of an area of vegetation, typically surrounding a water source that occurs where groundwater lies close to the desert surface, and where plant roots and wells can reach it. The oasis attracted and concentrated the Late Pastoral-Neolithic communities pushed out from the surrounding territories by drought. Moreover, the onset of new environmental conditions led to adaptive strategies to aridity that, starting from a simple exploitation of the land resources (such as agricultural activity), became more sophisticated after several centuries combining local land resource exploitation with long distance trade through the caravan routes (Liverani 2004).

It is not surprising, therefore, that the Wadi Tanezzuft was also densely inhabited during the Late Pastoral-Neolithic period (Fig. 5.6), when occupation in surrounding areas was considerably reduced and limited to specific movements related to nomadism. Configuration and distribution patterns of the sites dating from this period and found inside the perimeter of the oasis may suggest an interest in exploitation of the soil in a context of rising sedentarism. Individual sites are basically composed of fireplaces and storage pits, with vessels and faunal remains, but they include also a large number of grinding equipments, together with lithic hoes and gouges—possible indications of land management and crops processing.

In the central part of the alluvial plain of the Wadi Tanezzuft some phytogenic dunes related to tamarisk bushes are concentrated along the most depressed part of the alluvium. The dunes date from medieval times (ca. 600 year BP; Fig. 5.17), and they blanketed the alluvial plain in which Late Pastoral-Neolithic archaeological sites are buried (Cremaschi 2001). These sites were located close to the water resource despite the possibility of flooding. Micromorphology of a buried inceptisol developed at the top of the alluvium and connected to a Late Pastoral-Neolithic site shows strong bioturbation (root cast), fragmentation of the structure, and occurrence of coarse coatings associated with charcoal and phytolithes (Fig. 5.17). Micropedological evidence suggests some kind of soil management, including ploughing and the slashing and burning of the vegetal cover.

Clusters of tethering stones have been found in the alluvial plain at the northern end of the Tadrart Acacus massif (Fig. 5.18). They are distributed across the width of the valley, in the northern part of the Wadi Tanezzuft (Fig. 5.18). The significance of tethering stones in the Saharan archaeology is still under dis-

Fig. 5.17 Stratigraphy of a phytogenic dune (*A*) that covers a soil (*B*) developed at the top of the middle Holocene alluvium (*D*). A Late Pastoral-Neolithic site (*C*) is buried under the soil (site 96/267); uncalibrated radiocarbon dates are also indicated (After Cremaschi 2001). The insert (*E*) is a microphotograph of a thin section (plane polarized light) of the soil showing strong bioturbation, fragmentation of the structure, coarse coatings, small charcoal, and phytolithes

Fig. 5.18 **a** Distribution of the tethering stones at the northern fringe of the mid-Holocene Tanezzuft oasis (location in the insert). **b** A tethering stone from the northern Wadi Tanezzuft

cussion: even if a use in management of livestock cannot be excluded, the evidence of rock art suggests that they were certainly connected to hunting activity and used as a component of traps for wild animals (Pachur 1992; Cremaschi 2001; di Lernia et al. 2008). In this case the distribution of tethering stones corresponds with the extreme reach of the Wadi Tanezzuft, and therefore with the limits of the oasis in the late 5th millennium BP, as reconstructed on the basis of the geological evidence. Here the main concentrations of tethering stones, occurring at the interface between the oasis and the desert, would indicate the most probable access of wild game to the water in the oasis, and an intense exploitation of the marginal areas by hunters.

Radiocarbon dates indicate that the boundaries of the Tanezzuft oasis were stable in the 4th and 3rd millennia BP, but during the 2nd millennium BP the oasis contracted significantly (Fig. 5.14; Cremaschi 2003). At this time, archaeological evidence shows that the Tanezzuft oasis became the southern border of the Garamantian kingdom, and acted as a node on caravan routes connecting the central Sahara to sub-Saharan Africa (Liverani 2004). The climatic conditions at the beginning of the 2nd millennium BP and its effects on the Garamantian civilization are still under discussion. However, during their whole history the Garamantes tried to adjust to increasing aridity, for instance with water management (irrigation) for intensive agriculture within the oasis (Liverani 2005). For example, the foggara system (a traditional systems of water catchments and horizontal underground shafts that drain water and convey it by gravity) was used in the Germa (formerly Garama, capital of the Garamantian kingdom) oasis until medieval times (Mattingly 2003). In the southern Wadi Tanezzuft, where the main Garamantian settlements were located, there is no evidence of complex irrigation strategies apart from a few small canals now largely buried by sand dunes. It is likely that in the area of Ghat, El Barkat, and Fewet agriculture was sustained by shallow wells.

In the Tanezzuft valley the Garamantian settlement mainly consists of fortified villages and compounds located as a protection of the fringes of the oasis (Liverani 2005; Biagetti and di Lernia 2008). Information about actual land use is poor although archaeobotanical data confirm intensive agriculture (Mercuri et al. 2005). Several minor sites of this age, composed of clusters of fireplaces and scatterings of pottery, have been found buried by alluvial deposits (Cremaschi 2005). They indicate that the oasis was larger than today, and was probably sustained by the wet conditions evidenced by the presence of *Cupressus dupreziana,* which occurred between 2800 and 2200 years BP (Cremaschi et al. 2006). At that time Wadi Tanezzuft was still active and was a source of water for cultivation.

It is impossible to separate climate change and socio-political factors in explaining the collapse of the Garamantian kingdom. However, it was coincident with the onset of very dry conditions in the region and with the disaggregation of the boundary of Roman Empire in North Africa, and the consequent interruption of the commercial routes to the south (Liverani 2005). In any case the Garamantian occupation in the oasis ended about 1600 years BP. This coincides with the onset of very dry conditions, as indicated by the progradation of sand dunes into the oasis and by a sharp decrease of the tree-ring size in the dendroclimatic record of *Cupressus dupreziana* at 1573 year BP (Cremaschi 2005; Cremaschi et al. 2006).

The configuration of the oasis of the Wadi Tanezzuft at the end of the Garamantian kingdom was preserved by the exploitation of the residual water resource by means of shallow wells up to the beginning of the last century. These conditions lasted until a few decades ago, when the oasis slightly expanded again to a maximum extent in the 1980s thanks to the excavation of deep wells that exploited the fossil hydrological reserve. Today overexploitation of such waters is exhausting the reserve, and Ikonos satellite imagery shows recent plantations abandoned and obliterated by sand.

5.6 Conclusions

From the climatic history and cultural dynamics of the central Sahara a few conclusions can be reached.

a. A comparison of the geoarchaeological development throughout the Holocene of the Tadrart Acacus, Erg Uan Kasa, and Wadi Tanezzuft, strongly suggests that water availability was the main factor shaping the landscape and driving the evolution of settlement and the cultural dynamics of the population living there. For example, the early Holocene dry event (8.2 kiloyear BP event), though short, had a disruptive effect on the landscape. The most sensitive environment was the mountain where a discontinuity in anthropogenic sedimentation in caves and rockshelters marks a strong reduction in resources, likely followed by a change in survival strategies.
b. The modern period of hyperaridity began with the mid-late Holocene transition. The withdrawal of the African monsoon and the progressive fall in water resource mark the turning point of environmental conditions at ca. 5000 year BP. The response of fresh water systems seems to be instantaneous, while terrestrial environments have gradually adapted to increasing aridity. Finally, the water resources contracted to the oasis. The Pastoral-Neolithic communities which had settled all the landscape units were forced into the limits of the oasis, where they developed different subsistence strategies. The oases promoted a new, revolutionary model of territorial management, giving rise to different subsistence strategies evolving towards complex social systems.
c. The oases did not constitute stable geomorphologic system. They experienced a reduction in size following the decrease in the water resource. They could not halt the encroaching aridification, and crisis and collapse of societies such as the Garamantian was a common outcome. Moreover, the recent history of the Tanezzuft oasis shows that in a marginal environment the anthropic impact, even if very limited in time, would have quickened natural changes (as the recent lowering of the deep aquifers testifies), and thus accelerated desertification.
d. However, even after aridification, dry lands (as the Erg Uan Kasa and the northern part of the Wadi Tanezzuft) were never completely abandoned, and were subjected to reduced human exploitation. These areas, apparently remote and abandoned, were in reality frequented by nomads, hunters, and shepherds who exploited their scarce resources. These are fragile environments and the seemingly marginal but protracted human pressure may have ultimately contributed to accelerate landscape degradation. Large scale devastation may have resulted from the human-induced intensification of such interconnected processes as aeolian erosion, soil stripping, and the extreme reduction of vegetal cover. Marginal pastoralism might have enhanced the soil erosion and consequently the spread of desertification. Furthermore, the use of specific hunting artefacts in the desert areas, such as tethering stones in the Sahara region, or desert-kites in the Middle East (Helms and Bettis 1987), might have a long term effect on the local environment.
e. The steps toward aridity and the reduction in size of the oases appear to be connected to local conditions. In this sense, the present work should be intended as a confirmation of recent hypotheses concerning a differentiated Saharan mid-Holocene aridification and a slow rate of desertification in specific areas (Kröpelin et al. 2008). Care should be taken to avoid any regional generalization and future detailed studies should be encouraged in selected areas to better understand the effects on the landscape of the termination of the African Humid Period.
f. Although data archives of high resolution proxies (isotopes, tree-rings, CO_2, CH_4, as so on) give an idea of the intensity of climate change, a reliable interpretation of what happened in the physical

environment is offered only by combining geomorphologic evidence with the archaeological record into an integrated geoarchaeological approach. Holocene paleoenvironmental reconstructions in arid lands should be coupled with the study of the settlement pattern revealed by archaeology. The geoarchaeological approach to climate change offers an opportunity to understand the past and provides a key to solve present and future issues.

g. Finally, in the context of Quaternary changes, the modern Sahara seems to show anomalous behaviour. The dismantling of the Alpine glaciers, even at mid-latitudes, could be considered typical of an interglacial, but the modern expansion of the desert may correspond to environmental conditions more consistent with a glacial period. It is possible that desertification may have been intensified by human exploitation of the Sahara since the mid-late Holocene. This may be taken as one more case where humankind has significantly affected decrease in the size of a water reserve and a reduction of the sequestration of CO_2 in soils and sediments since early times. This would conform to the ideas proposed by Ruddiman (2003) concerning the early anthropogenic overprints on climate.

Acknowledgements The data exposed in this chapter were collected during the research conducted in the frame of the Italian-Libyan Archaeological Mission in the Acacus and Messak of the University of Rome, La Sapienza (at present leaded by S. di Lernia), from 1992 and 2003 under the direction of F. Mori and M. Liverani, with the supervision of the Department of Archaeology of Tripoli (DOA, Director Giuma Anag).

References

Alley RRB, Mayewski PA, Sowers T, et al. (1997) Holocene climatic instability: A prominent, widespread event 8200 years ago. Geology 25:483–486

Biagetti S, di Lernia S (2003) Vers un modèle ethnographique-écologique d'une société pastorale préhistorique saharienne. Sahara 14:7–30

Biagetti S, di Lernia S (2008) Combining intensive field survey and digital technologies: new data on the Garamantian castles of Wadi Awiss, Acacus Mts., Libyan Sahara. Journal of African Archaeology 6:57–85

Brooks N (2006) Cultural responses to aridity in the Middle Holocene and increased social complexity. Quaternary International 151:29–49

Brooks N, Chiapello I, di Lernia S, et al. (2005) The climate-environment-society nexus in the Sahara from prehistoric times to the present day. The Journal of North African Studies 10:253–292

Busche D, Erbe W (1987) Silicate karst landforms of the southern Sahara (Northeastern Niger and Southern Libya). Zeitschrift für Geomorphologie Supplement 64:55–72

Busche D, Hagedorn H (1980) Landform development in warm deserts – The central Sahara example. Zeitschrift für Geomorphologie Supplement 36:123–139

Carrara C, Cremaschi M, Quinif Y (1998) The travertine deposits in the Tadrart Acacus (Libyan Sahara) nature and age. In: Cremaschi M, di Lernia S (eds) Wadi Teshuinat – Palaeoenvironment and prehistory in south-western Fezzan (Libyan Sahara). CNR, Quaderni di Geodinamica Alpina e Quaternaria 7, Roma-Milano, Italy, pp 59–66

Cremaschi M (1998) Late Quaternary geological evidence for environmental changes in south-western Fezzan (Libyan Sahara). In: Cremaschi M, di Lernia S (eds) Wadi Teshuinat – Palaeoenvironment and prehistory in south-western Fezzan (Libyan Sahara). CNR, Quaderni di Geodinamica Alpina e Quaternaria 7, Roma-Milano, Italy, pp 13–47

Cremaschi M (2001) Holocene climatic changes in an archaeological landscape: the case study of Wadi Tanezzuft and its drainage basin (SW Fezzan, Libyan Sahara). Libyan Studies 32:3–27

Cremaschi M (2002) Late Pleistocene and Holocene climatic changes in the Central Sahara. The case of study of the south-western Fezzan, Libya. In: Hassan FA (ed) Droughts, Food and Culture, Kluver Academic/Plenum Publishers, New York, pp 65–81

Cremaschi M (2003) Steps and timing of the desertification during the Late Antiquity. The case study of the Tanezzuft oasis (Libyan Sahara). In: Liverani M (ed) Arid Lands in Roman Times, Arid Zone Archaeology vol 4, Edizioni All'Insegna del Giglio, Firenze, Italy, 2003, pp 1–14

Cremaschi M (2005) The Barkat oasis in the changing landscape of the Wadi Tanezzuft during the Holocene, in: Liverani M (ed) Aghram Nadharif. The Barkat oasis (Sha'abiya of Ghat, Libyan Sahara) in Garamantian times. Arid Zone Archaeology vol. 5, Edizioni All'Insegna del Giglio, Firenze, Italy, pp 13–20

Cremaschi M, di Lernia S (1996) Climatic changes and human adaptive strategies in the central Saharian massifs: the Tadrart Acacus and Messak Settafet perspective In: Pwiti G, Soper R (eds) Aspects of African Archaeology, Papers from the 10th Congress of the Pan-African Association of Prehistory, Harare, 1995, pp 39–51

Cremaschi M, di Lernia S (1998) Wadi Teshuinat – Palaeoenvironment and prehistory in south-western Fezzan (Libyan Sahara). Quaderni di Geodinamica Alpina e Quaternaria 7. CNR, Roma-Milano, Italy, 332 pp

Cremaschi M, di Lernia S (1999) Holocene climate change and cultural dynamics in the Libyan Sahara. African Archaeological Review 16:211–238

Cremaschi M, di Lernia S (2001) Environment and settlements in the Mid-Holocene palaeo oasis of wadi Tanezzuft (Libyan Sahara). Antiquity 75:815–825

Cremaschi M, Pelfini M, Santilli M (2006) Cupressus dupreziana: a dendroclimatic record for middle-late Holocene in the Central Sahara. The Holocene 16:293–303

Cremaschi M, Pizzi C, Zerboni A (2005) Evoluzione e scomparsa del sistema delta-lago di Garat Ouda (SW Fezzan,

Sahara Libico) tra l'antico e medio Olocene. In: Bondesan A, Fontana A (ed) Riassunti – Convegno Nazionale AIGEO: Montagne e Pianure, Materiali del Dipartimento di Geografia 28, Università di Padova, Italy, pp 82–83

Cremaschi M, Pizzi C, Zerboni A (2008) L'arte rupestre del Tadrart Acacus: testimone e vittima dei cambiamenti climatici. In: di Lernia S, Zampetti D (eds) La Memoria dell'Arte. Edizioni All'Insegna del Giglio, Firenze, Italy, pp 361–369

Cremaschi M, Trombino L (1999) Forming processes in the Uan Afuda sequence. Palaeoenvironment and human activities: a micromorphological approach. In: di Lernia S (ed) The Uan Afuda Cave (Tadrart Acacus, Libyan Sahara). Archaeological investigations of late Pleistocene and early Holocene human occupations. Arid Zone Archaeology vol. 1, Edizioni All'Insegna del Giglio, Firenze, Italy, pp 27–38

Cremaschi M, Trombino L (2001) The formation processes of the stratigraphic sequence of Uan Tabu and their palaeoenvironmental implications. In: Garcea EAA (ed) Uan Tabu in the settlement history of the Libyan Sahara. Arid Zone Archaeology vol. 2, Edizioni All'Insegna del Giglio, Firenze, Italy, pp 15–23

Cremaschi M, Zerboni A, Spötl C, et al. (2010) The calcareous tufa in the Tadrart Acacus Mt. (SW Fezzan, Libya). An early Holocene palaeoclimate archive in the central Sahara. Palaeogeography, Palaeoclimatology, Palaeoecology 287:81–94

Dagge J, Hamad O (2008) The long dry spell. Syria Today 43:20–23

deMenocal P, Ortiz PJ, Guilderson T, et al. (2000) Abrupt onset and termination of the African Humid Period: rapid climate responses to gradual insolation forcing. Quaternary Science Reviews 19:347–361

Desio A (1937) Geologia e Morfologia. In: Regia Società Geografica Italiana (ed) Il Sahara Italiano. Parte I: Fezzan e Oasi di Ghat, Società Italiana Arti Grafiche, Roma, Italy, pp 39–94.

Diamond J (2005) Collapse: How societies choose to fail or succeed. Viking Books, New York, 576 pp

di Lernia S ed. (1999a) The Uan Afuda Cave (Tadrart Acacus, Libyan Sahara). Hunter-gatherer societies of Central Sahara. Arid Zone Archaeology Monographs 3, Edizioni All'Insegna del Giglio, Firenze, Italy, 272 pp

di Lernia S (1999b) Discussing pastoralism. The case of the Acacus and surroundings (Libyan Sahara). Sahara 11: 7–20

di Lernia S. (1999c) The cultural sequence. In: di Lernia S (ed) The Uan Afuda Cave (Tadrart Acacus, Libyan Sahara). Hunter-gatherer societies of Central Sahara. Arid Zone Archaeology Monographs 3, Edizioni All'Insegna del Giglio, Firenze, Italy, pp 57–130

di Lernia S (2002) Dry climatic events and cultural trajectories: adjusting Middle Holocene pastoral economy of the Libyan Sahara. In: Hassan FA (ed) Droughts, Food and Culture, Kluver Academic/Plenum Publishers, New York, pp 225–250

di Lernia S, Mori L, Zerboni A (2008) Geo-Archaeological survey in the Kufra Eni Contract Area (Eastern Sahara, SE Libya). Sahara 19:7–26

Fantoli A (1937) Clima In: Regia Società Geografica Italiana (ed), Il Sahara Italiano. Parte I: Fezzan e Oasi di Ghat. Società Italiana Arti Grafiche, Roma, Italy, pp 97–119

Galeĉiĉ M (1984) Geological map of Libya, Explanatory booklet, sheet: Anay, NG 32-16, scale 1:250000. Socialist People's Libyan Arab Jamahiryah, Tripoli

Garcea EAA (2001) Uan Tabu in the settlement history of Libyan Sahara. Arid Zone Archaeology Monographs, 2, Edizioni All'Insegna del Giglio, Firenze, Italy, 256 pp

Gasse F (2000) Hydrological changes in the African tropics since the last glacial maximum. Quaternary Science Reviews 19:189–211

Girod A (2005) New data on Quaternary freshwater and land molluscs in the Sahara. Triton 12:21–30

Goudarzi GH (1970) Geology and mineral resources of Libya – a reconnaissance. Geological Survey Professional Paper 660, Washington, 104 pp

Hassan FA (1986) Holocene lakes and prehistoric settlements of the Western Fayum, Egypt. Journal of Archaeological Science 13:483–501

Helms S, Bettis AVG (1987) The desert "kites" of the Badiyat Esh-Sham and North Arabia. Paléorient 13:41–67

Hoelzmann P, Gasse F, Dupont LM, et al. (2004) Palaeoenvironmental changes in the arid and subarid belt (Sahara-Sahel-Arabian Peninsula) from 150 Kyr to present. In: Battarbee RW, Gasse F, Stickley CE (eds) Past Climate Variability Through Europe and Africa, Developments in Palaeoenvironmental Research Series, 6, Springer, Dordrecht, pp 219–256

Kalefa El-Gahali MA, (2005) Depositional environments and sequence stratigraphy of paralic glacial, paraglacial and postglacial Upper Ordovician siliciclastic deposits in the Murzuq Basin, SW Libya. Sedimentary Geology 177:145–173

Kröpelin S, Verschuren D, Lézine A-M, et al. (2008) Climate-Driven ecosystem succession in the Sahara: The past 6000 years. Science 320:765–768

Kuper R, Kröpelin S (2006) Climate-controlled Holocene occupation in the Sahara: motor of Africa's evolution. Science 313:803–807

Kutzback JE, Liu Z (2007) Response of the African monsoon to orbital forcing and ocean feedbacks in the Middle Holocene. Science 278:440–443

Liverani M (2004) Rediscovering the Garamantes: archaeology and history. Libyan Studies 35:191–200

Liverani M (2005) Aghram Nadharif. The Barkat Oasis. (Sha'abiya of Ghat, Libyan Sahara) in Garamantian Times. Arid Zone Archaeology Monographs 5, Edizioni All'Insegna del Giglio, Firenze, Italy, 520 pp

Martini M, Sibillia E, Zelaschi C, et al. (1998) TL and OSL dating of fossil dune sand in the Uan Afuda and Uan Tabu rockshelters, Tadrart Acacus (Libyan Sahara). In: Cremaschi M, di Lernia S (eds) Wadi Teshuinat – Palaeoenvironment and Prehistory in South-western Fezzan (Libyan Sahara). CNR, Quaderni di Geodinamica Alpina e Quaternaria 7, Roma-Milano, pp 107–122.

Mattingly DJ (2003) The archaeology of Fazzan. Volume 1, synthesis. The Society for Libyan Studies, London, 522 pp

Mayewski PA, Rohling EE, Stager JC, et al. (2004) Holocene climate variability. Quaternary Research 62:243–255

Mercuri AM (2008) Human influence, plant landscape evolution and climate inferences from the archaeobotanical records of the Wadi Teshuinat area (Libyan Sahara). Journal of Arid Environments 72:1950–1967

Mercuri AM, Trevisan Grandi G, Bosi G, et al. (2005) The archaeobotanical remains (pollen, seeds/fruits, and charcoal). In: Liverani M (ed), Aghram Nadharif. The Barkat oasis (Sha'abiya of Ghat, Libyan Sahara) in Garamantian times. Arid Zone Archaeology, vol. 5, Edizioni All'Insegna del Giglio, Firenze, Italy, pp 335–348

Pachur H-J (1992) Tethering stones as palaeoenvironmental indicators. Sahara 4:13–32

Pasa A, Pasa Durante MV (1962) Analisi paleoclimatiche nel deposito di Uan Muhuggiag, nel massiccio del Tadrart Acacus (Fezzan Meridionale). Memorie del Museo Civico di Storia Naturale 10:251–255

Perego A, Cremaschi M, Zerboni A (2007) Il telerilevamento nella ricostruzione della paleoidrografia olocenica in zone aride. Il caso di studio di Wadi Tanezzuft, Libia SO. Rendiconti online Società Geologica Italiana 4:97–99

Rognon P (1989) Biographie d'un Desert. Plon, Paris, 347 pp

Ruddiman WF (2003) The anthropogenic greenhouse era began thousands of years ago. Climatic Change 61:261–293

Walther H, Lieth H (1960) Klimadiagramm. Weltatlas. G. Fisher, Jena

Wendorf F, Karlen W, Schild R (2007) Middle Holocene environments of north and east Africa, with special emphasis on the African Sahara. In: Anderson DG, Maasch KA, Sandweiss DH (eds) Climate Change and Cultural Dynamics, Academic Press, London, pp. 189–227

Zerboni A (2006) Cambiamenti climatici olocenici nel Sahara centrale: nuovi archivi paleoambientali. Ph.D. Thesis, Università degli Studi di Milano, Milan, Italy, 232 pp

Zerboni A (2008) Holocene rock varnish on the Messak plateau (Libyan Sahara): Chronology of weathering processes. Geomorphology 102:640–651

Chapter 6
The Desertification of the Egyptian Sahara during the Holocene (the Last 10,000 years) and Its Influence on the Rise of Egyptian Civilization

Michael Brookfield

6.1 Introduction

Egyptian civilization is the gift of the Nile. But it is also the result of climatic change since the last ice age (Fagan 2004). During the last 10,000 years or so, desertification, punctuated by temporary reversals, has driven people out of what was once a well-watered savannah covering vast areas of the present Sahara into smaller areas fed by rivers and near-surface groundwaters. In Egypt, the result was a relatively large population in a relatively small area, the Nile Valley (Barich 1998). This has led to ideas that this population required organization in order for it to survive and allow complex regimented societies to develop; exemplified by the early Pharaonic period with its huge public works projects—the pyramids and associated monuments (Wilkinson 2003). But did complex societies evolve during this organization or was the organization a pre-requisite for the complex societies; did it take place before the constriction of society into the Nile Valley; and did it begin in one area and then spread? A popular view is that increasing social complexity occurred during the development of agriculture, made possible by a benign climate while severe climatic change interrupted this development (Burroughs 2005). On the other hand, there is increasing evidence that social complexity developed during climatic deterioration (Brooks 2006). Such questions require objective analysis of the timing of development of different social systems related to climatic change in different parts of Egypt and adjacent areas.

The area covered in this study is primarily encompassed by the present country of Egypt with some data from adjacent areas of Libya, Chad, Sudan, and even further away where necessary (Fig. 6.1). And because of the lack of evidence in the Nile delta and adjacent areas, most of this study deals with an area between Gilf Kebir and the Red Sea coast, south of Cairo (30°N). This band contains four rather separate areas: (a) the westernmost desert (around Gilf Kebir, Gebel Uweinat, and the Great Sand Sea), characterized by abundant cave engravings and paintings. There are few studies from this area and the few Bedouin inhabitants of the area left when the last wells dried out in the early twentieth Century; (b) the oases belt between the westernmost desert and the Nile, with few engravings or paintings but with many stratified sites; (c) the Nile Valley (including the Faiyum), with the best stratified sites and dates; and (d) the eastern desert between the Nile and the Red Sea, with mostly engravings and paintings.

This chapter attempts to: first, summarize the climatic changes from physical evidence; second to summarize the changes in human societies in different parts of Egypt as they moved from hunter-gathering in the Sahara savannah of the early Holocene, to the complex civilization of the early Pharoahs in the Nile Valley, and to relate these to climatic changes during this period. For the first we require fossils, sedimentary and landscape evidence that allows us to determine environment. For the second we require evidence of societal change that can be preserved: these are such structures as refuges and buildings, tools and other artefacts, pre-

M. Brookfield (✉)
Institute of Earth Sciences, Academia Sinica, Nankang
Taipei 11529, Taiwan. Mailing Address: 45 Dean Avenue
Guelph, ON N1G 1K6, Canada
e-mail: mbrookfi@hotmail.com

See Plate 5 in the Color Plate Section; also available at: extras.springer.com

I.P. Martini, W. Chesworth (eds.), *Landscapes and Societies*,
DOI 10.1007/978-90-481-9413-1_6, © Springer Science+Business Media B.V. 2010

Fig. 6.1 Location map of Egypt and Sudan with cited localities. Dashed line is current boundary between western desert and oases divisions. Nile river division separates oases and eastern desert division. Present oases are in bold italics

served art, sculpture and painting for example, and a stratigraphy or relative age which allows their placement in a historical context. For both we require good dating, not only of the societal changes but also the climatic changes, and this is difficult to get.

Relative dating comes from standard stratigraphic methods such as superposition and cross-cutting relationships. The main problems are: that time breaks of greater or lesser magnitude may separate the strata themselves and any incised pits; and the different artefacts used in different cultures at the same time obscure time equivalence.

Absolute dating (in years) comes from a variety of methods. Fortunately, artefacts from about 5000 BP (in Egypt) can sometimes be calibrated with written historical records. Prior to that, ^{14}C dates require calibration to calendar years with tree rings, but only up to about 7000 BP. Prior to this other methods are needed: this paper uses the Thorium/Uranium calibration with fossils corals back to 50,000 BP (Before Present, present being 1950) (Fairbanks et al. 2005); and all ^{14}C dates in this paper are calibrated and are given both in BP and BC: to convert add or subtract 1950 years respectively.

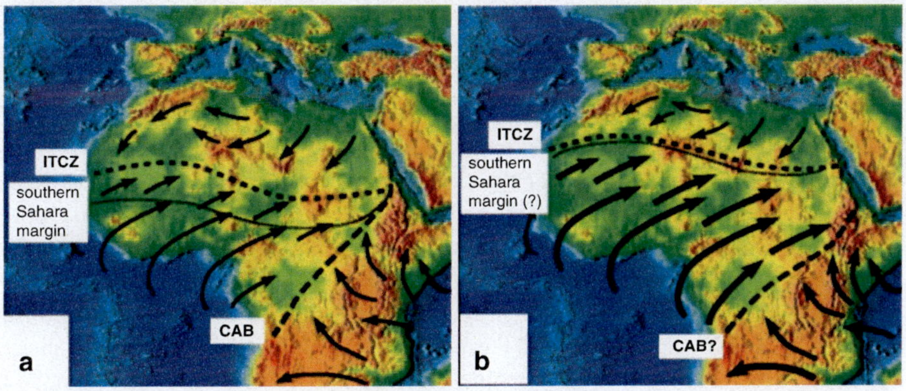

Fig. 6.2 **a** Present-day August positions of Intertropical Convergence Zone (ITCZ) and Congo Air Boundary (CAB) with southern Sahara margin and wind directions. **b** Inferred early Holocene position of ITCZ and winds. Note that more southern January ITCZ position would also imply penetration by northeasterly Mediterranean winds at that time

6.2 The Present Situation

The first necessity in interpreting climate and culture in the past is to determine existing relationships and interactions in comparable modern situations.

Tropical to subtropical climates are currently dominated by the two Hadley circulation cells. Hot air rises at the equator at the Intertropical Convergence Zone (ITCZ) and cooler air descends around 30°N & S and flows back towards the ITCZ. The ITCZ migrates north and south with the seasons, bringing summer rains to the Sahel and Sahara and winter rains to the Mediterranean coast. The northward movement of the (ITCZ) in the summer (Fig. 6.2a) determines the quantity of rainfall in a particular year: if its circulations penetrate far to the north there will be a long rainy season and good rains; if they do not penetrate sufficiently far north, then the rains may fail totally. During the winter, hot dry winds blow from the northeast. In spring, Mediterranean cyclones may rarely penetrate the eastern Sahara bringing rare downpours to hyperarid areas such as the Gilf Kebir. Rainfall decreases northwards from the tropical rainforests to the Sahara: and rainfall variability also decreases with increasing annual rainfall. A dry year with a 10% probability (9 out of 10 years will be wetter) has only half the rainfall of an average year at 200 mm, 65% at 400 mm and 80% at 1000 mm. During the early Holocene, the ITCZ penetrated much further north (Fig. 6.2b). Climate determines both the nature and location of the major ecotype belts and their land use (Fig. 6.3).

At present, there is a sharp division between the arid to hyperarid western and eastern deserts with their sparse nomadic Bedouin populations and the Nile Valley and oases with their dense settled agricultural populations. This sharp division did not occur in prehistoric times when the climate was generally wetter and the more gradual changes now seen south of the Sahara were often prevalent. The desert to savannah habitats and societies of northern Africa are thus particularly relevant for interpreting the pre-historic Holocene societies of Egypt (Midant-Reynes 2000; Thurston et al. 1993).

But some definitions of the type of societies present are needed. Foraging societies collect food available in nature by gathering wild plants, hunting wild animals and fishing. Pastoral societies raise livestock such as goats, cattle and sheep as food. Specialized pastoralists rely on such livestock, periodically moving the herds in search of fresh pasture and water. Multisource pastoralists mix herding with foraging and even farming. Farming is the growing of domesticated plants and raising of domesticated animals. None of these, of course are mutually exclusive which makes archaeological

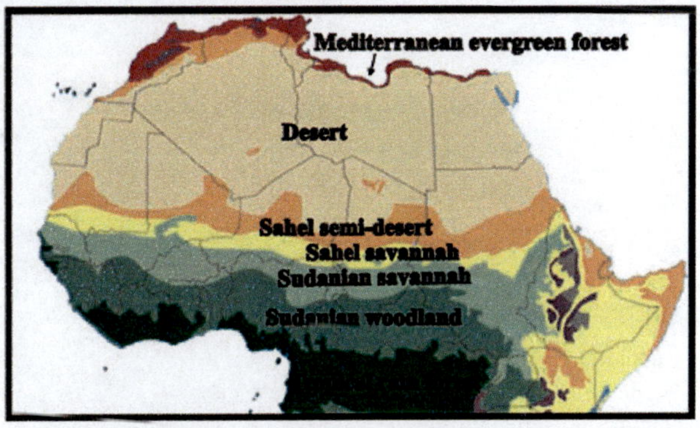

Fig. 6.3 Current biomes of northern Africa; the highest cattle concentrations are in the northern Sudanian savannah and southern Sahel biomes

interpretations very difficult. For example, among pastoralists, at one extreme the nomadic Fulani people herd cattle, goats and sheep across vast dry areas from west Africa to Sudan, keeping somewhat separate from the local agricultural populations, though they trade with them (Bonte 1999): at the other extreme the Raikas of Baluchistan often exchange milk, meat, wool, and manure, for grain, vegetables, and other goods produced by their settled neighbors. Also the cultivators compete and pay for having herds manure their fields and in exchange allow the herd to graze in their forests (Agrawal 1999). Foraging societies are now very limited, being mostly replaced by pastoralists and/or farmers.

The Sudanian savannah is divided into the East Sudanian savannah, east of the Cameroon highlands and the West Sudanian savannah to the west and these have slight differences (Fig. 6.4a). Rainfall (from April to October only), varies from 600 mm in the north to 1000 mm in the south. Typical species are trees and shrubs (*Combretum*, *Terminalia, Acacia*), and tall elephant grass (*P. purpureum*), with large mammals like the African elephant, giraffe, cheetah, leopard, lion, giant eland, black rhinoceras, and white rhinoceros, though many have been eliminated through over-hunting. During the dry season, most of the trees lose their leaves, and the grasses dry up and often burn. Human population density is high (50–100 people/km^2) with land use varying from farming in settlements to multi-resource and specialized pastoralism (Fig. 6.4b, c). The Sudanian savannah is the main area of arable farming because: the rainfall is high enough, yet leaching is still modest; the variability of rainfall is rather limited; and the availability of nutrients for growth of annuals is at maximum. The distribution of rainfall in a single season also promotes the growth of annual plant species and thus arable subsistence farming, where cereals are the obvious favored crops in view of the potential yield in both quantity and quality per unit area.

The Sahel is mostly covered in grassland with areas of savannah woodland and shrubland. At present, the Sahelian *Acacia* savannas form an unstable transition zone between the wooded Sudanian savannahs on the south, composed mainly of shrub and tree species and tall elephant grass, and the true Sahara Desert on the north (Fig. 6.4d). Grass cover is fairly continuous and dominated by annual grass species with *Acacia* species as the dominant trees. The Sudanian and Sahelian savannhs are the dominant areas of specialized pastoralism as shown by the density of cattle. In the northern Sahel, areas of desert shrub, alternate with areas of grassland and savannah. During the long dry season (November to May), many trees lose their leaves, and the predominantly annual grasses die. Before over-hunting, the Sahel had large populations of grazing mammals, including the scimitar-horned oryx, dama gazelle, dorcas gazelle, red-fronted gazelle, and bubal hartebeest, along with the African wild dog, cheetah, lion, and other large predators. Traditionally, most people in the Sahel are specialized pastoralists. Herds graze on high quality feed in the north during the wet season, and then trek several hundred kilometers to the south,

Fig. 6.4 Biomes and land use. **a** East Sudanian savannah. **b** Agriculture in Sudanian savannah. **c** Cattle herding in northern Sudanian savannah. **d** Sahelian Acacia grasslands in winter

to graze on more abundant, but less nutritious feed during the dry period. Before the domestication of animals, however, foraging would be the only viable strategy. In historic times, a series of decentralized empires (multiresource pastoralism) emerged after 750 AD, and supported several large independent trading cities in the Niger Bend region, including Timbuktu, Gao, and Djenné. Expansion south into the forest zone was prohibited as their horses and camels could not survive the heat and diseases of that region.

In the twentieth century, great variations in annual rainfall have caused relatively favorable times to alternate with killing droughts: and such alternations occurred in the past. The nature of these alternations requires investigation as they are critical to understanding Holocene changes. For example, perennial grasses, shrubs and trees extend their habitat to the north during long periods of higher rainfall and relative low exploitation of the vegetation. But a few years of drought are enough to push them back again, far to the south. This is a much faster process than the extension of habitats to the north, which is counteracted by the predominantly southward seed dispersion mechanisms by the Harmattan, the dry wind from the Sahara, and by migrating herds (Bremen 1992). We would expect that any northward migration of cultures would be slower than southward retreat. The biomes and their associated human cultures can be used to interpret the archaeological record in terms of climate change.

6.3 Climatic Change

Evidence of climatic changes comes from: (a) sedimentary evidence of sand dune movement and accumulation (dry) and active fluvial channeling and sediment accumulation in both fluvial channels and lakes (wet), though care has to be taken to try and separate arid and hyperarid stream deposits from semi-arid stream deposition (a very difficult thing to do), many wetter phase deposits, such as lake deposits have been, and are being, removed by aeolian deflation during dry phases (Kröpelein et al. 2008), and there can be significant lags between climatic change and sedimentary response (Kocurek 1999); (b) fossil evidence of the presence or absence of a biota indicating grassland, savannah, lake and desert environments; (c) artefacts indicating human activities in drier, savannah climates (such as hunting tools without other artefacts and without permanent settlements) and wetter climates where crops can be grown (such as carbonized cereal grains and grinding querns); (d) sculptures, engravings and paintings indicating the animals living in the area, such as giraffes (drier savannah) and hippopotamus (at least locally wetter periods) (Muzzolini 1992).

Contrary to earlier concepts of climatic stability during the Holocene, studies during the last decade indicate marked climatic changes in the Sahara which do not correlate with higher latitude climatic changes (Kuper and Kröpelein 2006). Astronomical forcing is one of the causes of these changes (Crucifix et al. 2002): for example, the maximum Earth obliquity was reached at 10,000 BP and has declined by about 1° of latitude since then: this affects the insolation of both hemispheres as well as the latitudinal migration of the Intertropical Convergence Zone (ITCZ), which would reach one degree further both north and south in the earliest Holocene. The present biotope distribution is related to the annual rainfall partly controlled by the summer position of the ITCZ (Fig. 6.2a). During the early Holocene maximum ITCZ displacement, the biotope distributions show increased summer-early autumn rainfall in the Sahara centered on about 20°N, extending southwards to the Sudanian woodland as well as northwards to the Mediterranean and southern Europe (Fig. 6.5), and with slightly cooler tropical and subtropical temperatures which would reduce evaporation (Salzmann et al. 2006).

Current ideas on climatic change in the Sahara have been summarized by Brooks (2006). After a long late glacial hyperarid period, including the Younger Dryas (from 12,900 to 11,500 BP), when the Nile underwent enormous fluctuations in discharge (Said 1993) and the huge sand draa of the Great Sand Sea developed (Bessler 2002), the entire Sahara became wetter as tropical rainfall belts shifted northwards by as much as 800 km (Nicholson and Flohn 1980). From around 10,500–6000 BP, the Sahara had many lakes with an abundant Sudanian- and Sahelian-type savannah climate flora and fauna (Damnati 2000). This period was interrupted by colder episodes of increased aridity, as shown by Nile discharges: the most severe was around 8000 BP which is inferred from a widespread cooling in the oxygen isotope signal between 8400 and 8000 BP and evidence of an abrupt climatic reorganization between 8200 and 7800 BP as circulations changed to full postglacial conditions (Brooks 2006).

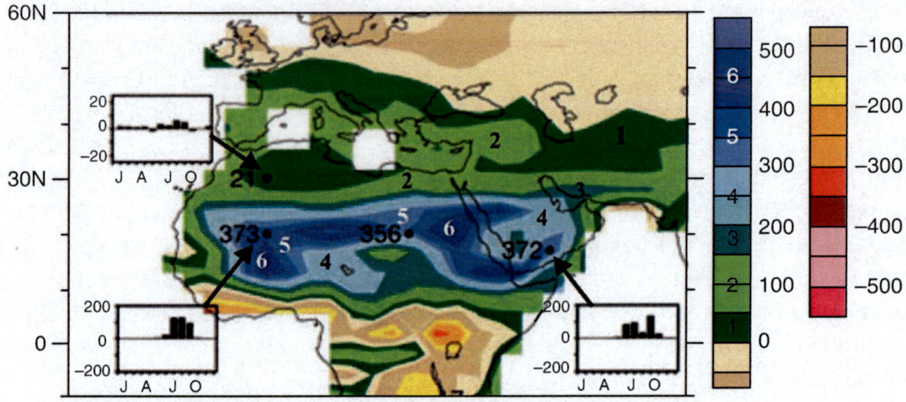

Fig. 6.5 Annual rainfall in northern Africa: early Holocene deviations from current annual rainfall; note max positive changes are in Sudanian, Sahelian and Desert areas

Though wetter conditions recurred after 8000 BP, the northern playas began to dry as early as 7700 BP (see oases belt below): most rainwater-fed playas began to wane around 7000 BP and became desiccated by 5500 BP. About 6200 BP the modern phase of hyperaridity had begun in southern Egypt (Wendorf and Schild 1998).

The Sahelian vegetation zones were 500–600 km north of their present range around 8000 BP, and 300–400 km north around 6400 BP. With increasing desiccation from 6000 BP onwards, the savannah formations retreated to the south until their present position was reached by 3800 BP (Neumann 1989). Some climate modelling results and paleoclimate data have indicated that the change from a semi-arid climate with about 250 mm of rainfall to a hyperarid climate with less than 50 mm/year of rainfall occurred over a relatively short period of time, on the order of hundreds of years (deMenocal et al. 2000). But recent evidence from sediment cores from Lake Yoa in the east-central Sahara (Fig. 6.1) indicates that a slower transition over several thousand years may have taken place, at least in that location (Kröpelein et al. 2008), while evidence from a number of localities suggest that transition times may have varied from place to place by as much as 2000 years (Salzmann et al. 2006). The rate of climatic change is important in understanding the feedback mechanisms involved, but the evidence is still inconclusive, especially as feedback and lag effects have to be considered (Kocurek 1999). For example, a change from desert to grassland in the southern Sahara would increase rainfall by 12% (Kutzbach et al. 1996). Abrupt climate changes over a few hundred years have a greater impact on human societies than changes over thousands of years.

6.4 Societal Changes

The best evidence of societal change comes from habitation structures with associated artefacts arranged in a stratigraphic order of change. These are very rare and mostly limited to the later stages of societal development which, being relatively young, have a better chance of preservation. Sculpture and paintings can also help determine such changes. The type of artefacts can also help in determining the activities of the contemporary population and, by comparison with existing societies help in determining the structure and complexity of those ancient societies (Diamond 1997). For example, religious structures suggest a priesthood specialization, while variations in contemporary grave goods suggest social stratification.

Our present knowledge of societal changes are summarized in numerous books and articles (Grimal 1992; Phillipson 2005; Midant-Reynes 2000; Wendorf and Schild 2001; Edwards 2004; Burroughs 2005) and four successive cultural phases have been recognized (Kuper and Kröpelein 2006). Although these phases will be followed in the summary, they need to be backed up by evidence from individual sites discussed below (all locations are on Fig. 6.1).

6.4.1 *Westernmost Desert*

In the westernmost desert, information is mostly limited to surface finds and the abundant paintings in caves, alcoves and rock outcrops at Uweinat, Arkenau and Gilf Kebir (Fig. 6.1), where numerous engravings and

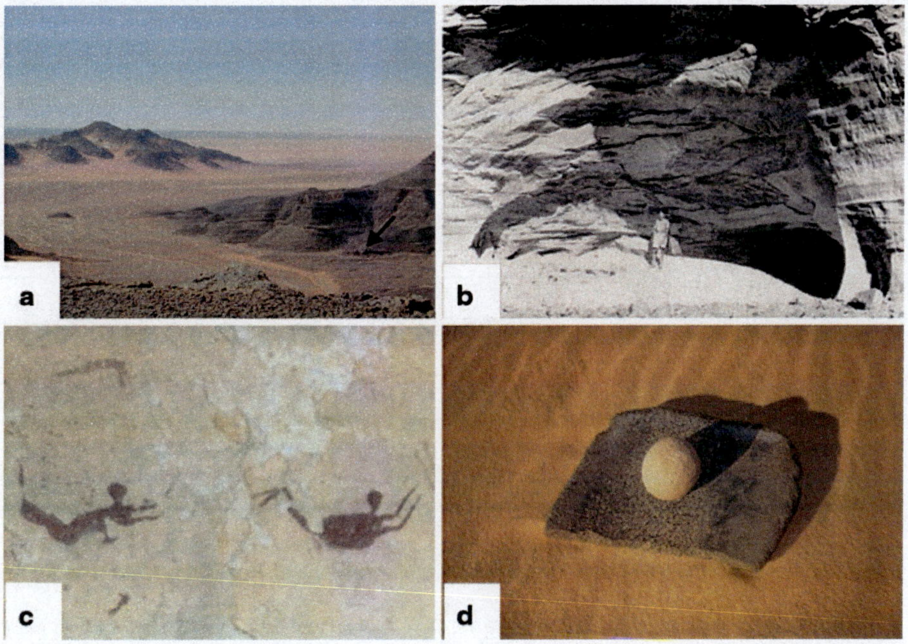

Fig. 6.6 Gilf Kebir. **a** Gilf Kebir looking west to Libya, arrow shows typical location of groundwater sapping that forms 'caves'. **b** Cave of swimmers alcove, with Count Almasy around 1933. **c** Swimmers. **d** Grinding quern and stone grinder

paintings, together with stone tools, show the activities of the inhabitants and their animals (Fig. 6.6).

The best stratified site is Wadi Bakht on the south-eastern side of the Gilf Kebir (Fig. 6.1). Here, an early Holocene dune at the end of the terminal Pleistocene hyperarid phase (around 10,000 BP) blocked a wadi, ponding back an ephemeral lake which was occupied until 5250 BP (Fig. 6.7a,b; Linstadter and Kröpelein 2004). Four cultural phases were distinguished in this and adjacent areas (Fig. 6.7c; Gehlen et al. 2002). Gilf A (10,000–8500 BP) has only limited small lithic tools and simple pottery. Gilf B (8500–6500 BP) associated with stone circles (probably dwelling sites) has a more extensive tool kit with evidence of wild cereal gathering (grinding stones) and hunting (foraging). Gilf C (6500–5500 BP) has evidence not only of foraging, but also of goat and possibly cattle husbandry (foraging- specialized pastoralism), whereas Gilf D (5500–5000 BP) is limited and represents a decline in populations, but seems to be cattle based (specialized pastoralism).

In the Great Sand Sea of alternating dunes and dune corridors, Neolithic stone tool sites are abundant on the stone pavements of the corridors (Fig. 6.8a, b) and hut circles and tools, marginal to eroding silts marking temporary playas or lakes indicate foraging before 5500 BP (Fig. 6.8c, d). But only a few ^{14}C dates have come from scattered playas; tools and pottery are dated only by cultural comparison and very little is known of the history of this vast area (Haynes 1985).

In the Abu Ballas area (Fig. 6.9a), stone circles with fire hearths (Pachur and Braun 1980; Kuper 1993) range from 9000 to 3500 BP, but with a concentration around 5800–5000 BP (Gabriel 1987). Storage jars of the New Kingdom (3500 BP) (Fig. 6.9b) record a trade route still used in Pharaonic times while the red-painted figure of a pharaoh (none other than Khufra, the builder of the great pyramid at Giza, reigned 4539–4516 BP) smiting his enemies was recently discovered on this route (Fig. 6.9c; Bergmann et al. 2001).

6.4.2 *Oases Belt*

Most data comes from the existing oases and playas, Siwa, Bahariya, Farafra, Dakhleh, Kharga, Nabta Playa, Selima, Wadi Howar and Laqiya (Fig. 6.1). Most are deflation basins that were periodically inundated throughout the Holocene (Brookes 1993).

Dakhleh Oasis (70 × 20 km) was fairly humid through the early Holocene, drying through the mid-Holocene, though short-lived arid episodes occurred even during the wetter phases (McDonald 2002, 2003). The mid-Holocene sequence in Dakhleh (Fig. 6.10)

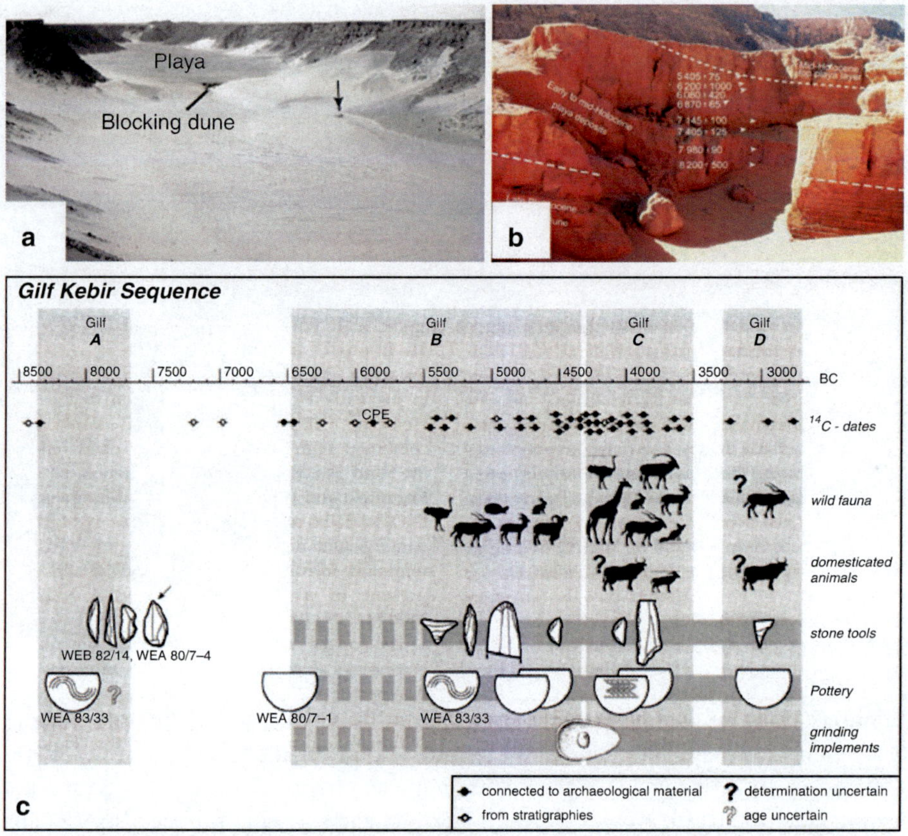

Fig. 6.7 Wadi Bakht. **a** General view with blocking dune and truck for scale arrowed. **b** Detail of blocking dune, playa sediments and ^{14}C dates (both from Linstadter and Kröpelein 2004). **c** Gilf Kebir sequence (from Gehlen et al. 2002)

begins with the Masara culture of foragers around 8800 BP. The Bashendi A culture (about 6400–5500 BP) has slab enclosure structures which seem large enough for cattle corrals (Fig. 6.11b), both plain and impressed pottery, and evidence of the use of wild cereals, sorghum and millets (multiresource pastoralism). In Bashendi B, from about 5500–4000 BP, correlated with increasing aridity, there are no slab structures but quite

Fig. 6.8 Great Sand Sea. **a** Dunes and dune corridors just north of Gilf Kebir. **b** Neolithic tool-making site in dune corridor (Plate 5b). **c** Camp sites around lake (yardangs in background). **d** Lake silts now eroded into yardangs with scattered flake tools at margins—detail of background to **c**

Fig. 6.9 Abu Ballas. **a** Abu Ballas monadnock. **b** Storage jars. **c** Cartouche of Khufru incised on rock

extensive scatters of hearth mounds with common grinding equipment (Fig. 6.11c). Prestige items such as bracelets, toggles, stone beads and small polished stone axes may indicate the mobile elites of pastoral groups (McDonald 2002, 2003). The impoverished Sheikh Muftah culture (about 4000–2800 BP) is confined to the central lowlands of the oasis where they are often associated with marshy areas: there is no sign of structures or permanent settlement and no evidence of cultivation. The same cultural units also occur in Kharga Oasis just to the east (Mandel and Simmons 2001; Warfe 2003).

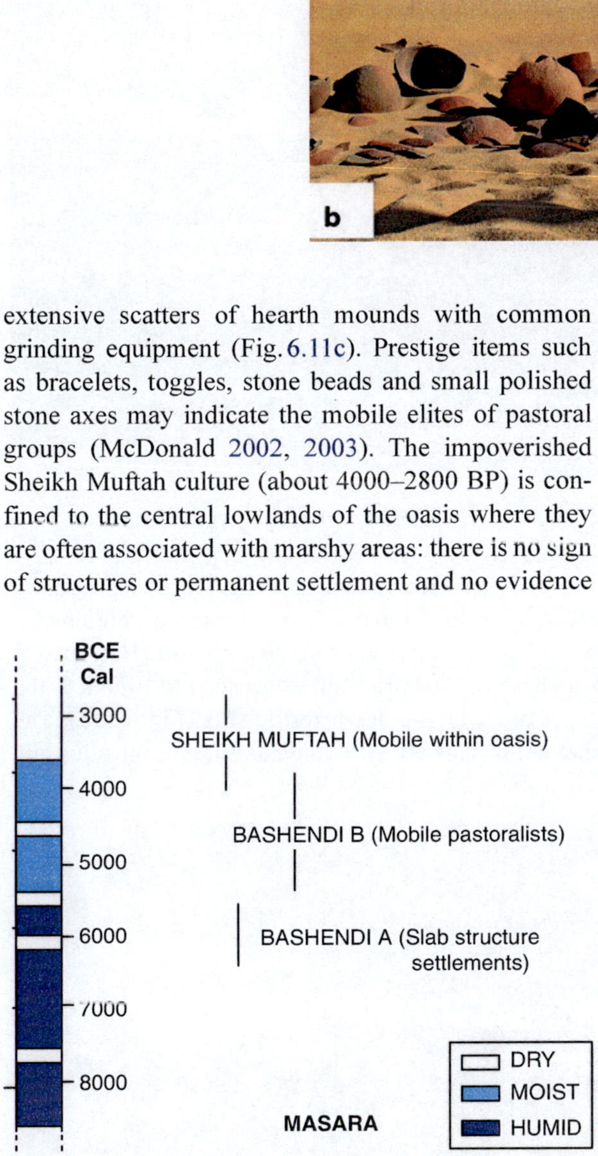

Fig. 6.10 Dakhleh Oasis: climate and cultural succession. (Courtesy of McDonald, modified)

Nabta Playa forms a natural depression of about 5000 km^2 (Fig. 6.1). Around 11,000 BP, temporary lakes or playas formed. Early cattle pastoralism, as well as the elaboration of religious beliefs and cultural practices was followed, beginning about 9000 BP, by larger permanent settlements which relied on wells and had sheep, cattle, and goat herding. After a period of intense drought between 8000 and 7000 BP, renewed permanent settlement recommenced, and by about 6500 BP, the inhabitants had developed a sophisticated, accurate way of marking time and seasons with a stone henge, using the stars as their guides (Fig. 6.12).

At Selima Oasis (Sudan), an extensive deep stratified lake formed around 9700 BP and lasted until about 6500 BP (Haynes et al. 1989). Its fossil plant and animal remains indicate a savannah climate, and many lake level and climatic fluctuations with a high lake level around 8000 BP (Haynes et al. 1989; Pachur and Holzmann 1991).

Wadi Howar (Sudan), the largest dry river system in the presently hyperarid Eastern Sahara, stretches over 1100 km from eastern Chad to the Nile (Figs. 6.1, 6.13a). But between 9500 to 4500 BP, lower Wadi Howar flowed through an environment with

Fig. 6.11 Dakhleh Oasis. **a** Present oasis. **b** Bashendi A dwelling area. **c** Bashendi B grinding querns. (**b, c** Courtesy of MMA McDonald)

numerous ground water outlets and freshwater lakes (Pachur and Kröpelein 1987). From 8000 to 7000 BP, the first inhabitants practiced fishing, hunting and gathering and produced ceramics decorated with Dotted Wavy Line and Laqiya-type patterns. From 6000 BP, intensive cattle herding began, and around 4200 BP, in response to increasing aridity, small livestock, sheep and goat, were added to the herds (Fig. 6.13b).

6.4.3 Nile Valley and Adjacent Areas

The Faiyum Depression is encircled by a northern escarpment and for much of its past it contained a lake (Lake Moeris) fed by a Nile run-off (Bar Yussef) which in prehistoric times seasonally flooded the depression, laying down fertile silts (Fig. 6.14a). The fluctuations in Nile flooding directly controlled the

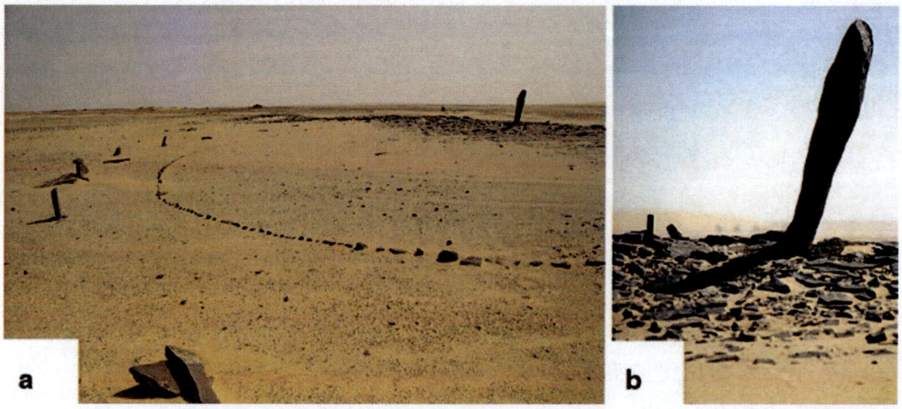

Fig. 6.12 Nabta playa. **a** Stone circle. **b** Standing stone

Fig. 6.13 Wadi Howar. **a** View across dry plain. **b** Cultural site with artefacts

levels of the lake which, in turn affected the settlements of the Faiyum (Fig. 6.14b) (Byrnes 2008). Although settlement of Faiyum began in Paleolithic times, only the latest Neolithic Qarunian is well represented (Wenke et al. 1988).

Qarunian settlements (ca. 9600–8200 BP) were generally on high-ground overlooking Lake Moeris (Wendorf and Schild 1976) and had a lithic industry, associated with a diet heavily based on fishing and hunting and an increasing reliance on wild grain: there is no sign of domestication. Though similar to contemporary Nabta Playa, there are major differences between the lithic tool kits which suggest little contact (Wenke and Brewer 1992). The Qarunian and the succeeding Faiyumian are separated by a long time gap of up to 1000 years, suggesting the Faiyum was abandoned for at least several centuries (Wenke 1999).

The Faiyumian (7200–6200 BP) represents the earliest known fully agricultural economy in Egypt, with hearths, lithic tools, pottery shards and some grinding stones (Fig. 6.14c). The economic basis for settlement was cereals (barley, emmer-wheat and flax) and domesticated animals (cattle, sheep, goats and pigs) derived from the Fertile Crescent of Arabia to the northeast (Zohary and Hopf 1993), though hunting and fishing continued to be practiced (Phillipson 2005). There may be a time gap between the Fayumian and the succeeding Moerian when Lake Qarun may have come close to drying up (Hassan 1986).

The Moerian (6200–5050 BP) has: more complex settlements with several hearths and light shelters; different ceramic styles and methods of producing and modifying stone tools. Faunal assemblages indicate a heavy reliance on fish, with only a few sheep and goat

Fig. 6.14 Faiyum. **a** Satellite view showing connection to Nile. **b** Present-day Lake Faiyum. **c** Faiyum A pot

remains, together with infrequent gazelle and waterfowl remains

In Lower Egypt (basically the delta), much has been buried by ongoing sedimentation (Fig. 6.1). The earliest known settlement is at Merimde 50 km northwest of Cairo on a mound above Nile flood levels, and gives calibrated ^{14}C dates of 6830–6200 BP. The house styles and street patterns reflects a growing level of urban organization. Like the Faiyumians, the Merimdian agriculturalists cultivated cereals, reared cattle, goats, and pigs, and hunted animals such as antelope, but their pottery is plain and simple. The dead were buried inside the unoccupied sections of the settlement, with few grave offerings, apart from the occasional beads, amulets or reed mats, suggesting little social stratification. El–Omari (around 5500 BP) is transitional to the Maadi culture (Mortensen 1999).

The Maadian (about 5800–5100 BP) is based on a center about 10 km south of Cairo, and is the most important of the local delta predynastic cultures. There are few grave goods and tombs in general, and the Maadi culture retained strong Neolithic characteristics, but with disk-shaped mace-heads, cosmetic palettes, stone vessels, and abundant copper (the first introduction of metal in this area). Significant trade is shown by imported Sinai copper, flint tools and lance blades (Midant-Reynes 2000). There is, however, no sign of anything more than basic social differentiation, certainly no sign of hierarchies, nor craft specialization based on function. And yet, contrast the rapid changes during this time at Naqada in Upper Egypt (Bard 1994).

In Upper Egypt (the upper Nile River valley south of Cairo), the Elkabian (around 9000 BP) is known from only one site where it is characterized by small lithic tools of a hunting/fishing culture. After a long gap of about 2000 years, the Tarifian (around 7000 BP) has hearth sites with lithic tools, no signs of agriculture, and only a few scattered pot fragments: it is little advanced from the Elkabian (Wetterstrom 1995). After a 500-year-old gap, the Badarian (about 6400–6000 BP) is the first farming culture in Upper Egypt. Small village settlements were occupied for only a short time and their huts seem to be built from very lightweight material (Hendrickx and Vermeersch 2000). They did, however, use metal and produced the first glazed pottery. Occasional characteristic Badarian finds further south and in the Wadi Hammamat show that the culture was established also in other parts of Upper Egypt.

The succeeding Naqada culture (6400–5000 BP), defined at Naqada about 25 km north of Luxor, shows more rapid change, especially during Naqada 3 (Phillipson 2005).

Naqada 1 (Amratian) (6400–5500 BP) is noted for its black-topped and painted pottery, with war and hunting represented as a dual theme. There are only few and very poor remains of dwellings and these were built from a mixture of mud, wood and reed.

Naqada 2 (Gerzean) (5500–5200 BP) is noted for the first marl pottery, usually with ochre-brown paintings on beige background. Graves become better equipped: often multiple burials sheltered up to 5 individuals. Through Naqada 2, copper tools replaced stone tools, silver and gold use increased, and the macehead changed from disc-shaped to pear-shaped and developed into a symbol of kingly power. Also, soft and hard stone-working techniques developed, which would be of central importance during younger Egyptian civilizations. This culture advanced both south and north from the area of Naqada 1, reaching the eastern edge of the Nile Delta and Nubia (where it is referred to as the Nubian A Group). The South Town of Naqada became the most advanced of Egypt's towns, fortified by walls, and a large mudbrick structure (30 × 50 m), possibly a royal palace, was built.

Naqada 3 (around 5200–5000 BP) marks the emergence of the early dynastic civilization, state building, and the development of a complex culture. In this period, graves become richer and contain more elaborate grave goods. Hierokonpolis, the main city, lies about 100 km below Naqada, on the west bank of the Nile. Naqada 3 extends all over Egypt and is characterized by some sensational firsts: the first hieroglyphs, the first graphical narratives on palettes, the first regular use of serekhs, the first truly royal cemeteries, and possibly, the first irrigation.

In the Sudan, around Wadi El-Khowi, societies may have been more stratified earlier, where settlements and cemetery sites, range from about 7000–5500 BP. In the cemeteries, the most important individuals were men buried at the highest point of each cemetery mound, whose graves contained the most important objects, such as elaborately decorated pottery, fine tools and weapons, and strange stone female figurines.

Around 5000 BP, a larger settlement had an advanced level of planning and an elaborate defense system, with centers of administration, private residences, storage huts, workshops, and cattle enclosures (Bonnet and Valabelle 2006).

From about 4500 BP, Egyptian written records supplement the archaeology. About 4500 BP the Egyptian pharaoh Snofru invaded and captured Lower Nubia and built forts and towns there. About 4250 BP, groups of cattle herders moved into Lower Nubia and the Egyptians withdrew. These people were experts in the use of the bow and arrow and formed three small chiefdoms or kingdoms. Their rulers were shown in rock drawings just like kings of Egypt, but with names in hieroglyphics unlike the names of any known Egyptian kings. Their most interesting remains are their cemeteries of round graves, their beautiful red or black pottery covered with white designs, the tiny cattle and animal figures buried in their tombs, and the plump, tattooed women drawn on pottery or made into figurines. About 4150 BP Egypt lost power and civil wars started, but by 4040 BP the pharaohs of the Middle Kingdom had reunited all of Egypt and by 3950 BP they had retaken Lower Nubia, built a series of forts, and began to attack Upper Nubia.

6.4.4 Eastern Desert

The prehistory of the Eastern desert is poorly known. The Tree Shelter site near Quseir, Egypt, is one of the rare stratified sites (Marinova et al. 2008) which starts around 8000 BP and continues until about 5000 BP. The archaeological finds show clear connections with the Nile Valley and the Western Desert during the wet Holocene period. The lowest level (about 8100–7800 BP) contained hunting and hide working lithic tools of nomadic hunters, similar to the Elkabian. The higher level has numerous hearths (about 6600–5000 BP) with animal and fish bones, Red Sea molluscs, and lithic tools similar to the Tarifian. Many of the abundant engravings (petroglyphs) on the wadi walls in the Eastern desert, can be matched with paintings on grave goods from the Naqada I (6400–5500 BP) cemeteries in the Nile Valley (Wilkinson 2003).

6.4.5 Cultural Summary

The time relationships of the oases and Nile cultures are shown in Fig. 6.15. The early cultures like the Qarunian (9600–8200 BP), Elkabian (ca. 9000 BP) and Masarian (ca. 8800 BP) were hunter-gatherers and fishers (foragers), exploiting savannah, lake and wetland environments, and living a existence governed by the cycle of the seasons. Their origins are unknown. The slightly younger Bashendi A culture (8000–7500 BP) of Dakhleh was a cattle-based culture (multiresource pastoralists) developed during increasing aridity.

The succeeding Faiyumian (7600–6200 BP) marks the first agricultural society in Egypt with a fully evolved, albeit simple, cultural and economic identity. Though their origins are disputed, they may have brought their agriculture with them from the Near East, as the contemporary Nile and Oases cattle-herding based cultures, like the Merimdian and Bashendi B (specialized pastoralists), show a decline from Bashendi A. That societal differentiation and social systems began to develop is shown by the differentiation of burials, with grave goods found in some (presumably higher status) graves but not others (Byrnes 2008). This indicates the development of a hierarchy, the existence

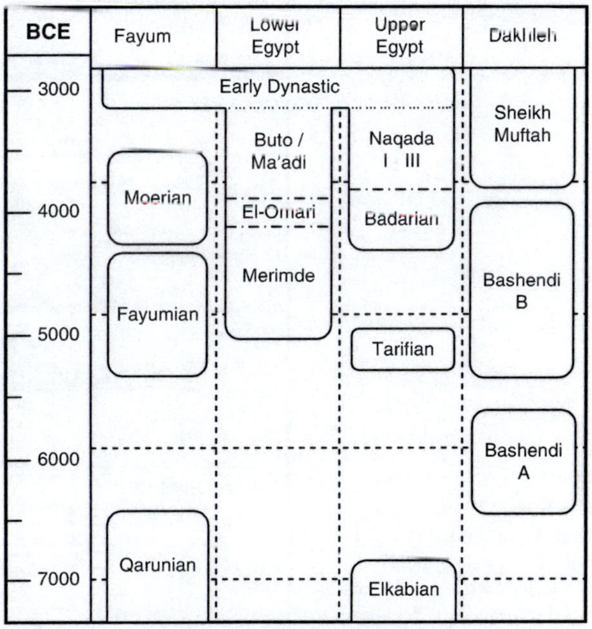

Fig. 6.15 Comparative chart of cultures from Dakhleh, Faiyum, and Lower and Upper Egypt

of skilled workers and an affiliation to a single given area. Thus, by about 7000 BP, social structures had advanced far enough to provide a springboard for their further elaboration in the Upper Egyptian Badarian and especially the Naquada cultures whose rapid evolution is continuous into the first dynastic societies.

6.5 Integration of Climatic and Societal Changes

During the end Pleistocene hyperarid phase, the Sahara extended about 400 km further south than it now does (Nichol 2004). Following this, after 11,000 BP, there are four distinct phases of human occupation of the Eastern Sahara (Kuper and Kröpelein 2006; Bubenizer and Riemer 2007): the Reoccupation phase (10,500–9000 BP); the Formation phase (9000–7300 BP) ending abruptly in areas without permanent water; the Regionalization phase (7300–5500 BP) featuring retreat to highland and Nile refuges; and the Marginalization phase (5500–3500 BP) (Fig. 6.16).

During the Reoccupation phase (10,500–9000 BP), the northward advance of monsoon rains at 10,500 BP transformed the Sahara into a savannah, allowed hunter gatherers from the south (defined by their tool kit and animal bones and already adapted to a savannah life-style) to migrate northwards into the present desert. Camp sites (10,000–9000 BP) around former lakes show migrations of over several hundred kilometers into the Great Sand Sea, now the most inhospitable area of the entire desert (Fig. 6.8). The dunes and interdunes must then have provided enough wild grains and other plants to feed the hunter gatherers and their game. At this time, there are few archaeological sites either along the Nile or in the Wadi Howar region. Their absence has been attributed to marshy, unhealthy conditions, though this is disputable.

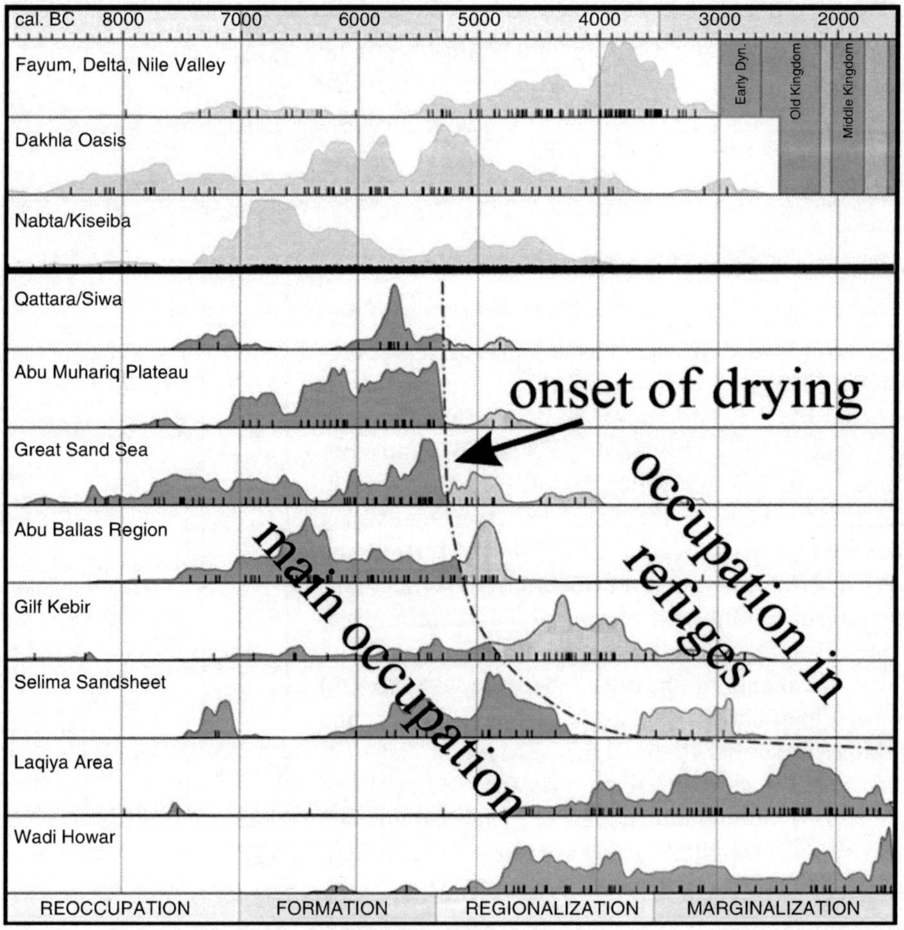

Fig. 6.16 Cultural sequence chart of calibrated ^{14}C dates for representative areas of western desert showing southward migration of drying up. At top are dates from Faiyum + Delta + Nile Valley, Dakhla, and Nabta Playa/Kiseiba. (After Kuper 2006)

During the Formation phase (9000–7300 BP) human populations adapted to regionally different ecological bases with multi-resource pastoralism analogous to those that persist in some existing African cultures. On the Abu Muhariq plateau, bifacial tool technology derived from the Levant completely changed the tool kit which was then transmitted to the younger cultures of the Nile Valley. By contrast, Sudan-type pottery is found as far north as the Great Sand Sea and central Oases belt. Domestic livestock was introduced, sheep and goats from the Middle East, cattle probably from local domestication. Nabta playa has the earliest documented domestic cattle (Hassan 2002), and these, together with goats, dominate the paintings in the westernmost desert, though the eastern desert has rather different and less detailed art. By the end of the formation phase, multi-resource pastoralism may have been the basis of human subsistence in all areas. Between 8000 BP and 7000 BP, the merging of the existing inhabitants and new arrivals is evident in the archaeological record by evidence of the fusion of the Saharan Neolithic herding and cultivation and Nilotic fishing with the first proto-agriculture from Near Eastern communities. The annual natural inundation and draining of the Nile floodplain, promoted a single crop of wild annual grasses.

During the Regionalization phase (7300–5500 BP), populations retreated from increasing desertification into refuges like the Gilf Kebir and plains further south, where rainfall was still sufficient. And this isolation may have fostered an even more regionalized cultural development. The old widespread wavy pottery was replaced by regional varieties. Only sporadic occupation occurred in the westernmost desert outside of the mountain refuges: for example, Abu Ballas was only sporadically occupied (or visited) by people from the oases to the east (McDonald 1998). The rise of specialized cattle pastoralism is reflected in the rock art of Uweinat and Gilf Kebir (if these paintings truly belong to this phase). Unlike the Near East, farming played no part in these developments. In Egypt, instead of the transition from nomadic hunter/gatherers to sedentary pottery producing farmers and stock keepers, we see pottery producing hunter/fisher groups replaced by nomadic cattle herders (Kuper 2006). In the Faiyum, Badarian occupation sites of ash layers, cultural debris and animal droppings resembling current African stock pans rather than any permanent dwellings.

During the marginalization phase (5500–3500 BP) permanent occupation of the desert was restricted to the northern Sudan (Kuper 2006). Even in ecological refuges, like the Gilf Kebir, permanent occupation ceased. By the beginning of the Early Predynastic Period, around 5000 BP, the inhabitants of Upper Egypt (upper Nile Valley) depended little on hunting for survival, having adopted an agricultural way of life, though field work has revealed no evidence of artificial irrigation dating to this period. The earliest currently available evidence for artificial irrigation is the macehead of the predynastic ruler King Scorpion (ruled around 5100 BP) (Fig. 6.17), which may not be the physical excavation of an irrigation canal but rather the symbolic cutting of a levee. Either way, a form of water control is evident, proving that the transition from a natural to an artificial irrigation system that was regulated (by the locals) had already been accomplished before Dynasty 1. The level of the Nile floods declined in the late predynastic (Said 1993), yet the response of the Ancient Egyptians was to cultivate new fields and construct new dwellings close to the new floodplain, instead of irrigating the existing fields by means of extended canals.

For the Pharaonic empire, well-established along the Nile after 5000 BP, the westernmost desert and oases, such as Abu Ballas in the western desert and Laqiya and Wadi Howar in the Sudan, played only a marginal role except as mineral exploration and sporadic trade routes to more fertile areas to the west, south and east (Fig. 6.1). Thus, at the beginning of the First Dynasty of Egypt (5050 BP) a united state had formed, with

Fig. 6.17 Scorpion King macehead, with king supervising agricultural digging activities

Memphis as probably the largest urban center, judging from the extent of the nearby cemetery fields on both sides of the Nile, at Saqqara (West Bank) and Helwan (East Bank). The development, however, of the requisite social specialization and hierarchies had begun much earlier, during the increasing aridity, in predynastic Naqada times at the latest.

6.6 Conclusions

The rise of Egyptian civilization followed the development (or immigation) of agricultural societies in the Faiyum (Faiyumian) at the end of the optimum formation phase and the start of increasing aridity of the regionalization phase around 7000 BP During the next thousand years, these societies became increasingly fragmented as conditions worsened, with groups retreating to refuges like the Gilf Kebir and Nile Valley, returning to specialized pastoralism or migrating southward back to the Sudan. In Egypt, the onset of hyperaridity around 5500 BP, forced migration of already structured agricultural societies into the Nile Valley refuge where they developed greater innovation and organization rapidly over the next 500 years so that, by 5000 BP a complex Pharaonic society dominated the entire Egyptian Nile Valley. Though a broad oversimplified outline of the interaction of climate and the development of Egyptian societies is now possible (though liable to be modified by new findings in Sudan), accurate dating of the various phases in local areas is still inadequate and too inaccurate to separate specifically local from broader climatic and cultural changes (Vernet 2002). Such accurate dating to within one hundred years, or better one human generation (about 20 years), is one of the most important objectives of future research, together with scientific study of field areas, now rapidly disappearing both with the growth of existing human settlements (in the Nile areas) and by the thoughtless unrecorded removal of artefacts' everywhere (especially in the now accessible desert areas).

References

Agrawal A (1999) Greener Pastures. Duke University Press, Durham, North Carolina, USA, 219 pp

Bard KA (1994) The Egyptian Predynastic: a review of the evidence. Journal of Field Archaeology 21:265–288.

Barich BE (1998) People, water and grain: The beginnings of domestication in the Sahara and the Nile Valley. L'Erma di Bretschneider, Rome, 155 pp

Bergmann C, Kuhlmann KP (2001) Die Expedition des Cheops. GEO Special 5:120–127

Bessler H (2002) The Great Sand Sea (Egypt) during the late Pleistocene and the Holocene. Zeitschrift fur Geomorphologie NF 127:1–19

Bonnet C, Valbelle D (2006) The Nubian Pharaohs: Black Kings on the Nile. The American University in Cairo Press, Cairo and New York, 215 pp

Bonte P (1999) La vache ou le mil: Peuls et Dogons au Séno (Mali) In: Botte R, Schmitz J, Boutrais J (eds) Figures peules. Collection Hommes et Sociétés, Karthala, Paris, pp 385–404

Bremen H (1992) The sustainability concept in relation to rural development: Offering stones for bread. Proceedings of the 4th Sahel Workshop, Sustainable Development in the Sahel 6–8 January 1992, Gilleleje, Denmark, AAU Reports 29:37–52

Brookes IA (1993) Geomorphology and Quaternary geology of the Dakhla oasis region, Egypt. Quaternary Science Reviews 12:529–552

Brooks N (2006) Cultural responses to aridity in the Middle Holocene and increased social complexity. Quaternary International 151:29–49

Bubenizer O, Reimer H (2007) Holocene climatic change and human settlement between the central Sahara and the Nile Valley: archaeological and geomorphological results. Geoarchaeology 22:607–620

Burroughs WJ (2005) Climate change in Prehistory: the end of the reign of chaos. Cambridge University Press, Cambridge, 376 pp

Byrnes A (2008) Predynastic Faiyum, S. Cairo and Western Delta. http://www.faiyum.com/index.html.

Crucifix M, Loutre M-F, Tulkens P, et al. (2002) Climate evolution during the Holocene: a study with an Earth system model of intermediate complexity. Climate Dynamics 19:43–60

Damnati B (2000) Holocene lake records in the northern hemisphere of Africa. Journal of African Earth Sciences 31:253–262

deMenocal P, Ortiz J, Guilderson T, et al. (2000) Abrupt onset and termination of the African humid period: rapid climate responses to gradual insolation forcing. Quaternary Science Reviews 19:347–361

Diamond J (1997) Guns, Germs, and Steel: the fates of human societies. Norton & Co., New York, 494 pp

Edwards DN (2004) The Nubian past. Routledge, London, 348 pp

Fagan B (2004) The long summer: how climate changed civilization. Grant, London, 384 pp

Fairbanks RG, Mortlock RA, Chiu TC, et al. (2005) Radiocarbon calibration curve spanning 0 to 50,000 BP based on paired $^{230}Th/^{234}U/^{238}U$ and ^{14}C dates on pristine corals. Quaternary Science Reviews 24:1781–1796

Gabriel B (1987) Palaeoecological evidence from Neolithic fireplaces in the Sahara. African Archaeological Review 5:93–103.

Gehlen B, Kindermann K, Linstädter J, Reimer H (2002) The Holocene occupation of the eastern Sahara: regional chronologies and supra-regional developments in four areas of the absolute desert. In: Lenssen-Erz T, Tegtmeier L, Kröpelin S, et al. (eds) Tides of the Desert-Gezeiten der Wüste. Heinrich-Barth-Institut, Köln, pp 85–116.

Grimal N (1992) A history of Ancient Egypt. Blackwell Publishers Ltd., Oxford, 534 pp

Hassan F 1986 Holocene Lakes and Prehistoric Settlements of the Western Faiyum, Egypt. Journal of Archaeological Science 13:483–501

Hassan F (2002) Palaeoclimate, food and cultural change in Africa: an overview. In: Hassan FA (ed) Droughts, food and culture. Kluwer Academic Press, New York, pp 11–26.

Haynes CV (1985) Quaternary studies, western desert, Egypt and Sudan 1975-1978. National Geographic Society Research Reports 15: 257-293.

Haynes CV, Eyles CH, Pavlish LA, et al. (1989) Holocene palaeoecology of the eastern Sahara: Selima oasis. Quaternary Science Reviews 8:109–136

Hendrickx S, Vermeersch P (2000) From the Palaeolothic to the Badarian Culture (c.700,000–4000 BC). In: Shaw A (ed) The Oxford History of Ancient Egypt. Oxford Univeristy Press, Oxford, pp 17–43

Kocurek G (1999) The aeolian rock record (Yes, Virginia, it exists, but it really is rather special to create one). In: Goudie A, Livingstone I (eds) Aeolian Environments, Sediments and Landforms. John Wiley, London, pp 239–259

Kröpelein S, Verschuren D, Lézine A-M, et al. (2008) Climate-driven ecosystem succession in the Sahara: the past 6000 years. Science 32:765–768.

Kuper R (1993) Sahel in Egypt: environmental change and cultural development in the Abu Ballas area, Libyan desert. Studies in African Archaeology, Poznan, Poland 4:213–223

Kuper R (2006) After 5000 BC: the Libyan desert in transition. Comptes Rendus Paleovol 5:409–419

Kuper R, Kröpelein S (2006) Climate-controlled Holocene occupation in the Sahara: motor of African evolution. Science 313:803–807

Kutzbach JE, Bonan G, Foley J, Harrison SP (1996) Vegetation and soil feedbacks on the response of the African monsoon to orbital forcing in the early to middle Holocene. Nature 384:623–626

Linstadter J, Kröpelein S (2004) Wadi Bakht revisited: Holocene climate change and prehistoric occupation in the Gilf Kebir region of the Eastern Sahara, SW Egypt. Geoarchaeology 19:753–778

Mandel RD, Simmons AH (2001) Prehistoric occupation of late Quaternary landscapes near Kharga oasis, western desert of Egypt. Geoarchaeology 16:95–117

Marinova E, Linseele V, Vermeersch P (2008) Holocene environment and subsistence patterns near the Tree Shelter, Red Sea mountains, Egypt. Quaternary Research 70:392–397

McDonald MMA (1998) Early African pastoralism: view from Dakhleh oasis (south central Egypt). Journal of Anthropological Archaeology 17:124–142

McDonald MMA (2002) Dakhleh oasis in predynastic and early dynastic times: Bashendi B and the Sheikh Muftah cultural units. Archeo-Nil 12:109–120.

McDonald MMA (2003) The early Holocene Masara A and Masara C cultural sub-units of Dakhleh oasis within a wider cultural setting. In: Bowen GE, Hope CA (eds) The Oasis Papers 3, Proceedings of the Third International Conference of the Dakhleh Oasis Project. Oxbow Books, Oxford, pp 4–69

McDonald MMA (2006) Kharga Oasis, Egypt: key to timing transdesert contacts in the mid-Holocene. Proceedings of the 18th biennial meeting of SAfA, Calgary, www.safa.rice.edu

Midant-Reynes B (2000) The Prehistory of Egypt. Blackwell Pub, Oxford, 328 pp

Mortensen B (1999) El-Omari. In: Bard KA (ed) Encyclopedia of the Archaeology of Ancient Egypt. Routledge, London, pp 592–594

Muzzolini A (1992) Dating the earliest central Sahara rock art: archaeological and linguistic data. In: Friedman R, Adams B (eds) The Followers of Horus. Egyptian Studies Association Publication No. 2: Oxbow Monograph 20. Oxbow Books, Oxford, pp 147–154

Neumann K (1989) Holocene vegetation of the Eastern Sahara: charcoal from prehistoric sites. African Archaeological Review 7:97–116

Nichol K (2004) Recent environmental change and prehistoric human activity in Egypt and northern Sudan. Quaternary Science Reviews 23:561–580

Nicholson SE, Flohn H (1980) African environmental and climatic changes and the general atmospheric circulation in Late Pleistocene and Holocene. Climatic change 2:313–348

Pachur HJ, Braun G (1980) The paleoclimate of the central Sudan, Libya and the Libyan desert. Palaeoecology of Africa 12:351–363

Pachur HJ, Hoelzmann P (1991) Paleoclimatic implications of late Quaternary lacustrine sediments in western Nubia, Sudan. Quaternary Research 36:257–276

Pachur HJ, Kröpelein S (1987) Wadi Howar paleoclimatic evidence from an extinct river system in the southeastern Sahara. Science 237:298–300

Phillipson DW (2005) African archaeology, 3rd edn. Cambridge University Press, Cambridge, 398 pp

Said R (1993) The River Nile. Pergamon, Oxford, 320 pp

Salzmann U, Hoelzmann P, Waller MP (2006) The termination of the Holocene humid period in west Africa: abrupt climatic changes, local fluctuations and regional trends. Geophysical Research Abstracts 8:2304.

Thurston S, Sinclair P, Andah B, Okpoko, A (eds) (1993) The archaeology of Africa: food, metal and towns. Routledge, London, 857 pp

Vernet R (2002) Climate during the late Holocene in the Sahara and the Sahel: evolution and consequences on human settlement. In: Hassan FA (ed) Droughts, food and culture. Kluwer Academic Press, New York, pp 47–63

Warfe AR (2003) Cultural origins of the Egyptian Neolithic and Predynastic: an evaluation of the evidence from the Dakhleh oasis (south central Egypt). African Archaeological Review 20:175–202

Wendorf F, Schild R (1976) Prehistory of the Nile Valley. Academic Press, New York, 708 pp

Wendorf F, Schild R (1998) Nabta playa and its role in northeastern African prehistory. Journal of Anthropological Archaeology 17:97–123

Wendorf F, Schild R, Nelson K, et al. (2001) Holocene settlement of the Egyptian Sahara. Kluwer Academic/Plenum Publishers, New York, 708 pp

Wenke RJ (1999) Patterns in Prehistory. Oxford University Press, Oxford, 712 pp

Wenke RJ, Brewer (1992) The Neolithic-Predynastic Transition in the Faiyum Depression. In: Friedman R, Adams B (eds) Egyptian Studies Association Publication, Followers of Horus Oxbow Monograph 20:175–184

Wenke RJ, Long JE, Buck, PE (1988) Epipalaeolithic and Neolithic subsistence and settlement in the Fayyum oasis of Egypt. Journal of Field Archaeology 15:29–51

Wetterstrom W (1995) Foraging and farming in Egypt: the transition from hunting and gathering to horticulture in the Nile valley. In: Shaw T, Sinclair P, Andah B, et al. (eds) The archaeology of Africa. Routledge, London, pp 165–226

Wilkinson T (2003) Genesis of the Pharoahs. Thames & Hudson, London, 208 pp

Zohary D, Hopf M (1993) Domestication of plants in the Old World. Clarendon Press, Oxford, 278 pp

Chapter 7
Paleoenvironments and Prehistory in the Holocene of SE Arabia

Andrew S. Goudie and Adrian G. Parker

7.1 Introduction

There are many examples of the effects of climate change on human societies evident during the Holocene, not least in West Asia (Staubwasser and Weiss 2006). For example, the 8200 cal. year climate event may have forced abandonment of agricultural settlements in northern Mesopotamia and the Levant (Anderson et al. 2007). Conversely, the 'Greening of the Sahara' in the early to mid-Holocene, may have led to an explosion of activity by Neolithic peoples (Petit-Maire et al. 1999). From around 6000 cal. year BP a reduction of rainfall and of monsoonal strength in North Africa, the Near East and Arabia could have forced people out of the deserts into more favorable environments. Around 5200 cal. year BP (Parker et al. 2006a, b; Staubwasser and Weiss 2006) a rapid drying and cooling event in the Middle East may have led to the collapse of the Uruk Culture in southern Mesopotamia. Wright (2001, pp. 145–146) sees this as a period of "differential growth, accelerated inter-regional conflict, the emergence of large polities and their collapse". Around 4200–4100 BP another sharp climatic deterioration may also have caused severe problems for many urban centers (Staubwasser et al. 2003). In general, agricultural intensification and domestication may have been stimulated by episodes of increased aridity (Sherratt 1997) and there was an association in the mid-Holocene between desiccation and increasing social complexity in the central Sahara and Egypt (Brooks 2006).

Enhanced aridity, Brooks argues, caused population agglomeration in environmental refugia characterized by the presence of surface water (such as the Nile Valley). Curiously, whereas hunter gatherers and foragers were able to move in response to changes in climate, more advanced peoples have become increasingly anchored to particular locations and have been less able to adapt to changing conditions by means of migration (Linder 2006).

This chapter examines the evidence for Holocene environmental changes in SE Arabia and relates these to the archaeological evidence for changes in human cultures. It concentrates in particular on some of the abrupt drying events that took place at around 8200, 5200 and 4200 cal. year BP. All ^{14}C ages reported here were calibrated using OxCal 4.05 (Bronk Ramsey 2001).

7.2 The Area

7.2.1 Introduction

Southeast Arabia (Fig. 7.1) lies between the Rub'al-Khali desert and the Oman Mountains (Ru'us al Jibal and Hajar ranges) (which rise over 2000 m asl) (Parker et al. 2004). The region is arid to hyperarid and is currently located outside the range of the Indian Ocean Monsoon (IOM) summer rainfall. Precipitation, enhanced by the orographic effects of the Oman Mountains, occurs in the winter and is associated with westerly depressions that originate in the eastern Mediterranean and then penetrate into the Arabian Gulf. Rainfall is low (ca. 120 mm), but there is a strong gradient across the region between the Ru'us al Jibal (ca. 140 mm) and Dubai (ca. 80 mm). The wind regime

A. S. Goudie (✉)
St Cross College, St Giles, Oxford, OX1 3LZ, UK
e-mail: andrew.goudie@stx.ox.ac.uk

See Plate 6 in the Color Plate Section; also available at: extras.springer.com

I. P. Martini, W. Chesworth (eds.), *Landscapes and Societies*,
DOI 10.1007/978-90-481-9413-1_7, © Springer Science+Business Media B.V. 2010

Fig. 7.1 Location map of the study area with sites mentioned in the text. (*1* Awafi; *2* al-Daith; *3* Jazirat al-Hamra; *4* Wahalah; *5* UAQ2; *6* Akab island; *7* Point 69; *8* Sharjah; *9* RAK loess deposit; *10* Idhn; *11* Wadi Dhaid; *12* Mleiha; *13* Madam Plain. UAQ = Umm al-Qawain; RAK = Ras al-Khaimah)

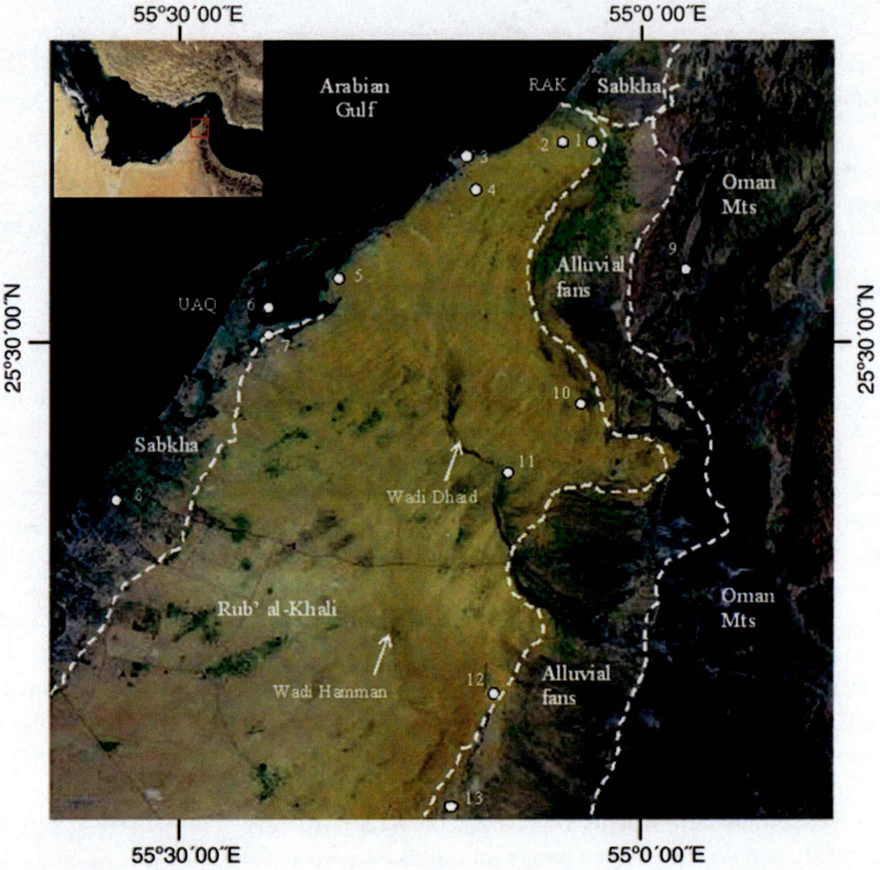

is complex with 52% of sand moving winds blowing from the west and north west, driven by the Shamal, and 28% of sand moving winds blowing from the south east (Goudie et al. 2000a, c; White et al. 2001).

7.2.2 The Oman Mountains

Limestones and dolomitic limestones dominate the northern part of the Oman Mountains, whilst argillaceous limestones, shales and cherts play an important but subsidiary role. Further south, the metamorphic cherts, ophiolites and limestones predominate with some ultrabasic igneous rocks in the far south (Cox et al. 1999; Nasir and Musallam 1998). The Oman Mountains have provided humans with raw materials for building, the production of metals, especially copper and iron (Weeks 2004), chlorite for the production of stone vessels (David et al. 1990) and chert for the production of lithics (Calley et al. 1997).

The Oman Mountains were formed during Paleocene to Miocene times and became faulted and uplifted when the Tethys Ocean closed as a result of the interaction between the Asian and Arabian plates (Glennie et al. 1974). Goudie et al. (2000b) mapped ramps of peridesert loess plastered against the lower slopes of the mountain front near Wadi Haqil and Wadi al Bih. These aeolian landforms and sediments are derived by deflation from the large alluvial fan surfaces and neighboring coastal plain. Preliminary Optically stimulated luminescence (OSL) dates for the main loess bed range from ca. 20,800 to 10,540 year BP.

7.2.3 Sand Deserts

Sandy deserts cover around round one third of Arabia (Edgell 2006). The largest of these is the Rub' al Khali, the world's most extensive sand sea. It covers approximately 560,000 km² (Goudie 2002) and its eastern extremity extends into the study area (Fig. 7.1). The main dune field is composed of large (up to 100 m high), red, partially vegetated mega-linear dunes. Their crests trend SSW-NNE and are separated by interdunal corridors that are approximately 2 km apart. In the south of the study area, where rainfall is less, the dunes have less vegetation and are more active. In places, their crests have been re-worked by more recent activity into smaller secondary forms. In the sector within Ras al-Khaimah emirate, the secondary dune pattern consists of branching, relatively low (5–15 m high) linear dunes with tuning-fork junctions, which are orientated NW to SW (Goudie et al. 2000a), though in the coastal region, between Sharjah and Ras al-Khaimah, the dunes also comprise parabolic dunes.

The morphology, provenance and origin of the dunes are complex. Using LANDSAT imagery, Breed et al. (1979) described them as compound feathered linear dunes. Glennie (1991, 1998) mapped these dunes as deflated megalinear dunes, formed initially by winds blowing from SW to NE. Under current wind conditions, however, the megadunes are transverse to the NW Shamal winds, and so, as discussed above, are overlain by secondary patterns.

The provenance of sand has been studied by various workers (Besler 1982; Anton and Ince 1986; Ahmed et al. 1998; Alsharhan et al. 1998; El-Sayed 1999; Abu-Zeid et al. 2001; Howari et al. 2007). The dunes become redder to the north and northeast as they approach the mountains (owing to a reduced carbonate content and the input of iron-bearing minerals from the ophiolites of the Oman Mountains), whilst closer to the coast they are carbonate-rich (White et al. 2001; Teller et al. 2000). Some of this carbonate sand may have been derived from the Zagros Mountains of Iran at times of low sea-levels in the Gulf (Garzanti et al. 2003). Even mega-dunes at some distance from the sea have carbonate contents that are generally 30–60%.

Dates of dune-building activity in the United Arab Emirates (UAE) (reviewed in Teller et al. 2000; Glennie and Singhvi 2002) reveal a range from the Middle Pleistocene (ca. 800,000 year BP; Hadley et al. 1998) through to the late Holocene (Goudie et al. 2000a; Stokes et al. 2003), with most dune deposits dating to between 100,000 and 10,000 yearr BP. Lancaster (2004) found that dates for major linear dunes clustered between 15.1 and 18.3 year BP. In Sharjah, carbonate nodules from boreholes through dune sands underlying the present coastline were dated to the period corresponding to the Last Glacial Maximum (LGM) (Dalongeville et al. 1991). Dates collated from dune sequences in the study area are listed in Table 7.1.

Quarries have yielded some important sections through the dune systems. At al-Daith, Ras al-Khaimah, Late Pleistocene dune sands (18,000–14,700 BP) were dated from beneath a Holocene shell midden (Parker and Goudie 2007). Dune sediments, exposed beneath the lake sediments at Awafi, Ras al-Khaimah, were dated to the LGM (ca. 17,000 BP), though the main mega-linear dune sequence was dated to between 12,000 and 9000 BP (Table 7.1; Goudie et al. 2000a). Within the lake sequence at Awafi a 15 cm thick sand layer represents a phase of lake desiccation that was accompanied by inundation of the basin by aeolian sand. A date of 4100 cal. year BP was obtained for this dry event (Parker et al. 2004, 2006a). A near identical date of 4180 cal. year BP was obtained from dune sand overlying a buried Hafit Period Bronze Age shell midden at al-Daith, Ras al Khaimah, indicating dune reactivation at this time (Parker and Goudie 2007). To the south at Idhn, 40 m of dune, resting upon gravel veneered pediment, accumulated at ~1000 BP, exhibiting a phase of very rapid accretion (Stokes et al. 2003).

7.2.4 *Alluvial Fans and Gravels*

Alluvial fans coalesce to from a continuous belt along the mountain front (Fig. 7.1; al-Farraj 1995). The fan material comprises boulders, cobbles, gravels and sands made up largely of limestone and dolomite with important but lesser amounts of chert. The width of the fans varies between 2 and 16 km and relates to the size of the drainage basins supplying sediment (al-Farraj and Harvey 2004). Near Ras al-Khaimah town, the fan gravels, often cemented by calcrete, are at least 50 m thick. The fan surface has a gently sloping surface (generally less than 3°) and declines westwards to disappear beneath the almost flat surface of the coastal sabkhas and barriers (Schmidt 1989).

Coarse fan gravels pre-date the linear dunes and can be found in exposures tens of kilometers from the mountain front (at the Camel Pit, Umm al-Qawain, 25°26′N 055°47′E). Ancient river systems can be

Table 7.1 Dates for aeolian sediments

Site	Lab code	Depth (m)	Age (ka)	±1σ	Reference
Awafi	RAK 7	0.5	0.19	0.04	Goudie et al. (2000a)
Awafi	RAK 1	5.0	9.10	0.30	
Awafi	RAK 3	7.5	9.70	0.30	
Awafi	RAK 4	9.5	10.30	0.60	
Awafi	RAK 6	14.0	10.70	0.30	
Awafi	RAK 2	6.0	10.90	0.30	
Awafi	RAK 5	12.0	13.50	0.70	
Awafi	AWAFI1	0.6	4.10	0.24	Parker et al. (2004)
Awafi	AWAFI2	3.7	17.65	1.79	
UAQ2	RAK 00/2	8.2	16.15	0.91	
UAQ2	RAK 00/5	2.0	10.91	1.02	
Al-Daith	RAK/0014	1.2	0.54	0.17	Parker and Goudie (2007)
Al-Daith	RAK/00/13	2.6	4.18	0.72	
Al-Daith	RAK/00/12	3.8	7.58	0.59	
Al-Daith	RAK/00/11	4.6	14.77	1.33	
Al-Daith	RAK/00/10	5.5	15.51	1.00	
Al-Daith	RAK/00/A1	10.0	18.12	1.66	
Wadi-Bih	RAK Loess 01	1.65	10.54	1.25	
Wadi-Bih	RAK Loess 02	2.35	20.80	1.81	
IDHN	RAK13	6.0	0.80	0.02	Goudie et al. 2000a
IDHN	RAK12	10.0	0.30	0.30	
IDHN	RAK11	15.0	0.50	0.40	
IDHN	RAK10	20.0	0.95	0.30	
IDHN	RAK9	25.0	1.09	0.60	
IDHN	RAK8	30.0	1.03	0.40	
IDHN	RAK14	35.0	0.96	0.80	

traced through the dune field (Fig. 7.1) including the relict courses of the Wadis Dhaid and Hamman, which formerly flowed into the Arabian Gulf. These are indicative of markedly more vigorous fluvial activity than occurs today.

There may be at least three district wadi terrace sequences in Southeast Arabia (Dalongeville et al. 1991; al Farraj 1995). The oldest one floors the wadis and is consolidated by calcrete. In southeast Arabia these deposits have not been dated directly but are thought to date to ca. 120,000 year BP and thus to the warm, moist period corresponding to the last interglacial period (Dalongeville et al. 1991). OSL and U-Th dates elsewhere in Arabia relating to a humid phase during this period (Burns et al. 1998; Glennie and Singhvi 2002) support this assumption.

Dalongeville et al. (1991) suggest that the next major phase of aggradation, based on evidence from terraces in Wadi Dhaid, occurred between 30,000 and 21,000 year BP. This also corresponds with dates for a wet phase from elsewhere in Arabia (McClure 1976; Schulz and Whitney 1986; Clark and Fontes 1990; Wood and Imes 1995).

The youngest terraces correspond to the early-middle Holocene period. Within the Wadi Dhaid catchment, dates of 11,100–10,500 cal. year BP (9470 ± 170 ^{14}C year BP) from the base of the terrace at Fallaj al Moalla and of 7870–7670 cal. year BP (6950 ± 120 ^{14}C year BP) at al-Madam have been recorded (Dalongeville and Bescançon 1997; Dalongeville 1999). In addition to the terrace gravels, fluvial silts at Mleiha were dated between 6170–4540 cal. year BP (5190–5194 ± 175 ^{14}C year BP) (Dalongeville 1999). Dates from the region are collated in Table 7.2.

7.2.5 Lakes

Former Holocene lakes are evident in the eastern sector of the Rub' al-Khali dune field (Parker et al. 2004). The deposits at Awafi (25°42′57″ N, 057°55′57″ E) accumulated within a closed basin extending over some 2 km^2. This is bounded by 50 m high mega-linear dunes of LGM, Younger Dryas (YD), and earli-

Table 7.2 Dates for lacustrine and fluvial sediments

Site	Lab code	Environment	Age ^{14}C	Error (1σ)	Age cal. BP (1σ)	Age cal. BP (2σ)	Source
Jebel Hafit	Hv-14592	Fluvial	155	50	290–0	290–0	Gebel et al. (1989)
Al Ain	HD-8095-8137	Fluvial	200	40	300–0	310–0	Gebel et al. (1989)
Al Ain	HD-8096-8136	Fluvial	230	40	310–0	430–0	Gebel et al. (1989)
Mleiha	Ly-7319	Fluvial	4175	110	4840–4540	5000–4470	Dalongeville (1999)
Awafi	Poz-3932	Lacustrine	4450	40	5280–4970	5290–4870	Parker et al. (2004)
Awafi	Poz-3933	Lacustrine	4670	40	5470–5320	5580–5310	Parker et al. (2004)
Mleiha	Ly-6723	Fluvial	4990	205	6000–5450	6300–5300	Dalongeville (1999)
Mleiha	OxA-5367	Fluvial	5090	75	5920–5740	5990–5650	Dalongeville (1999)
Awafi	Poz-3881	Lacustrine	5190	50	6170–5900	6170–5750	Parker et al. (2004)
Al Ain	Hv-10917	Lacustrine	5795	100	6730–6450	6900–6300	Gebel et al. (1989)
Awafi	Poz-3882	Lacustrine	6800	50	7675–7590	7740–7570	Parker et al. (2004)
Al Ain	Hv-12709	Lacustrine	6930	75	7840–7670	7940–7610	Gebel et al. (1989)
Al-Madam	OxA-5369	Fluvial	6950	120	7930–7670	7980–7570	Dalongeville (1999)
Awafi	Poz-3934	Lacustrine	7040	40	7940–7790	7950–7570	Parker et al. (2004)
Awafi	Poz-3529	Lacustrine	7170	50	8110–7880	8120–7860	Parker et al. (2004)
Awafi	Poz-3685	Lacustrine	7240	40	8160–7970	8170–7960	Parker et al. (2004)
Falaj Al-Moalla	Ly-3924	Fluvial	9470	170	11,100–10,500	11,200–10,250	Dalongeville (1999)
Al Ain	Hv-12708	Lacustrine	9700	90	11,210–10,790	11,250–10,700	Gebel et al. (1989)

est Holocene age (Goudie et al. 2000a). Their crests have been reworked by a later phase of aeolian activity in the mid-late Holocene (Parker et al. 2004). The sediment sequence at Awafi was examined for pollen, phytoliths and selected physical and geochemical sediment properties (Parker et al. 2004, 2006a).

Four main units (Fig. 7.2) were identified in the field. Unit 1 is a basal gravel lag deposit overlain by yellow-orange mottled sand dated to the LGM. Unit 2 is a lower Holocene carbonate-rich marl with laminations. Pollen and phytolith analyses show that the surrounding vegetation consisted of C3 savannah grassland with woody elements (8500–6000 cal. year BP). Middle Holocene marls and sands compose unit 3. The surrounding vegetation included C4 (arid adapted) grassland, which becomes sparse up profile (6000–4000 cal year BP). Unit 4 consists of red aeolian sands with fine carbonate laminae, indicating a dry lake basin with periodic inundation and frequent episodes of windblown sand in the last 4000 years. Six radiocarbon (Table 7.2) and two OSL dates (Table 7.1) constrain the age of the deposit.

By dating the dunes, fans and terrace gravels, loess deposits and lake beds described above it has proved possible to establish a sequence of environmental changes in the region, which can be related to societal changes.

7.3 Environmental and Societal Changes

7.3.1 *The Late Pleistocene and Early Holocene*

7.3.1.1 Environmental Changes

Dates from Sharjah (Dalongeville et al. 1991) al-Daith (Parker and Goudie 2007) and Awafi (Parker et al. 2004) fall within the period corresponding to the Last Glacial Maximum (LGM). At that time there was widespread aeolian deposition with the development of mega-linear dunes (Glennie and Singhvi 2002). These were formed initially by intensified Shamal winds blowing from SW to NE and their orientation is transverse to the current Shamal. A second, younger, phase of rapid dune emplacement occurred between 12,000 and 9000 BP (Goudie et al. 2000a). This period encompasses the European Younger Dryas (YD) and is close to the period of maximum rate of sea level rise (Lambeck 1996) supporting the idea that sea level rise released sediment by coastal erosion, and formed transgressive dunes that carried carbonate sand inland, driven by the Shamal winds (Hadley et al. 1998).

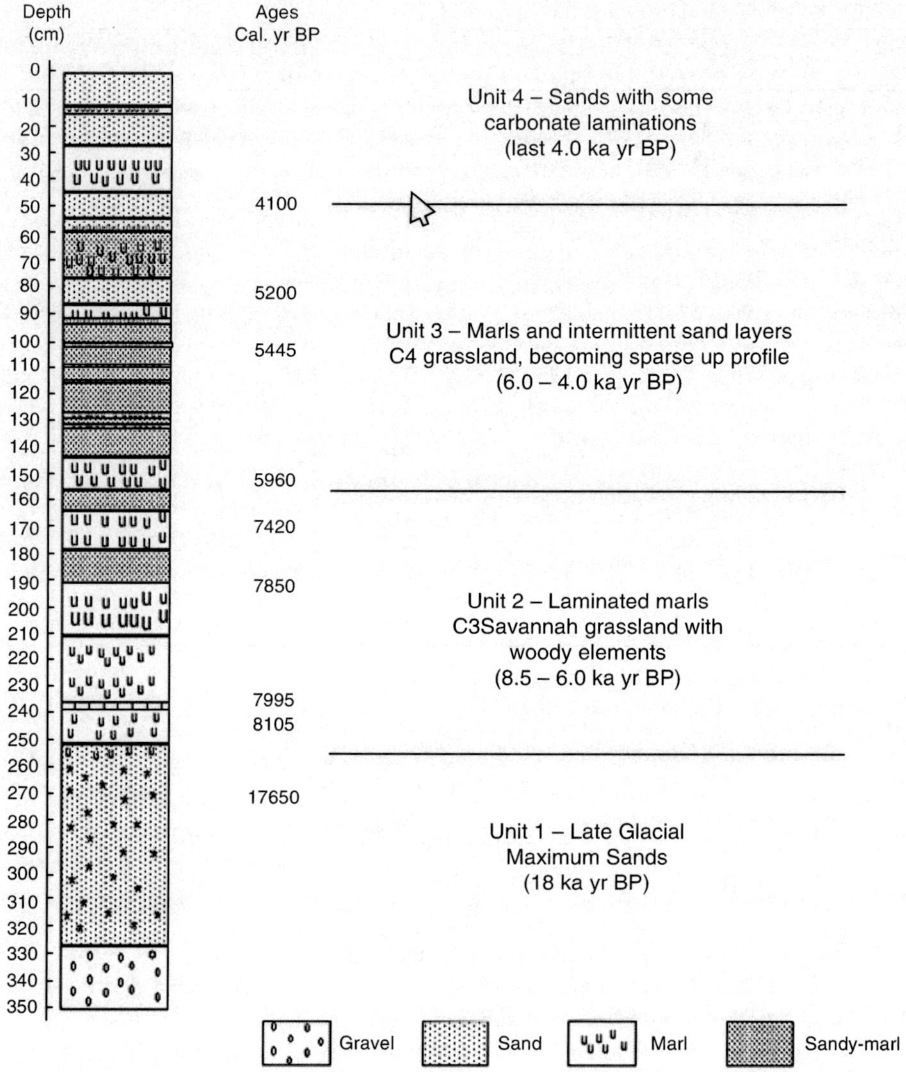

Fig. 7.2 Sedimentary sequence at Awafi

Enhanced aridity has been noted during the YD and pre-Boreal in the North Arabian Sea (Von Rad et al. 1999). Further to the south, shifts in $\delta^{18}O$ from Foraminifera in cores off the Oman coast (Gupta et al. 2003) imply that there was an arid phase between 11,000 and 9000 BP separated by a brief wet phase. This may correspond with the fluvial and lacustrine deposits from Al Ain (Gebel et al. 1989) and Fallaj al-Moalla (Dalongeville and Bescancon 1997).

In contrast to the latter phase of intense aeolian transport and deposition in the Arabian Gulf region, southern Arabia was under the influence of the IOM at this time. Speleothem data from southern Oman and Socotra (Fleitmann et al. 2007) record the northwards movement of the Intertropical Convergence Zone (ITCZ) and incursion of the IOM by 10,300 BP and into northern Oman by 9600 BP (Neff et al. 2001). At Dhamar in Yemen a peat deposit (dated at 10,253–10,560 BP) and lake beds document the arrival of high moisture conditions (Davies 2006). At Awafi, cessation of dune emplacement occurred at 9000 cal. year BP (Goudie et al. 2000a) and the onset of lacustrine sedimentation took place after 8500 cal. year BP (Parker et al. 2004). Thus it took approximately 1800 years for the IOM to move from southern (15°N) to northern Arabia (25°N). This is important as it provides information on the northwards migration and latitudinal position of the summer ITCZ and the incursion of monsoonal rainfall across eastern Arabia during the early Holocene. Saher et al. (2007) classify such environmental changes into a glacial monsoon mode from 20,000 to 13,000 BP, with a weak SW monsoon and strong cooling by glacial NE monsoon winds compared to the modern situation. A transitional mode occurred between 13,000 and 8000 BP, and then the modern monsoon mode started at ca. 8000 BP.

7.3.1.2 Societal Changes

During the period corresponding to the Younger Dryas, Natufian peoples, who had occupied the Euphrates valley of modern day Syria, deserted much of the area as the climate became colder and drier. Conditions were no longer conducive to their sedentary way of life based on the harvesting of wild cereals and the hunting of gazelle. However, in some areas (the Levant and Zagros foothills) early settlements survived during this period of environmental stress through cultivation of cereal crops (Weiss 2000).

The onset of wet conditions in the early Holocene of southeast Arabia closely parallels the earliest evidence for Holocene human occupation in the region. At Nad al-Thamum, in the Emirate of Sharjah, a fragment of *Fasciolaria* shell, probably used as a water vessel, has been dated to 8434 ± 40 ^{14}C BP (9530–9330 cal. year BP, Hd-24356). This fragment was found in a multi-period occupation site on a dune overlooking an as yet undated paleolake 65 km from the coast. The position of this site in the landscape may suggest that the onset of lacustrine conditions here was earlier than at Awafi.

From the capital area of Oman a date of 9615 ± 65 BP (10,791–11,138 cal. year BP, Hv-12964) was obtained from hearth charcoal (Uerpmann and Uerpmann 2003). In the Dhofar region of Oman, Cremaschi and Negrino (2005) dated charcoal in a layer containing many Epipaleolithic flint artefacts from a rockshelter in Gebel Qara to 9130 ± 290 BP (10,700–9750 cal. year BP, GX-24065). This age lies within the range of the onset of wet conditions at 10,300 BP recorded in the Qunf cave speleothem by Fleitmann et al. (2003) in southern Oman.

There was an abrupt climate event at 8200 cal. BP (the 8.2 ka event) that interrupted an otherwise humid period (Brooks 2006; Fleitmann et al. 2003, 2007). In southeastern Arabia this event is recorded in the Hoti Cave speleothem record, where a positive shift in the δ^{18}O of calcite records a precipitation minimum (Neff et al. 2001; Fleitmann et al. 2007), and the Awafi Lake sediment record, where a large influx of aeolian material entered the lake system (Parker et al. 2006a, b). The 8.2 ka event may represent the final collapse of the Laurentide Ice Sheet into the North Atlantic and a reduction in salinity and sea-surface temperatures (Alley et al. 1997; Bond et al. 1997). The impact on the climate system was global in magnitude and extent and has been observed in North Africa (Gasse and Van Campo 1994; deMenocal et al. 2000; Kuper and Kröpelin 2006), the Near East (Bar-Matthews et al. 1997), and the Arabian Sea (Gupta et al. 2003) and Oman stalactite records (Neff et al. 2001; Fleitmann et al. 2003, 2007).

Across the Near and Middle East it has been suggested that this event led to a drought that lasted at least 200 years and which forced the abandonment of agricultural settlements in northern Mesopotamia and the Levant. In Arabia, after this 8.2 ka event, the monsoon rains returned and led to the development of savannah grassland with interspersed *Acacia* trees (Parker et al. 2004). The Neolithic herder-gatherers from the Near East migrated into this landscape with sheep, goats, and cattle. In southeastern Arabia it has been suggested that transhumance was practiced between the mountains in the summer and the Arabian Gulf coast in winter. On the coast, large Neolithic shell middens developed as part of this cycle. In the central and western regions of the Sahara cattle herding occurred from ca. 7.0 ka, having been encouraged by climatic deterioration (Nicoll 2004). The move to cattle herding is thought to be related to having predictability of food availability which hunting and gathering does not guarantee. In addition, as conditions became progressively drier, pastoralism based on sheep and goats, which are more tolerant of aridity and can tolerate partially saline water sources, became more established (Brooks 2006).

7.3.2 The Early Holocene (8000–5500 cal. year BP)

7.3.2.1 Environmental change

The early to middle Holocene in this region experienced wetter conditions owing to the incursion of the Indian Ocean Monsoon (Parker et al. 2006a) and supporting evidence for more humid conditions comes from Mleiha in the Wadi Dhaid catchment (Dalongeville and Besancon 1997). At Awafi a lake formed about 8500 year BP and the dune field became stabilised and vegetated during the early Holocene with C3 grasslands and scatters of woody elements, including *Acacia*, *Prosopis* and *Tamarix* (Parker et al. 2004). The evidence for a rich grassland cover supports archaeological evidence for herding between the mountains, desert and coast during the period of maximum monsoonal rainfall (Uerpmann 2002).

At Awafi the accumulation of Unit 3 coincided with the onset of regional aridity, an influx of aeolian material, and dune reactivation and accretion. The IOM weakened and retreated southwards around 5900 cal year BP (Neff et al. 2001; Uerpmann 2002; Parker et al. 2004, 2006a). This event is clearly marked in the Hoti cave speleothem, which records a large reduction in precipitation immediately prior to this date and a marked change in isotopic signature (Neff et al. 2001; Fleitmann et al. 2007). The lakes of the central Rub al Khali also ceased to exist beyond this time (McClure 1976).

7.3.2.2 Societal Change

The reduction in rainfall at ca. 6.0 ka occurred across North Africa, the Near East, and Arabia owing to a weakening of the monsoon systems because of a reduction in Northern Hemispheric insolation (Fleitmann et al. 2003, 2007). Steig (1999) describes 6.0 ka as the time of steepest decline in Northern Hemispheric radiation. The rainfall reduction ended the practices of Neolithic herding in the desert interiors of Arabia and North Africa and forced these peoples to move to the hydrologically more favored mountains, coasts, oases or riverine areas.

The southeastern Arabian archaeological record indicates that the ABT/Ubaid period came to an abrupt end in eastern Arabia and the Oman peninsula at 5800 cal year BP and there is no evidence of human presence in the area between Kuwait and the Oman Mountains for approximately 600 years (Uerpmann 2002). Limited pockets of human occupation appear to have persisted in deeply incised wadis on the Indian Ocean coast that provided shelter and limited water supplies. This period has been described as the Dark Millennium in the eastern Arabian Gulf region owing to the lack of known archaeological sites (Vogt 1994). Uerpmann (1992, 2002) has suggested that climatic deterioration caused dramatic changes in semi-desert nomadism, subsistence, and settlement patterns around 5800 cal. year BP. The number of known sites suggests that the population shrank considerably at the time and became concentrated in the few parts of Arabia that offered greater ecological diversity (Uerpmann and Uerpmann 1996).

7.3.3 Mid-late Holocene (5500–4000 cal. year BP)

7.3.3.1 Environmental Change

The last 4000 years have been predominantly dry and until the last few hundred years there are no dates supporting wet conditions from either the lacustrine or wadi systems. Evidence from Awafi revealed a short-lived, intense arid event leading to lake desiccation and infilling of the basin with sand (Parker et al. 2004). The vegetation cover was incomplete and dominated by C4 grassland and *Cyperaceae*. Bray and Stokes (2003) showed a late Holocene peak in aeolian activity and dune accretion at about 3000 BP in the Liwa region, causing the accumulation of the megabarchan ridges. At Idhn, Ras al-Khaimah, up to 40 m of dune sand accumulation occurred at about 1000 BP (Goudie et al. 2000a; Stokes et al. 2003). This coincides with a proposed precipitation minimum across the region as recorded in the deep-sea cores in the northeastern Arabian Sea off Pakistan (von Rad et al. 1999). In addition, Stokes et al. (2003) suggested that the reactivation may have been triggered by overgrazing by domestic stock and intensive removal of trees for the thriving copper smelting tradition during the Abbassid Period (750–1220 AD).

7.3.3.2 Societal Change

Around 5200 cal year BP a century-scale rapid drying and cooling event took place across the Middle East, resulting in the migration of people in the Anatolian and Iranian plateaus. In southern Mesopotamia, intense drought led to the collapse of the urban centered Uruk culture, which had been dependent on irrigation for agriculture (Brooks 2006). In Arabia, as the rains failed and the grassland cover disappeared, the Neolithic herders of the Rub' al-Khali abandoned the desert interior in favor of upland and coastal sites (Parker et al. 2006b).

A return to wetter conditions after 5200 BP coincided with Bronze Age human resettlement in the region, at first during the Hafit Period (Potts 1990; Parker and Goudie 2007) followed by a much larger expansion of settlement during the Umm al-Nar Period. This latter period is the most important concerning the development of civilization in southeastern Arabia, with expansion of settlement and increasing overseas trade. In particular the trade of copper to Mesopotamia and the Indus Valley is thought to have made this area considerably wealthy. The settlements of the Umm al-Nar Period were essentially heavily fortified, oasis towns built around a single underground well. The building of fortifications is of particular significance as it suggests that the early inhabitants felt it necessary to safeguard their water supply. Indeed, Potts (1997) suggests that these fortresses acted almost as a lock placed upon the vital water source. It is argued that rainfall was insufficient to allow the Bronze Age inhabitants of the region to adopt a semi-nomadic lifestyle as earlier cultures had (Arabian Bifacial/Ubaid).

Around 4200–4100 cal. year BP evidence for a major shift in climate from a number of regions of the globe has been recognized. A pronounced dry event is recorded from Red Sea sediments at around this time (Arz et al. 2006), and also from a core in the Gulf of Oman (Cullen et al. 2000), where mineralogical and geochemical analyses of the sediments revealed a large increase in windblown dust derived from Mesopotamia. These indicated a 300-year aridity event commencing at 4100 cal year BP. This may have contributed to the collapse of the Akkadian Empire in Mesopotamia (Weiss et al. 1993). Kuzucuoğlu and Marro (2007) described a decline in the Khabur basin but also noted a settlement increase along parts of the middle Euphrates valley at this time. This culture was established on the broad, alluvial plain between the Tigris and Euphrates ca. 4300–4200 cal. year BP. The empire was reliant on rain-fed agriculture in the upper reaches of river catchments along with irrigation agriculture in southern Mesopotamia. Archaeological evidence suggests that there was a large influx of refugees into southern Mesopotamia from the north, which led to the building of a 180 km long wall to stem the influx of people. At Tell Leilan in northeastern Syria the occupation layering representing the social collapse was overlain by 1 m of windblown silt that was sterile of any archaeological artefacts (Weiss et al. 1993). Further west, in the Arabian Gulf region, there was a marked cultural change at 4100 cal year BP between the Umm an-Nar and Wadi Suq periods.

7.4 Conclusions

During the late Quaternary, the climate of Arabia has fluctuated between periods of relative aridity and periods of higher rainfall and fluvial activity, dominated by the influence of the IMO and westerlies. This has left a rich legacy of landforms from which temporal and spatial patterns of environmental change are reconstructed. Such changes in the evolution and modification of this landscape will have influenced humans living in and exploiting this landscape for food and water, raw materials and trade routes. The period corresponding to the Late Glacial was characterized by intense aridity and accumulation of mega-linear dunes driven by the Shamal winds. In the Arabian Gulf region, this continued into the earliest part of the Holocene, whilst southern Arabia was under the influence of the IOM. The monsoon rains migrated into the Gulf region between 8500–6000 cal year BP. During this time, semi-nomadic herders, grazing their animals in a landscape covered with C3 savannah grasslands, occupied the region. The Neolithic peoples also practiced hunting and fishing and the collection of shellfish was an important activity. Pottery shows links with Mesopotamia at this time. From 6000 cal year BP the IOM retreated south and rainfall was derived from winter westerly sources. Under drier conditions there was a switch to a sparser cover of C4 grasses. From 4500 cal year BP the climate became much drier with the development of stronger westerly summer Shamal winds and the reactivation of dunes across the region. An intense arid period occurred at 4100 cal year BP, corresponding with major drought conditions in Mesopotamia and the Indus region, and leading to major changes in society. This event occurs at a major transition in the Arabian Gulf Bronze Age archaeological record and with a decline in population or increased population mobility. The last 4000 years have largely been characterized by arid conditions similar to those found in the region today.

References

Ahmed EA, Soliman MA, Alsharhan AS, Tamer S (1998) Minerological characteristics of the Quaternary sand dunes in the eastern province in Abu Dhabi, United Arab Emirates. In: Alsharhan AS, Glennie KW, Whittle GL, Kendall CGStC (eds) Quaternary Deserts and Climatic Change. Balkema, Rotterdam, pp 85–90

Abu-Zeid MM, Baghdady AR, El-Etr HA (2001) Textural attributes, mineralogy and provenance of sand dune fields in the greater Al Ain area, United Arab Emirates. Journal of Arid Environments 48:475–499

Alley RE, Mayewski PA, Sowers T, et al. (1997) Holocene climate instability: a prominent, widespread event 8200 yr ago. Geology 25:483–486

al-Farraj A (1995) Geomorphology of the Wadi Bih, Ras al-Khaimah, UAE. Ph.D thesis, University of Liverpool, Great Britain

al-Farraj A, Harvey A (2004) Late Quaternary interactions between aeolian and fluvial processes: a case study in the northern UAE. Journal of Arid Environments 56:235–248

Alsharhan AS, Ahmed EA, Tamer S (1998) Textural characteristics of Quaternary sand dunes in the eastern province of Abu Dhabi, United Arab Emirates. In: Alsharhan AS, Glennie KW, Whittle GL, Kendall CGStC (eds) Quaternary Deserts and Climatic Change. A.A. Balkema, Rotterdam, pp 91–108

Anderson D, Goudie AS., Parker AG (2007) Global environments through the Quaternary. Oxford, Oxford University Press, 359 pp

Anton D, Ince F (1986) A study of sand colour and maturity in Saudi Arabia. Zeitschrift für Geomorphologie NF 39: 339–356

Arz HW, Lamy F, Patzold J (2006) A pronounced dry event recorded around 4.2 ka in brine sediments from the northern Red Sea. Quaternary Research 66:432–441

Bar-Matthews M, Ayalon A, Kaufman A (1997) Late Quaternary paleoclimate in the eastern Mediterranean region from stable isotope analysis of speleothems at Soreq Cave, Israel. Quaternary Research 47:155–168

Besler H (1982) The north-eastern Rub' al-Khali, within the borders of the United Arab Emirates. Zeitschrift für Geomorphologie NF 26:495–504

Bond G, Showers W, Cheseby M et al (1997) A pervasive millennial-scale cycle in North Atlantic Holocene and glacial climates. Science 278:1257–1265

Bray HE, Stokes S (2003) Chronologies for Late Quaternary barchans dune reactivation in the southeastern Arabian Peninsula. Quaternary Science Reviews 22:1027–1033

Breed CS, Fryberger SG, Andrews S et al (1979) Regional studies of sand seas using Landsat (ERTS) imagery. In: McKee ED (ed) A Study of Global Sand Seas. US Geological Survey Professional Paper 1052:305–397

Bronk Ramsay C (2001) Development of the radiocarbon calibration program OxCal. Radiocarbon 43:355–363

Brooks N (2006) Cultural responses to aridity in the Middle Holocene and increased social complexity, Quaternary International 151:29–49

Burns SJ, Matter A, Frank N, Mangini A (1998) Spleleothem-based palaeoclimate record from northern Oman. Geology 26:499–502

Calley S, Dalongeville R, Sanlaville P, Santoni MA (1997) The Dhaid-fill plain–geomorphology and prehistory. In: Archaeological Surveys in Sharjah Emirate (U.A.E.) Third Report (1986). Maison del'Orient, CNRS France, pp 16–22

Clark I, Fontes J-C (1990) Paleoclimatic reconstruction in Northern Oman based on carbonates from hyperalkaline ground-waters. Quaternary Research 33:320–336

Cox J, Searle M, Pedersen R (1999) The petrogenesis of leucogranitic dykes intruding into the northern Semail amphibolites, United Arab Emirates: field relationships, geochemistry and Sr/Nd isotope systematics. Contributions to Mineralogy and Petrology 137:267–287

Cremaschi M, Negrino F (2005) Evidence for an abrupt climatic change at 8700 14C yr BP in roackshelters and caves of Gebel Qara (Dhofar-Oman): palaeoenvironmental implications. Geoarchaeology 20:559–579

Cullen HM, deMenocal PB, Hemming S, Hemming G et al (2000) Climate change and the collapse of the Akkadian empire: evidence from the deep sea. Geology 28:379–382

Dalongeville R. (1999) Mleiha: étude physique et paléoenvironment. In: Mouton M (ed) Mleiha I; Environment, Statégies de Subsistance et Artisanats. Travaux Maison de l'Orient 29, Lyon, pp. 33–54

Dalongeville R, Bescançon J (1997) Morphological study of the inner basin of Sharjah. Archaeological Surveys in Sharjah Emirate (UAE) Fifth Report (1989), pp 24–25

Dalongeville R, Medwecki de V, Sanlaville P (1991) Évolution du piédmont occidental de l'Oman depuis le Pléistocène supérieur. 116e Congress National Société.Savants., Chambrey, Déserts, PICG 252:93–115

David H, Tegyey M, Le Metour J, Wyns R (1990) Les vases en chlorite dans la péninsula d'Oman: une étude pétrographique appliqué á l'archéologie. Comptes Rendus de l'Académie des Sciences de Paris 311:951–958

Davies CP (2006) Holocene paleoclimates of southern Arabia from lacustrine deposits of the Dhamar highlands, Yemen. Quaternary Research 66:454–464

deMenocal P, Ortiz J, Guilderson T, Sarnthein M, et al (2000) Abrupt onset and termination of the African Humid Period: rapid climate responses to gradual insolation forcing. Quaternary Science Reviews 19:347–361

Edgell HS (2006) Arabian Deserts: Nature, Origin and Evolution. Berlin, Springer, 592 pp

El-Sayed MI (1999) Sedimentological characteristics and morphology of the aeolian sand dunes in the eastern part of the UAE: a case study from Rub' al Khali. Sedimentary Geology 123:219–238

Fleitmann D, Burns SJ, Mudelsee M, Neff U, et al. (2003) Holocene forcing of the Indian monsoon recorded in a stalactite from southern Oman. Science 300:1737–1739

Fleitmann D, Burns SJ, Mangini A, et al. (2007) Holocene ITCZ and Indian monsoon dynamics recorded in stalagmites from Oman and Yemen (Socotra). Quaternary Science Reviews 26:170–188

Garzanti E, Ando S, Vezzoli G, Dell'era, D (2003) From rifted margins to foreland Basins: investigating provenance and sediment dispersal across desert Arabia (Oman, U.A.E.). Journal of Sedimentary Research 73:572–588

Gasse F, Van Campo E (1994) Abrupt post-glacial climatic events in West Africa and North Africa monsoon domains. Earth and Planetary Science Letters 126:435–456

Gebel HG, Hannss C, Liebau A, Raehle W (1989) The Late Quaternary environments of 'Ain al-Faida/Al Ain, Abu Dhabi Emirate. Archaeology in the United Arab Emirates 5:9–48

Glennie KW (1991) Sand dunes in the Emirates. Tribulus 1:14–18

Glennie KW (1998) The desert of southeast Arabia: a product of Quaternary climatic change. In: Alsharhan AS, Glennie KW, Whittle GL, Kendall GGStC (eds) Quaternary Deserts and Climatic Change. Balkema, Rotterdam, pp. 279–291

Glennie KW, Boeuf MGA, Hughes Clarke MW, et al. (1974) Geology of the Oman Mountains. Verhandelingen Koninklijk Nederlands Geologisch Mijnbouwkundig Genoostschap 31:1–423

Glennie KW, Singhvi AK (2002) Event stratigraphy, palaeoenvironment and chronology of SE Arabian deserts. Quaternary Science Reviews 21:853–869

Goudie AS (2002) Great Warm Deserts of the World: Landscapes and Evolution. Oxford University Press, Oxford, 444 pp

Goudie AS, Colls A, Stokes S, Parker AG, et al. (2000a) Latest Pleistocene dune construction at the north-eastern edge of the Rub al Khali, United Arab Emirates. Sedimentology 47:1011–1021

Goudie AS, Parker AG, White K, Al-Farraj A, et al. (2000b) Desert Loess in Ras al Khaimah, United Arab Emirates. Journal of Arid Environments 46:123–135

Goudie AS, Parker AG, Al-Farraj A (2000c) Coastal change in Ras al-Khaimah (United Arab Emirates): a cartographic analysis. Geographical Journal 166:14–25

Gupta A, Anderson DM, Overpeck JT (2003) Abrupt changes in the Asian southwest monsoon during the Holocene and their links to the North Atlantic Ocean. Nature 421:354–356

Hadley DG, Brouwers EM, Bown TM (1998) Quaternary paleodunes, Arabian Gulf coast, Abu Dhabi Emirate: age and palaeoenvironmental evolution. In: Alsharhan, AS, Glennie KW, Whittle GL, Kendall CGStC (eds). Quaternary Deserts and Climatic Change. Balkema, Rotterdam, pp 123–133

Howari FM, Baghdady A, Goodell PC (2007) Mineralogical and geomorphological characterization of sand dunes in the eastern part of United Arab Emirates using orbital remote sensing integrated with field investigations. Geomorphology 83:67–81

Kuper R, Kröpelin S (2006) Climate-controlled Holocene occupation in the Sahara: motor of Africa's evolution. Science 313:803–807

Kuzucuoğlu C and Marro C (2007) Sociétés humaines et changement climatique à la fin du troisième millénaire: une crise a-t-elle eu lieu en haute Mésopotamie? Paris, De Boccard

Lambeck K (1996) Shoreline reconstruction for the Persian Gulf since the Last Glacial Maximum. Earth and Planetary Science Letters 142:43–57

Lancaster N et al (2004) Age and paleowind regime of linear dunes in the northern Rub Al Khali. Geological Society of America Abstracts with Programs 36(5), p 122

Linder E (2006) The winds of change. Simon and Schuster, New York, 302 pp

McClure HA (1976) Radiocarbon chronology of late Quaternary lakes in the Arabian Desert. Nature 263:755

Nasir S Musallam AK (1998) Ferrous tourmaline (Scholr) in pegmatite granite from Khor Fakkan massif, Semail ophoiolite. Chemie der Erde–Geochemistry 58:187–195

Neff U, Burns SJ, Mangini A, Mudelsee M, et al. (2001) Strong coherence between solar variability and the monsoon in Oman between 9 and 6 kyr ago. Nature 411:290–293

Nicoll K (2004) Recent environmental change and prehistoric human activity in Egypt and Northern Sudan, Quaternary Science Reviews 23:561–580

Parker AG, Goudie AS (2007) Development of the Bronze Age landscape in the southeastern Arabian Gulf: new evidence from a buried shell midden in the eastern extremity of the Rub' al-Khali desert, Emirate of Ras al-Khaimah, UAE. Arabian Archaeology and Epigraphy 18:132–138

Parker AG, Eckersley L, Smith MM, et al. (2004) Holocene vegetation dynamics in the northeastern Rub' al- Khali desert, Arabian Peninsula: a pollen, phytolith and carbon isotope study. Journal of Quaternary Science 19:665–676

Parker AG, Goudie AS, Stokes S, White K, et al. (2006a) A record of Holocene Climate Change from lake geochemical analyses in southeastern Arabia. Quaternary Research 66:465–476

Parker AG, Wilkinson TJ, Davies C (2006b) The early-mid Holocene period in Arabia: some recent evidence from lacustrine sequences in eastern and southwestern Arabia. Proceedings of the Seminar for Arabian Studies 36:243–255

Petit-Maire N, Burollet PF, Ballais J-L, et al. (1999) Paléoclimats Holocènes du Sahara septentionale, dépôts lacustres et terrasses alluviales en bordure du Grand Erg Oriental à l'extrême Sud de la Tunisie. Comptes Rendus Académie des Sciences, Series 2, 312:1661–1666

Potts D (1990) The Arabian Gulf in Antiquity. Volume 1. From Prehistory to the fall of the Achaemenid Empire. Oxford University Press, Oxford, 419 pp

Potts DT (1997) Before the Emirates: an archaeological and historical account of developments in the region c.5000 BC to 676 AD. In: Ghareeb E, Al-Abed I (eds) Perspectives on the United Arab Emirates. Trident Press, London, pp 36–73

Saher MH, Peeters FJC, Kroon D (2007) Sea surface temperatures during the SW and NE monsoon seasons in the western Arabian Sea over the past 20,000 years. Palaeogeography, Palaeoclimatology, Palaeoecology 249:216–228

Schmidt J (1989) Untersuchungen zur morphogenese der Oman Mountains im Emirat Ras Al-Khaimah. Zeitschrift für Geomorphologie Supplementband 74.45–55

Schulz E, Whitney J (1986) Upper Pleistocene and Holocene lakes in the An Nafud, Saudi Arabia. Hydrobiologie 28:175–190

Sherratt A (1997) Climatic cycles and behavioural revolutions: the emergence of modern humans and the beginning of farming. Antiquity 71:271–287

Staubwasser M, Sirocko F, Grootes PM, Segl M (2003) Climate change at the 4.2 ka BP termination of the Indus valley civilisation and Holocene south Asian monsoon variability. Geophysical Research Letters 30, 1425, doi:1029/2002GLO016822

Staubwasser W, Weiss H (2006) Holocene climatic and cultural evolution in late prehistoric-early historic West Asia. Quaternary Research 66:372–387

Steig EJ (1999) Mid-Holocene climate change. Science 286:1485–1487

Stokes S, Goudie AS, Colls A, al-Farraj A (2003) Optical dating as a tool for studying dune reactivation, accretion rates and desertification over decadal, centennial and millennial timescales. In: Alsharhan AS, Wood WW, Goudie AS, et al. (eds) Desertification in the Third Millennium. Swets and Zeitlinger, Lisse, pp 53–60

Teller JT, Glennie KW, Lancaster N, Singhvi AK (2000) Calcareous dunes of the United Arab Emirates and Noah's Flood: the post glacial reflooding of the Persian (Arabian) Gulf. Quaternary International 68–71:297–308

Uerpmann M (1992) Structuring the Late Stone Age of southeastern Arabia. Arabian Archaeology and Epigraphy 3:65–109

Uerpmann M (2002) The Dark Millennium- Remarks on the final Stone Age in the Emirates and Oman. In: Potts D, al-Naboodah H, Hellyer P. (eds) Archaeology of the United Arab Emirates. Proceedings of the first international conference on the archaeology of the U.A.E. Trident Press, London, pp 74–8

Uerpmann M, Uerpmann H-P (1996) 'Ubaid pottery in the eastern Gulf - new evidence from Umm al-Qaiwain (UAE). Arabian Archaeology and Epigraphy 7:125–129

Uerpmann H-P, Uerpmann M (2003) Stone Age sites and their natural environment. The Capital Area of Northern Oman, pt.III. Beihefte zum Tübinger Atlas des Vorderen Orients Reihe A (Naturwissenschaften) Nr.31/3. Dr Ludwig Reichert: Wiesbaden

Vogt B (1994) In search for coastal sites in prehistoric Makkan: mid-Holocene "shell eaters" in the coastal desert of Ras al Khaimah, UAE. In: Kenoyer, J.M. (ed) From Summer to Meluhha. Wisconsin Archaeological Report No 3, pp 113–128

Von Rad U, Schulz H, Reich V, et al. (1999) Multiple-monsoon-controlled breakdown of oxygen-minimum conditions during the past 30,000 years documented in laminated sediments off Pakistan. Palaeogeography, Palaeoclimatology, Palaeoecology 152:129–161

Weeks L (2004) Early Metallurgy of the Persian Gulf: Technology, Trade, and the Bronze Age World. American School of Prehistoric Research Monograph Series. Brill Academic Publishers, Boston, 254 pp

Weiss H (2000) Beyond the Younger Dryas: collapse as adaptation to abrupt climate change in ancient west Asia and the eastern Mediterranean. In: Bawden G, Reycraft RW (ed) Environmental Disaster and the Archaeology of Human Response. Maxwell Museum of Anthropology, Anthropological Papers, vol. 7. pp 75–98. University of New Mexico Press

Weiss H, Courty M-A, Wetterstrom W, et al. (1993) The genesis and collapse of Third Millennium north Mesopotamian civilization. Science 261:995–1004

White K, Goudie AS, Parker AG, al-Farraj A (2001) Mapping the geography of the northern Rub' al Khali using multispectral remote sensing techniques. Earth Surface Processes and Landforms 26:735–748

Wood W, Imes JL (1995) How wet is wet? Precipitation constraints on late Quaternary climate in the southern Arabian peninsula, Journal of Hydrology 164:263–268

Wright HT (2001) Cultural action in the Uruk world. In: MS Rothman (ed) Uruk Mesopotamia and its neighbours: cross-cultural interactions in the era of state formation. School of American Research Advanced Seminar Series, 123–147

Chapter 8
Human Paleoecology in the Ancient Metal-Smelting and Farming Complex in the Wadi Faynan, SW Jordan, at the Desert Margin in the Middle East

Chris Hunt and Hwedi el-Rishi

8.1 Introduction

This chapter reviews existing information, describes new geoarchaeological evidence and from this infers aspects of the human paleoecology and land use in a landscape heavily affected by millennia of metal-winning and metal-processing in the Wadi Faynan and its tributaries in southwest Jordan. The Wadi Faynan lies in an ecotonal position on the margins between the warm desertic Wadi 'Araba and the Jordanian uplands, which are characterized by steppe and at high altitude, Mediterranean dry forest. It was a key Middle Eastern industrial center from the early 3rd millennium BC to the Byzantine period (Barker et al. 2007a; Hauptmann 2007). The metal industry in the Faynan has considerable time depth, since in the Neolithic, before smelting started; the brightly-colored copper ores were extracted for ornamental purposes and cosmetics. The environment in the Wadi Faynan was harsh by any standards, and resources were difficult to access and extract. Metal-winning and metal-processing causes multiple, more or less severe impacts on the environment, which vary depending on the location and nature of the site. The highly-polluted environment is known to have caused severe health impacts (Grattan et al. 2002). Yet in spite of these difficulties, during several millennia people lived and extracted copper in the Wadi Faynan. This chapter describes and examines evidence of how these ancient miners, metal workers and farmers lived at the desert margin and amidst the industrial pollution.

C. Hunt (✉)
School of Geography, Archaeology & Palaeoecology
Queen's University Belfast, Belfast BT7 1NN, UK
e-mail: c.hunt@qub.ac.uk

8.2 Materials and Methods

The Wadi Faynan (Fig. 8.1) lies on the margins of the Dead Sea—'Arabah rift in southwest Jordan and drains a landscape deeply incised into Precambrian, lower Paleozoic, and upper Cretaceous to Paleogene successions in the mountains of Edom, some 30 km north of Petra. Among the Paleozoic succession, intense sedimentary copper mineralization occurs in the Lower Cambrian Dolomite-Limestone Shale Unit, with secondary mineralization occurring in fracture-fills in the Lower Brown Sandstones, which lie later in the Cambrian (Hauptmann 2007).

Hunter-gatherer activity in the area can be traced back to the Lower Paleolithic (McLaren et al. 2007). In the Holocene, Pre-Pottery Neolithic A (PPNA) settlements appear about 9600 BC. The subsistence pattern of the PPNA appears to have been primarily hunting and gathering in a partly-wooded steppe landscape (Barker et al. 2007c). Intensive cereal use and complex settlements with substantial architecture are evident in the Pre-Pottery Neolithic B about 8500 BC (Simmons and Najjar 1996) . During the Pottery Neolithic (about 6000 BC), the landscape began to aridify, so the tree cover was markedly reduced and farming became sufficiently widespread that soil erosion caused major valley alluviation (Hunt et al. 2007a). Pottery Neolithic waterside settlements are known in the Faynan and its tributary wadis. Generalized metal pollution was minimal and episodes appear to have been short-lived and probably related mostly to the production of brightly-colored copper ore powders and beads (Grattan et al. 2007). There are raised levels of heavy metals in late

See Plate 7 in the Color Plate Section; also available at: extras.springer.com

I. P. Martini, W. Chesworth (eds.), *Landscapes and Societies,*
DOI 10.1007/978-90-481-9413-1_8, © Springer Science+Business Media B.V. 2010

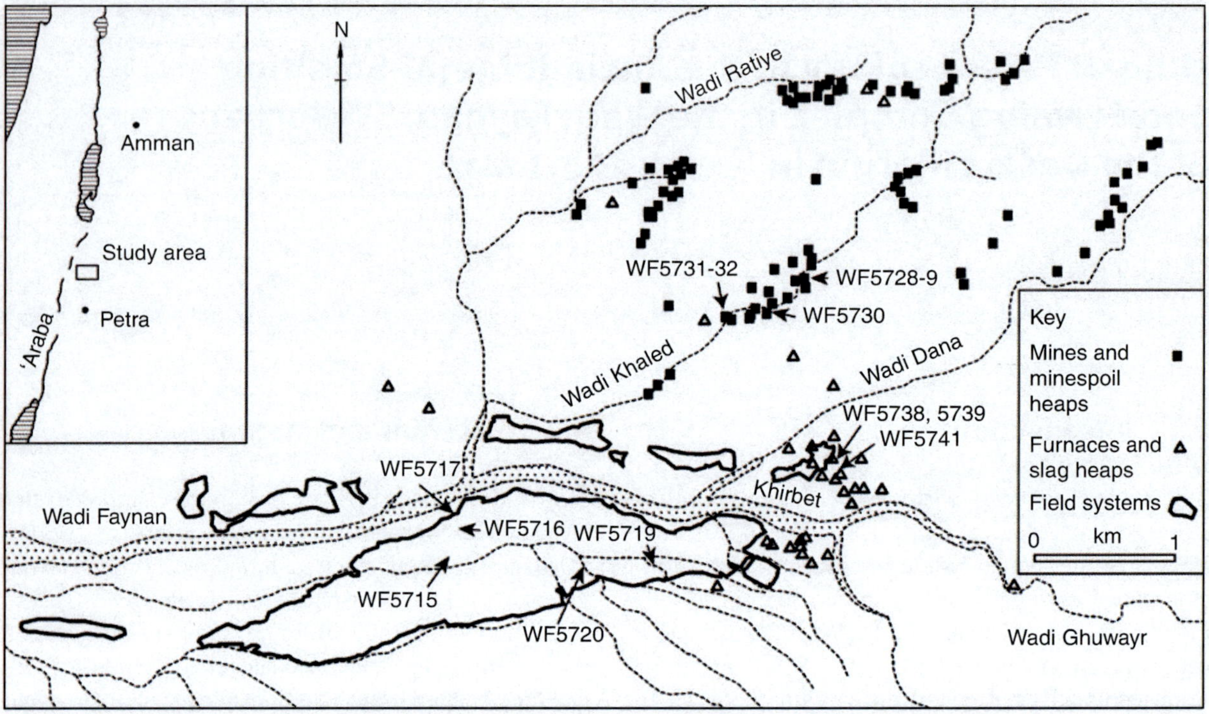

Fig.8.1 The Wadi Faynan farming and metal-production complex showing sites mentioned in the text

Neolithic deposits consistent with the heating of copper ores, from about 5500 BC (Grattan et al. 2007).

Early copper metallurgy can be deduced from the occurrence of extremely polluted sediments and slags associated with dates as early as 4500 BC in the Chalcolithic (as documented below) but archaeological evidence of metal-smelting technology is not yet forthcoming (Adams 1999; Hauptmann 2007). More substantial, extensive and well-attested metal-producing activity occurred in the Early Bronze Age, from about 3600 BC (Adams 1999, 2002; Levy et al. 2002). Very intense metal production activity occurred during the later Early Bronze Age (about 2950–1950 BC). There was then something of a hiatus for nearly a millennium, during which the area had only transient populations (Adams 1999). Dated slag deposits suggest (Grattan et al. 2007) that metal production seems to have started again during the later Bronze Age 1 (about 1700 BC), with small scale activity episodically into Iron Age I (about 1300 BC). Activity intensified during Iron Age II (about 1000 BC), during the rise of what became the Edomite kingdom (Levy et al. 2004). The Nabateans, who seem to have come into the region from NW Arabia, seem to have been less engaged with metal-production, though there was some activity in the Nabatean period (312–106 BC) (Mattingly et al. 2007b). From shortly after the annexation of the Nabatean state by Imperial Rome, the Faynan orefield became one of the industrial powerhouses of the classical world, with large-scale mining and smelting during the Roman and Byzantine periods (106 BC–650 AD) with a cadre of professionals overseeing labor by convicts (Hauptmann 2007; Grattan et al. 2007; Mattingly 2007a).

The legacy of these episodes is a landscape which is still highly polluted. Grattan et al. (2007) argue that the Faynan metal industry made a notable contribution to global environmental levels of heavy metals during the Prehistoric and classical phases and that the intensity of pollution in the Faynan orefield was rarely surpassed elsewhere before the nineteenth century AD.

Today, the Wadi Faynan has an annual average rainfall estimated at between 70 and 150 mm/a (Hunt et al. 2007a). The landscape in the lower parts of the wadi is largely very-degraded steppe, dominated by annual grasses, sedges, knapweeds, daisies and thistles (Asteraceae) and patches of sandwort group (Caryophyllaceae) and chenopod scrub. The wadi floors show classic braided bedform morphology, are occasionally flooded by runoff from winter rainfall, but are usually dry and dotted with sparse oleander bushes except near rare mountain-front springs, where reed-dominated vegetation with palms, tamarisks and willows is present. The Edom Mountains above the Wadi Faynan are characterized by a zone of *Artemisia* steppe associated with annual rainfall of 150–200 mm at intermediate altitudes, a zone of juniper woodland associated with annual rainfall of 200–250 mm at higher altitudes and sparse remnants of mixed Mediterranean woodland at very high altitudes, associated with annual rainfall of 250–450 mm (Engel and Frey 1996; Hunt et al. 2007a). All of the vegetation belts have become extremely degraded over the last 40 years as the result of increasingly-intensive grazing by Bedouin flocks.

Research on environmental and landscape change in the Wadi Faynan suggests a relatively wet early Holocene, with Mediterranean woodland at low altitude and settlements alongside perennial rivers until the Late Neolithic (Hunt et al. 2004, 2007a; Barker et al. 2007c). A relatively diverse but treeless steppe was still present until the Chalcolithic/Early Bronze Age, when small water catchment systems appear (Hunt and Gilbertson 1998; Barker et al. 2007b), but thereafter aridification seems to have set in and floodwater-farming was apparently needed to sustain large-scale Iron Age II to Roman agriculture, resulting in the construction of the enormous Wadi Faynan field system (Fig. 8.1; Mattingly et al. 2007a, b). In vegetational terms, the effects of the wetter Roman period climate in the region (Heim et al. 1997) appear to have been negated by human impact, most notably the intense regional wood-gathering required to produce fuel for the smelting industry in the wadi, with the vegetation remaining highly degraded (Hunt et al. 2007a; Mattingly et al. 2007b).

The sites discussed here were identified and assigned to their landscape context through a combination of air-photo interpretation and ground survey. The air-photos derived from wartime British coverage held in the archives of the University of Western Australia (El-Rishi et al. 2007). Air-photo interpretation narrowed the areas for ground survey. The research area is sparsely vegetated because of aridity and heavy grazing so structures and features such as slag heaps and field walls were easily visible to ground survey. Sites were identified in the field for sampling of natural exposures (for instance in gully walls or river cutbanks), or for test excavation, if no natural exposures existed (El-Rishi et al. 2007). Natural sections or test-pit faces were cleaned, drawn, photographed and sampled. Samples were dried in the field-camp under clean newspaper to minimize contamination, then returned to the laboratory in sealed polythene bags. The normal sample size was 1 kg. In the laboratory, analysis included visual characterization, sieving for macroscopic remains, palynological analysis, analysis of heavy metal content using X-ray fluorescence (XRF) on a Spectro X-Lab, magnetic susceptibility using a Bartington MS2b meter and organic carbon content by loss on ignition. Samples were submitted for radiocarbon analysis using accelerator mass spectrometry and calibrated using Calib 501. Radiocarbon dates are displayed in calendar years BC or AD (Hunt et al. 2007b). Pollen and metal data were displayed graphically using Tilia and TGView.

In a general way, undated sites can be placed in their chronological context using chemostratigraphy. The pollution was characteristically copper-dominated and relatively low in lead during the Early Bronze Age to Iron Age (2000–750 BC), then much higher in lead through the Classical-Byzantine periods (750 BC–620 AD) as a result of changes in the ore sources exploited (Grattan et al. 2007). Other lines of dating evidence, such as radiocarbon dates (Hunt et al. 2007b) and archaeological seriation are also sometimes available.

8.3 The Farmscape

One of the most prominent parts of the Faynan complex of sites are the field systems, which in broad terms date to a series of episodes between the early Bronze Age and the Late Byzantine period. The farmscape around the metal-processing sites was sampled in a series of sites in and on the edge of the Wadi Faynan farming system. Some of these sites, such as WF5715 (Fig. 8.2) and WF5717 (Fig. 8.3) are field-fills, which accreted vertically as sediment-laden irrigation water was distributed on the fields of the irrigation system. Most, such as at WF5717, show sedimentary and pedogenic structures consistent with vertical accretion from relatively quiet waters interspersed with welded soil development. Signs of plough cultivation are unusual in the field system—Site WF5715 is the only site at which a plough soil was recorded. It is possible, how-

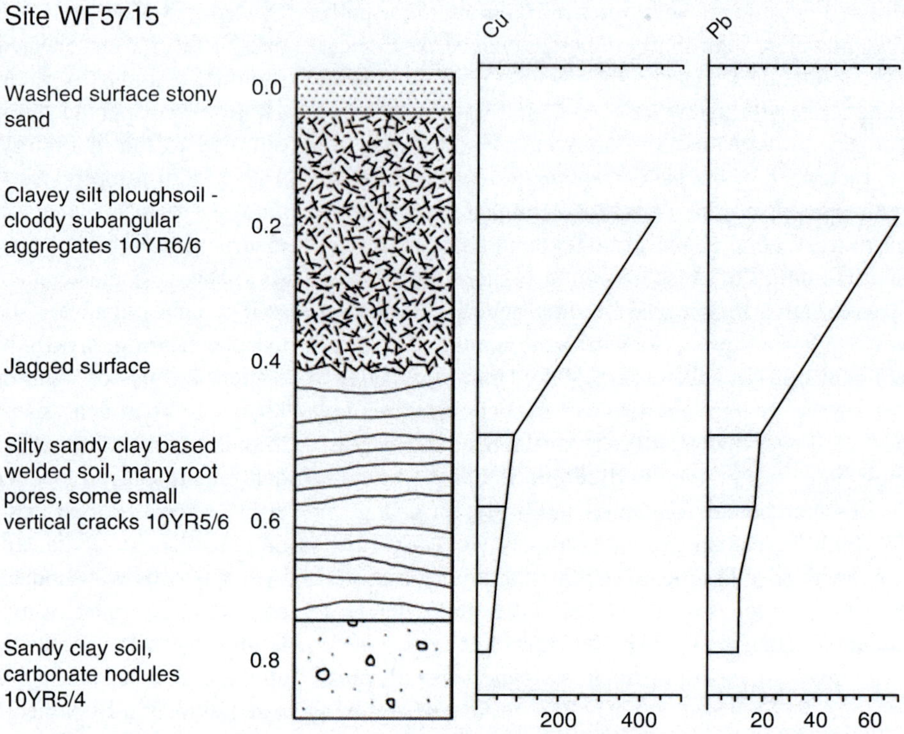

Fig. 8.2 Summary of stratigraphy and geochemistry of a field-fill at WF5715. Depths in meters, Cu and Pb concentrations in ppm. The very low lead figures in the welded soil suggest that this accumulated before the Iron-Age/Classical metal production episodes when lead pollution was relatively high, and thus that it relates to the Early Bronze Age episode

ever, since parts of the field system are highly deflated, that plough soils were more prevalent in the past, but have been removed subsequently by erosion. The presence of *Melanopsis* at site WF5717 suggests irrigation from a perennial water source, while *Theba*, which is common in field system soils and watercourse deposits, is associated with steppe vegetation in the study region, but not with arable fields.

Other sites, such as WF5716, contain waterlain sediments, often sandy and gravelly, laid down in the channels of the irrigation system (Fig. 8.4). Others, such as 5719 (Table 8.1) and 5720 (Figs. 8.5, 8.6, 8.7) are buried water-control features within or marginal to these channels. Site 5719 contained two generations of water-control structures and stratified between them several Late Iron Age shards. A pollen assemblage was recovered from a silty horizon stratified beneath the stonework of the upper water-control structure (Table 8.1). The assemblage is dominated by steppe taxa—Caryophyllaceae, *Artemisia*, *Asphodelus*, *Bidens* type, Lactucae. Pollen derived from woodlands on the Edom Mountains (*Pinus*, *Ostrya*, *Juniperus*) is fairly prominent. Pollen of cereals is frequent enough in the assemblage to suggest nearby cultivation. The palynofacies assemblage contains abundant burnt woody matter and some spherules, consistent with metal-processing nearby.

Site 5720 is broadly Nabatean-Late Roman in age (Figs. 8.5, 8.6, 8.7). The pollen assemblages (Fig. 8.6) are again dominated by steppe taxa—Caryophyllaceae, *Artemisia*, *Asphodelus*, *Trifolium*, *Plantago*; although the youngest sample, which may be Late Roman in age, shows a decline in steppe species and higher percentages of taxa typical of degraded (*Centaurea*, *Helichrysum*, Lactucae, *Malva*) and desertic (*Ephedra*) landscapes. Pollen of cultivated taxa (cereals and date palm) is very sparse. Algae (mostly Zygnemataceae) suggest standing water in the system for several months, as a minimum. This contrasts with the assemblage from WF5719, where algae are not present, suggesting ephemeral water-supply in the Iron Age. This observation chimes with the work of Heim et al. (1997), who suggest higher regional rainfall in the Roman period. The palynofacies analysis shows high percentages of thermally mature woody matter and some spherules, most probably derived from nearby metal production areas (Fig. 8.7).

While cereal cultivation seems to have taken place locally until the Late Iron Age, the surviving soils, mollusc and pollen evidence for the Classical period together suggests that much of the field system was used primarily for grazing, rather than for arable agriculture as suggested by Mattingly et al. (2007a).

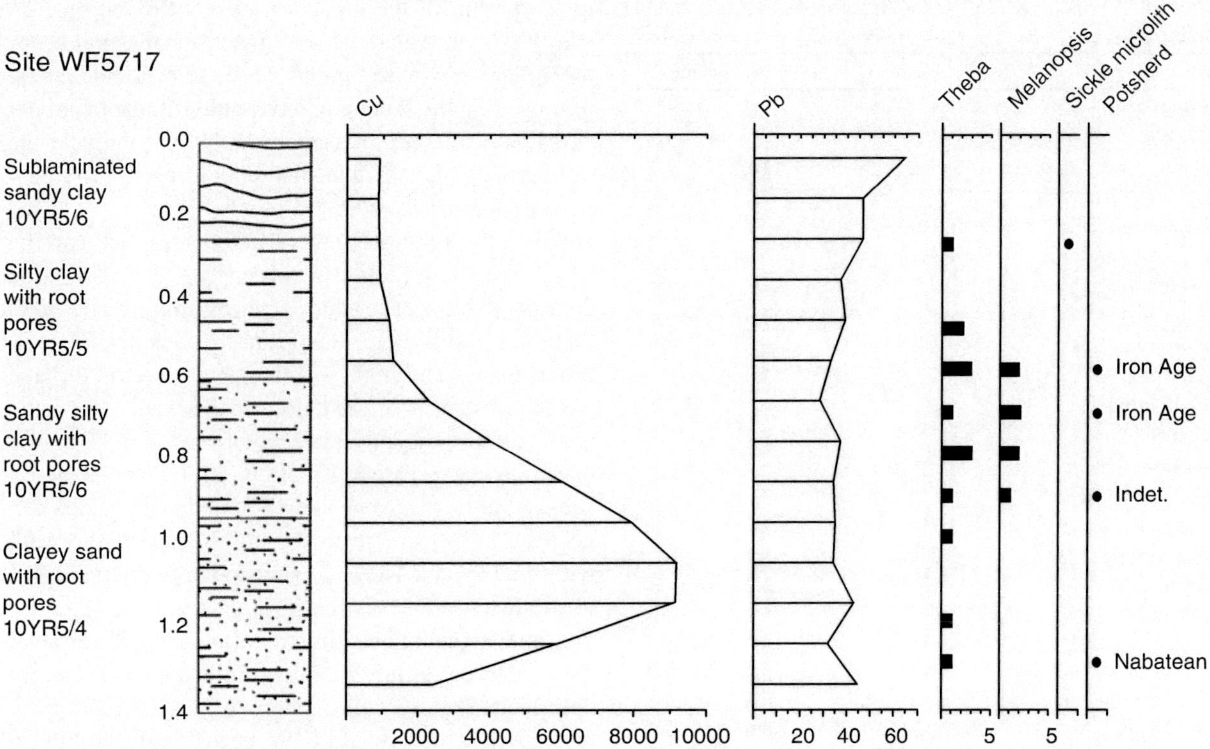

Fig. 8.3 Summary of stratigraphy, geochemistry, molluscs and artefacts from a field fill at WF5717. Depths in meters, Cu and Pb concentrations are in ppm. It would appear that this sequence was initiated in Nabatean (early Classical) times, on the pottery evidence, and the high lead levels in the upper part of the deposits can be used to suggest that sedimentation continued during Roman-Byzantine times. The Iron Age potsherds are thus recycled

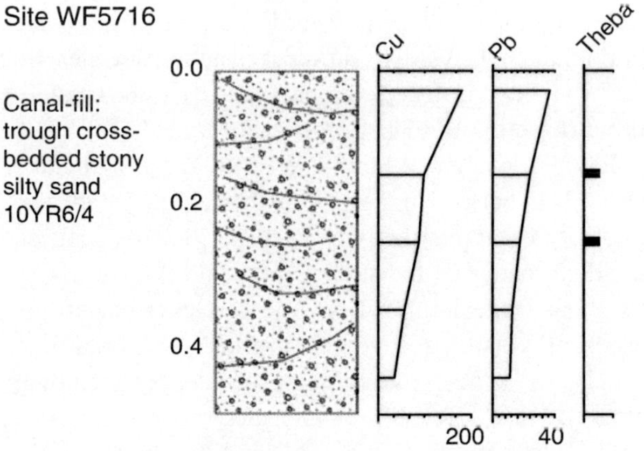

Fig. 8.4 Summary of stratigraphy, geochemistry and molluscs from an irrigation canal fill at WF5716. Depths in meters, Cu and Pb concentrations are in ppm. The low, but rising figures for heavy metals, especially the low lead level, may suggest that this is a relatively early feature, perhaps broadly contemporaneous with the nearby field fill WF5715

It is clear from the chemical analyses that the field fills are highly polluted, with pollution levels during Nabatean and later periods certainly high enough to impact severely on plant productivity (cf. Pyatt et al. 1999, for modern comparative studies, which show that grain yields in highly polluted areas in the region diminish to half those in relatively unpolluted locations). Thus it could be argued that the field system was very large because yields were low. It could further be argued that the Roman and Byzantine managers of the field system were aware that yields were low, since field-walking has disclosed evidence of intensive Roman/Byzantine manuring of the fields using the contents of domestic middens (Mattingly et al. 2007a). These, of course, were likely also polluted with heavy metals and thus will have increased the pollution load in the field-soils. Animals grazed on the field system and people eating their meat or eating grain grown there will have bio-accumulated heavy metals, particularly lead.

Table 8.1 Pollen and palynofacies assemblages from Sample WF5719A

Species	Number	Percentage
Plateau		
Cedrus	1	0.3
Pinus	35	10.7
Ostrya	7	2.1
Juniperus	1	0.3
Daphne	1	0.3
Waterside		
Palmae	5	1.5
Pteropsida	1	0.3
Montia	1	0.3
Cultivated		
Cereal	7	2.1
Steppe		
Caryophyllaceae	110	33.5
Artemisia	71	21.6
Bidens type	20	6.1
Asphodelus	17	5.2
Lactuceae	15	4.6
Poaceae	11	3.4
Cyperaceae	3	0.9
Helichrysum type	2	0.6
Bellis type	2	0.6
Acacia	2	0.6
Hippophae	1	0.3
Anenome type	1	0.3
Trifolium type	1	0.3
Desertic		
Chenopodiaceae	9	2.7
Haloxylon type	2	0.6
Indeterminate	2	0.6
Total	**328**	**100.0**
Palynofacies		
Thermally mature	72	43.1
Spherules	2	1.2
Plant cell walls etc.	32	19.2
Pollen	56	33.5
Insect	1	0.6
Fungal spores	1	0.6
Fungal hyphae	3	1.8
Total	**167**	**100.0**

8.4 Metal-Extraction Sites

The landscape around Wadi Faynan is littered with the remains of metal-winning sites (Hauptmann and Weisgerber 1987, 1992; Hauptmann 1989, 2000). These include surface trenches, shaft and adit mines and spoil heaps, but few provide depositional sites in which environmental evidence has been preserved relating to the period of the metal-winning, often because of their position, high on steep hillsides. Minespoil heaps and related fluvial sediments were sampled in the main mining area, the Wadi Khaled (Table 8.2). Both copper and lead levels are extremely high, especially in the minespoil, which is often silty in texture and extremely friable. Runoff from the minespoil heaps introduced significant pollution into watercourses, as can be seen from the pollution levels in the related riverine deposits. The dust produced during mining and spoil dumping would also have carried considerable pollution loads. The high sediment input seems to have led to substantial aggradation of the wadi floor—the wadi has subsequently incised between 2 and 3 m into these deposits of highly polluted alluvium, leaving a substantial river terrace at the studied sites, which can be traced downstream to the confluence of the wadi with the Faynan. These deposits do not contain pollen or molluscs, but it can be inferred from the aggradation and its contained fluvial/colluvial bedforms that the landscape during the mining episodes was largely devegetated as a result of the extreme toxicity of the minespoil, which would have been highly mobile in the environment.

8.5 Metal-Working Sites

Three sites near Khirbet Faynan were sampled (Figs. 8.8, 8.9, 8.10). These are localities where metal-rich smelting spoil was discarded on the edge of the braidplain of the Wadi Dana, close to its confluence with the Wadi Faynan. The sediments date from the Chalcolithic (base of WF5741), Iron Age I and II (WF5738, WF5739), Roman (WF5738, WF5739) and Mameluke (WF5741) periods and all are extremely polluted. The intensity of human activity in these locations caused the disturbance of the stratigraphy, and thus to dating reversals, especially in Iron Age II. Nevertheless, the ashes, silty ashes and gravels preserve abundant food debris and other environmental evidence, including charred barley, grape pips and date stones, bones of sheep/goats and fish, shells of land and marine molluscs. The sheep/goat bones are mostly rib and long-bone fragments, suggesting butchery elsewhere of sheep and/or goats and consumption of high-quality meat on site, while the fishbones must represent dried or salt fish imported from coastal areas or the Jordan

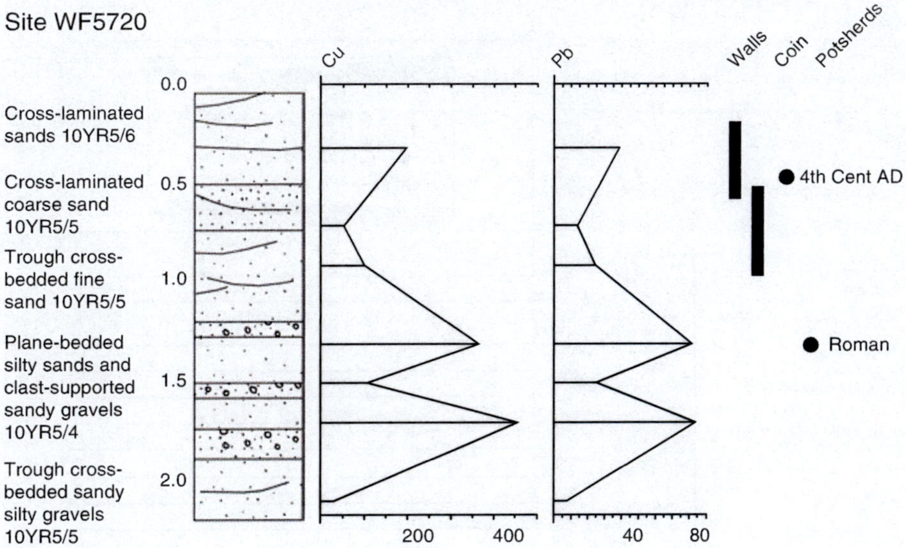

Fig. 8.5 Summary of stratigraphy, geochemistry and archaeology associated with a water-control structure at WF5720. Depths in meters, Cu and Pb concentrations are in ppm. The metal concentrations and artefacts are consistent with this site being broadly of Roman age

Valley, as there are no local sources. Given the highly polluted and active environments represented by these sites, it is possible that the land snail fragments also represent food items, rather than animals which were living on site—there are occasional records of land molluscs being a regular part of the prehistoric diet elsewhere in the arid zone, although these are not usually seen as high-status food. The marine molluscs are derived from the Mediterranean or Red Sea coasts and are extremely unlikely to have been food items, partly because of the difficulty of rapid transportation from the coast to ensure freshness, partly because the specimens are all very small, and partly because the taxa concerned—cowries and *Conus* sp.—were rarely eaten but were prized for ornamental purposes and sometimes currency. It is thus more probable that these had some non-dietary significance for their owners. Artefacts are also abundant, mostly coarseware potsherds, but also substantial numbers of lithic artefacts. These latter are mostly irregular, but sharp, chert flakes and it is probable that these reflect the regular use of stone tools as late as the Iron Age.

8.6 Discussion

The Faynan was challenging climatically—cold in winter, unbearably hot in summer and with a seasonal, scanty and irregular rainfall regime. By the Bronze Age, the Faynan landscape was already too dry for rainfed farming, so irrigated agriculture was necessary from then on (Hunt et al. 2007a), and was certainly still in operation as late as the Roman period. Similarly, it is likely that the demand for wood for fuel for smelting could only be supplied from regional sources (Hunt et al. 2007a).

Cereal pollen and traces of agricultural soil erosion are known in Neolithic (Hunt et al. 2007a), Chalcolithic/Early Bronze Age (Barker et al. 2007b) and Iron Age contexts. The evidence presented here is, however, inconsistent with the field system being the breadbasket of the Faynan metal industry during the Classical period. The soils, mollusc and pollen evidence would instead be consistent with the field system being used mostly for grazing rather than for arable agriculture at this time. Certainly, the levels of cereal pollen during Classical times are significantly less than those present in the Chalcolithic/Early Bronze Age (Hunt et al. 2007a; Barker et al. 2007b) or Iron Age farmscapes when grain was almost certainly produced in the Faynan. Thus, it is extremely likely that a large proportion of grain was imported into the Faynan during Roman and most probably Byzantine times. There is no trace of grape or olive and only very occasional grains of date pollen from the Faynan for any period, so it is highly likely that the seeds found in the present study reflect imported produce, from the Chalcolithic onward.

Pastoral agriculture occurred in the Faynan from the Pottery Neolithic, and possibly earlier. Hunt et al. (2007) suggest that vegetational change and soil ero-

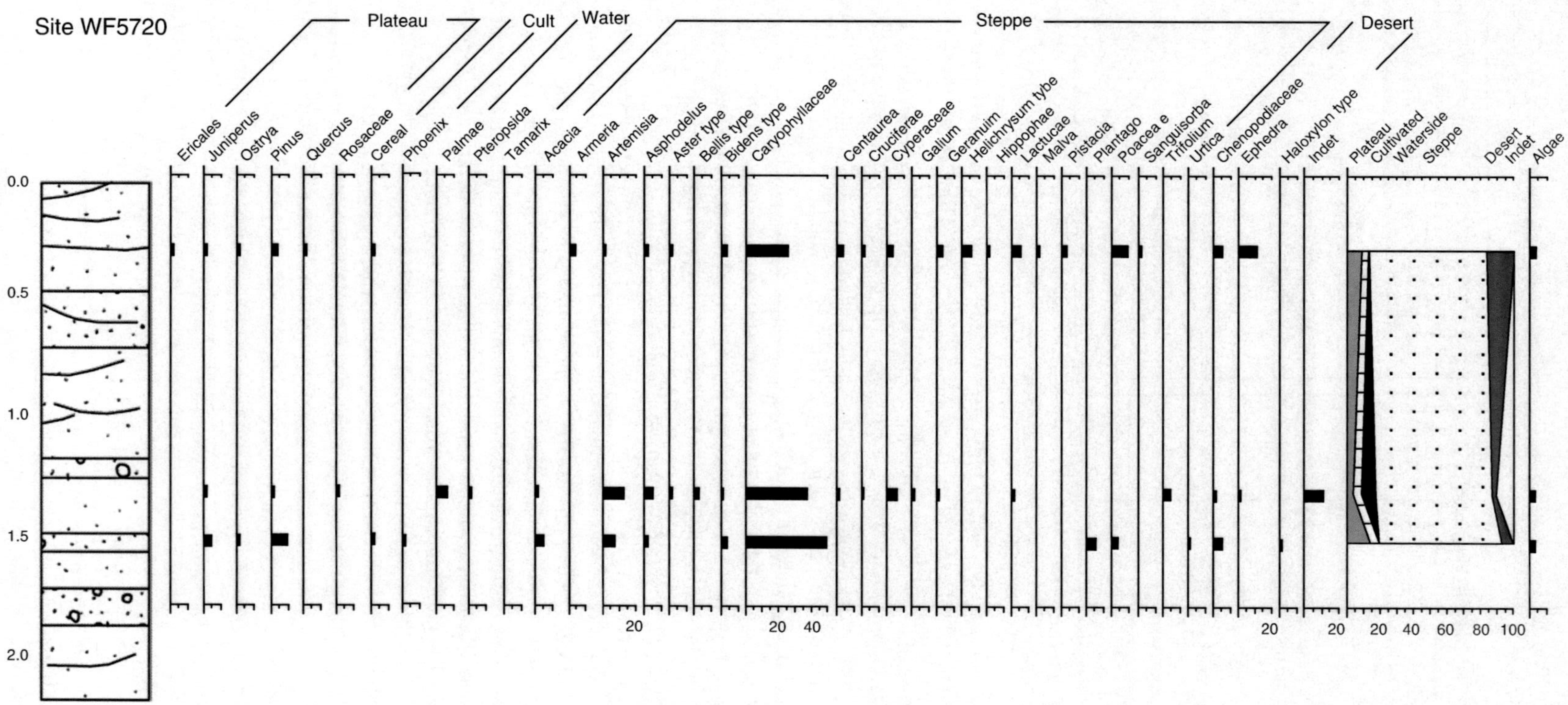

Fig. 8.6 Pollen percentage diagram from site WF5720. Depths in meters. Presences of pollen taxa in samples at 0.7, 0.9 and 1.7 m, which were too sparse to count, are shown as dots

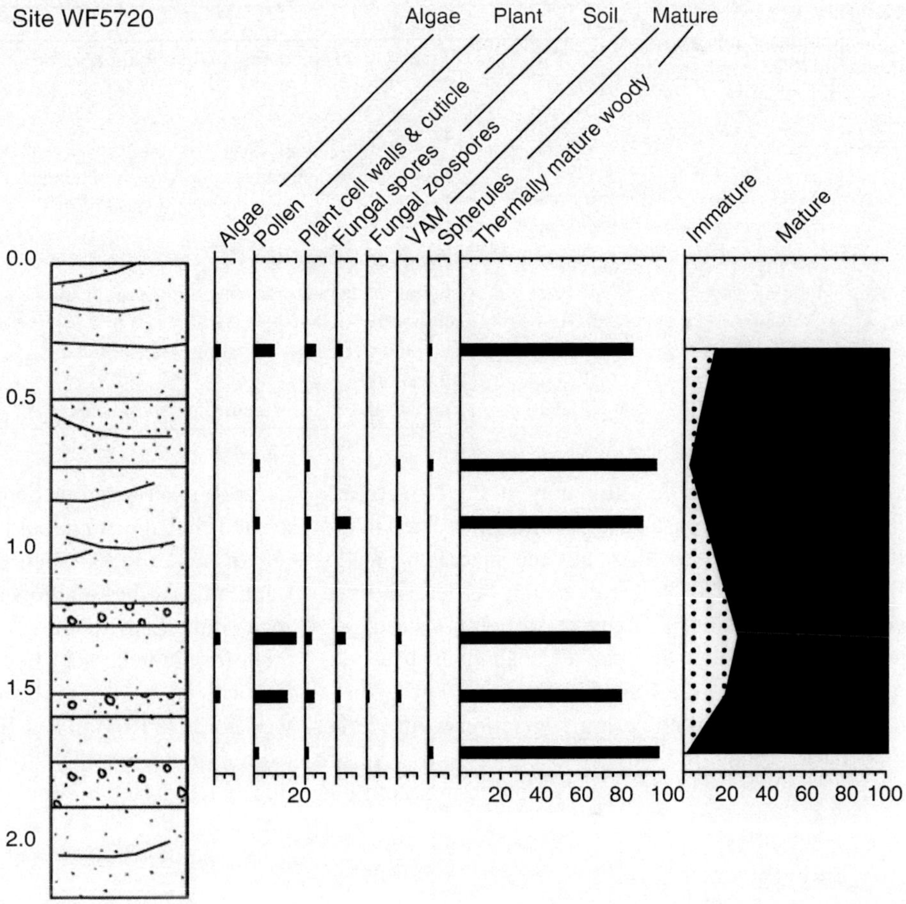

Fig. 8.7 Palynofacies analysis from site WF5720. Depths in meters

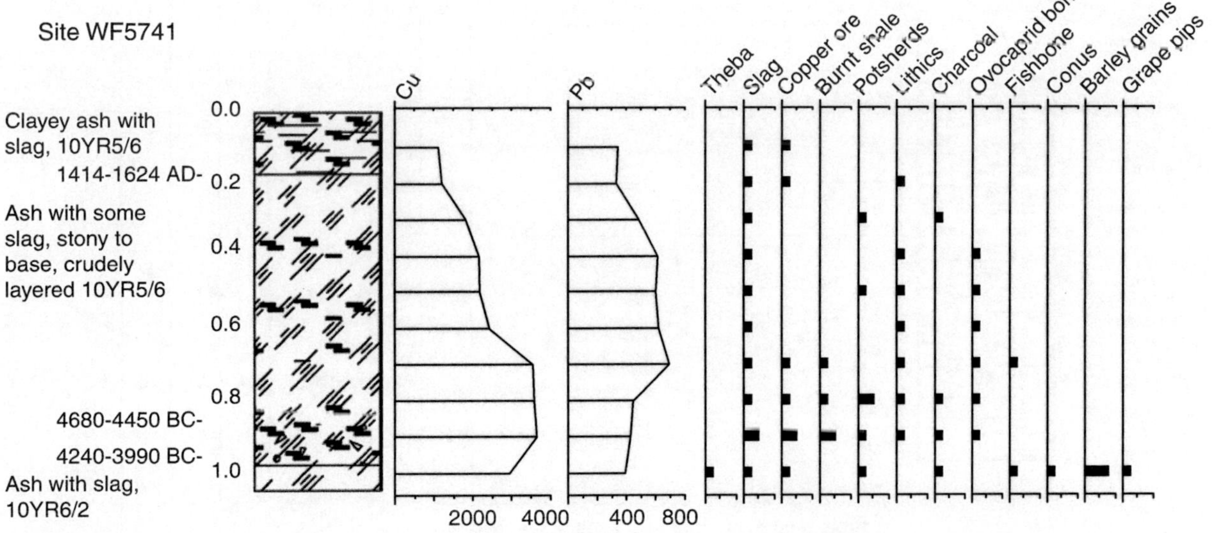

Fig. 8.8 Summary of dating, stratigraphy, heavy metals, molluscs, artefacts and environmental evidence from site WF5741. (Cu and Pb values are in ppm)

Table 8.2 Heavy metal concentrations in minespoil heaps and related sediment in the Wadi Khaled

Sample	Description	Cu (ppm)	Pb (ppm)
5728/1	Fluvial gravels predating main mining episode	224	36
5728/2		201	30
5728/3		232	31
5728/4	Minespoil interbedded with fluvial gravels	7485	734
5728/5	Fluvial gravels interdigitating with spoil heap	4331	415
5729/1	Fluvial gravels downstream from spoil heap	812	91
5730/1	Fluvial sands predating mining	261	41
5730/2	Minespoil	3388	362
5730/3	Fluvial sands interdigitating with spoil heap	2821	251
5730/4	Minespoil	9526	3127
5731/1	Fluvial gravels interdigitating with spoil heap	938	85
5731/2	Minespoil	5550	438
5732/1	Fluvial sands downstream from spoil heap	505	62

sion during the later Neolithic may relate to overgrazing. It is likely that pastoral agriculture persisted in the Faynan after the Neolithic, but the increasing aridification is likely to have forced the herders to have adopted increasingly-extensive foraging strategies and possibly to have adopted a transhumant lifestyle, taking stock up to the Edom Plateau during the summers, when the rainfed spring vegetation would have died back. Today, the current stocking density operated by the local Bedouin is sufficient to maintain the vegetation in a rather degraded state and it could be argued that the rangeland is near its carrying capacity. In those times when there was a large industrial population to be fed, however, it is very probable that local rangeland resources would have been insufficient for the necessary herds. Thus, it is likely that the majority of the animal protein eaten by the Faynan workforce in Roman and Byzantine times was derived from more or less distant sources. Part of the logic for this statement is that most of the Faynan field system

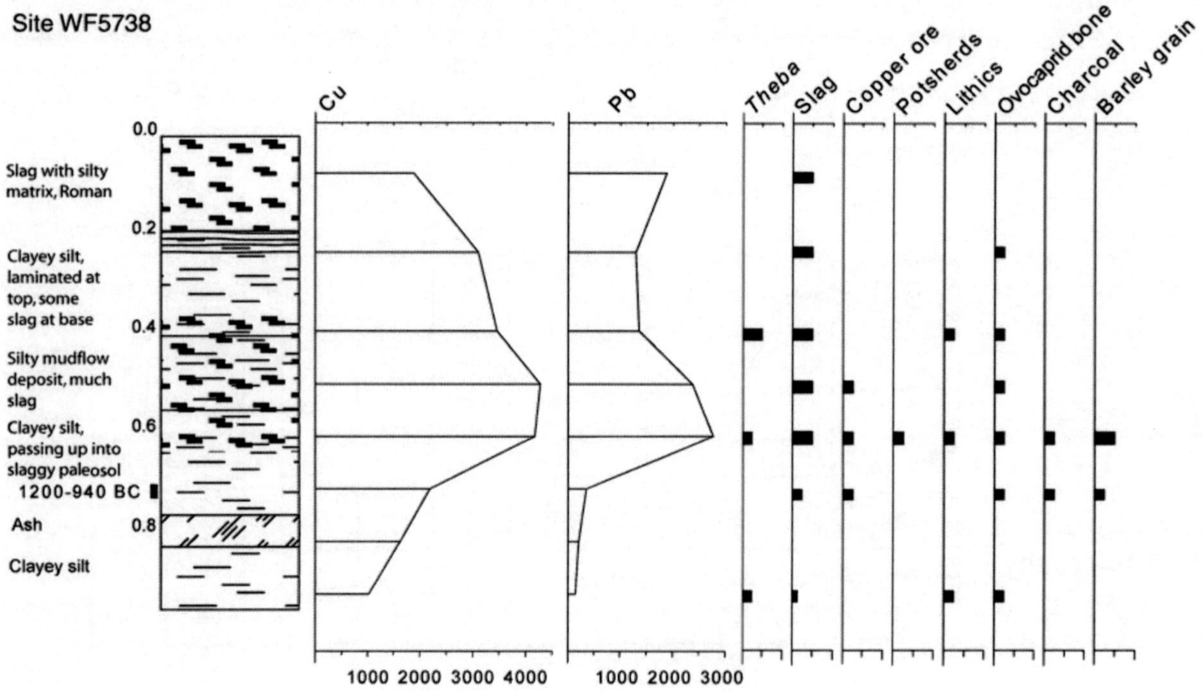

Fig. 8.9 Summary of dating, stratigraphy, heavy metals, molluscs, artefacts and environmental evidence from site WF5738. (Cu and Pb values are in ppm)

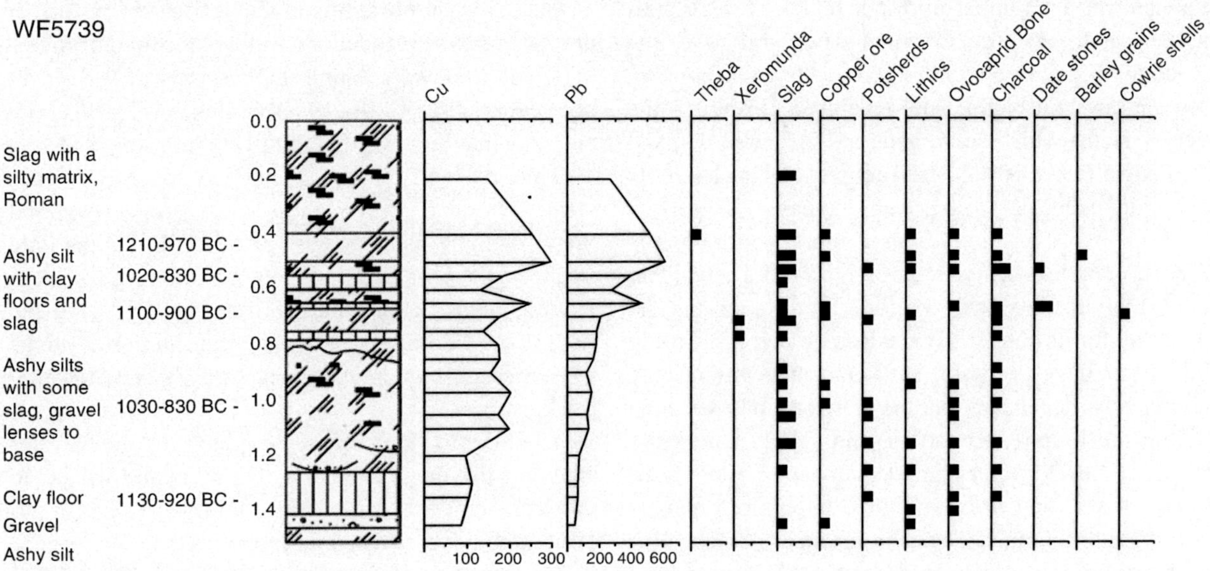

Fig. 8.10 Summary of dating, stratigraphy, heavy metals, molluscs, artefacts and environmental evidence from site WF5739. (Cu and Pb values are in ppm)

can only have functioned seasonally and inefficiently (Crook 2009). The field system was fed by run-off, either from adjacent slopes, or from diversions from the main wadi channel. Although there is occasional evidence for perennial water in parts of the irrigation system, it is likely that there was, at most, a seasonal flush of forage in the largest part of the system: it would have been necessary to take stock elsewhere to graze at other times, possibly onto the upland plateau of the Edom Mountains, as some local Bedouin groups do during summer today. Indeed Mattingly et al. (2007a) document Classical-period pastoral sites at this time in the hills overlooking the Faynan. Thus, for much of the year, the stock would have been feeding in areas where the level of metal pollution was relatively low.

Imported food (and other items) was thus present in the Faynan from the Chalcolithic, when the import of dried fish and raisins can be substantiated (Fig. 8.8). There is also good evidence for the import of food items in the Iron Age. Mattingley et al. (2007a) document imported fish for the Roman period and it can be argued that by Classical times, it is very likely that the overwhelming proportion of food consumed by the Faynan workforce originated elsewhere. This observation has profound significance for the health of the industrial population, since it means that the dietary ingestion of heavy metal pollution would have been minimized. The metal workers evidently had the resources to be able to import exotic food items and other non-essentials, such as seashell, even in the earliest phases. It is notable also that the sheep/goat bones associated with the smelting sites are derived from the high-quality parts of the carcass. These were not low-status individuals, but seem, rather, to have been well-rewarded professionals in each of the key metal-winning phases: the Chalcolithic/Early Bronze Age, the Iron Age and the Roman/Byzantine periods.

In addition to the climatic problems, the metal extraction and smelting which characterized the Faynan orefield led to considerable risks for the workforce. Metals would have been ingested through the breathing in of vapours and from polluted dust, through polluted water, through contamination of food on dusty surfaces during preparation, and from the metal load contained in locally-grown food, as both cereals and grazing animals are known to bioaccumulate (Grattan et al. 2002, 2007; Pyatt and Grattan 2001; Pyatt et al. 2005). The only mitigating factor is the strong probability, discussed above, that much of the food was imported and would thus have been relatively low in metal pollution. Nevertheless, life expectancy among the professional groups who operated the system would have been low and it is thus likely that the metal industry in the

Faynan was a net importer of personnel: professionals as well as the well-documented slave laborers of the Roman Period. This could only have been possible in regional systems with well-developed linkages—the very systems which could generate sufficient demand for copper to justify industrial activity in such a remote and arduous locality as the Wadi Faynan.

As it expanded and became more polluting, the Faynan complex became progressively more dependent upon the outside world, for fuel, food, labor and for a market for its copper. By the Roman period, it could no longer have been self-sufficient, but was instead completely dependent on the network of exchange within the Empire. Export to, and supply from, distant localities, together with local transport of ore from mines to smelting sites, would have required numerous draught animals. The maintenance of these animals would have generated considerable demand for forage in the Faynan. Forage and holding areas would have also been necessary for animals imported for meat, while they awaited slaughter. We therefore reinterpret the great WF4 field system as primarily operated for animal forage and stock-penning, rather than for arable agriculture. This interpretation would conform to the hydrology of the system (Crook 2009) and to the geoarchaeological evidence described above.

8.7 Conclusions

This chapter has explored the human Paleoecology of the Faynan orefield over the Holocene and has reinterpreted the function of the WF4 field system. During the Pre-Pottery Neolithic, early inhabitants adopted stock-rearing and cereal cultivation in a relatively benevolent wooded steppe landscape. In the later Neolithic, aridification set in, the steppe became devoid of trees and soil erosion became marked. The exploitation of copper ores started, at first for powders and ornamental stone, although there are indications that ores were being heated by the end of the Neolithic, whether purposely or by mistake is unclear.

Aridification became more marked during the Chalcolithic and Early Bronze Age. Cereal farming was still occurring locally, but now irrigated using small-scale floodwater farming. Sizeable settlements are known at this time. The earliest dated slags point to the inception of copper extraction during the Chalcolithic. Already, at that stage, imported items including seashells, fish and raisins were present in the diet of the metal-processors, and they were eating the best cuts of the sheep/goat carcass, suggesting external linkages by this time and also that they were of relatively high status.

The area was largely abandoned apart from transient pastoralists during the Middle Bronze Age and metal-winning seems to have restarted during the Late Bronze Age. Some activity continued during Iron Age I, with intensification during Iron Age II during the rise of the Edomite kingdom. Again, imported items were present in the metal processors' diet and they ate the 'best cuts'.

The Edomite kingdom was replaced by the Nabateans, and the metal industry of the Faynan waned in importance at this time. Under Imperial Rome, however, the Faynan reached its apogee. At this stage, convicts were at times imported as laborers. The skilled work was done by professionals, however, and again, they seem to have been well-fed and well-rewarded, with most food coming from distant sources. The fact that their food was largely imported and thus low in bioaccumulated heavy metals would have been critical in prolonging the lives of the metal workers. Miners and metalworkers on the Faynan orefield confronted a harsh environment, which their activities seem to have made even more hostile. They ran considerable risks, but it seems that they were relatively well-rewarded, by the standards of the ancient world.

The orefield was exploited only to satisfy the needs of the outside world for copper during phases where there were well-integrated economic systems, and there seems to have been tangible rewards for so doing. In many ways, the exploitation of copper ores in this remote and hostile landscape seems only to have happened and been possible because of its linkages with the outside world. The geoarchaeological evidence points to the overwhelming importance of the instrument of these linkages—pack animals—as the object of the great field system of the Faynan. As a metal-winning center Faynan was nothing without its linkages—it could only exist as a node in the local and regional networks.

Acknowledgements We thank all the members of the Faynan Project for their cooperation, particularly Professors Graeme Barker, David Gilbertson, John Grattan and David Mattingley. We also thank the directors and employees of the Department of Antiquities, Jordan, the RSCN, Jordan and the CBRL for permissions and logistical support and the many funding bodies, including the Libyan Government's Postgraduate Studentship programme, the British Academy, the CBRL and the Society of Antiquaries.

References

Adams RB (1999) The development of copper metallurgy during the Early Bronze Age of the southern Levant: Evidence from the Faynan Region, southern Jordan. Unpublished PhD Thesis, Department of Archaeology and Prehistory, Sheffield University. [available online at http://russellbadams.brinkster.net/PhDThesis.htm]

Adams RB (2002) From farms to factories: The development of copper production at Faynan, southern Jordan, during the Early Bronze Age. In: Ottaway BS, Wagner EC (eds), Metals and society. British Archaeological Reports, International Series S1061. Archaeopress, Oxford, pp 21–32

Barker G, Gilbertson D, Mattingly D (eds) (2007a) Archaeology and desertification: The Wadi Faynan landscape survey, Southern Jordan. Oxford, Oxbow Books, CBRL Levant Series vol 6, 510 pp

Barker G, Adams R, Creighton O, et al. (2007b) Chalcolithic (c.5000–3600 cal. BC) and Bronze Age (c.3600–1200 cal. BC) settlement in Wadi Faynan. In: Barker G, Gilbertson D, Mattingly D (eds) Archaeology and desertification: The Wadi Faynan landscape survey, Southern Jordan. Oxford, Oxbow Books, CBRL Levant Series 6:227–270

Barker G, Hunt C, McLaren S, et al. (2007c) Early Holocene environments and early farming. In: Barker G, Gilbertson D, Mattingly D (eds) Archaeology and desertification: The Wadi Faynan landscape survey, Southern Jordan. Oxford, Oxbow Books, CBRL Levant Series vol 6:199–226

Crook D (2009) Hydrology of the combination irrigation system in the Wadi Faynan, Jordan. Journal of Archaeological Science 36: 2427–2436

El-Rishi H, Hunt C, Gilbertson D, et al. (2007) The past and present landscapes of the Wadi Faynan: geoarchaeological approaches and frameworks. In: Barker G, Gilbertson D, Mattingly D (eds) Archaeology and desertification: The Wadi Faynan landscape survey, Southern Jordan. Oxford, Oxbow Books, CBRL Levant Series vol 6: 59–96

Engel T, Frey W (1996) Fuel resources for copper smelting in antiquity in selected woodlands in the Edom highlands to the Wadi Araba, Jordan. Flora 191:29–39

Grattan J, Huxley SN, Karaki LA, et al. (2002) Death more desirable than life? The human skeletal record and toxicological implications of ancient copper mining and smelting in Wadi Faynan, southwest Jordan. Ecotoxicology and Industrial Health 18:279–307

Grattan JP, Gilbertson DD, Hunt CO (2007) The local and global dimensions of metalliferous pollution derived from a reconstruction of an eight thousand year record of copper smelting and mining at a desert-mountain frontier in southern Jordan. Journal of Archaeological Science 34:83–110

Hauptmann A (1989) The earliest periods of copper metallurgy in Feinan, Jordan. In: Hauptmann A, Pernicka E, Wagner GA (eds), Archaometallurgie der Alten Welt e Der Anschnitt. Beiheft 7:119–136

Hauptmann A (2000) Zur Fruhen Metallurgie des Kupfers in Fenan/Jordanien. Der Anschitt 1. Deutsch Bergbau Museum, Bochum, 238 pp

Hauptmann A (2007) The archaeometallurgy of copper: Evidence from Faynan, Jordan. Berlin: Springer. (available online at http://www.springerlink.com/content/h62404/?p=16e50d8fb581419488683e29000a94ed&pi=0)

Hauptmann A, Weisgerber G (1987) Archaeometallurgical and mining archaeological investigations in the Feinan, Wadi Arabah Jordan. Annual of the Department of Antiquities Jordan 31:419–437

Hauptmann A, Weisgerber G (1992) Periods of ore exploitation and metal production in the area of Feinan, Wadi 'Arabah, Jordan. In Hadidi A (ed), Studies in the history and archaeology of Jordan, vol 4. Department of Antiquities of Jordan, Amman pp 61–66

Hauptmann A, Begemann F, Heitkemper E, et al. (1992) Early copper produced at Feinan, Wadi Araba, Jordan: the composition of ores and copper. Archaeomaterials 6:1–33

Heim C, Nowaczyk NR, Negendank JFW (1997) Near East desertification: Evidence from the Dead Sea. Naturwissenschaften 84:398–401

Hunt CO, Gilbertson DD (1998) Context and impacts of ancient catchment management in Mediterranean countries: implications for sustainable resource use. In: Wheater D, Kirby C (eds) Hydrology in a changing environment. Volume II. John Wiley, Chichester, pp 473–483

Hunt CO, El Rishi HA, Gilbertson DD, et al. (2004) Early Holocene palaeoenvironments in the Wadi Faynan, Jordan. The Holocene 14, 6:921–930

Hunt CO, Gilbertson DD, El-Rishi HA (2007a) An 8000-year history of landscape, climate and copper exploitation in the Middle East: the Wadi Faynan and the Wadi Dana National Reserve in southern Jordan. Journal of Archaeological Science 34:1306–1338

Hunt C, Mattingly D, Barker G, et al. (2007b) Published radiocarbon dates from the Faynan region. In: Barker G, Gilbertson D, Mattingly D (eds) Archaeology and desertification: The Wadi Faynan landscape survey, Southern Jordan. Oxford, Oxbow Books, CBRL Levant Series 6:445–464

Levy TE, Adams RB, Hauptmann A, et al. (2002) Early Bronze Age metallurgy: A newly discovered copper manufactory in southern Jordan. Antiquity 76:425–437

Levy TE, Adams RB, Najjar M, et al. (2004) Reassessing the chronology of Biblical Edom: New excavations and 14C dates from Khirbat en Nahas (Jordan). Antiquity 78:863–876

Mattingly D, Newson P, Creighton O, et al. (2007a) A landscape of imperial power: Roman and Byzantine *Phaino*. In: Barker G, Gilbertson D, Mattingly D (eds) Archaeology and desertification: The Wadi Faynan landscape survey, Southern Jordan. Oxford, Oxbow Books, CBRL Levant Series 6:305–348

Mattingly D, Newson P, Grattan J, et al. (2007b) The making of early states: the Iron age and Nabatean periods. In: Barker G, Gilbertson D, Mattingly D (eds) Archaeology and desertification: The Wadi Faynan landscape survey, Southern Jordan. Oxford, Oxbow Books, CBRL Levant Series 6:271–304

McLaren S, Reynolds T, Gilbertson D, et al. (2007) Pleistocene environments and human settlement. In: Barker G, Gilbertson D, Mattingly D (eds) Archaeology and desertification: The Wadi Faynan landscape survey, Southern Jordan. Oxford, Oxbow Books, CBRL Levant Series 6:177–198

Pyatt FB, Grattan JP (2001) Consequences of ancient mining activities on the health of ancient and modern human populations, Journal of Public Health Medicine 23:235–236

Pyatt FB, Barker GW, Birch P, et al. (1999) King Solomon's miners - starvation and bioaccumulation? An environmental archaeological investigation in southern Jordan. Ecotoxicology and Environmental Safety 43 (3):305–308

Pyatt FB, Walker C, Sheen T, et al. (2005) Environmental toxicology: heavy metal content of skeletons from an ancient metalliferous polluted area of Southern Jordan with particular reference to bioaccumulation and human health. Ecotoxicology and Environmental Safety 60:295–300

Simmons A, Najjar M (1996) Test excavations at Ghuwair I, a Neolithic settlement in the Wadi Feinan. American Center of Oriental Research Newsletter 8.2:7–8

Chapter 9
Empire and Environment in the Northern Fertile Crescent

Tony J. Wilkinson

9.1 Introduction

Because of their huge size, empires are daunting for archaeological study. Although some features of early Near Eastern empires have been studied since the very earliest trenches were sunk into the Assyrian capital cities of Nimrud and Nineveh (Fig. 9.1), the implications of the development of territorial empires have not been fully absorbed into the study of human-environment relations. The later territorial empires of the first millennium BC and AD fundamentally changed the landscapes of the Near East in ways that did not previously obtain. For example, features of monumental scale (which are often associated with empires), and which include huge canal systems such as those of the Sasanians, necessarily had massive impacts on the environment, but more widespread, and ultimately perhaps more significant in terms of human impacts on the environment, are the smaller scale features that are often under-represented by archaeologists. This chapter relates the signatures of the cultural landscapes of the later territorial empires of the Near East to the local environment and landscape degradation.

Empires provide the opportunity to relate socio-political processes explicitly to environmental conditions and more specifically to evaluate the links between humans and the environment. Nevertheless, because of their size and administrative complexity the recognition of causal links may be difficult. Territorial empires can be defined as territorially expansive polities in which a ruling power effectively controls and dominates a number of smaller and often weaker subordinate societies and their territories (Doyle 1986). In terms of the landscape, the royal household, ruling out of an imperial capital, has control of vast areas of terrain that can be transformed by the imposition of new patterns of settlement or by the introduction of technological innovations such as irrigation. By so doing, not only do the changes of settlement impact the landscape (by degrading it), innovations such as the introduction of irrigation, can result in the limitations of the environment being ignored or over-ridden.

Examples are drawn primarily from the Neo-Assyrian Empire, the Sasanian Empire in Iran, and the Seleucid, Roman and Byzantine empires in southern Turkey and northwest Syria. The chronological range is from around 900 BC through to the demise of the Abbasid caliphate in the tenth century AD (Table 9.1).

The area dealt with in this chapter occupies that part of the Fertile Crescent in northern Syria, northern Iraq, and southern Turkey where agriculture is primarily dependent upon rainfall. Nevertheless, irrigation became increasingly important from the first millennium BC, and is common today, although frequently today this is at the expense of local water tables that are rapidly falling. Rainfall exceeds 700 mm/annum in the Amanus Mountains flanking the Amuq Valley in southern Turkey (Fig. 9.1), but falls to 300 mm throughout much of the Jazira located between the Tigris and Euphrates rivers, and even less in the semi-desert steppe that fringes the Syrian Desert to the south. Between the Tigris and Euphrates rivers the landscape consists of rolling uplands on Tertiary limestone, sandstone and occasional basaltic outcrops, separated by broad alluvial basins and river valleys, some of which are tributary to the Tigris and Euphrates rivers. Further

T. J. Wilkinson (✉)
Dept. of Archaeology, Durham University
South Road, Durham, DH1 3LE, UK
e-mail: t.j.wilkinson@durham.ac.uk

See Plate 8 in the Color Plate Section; also available at: extras.springer.com

I. P. Martini, W. Chesworth (eds.), *Landscapes and Societies*,
DOI 10.1007/978-90-481-9413-1_9, © Springer Science+Business Media B.V. 2010

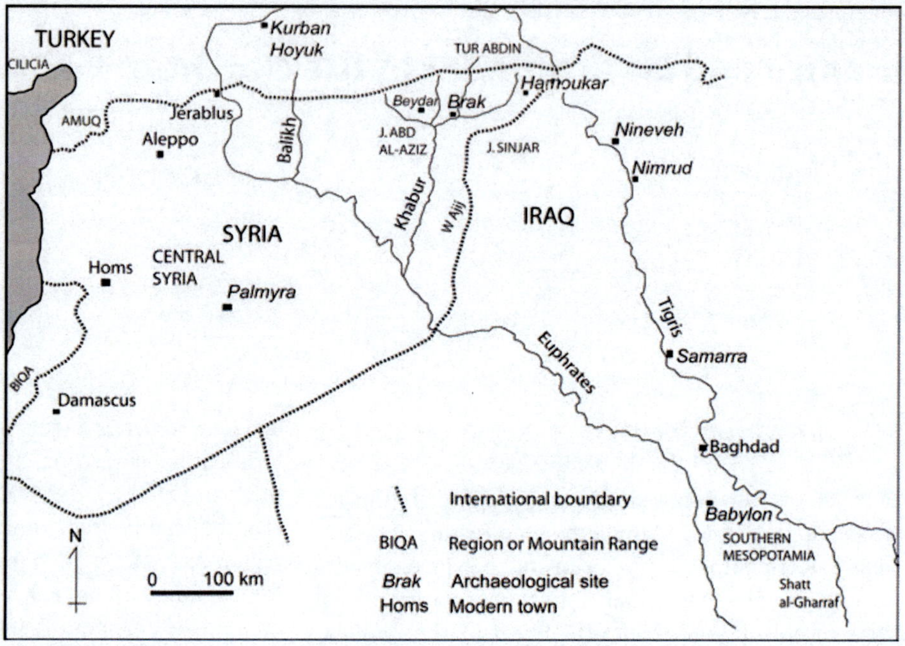

Fig. 9.1 Main sites and regions discussed in text

west near the Mediterranean, the Amanus Mountains in Turkey and neighboring ranges in northern Syria are characterized by higher-energy erosional regimes than the steppe lands to the east. During the early and mid-Holocene (Neolithic, Chalcolithic, and Bronze Age), settlement was primarily focused on the broader alluvial plains and valleys, but during the later territorial empires of the first millennium BC and later, settlement spread beyond these core areas into the uplands and out into the more climatically marginal areas of the steppe. Occasional reference is also made to southern Mesopotamia and SW Iran (Khuzestan) where irrigation is essential on the otherwise arid alluvial plains of the Tigris, Euphrates, Karun, and Dez rivers.

The excavation of sites belonging to the ancient empires stretches back to the very beginnings of archaeology and Assyriology, with the excavations by Layard and Botta in the Assyrian capitals of Nimrud, Khorsabad, and Nineveh (Larsen 1994). Only in more recent years have smaller settlements and their landscape context become the focus of investigations. For example, the pioneering studies of David Oates in northern Iraq and Syria focused on both the Neo-Assyrian remains of the first half of the first millennium BC, and also on the Parthian, Sasanian and Roman sites that followed (Oates 1968). Working during the initial years of aerial survey, pioneers such as Mouterde and Poidebard (1945), made the first strides in the mapping of central Syria, subsequently continued on the ground by Bernard Geyer and colleagues (Geyer and Rousset 2001; Geyer et al. 2007). In northern Syria a range of archaeological surveys and geoarchaeological investigations conducted since the 1980s have extended the range of data so that it is now possible to make generalizations from them (Ergenzinger et al. 1988; Wilkinson et al. 2005; Morandi Bonacassi 2000).

Table 9.1 The chronology of the main later empires in the northern Fertile Crescent. Note that gaps in the chronology indicate periods of transition

Period	Approximate dates
Neo-Assyrian empire	911–612 BC
Persian/Achaemenid	539–333 BC
Hellenistic/Seleucid (to west)	333–64 BC
Parthian/Arsacid (to east)	247 BC–224 AD
Sasanian (to east)	224–651 AD
Roman (to west)	64 BC–395 AD
Byzantine (to west)	395 AD–636 AD
Umayyad (early Islamic)	661–750 AD
Abbasid (early Islamic)	750–968 AD

During the early days of geoarchaeology, and even before the term had been coined, both Butzer (1964) and Vita-Finzi (1969) had emphasized the genesis of alluvial fills, albeit with rather different conclusions. Thus Butzer saw a more fine grained temporal record and more complex inter-relations between humans and the environment, whereas Vita-Finzi focused on region-wide climatic change as a causal factor. In the Near East, Rosen (1986), Brückner (1986), Goldberg (1998), Cordova (2000) and others have taken an approach similar to Butzer, and in recent years there has been an increasing tendency to focus upon the complexity of the relationship between humans and the environment and the role of humans in mediating some of the affects of climatic change (Rosen 2007). Similarly, Bintliff, whose approach originally closely followed that of Vita-Finzi, later adopted an approach that emphasized both human and climatic factors (Bintliff 1992, 2002). Although these pioneering studies were fundamental to the development of the field, because they focused upon the geomorphologic evidence and processes they frequently under represented the record of settlement, which in most studies remained unquantified. Moreover, for a variety of reasons some studies have only been able to focus on valley floor records at the expense of the valley slopes and interfluves that supplied the sediments for the valley floors (Beach and Luzzadder-Beach 2008). Not until the geoarchaeological record was studied in tandem with quantitative studies of settlement during the surveys of the 1980s and later has it become possible to gain a clearer cause and effect relationship between human settlement and associated physical responses (Casana 2008). The primary focus of this chapter is to bring together the results of such investigations over the past 30 years, with particular emphasis upon how the spread of settlement and water supply technology during the later empires contributed to our understanding of interactions between humans and the environment. Nevertheless, it is important to appreciate that settlement and land use configurations respond in many different ways to variations in rainfall, so that it is frequently difficult to separate the effects of climate and humans. In the narrative that follows, the primary attention is paid to the human circumstances that create the conditions for climatic change to operate.

9.2 Case Studies: Settlement Dispersal and Soil Erosion

Although ancient territorial empires are best known for archaeological features of monumental scale, such as the palaces and associated zigurrats of Nimrud and Khorsabad, or the temple-palace complexes of Babylon, Rome, and Ctesiphon, in terms of the landscape, it is often the smaller features that cumulatively have had greatest impact. Whereas nineteenth century archaeological investigations of the Assyrian and Neo-Babylonian empires tended to focus upon the more massive constructions, archaeology in the twentieth century slowly shifted its emphasis to a range of features that were more representative of the entire empire in question. The early Middle Eastern empires were highly centralized and administered by the king, who usually resided in his imperial city, or went out on military campaigns to consolidate and maintain his rule of the outlying provinces. In order to fund the lavish imperial lifestyles, to supply food for their massive populations and to support an army, it was necessary to strengthen the economy and invest in agriculture, often in more far flung parts of the empire. One policy that was particularly common during the Neo-Assyrian Empire was the re-settlement of deportees into marginal steppe lands, many of which were located in the Jazira between the Tigris and Euphrates rivers (Oded 1979). As a result of centralized decisions on geographically marginal areas, local impacts on the landscape (such as deforestation and soil erosion) were often a result of decisions taken in the physically remote imperial capitals. This is very different from the situation that prevailed during the city states of the Bronze Age, when local agricultural decisions were probably most often made by the local king housed in a local settlement (a 'tell').

Neo-Assyrian cuneiform texts provide us with numerous examples of such agricultural policies, many of which appear to be statements of imperial bombast. Some for example, boast of magnificent achievements such as the settling of entire regions by the relocation of large numbers of displaced populations, whereas others describe large land-reclamation schemes and their attended settlements (Radner 2000). Such schemes were difficult to identify during the early stages of archaeological surveys, when techniques were rather coarse, but in recent years field surveys at increasing levels of intensity have demonstrated what appears to be the signature of resettlement schemes. These mainly took the form of areas of small dispersed settlements that dot the landscape in between the earlier and more recognizable pattern of tells. Exemplar landscapes include those north and south of the Jebel Sinjar (NW Iraq), as well as in the Khabur and Balikh Valleys and in the Wadi Ajij near the Iraq/Syrian border (Fig. 9.1; Oates 1968; Bernbeck 1993; Wilkinson 1995; Morandi Bonacassi 2000; Wilkinson et al. 2005).

The settlement landscape of Iron Age Upper Mesopotamia consisted of a broad dispersal of farmsteads,

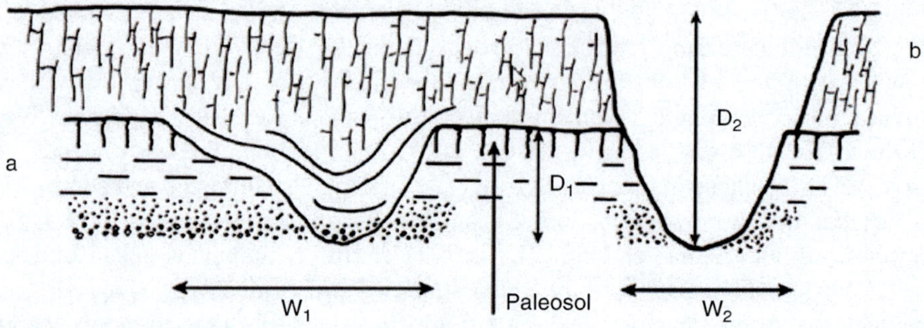

Fig. 9.2 Sketch section through wadi fills in the Wadi Jaghjagh near Tell Brak (D_2 is roughly 6 m, D_1 3 m; W_2 is 10–15 m; W_1 > 15 m). (After Wilkinson 2003)

hamlets and villages with occasional large towns dominated by one massive imperial capital. By no means were all of these settlement schemes imperial foundations: many may have simply resulted from the spontaneous sedentarization of Aramaean nomads who chose to settle under the administrative umbrella of the Neo-Assyrian kings. Nevertheless, this phase of settlement represented a dramatic departure from the characteristic signature of the Bronze Age which consisted of a hierarchical pattern dominated by large tells up to 100 ha area, with progressively smaller centers of power (again 'tells') which formed prominent mounds ranging from 1 to 50 ha in area and 5–30 m in height. The larger Bronze Age tells were distributed between 10 and 30 km apart, whereas the smaller subordinate settlements were usually distributed around the centers. One should not, however, assume that necessarily the largest sites were the most important politically because power relationships between city states could be remarkably fluid. Moreover, the later empires cannot be categorized as homogenous. Rather they comprised a mosaic of administrative entities, satraps, provinces and vassals each of which must have had its own signature on the landscape. Despite these variations, both textual sources and settlement studies demonstrate that there was a shift in the spatial imposition of power between the era of Bronze Age city states and the later territorial empires.

Evidence for the impact of Iron Age settlement on the degradation of the Jazira landscape is ambiguous. Surveys in the Khabur basin around tells Brak, Beydar, and Hamoukar all demonstrate that the dispersal of populations into farmsteads and small villages and lower towns (beneath tells) was in place by the first quarter of the first millennium BC (Morandi Bonacassi 2000; Wilkinson and Barbanes 2000). After the end of the second millennium BC, that is around the onset of the Iron Age or slightly later, the Wadi Jaghjagh near Brak underwent a change of regime from a broader sinuous channel with a grayish silty fill to an entrenched highly meandering channel incised into a brown, blocky clay-rich alluvial fill (Fig. 9.2; Wilkinson 2003). Optically stimulated luminescence (OSL), TL screening, and radiocarbon dating of sediments and carbonized material from the wadis Jaghjagh, Khanzir, and Jarrah provide age estimates on this upper clay fill in the region of 2600 ± 400 BP (OSL), 2600 ± 800 BP (TL screening), 2300 ± 200 BP (OSL) and 2200 ± 900 (TL screening) (Deckers and Riehl 2007). In addition, around AD 244 (1780 ± 30 BP), fragments of wood charcoal in clayey sand in the Wadi Jaghjagh (Deckers and Riehl 2007) suggest that there was localized forest destruction during the Roman/Parthian period. Despite their very broad standard deviations, the OSL and TL estimates broadly support field stratigraphic and artifactual chronologies for these fills (Wilkinson 2003). The upper brown blocky clay fill therefore approximates to the time when the Jazira was becoming heavily re-settled, following a phase of sparse settlement and partial desertion during the late third and part of the second millennium BC. Archaeological surveys and satellite images indicate that starting in the Late Bronze Age, and gaining momentum in the Iron Age, settlements developed on land between tells, which often continued to be settled in the form of lower towns. This dispersed pattern of settlement in the Khabur basin was maintained during the Hellenistic, Roman/Parthian and Byzantine/Sasanian periods and it was during the early Christian period that significant settlement occurred on the hills of the Tur Abdin to the north. In other words not only was the total area of cultivated land increased by the Iron Age dispersal, but also the settlement and cultivation of the hills, (also evident on satellite images), must have resulted in significant erosion of soil into the fluvial system. This erosion, combined with evidence for increased abstraction of water from the Jaghjagh by Partho-Sasanian canal systems, suggests that the main rivers such as the Jaghjagh were receiving both increased sediment loads as well as lower discharges due to abstraction. When combined these factors appear to have contributed to the aggradation of the upper clay fills.

During the first millennium BC, Iron Age settlements extended into parts of the climatically marginal steppe around the Jebel Abd al-Aziz and to the east of the Khabur River (Bernbeck 1993; Hole and Kouchoukos in press), despite the fact that climatic conditions were significantly drier than during the optimum phase of Bronze Age settlement. This is evident from the proxy record of climate from Lake Van, located in eastern Turkey some 100–200 km north of the area in question, which shows that this phase of settlement extension occurred when the climate either experienced significant inter-annual variability, or was somewhat drier than today (Fig. 9.3).

Similar dispersed patterns of settlement prevailed during other phases of territorial empires in the Near East, namely during the Hellenistic, Parthian-Roman, Sasanian-Byzantine, and early Islamic empires, but in these cases the relationship between settlement and valley fills is much clearer. Particularly distinctive landscapes of dispersed settlement are evident along the Euphrates Valley in Turkey and Syria as well as further west in southern Turkey and NW Syria. If aggregate settlement area is used as a proxy indication for population, it appears that in many areas population increased to levels that were significantly above those of the Bronze and Iron Ages. This trend of settlement increase has been observed along the Turkish Euphrates around Kurban and Lidar Höyüks (Wilkinson 1990; Gerber 1996), along the Syrian Euphrates near Carchemish (Fig. 9.4a) (Wilkinson et al. 2007), further west in the Amuq Valley of southern Turkey (Casana and Wilkinson 2005; Casana 2008), in the basalt plateaus north of Homs (Philip et al. 2002), as well as in the Biqa valley of Lebanon (Marfoe 1979). Perhaps more important in terms of landscape degradation, however, is that the number of sites increased even more than overall aggregate settlement area. This meant that areas that had formerly been open steppe, pasture, upland or woodland, received the imprint of very many small settlements, which in aggregate probably had greater impact than around constrained nucleated settlements located within long-settled lowlands. This is especially the case because the Bronze Age phase of nucleated tell-type settlement developed on silt-clay plains that generally were less susceptible to erosion.

Unlike the phase of Iron Age settlement noted above, the Classical to Early Islamic phases of settlement frequently spread into the uplands of limestone, sandstone and basalt that appear, during the Bronze and Iron Ages, to have been under-settled marginal lands or pastures. Consequently, their impact on the environment was considerable. Not only did the extension of settlement apparently result in the acceleration of the removal of the woodland cover (Rosen 1997; Deckers and Riehl 2007), it also was associated with

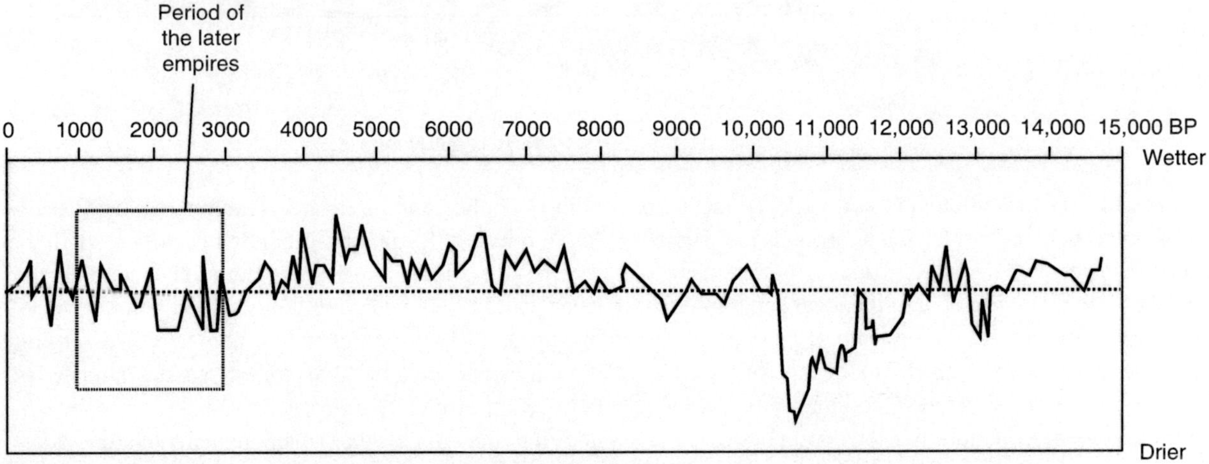

Fig. 9.3 Proxy isotopic record from Lake Van showing the moist phases (*above the dotted line*) and dry phases (*below*). The period of the later empires of the Iron Age and Roman-Byzantine period is indicated within the box. (After Lemcke and Sturm 1997)

Fig. 9.4 Long-term settlement curves. **a** Kurban Hoyuk area on the Turkish Euphrates. (After Wilkinson 1990) **b** Jerablus area near Carchemish: Note how the non-tells increase in number after the Iron Age

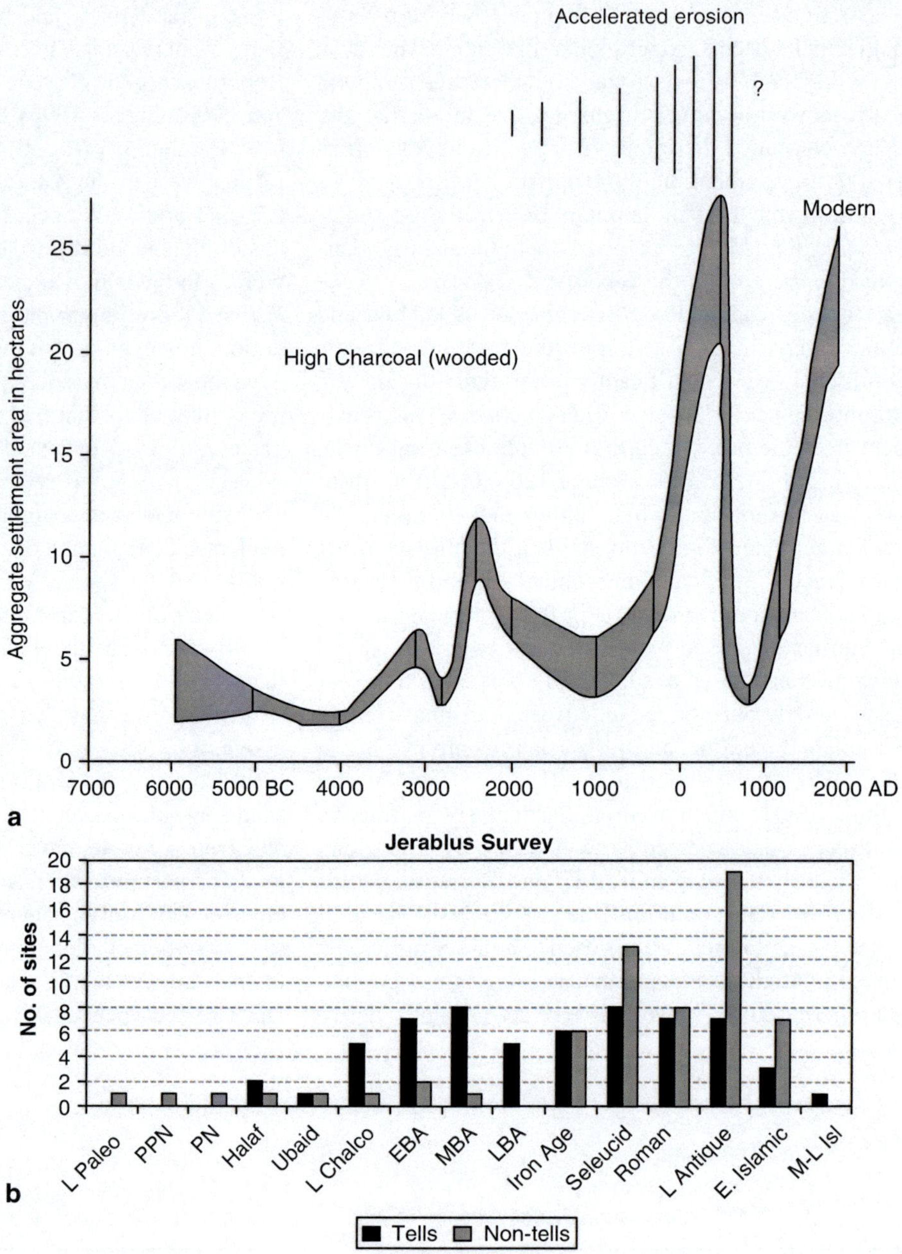

a phase of accelerated erosion and valley floor deposition that became widespread around the eastern Mediterranean.

Thanks to diversification of the local geography, the extension of settlement varied from place to place. For example, in the area of Kurban Höyük (Turkey), Late Roman–Byzantine settlement colonized all available landscape niches, namely the lower Euphrates Valley that was the long term locus of settlement, the episodically settled upper terrace, as well as the higher ridges beyond, the latter becoming pastures equipped with animal pounds and associated features. Around Jerablus in northern Syria, non-mounded but often spatially extensive settlements of short duration, occupied the river valleys between the long-settled tells, extending up previously uninhabited wadis and also settling the limestone uplands that comprised the interfluves between the main long-settled wadis (Fig. 9.4b). In addition, the uplands were also the locus of stone quarrying which occurred not only near the settlements, but again extended on to the uplands, areas that under previous settlement regimes must have been upland pastures presumably held in some form of common arrangement by local communities based in the main valleys.

At the same time, as settlement dispersal occurred, the river valleys became partly infilled with sedimentary accumulations that included post-Roman colluvia and thick deposits of high-energy alluvial deposits. Thus along the River Amarna, near Jerablus, a 2.5 m thickness of alternating beds of colluvia, and weakly developed paleosols overlay channel gravels of the river, whereas in the nearby Wadi Seraisat, thick cobble and gravel beds of high-energy channels planed across and destroyed a pre-existing Roman–Byzantine industrial area (Fig. 9.5a, b). In the Kurban Höyük area, the best examples of landscape degradation associated with settlement extension took the form of the rapid growth of alluvial fans. Although these were apparently initiated at the beginning of the second millennium BC following a local settlement peak of the early Bronze Age date, arguably they attained their maximum extent during the Late Roman and Byzantine phases when erosion had eventually bitten into the chalky bedrock and deposited an upper alluvial fan fill (Wilkinson 1999). Overall, it is common to find Hellenistic, Roman, and Byzantine buildings or other

Fig. 9.5 **a** A colluvial/alluvial sequence on the south bank of the Amarna River south of Carchemish in Syria. Sediments above the figures' heads are dated to the post-Byzantine period. **b** High-energy wadi gravels overlying Late Roman/Byzantine kilns in the Wadi Seraisat south of Carchemish, Syria

Fig. 9.6 Settlement zones inferred for the Amanus Mountains during the Roman/Byzantine period

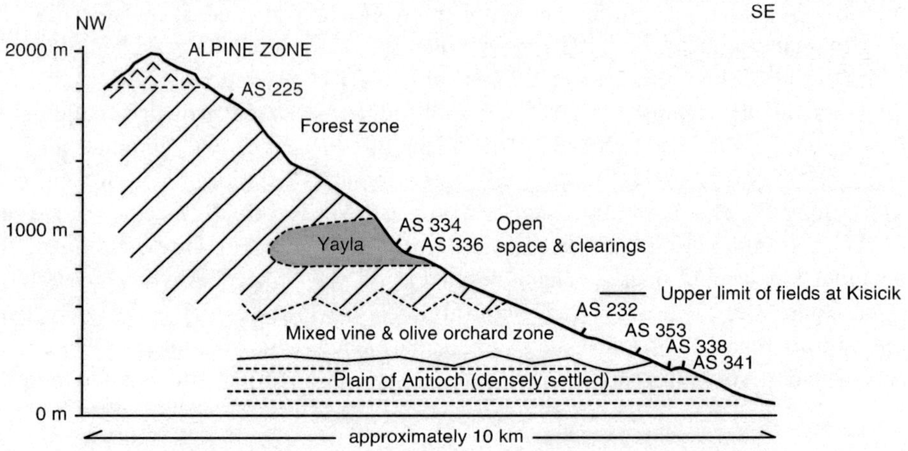

structures sealed below thick accumulations of colluvial, alluvial fan or alluvial deposits, examples having been found in the western Amuq Valley (Casana and Wilkinson 2005), the Cilician coast (Beach and Luzzadder-Beach 2008), the Jebel al-Aqra, near the Amuq (Casana 2008), along the Syrian Euphrates near Meskene (Harper 1975), as well as in numerous locations around the Mediterranean (Vita-Finzi 1969).

Further west, the so-called Massif Calcaire to the NW of Aleppo, provides the clearest indication of settlement encroachment on to the uplands. In this area of denuded limestone uplands, some 700 villages and monastic communities remain today as upstanding remains of one of the most remarkable ancient landscapes of the Middle East. Settlements are surrounded by their relict fields and track-ways to form a relatively brief phase of cultural landscape that was initiated during the Seleucid Empire, but reached its climax during the eastern Roman–Byzantine Empire between the fourth and sixth centuries AD (Tchalenko 1953; Tate 1992). Although no geoarchaeological studies have been published for the best known part of this terrain, neighboring areas in southern Turkey, which show a more subtle manifestation of the same process, demonstrate that the extension of settlement into the uplands was associated with erosion of the same uplands and the complementary accumulation of colluvial fills, growth of alluvial fans and valley floor alluviation (Casana and Wilkinson 2005; Casana 2008). This area, which fell within the immediate hinterland of Antioch, experienced the spread of a rash of minor settlements into erodable sandstone and shale uplands south east of the main city, on to limestone uplands further to the east, and to elevations of 1000 m above sea level in the Amanus Mountains in the north (Fig. 9.6). In the Amanus Mountains this upper extension of settlement was probably associated with vine and olive husbandry which is still practiced up to similar elevations today. Most of these settlements appear to have been the farmsteads of individual households or small villages. Nevertheless, the existence of temples on hill tops and mining settlements within the metalliferous zone of the Amanus Mountains demonstrate that both secular and religious activities must have contributed to the loss of vegetation and soil cover. In the most vulnerable areas of sandstone and shale, Casana has demonstrated that massive aggradation can result where settlement extended rapidly into what had previously been wooded uplands. The resultant erosion resulted in a burst of aggradation in the local valleys that ranged from 12 to 30 times the long-term mean for the region (Casana 2008). On the other hand the encroachment of settlement into the nearby Amanus Mountains appears to have resulted in the growth of alluvial fans in the neighboring lowlands (Casana and Wilkinson 2005). Similar aggradation is also evident in Cilicia, in the vicinity of Kinet Höyük, where some 1.0–5.0 m of post-Hellenistic aggradation has been recorded. This included 1.00 and 1.8 m of aggradation over successive re-buildings of Roman roads (Beach and Luzzadder-Beach 2008). As in the Jebel al-Aqra area, rates of sediment aggradation peaked during the later empires and between the Hellenistic and Later Roman periods accumulation rates were 2.5 times greater than both before or after this period (Beach and Luzzadder-Beach 2008).

However, probably the most integrated record of settlement and landscape impacts derives from the

Taurus Mountains of western Turkey, where Waelkens and colleagues (Vermoere et al. 2000) have used multi-proxy records of settlement, biotic indicators and geo-archaeology to demonstrate that settlement growth, woodland clearance, fire-induced erosion and accumulation of valley fills all coincided to form a single horizon, the Beyşehir Occupation Phase. Broadly dated between 3265 and 1300 BP, this horizon includes much more specific events of fire impacts between 2320 and 1820 cal year BP (Kaniewski et al. 2008).

9.3 Case Studies: The Extension of Settlement into Climatically Marginal Areas

Because the later empires were powerful polities that exercised their administrative clout over large areas, and often moved entire communities from place to place, the relationship of empires to the environment was often different from that of earlier political regimes. In some cases, colonization of land extended into the semi-desert or uplands on to terrain that would have been regarded as environmentally marginal by earlier regimes. Colonization often apparently disregarded the limitations of such environments.

Written sources from the Ottoman Empire demonstrate that settlement processes in climatically marginal areas were complex and often did not simply consist of the colonization of marginal areas during moister intervals and retreat during dry spells. Rather, Norman Lewis has demonstrated that during the eighteenth and nineteenth centuries AD, settlement in Syria and Jordan depended mainly upon the administrative strength of the Ottoman Empire (Lewis 1987). During times of weakness, or when the local administrative apparatus of the empire was weak, Bedouin tribes from the deserts became the scourge of the marginal lands, even threatening the city of Aleppo. At other times, under strengthened administrations, the area witnessed the sedentarization of Bedouin tribes, or optimum relations between the sedentary and mobile communities. Although this may not be the entire story, again it indicates that even in areas where rainfall was crucial for successful agriculture, in sub-marginal areas where agriculture is possible, human administrative factors can lead to areas of high settlement potential becoming deserted.

The Neo-Assyrian Empire expanded into the climatically marginal steppe of the Jazira of northern Syria, an area that had been sparsely populated during the Late Bronze Age and Early Iron Age. On the other hand, under the later Roman empires in the deserts and uplands of the Levant, there was conspicuous settlement expansion under conditions of economic growth (Rosen 2007).

Perhaps one of the best documented examples of such settlement expansion into climatically marginal areas comes from central Syria where the record for the later empires is, however, just one of a series of settlement cycles. These start with sparse Pre-Pottery Neolithic B (PPNB) occupation followed by a retreat or even absence of settlement during the Halaf, Ubaid and Late Chalcolithic settlement (Fig. 9.7, ca. 6500 BP). There followed a rapid expansion of settlement during the Early Bronze Age, which reached some 57 km east of the Chalcolithic limit during Early Bronze IV (Fig. 9.7, ca. 4300 BP). After this there was a thinning of population during the Middle Bronze Age, and virtually a desertion during the Late Bronze Age, as settlement retreated towards the long-settled lands in the west. During the Iron Age (ca. 3000–2600 BP) settlement continued to be confined primarily to the west, a situation that contrasted with that between the Tigris and Euphrates where the Iron Age witnessed colonization and the establishment of settlement schemes. At this time, central Syria showed minimal evidence for sedentary occupation (Geyer and Rousset 2001; Geyer et al. 2007), probably because it was the domain of Aramaean and affiliated mobile groups and was beyond the effective reach of Neo-Assyrian imperial policy. By the Hellenistic period (starting in the third century BC) there commenced a rapid extension of settlement eastwards with settlement extending 15, 47, and 87 km east of the Chalcolithic minimum in the Hellenistic (2300 BP), Roman (2000 BP) and Byzantine (1500 BP) periods respectively (Geyer et al. 2007). After this, there followed progressive retreat through the early Islamic, Ayyubid and Mamluk periods until the recovery of settlement during the Ottoman revival of the seventeenth and eighteenth centuries and the twentieth century AD.

Although, it has been suggested that the dramatic Roman-Byzantine expansion of settlement occurred during a period when the climate was relatively favorable to agriculture (Geyer and Rousset 2001), the isotopic proxy records from Soreq Cave (Israel) and Lake

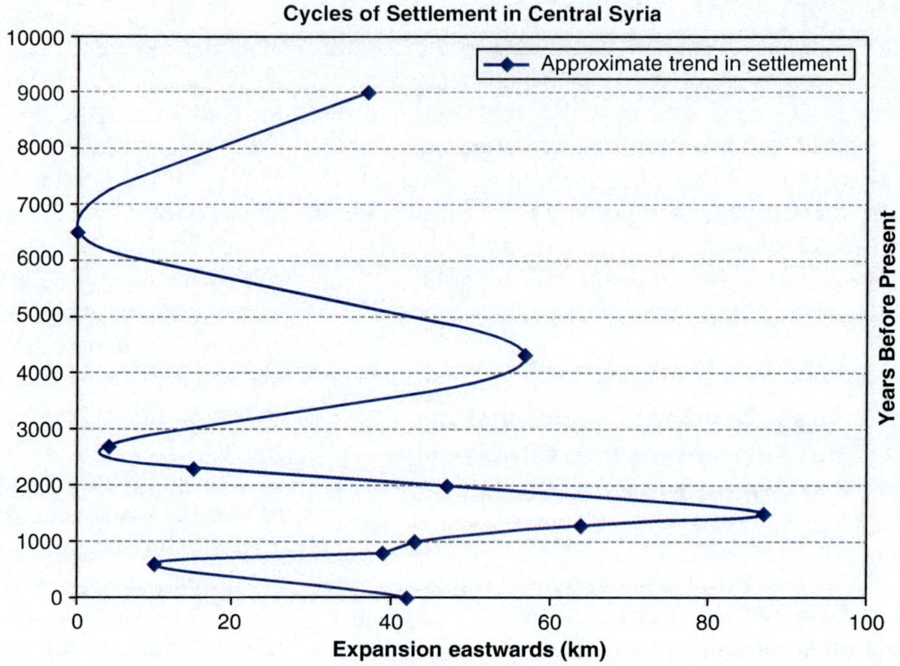

Fig. 9.7 Diagram showing cycles of settlement expansion and decline in central Syria. (After Geyer and Rousset 2001)

Van (Turkey), indicate the climate was rather drier than obtained during the Chalcolithic and Early Bronze Age. More significantly, during these six to eight centuries, the isotopic records indicate that climate varied around present day levels with both wetter and drier conditions prevailing (Fig. 9.3). In other words, there was no single wetting trend that could account for such colonization, although there were wetter phases when agricultural production would have been more successful. A similar point has been made by Arlene Rosen for the southern Levant, but using the climate proxy curve from Soreq Cave (Rosen 2007). For this period of territorial empires such settlement would have been encouraged within a relatively secure imperial administration, so that any shortfalls in production could be (within reason) met by the wider network of supply that prevailed (Rosen 2007), and most important, any climatic limitations could be overridden by hydraulic technology. Although this administrative hold was relatively light or absent during the Seleucid Empire, by the Roman period the extension of settlement was contained within and west of the Roman frontier system and the frontier road, the 'Strata Diocletiana' which ran from Damascus to Palmyra (Tadmor), and then Resafa, reaching the Euphrates near Sura. The emplacement of a Roman military frontier with military roads and forts was associated with a long step-by-step advance of the Roman military presence and a settled population into the dry steppe (Millar 1993). That protection from the Saracens ('bedu') was part of this programme is emphasized by inscriptions that record the construction of reservoirs in the Jordanian steppe to afford protection to the population away from the Saracens (Millar 1993). Within central Syria, although the *Pax Romana*, may have contributed the overall administrative framework, in reality, on the ground many of the settlements had semitic, that is local names and were under the administration of local governors and/or cities such as Palmyra (in the desert), Andarin (in the steppe), and Apamea (in the verdant west; Millar 1993; Butcher 2003).

Significant settlement expansion in the fourth century AD may have been encouraged by the reorganization of the Roman tax system under Diocletian, and yet further by Constantine who provided tax incentives for military veterans who farmed otherwise deserted lands (Butcher 2003). By the Byzantine period, in the fifth and sixth centuries AD, it is clear that settlement extension occurred both under the administration of a territorial empire as well as under conditions where more sophisticated water systems were being developed and built, specifically 'qanats' (underground irrigatin canals), wells, and cisterns (Geyer et al. 2007). Settlement expansion in central Syria was especially associated with the construction of major systems of 'qanat Romani' (Lightfoot 1996; Geyer and Rousset

2001), and their attendant monumental cisterns which not only enabled the land to be cultivated, but even enabled communities such as that at the town of Andarin (ancient Androna) to farm fish (Mango 2002).

Although subject to crises of drought, cold and plague, the climate of the fourth to sixth centuries AD was little different from that of the present day, and certainly was drier than during the optimum conditions of the sixth, fifth, and fourth millennia BC, when settlement was either rare, or limited to very few sites. Perhaps what is most telling about the cycles of expansion and contraction of settlements that occur in this region is how little they correlated with phases of wetter and drier conditions. In fact, for the southern Levant the later empires were in place during a period of less variable climatic conditions (Rosen 2007), although such stability is less evident in the Lake Van curve (Fig. 9.3).

9.4 Case Studies: Water Supply and the Spread of Irrigation Technologies

A key factor that enabled territorial expansion to take place into the marginal lands of empires was the adoption of hydraulic systems for irrigation and water supply which were harnessed to over ride any environmental limitations. These technologies, which had been developed during the Neo-Assyrian Empire and earlier, were refined and elaborated during the later empires and not only enabled dry-lands to be colonized, but also contributed to increased agricultural productivity in climatically marginal areas. Such developments also resulted in changes in the hydrological regimes of rivers, or even their diversion.

During the expansive later territorial empires, administrative control extended over large areas, either directly or via local governors or city administrations. This enabled larger and longer systems of water supply to be constructed with the result that the Middle East witnessed the rapid expansion of irrigation systems and their associated settlements over areas that, hitherto, had been only sparsely settled. This is particularly evident in the zone that separates irrigated southern Mesopotamia from the rain-fed northern regions of upper Mesopotamia. Here large areas of land were settled (Fig. 9.1): to the west of the Tigris near Samarra (Adams 1972); along the lower Khabur River (Ergenzinger et al. 1988); along the lower Balikh in Syria (Wilkinson 1998); in the Euphrates valley south of Jerablus (Carchemish) on the Turkish border (Wilkinson et al. 2007).

The withdrawal of substantial amounts of water from rivers for irrigation appears to have had significant impacts on the discharge of some of the Euphrates tributaries. This was especially apparent along the modest sized Balikh River whose discharge of around $6\,m^3/sec$ ($=15.77$ million $m^3/month$) appears to have been significantly depleted by the withdrawal of water (Kolars and Mitchell 1991; Wilkinson 1998). Textual records indicate that there was some degree of water competition between upstream and downstream communities during the second millennium BC (Dossin 1974; Villard 1987). By the late first millennium BC and early first millennium AD when canals up to 6 m wide were in operation, sections through paleochannels of the Balikh demonstrate a progressive diminution of flow through the first millennia BC and AD (Wilkinson 1998). Similar levels of abstraction probably occurred along the Khabur River in NE Syria where large canal systems were in use from probably the first millennium BC (Ergenzinger et al. 1988; Morandi Bonacassi 2000), as well as along the west bank tributary of the River Amarna near Carchemish where Roman–Byzantine conduits appear to have replaced an earlier earthen canal (Wilkinson et al. 2007).

Not only did the abstraction of river waters for irrigation result in depletion of river flow, but also the discharge of excess water from the downstream (outfall) ends of canals and irrigation systems frequently contributed to the development of marshes. This process has been well documented for southern Mesopotamia where cuneiform texts, Islamic histories and archaeological surveys indicate the formation of marshes during the Neo-Babylonian, Sasanian, and early Islamic empires. Such inundation appears to have developed, in part, as a result of the excess outflow of irrigation waters into large sumps or flood basins, the deterioration of canals, major floods or even deliberate human river diversion (Cole 1994; Adams 1981).

Such processes also had significant impacts on the local environment in NW Syria and southern Turkey where extensive marshes and shallow lakes developed during the late first millennium BC and first millennium AD (Wilkinson 1997; Casana 2004; Eger 2008b). The history of the lake of Antioch within the Amuq Valley of southern Turkey is not known in detail, but

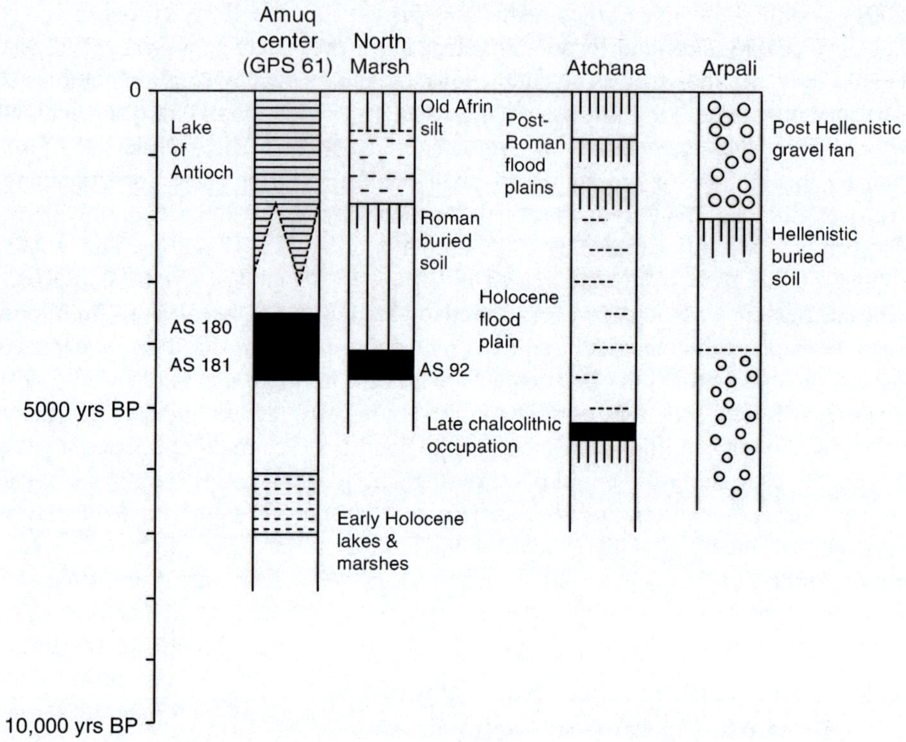

Fig. 9.8 Sections through the Amuq Plain showing lake beds, alluvial and fan deposits in relation to buried settlement horizons. (After Yener et al. 2000)

it is clearly post Bronze Age in date (Fig. 9.8). According to the Late Roman writers Libanius and Malalas, a lake was present in the early centuries AD and appears to have further expanded in early Islamic times (Eger 2008a). Ceramics on the associated sites date the construction of canals to the early Islamic period (seventh and early eighth centuries AD, Eger 2008a, b), at which time the lake and its reed-fringed marshes appear to have expanded eastwards. Two major canal systems diverted flow off the Afrin River and were used to supply settlements and presumably their irrigation systems with water in the eastern plains. Although it is likely that no single factor contributed to the expansion of the Amuq Lake and its marshes, it appears that a combination of accelerated deposition in the Amuq, flooding and perhaps avulsion of rivers, increased runoff from the surrounding uplands, and the outfall of excess irrigation water from canals contributed to the development and growth of the lakes and marshes (Wilkinson 1997; Casana 2004; Eger 2008a; Gerritsen et al. 2008). Overall, these canals not only reduced the flow of the ancient Afrin River, but also diverted some of that flow into marshes, thereby transforming a riverine plain drained by a vigorous trunk river into a plain drained by a diminished river with extensive marshes.

Elsewhere, however, other marshes shrunk during the late Roman period. For example the Berekat marsh in the western Taurus Mountains of Turkey was reduced as a result of the increased discharge of sediment from neighboring slopes caused by succession of fire events (Kaniewski et al. 2008).

Because the construction of canal off-takes introduced weaknesses into river banks and levees, it appears that in lowland Mesopotamia and southwest Iran, the entire flow of rivers could be diverted along canals. A well-documented recent avulsion has been recorded to the SW of Baghdad where during the final stages of Ottoman administration, the Euphrates started to flow along a canal, the Hindiyeh branch, leaving the Hilla branch of the river empty (Cadoux 1906; Gibson 1972). Despite the best efforts of the Turkish government to construct a barrage, this only diverted 1/3 of the flow back down the Hilla branch so that the flow down the Hindiyeh branch predominated. Then in 1903 the barrage broke leaving the Hilla branch dry so that the entire flow adopted the Hindiyeh branch. In terms of human impacts on the environment, it is significant that what is frequently regarded as a natural process, namely avulsion, in this case resulted from a major river adopting a canal (Hindiyeh branch) for its course. This then led Cadoux to suggest that the Shatt al-Gharraf in the southern plains of Mesopotamia was also originally a canal, before it became a branch of the Tigris. Cadoux also blamed the increased diversion of water into irrigation canals, as well as the presence of temporary dams for water diversion, for the increased amount of sedimentation within the channels. By providing alternative flow paths for the rivers and trapping sediment that tended to clog the flow, these two factors contributed to the tendency of the channel to shift to a more advantageous course (Cadoux 1906).

A particularly good historical example of the role of human agency in the diversion of rivers is provided by the Gargar River in Khuzestan, SW Iran, where a canal appears to have initiated a major river avulsion. According to Islamic geographers, a massive gorge was dug at Shushtar under the administration of the Sasanian King Chosroe. Later this canal adopted many of the flow characteristics of a natural river, such as sinuosity and fluvial terraces, a situation which prevails to the present day. This diversion, which is attested by both geoarchaeological investigations and early Islamic geographers (Alizadeh et al. 2004; Moghaddam and Miri 2007) not only resulted in a massive diversion of the waters of the Karun River, but also initiated a phase of landscape incision and badland development that itself exhumed a prehistoric archaeological landscape along the east bank of the Gargar River. Similarly along the Tigris River, the adoption of a more easterly course for the Tigris near Samarra, arguably represents the avulsion of much of the river's flow along a series of canals dug to capture irrigation water during the Sasanian and early Islamic empires (Adams 1965; Northedge et al. 1990).

9.5 Discussion

Because archaeology, and indeed the ancient world, is often subdivided into chronological specialties such as Assyriology, Classics, Roman, Byzantine, and so on, there has been a tendency to ignore the broad trends in settlement that prevail over longer periods of time. Instead scholars tend to compare relatively brief slices of contiguous time, over which the broad trends are hardly visible. In reality, it is over cycles such as the Early Bronze Age to Roman or Byzantine periods that truly significant changes in the landscape are evident. Despite this relative myopia, one of the key trends in Near Eastern Settlement has not escaped Roman archaeologists: "In the midst of all this uncertainty one rural trend is absolutely clear. During the period of Roman rule the settled rural population increased enormously in Syria and the Near East. Good land was densely settled, and marginal lands such as the dry steppe or stony highlands were occupied more intensively than at any time before." (Butcher 2003, p. 140).

As argued above, the phase of Roman settlement was just part of a broader trend that started with the Neo-Assyrian Empire and continued at least in places into the Abbasid Caliphate. Nevertheless it should not be assumed that empires were uniform entities with consistent administration. They were usually composed of a mosaic of provinces and cities, together with client states and other buffer states beyond (Butcher 2003) and the patterns of landholding and rural settlement within them must have been equally complex. In the Byzantine Empire, estates and other forms of extensive land holdings grew and developed in association with the church and monastic communities (Ward-Perkins 2000). On the other hand at the village level of organization, the patterns of Roman and Byzantine rural settlement in areas such as Kurban Höyük in Turkey did not develop within the core of imperial settlement but rather were administratively located within Osroehene, a small province that was not established until 194 or 195 AD (Wilkinson 1990; Butcher 2003).

If the political administration of the later empires was complex, the associated human-environment interactions were even more so, and included (at risk of oversimplification) such as extension of settlement into climatically marginal zones and uplands; associated loss of tree cover; increased erosion and associated valley floor aggradation; the spread of irrigation technologies; depletion of river discharges by diversion of water into canals; canal construction combined with high flood peaks resulted in avulsion and diversion of river channels, especially in lowland riverine plains; the inadvertent but sometimes deliberate creation of swamps and marshes as a by-product of the construction of canals; and the diminution of marshes as a result of the accumulation of eroded sediments.

In addition, landscape degradation in upland basins exacerbated flood levels downstream and the loss of forest cover may have contributed a feedback affect to local climates and atmospheric circulation. When combined, the extension of settlement into marginal lands, loss of tree cover, enhanced runoff and abstraction of water for irrigation all appear to have increased peak flood discharges and lowered the dry season flows. Together these contributed to increased variations of stream discharge and more flashy flow regimes.

In the NW Levant, the dispersal of villages and farmsteads away from pre-existing tells frequently resulted in terrain that had previously been fairly stable under less erosive regimes becoming eroded. This contributed increasing amounts of sediment to the lowlands as valley fills. Each area, however, was affected by these processes differently. Roman-Byzantine settlement along the Turkish Euphrates included the extension of hamlets and farmsteads onto higher river terraces, whereas around the Amuq Valley such extension occurred into the Amanus Mountains or the readily eroded hills south of Antioch. In NW Syria and near Carchemish, extension occurred into the limestone hills which appear to have been partially stripped of their soil cover, whereas in central Syria, the equivalent settlement was into climatically marginal areas of semi-desert. In contrast, for the earlier Neo-Assyrian expansion, most settlement infilled the voids between pre-existing lowland settlements, although around Tell Beydar this did include settlement on to low interfluves and basaltic uplands. In addition, occasional forts that were constructed on low hills represented a minor extension on to higher land, but settlement expansion into the uplands was never as pervasive as during the Roman-Byzantine periods.

Significantly, landscape degradation was not simply caused by local population pressure or for the satisfaction of local needs. Rather, erosion was caused by the demands of the greater empire, so that erosion in one area reflected the demand for economic crops and products either in distant locations or within the core of empires. This is best exemplified by the massive increase in the demand for olive oil around the greater Mediterranean that stimulated the growth of orchards and settlement in the uplands of the NW Levant (Mattingly 1996; Ward-Perkins 2000). In addition, the presence of monastic communities and centers of pilgrimage at sites such as Qalat Siman northwest of Aleppo demonstrate that expansion was not simply a result of economic growth; rather religion and economic growth were intertwined. Although many features of the spatial economy of the Roman and Byzantine empires continue to be debated, as Chew (2001) has argued for the development of World System economies, landscape degradation is often exported as a result of needs in one part of a trading or administrative system being met by production elsewhere.

Such settlement (and associated erosion) did not simply occur under the macroscopic conditions that prevailed when imperial powers were able to effectively control or administer vast areas. They were also scaled down through a hierarchical system of human action that prevailed under such empires. At the level of the individual, land holdings that had previously been held by a corporate group or community for episodic pasture, might have been regarded by the imperial powers as underused or deserted. As a result peasants farming that land and maintaining its productivity would have been granted ownership (Butcher 2003), with the result that episodic pasture or forage land would have shifted to cultivation with a significant increase of sediment yield and soil erosion. This settlement and the associated landscape degradation appear to have had their origins in a number of factors that include extension of private land holdings into what presumably have been common property, allocation of land to veterans under Roman administration, growth of intensive agriculture around growing urban areas (Casana 2003), settlement within estates, or forced displacements of populations from other regions. In addition, the spread of technologies such as iron tools, which was easier when vast areas were controlled by a single administration, may also have contributed to the extension of cultivated land use. When the new farmers were drawn from non-local populations, they would have lacked an intimate knowledge of the local soils, with the result that landscape degradation may have been exacerbated by inappropriate husbandry practices. All such processes of land allocation appear to have contributed to this phase of settlement and its associated degradation.

In the Jerablus, Amuq, Biqa, and upper Euphrates areas, estimates of aggregate settlement area suggests that there was a significant population increase during the Roman-Byzantine empires, and this combined with the expansion of settlement on to erodable uplands and the abstraction of water from the main wadis resulted in the increased accumulation of valley fills and more flashy flow conditions. Such circumstances appear to have been exacerbated by the diversion of outflow water from canals into former flood basins that subsequently became persistent marshes or lakes.

Human activities and population extension under growing territorial empires indeed provide a compelling cause for increased soil erosion and degradation, over and above the levels that would have been expected under the prevailing natural regime. Nevertheless, by focusing on cultural process, the above discussion may give the impression that the aggressive spread of human settlement and cultural landscapes was the only force driving soil erosion and accelerated aggradation. This is not the case, however, because the spread of human activity may simply have created the pre-conditions for soil erosion in the form of bare land and conditions favorable for increased runoff. Consequently, when storms or deluges occurred, the resultant degradation would have been greater than before such activity (Casana 2008; Bintliff 2002; Grove and Rackham 2001). Whether such erosion coincided with a period of extension of cultivation and increased rainfall or simply took place under normal rainfall conditions is unclear. Although long-term proxy records such as that from Saroq Cave in Israel suggest there was no obvious increase in rainfall during the phase of erosion in the Amuq, or indeed elsewhere in the Near East during the later empires, our climate sources are not sufficiently high resolution for us to be certain (Casana 2008).

In conclusion, the relationship between human action and the environment during the later empires was a tangled one. Not only did individual humans, provincial and city administrations, as well as imperial rulers, initiate significant amounts of settlement that triggered erosion and aggradation, they also had a significant impact on certain proxy indicators of environmental change by diminishing river flow or by creating marshes. In addition, the widespread adoption of technologies of water distribution in marginal or sub-optimal areas enabled dry climatic cycles to be over-ridden, or for locally verdant but geographically remote environments to be utilized for cultivation for the first time. Moreover, by extending their reach into climatically marginal lands, imperial administrations may have enabled the maximum opportunity to be taken of short-lived moist phases. In addition, the increased impact of humans on the landscape itself changed the impact of weather and climate on the land. Although such inter-relations may appear bewilderingly complex, the careful reading of historical records, geoarchaeological evidence, proxy climate records and archaeological surveys demonstrates that the individual strands of such inter-relationships can be unravelled to produce a more nuanced understanding of past landscape change.

Acknowledgements Many thanks to Claudio Vita-Finzi and John Bintliff for discussions at various times on this subject.

References

Adams R Mc (1965) Land behind Baghdad. University of Chicago Press, Chicago, 187 pp

Adams R Mc (1972) Settlement and irrigation patterns in ancient Akkad. In: Gibson McG (ed) The city and area of Kish. Field Research Projects, Miami, pp 182–208

Adams R Mc (1981) Heartland of cities. University of Chicago Press, Chicago, 362 pp

Alizadeh A, Kouchoukos N, Wilkinson, TJ, et al. (2004) Human-environment interactions on the Upper Khuzestan Plains, southwest Iran. Recent investigations. PaleOrient 30:69–88

Beach TP, Luzzadder-Beach S (2008) Geoarchaeology and aggradation around Kinet Höyük, an archaeological mound in the Eastern Mediterranean, Turkey. Geomorphology 101:416–428

Bernbeck R (1993) Steppe als Kulturlandschaft. Dietrich Reimer, Berlin, 210 pp

Bintliff J (1992) Erosion in the Mediterranean lands: a reconsideration of pattern, process and methodology. In: Bell M, Boardman J (eds) Past and present soil erosion. Archaeological and Geographical Perspectives, Oxbow Books, Oxford, pp 125–31

Bintliff J (2002) Time, process and catastrophe in the study of Mediterranean alluvial history: a review. World Archaeology 33:417–435

Brückner H (1986) Man's impact on the evolution of the physical environment in the Mediterranean region in historical times. GeoJournal 13:7–17

Butcher K (2003) Roman Syria and the Near East. British Museum Press, London, 472 pp

Butzer KW (1964) Environment and archaeology. An introduction to Pleistocene Geography. Aldine, Chicago, 524 pp

Cadoux HW (1906) Recent changes in the course of the lower Euphrates. Geographical Journal 28:266–77

Casana JJ (2003) From Alalakh to Antioch: settlement, land use and environmental change in the Amuq Valley of Southern Turkey. PhD. Dissertation, NELC, University of Chicago

Casana JJ (2004) The archaeological landscape of Late Roman Antioch. In: Sandwell I, Huskinson J (eds) Culture and society in Late Roman Antioch, Oxbow, Oxford, pp 102–25

Casana, JJ (2008) Mediterranean valleys revisited: linking soil erosion, land use and climate variability in the Northern Levant. Geomorphology 101:429–442

Casana JJ, Wilkinson TJ (2005) Settlement and landscapes in the Amuq Region. In: Yener KA (ed) The archaeology of the Amuq Plain. Oriental Institute Publications 131, Chicago, 25–65 and 203–280

Chew SC (2001) World ecological degradation: Accumulation, urbanization and deforestation 3000 BC–AD 2000. Alta Mira, Lanham, New York, 217 pp

Cole SW (1994) Marsh formation in the Borsippa region and the course of the lower Euphrates. Journal of Near Eastern Studies 53:81–109

Cordova C (2000) Geomorphological evidence of intense prehistoric soil erosion in the highlands of central Jordan. Physical Geography 21:538–567

Deckers K, Riehl S (2007) Fluvial environmental contexts for archaeological sites in the Upper Khabur basin (northeastern Syria). Quaternary Research 67:337–348

Dossin G (1974) Le site de Tuttul-sur-Balikh. Revue Assyriologique 68:25–34

Doyle MW (1986) Empires. Cornell University Press, Ithaca, 407 pp

Eger A (2008a) The spaces between the teeth. Environment, settlement and interaction on the Islamic-Byzantine frontier. PhD dissertation, NELC, University of Chicago, 567 pp

Eger A (2008b) The early Islamic period (mid 7th to mid 10th centuries). In: Gerritsen F, De Giorgi A, Eger A, et al. (eds) Settlement and landscape transformations in the Amuq Valley, Hatay. A Long-term perspective. Anatolica 34:267–274

Ergenzinger PJ, Frey W, Kühne H, Kurschner H (1988) The reconstruction of environment, irrigation and development of settlement on the Habur in North-east Syria. In: Bintliff J, Davidson DA, Grant EG (eds) Conceptual Issues in environmental archaeology, University Press, Edinburgh, pp 108–128

Gerritsen F, De Giorgi A, Eger A, et al. (2008) Settlement and landscape transformations in the Amuq Valley, Hatay. A Long-term perspective. Anatolica 34:241–314

Gerber C (1996) Die Umgebung des Lidar Höyük von Hellenistischer bis FrühIslamischer Zeit. In: Bartl K, Hauser SR (eds) Continuity and Change in Northern Mesopotamia from the Hellenistic to the Early Islamic Period. Dietrich Reimer, Berliner Beitrage zum Vorderen Orient Band 17, Berlin, pp 303–332

Geyer B, Rousset MO (2001) Les steppe arides de la Syrie du nord à l' époque Byzantine ou "la ruée vers l' est". In: Geyer B (ed), Conquête de la Steppe: et appropriation des terres sur les marges arides du Croissant fertile. Travaux de la Maison de l'Orient Méditerranéen no. 36, de Boccard, Paris, pp 111–121

Geyer B, al-Dbiyat M, Awad N, et al. (2007) The arid margins of Northern Syria: occupation of the land and modes of exploitation in the Bronze Age. In: Morandi Bonacossi D (ed) Urban and natural landscapes of an ancient Syrian capital. Settlement and environment at Tell Mishrefeh/Qatna and in central-western Syria. Studi Archeologici su Qatna 01. Forum, University of Udine, Italy, pp 269–281

Gibson M (1972) The city and area of Kish. Field Research Projects, Miami, Coconut Grove, 316 pp

Grove AT, Rackham O (2001) The nature of Mediterranean Europe. An ecological history. Yale University Press, New Haven, 384 pp

Goldberg P (1998) The changing landscape. In: Levy TE (ed) The archaeology of society in the Holy Land. Leicester University Press, Leicester UK, pp 40–57

Harper RP (1975) Excavations at Dibsi Faraj, northern Syria, 1972–1974: a preliminary note on the site and its monuments. Dumbarton Oaks Papers 29:319–338

Hole F, Kouchoukos N (in press) Preliminary report on an archaeological survey in the western Khabur basin, 1994. Annales Archéologiques Arabes Syriennes

Kaniewski D, Paulissen E, De Laet V, Waelkens M (2008) Late Holocene fire impact and post-fire regeneration from the Bereket basin, Taurus Mountains, southwest Turkey. Quaternary Research 70:228–239

Kolars JF, Mitchell WA (1991) The Euphrates River and SE Anatolia development project. Carbondale: Southern Illinois University Press, 390 pp

Larsen MT (1994) The conquest of Assyria. Excavations in an antique land 1840–1860. Routledge, London, 390 pp

Lemcke G, Sturm M (1997) ^{18}O and trace element measurements as proxy for the reconstruction of climate changes at Lake Van (Turkey): Preliminary results. In: Dalfes HN, Kukla G, Weiss H (eds) Third millennium BC climate change and Old World collapse. NATO ASI Series, Global Environmental Change 49. Springer, Berlin, pp 653–678

Lewis N (1987) Nomads and settlers in Syria and Jordan 1800–1980. Cambridge University Press, Cambridge, 268 pp

Lightfoot DR (1996) Syrian qanat Romani: history, ecology, abandonment. Journal of Arid Environments 33:321–336

Mango M (2002) Fishing in the desert. Palaeoslavica 10:323–330

Marfoe L (1979) The integrative transformation: patterns of sociopolitical organization in southern Syria. Bulletin of the American Schools of Oriental Research 234:1–42

Mattingly DJ (1996) First Fruit? The olive in the Roman world. In: Shipley G, Salmon J (eds) Human Landscapes in Classical Antiquity. Environment and Culture, Routledge, London, pp 213–253.

Millar F (1993) The Roman Near East 31–AD 337. Harvard University Press, Boston, 587 pp

Moghaddam A, Miri N (2007) Archaeological surveys in the "eastern corridor", south-western Iran. Iran 45:23–55

Morandi Bonacassi D (2000) The Syrian Jazireh in the Late Assyrian period: A view from the countryside. In: Bunnens G (ed), Essays on Syria in the Iron Age. Peeters Press, Louvain, pp 349–396

Mouterde R, Poidebard A (1945) Le Limes de Chalcis. Organisation de la steppe en Haute Syrie Romaine. Geuthner P, B.A.H. 38, Beyrouth-Paris

Northedge A, Wilkinson TJ, Falkner R (1990) Survey and excavations at Samarra. Iraq 52:121–148

Oates D (1968) Studies in the ancient history of northern Iraq. British Academy, London, 176

Oded B (1979) Mass deportations and deportees in the Neo-Assyrian Empire. Ludwig Reichert, Wiesbaden, 142 pp

Philip G, Jabour F, Beck A, (2002) Settlement and landscape development in the Homs region, Syria: research questions, preliminary results 1999-2000 and future potential. Levant 34:1–23

Radner K (2000) How did the Neo-Assyrian king perceive his land and resources. In: Jas R (ed), Rainfall and Agriculture in Northern Mesopotamia. Nederlands Historisch-Archaeologisch Instituut te Istanbul 88, Leiden, pp 233–246

Rosen AM (1986) Cities of Clay. The geoarchaeology of tells. University of Chicago Press, Chicago, 167 pp

Rosen AM (1997) The geoarchaeology of Holocene environments and land use at Kazane Höyük, S.E. Turkey. Geoarchaeology 12:395–416

Rosen AM (2007) Civilizing climate: Social responses to climate change in the ancient Near East. Alta Mira Press, Lanham, New York, 2007 pp

Tate G (1992) Les campagnes de la Syrie du Nord (Tome 1). Librairie Orientaliste Paul Geuthner, Paris, 364 pp

Tchalenko G (1953) Villages antique de la Syrie du Nord II. Vol 1. Librairie Orientaliste Paul Geuthner, Paris, 442 pp

Vermoere M, Smets E, Waelkens M, et al. (2000) Late Holocene environmental change and the record of human impact at Gravgaz near Sagalassos, Southwest Turkey. Journal of Archaeological Science 27:571–595

Villard P (1987) Un conflt d' autorités à propos des eaux du Balih. MARI 5:59–96

Vita-Finzi C (1969) The Mediterranean valleys: geological changes in historical times. Cambridge University Press, Cambridge, 140 pp

Ward-Perkins B (2000) Specialized production and exchange. In: Cameron A, Ward-Perkins B, Whitby M (eds) The Cambridge ancient history vol. XIV, Late Antiquity: empires and successors, A.D. 425–600. Cambridge University Press, Cambridge, pp 346–391

Wilkinson TJ (1990) Town and country in SE Anatolia vol.1: Settlement and land use at Kurban Hoyuk and other sites in the lower Karababa Basin. Oriental Institute Publications 109, Chicago, 315 pp

Wilkinson TJ (1995) Late-Assyrian settlement geography in Upper Mesopotamia. In: Liverani M (ed) Neo-Assyrian geography. University di Roma, Quarderni di Geografia Storica, 5, Rome, pp 139–159

Wilkinson TJ (1997) The history of the Lake of Antioch: A Preliminary note. In: Young G, Chavalas M, Averbeck R (eds), Crossing boundaries and linking horizons: Studies in honor of Michael C. Astour on His 80th Birthday. CDL Press, Bethesda, Maryland, pp 557–576

Wilkinson TJ (1998) Water and human settlement in the Balikh Valley, Syria: Investigations from 1992–1995. Journal of Field Archaeology 25:63–87

Wilkinson TJ (1999) Holocene valley fills of southern Turkey and NW Syria. Recent geoarchaeological contributions. Quaternary Science Reviews 18:555–572

Wilkinson TJ (2003) Archaeological landscapes of the Near East. University of Arizona Press, Tucson, Arizona, 260 pp

Wilkinson TJ, Barbanes E (2000) Settlement patterns in the Syrian Jazirah during the Iron Age. In: Guy Bunnens (ed) Essays on Syria in the Iron Age. Peeters Press, Louvain, pp 397–422

Wilkinson TJ, Wilkinson EB, Ur J, Altaweel M (2005) Landscape and settlement in the Neo-Assyrian Empire. Bulletin of the American Schools of Oriental Research 340:23–56

Wilkinson TJ, Peltenburg E, McCarthy A, et al. (2007) Archaeology in the land of Carchemish: landscape surveys in the area of Jerablus Tahtani, 2006. Levant 39:213–247

Yener KA, Edens C, Harrison TP, et al. (2000) The Amuq Valley Regional Project 1995–1998. American Journal of Archaeology 104:163–220

Part III
The Mediterranean and European World–Warm-Temperate Mediterranean Lands

Chapter 10
The Interplay between Environment and People from Neolithic to Classical Times in Greece and Albania

Eric Fouache and Kosmas Pavlopoulos

10.1 Introduction

The objective of this chapter is to examine environmental and societal changes in the southern part of the Balkan Peninsula. By virtue of its position facing Asia Minor (Anatolia), the southern part of the Balkans, comprised of Greece (here defined to include Attica, Boetia and the Peloponnesus) and Albania, played an important role in the transmission of agriculture practices and civilization to the European continent, a process commonly labelled Neolithization (Van Andel et al. 1980; Perlès 2001; Touchais and Renard 2002). The Near East contributions are undeniable, but it also appears that indigenous civilizations began very quickly to differentiate themselves. This led to a dense agricultural development of the region at the start of the Bronze Age, occasionally with an impressive management of hydraulic constraints, culminating notably in the management of 'poljes' (areas of coalescent karst sinkhole or dolines) during the Mycenaean period in Boetia, Thessaly, as well as the Peloponnesus (Knauss 1991). The agricultural cultivation and the agro-pastoral activities that followed, combined with the role of climatic fluctuations, resulted in the intense clearing of land that at times led to significant erosion. The agricultural development was eventually accompanied by the development of cities. The Minoan period, limited to Crete, and the following Mycenaean period that extended to the whole of ancient Greece provide the most impressive examples. This substratum largely prepared for the advent of Greek cities, the 'Athenian miracle' and the expansion of Hellenism across the whole of the Mediterranean Basin and the Black Sea.

E. Fouache (✉)
Département de Géographie
Université Paris Ouest Nanterre la Défense
Paris, France
e-mail: eric.g.fouache@wanadoo.fr

38 rue Louis Ricard, 76000 Rouen, France

10.2 Natural Development of the Postglacial Environment

Geological changes just before human beings entered the landscape as farmers was immense, and it is necessary to summarize what went on because this was the landscape that humanity inherited.

The first Neolithic sites appeared in Greece around 9000 years BP. At that time, the hunting territories and trade networks were much more extensive than today due to the fact that postglacial sea-level rise, which began after 15,000 years BP was still far from complete (Fig. 10.1). Neolithic people coming from the Near East settled in dry Mediterranean environments like Beotia and Thessaly (Perlès 2001). The limited number of Neolithic sites, and their fragmentary distribution, identified in western Greece and around the Gulf of Corinth for example could also be explained by the subsequent submersion, sometime to the point of drowning, of coastal zones. This is attested in the Gulf of Corinth where a paleolake was very quickly submerged starting in 13,000 BP (80 m below psl (psl = relative present sea level)) (Moretti et al. 2004). The same phenomenon is likely to have occurred between Corfu and the continent. At about 6000 years BP the postglacial sea had risen to about 6 m below psl (Lambeck and Purcell 2005). Deltaic progradation,

See Plate 9a in the Color Plate Section; also available at: extras.springer.com

I. P. Martini, W. Chesworth (eds.), *Landscapes and Societies*,
DOI 10.1007/978-90-481-9413-1_10, © Springer Science+Business Media B.V. 2010

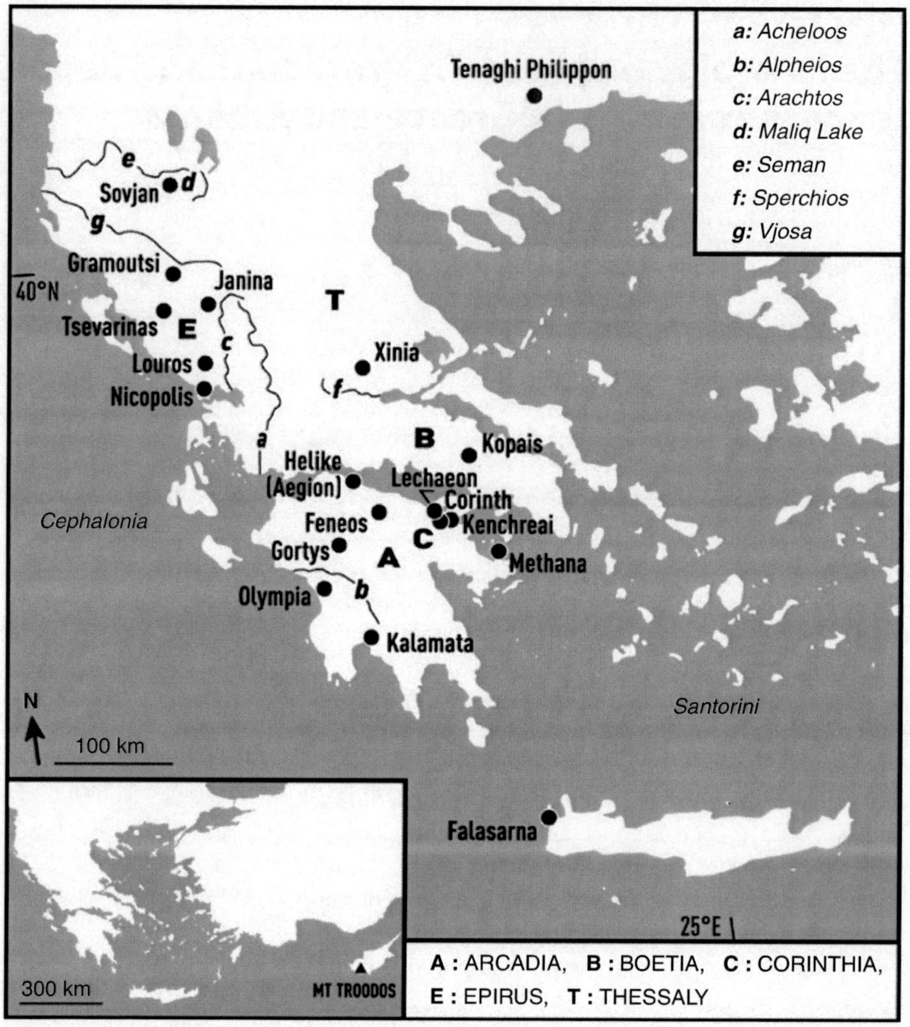

Fig. 10.1 Location map

alleviation, and formation of coastal strips favored by a slow sea-level rise prevailed after this time (Fouache and Pavlopoulos 2005).

Globally, at the end of the last glacial period 18,000–20,000 years BP the sea level was almost 120 m lower than it is today. By the end of the period and the retreat of the glaciers, the sea level was initially rising at the fast rate of 1 m/year until almost 6000 years BP, after which it reduced to almost 2 mm/year (Fouache 2006).

For the Mediterranean Sea, Lambeck and Purcell (2005) have stated that an eustatic component of sea-level rise since the Roman Period along the Tyrrhenian Sea coast of Italy central Mediterranean, has been 1.25 ± 0.25 m and that the onset of modern relative sea-level rise (0.13 ± 0.09 m) occurred approximately from the late nineteenth century or early twentieth century.

Local rapid tectonic movements have occurred as well, for example, the up rise or the collapse by about one meter along submerged fault scarps in Perachora (Pirazzoli et al. 1982). Thus, the two harbors of ancient Corinth find themselves in very different situations (Fouache and Pavlopoulos 2005). The ancient harbor of Lechaeon, the western harbor of ancient Corinth, was raised above sea level between 500 and 200 BC. By contrast the port of Kenchreai located on the Saronic Gulf was instead submerged after a series of earthquakes (including those of 77 BC, 365 BC, and during the sixth century AD). In all the known localities, such tectonic displacements progressively damaged the structures while the harbors were still active and disrupted them further after the sites were abandoned.

The postglacial transgression may have given birth to a number of myths. This is particularly true when it affected lake basins that became connected with the Mediterranean Sea, and where the inundations were rapid enough to adversely affect human societies. Thus, according to Ryan and Pitman (1998), the filling of the Black Sea around 5600 years BC could be the source of the myth of Noah's flood.

10.2.1 Holocene Climatic Variability and the Evolution of the Vegetation Cover

A major consequence of the postglacial warming was the evolution of the vegetation cover of the landscape. Pollen analysis best allows follow the evolution of vegetation cover, a thereby understand the effects of global warming, of short-term climatic pulsations, and of anthropic changes on environments. In Greece two types of Holocene sediment have been examined: lacustrine or peat sediment (Willis 1994; Denèfle et al. 2000) and marine sediment, in particular sapropels, showing particularly strong accumulations from 9000 BP to 6000 years BP (Rossignol-Strick 1995).

Pollen analysis indicates that during the last cold period the climate was arid, cold in winter, briefly hot in summer (Camerlenghi and McCoy 1989; Cheddadi et al. 1991; Petit-Maire 1999). These climatic conditions permitted the development of a cold steppe vegetation, *Artemisia* (sage brush), in the lower-lying zones of the western Mediterranean areas as well as in the Balkans (Tzedakis 1993). Continental pollen diagrams also identify the vegetation of the mid-altitude zones for the southern Balkan Peninsula.

The sites in question are Gramousti (285 m asl; Willis 1992a), Tsevarinas (450 m asl; Turner and Sanchez-Goni 1997), Janina (500 m asl; Bottema 1974) and Xinias (450 m; Bottema 1979). Just one site is located at 1800 m altitude in Rezina (Willis 1992b, 1994). With the exception of two typically Mediterranean taxa, *Olea* and *Pistacia*, all the other elements making up the mixed forest, which developed during the early Holocene epoch, were already present during the last Pleistocene cold period. Everything indicates that the region played an important role as a refuge for vegetation, in particular the inter-mountainous basins situated at mid-level altitudes (between 500 and 800 m). The Korçë Basin, where Lake Maliq is located, has been identified as one of these vegetation refuges (Denèfle et al. 2000). The presence of these species and local edaphic conditions led palynologists to consider that at altitudes between 500 and 800 m the summer was cool and dry, the winter cool and damp, and in summer the climatic environment allowed for the maintaining of a forest ecosystem, with no consensus as to its true extent. Cold steppe vegetation (*Artemisia*) prevailed below these mid-level altitudes.

10.2.2 Postglacial Evolution of the Vegetation Cover

Deglaciation is characterized by the decline in the cold steppe sage-brush. Globally, from 18,000 to 11,000 years BP aridity increased (Cheddadi et al. 1991), interrupted briefly by a more humid period as indicated by a relative increase in *Quercus* pollen, which correlates with the so-called Bölling–Alleröd period (13,000–11,000 years BP). Maximum aridity was attained during the Younger Dryas (11,000–10,000 years BP). According to Rossignol-Strick (1995), in the southeastern Balkans and the western Mediterranean the Holocene climate change took place around 9800 years BP.

The global image revealed by data obtained from on the core from Tenaghi-Philippon, shows that since 8000 years BP tree pollen gains constitute 90% of the pollen spectrum. In other words the region was covered by forest (Wijmstra 1969; Greig and Turner 1974; Lespez 1999). The plain and low hills of the area were occupied by a mixed oak grove that included, in addition to deciduous oaks, lime or linden trees (genus *Tilia*) and elms, ash and hazels in the clearings. The mountainsides were covered by more sparse oak groves that included hornbeams (genus *Carpinus*). Around 5000 years BP a forest of firs developed on the upper portions of the mountainsides.

In brief, the first human settlers inherited a landscape that had been profoundly changed over millions of years by the deep-seated processes of plate tectonics—still manifested by earthquakes and volcanic activity in the region – by the steady workings of the erosional cycle at the surface, and by the cover of vegetation that evolved in concert with climatic changes at the end of the last glaciation. All of these changes formed the landscape that dominated the lives of its human inhabitants from the beginning of the Holocene, until agriculture began to turn *Homo sapiens* into a geomorphologic force.

10.3 The Coming of Agriculture and its Impact

Agriculture entered the peninsula from Anatolia, about 9000 years ago. The existing Mesolithic population, living as hunters and gatherers, had only a small ecological footprint compared to the Neolithic farmers who were to follow them. The latter first settled in lowland, riverine sites, with humid, easily worked soils. Later, possibly as a result of population pressure, they expanded onto the hills. By about 4500 years ago agriculture had expanded throughout the peninsula.

10.3.1 Land Clearing, an Agricultural Practice Especially Visible after 4500 Years BP

After 4500 years BP, during the Bronze Age, at Tenaghi-Philippon and Lake Maliq in the Korçë Basin, there was a continuous growth of *Fagus* at the expense of *Abies*, while at the same time cultivated species such as grasses or *Juglans* developed as well as typically Mediterranean species like *Quercus ilex* and *Olea*. This is the result of the increasing influence of agriculture on the plains of Drama, as well as the influence of a slight lowering of winter temperatures and more marked summer dryness (Huntley and Prentice 1989). The gradual passage towards dryer conditions, in particular in summer, is recorded after 5000 years BP in the north of the Mediterranean basin (Harrison and Digerfeld 1993). It is at this time that a climate closer to the current one would have developed.

The fact that the impact of the anthropogenic land clearings is not evident in the pollen diagrams in any marked way before 4500 years BP indicates that the Neolithic agricultural diffusion and the commencement of land clearing were not strictly contemporaneous. This has led certain authors (Willis and Bennett 1994) to suggest that Neolithization initially took place in the east of the Pindus region in limited naturally-open spaces such as forest clearings or lake borders. Two good examples are the Sovjan site and Lake Maliq in the Korçë Basin (Fouache et al. 2001; Fouache 2006). It is only during the Bronze Age that significant land clearings occurred. Between 2500 BC and 1900 BC (end of the Early Bronze Age and beginning of the Middle Bronze Age) vegetation changes were still slight.

The growth of garrigue (or maquis) scrubland made-up of evergreen species of Mediterranean affinity probably marks the first significant anthropic transformations of the vegetation cover in an environment of dry summers. The first clear signs of transformation of vegetation by humans occur between 1900 BC and 1300 BC. The development of olive trees is attested in the region. After 1300 BC and up to the end of Roman times, periods of forest retreat alternated with those of re-advance: the still-elevated percentage of arborescent (inter-forest ground cover) species indicates an environment still largely forest-like but with secondary formations becoming increasingly important.

Today the maquis shrubland is undergoing a process of re-establishment in the southern Balkans, except for Albania. This is due to the massive rural depopulation and quasi-disappearance of extensive pastureland. The Albanian exception is explained by high rural densities and persistent auto-subsistence practices such as pasturing of small family herds and cutting of trees for firewood.

10.3.2 Holocene Climatic Variability

Paleoclimatic reconstitutions for the past 15,000 years (Fig. 10.2) based on palynological data from Lake Maliq in the Korçë Basin (Bordon et al. 2009) show some, but not extreme, variations in Holocene climate, the most important being a distinct cooling accompanied by dryness between 8300 and 8100 cal years BP. This event is coeval with the so-called '8.2 ka event', widely recorded in ice cores and in both marine and terrestrial archives in the Northern Hemisphere (Mayewski et al. 2004; Alley and Agustsdottir 2005). Otherwise, the evidence from Lake Maliq indicates the Holocene has been relatively stable and that the major features of the current Mediterranean climate was established after 5000 years BP. These reconstructions suggest cold and dry conditions especially in wintertime (a prominent characteristic of the Mediterranean biome) are consistent with previous observations in the Northern Hemisphere (Alley and Agustsdottir 2005). Magny et al. (2003) show that drier conditions occurred in both northern and southern Europe in response to the 8.2 ka cooling. By contrast the climate of central Europe was

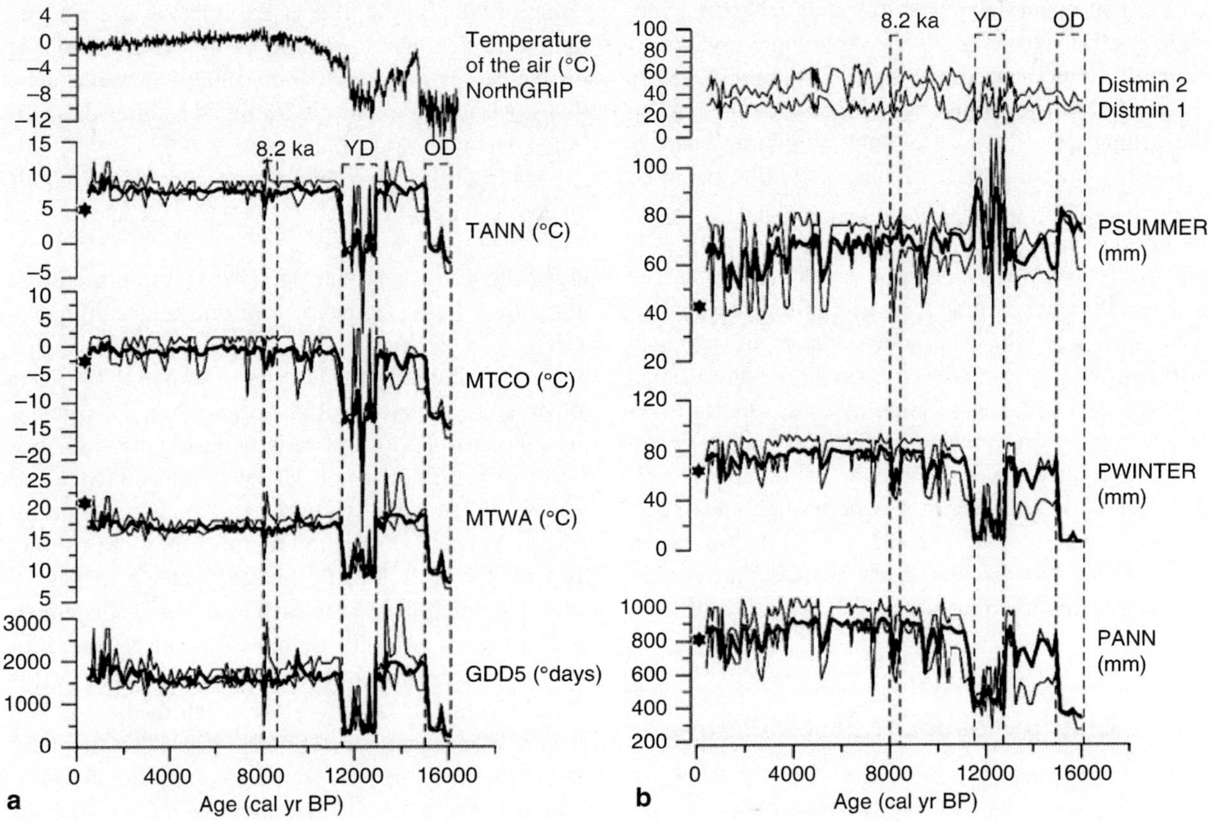

Fig. 10.2 **a** Pollen-inferred quantitative climate reconstruction at Lake Maliq pollen using MAT with biome constraint of temperature parameters: mean annual temperature (*TANN*), mean temperature of the coldest month (*MTCO*) and the warmest month (*MTWA*), annual growing degree days above 5°C (*GDD5*). Mean values are plotted (*line*) together with the error bars which are plotted in *gray*. The major Holocene and Lateglacial events (8.2 ka, Younger Dryas, Oldest Dryas) are indicated. The modern climate values are indicated with a *black star*. Values of air temperature estimated at NorthGRIP from delta ^{18}O measurements of Johnsen et al. (2001) are given at the *top* of the figure. **b** Pollen-inferred quantitative climate reconstruction at Lake Maliq pollen using MAT with biome constraint of hydrological parameters: summer precipitations (*Psummer*) calculated as the mean of the precipitations during June–September; winter precipitations (*Pwinter*) calculated as the mean of the precipitations during October–May, annual precipitation (*PANN*). The Euclidian distances calculated between the eight modern pollen assemblages considered as the best analogues and the fossil assemblage (nearest and furthest) are indicated at the top of the figure as *Distmin 1* and *2*. (After Bordon et al. 2009)

wetter. Reconstruction of the precipitation history at Lake Maliq supports this result. The cooling recorded at Lake Maliq (−2°C for TANN) is similar to the cooling of −2°C in central Europe from the Ammersee ^{18}O record (von Grafenstein et al. 1999) and a simulated cooling of between −1° and 0°C in the Balkans, from ECBilt-CLIO model results (Wiersma and Renssen 2006). Climatic variations that have occurred in time and space seem primarily related to the repartition of precipitation between summer and winter. This factor is of extreme importance for agriculture and perhaps the principal cause of the large variations observed in lake levels (in Lake Maliq for example) at different periods.

The best way to remain protected from exceptional climatic phenomena was to be located near a permanent lake area. This explains why not all environments played roles of equal importance in regards to the appearance of new dynamics in the landscape, as with the case of the development of agriculture. Intermontane basins in the south of the Balkan peninsula, which have lake areas and swamplands as is the case of the poljes of Thessaly (Kopais Lake notably) and Boetia, played a considerable role in sheltering and promoting re-established forest growth during the first half of the Holocene as well as the process of Neolithization as a whole.

The consequences of Neolithization have been very significant in terms of geomorphologic dynamics. Wood clearance and overgrazing very rapidly led to the disappearance of the original forests dating back to the Climatic Optimum. It is unlikely that any natural landscape, uninfluenced by human activities, persisted in the Peloponnesus after the Bronze Age. At the same time, alluvial deposits in valley bottoms and deltas tell us that soil weathering and erosion never ceased to intensify, with complete erosion of forest paleosols from limestone slopes. Large volumes of sediment also reached the sea between Bronze Age and Roman times. This led to the siltation of some coastal areas and harbors, such as in the Thessaloniki coastal plain in north-central Greece (Fouache et al. 2008). Evidence of all these events is present throughout Greece. Harbors were silted up and enclosed in the middle of the plain far away from the sea, such as the ports of Pela and Oeniades in the delta of the Acheloos River.

10.4 Changes over the Last Millennium

In addition to the relatively uniformitarian geomorphologic dynamics of the Holocene, there are a number of catastrophic elements of a marginal but spectacular nature that need to be taken into account. These include volcanic eruptions, tsunamis resulting from undersea earthquakes, and particularly the direct and indirect effects of earthquakes.

10.4.1 Catastrophic Events

Impressive though they may be, we must guard against privileging catastrophic events to explain major historical changes (Helly 1987). A case in point is the current reinterpretation of the hypothesis of Marinatos (1939) that the demise of the Minoan civilization resulted from the catastrophic caldera-type volcanic eruption of Santorini in the Aegean. The chronologies established by archaeological evidence and by radiometric dating, support a different hypothesis. The destruction of Minoan palaces is placed around 1450 BC whereas the most recent dating of the Santorini explosion ranges between 1635 BC and 1625 BC (Michael 1980; Aitken 1988; Baillie 1989). Earthquakes are also no longer considered responsible for the abandonment of the Minoan palaces (Poursoulis 1999; Poursoulis et al. 2000), which were instead destroyed by fires or abandoned during periods of troubles.

Nevertheless, the entire region considered in this chapter is situated within a tectonically active area associated with the subduction of the African Plate under the Eurasian Plate (Fig. 10.3). It is therefore prone to a high degree of volcanic and earthquake activity.

The distribution of volcanoes is limited to the inner volcanic arc situated in the Aegean Sea. Two volcanoes have had periods of activity during the Holocene and historical periods, the Methana Volcano (Mee and Forbes 1997) and the aforesaid, more famous Santorini (Thera) Volcano. Volcanic eruptions may have direct disastrous effects locally, but they can also trigger other damaging processes such as tsunamis generated by collapse of volcanic structures such (the Santorini caldera for example) or by undersea slumps or volcanic related undersea earthquakes such as the 365 AD tsunami (Papazachos and Dimitriu 1991). However the low frequency of local volcanic eruptions is too low to permanently modify the geomorphologic dynamics of a region.

Seismic activity is instead widespread in the region, and is both frequent and enduring and can have a significant influence on the landscape. Seismic events act

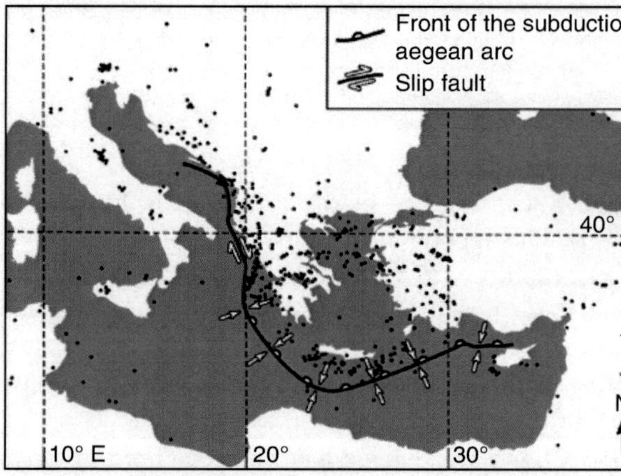

Fig. 10.3 Distribution of major earthquakes (above 5 on the Richter scale) between 1964 and 1977. (Information from Center National de la Recherche Scientifique, Institut de Physique du Globe de Paris)

on geomorphologic dynamics in two ways, the instantaneous manifestations of an earthquake and deferred effects. The latter often lead to a substitution of rapid processes for slow ones particularly on mountainsides (Dufaure 1984).

The zones closest to the epicenters are the most affected by the seismic events. These are located along fault lines and, depending on the type of substrate, may generate a variety of landscape modifications. For example, fissures are open, fault-scarps formed, rock and sediment slumps are generated in steeper slopes, and mudflows can also develop in relatively flat areas. Locally this greatly enhances erosion from mountain sides leading to siltation of streams and alluvial valleys. Furthermore spectacular variations in the landscape have occurred as well, examples being the deflection of water courses as in the case for the Sperchios River in 426 BC, or the sudden sinking of the city of ancient Helike in 373 BC near the Gulf of Corinth (Dufaure 1976a). On the whole, however, the effects of punctuated catastrophic phenomena associated with volcanism and earthquakes remain relatively limited compared with those induced by climatic effects and human activities. Nevertheless, the combined effect of all these mechanisms led to much erosion and numerous landslides throughout Greece (Fig. 10.4).

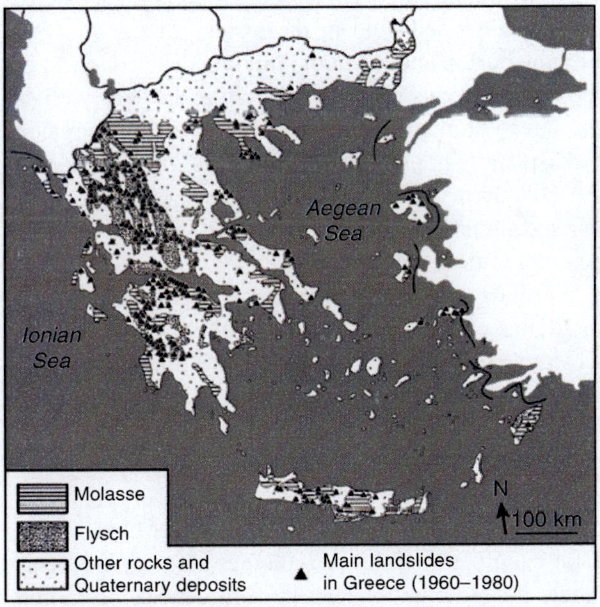

Fig. 10.4 Main landslides in Greece between 1960 and 1980. (After Flageollet 1989)

10.4.2 Climatic Changes

Climate affects erosion through changes in temperature and precipitation. Climatic conditions favored post-Pliocene reforestation after 8000 years BP. The slopes became protected against wholesale erosion and predominantly fine sediments were removed from the slopes and transported seaward. Only after 5000 years BP did summer dryness become clearly evident, and this favored natural fires that led to temporary openings of the forest and to local intense erosion during the autumnal precipitations. The loss of protective vegetation from the slope continued and increased significantly on account of human activities, starting in the Bronze Age. Deforestation resulted in a reduced strength of the slope sediments and increased the number of landslides, less retention of water by the soils with increased mass flows and overland flow erosion, and greater production of coarse-grained sediments. Increased seasonal runoff also led to greater but variable stream flows and the alternate development of braided and meandering channels in the same river systems. The increased erosional power of streams is well illustrated by the terraces of the Gortynios River, a tributary of the Alpheios River, in Arcadia (Dufaure 1975; Bousquet et al. 1983; Fouache 1999). There the river has entrenched into a bedrock channel cutting through the uppermost Pleistocene alluvial and colluvial deposits and an early Holocene alluvial terrace (Fig. 10.5).

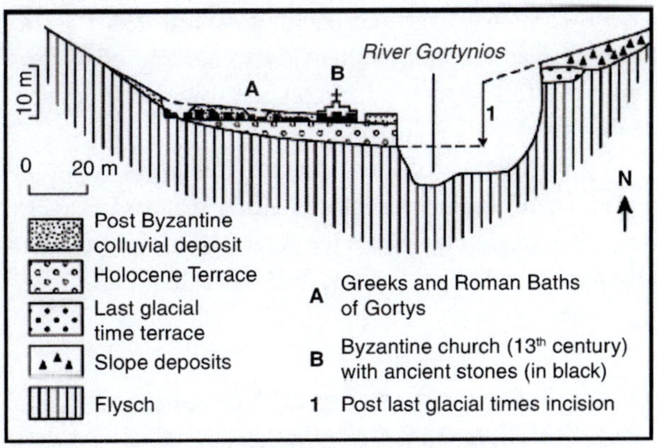

Fig. 10.5 Holocene erosion and sedimentation on the site of the Asclepieiion (healing temple) of Gortys (Arcadia). (After Fouache 1999)

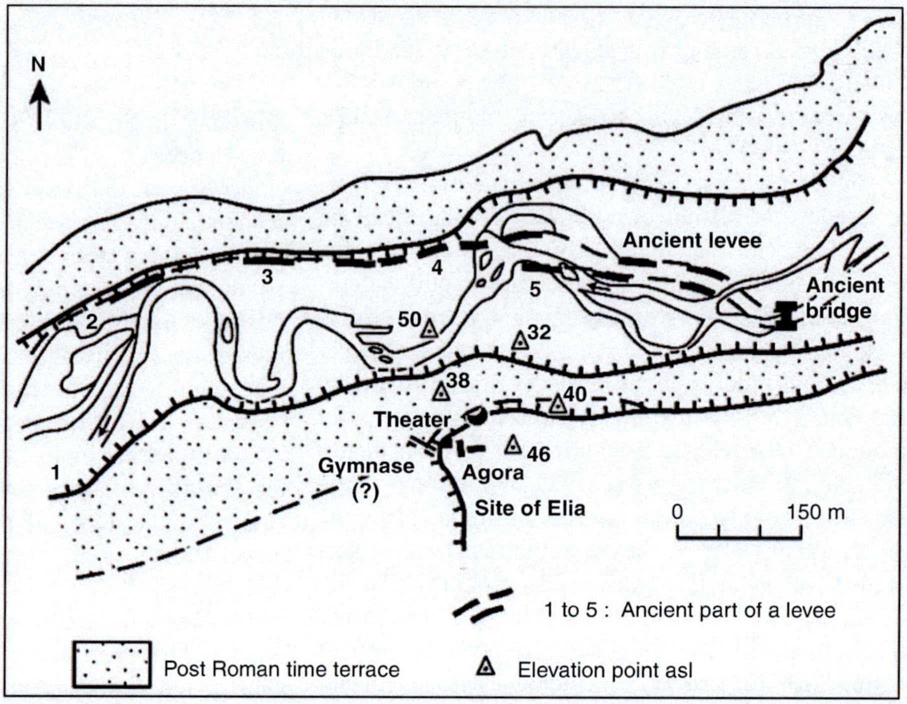

Fig. 10.6 Changes in streambed pattern of the Pinios River near the site of Elia between antiquity and nineteenth century. (After Fouache 1999)

The Pineios (Fig. 10.6) and Alpheios rivers (Fig. 10.7) in the Elia prefecture, record a transition from a meandering to a braided course during the Holocene. This has been reported from other rivers across Greece and adjacent Albania (Fouache 1999). Conversely, nowadays in these two rivers and everywhere else in the piedmont zone, active riverbeds are located in the middle of ancient braided systems. We interpret the formation of these braided river systems to have primarily occurred, as in other places such as in the French alpine region, during the Little Ice Age (1550–1850 AD) (Grove 1988; Bravard 1989). The braiding was due not exclusively to a surge of exceptional rainfall events, but rather to a higher frequency of flooding.

A consequence of these dynamic changes was that a number of ancient structures along waterways were destroyed during the Little Ice Age. The hippodrome at Olympia (see below) was destroyed at this time.

10.4.3 Consequences of Socioeconomic and Historic Change

In the southern Balkans, most of the populated sites were abandoned after of the collapse/downfall of the western Roman Empire. Byzantine society reorganized by retreating to major urban centers, and Slavic populations settled down in the area between the sixth and the ninth centuries AD. The agro-pastoral practices of the Slavic people in regions with a lithology that is sensitive to weathering such as the Olympia area, caused erosion unprecedented in the history of the site that became buried under 6 m of alluvium. The conquest of the Peloponnesus by Villehardouin in 1204 as well as the later Ottoman conquest further increased erosion by displacing populations that were reluctant to accept the conqueror's authority. Orthodox monasteries, for instance, retreated further inland into refuge areas like Arcadia in the Peloponnesus.

Similarly there has been strong progradation of the deltas of the Acheloos and Arachtos rivers on the western coast of Greece during the nineteenth century (Fouache 1999). The same thing has occurred on deltas of the Albanian coast, in particular in the conjoined delta of the Seman and Vjosa rivers. There are of course complications arising from climatic changes, especially during the Little Ice Age, but a clear example of anthropic changes is found in Epirus where the systematic clearing of forest was promoted by Ali Pacha of Janina to supply timber for shipbuilding to French and English arsenals of Toulon and Malta, as well as for providing wood and charcoal for

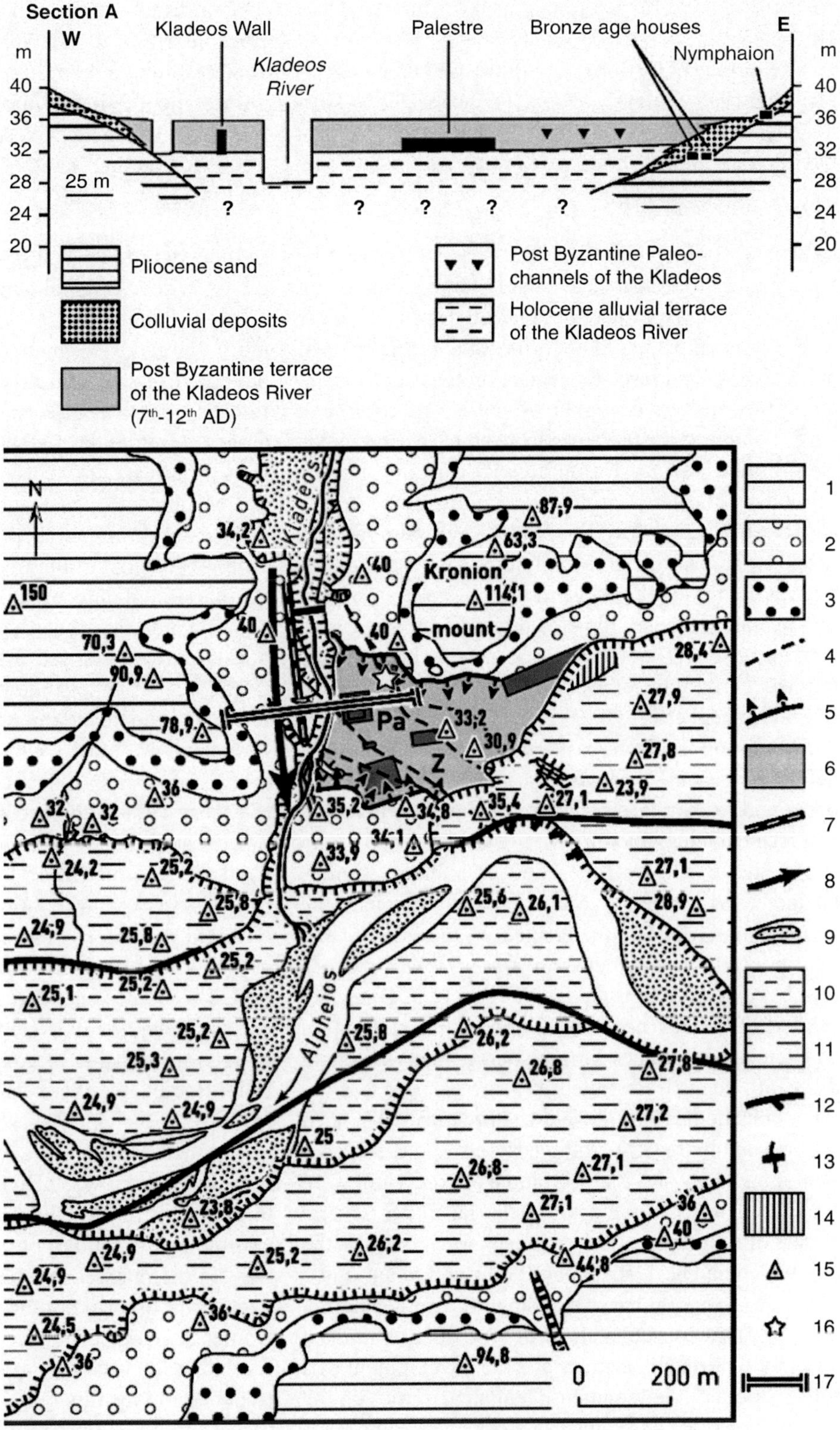

Fig. 10.7 Geomorphic map showing in fluvial and sediment patterns at Olympia, Greece (After Fouache 1999) (*1* Pliocene hill; *2* Olympia terrace; *3* Post Byzantine colluviations; *4* Post Byzantine paleochannels of the Kladeos River; *5* Limit of archaeological excavations; *6* Archaeological excavation (Olympie); *7* Kladeos wall (artificial levee); *8* Kladeos River during Greek and Roman times; *9* Channels of the Alpheios River; *10* Present times floodplain; *11* Floodplain during the Little Ice Age; *12* Present times levee; *13* Bridge on the Kladeos River; *14* Area artificially filled with sediments; *15* Elevation point in m asl; *16* Ancient bridge (second century BC); *17* Section A, ***PA*** Palestre, ***Z*** Temple of Zeus)

the cities that were developing at this time (McNeill 1992).

The terrace of Olympia is another interesting example (Fig. 10.7; Dufaure 1976b; Fouache 1999). During the Greco-Roman antiquity, games were held at Olympia and many monuments were built at the site. Olympia is located on a mid Holocene alluvial terrace at the confluence of the Alpheios River and a small affluent from the right bank, the Kladeos River. The left bank of the Kladeos River had been equipped with a small levee for protection against floods, however levees were rare and did not protect the central part of the Altis, which housed the temple of Zeus at Olympia, while the site was occupied. However, after the area was abandoned in the seventh century AD, remains were buried under a loamy alluvial 6 m high terrace formed by the Kladeos River throughout the eighth and twelfth centuries (Fouache 1999). Similar terraces developed along several other tributaries of the middle course of the Alpheios River, an area underlain by easily eroded Plio-Calabrian marine sands and conglomerates. The siltation and development of the terraces chronologically corresponds with the arrival of Slavic shepherds in the region (Fouache 1999). It is thus tempting to blame pastoral practices, such as the deliberate burning of vegetation for regenerative purposes and for land clearance, for the increased erosion and local sedimentation.

Another example of the effect of human activities on increased erosion and siltation is the ancient site of Asclepieion at Gortys, erected on the bank of the Gortynios River, a tributary of the upstream reaches of the Alpheios River in Arcadia (Fig. 10.5; Bousquet et al. 1983; Fouache 1999). Agricultural cultivation and extensive livestock farming during the twelfth and thirteenth centuries resulted in some clearings of the mountainside near the ancient site of Gortys. The area is underlain by readily erodible flysch, and the formation of badlands and a large volume of colluvium were a consequence. This led to the burial under several meters of sediments of the antique thermal baths of the city which had been abandoned at the end of antiquity, as well as to partial burial of a small Byzantine church (Fig. 10.5). The establishment of an Orthodox monastery in the thirteenth century and the development of a village around it led to intense human activities on the countryside that caused intense erosion and the formation of badlands.

Conversely, there are many examples of anthropogenic work that have been beneficial to human societies while being minimally disruptive to the environment. This is the case of hydraulic control practiced during the Mycenaean period, around 1600–1200 BC, predominantly in the poljes of Arcadia and Boetia (Knauss 1991). In these karstic dissolution mega-forms there were thick accumulations of potentially very fertile clay. For the Mycenaean small settlements to develop on the hums, limestone pinnacles in a naturally defensive location safe from flooding, the agricultural development of the plains or poljes would have had to overcome three inter-related difficulties in order to succeed. To start, it would have been necessary to maintain the 'ponors' (karstic orifices), as is still the case this day, in order to ensure they did not become blocked or obstructed and thereby form a lake of significant depth. This happened with the polje of Feneos, in the north of Peloponnesus, where the careful maintenance of the underground drainage channels had been assured during the Ottoman period but soon lapsed following disturbances of the Great War of Independence of 1821 and 1829. In a few years the abandonment led to the formation of a lake some 40 m deep, which eventually drained when the system finally managed to unblock itself in 1892 (Dufaure 1975). The second problem requiring solution was that of winter flooding provoked by water from the saturated deep karst systems, which would rise backing up through the same underground channels. On the other hand, in summer it was necessary to find a way to gain access to the water for irrigation. The Mycenaean civilization was able to develop an unconventional system of maintenance of the karstic orifices and of control of the network of surficial floodwater channels by constructing a series of low walls. In winter these functioned as levees and in summer as dams. The water was thus available for irrigation, and it was also trained into artificial canals to power mills. All these adjustments were done by exploiting the topography in the most advantageous way, in the manner in which the Romans would later excel with their hydraulic systems of aqueducts. The Roman aqueduct of Nicopolis, located in Epirus, for example was constructed primarily by digging trenches and locally by building tunnels in such a way that it did not significantly destabilize the mountainsides (Doukellis et al. 1995).

10.5 Conclusion

The hunters and gatherers of Mesolithic societies had little impact on the vegetation of the southern Balkans. Conversely, from 9000 BP and in less than 5000 years, the Neolithization led to massive wood clearance which generated soil erosion that continued to increase until our era. Numerous bays silted up and deltas prograded rapidly terrestrializing numerous harbors. In the meantime, valley bottoms experienced intense alluviation. From the Bronze Age, towards 4500 years BP, the erosion dynamics were mainly directed by a combination of climatic, political, and socio-economic factors that account for the fact that each main basin has its own history. In terms of soil erosion, rural exodus after World War II was a positive factor in the sense that it encouraged the renewal of the vegetal cover. But today, the latter is not maintained and is consequently threatened by fires. The disastrous fire that broke out in Peloponnesus in the summer of 2007 may be a warning of things to come.

Acknowledgements We wish to thank the French School of Athens and the French Ministry of Foreign Affairs for the support to the MALRHO program directed by Eric Fouache.

References

Aitken MJ (1988) The Minoan eruption of Thera, Santorini: A re–assessment of the radiocarbon dates. In: Jones RE, Catling HW (eds) New aspects of archaeological science in Greece. British School at Athens, Occasional Paper 3 of the Fitch Laboratory, Athens, pp 19–24

Alley RB, Agustsdottir AM (2005) The 8k event: cause and consequences of a major Holocene abrupt climate change. Quaternary Science Reviews 24:1123–1149

Bordon A, Peyron O, Lézine AM, et al. (2009) Pollen inferred Late-Glacial and Holocene Climate in southerns Balkans (Lake Maliq). Quaternary International, 200:19–30

Bottema S (1974) Late Quaternary vegetation history of northwestern Greece. Thesis, State University, Groningen, 190 pp

Bottema S (1979) Pollen analytical investigations in Thessaly (Greece). Paleohistoria 21:19–40

Bousquet B, Dufaure JJ, Péchoux PY (1983) Temps historiques et évolution des paysages égéens. Méditerranée 2:3–25

Bravard JP (1989) La métamorphose des rivières des Alpes françaises à la fin du Moyen Age et à l'époque moderne. Bulletin de la Société Géographique de Liège 25:145–157

Camerlenghi A, McCoy FW (1989) Physiography and structure of Bacino Bannock (eastern Mediterranean). Geo-Marine Letters 10:23–30

Cheddadi R, Rossignol–Strick M, Fontugne M (1991) Eastern Mediterranean paleoclimates from 26 to 5ka BP documented by pollen and isotopic analysis of a core in the anoxic Bannock basin. Marine Geology 100:53–66

Denèfle M, Lézine AM, Fouache E, Dufaure, JJ (2000) A 12000 years pollen record of Lake Maliq (Albania). Quaternary Research 54:423–432

Doukellis P, Dufaure JJ, Fouache E (1995) Le contexte géomorphologique et historique de l'aqueduc de Nicopolis. Bulletin de Correspondance Hellénique 119:209–233

Dufaure JJ (1975) Le relief du Péloponnèse Thèse d'Etat, Paris, Université de Paris-Sorbonne, 1422 pp

Dufaure JJ (1976a) Contraintes naturelles et historiques dans la mise en valeur des plaines grecques. Cahiers de Géographie de Rouen 6:5–27

Dufaure JJ (1976b) La terrasse holocène d'Olympie et ses équivalents méditerranéens. Bulletin de l'Association des Géographes Français 433:85–94

Dufaure JJ (ed) (1984) La mobilité des paysages méditerranéens: hommage à Pierre Birot. Université de Toulouse le Mirail: revue géographique des Pyrénées et du Sud-Ouest, 387 pp

Flageollet, JC (1989) Les mouvements de terrain et leur prevention. Masson, Paris, 22 pp

Fouache E (1999) L'alluvionnement historique en Grèce Occidentale et au Péloponnèse: géomorphologie, archéologie, histoire. Bulletin de Correspondance Hellénique, supplément 33, Athènes: École Française d'Athènes, Paris: De Boccard, 235 pp

Fouache E (2006) 10000 d'évolution des paysages en Adriatique et en Méditerranée Orientale. Travaux de la Maison de l'Orient Méditerranéen (TOM), Volume 45, Lyon, Paris: De Boccard, 225 pp

Fouache E, Pavlopoulos K (2005) Sea level changes in Eastern Mediterranean during Holocene: Indicators and human impacts. Zeitschrift fur Geomorphologie, Supplementary volume 137, 193 pp

Fouache E, Dufaure JJ, Denèfle M, et al. (2001) Man and environment around Lake Maliq (southern Albania) during the late Holocene. Vegetation History and Archaebotany 10:79–86

Fouache E, Ghilardi M, Vouvalidis K, et al. (2008). Contribution on the Holocene reconstruction of Thessaloniki coastal plain, Greece. Journal of Coastal Research, 24:1161–1173

Greig JRA, Turner J (1974) Some pollen diagrams from Greece with special reference to pollen as an indicator of human activity. Paleohistoria 24:258–289

Grove JM (1988) The Little Ice Age. Methuen, New York, 498 pp

Harrison SP, Digerfeld G (1993) European lakes as palaeohydrological and paleoclimatic indicators. Quaternary Science Reviews 12:233–248

Helly B (1987) La Grèce antique face aux phénomènes sismiques. PACT 18 (III.2):143–160

Huntley B, Prentice C (1989) Holocene vegetation and climates of Europe. In: Huntley B, Webb T (eds) Vegetation history, Handbook of vegetation science 7. Kluwer Academic Publishers, Dordrecht, pp 633–672

Johnsen SJ, Dahl-Jensen D, Gundestrup N, et al. (2001) Oxygen isotope and palaeotemperature records from six Greenland ice-core stations: Camp Century, Dye-3, GRIP, GISP2,

Renland and NorthGRIP. Journal of Quaternary Science, 16:229–307

Knauss J (1991) Arkadian and Boiotian Orchomenos, centres of Mycenaean hydraulic engineering. Irrigation and Drainage Systems 5:363–381

Lambeck K, Purcell A (2005) Sea-level change in the Mediterranean Sea since the LGM: model predictions for tectonically stable areas. Quaternary Science Reviews 24:1969–1988

Lespez L (1999) L'évolution des modelés et des paysages de la plaine de Drama et de ses bordures au cours de l'Holocène (Macédoine Orientale, Grèce). Thèse Université de Clermont II, 539 pp

Magny M, Begeot C, Guiot J, Peyron O (2003) Contrasting patterns of hydrological changes in Europe in response to Holocene climate cooling phases. Quaternary Science Reviews 22:1589–1596

Marinatos SP (1939) The volcanic destruction of Minoan Crete. Antiquity 13:425–439

Mayewski PA, Rohling E, Stager C, et al. (2004) Holocene climate variability. Quaternary Research 62:243–255

McNeill JR (1992) The mountains of the Mediterranean World: An environmental History. Cambridge University Press, Cambridge, 423 pp

Mee C, Forbes H (1997) A rough and rocky place: the landscape and settlement history of the Methana peninsula, Greece. Liverpool University Press, Liverpool, 370 pp

Michael HN (1980) Radiocarbon dates from the site of Akrotiki Thera 1967–1977. In: Doumas C (ed) Thera and the Aegean World, Proceedings 2nd international scientific congress, Santorini, Greece, 1978. Thera Foundation, London, pp 791–795

Moretti I, Lykousis V, Sakellariou D, et al. (2004) Sedimentation and subsidence rate in the Gulf of Corinth: what we learn from the Marion Dufresnes long-piston coring, Comptes rendus Geosciences, 336:291–299

Papazachos BC, Dimitriu PP (1991) Tsunamis in and near Greece and their relation to the eartquake focal mechanisms. Natural Hazards 4:161–170

Perlès C (2001) The early Neolithic in Greece. Cambridge World Archaeology, 350 pp

Petit-Maire N (1999) Variabilité naturelle des environnements terrestres : les deux derniers extrêmes climatiques (18000 +/− 2000 et 8000 +/− 1000 ans BP). Comptes rendus de l'Académie des Sciences, Sciences de la terre et des planètes 328:273–279

Pirazzoli P, Thommeret J, Thommeret Y, et al. (1982) Crustal block movements from Holocene shorelines: Crete and Antikythira (Greece). Tectonophysics 86:27–43

Poursoulis G (1999) La destruction des palais Minoens. Thèse Université Paris 1 Panthéon-Sorbonne, 2 vol, 482 pp and 147 pp respectively

Poursoulis G, Dalongeville R, Helly B (2000) Destruction des édifices minoens et sismicité récurrente en Crète (Grèce). Géomorphologie 4:253–266

Rossignol–Strick M (1995) Sea–land correlation of pollen records in the eastern Mediterranean for the glacial–interglacial transition – biostratigraphy, versus radiometric time-scale. Quaternary Science Reviews 14:893–915

Ryan W, Pitman W (1998) Noahs flood: the new scientific discoveries about the event that changed history. Simon and Schuster, London, 337 pp

Touchais G, Renard J (2002) L'Albanie dans l'Europe Préhistorique. Actes du colloque de Lorient, 8-10 juin 2000. Bulletin de Correspondance Hellénique, Supplément 42, 264 pp

Turner C, Sanchez Goni MF (1997) Late glacial landscape and vegetation in Epirus. In: Bailey G (ed) Klithi: Palaeolithic settlement and Quaternary landscapes in northwest Greece. Vol. 2: Klithi in its local and regional setting, chapter 29, pp 559–585

Tzedakis PC (1993) Long term tree populations in northwest Greece throught multiple Quaternary climatic cycles. Nature 364:437–40

Van Andel TH, Jacobsen TW, Jolly J.B, Lianos N (1980) Late Quaternary history of the coastal zone near Franchti cave, southern Argolid, Greece. Journal of Field Archaeology 7:389–402

von Grafenstein U, Erlenkeuser H, Brauer A, et al. (1999) A mid-European decadal isotope-climate record from 15,500 to 5000 years BP. Science 284:1654–1657

Wiersma AP, Renssen H (2006) Model–data comparison for the 8.2ka BP event: confirmation of a forcing mechanism by catastrophic drainage of Laurentide lakes. Quaternary Science Reviews 25:63–88

Wijmstra TA (1969) Palynology of the first 30 metres of a deep section in northern Greece. Acta botanica neerlandica 18:511–527

Willis KJ (1992a) The late Quaternary vegetational history of northwest Greece. I Lake Gramousti. The New Phytologist 121:101–117

Willis KJ (1992b) The late Quatemary vegetational history of northwest Greece. II Rezina marsh. The New Phytologist 121:119–138

Willis KJ (1994) The vegetational history of the Balkans. Quaternary Science Reviews 13:769–788

Willis KJ, Bennet KD (1994) The Neolithic transition – fact or fiction? Palaeoecological evidence from the Balkans. The Holocene 4:326–330

Chapter 11
The Nuragic People: Their Settlements, Economic Activities and Use of the Land, Sardinia, Italy[1]

Anna Depalmas and Rita T. Melis

11.1 Introduction

Sardinia is the second largest island in the Mediterranean (Fig. 11.1) and is well-known for its Nuragic society, which developed in the Bronze and Early Iron Ages from 1900 to 730 BC (Nuragic Age, Table 11.1).

Past studies of this civilization were based primarily on the analysis of buildings and artefacts. The distribution of the various settlements was analyzed in detail only in relatively limited areas, and generally with particular interest being paid to their use for military purposes. The relationships between the human societies and the terrain have been mainly overlooked, in part because of the absence of collaboration between archaeologists and Earth scientists. The aim of this chapter is, therefore, to explore and partly remedy this lack of knowledge of these relationships.

11.2 Geology and Geomorphology of Sardinia

Sardinia is approximately 240 km long from north to south and 140 km wide, and has a varied landscape. The island has had a complex geological history since Paleozoic times (Fig. 11.2). It is characterized by rugged mountain ranges and massifs which are up to 1834 m high, with Paleozoic metamorphic and intrusive rocks mainly in the center and Northeast of the island, and localities of deformed Mesozoic carbonate, particularly along the East coast. The latter have been intensely affected by karst. Several major volcanic eruptions have occurred, primarily during the Oligocene–Miocene (primarily andesite, rhyolites and pyroclastic materials) and the Pliocene–Pleistocene (primarily basalt) periods in the central-western part of the island. Due to erosion, the ancient volcanic rocks now form several high wide plateaus with steep sides. Extensional and transtensional tectonic movements during the Pliocene-Quaternary reactivated older faults systems and dissected the island in NE-SW and NW-SE directions. These led to the formation of several graben including the Campidano Graben in the central-south part of the island (Fig. 11.3) partially filled with more than 600 m of syntectonic deposits. Most valleys of Sardinia have relatively flat bottoms and relatively steep sides. There are Quaternary alluvial deposits, travertines, aeolianites, coastal deposits, and lagoonal sediments in several areas. Flat lowlands are few and small in size, the larger ones being in the lower areas of the Campidano Graben and in some coastal zones, primarily in the northwest where they are often covered by aeolian sands. There are a few torrential rivers in the island. One of the largest is the Tirso River in the central west. It is 151 km long and there is a small fertile alluvial plain in the lower reaches.

Sardinia has a Mediterranean climate with dry, hot summers and mild winters. The mountain areas are colder, with annual snowfalls in the higher areas and occasional ones elsewhere. The island is swept by strong winds, particularly the cold Mistral from the

A. Depalmas (✉)
Dipartimento di Scienze Umanistiche e dell'Antichità
Università di Sassari, Piazza Conte di Moriana 8
07100 Sassari, Italy
e-mail: depalmas@uniss.it

[1] A Depalmas is principally responsible for the archaeological part of the chapter, and RT Melis for the landscape.

See Plates 9b, 10 in the Color Plate Section; also available at: extras.springer.com

I. P. Martini, W. Chesworth (eds.), *Landscapes and Societies*,
DOI 10.1007/978-90-481-9413-1_11, © Springer Science+Business Media B.V. 2010

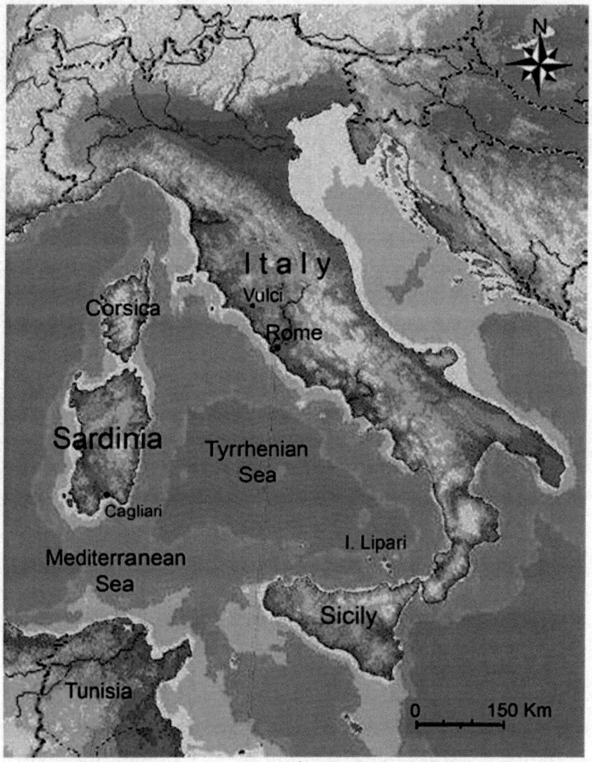

Fig. 11.1 Location of Sardinia in the West Mediterranean

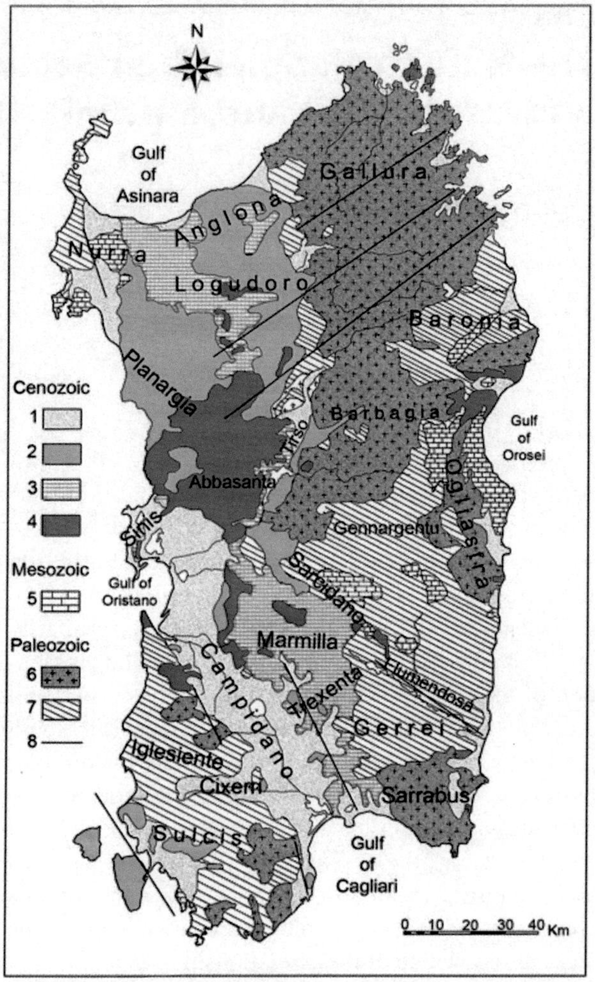

Fig. 11.2 Schematic geological map of Sardinia. **Cenozoic:** *1* Continental and marine deposits (alluvial, colluvial, aeolian and littoral gravels, sands, and silts) (Holocene–Pleistocene), *2* alkaline, transitional and subalkaline volcanic cycle (Plio-Pleistocene), *3* marine succession and continental deposits (marls, sandstones, marly sandstones, silstones, calcarenites, silty, conglomerates, and sandy marls) (Upper Miocene, Lower Middle Miocene, and Lower Miocene), *4* calkaline volcanic cycle (andesites, andesitic basalt, ignimbrites, and epiclastic deposits) (Oligocene–Miocene). **Mesozoic:** *5* marine and transitional successions dolomitic limestones, limestones, dolostones, marly dolostones (Upper Cretaceous, Middle Triassic, Lower Cretaceous). **Paleozoic:** *6* intrusive complex (equigranular leucogranites, equigranular monzogranites, and tonalites) (Upper Carboniferous, Permian), *7* Hercynian metamorphic complex (shales, micashistes, metasilstones, metasandstones, metalimestones, and limestones) (Carboniferous, Devonian, Ordovician, Silurian, Cambrian), *8* faults

northwest and the hot Scirocco from the south, which occasionally causes dust storms. At present there are some local cases of desertification due to global climatic changes and intense human exploitation of the land. Similar conditions have occurred in the past, when they were due, for example, to the exploitation and cutting down of the ancient forests of ilex, oak, chestnut, and carob, partly to create more pastureland for grazing. This contributed to the intense erosion of soil in the highlands such as the many currently barren carbonate hills. At the moment only one sixth of the total land area (approximately 475,000 ha) is forested (mainly *Quercus ilex*, *Quercus suber*, but also *Quercus pubescens* and *Juniperus phoenicea*).

11.3 The Nuragic Civilization

The Nuragic civilization developed in Sardinia in the 2nd millennium BC and was primarily characterized throughout most of its existence (until about 1150 BC) by the round stone towers called nuraghe, many of which are still standing (Table 11.1). The structures changed over time. These changes, and also other artefacts, allow us to identify different societies and their habits (Table 11.2).

The nuraghe are essentially conical, several-storied, dry-stone towers. There are two main types, the early 'corridor' nuraghe or proto-nuraghe (Fig. 11.4),

Table 11.1 Chronology of Sardinian archaeology from the Lower Paleolithic to the end of the Roman Empire based on calibrated datations. The definition of the Nuragic civilization in five phases (Nuragic I-V) is after Lilliu (1999)

Paleolithic	Lower		Rio Altana (?)				450000-200000 B.P. (?)
			Codrovulos (?)				200000-150000 B.P. (?)
	Upper		Grotta Corbeddu				20000-12000 B.P.
Mesolithic			Grotta Corbeddu				9000-6500 B.C.
			Grotta Su Coloru				7000-6200 B.C.
Neolithic	Early		Su Carroppu				5900-5500 B.C.
			Filiestru - Grotta Verde				5500-4900 B.C.
			Filiestru B				
	Middle		Bonu Ighinu				4900-4400 B.C.
	Late		San Ciriaco				4400-4100 B.C.
	Final		San Michele di Ozieri				4100-3500 B.C.
Eneolithic (Copper Age)	Early		Sub Ozieri				3600-2800 B.C.
				Filigosa			3000-2400 B.C.
					Abealzu		2600-2300 B.C.
	Full		Monte Claro				2900-2200 B.C
	Final				Beaker A_1		2500-2100 B.C.
					Beaker A_2 (*Facies*, Sulcitana)		
	Early		Bonnannaro A_1		Beaker B		2300-1700 B.C.
			Bonnannaro A_2 Sant'Iroxi				
Bronze Age	Middle	Initial	Sa Turricula			Nuragic I	1700-1600 B.C.
		Full	Muru Mannu, San Cosimo				
			Tamuli (?)			Nuragic II	1600-1350 B.C.
Late	Recent		- *facies* "a pettine" (North Sardinia) - Antigori – Su Mulinu (South Sardinia)			Nuragic III	1350-1200 B.C.
	Final		Proto-geometric			Nuragic IV	1200-1000 B.C.
Iron Age	Early		I	Geometric		Nuragic IV Phoenician	1000-730 B.C.
				Orientalizing			730-580 B.C.
			II	Archaic		Nuragic V	580-500 B.C.
	Late			Punic			525-238 B.C.
				Roman		Republican	238-1 b.C
				Roman		Imperial	1 d.C-476 d.C

and the later 'tholos' nuraghe. The latter can be further subdivided into nuraghe with either a single tower (mono-tower) (Fig. 11.5) or multiple towers (complex) (Fig. 11.6). In later times the use of the nuraghe diminished as increasingly complex villages developed, either around the ancient structures or in other areas in the countryside (Fig. 11.7; Depalmas 2003; Dyson and Rowland 2007). The ancient sites were rarely abandoned and the structures may have been used for different purposes, in different areas. The dry-stone constructions for civil purposes in the rest of the Mediterranean (Corsica and the Balearic Islands) are markedly different in their construction techniques, despite apparent similarities (Contu 1997).

11.3.1 Early Middle Bronze Age

The early stages of the Middle Bronze Age were characterized by the corridor nuraghe. These are buildings with strong rough stone walls and smaller internal areas (Fig. 11.4). There is no typical floor plan and some are elliptical, some quadrilateral, and some circular. All of the monuments have an internal corridor which is either straight or elbow ('a gomito'). The buildings may have two entrances. Sometimes, apart from the corridor, there are other small spaces. It is not rare to find a stone stairway in the corridor, which leads up to an upper terrace. We hypothesize that the original buildings were between eight and fifteen meters high.

Village construction began during this first phase. These consisted of groups of stone huts, or, in the plains where no stone was available, perishable material and/or crude mud bricks. Some of the huts were half buried in the earth. In the earliest phase while rectangular huts are found, such as Sa Turricola, Muros (Fig. 11.3), circular huts were most common.

The greatest density of corridor nuraghe occurs in central Sardinia on basalt plateaus. They are less com-

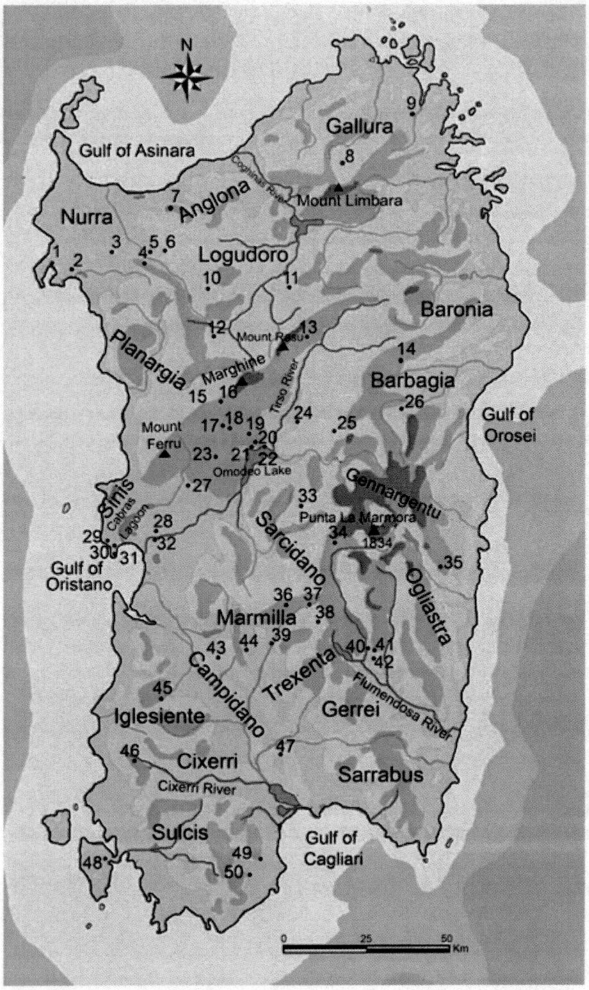

Fig. 11.3 Map of Sardinia, showing major regions, rivers, mountains and principals sites mentioned: *1* Sant'Imbenia, Alghero; *2* Palmavera, Alghero; *3* Olmedo; *4* Chessedu, Uri; *5* S'Iscia 'e sas Piras, Usini; *6* Sa Turricola, Muros; *7* Serra Niedda, Sorso; *8* Nuchis; *9* Li Lolghi, Arzachena; *10* Sant'Antonio, Siligo; *11* Ozieri; *12* Giave; *13* Bonotta, Bultei; *14* Su Tempiesu, Orune; *15* Tamuli, Macomer; *16* Santa Barbara, Macomer; *17* Toscono, Borore; *18* Duos Nuraghes, Borore; *19* Ulinu, Sedilo; *20* Lighei, Sedilo; *21* Iloi, Sedilo; *22* Talasai, Sedilo; *23* Abbasanta; *24* Ottana; *25* Sarule; *26* Predu Zedda, Oliena; *27* Crabia, Paulilatino; *28* Santa Vittoria, Nuraxinieddu; *29* Funtana Meiga; *30* Tharros; *31* Su Murru Mannu, Cabras; *32* Oristano; *33* Talei, Sorgono; *34* Funtana Raminosa, Gadoni; *35* Ilbono; *36* Brunku Madugui, Gesturi; *37* Is Paras, Isili; *38* Santa Vittoria, Serri; *39* Su Mulinu, Villanovafranca; *40* Gasoru, Orroli; *41* Arrubiu, Orroli; *42* Su Putzu, Orroli; *43* Sant'Anastasia, Sardara; *44* Genna Maria, Villanovaforru; *45* San Cosimo, Gonnosfanadiga; *46* Iglesias; *47* Su Stradoni di Deximu, San Sperate; *48* Sulci-Sant'Antioco; *49* Antigori, Sarroch; *50* Perda 'e Accuzzai, Villa San Pietro

mon in other landscapes, such as granite or volcanic rock areas. These nuraghe are not found in valley bottoms or on plains.

During these times the deceased persons were buried in specially constructed tombs a short distance away from the nuraghe and villages. The relationship between the number of nuraghe and tombs has not been calculated, and probably cannot be estimated precisely. One to five tombs (funerary constructions) have been so far found in the areas of highest density of nuraghe. These buildings, called Giant's tombs, have funerary chambers and concave stone fronts ('exedrae') (Fig. 11.8).

The oldest tombs (Early Middle Bronze Age) date from the same period as the corridor nuraghe, and are made of vertical slabs of stone with flat roofs. The exedrae are delimited by stone slabs fixed in the soil and at the center there is a large and imposing stele with a rounded top (Fig. 11.8a). This type of tomb is often imposing and had a collective or communal structure.

In the Early Middle Bronze Age (Sa Turricola facies) the most common artefacts were pans with low sides and flat bottoms (Fig. 11.9a) although milk-boilers—cylindrical containers with inside-listel for inserting perforated diaphragms (Fig. 11.9e)—were also widely used. The cooking dish—a type of bell-shaped pan with a convex top—also appears, and this is found throughout the whole Nuragic period (Fig. 11.9f–g). Apart from these various open (dishes and bowls) and closed (clay jars and vases with necks) forms, other different types are found throughout the whole Nuragic period (Fig. 11.9b–d).

There are insufficient data to accurately reconstruct the economy of this phase. Analysis of the remains found inside the corridor nuraghe Brunku Madugui shows that goats and sheep (sheep, *Ovis aries* L.; muflon, *Ovis musimon* Pallas; goat, *Capras hircus* L.) were the most frequently found species (58.5% of the total, with 33 animals identified). These were followed by swine (pigs and wild boar, *Sus scrofa* ssp. L.) (20.2%, 8 animals), deer (*Cervus elaphus* L.) (6.4%, 6 animals), and oxen (*Bos taurus* L.) (15%, 5 animals). The sheep and goats were slaughtered still young, usually between one and three years of age as were most of the swine (less than two years of age) whereas the age of the slaughtered oxen varied more, from one to eleven years of age (Fonzo 1987).

Table 11.2 A summary of the main aspects of Nuragic civilization

	← Nuragic Age →				
PERIOD	Early Middle Bronze Nuragic I	Late Middle Bronze Nuragic II	Recent Bronze Nuragic III	Final Bronze Nuragic IV	Early Iron age Nuragic IV (Geometric/Phoenician) End of Nuragic Age Nuragic V
Time (1700 ~ 580 BC)	1700–1600	1600–1350	1350–1200	1200–1000	1000–580 (510 BC Phoenician occupation)
Climate	Wet and cool (?)	Wet, cold and xeric (?)	Wet, cold and xeric (?)	Drier (?)	Drier (?)
Building Nuraghe	corridor nuraghe *used for family habitation*	Single tower tholos nuraghe Local concentrations of nuraghe *indicating agglomeration of few families*	Development of some complex nuraghe with multiple towers in selected sites. *Development of tribes*	Intense development of hut villages; decreased importance of nuraghe (*used for religious purposes or storage*)	Intense development of hut villages; decreased importance of nuraghe (*used for religious purposes or storage*)
Villages	• huts without stone walls • huts with stone walls (rarer), sometimes rectangular	• circular huts with stone walls • huts without stone walls	• circular huts with stone walls • huts without stone walls	• villages of stone walled huts: – isolated, – around the nuraghe, – around a cult site • multi roomed buildings	• villages of stone walled huts: – isolated, – around the nuraghe, – around a cult site • multi roomed buildings
Graves structures	'Giants' tombs with slabs set edgewise and "stele centinata"	'Giants' tombs' with blocks in rows and monumental front in rughs or squared blocks.	'Giants' tombs' 'corridor tombs' without exedra	'Giants' tombs' and corridor tombs till used but also personal tombs as trench or pits ('a fossa or 'a pozzetto')	'Giant's Tombs' and corridor tombs still used but also personal tombs as trench or pits
Sacred structures	–	• Square temples with central hall ("megaron") • external area of tombs (exedra)	• Sacred springs (?) • Square temples with central hall ("megaron")	• Sacred springs • Sacred wells • Square temples with central hall ("megaron")	• Sacred springs • Sacred wells • Square temples with central hall ("megaron")
Manufactured items	Bread pan, cooking dish, milk-boiling, bronze ax	Bread pan, pots, milk-boiling, bronze ax	Bread pan with comb decoration; storage jars	Metallurgical activity; specialized workshops, bronze figurines; Jugs, storage jars	Metallurgical activity; specialized workshops, bronze figurines Jugs, storage jars
Demography	Low density	Rapid increase in population	High density	High density	Reduction in density
Forest	Extensive cover, low human impact	Still extensive, small human impact	Reduction in forested areas due to fires and human use of land for agriculture	More intense deforesting	More intense deforesting, degradation of the environment
Agriculture	Cereals, grazing	Cereals, legumes (*Vicia faba*)	Cereals, wild fruit animal husbandry	Cereals (wheat, barley), wild grapes, animal husbandry	Cereals, cultivated grapes
Animal Husbandry	Goats, pigs	Cattle, pigs, goats/sheep	Pigs	Pigs, goats/sheep	Cattle, goats/sheep, pigs
Hunting	Deer		Deer	Prolagus	Deer, Prolagus, Birds
Site Distribution Landscapes	Generally low, with concentrations on high basalt plateaus, near water courses	Widespread all over the island, with intense occupation of high basalt plateaus	Widespread all over the island, with intense occupation of high basalt plateaus	Widespread all over the island, with intense occupation of high basalt plateaus	Widespread all over the island, with intense occupation of high basalt plateaus (?)

The five Nuragic phases (Nuragic I–V) are after Lilliu (1999).

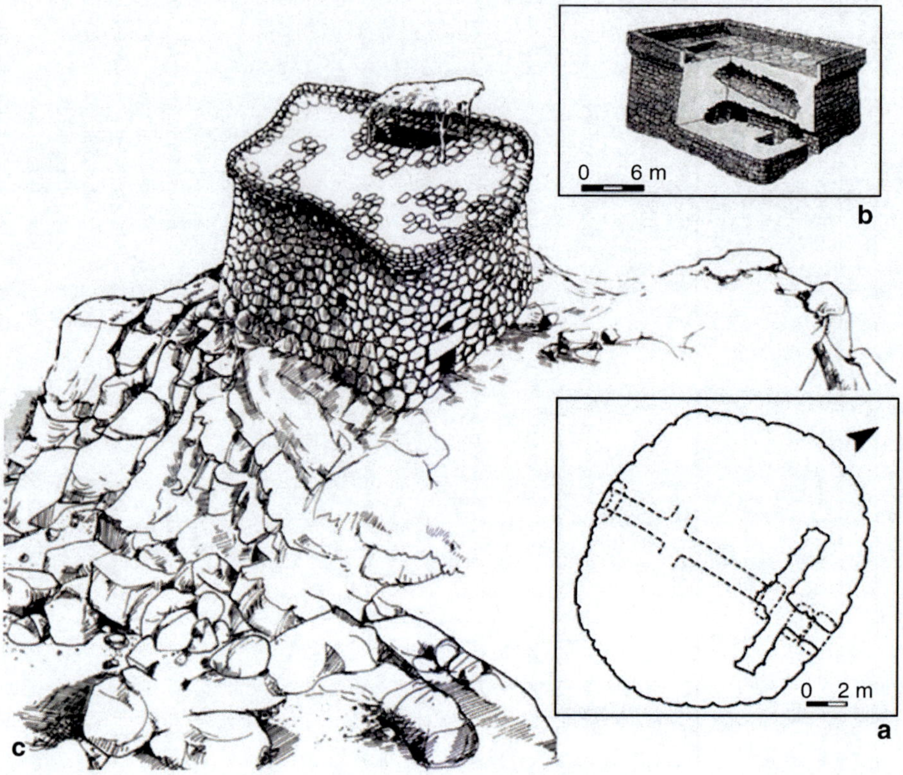

Fig. 11.4 The corridor nuraghe of Ulinu (Sedilo). **a** Plan. **b** Idealised section. **c** Idealised reconstruction of a corridor nuraghe. (After Tanda 1990, 1996)

11.3.2 Late Middle to Recent Bronze Age

In the late Middle Bronze Age and the Recent Bronze Age, single towered tholos nuraghe predominated (Fig. 11.5). This type of nuraghe has an entrance corridor, an opening on a side for a stairway which leads up to the upper floor(s), and a niche on the opposite side. The corridor leads into a circular chamber covered with a false cupola (tholos), with one to three niches in the walls. The upper floor may contain a similar chamber to that on the ground floor or else an open terrace. This floor originally had a wooden structure supported by stone projections which extended upwards from the last layer of stones. It is hypothesized that these towers were originally 20 m high or so.

This type of building was constructed first as single tower in the Late Middle Bronze Age and later in Recent Bronze Age as complex nuraghe with two and five towers (Fig. 11.6). Their use may have varied, but on the whole they can be considered to have always been principally dwelling places. Villages may also have grown around the nuraghe even though one also finds villages not associated with nuraghe during this period.

The village huts were circular with basal stone walls 0.80–1.70 m high (Fig. 11.7a). The huts usually had beams and small cross beams supported by the stone walls and wooden roofs or, on rare occasions, flat stone slab roofs; at the center of the hut there was a fireplace or a nether millstone. Sometimes there was an outer stone wall surrounding the settlement.

In this period there was a marked increase in the number of single tower tholos nuraghe. A number of about 9000 units was suggested (Contu 1997). They were built everywhere including valley bottoms, although they were more common on basalt plateaus and in areas of volcanic rock. There were fewer complex tholos nuraghe. Where present they are found on the same landscapes as the single tower nuraghe.

The discovery of ceramic materials of this period in rectangular structures ('megaron' temples) and in springs and wells, which have been identified as cult ritual sites, may indicate that sacred architecture may have already started in this period. During the Middle and Recent Bronze Age the places used for religious ceremonies seem to coincide with burial places. Communal ceremonies also took place in the

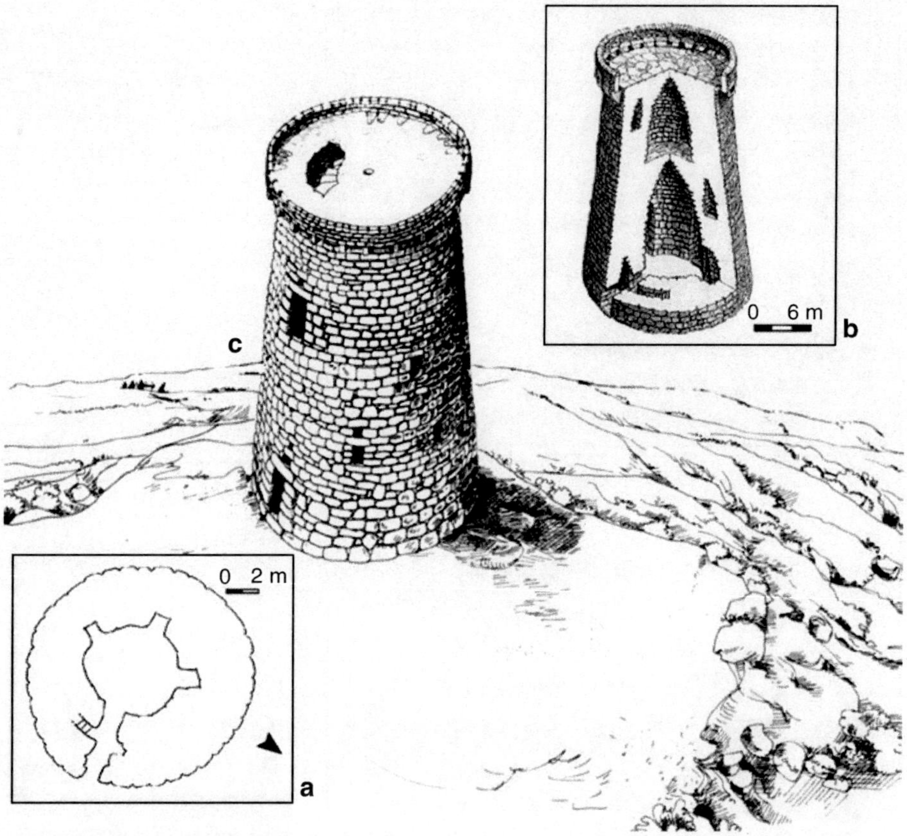

Fig. 11.5 The single tower tholos nuraghe. **a** Plan of nuraghe Lighei (Sedilo). **b** Idealized section. **c** Idealized reconstruction of a single tower two-storey tholos nuraghe. (After Tanda 1990, 1996)

exedra of the tombs and the spaces where offerings were left. The presence of conic stones ('betils') near the tombs suggest that ancestor worship may have been practiced.

The Giant's tombs of this period have exedra made with courses of rough or squared stones and the funerary corridors are roofed with projecting courses and ogival sections (Figs 11.8b, c).

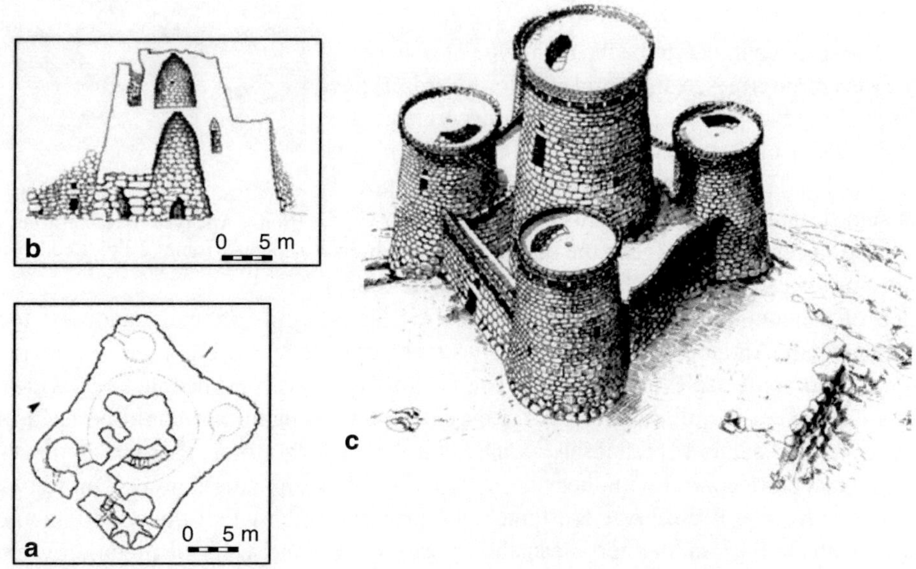

Fig. 11.6 The complex tholos nuraghe. **a** Plan. **b** Section of nuraghe Santa Barbara, Macomer (After Moravetti 1998b). **c** Idealized reconstruction of a complex three-storey tholos nuraghe. (After Tanda 1990)

Fig. 11.7 The Nuragic village. **a** Hut 3 of Iloi village (Sedilo). **b** Idealised reconstruction of village huts of Serra Orrios (Dorgali). (After Moravetti 1998a)

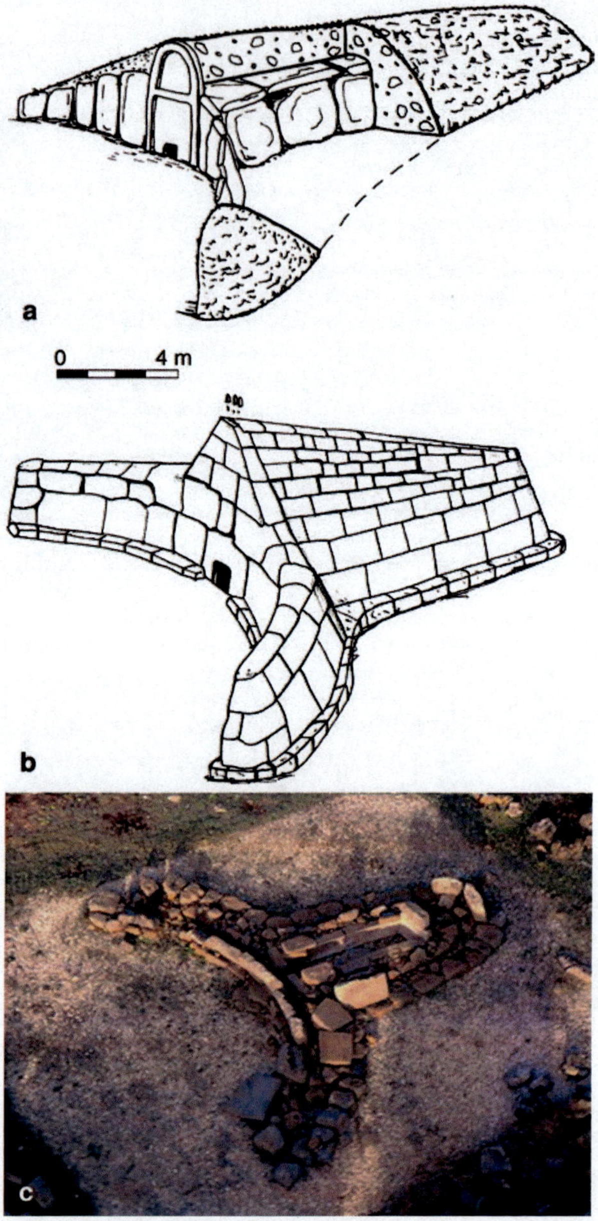

Fig. 11.8 Giant's Tombs. **a** Reconstruction of Early Middle Bronze Age type. **b** Reconstruction of Middle Bronze Age Type. **c** Giant's Tomb of Iloi (Sedilo). (After Tanda 2003)

The 'pan' continues to be the main type of pottery in the Recent Bronze Age. In central and North Sardinia the internal base of the vases of this age is decorated with impressions of a comb determining the 'comb phase facies' (Fig. 11.9h). Large vases ('dolii') start to appear in significant numbers during this phase (Fig. 11.9i).

The Recent Bronze Age is the only period in which differences can be established in the ceramic production of the southern and northern parts of the island. In the south, Antigori (undecorated) facies were produced and comb facies in the north. The reasons for this difference are still not clear, but it may be due to the influence of the Mycenaeans, as the south of the island had more contact with them.

In the Middle Bronze Age Sardinia had more contacts with the Eastern Mediterranean and Cypriot style arms, utensils, and new technologies that influenced the development of metallurgy in the island. The Nuragic people had a wide range of tools at their command that could be used to smelt bronze and manufact utensils such as pincers, coal shovels, hammers and small anvils. These types of manufactured goods show that there were many connections with Aegean and Cypriot metallurgy and perhaps the rest of the

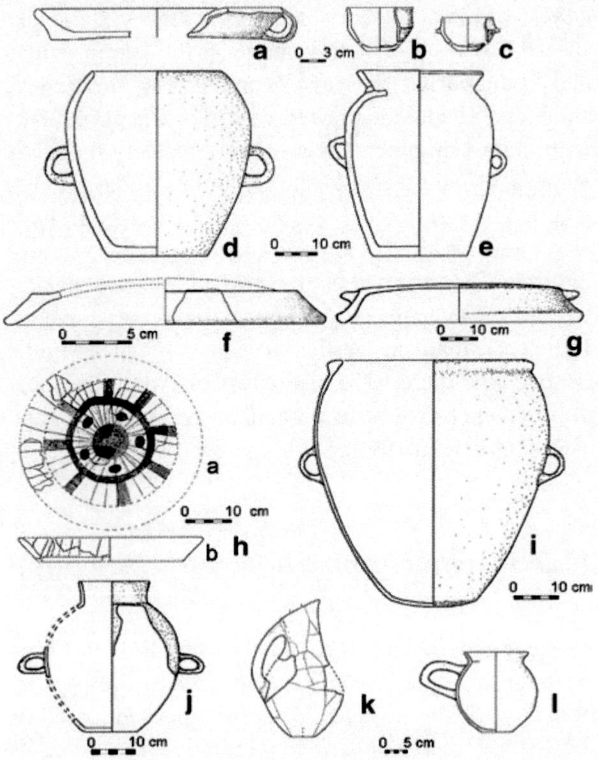

Fig. 11.9 Ceramic vessel forms of Middle Bronze Age: bread pan (**a**), carenated bowls (**b–c**), storage jar (**d**), milk-boiling vessel (**e**), "cooking dish" (**f**); Ceramic vessel forms of Final Bronze Age-early Iron Age: "cooking dish" (**g**); Ceramic vessel forms of Recent Bronze Age: bread pan with comb decorations (**ha** top view, **hb** profile), storage jar (**i**); storage jar (**j**), askoid jug (**k**), large handled milk-boiling vessel (**l**). [From nuraghe Talei, Sorgono (**a–c**), Giant's tomb of Li Lolghi, Arzachena (**d**), village of Santa Vittoria, Nuraxinieddu (**e**), nuraghe Brunku Madugui, Gesturi (**f**), village of Palmavera, Alghero (**g**, **l**), nuraghe Chessedu, Uri (**h**), tomb of Perda 'e Accuzzai, Villa San Pietro (**i**), village of Brunku Madugui (**j**), village of Castello, Lipari (**k**). (After Fadda 1998 (**a–c**); Badas 1993 (**f**); Moravetti 1992 (**g**, **l**); Bagella et al. 2000) (**d–e**, **h–j**); drawing by Depalmas (**k**)]

Mediterranean world. It is therefore hypothesized that at first the mining and smelting was supervised by people from the Aegean and later it was continued by locals. In any case objects continued to be imported from the Eastern Mediterranean and then imitated until the Iron Age.

We have no precise information on the organization of Nuragic societies particularly during the early stages of their development. Some deduction, however, can be arrived at by analysing the distribution of the settlements. The absence of firm evidence indicating a hierarchical system leads us to hypothesize that it was primarily a family based society, although after the great increase in population during the Late Bronze Age these family groups developed into territorially based tribes.

We do not have enough data to reconstruct accurately the climate and the agricultural practice of the Nuragic people. However on the basis of the scanty botanic data available, and after comparison with other Mediterranean areas (Van Joolen 2003), we can hypothesize that during the Bronze Age Sardinia had a Sub-Boreal temperate and wet climate, wetter than the present.

The information we have about the agricultural produce, such as the carbonised remains of grains found in some nuraghe, indicates that it was varied and influenced by the geological and geomorphologic settings. Few grain species were found in the Middle Bronze Age Duos Nuraghes complex on the basalt plateau of Abbasanta. They consist of remains of herbaceous plants, such as *Triticum dicoccum, Vicia faba* and *Olea europaea* (olive) that could not be determined whether cultivated or wild. However cultivated barley is found but not the naked barley variety whose cultivation, according to the evidence from the rest of the Western Mediterranean, ended during the Early Bronze Age just before the Middle Bronze Age (Bakels 2002).

In the Recent stage of Nuragic culture more species are found. These include *Hordeum vulgare* and *Triticum durum*. The oats may have been wild or domesticated. No *Vicia faba* was found, but very few samples have been analyzed. On the other hand four types of fruit have been identified. These are *Prunus spinosa* L. (sloe) and *Rubus sp.* (blackberry), which were collected in the wild, *Ficus carica* L. and *Vitis vinifera* L., which may have been either wild or domesticated (Bakels 2002). There are also numerous herbaceous species with a marked abundance of clover and forage grasses.

The paleo-palynological analyses carried out in central south Sardinia at the nuraghe of Arrubiu, Orroli (Sarcidano) (López et al. 2005) on the basalt plateau (550 m above sea level) that is bordered by the deep valley of the Flumendosa River, indicate that at the end of the Middle Bronze Age there was a heavily wooded environment (80% of total pollen is from trees/bushes). This was mainly oak trees and, to a greater or lesser extent, olive trees. However there were also alders, ash, elms and poplars. There are few rockroses, which are indicators of forest degradation. Grasses cover only 13% of the area and among these graminaceae predominate. The corresponding climate must have been cold and dry.

The situation continued to change with an increase in the amount of heather and a reduction in the number of oaks and olive trees. This indicates that deforestation was taking place, as can be seen from the presence of *Glomus* cf. *fasciculatum*, and this is also confirmed by the increase in the number of grass species and cereals. Among the non-pollen microfossils there are some which indicate that fires had taken place (*Chaetomium* sp.). This phenomenon increased with time and in the Recent Bronze Age the percentage of tree/bush pollens fell still further to 10% leaving more space to the Mediterranean maquis. There was a fall in the cultivation of cereal crops and an increase in pasture.

Other research in the same area at the Gasoru nuraghe, Orroli, about 2.3 km northeast of Arrubiu, has found that in the Recent Bronze Age the countryside was characterized by open oak forests with extensive cultivation of cereals around the monuments. The high percentage of *Chaetomium* sp. suggests that fires were used to increase the area available for cultivation.

Coprophilous fungi have been found among the microfossils, which leads us to hypothesize that animal husbandry took place near the building.

During the Bronze Age domestic animals became smaller in size and this was also the case in Sardinia (Wilkens 2003). This was due to changes in the climate and perhaps also to more intensive grazing (Fonzo 1987). The cattle at the Arrubiu nuraghe are estimated to have had an average height of 95 cm (Fonzo 2003). Cattle were the most important source of protein at the Arrubiu nuraghe of Orroli, followed by swine, sheep, and goat. In the Late Bronze Age swine became more important and very young animals were eaten. Hunting was of great importance as can be seen from the large quantity of deer remains found on the site.

Studies of human bones from the Nuragic period have discovered that the population suffered from various diseases. Hyperostosis porosity appears to have been very common. This disease is closely linked to chronic forms of anaemia, such as those linked to malaria, or to severe malnutrition. Malaria was reported to be present in the island in ancient times (it was already there in the Late Neolithic period) and traces of this are still evident today from the large number of patients who suffer from lack of glucose-6-phosphate dehydrogenase (G-6-PD). This pathology originated in persons who had successfully overcome attacks of malaria (Floris and Sanna 1999). The widespread diffusion of thalassaemia in the island is also due to the earlier presence of malaria, as the dimensions and shapes of the red cell globules in patients with thalassemia provide a natural protection against malaria. Bone tuberculosis (at Predu Zedda, Oliena) and rickets (such as at S'Iscia 'e sas Piras, Usini) have also been reported. General analysis of the remains indicates, however, that the population was healthier and on average taller (males: 165.2 cm; females: 153.5 cm) than in earlier periods. In the Middle Bronze Age there were also cases of cranial trepanation, degenerative arthritis and, in rare cases, dental caries (Germanà 1999).

11.3.3 Final Bronze Age–Early Iron Age

In the Final Bronze Age (1200–1000 BC) the construction of nuraghe slowly stopped and the population moved into villages. The nuraghe were used as food stores or for cult rituals, although in rare cases they continued to be inhabited. When the walls of the nuraghe deteriorated and the upper parts of the walls fell down, rather than rebuild them the people used the stones for building their huts. These were built around the nuraghe and sometimes even inside the courtyard or bastion of the nuraghe. There are also many villages which grew up far away from a nuraghe.

In this phase the ground plan of the huts were not only circular in form but also trapezoid, rectangular, elliptical and other shapes. There were central courtyards in the groups of huts that formed the villages, as mentioned earlier (Fig. 11.7b). In the villages there is generally also a large circular hut with seats around the inside walls. These are called the 'meeting huts'. Inside them stone models of nuraghe are often found, which evidence of the cult of the nuraghe as works of their ancestors (Fig. 11.10e–f). This is the phase when the sacred architecture, represented by megaron temples, circular structures, and sacred wells and springs, was fully developed (Fig. 11.11). Around the temples small groups of huts were built. These were probably only used during festivals.

In the Final Bronze and Early Iron Age the Giant's Tombs were still used, but different types of sepulchre were also built as well as collective (corridor) or individual (trench or pit) tombs.

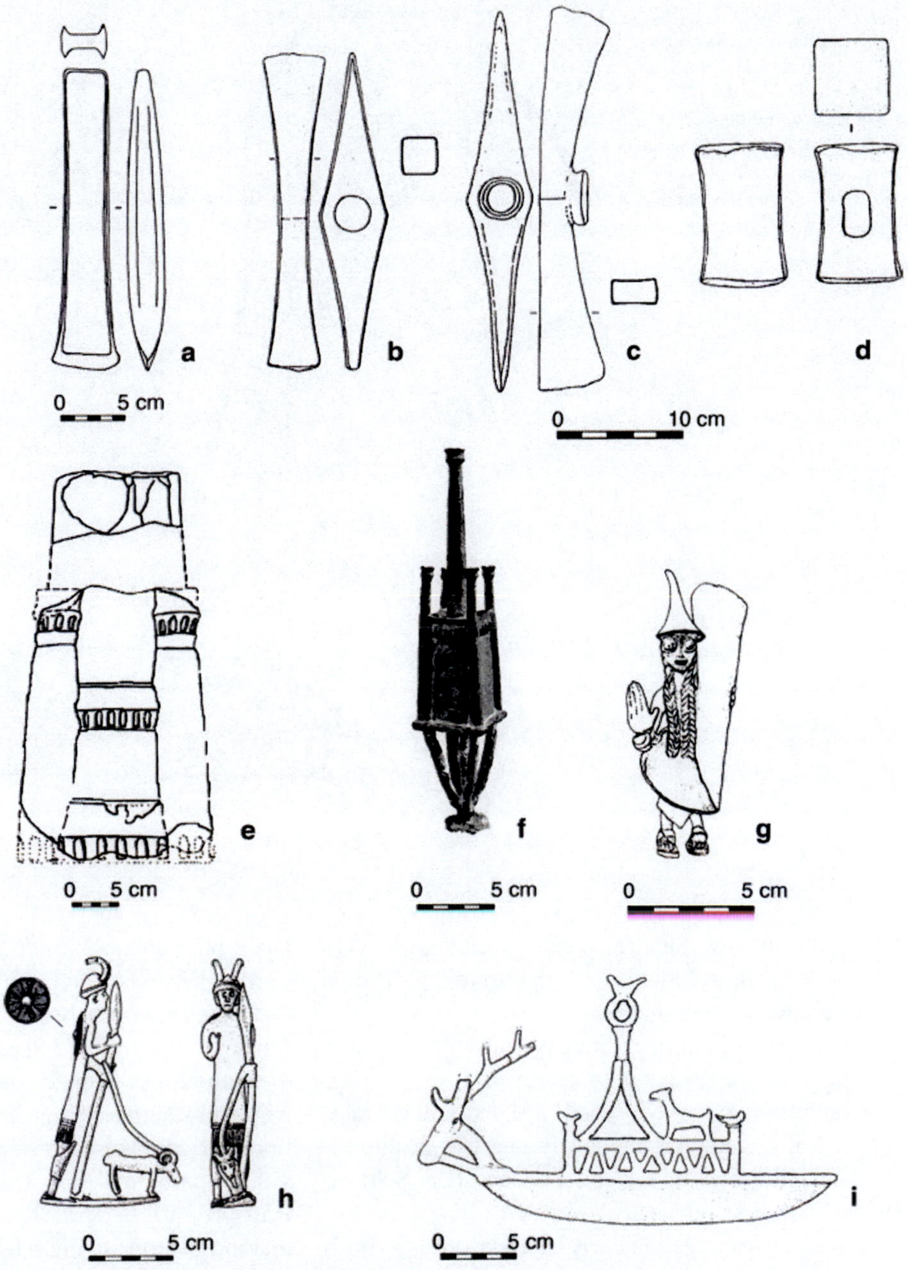

Fig. 11.10 Bronze ax (**a**), double axes (**b-c**), hammer (**d**); carved-stone model (**e**) and bronze model (**f**) of a complex nuraghe, bronze figurines of a pugilist/priest (**g**) and a warrior/shepherd (**h**), bronze-boat lamp with figure-head and dog/bird gunwale figures (**i**) from Ilbono (**a**), Sarule (**b**), unknown provenance (**c**), Nuchis (**d**), Su Stradoni di Deximu, San Sperate (**e**), Olmedo (**f**), Ponte Rotto/Cavalupo, Vulci (**g**), Serra Niedda, Sorso (**h**), Bonotta, Bultei (**i**). After Lo Schiavo 2005 (1–4); Lilliu 1999 (5–6); Pacciarelli 2000 (7); Rovina 2002 (8); Depalmas 2005 (9)

In this phase large containers (dolii) were widely used. These were in the form of jugs with large handles and askoid jugs, closed vases with narrow asymmetrical necks (Fig. 11.10m). These were often finely decorated.

Metallurgy also developed greatly during this period, and this provides us with important information on the technology used in the economic activities of the Nuragic people (Lo Schiavo 2005). Axes (Fig. 11.10a–c), chisels, awls, wedges, drills, files and saws have been found which were used for working wood, while scrapers, blades, borers and punches were used for working leather.

In the Final Bronze and Early Iron ages there was a great increase in the tools used in agriculture due to the developments in metallurgy. This can be seen from the advances in crop cultivation, which seems, together with animal husbandry, to have been the basis of the

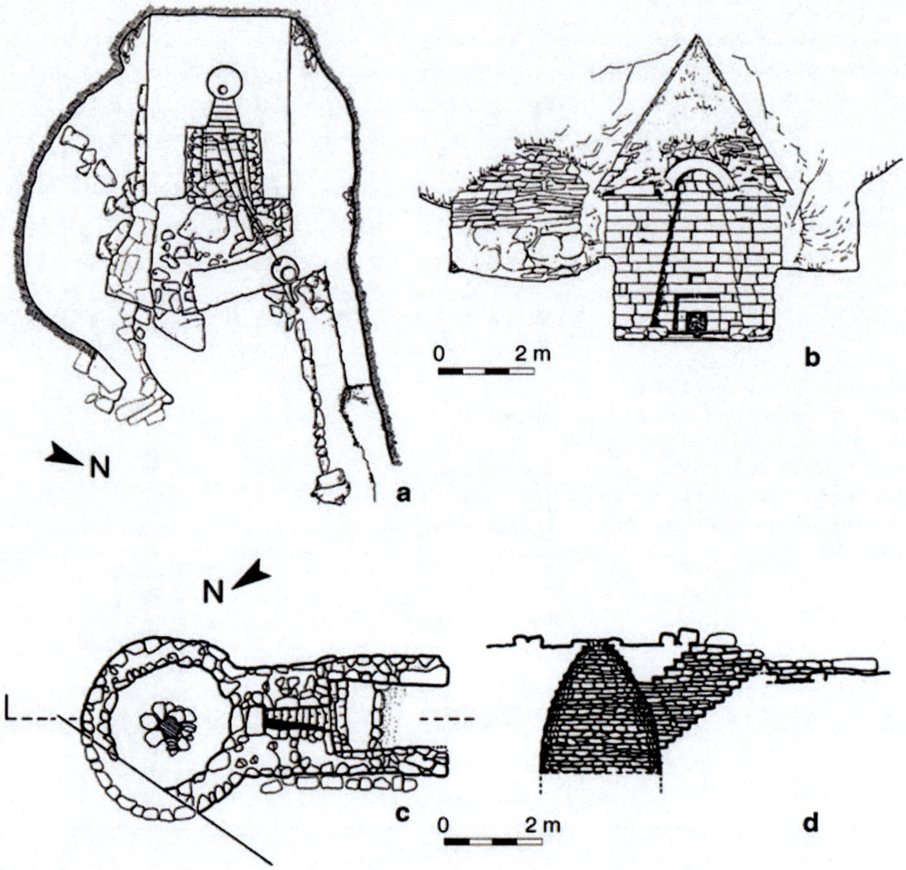

Fig. 11.11 Sacred constructions. **a** Plan of sacred spring of Su Tempiesu di Orune. **b** Prospect of the same (After Fadda 1988). **c** Plans of sacred well of Santa Vittoria, Serri. **d** Longitudinal profile of the same. (After Santoni 1985)

Nuragic economy. Sickles, pickaxes and hoes were used in agriculture. Sickles are the agricultural tools which are most often found.

The use of animals and ploughs is also documented in certain bronze figures of yoked oxen.

Examples of carved wood are extremely rare, although not unknown, obviously because of the climate and the fact that wood perishes (Lo Schiavo 1981).

In the Final Bronze and Early Iron Age one also finds bronze figures of men, women, animals, and objects. These are often found in places where cult rituals took place which indicate that they were particularly valuable gifts and offerings. The figures provide valuable information about the clothing, arms and equipment in the Nuragic period (Fig. 11.10g–h). The representation of wheels, carts and boats provide information on the means of transport used (Fig. 11.10i).

The location of Sardinia in the center of the Mediterranean and its resources of valuable metals favored considerable commerce. Some of this was short distance but some was medium distance such as to the Tyrrhenian coast, the Eolic Islands, and Sicily, but also to farther areas such and to Greece, Cyprus and the Levant to the east and Spanish peninsula to the west.

In the Final Bronze Age and Early Iron Age there were substantial changes in society. Social stratifications appear. This can be seen from the use of individual tombs and also from the bronze statuettes of warriors, priests, priestesses, craftsmen, and people making offerings (Fig. 11.10g–h). Direct evidence of how the land was used has been found at the Gasoru nuraghe at Orroli. Here the passage from the Final Bronze Age to the Iron Age seems to have been characterized by increased deforestation and a marked increase in pasture of graminacea and other nitrophile species. The high incidence of *Chaetomium* sp. suggests that the woods were burnt. A similar situation for the Final Bronze Age is found at Su Putzu (López et al. 2005). The deforested land still had some remaining trees and the pasture consisted of anthropic nitrophile grasses and cultivated graminacea.

Studies in other Mediterranean areas indicate that the climate at the beginning of the Iron Age (Early Sub-Atlantic Iron Age 1000–600 BC) was arid (Van Joolen 2003). Oak acorns were found in a large vase in the village of Genna Maria, Villanovaforru (Marmilla zone), which dates from the Early Iron Age. By an extraordinary chance the external coverings of these acorns had been preserved. Apart from *Hordeum vulgare* (hulled six-rowed barley) a single grain of *Triticum monococcum* L. (einkorn) has been recovered, and also *Triticum* sp. (naked wheat) which, according to Corrie Bakels, could be macaroni wheat (Bakels 2002). Other data on the cultivation of wheat, barley and, perhaps, legumes, have been found at the Early Iron Level of the Toscono di Borore nuraghe (Wetterstrom 1987). The grape seeds found at Duos Nuraghes at the Final Bronze Age level are the squat type with short stalks which are characteristic of *Vitis vinifera* L. var. *sylvestris*. This is a wild species which was often used in this period. Discoveries at Genna Maria indicate however that in the Early Iron Age there were cultivated as well as wild species.

At the Arrubiu nuraghe at Orroli in the Final Bronze Age swine were increasingly important while there was a marked reduction in the number of cattle as well as in the number of deer. The number of sheep and goats remained constant and made up half of the total livestock. Among the wild species that were hunted there were many examples of *Prolagus*. In the Final Bronze and Early Iron Age there was once again an increase in the number of cattle and wild animals (deer, *Prolagus*, birds). There were fewer sheep than goats. However the presence of adults indicates that they were used more for their skins, wool and milk than for their meat. Deer hunting also increased (Fonzo 2003).

Remains of paleofauna have been found at the Toscono di Borore nuraghe. These have allowed us to establish an economic framework in which muflons were used more and indeed made up 50% of the domestic and hunted animals. Cattle made up only 19% and swine 17%.

25% of the animal remains were deer and roebuck (25%) (Webster 1996). In general, from analysis of the paleofauna remains one can see that the number of domestic animals of different species was better balanced than in the previous period, although often there were more sheep and goats or cattle.

The paleofauna remains from the Early Iron Age at Genna Maria and Sant'Anastasia also indicate that there were more sheep and goats, with swine, cattle and deer the next most numerous. However it is clear that cattle followed by swine, deer and roebuck were the most valuable animals, as they provided more meat. The domestic animals were slaughtered young as can be seen from the low percentage of old animals. Their presence however indicates that they were used for other purposes than meat, such as skins, milk or motive power. The high percentage of wild animals is some sites such as Genna Maria and Sant'Anastasia shows that while animal husbandry was the main occupation, hunting was also of great importance.

The remains of shellfish and rare remains of fish show that the sea was also a resource that was exploited, and in some settlements near the sea such as Sant'Imbenia, Alghero they make up a high percentage of the animal remains. They were also found, albeit in smaller quantities, in internal or non-coastal villages such as Sant'Antonio, Siligo and Is Paras, Isili (Wilkens 2003). The rarity of such finds may however be due to the inadequate methods used for examining the sediments.

Analysis of the human remains from the Final Bronze Age in the Gallura region show that there were anthropological differences between the groups buried in tafoni, or natural caves, and those buried in Giant's Tombs. The former suffered from rickets, alveolar pyorrhea, tartar and severe dental wear, while the latter have traces of tumours and hyperostosis (Germanà 1999).

11.3.4 End of the Nuragic Age (Phoenician Influence and then Conquest of the Island)

The Nuragic age seems to have definitely ended with the establishment of Punic coastal settlements in the VI Century BC and then the Carthaginian total conquest of the island in 510 BC. Previous contacts with the Phoenicians in the IX century BC do not seem to have caused substantial changes in the culture, but rather, indeed, to have stimulated other contacts with the Eastern Mediterranean and the Etruscan world. An example of this is the centers of Sulci in the south and Sant'Imbenia in the north. Here there is evidence of both materials imported for the Aegean and the Levant and changes in local production influenced by foreign shapes and technology.

In the later stages of the Early Iron Age there was a marked reduction in the population evidenced by a smaller number of archaeological data. The indigenous community was unable to respond to this crisis, and in later stages it is no longer possible to distinguish with certainty between the processes and actions of the local population and those of people arriving from other places.

11.4 Distribution of the Different Types of Settlements in the Various Landscapes

The Nuragic settlements are found on all types of terrain, although the differences in their density do depend on the type of terrain.

11.4.1 Mountainous Landscape

This is the landscape of mountain ranges and massifs with crests and steep slopes, interrupted by large stretches of table land. The soils have suffered from leaching and erosion. In this terrain alluvial plains cannot be extensive. Here corridor and tholos nuraghe and villages are common. They are not found only on the summits of the reliefs but also along the slopes and in the valley bottoms.

On granite mountain ranges one finds mainly corridor nuraghe that blend into the landscape. Their walls are often supported by rock outcrops and natural forms originating from weathering (tafoni) are incorporated in the nuraghe structure. In addition the presence of broken granite blocks and tors (heaps of rocks) without doubt encouraged construction of the nuraghe by providing ideal building material. Natural shelters (such as caves and tafoni) were used both for living spaces and funeral purposes. The granite terrains with sandy soils have been particularly well adapted to a mixed activity combining agriculture and pastoralism. Evidence of pastoral activity is evidenced by the discovery of milk-boilers used to process it.

In the mountainous areas with a metamorphic substrate such as Gennargentu, Gerrei, and Sulcis Iglesiente there is not a very high density of nuraghe and villages. The softer rocks and clay-rich soils are not suitable for high buildings. Tholos and complex nuraghe are found near the tops of the highest plateaus. These may have been seasonal sites occupied during transhumance. Some of these areas, such as Gerrei and Sulcis-Iglesiente, are rich in minerals (copper, lead, silver, iron and some tin). The nuraghe may have been used to store the metals and even for the first processing of them as indicated by the finding of primary slag in three of the ten nuraghe in the mines areas (Arca and Tuveri 1993). In the mining basin of Iglesias, instead, the density of nuraghe is not particularly high despite the great richness in lead and zinc. This may be due to the low fertility of the soil and the very rough terrain (Giardino 1995).

Sarrabus is another area where there is a clear relationship between the Nuragic settlements and the mineral resources. In this area there is a marked density of complex nuraghe and villages near the silver and copper deposits. However here too, apart from the discovery of casting slag in one nuraghe, there is no definite evidence of the exploitation of the mineral resources in the proto-historic epoch (Usai 1991).

The Mesozoic and Paleozoic limestone rock areas, such as Barbagia and Baronia, are very rugged, with deep canyons and almost vertical slopes. There are many caves in these areas. Although few of the caves have been explored, the little information that we have supports the hypothesis that they were used for cult rituals and, more rarely, were inhabited. In these areas one finds tholos nuraghe and villages, although not at a particularly high density.

11.4.2 Plateau Landscapes

Numerous extensive plateaus exist in Sardinia underlined either by limestone or basalt. Every type of nuraghe and villages occur both on the summits and on their slopes.

The highest and most impressive plateaus are the limestone ones of Sarcidano and Ogliastra, affected deep karst. On the tops of the plateaus rocky outcrops predominate, and poor clay soil is mainly found in low-lying areas. The density of settlement is much lower than on the basalt plateaus.

The highest density of nuraghe is found on the basalt plateaus, with more than one per square kilometer. The plateaus have rocky outcrops and thin but rich soil (at the present the Andosoils are in the highest plateau, about 800 m). There is abundant water thanks to the many springs and marshes. The marshes provide water for the livestock and do not dry up even in the driest seasons. The basalt substrate is also broken up into blocks of various sizes which are suitable for use as building material. These plateaus are of different heights, ranging from 75 to 700 above sea level, but the intensity of distribution of settlements does not seem to be influenced by the altitude.

High, and sometimes very large and isolated, basalt plateaus are typical of the central-west Sardinian landscape. On these plateaus there is a high density of nuraghe, both on their edges and in their interior. More than 50% of the 350 corridor nuraghe of Sardinia are found on these plateaus. Abbasanta is an important example of a basalt plateau. It is 456 km^2 in area and overlooks the valley of the Tirso. During the start of the Middle Bronze Age 101 corridor nuraghe were built here. They are mainly on the flat top of the plateau near the edge, on the slopes, and near streams. There are fewer settlements in the center of the plateau.

Between the end of the Middle Bronze Age and Recent Bronze Age, 293 tholos single tower were built. These were also built in the central areas of the plateau which had previously been little used for settlement. Furthermore, in the Recent Bronze Age 48 single tower nuraghe on the edges and in the center of the plateau were modified into complex nuraghe by the addition of additional towers. There are 125 villages of various sizes on the plateau. They are almost all on the flat top. They may be from the Final Bronze Age as is also the case in other parts of Sardinia, but this cannot be confirmed because of lack of good chronological data (Depalmas 2003).

11.4.3 Hilly Landscapes

In Sardinia there are extensive hilly terrains especially in Nurra, Logudoro, Anglona, Planargia, Marmilla, and Trexenta. The landscape varies, depending on the geology. The areas of sedimentary or volcanic rock are characterized by rocky ridges and gentle hills with rounded to flat summits, of varying heights. Wide, sometimes swampy, valleys separate the hills. In these areas the distribution of the nuraghe often depends on the surface rocks and the soil. High densities of tholos single tower and complex nuraghe and villages are found in areas underlined by as sandstone, limestone, and particularly volcanic rocks with fertile soil.

11.4.4 Lowland Landscapes

Sardinia has a limited amount of lowlands, just 18.5% of the whole island. The largest plains in Sardinia, the Campidano and Cixerri, are of tectonic origin, whereas the smaller ones, such as the plains of Giave and Ozieri, result from the drying out of ancient lakes. Campidano is the largest plain. On its borders there are alluvial fans which extend into the center of the plain, creating wide piedmont slip faces. The western border is noteworthy for its alluvial fans that originate from Paleozoic mountains. Their soil is mainly acid, poor and coarse. The few tholos single tower and complex nuraghe are found near small volcanic outcrops such as those found in the north of Campidano. On the eastern border of the plain there are alluvial fans originating from limestone hills and once again the density of nuraghe is low. The central part of Campidano alluvial plain is characterized by soils with high fertility, flat morphology and wetlands reclaimed in the recent past. There are not many nuraghe but Middle, Recent and Final Bronze Age villages, built with remains of perishable material (logs, branches, packed clay) deep-set in the soil are common.

11.4.5 Coastal Landscape

The coast has a marked variety of landscapes. Most of the coast is rocky. There are high cliffs in the limestone areas and deep inlets and bays in the granite areas with long beaches bordered by wide areas of dunes on the west coasts. There are few coastal plains and they develop preferentially near river mouths and mainly on the east and north coasts of the island. These landscapes are characterized by lagoons, sandbars, and coastal dunes. There are not many nuraghe in coastal areas. Tholos complex nuraghe that are present are preferentially located on slightly higher lands near lagoons and river mouths. They rarely occur on rocky headland areas.

11.5 Case Studies

In order to evaluate the relationship between the Nuragic sites and the landscape we examine in some detail two areas with different environments and high density of settlements: the Tirso valley and the Sinis peninsula.

11.5.1 The Tirso Middle Valley

The Tirso middle valley is located in central Sardinia between the volcanic massif of Monti Ferru and the River Tirso. It is in the southern part of the Ottana Graben. This area is characterized by pyroclastic materials such as ignimbrite and tuffs, basaltic lava, and by fluvial sandstones. There are granite outcrops in the northeastern part of the area.

The morphologic elements of the region consist of a basalt plateau, the Tirso alluvial plain, granite hills, other hilly area, and escarpment areas (Fig. 11.12).

The basaltic plateau has mainly flat top and variable slopes. The plateau top is characterized by small furrowed paleovalleys, depressions, and swamps. The soil is rich and easy to work but very thin, and there are large areas of bare rock. The deeper soils are found in the swampy areas and are hydromorphic and clayey. There are many seasonal springs all over the plateau. There is a high density of nuraghe all over the top of the plateau, including groupings of three to seven nuraghe. The highest concentrations are found along the small paleovalleys and near the swamps. The corridor nuraghe are mainly found near the edges of the plateau. Paleobotanic studies of nearby nuraghe such as Toscono and Duos Nuraghes have found that the area was used for grazing and for the cultivation of cereals.

The upper parts of some slopes have rock scarps subject to rockfall. This has affected the nuraghe built near the edge during the Bronze Age, as indicated by the presence of supporting walls and repairs to some walls (Fig. 11.13). The middle and lower slopes have been subjected to water erosion. There are small flat areas of deep rich soil on which corridor nuraghe and villages were built, although less than on the plateau itself.

The plain of the River Tirso is separated into two parts by highlands (Fig. 11.12). The northern upper part is an alluvial plain subject to flooding terraces with generally poor acid soils. No nuraghe are present. The southern lower part is presently occupied by a water reservoir (Omodeo Lake), the rest has alluvial terraces with rich soils and many single tower tholos nuraghe have been built there.

Low granite hills with gently sloping sides and wide valleys with flat bottoms are found in the north-eastern sector of the area. There is widespread intense erosion in the hills and the soil is very thin with areas of bare

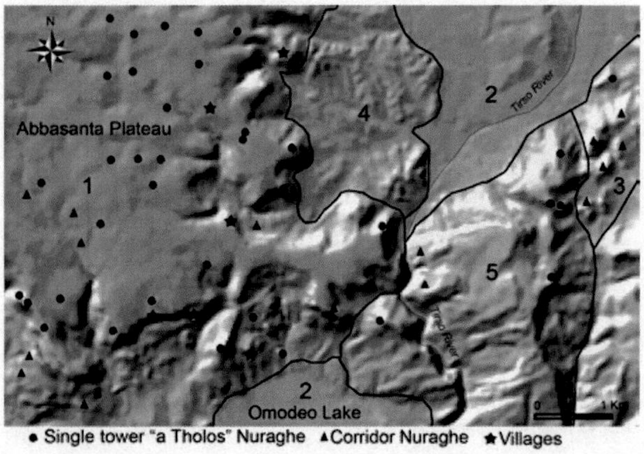

Fig. 11.12 Digital terrain model (DTM) of the central Tirso valley showing morphologic units and distribution of nuraghe in the Sedilo area: *1* basalt plateau, *2* alluvial plain, *3* granite hills, *4* gently hills, *5* cuestas

Fig. 11.13 The Talasai nuraghe of Sedilo with its supporting wall along the edge of the basalt plateau, suffering from subsidence. (After Tanda 1996)

rock. The only nuraghe present are of the corridor type and these are concentrated on the middle and upper slopes, near springs and small flat areas where the soil is deeper.

Between the basalt plateau and the valley of the Tirso River there are gently rolling hills with volcanoclastic and sandstone substrate. There is intense erosion and leaching and the soils are thin and of poor quality. There are no nuraghe in this area, also because there was no suitable building material.

The escarpment area is characterized by cuestas (asymmetric ridges, steep on one side and gently sloping on the other) of ignimbrite and tuffs. Rock falls and rill erosion affect the steepest slopes. The soils are very thin with areas of rocky outcrops. The nuraghe are almost exclusively of the corridor type and are found on the slopes facing the Tirso River, on small flat areas, and near springs (Melis 1998).

11.5.2 Sinis Peninsula

The Sinis peninsula is located at the north-west end of the Campidano plain, north of the Gulf of Oristano. It is underlain by a variety of Tertiary rocks and sediments including andesites and basalts, limestones, and Quaternary continental to costal and marine deposits, mainly calcareous sands.

The Sinis peninsula is mainly characterized by gently rolling hills with an average elevation of sixty meters above sea level. There are clear differences in the geomorphology of the northern and central-southern parts (Fig. 11.14). The northern part is a flat plain with large lagoons or freshwater marshes delimited to the north by high cliffs and aeolian dunes. The southern-central part by contrast has gently rolling hills with flat limestone tops, loamy in the central section. These are separated by wide paleovalleys. There is a small north–south oriented basaltic plateau in part of the southern area. The eastern side of this has a straight concave slope that faces toward the coastal Cabras lagoon. The western side slopes gradually to the coastline. The coastline is mainly low and sandy with rocky headlands. There is great variety of soils, the richest ones being developed on top and slopes of the basalt and on limestone terrains.

On the peninsula there are numerous single tower and complex tholos nuraghe, particularly in the southern part, but no corridor nuraghe (Fig. 11.14). The oldest example of the Nuragic village is that of Murru Mannu that dates from the early Middle Bronze Age. Most (45%) of the 69 single tower tholos nuraghe are found on top and slopes of the plateau. On the slopes the density is particularly high with more than two per square kilometer. Most of the remaining nuraghe (43%) occurs on hills near wetlands, and the rest near the low-lying coasts (9%) and on coastal headlands (3%). In the ratio is one complex nuraghe (total number: 37 units) for every three single tower nuraghe.

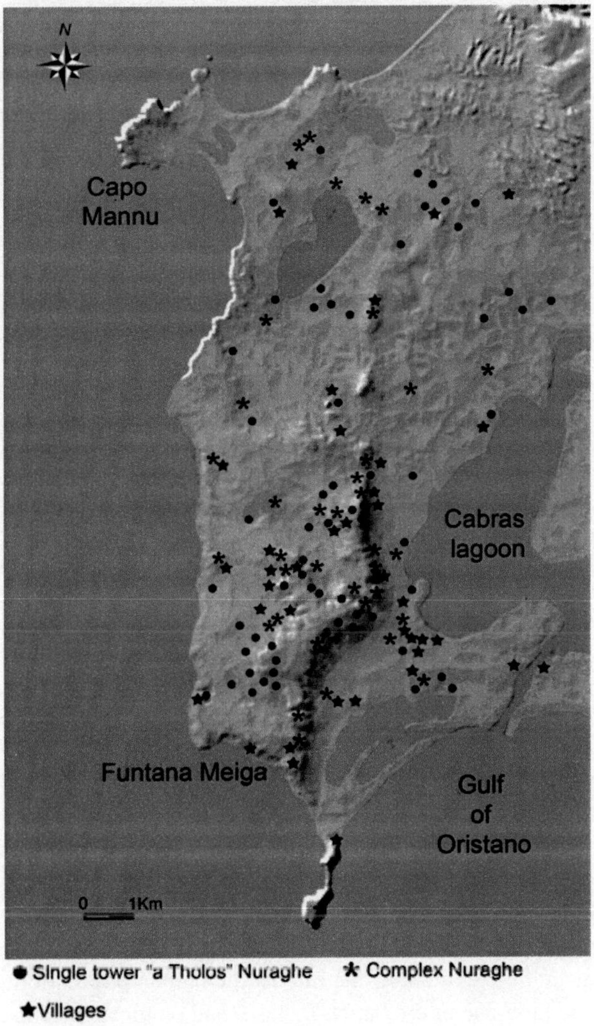

Fig. 11.14 Digital terrain model (DTM) of the Sinis area and distribution of nuraghe

In the Final Bronze Age and Early Iron Age there were two significant changes. One was the establishment of 31 new villages. Some of these were also built on the plains near the lagoons and the coast.

11.6 Discussion

At the beginning of the Middle Bronze Age the formative period of Nuragic culture was marked by the construction of corridor nuraghe. Data from the few excavations indicate that they were used as dwelling places for family groups. The low density of these nuraghe also indicates that the society was still not very cohesive and consisted of isolated, small family groups. Most of these nuraghe were located in places such as basalt plateaus where the land could be used for pasture and, to a lesser extent, agriculture.

During this phase, bronze axes become a very important weapon. One of the possible interpretations of this is their use for more intense deforestation to obtain open spaces for agriculture and grazing. Another indication of increasing agricultural activities is the widespread use of pans. These are particularly suitable for cooking and preparing cereal foods, as are the bell-shaped pans, which can be used for baking sheets of bread. In addition the many milk-boilers have been found that were used for processing milk and for preparing milk based products. The lack of paleobotanic and stratigraphic data from this period makes it difficult to evaluate what impact the human population and their use of the land may have had on the area. The prevalence of sheep and goats, but also swines and deer among the paleofauna may indicate that the economy was still linked to exploitation of the forests.

In the final phase of the Middle Bronze Age a more elaborate architecture developed with more advanced techniques. This was an important moment in the development of Nuragic civilization. The spread of single tower tholos nuraghe was associated with a great increase in the population. This can be seen from the way in which the buildings spread out all over the island and were built on different landscapes and in areas not inhabited in preceding periods. The occurrence of small groups of nuraghe may reflect the tendency of families to associate in larger communities.

The distribution of nuraghe suggests possible hypotheses about their functions. These monuments on highlands did not always occupy the optimal defensive position. They did not occupy the highest elevation in an area and did not have the best view of the surrounding land. In contrast to the settlements of the Italian Peninsula that were built for defensive purposes (di Gennaro 1996), the Nuragic community did not generally appear to be particularly worried about possible threats from outsiders. The nuraghe built along the coast are not planned in such a way as to form part of a defensive network (Depalmas 2002).

Although the paleobotanic and pollen data show that agriculture, in particular the cultivation of cereals and legumes (*Vicia faba*), was developing, the terrain was still heavily wooded. Thus agriculture did not significantly degrade the terrain and humans had little impact on the environment. Indeed the remains of game are consistently found, and confirm that the island was still covered with well-developed forests. Different factors played a part in creating the new situation, whose effects can be seen not only in the more advanced architecture but also in the growth in population, the spread of economic activities throughout the island, and the improvements in the economic conditions of the population and their quality of life. The result was the increase in the average stature of the individuals and an overall reduction in diseases, even if a widespread presence of porotic hyperostosis—which may also have been due to malaria—is recorded.

In the Recent Bronze Age, society was more complex as indicated by construction of complex nuraghe. These structures are often controversially interpreted as proof that there was an increased stratification of society, made evident by the building of what may have functioned as palaces and castles, the residences of chieftains. For the entire Nuragic period there is, however, no strong evidence of differences in social status in the community. Indeed no indicators have been found of role distinctions either in the artefacts connected to powerful figures, or in the single tombs, or in distinctive grave goods. On the other hand the complex towers were built using the same system of construction as the simple towers, but with the addition of service areas such as courtyards and staircases. In addition there are no areas particularly designed for administrative or ceremonial functions. Nuraghe of this type are found in zones where the surrounding resources were abundant. An example of this is the Sinis peninsula where there is a large number of complex nuraghe in areas rich in resources. Their distribution seems to be connected to a society that had moved beyond the family group and was already organized in territorially based tribes. The increased use of the land can be seen from the reduction in the size of the forests—including deliberate burning—and further increases in the area under cultivation or pasture.

Major changes in the society continued in the period between the Recent and Final Bronze Age. We do not know whether they were fuelled by any economic or environmental reason. The new society of the last part of the Bronze Age lived in villages, although they also continued to use, or rather reuse the nuraghe as storehouses or for religious rituals. It was a particularly flourishing society which also developed many specialised crafts, among which metal-working is a prominent example. In the Final Bronze and Early Iron Age villages there were places that can be considered specialised workshops as indicted by the way that they were equipped and the tools that were used in them. The new villages often grew up around the nuraghe. The occupation of new areas for farming and grazing was the result of increased deforestation by burning. There was more variety in the crops that were grown, and wild berries and fruit were also consumed. Although grapes were probably used for wine, as can be seen from the widespread discovery of askoid jugs, these may not have been domesticated grapevine species. The numerous agricultural and craftsmen's tools found in these phases confirm that productive activity developed, as does the establishment of grain stores and the widespread use of large storage vases.

In the Final Bronze Age and Early Iron Age not only was animal husbandry, especially of cattle, widely practiced, but hunting also acquired new vigour. This may have been connected to the habits of the elite. The discovery of the remains of adult domesticated animals may be due to increased use of wool and leather. The increase in the number of cattle, and especially adult animals that were thus used for all their working lives, perhaps as draught animals, may be another indication of the increase in agriculture. The human bones found in Gallura allow us to hypothesize that there were social differences between different groups in the area. Malnutrition was found mainly in those occupied in pastoral activities and buried in caves, while those buried in Giant's Tombs were better fed. The presence of bronze statuettes showing different social categories, such as warriors, priests and craftsmen supports this hypothesis.

In the advanced stages of the Early Iron Age the population decreased. It is uncertain whether this was caused by the climate becoming more arid, or by environmental degradation due to overuse of the land, or by combination of the two effects. At present there is insufficient data to fully explain the phenomenon.

11.7 Conclusions

The typical constructions and products of Nuragic civilization, such as nuraghe, Giant's tombs, cult sites, pottery and metal objects are homogeneous all over Sardinia. This cultural homogeneity is preserved even if some changes occurred during the different historical phases. Thus in general one can say that local conditions did not vary so greatly that they negated this homogeneity. By contrast strong local differences are noted in other islands such as Sicily or Corsica. Indeed, in the nearby Corsica, towers are found almost exclusively in the south of the island, and their almost complete absence in the north of the island suggests that there were great cultural differences.

Nuraghe are found all over Sardinia but a closer examination of the situation in different areas shows that there were differences in their concentration or scarcity. This was linked to the landscape. Analysis of the relationship between the monuments and the settled areas allows one to state that the presence of nuraghe was mainly determined by the environmental resources and the possible economic exploitation of the area, rather than any defensive consideration. Although paleoenvironmental data are limited, a trend emerges of the populations of local areas expanding and occupying—in regions with more resources—every available ecozone. The intensive exploitation for agriculture and grazing of these areas was achieved by reducing the forested area, both by cutting down trees and by burning. This was a gradual process which substantially modified the landscape. The strategy employed to use the land profitably involved using a wide range of bronze and stone tools and animal power (draught animals). This human impact may have caused a significant degradation of the environment, which occurred primarily in the final stages of Nuragic civilization.

Acknowledgements We like thank Peter Norton for traducing our original Italian manuscript into English.

References

Arca M, Tuveri C (1993) Nota sulle miniere di rame in Sardegna. In: Kirova TK (ed) L'uomo e le miniere in Sardegna. Edizioni Della Torre, Cagliari, pp 21–24

Badas U (1993) Il nuraghe Bruncu Madugui di Gesturi: un riesame del monumento e del corredo ceramico. Quaderni della Soprintendenza Archeologica per le province di Cagliari ed Oristano 9:31–76

Bagella S, Depalmas A, Manunza MR et al. (2000) Forme vascolari dell'età del Bronzo in Sardegna. In: Atti Congresso Nazionale Criteri di nomenclatura e di terminologia inerente alla definizione delle forme vascolari del Neolitico/Eneolitico e del Bronzo/Ferro, Octavo, Firenze 2000, pp 491–503

Bakels C (2002) Plant remains from Sardinia, Italy, with notes on barley and grape. Vegetation History and Archaeobotany 11:3–8

Contu E (1997) La Sardegna preistorica e nuragica. La Sardegna dei nuraghi, 2, Chiarella, Sassari, 520 pp

Depalmas A (2002) Approdi e insediamenti costieri nella Sardegna di età nuragica. In: Atti del Quinto Incontro di studi Paesaggi d'acque. Preistoria e Protostoria in Etruria, Milano, pp 391–402

Depalmas A (2003) Scelte insediative e aspetti del popolamento nella Sardegna di età nuragica. Histria Antiqua 11:13–21

Depalmas A (2005) Le navicelle di bronzo della Sardegna nuragica. Gasperini, Cagliari, 398 pp

di Gennaro F (1996) L'Italia centrale e meridionale (part of chapter by Cardarelli A, di Gennaro F: L'Italia). In: Belardelli C, Neugebauer JW, Novotná M, Novotny B, et al. (eds) The Bronze Age in Europe and the Mediterranean (XIII International Congress of Prehistoric and Protohistoric Sciences, Forlì), pp 259–266

Dyson SL, Rowland RJ (2007) Archaeology & history in Sardinia from the Stone Age to the Middle Ages: Sheperds, sailors, and onquerors. UPenn Museum of Archaeology, Philadelphia, 240 pp

Fadda MA (1988) La fonte sacra di Su Tempiesu. Guide e itinerari. Delfino, Roma, 41 pp

Fadda MA (1998) Nuovi elementi di datazione dell'Età del Bronzo Medio: Lo scavo del Nuraghe Talei di Sorgono e della tomba di giganti Sa Pattada di Macomer. In: Balmuth MS, Tykot RH (eds), Sardinian and Aegean Chronology. Studies in Sardinian Archaeology V. Oxbow Books, Oxford, 179–193

Floris G, Sanna E (eds) (1999) L'Uomo in Sardegna. Aspetti di Antropobiologia ed Ecologia umana. Zonza Editori, Quartu Sant'Elena, 219 pp

Fonzo O (1987) Reperti faunistici in Marmilla e Campidano nell'età del bronzo e nella prima età del ferro. In: Atti del II Convegno di studi La Sardegna nel Mediterraneo tra il secondo e il primo millennio A.C. Stef, Cagliari, pp 233–242

Fonzo O (2003) L'ambiente e le sue risorse: la caccia e l'allevamento del bestiame. In: Cossu T, Campus F, Leonelli V, et al. (eds) La vita nel nuraghe Arrubiu. Grafica del Parteolla, Dolianova, pp 113–133

Germanà F (1999) Malattie e culture della Sardegna preistorica. In: Floris G, Sanna E (eds) L'Uomo in Sardegna – Aspetti di antropobiologia ed ecologia umana. Zonza, Quartu Sant'Elena, pp 21–39

Giardino C (1995) Il Mediterraneo occidentale fra XIV ed VIII secolo A.C.: cerchie minerarie e metallurgiche. British Archaeological Reports, International Series 612, Oxford, 368 pp

Lilliu G (1999) La civiltà nuragica. Carlo Delfino, Roma, 242 pp

López P, López Sáez JA, Macías R (2005) Estudio de la paleovegetación de algunos yacimentos de la Edad del Bronce en el SE de Cerdeña. Ruiz-Gálvez M (ed) Territorio nurágico y paisaje antiguo la Meseta de Pranemuru (Cerdeña) en la Edad del Bronce. Fernández Ciudad, Pinto, pp 91–105

Lo Schiavo F (1981) Economia e società nell'età dei nuraghi. In: Atzeni E, Barreca F, Ferrarese Ceruti, et al. (authors) Ichnussa: la Sardegna dalle origini all'età classica. Garzanti-Scheiwiller, Milano, pp 255–347

Lo Schiavo F (2005) Archaeometallurgy in Sardinia: from the origins to the early iron Age. Monique Mergoil, Montagnac, 412 pp

Melis RT (1998) La ricostruzione paleo-ambientale come strumento di indagine archeologica. In: Tanda G (ed) I monumenti nel contesto territoriale. Antichità Sarde. Studi e Ricerche, 3/III, Soter, Villanova Monteleone, pp 9–20

Moravetti A (1992) Il complesso nuragico di Palmavera. Guide e Itinerari. Delfino, Roma, 139 pp

Moravetti A (1998a) Ricerche archeologiche nel Marghine-Planargia, I. Delfino, Roma, 714 pp

Moravetti A (1998b) Serra Orrios e i monumenti archeologici di Dorgali, Guide e Itinerari. Delfino, Roma, 123 pp

Pacciarelli M (2000) Dal villaggio alla città. La svolta protourbana del 1000 A.C. nell'Italia tirrenica. All'Insegna del Giglio, Firenze, 210 pp

Rovina D (2002) Il santuario nuragico di Serra Niedda a Sorso (SS). Beta Gamma, Viterbo, 31 pp

Santoni V (1985) I templi di età nuragica. In: Atzeni E, Contu E, Sardegna, et al. (authors) Sardegna preistorica. Nuraghi a Milano, Electa, pp 181–207

Tanda G (ed) (1990) Ottana: Archeologia e territorio. Studio Stampa, Nuoro, 232 pp

Tanda G (ed) (1996) Sedilo 1: I monumenti del progetto "Iloi" nel contesto territoriale comunale. Antichità Sarde. Studi e Ricerche, 1/III. Soter, Villanova Monteleone, 268 pp

Tanda G (ed) (2003) Sedilo 7: La tomba di giganti 2 di Iloi (Sedilo-OR). Antichità Sarde. Studi e Ricerche, 1/IV. Soter, Villanova Monteleone, 258 pp

Usai D (1991) Modelli d'insediamento nel Sarrabus dal Neolitico all'età del Bronzo. Quaderni della Soprintendenza Archeologica per le province di Cagliari ed Oristano 7:117–134

Van Joolen E (2003) Archaeological land evaluation, a reconstruction of the suitability of ancient landscapes for various land uses in Italy focused on the first millennium B.C. Phd Thesis, Groningen Institute of Archaology, Groningen, 291 pp

Webster GS (1996) A Prehistory of Sardinia 2300-500 B.C. Sheffield Accademic Press, Bath, 223 pp

Wetterstrom W (1987) A preliminary report on the plant remains from nuraghe Toscono. In: Michels J, Webster GS (eds) Studies in Nuragic Archaeology: Village Excavations at Nuraghe Urpes and Nuraghe Toscono. British Archaeological Reports, International Series 373, Oxford, pp 93–103

Wilkens B (2003) La fauna sarda durante l'Olocene: le conoscenze attuali. Sardinia, Corsica et Baleares Antiquae I:181–197

Chapter 12
Floods, Mudflows, Landslides: Adaptation of Etruscan–Roman Communities to Hydrogeological Hazards in the Arno River Catchment (Tuscany, Central Italy)

Marco Benvenuti, Cristina Bellini, Gianfranco Censini, Marta Mariotti-Lippi, Pasquino Pallecchi, and Mario Sagri

12.1 Introduction

Since the Neolithic, large fluvial catchments of the Middle East, North Africa, and the Mediterranean favored the onset and development of complex human communities that adopted significantly different socio-economic practices from those of the nomadic, hunter-gatherer cultures of early Prehistory (Redman 1978; Bellwood 2004). Alluvial soil fertility triggered the onset of agriculture that in turn favored permanent settlements, conservation and trade of excess harvest. The natural environment, thus, stimulated human creativity in exploiting, managing and reshaping the land surface. This interdependence between humankind and the environment has significantly changed over time.

In the early–middle Holocene, river flooding was perceived first of all as a chance to get better harvest from weatherable minerals and nutrients supplied by seasonal floods to the soils (Bellwood 2004). Hydraulic engineering and land reclamation drastically changed the physiography of alluvial plains in attempting to control their natural geomorphic and depositional dynamics regulated by climatic, tectonic and, in coastal plains, eustatic factors. Large towns, which flourished around fertile croplands, attracted trading and differentiated handicrafts. With expanding urban centers, alluvial plains were deeply transformed and progressively occupied by different and varied activities other than agriculture.

In historical times, spaces for accommodating river flooding and natural channel migration were progressively reduced, making river dynamics to be felt more as a threat than an opportunity. Since then, the early positive loop between man and the environment has changed into a difficult coexistence becoming dramatic during floods that, though not particularly large, can be felt as catastrophic events. Millennia to century-scale climatic fluctuations punctuating the Holocene (Mayewski et al. 2004), complicated such a coexistence forcing human adaptation to a rapidly changing environment. Alluvial plains characterized by regular and predictable floods may have been affected in few decades or centuries by higher magnitude and differently distributed floods or by droughts triggered by changed climatic conditions.

This instability may have made prompt reactions of human communities difficult, with negative consequences on their development. Geomorphic conditions and climatic changes may also have determined dangerous environments for human settling and activities at the footslopes of fluvial valleys or in mountain catchments. Gravity-driven processes, such as landslides and debris- or mud-flows, may have become serious menaces for ancient communities under changing climates and/or over-exploitation of mountain slopes, for instance through extensive deforestation.

Italy provides several examples of the importance of alluvial environments to the Neolithic establishment and later development of human settlements, most notably the Tiber, Po, and Arno watersheds. This chapter analyzes a geoarchaeological transect on the Arno River drainage basin in central and NE Tuscany (Italy), based on three archaeological sites spanning the Etruscan and Roman periods and distributed from the lower reaches of the river to its source area (Fig. 12.1). The aim of this paper is to show how ancient Italic

M. Benvenuti (✉)
Dipartimento di Scienze della Terra, Università di Firenze
Via G. La Pira 4, 50121 Florence, Italy
e-mail: oredep@unifi.it

See Plate 12 in the Color Plate Section; also available at: extras.springer.com

I. P. Martini, W. Chesworth (eds.), *Landscapes and Societies*,
DOI 10.1007/978-90-481-9413-1_12, © Springer Science+Business Media B.V. 2010

civilizations adapted and reacted to, eventually managing (albeit imperfectly), the hydro-geological risks on alluvial plains and mountain catchments exploited for commercial, residential, and religious purposes.

12.2 Regional Setting

The Arno River (Tuscany, Central Italy) is 241 km long, draining a watershed of about 8830 km² with a mean elevation of 353 m asl (Fig. 12.1). The river has a complex course which crosses different landscapes resulting from geologic processes active since the Miocene. The upper part of the catchment is located within the axial chain of the Northern Apennines, mostly composed of thick arenaceous-pelitic successions dating back to the early–late Miocene (Tuscan-Umbro-Marchean successions; Vai 2001). Locally, these are tectonically overlain by thick claystone, limestone, sandstone, and ophiolites, Jurassic–early Cenozoic in age (Ligurid succession; Vai 2001). On the whole, these different successions record a continental collision between the European and African plates. Initially, these events are attested by foredeep basins dominated by turbidites (Tuscan-Umbro-Marchean successions; Vai 2001). Since the late Oligocene, the depositional basins migrated toward ENE following a progressive crustal thrusting and folding due to the Africa–Europe continental collision (Vai 2001). Such a deformation brought the Ligurid succession, formed in an oceanic domain interposed between the continental landmasses, to thrust over and progressively suture the foredeep basins.

Since the middle Tortonian, the axial chain has emerged and has been affected by differential subsidence that determined, from the late Pliocene, the opening of several intermontane basins. The latter, considered traditionally to have developed in a domi-

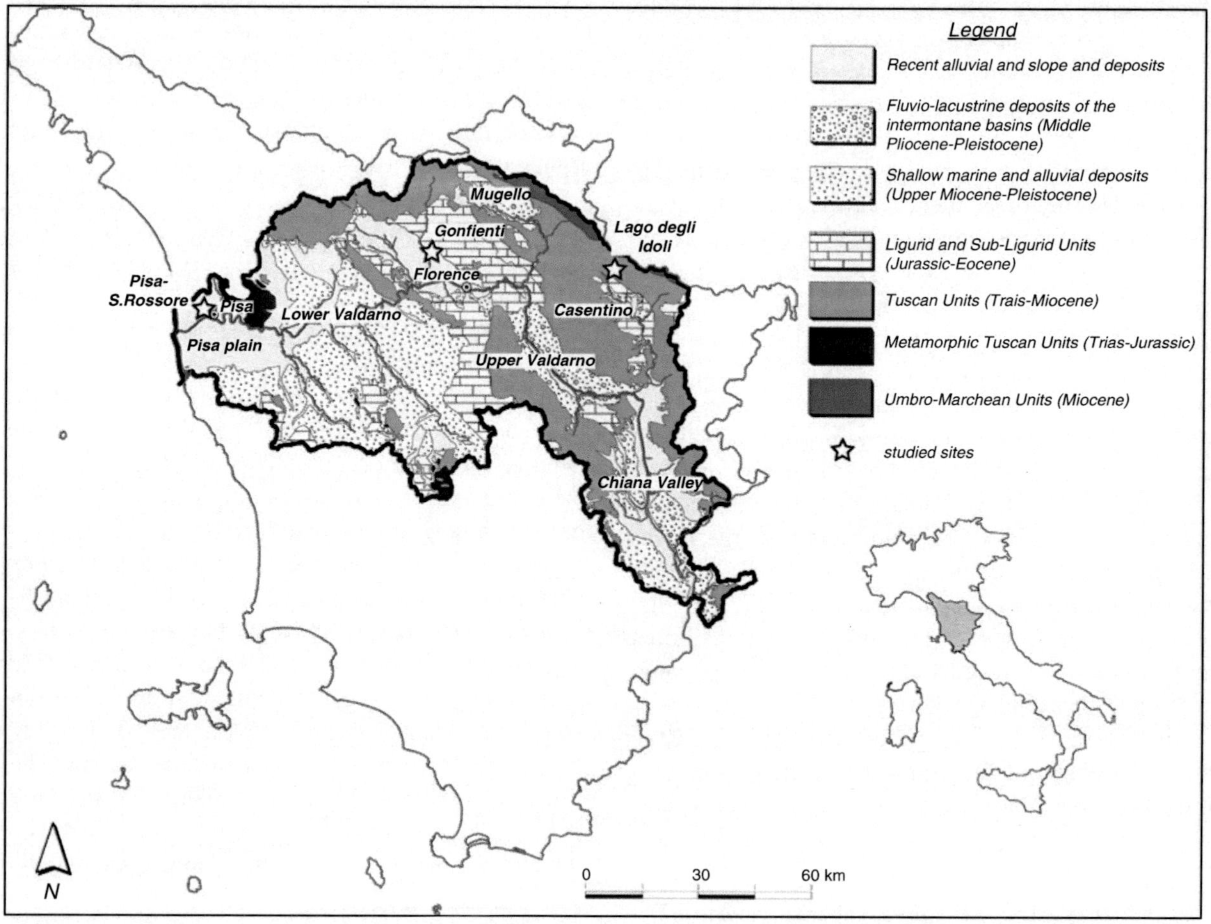

Fig. 12.1 Schematic geological map of the Arno River drainage basin and location of the studied sites. (After Carmignani and Lazzarotto 2004) (Administrative boundary of the Tuscan Region is outlined)

nant extensional regime (Martini and Sagri 1993; Martini et al. 2001), have been recently referred to a more complex tectonic setting including upper Pliocene–lower Pleistocene dominant compression (Sani et al. 2009). The tectonic activity was a main factor in creating space to accommodate up to 1000 m thick fluvio-lacustrine clastic successions in these basins (Martini and Sagri 1993; Martini et al. 2001; Sani et al. 2009). The upper-mid course of the Arno River crosses several of these basins, such as the Casentino, Upper Valdarno, and the Florence basins. The middle–lower reaches of the Arno River cross wide plains underlined by Pliocene and Quaternary fluvial and shallow marine terrigenous deposits up to 2500 m thick. Finally, the Arno River reaches the Tyrrhenian Sea coast and develops a cuspate delta (Pranzini 2001).

Human occupation of the Arno River drainage basin is recorded since the Middle Pleistocene when early–middle Paleolithic people moved onto this territory, mostly on a seasonal basis, for hunting and gathering activities. Early settlements on or near the alluvial plains are documented since the late Paleolithic becoming only in the late Neolithic and Bronze–Iron ages more complex and stable (Martini et al. 1999). Several Neolithic and Eneolithic (Italian Copper Age) settlements have been found in the Arno Valley, where widespread wetlands forced the construction of draining channels as documented in the western periphery of Florence. The Etruscan and Roman people induced significant morphological changes to the alluvial plains by building relatively large urban centers, reclaiming wetlands for extensive agriculture, and construction of road networks. Mountainous areas were largely exploited for wood and locally used as sacred places due to the high spiritual significance of the water springs and forests.

12.3 Methods

Methodology applied in this study benefited from multidisciplinary contributions from different branches of Geology and Paleobiology. Aerial and satellite image analysis and GIS analysis of high-resolution digital topographic maps were used to produce geomorphologic descriptions.

In two of the three studied sites (Gonfienti and Lago degli Idoli) geophysical surveys allowed recognition of archaeological remains in the shallow subsoil and a better paleoenvironmental understanding. Specifically, a survey with the Ground Penetrating Radar was carried out in the Gonfienti site by means of a low frequency antenna (100 MHz) system to explore the low resistivity alluvial soils. At the Lago degli Idoli, seismic data were collected with an OYO McSeis SX 24 channel Seismograph with geophones separation of 3 m along two explored sections. Acquired data were elaborated with specific software to obtain a WET (Wavepath Eikonal Traveltime) seismic tomography. Sedimentary facies description and interpretation have been performed on the archaeological excavations through detailed logging and correlation. On the exposed sections sampling for palynological and geochronological analyses was carried out. Palynological analyses were conducted through standardized chemo-physical laboratory treatment of small volumes of fine-grained sediments; the residuals were observed at light microscope at 400–1000× for pollen grain identification and counting for statistical elaboration (Faegri and Iversen 1989). Charcoal, wood fragments and mollusc shells were collected for standard and AMS C^{14} dating carried out at research and commercial laboratories.

12.4 Case Studies

12.4.1 *Terminal Reach of the Arno River: The Roman Fluvial Harbor of S. Rossore, Near Pisa*

12.4.1.1 General Setting

This site is located in the lower alluvial plain of the Arno and Serchio rivers at about 11 km from the Tyrrhenian Sea (Fig. 12.2a). This wide plain is bounded to the north by highlands including the Pisani Mountains that are composed of late Paleozoic–early Mesozoic metamorphic and mainly sedimentary clastic rocks (Rau and Tongiorgi 1974).

The overall morphology of the Arno and Serchio river floodplains and adjacent coastal plain was reached during the Late Pleistocene–early Holocene (Aguzzi et al. 2005, 2007), but superficial changes have occurred since then due to natural and anthropogenic factors. Historical chronicles from the geographer Strabo and the poet Rutilio Namaziano report a rather different hydrography of the plain from the present one (Della Rocca et al. 1987). For example, according to Rutilio Namaziano

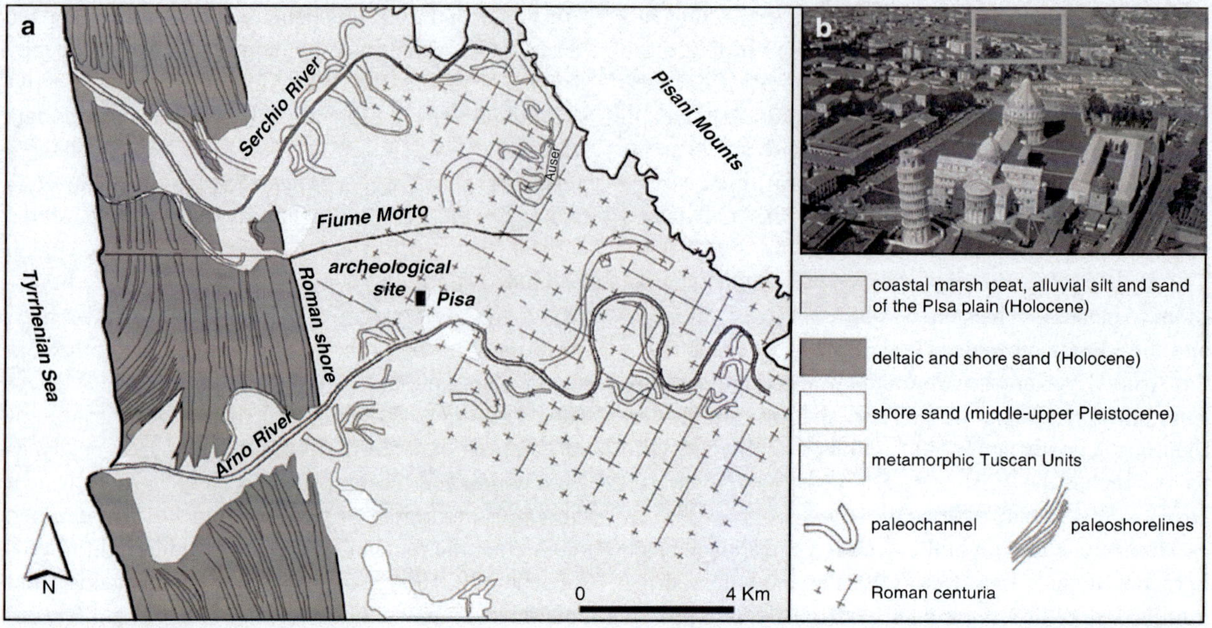

Fig. 12.2 **a** Schematic geomorphologic map of the Pisa plain with location of the Pisa-S. Rossore site (Distribution of paleoshorelines, paleochannels and of the Roman 'centuria' after Della Rocca et al. (1987). **b** Oblique aerial view of the Piazza dei Miracoli downtown Pisa, the squared area indicates the approximate location of the archaeological site

(*De Reditu Suo*, first book), the Ausur (more frequently cited as Auser) River—an ancient precursor of the Serchio River—flowed into the Arno River at the site of Roman Pisa. In the fifth century its course was diverted and the Serchio River now flows directly into the sea about 10 km NW of Pisa. Detailed reconstructions of the ancient hydrography around Pisa have been proposed among others by Bruni and Cosci (2003) for the late Roman age and by Carratori et al. (1994) for the Middle Age (Martini et al. this volume), basing on remote sensing, geomorphologic and historical data.

Furthermore, in Roman times the coastline was about 4–4.5 km west of the archaeological site of Pisa-S. Rossore whereas now it is 11 km away (Benvenuti et al. 2006).

12.4.1.2 Description, Stratigraphy, and Paleoenvironment of the Site

In late 1998, a 5 m-deep excavation on a surface of about 3000 m^2 located 2 km W of the Tower of Pisa (Fig. 12.2b), brought to light several shipwrecks and many other archaeological materials, mostly related to the ship cargoes, attesting the vestige of a previously unknown ancient urban wharf. At a first glance, the archaeological site resembled a chaotic and meaningless mix of ships, scattered wood planks deriving from the ship plating, thousands of complete and broken amphorae, and other various materials encased in a sandy matrix (Fig. 12.3; Camilli 2002; Bruni 2003). A detailed stratigraphic analysis performed on several trenches and extended through shallow cores (Fig. 12.4), allowed to re-order this exceptional record of an unexpected riverine urban wharf buried in the shallow subsoil of Pisa (Benvenuti et al. 2006).

The stratigraphic relations between archaeological materials and encasing sediments outline the occurrence of four main sand-rich, cross-stratified, units (units 1–4 in Fig. 12.4; Benvenuti et al. 2006). In the southern portion of the site, a fifth sand-rich deposit, unit 0, has been recognized. This poorly exposed unit, composed of intervening sands and muds in horizontal, lensing, and cross-stratified beds, represents a complex sedimentary body accumulated possibly since the late Etruscan period and throughout the Roman Age. This is indicated by radiocarbon dates of woods from an *in situ* wooden palisade (Fig. 12.3c) attesting a possible Etruscan docking site, and from poorly-preserved remains of a ship (Belluomini et al. 2002; Benvenuti et al. 2006).

Units 1–4 are composed of centimeter- to meter-thick planar inclined beds of medium- to coarse-grained sand encasing archaeological materials ranging in size from entire ships to pottery clasts (Fig. 12.3a, b, d, e).

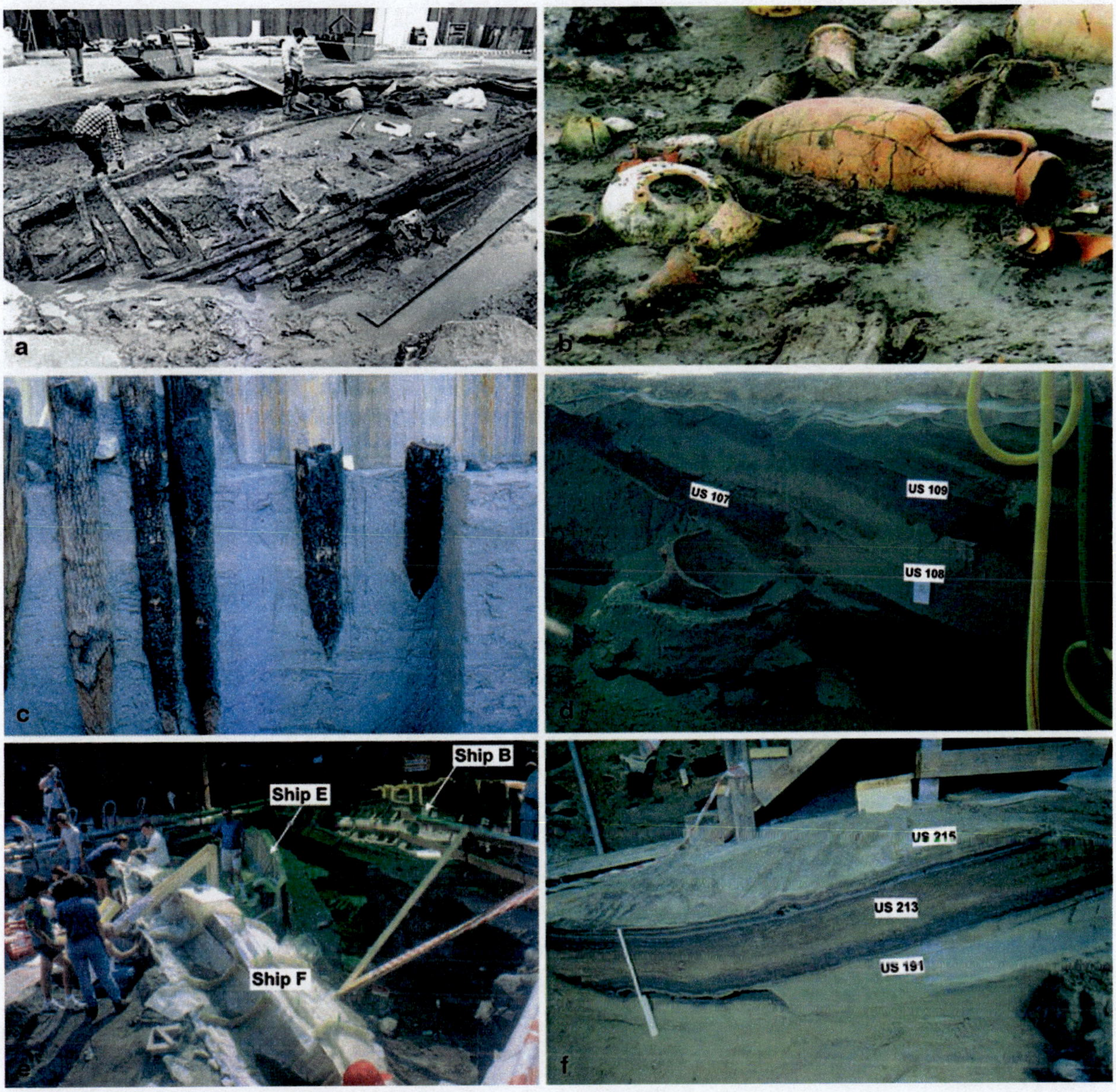

Fig. 12.3 Materials and sediments from the Pisa-S. Rossore excavation site. **a** Ship D originally buried in unit 4 (see Fig. 12.4 for location in the site). **b** 'Amphorae' dispersed in the sediment of unit 0. **c** The Etruscan wooden palisade fixed in deposits of unit 0; wood stakes are 0.4–1 m long. **d** NW-dipping cross-bedded sands of unit 1, encasing an amphora fragment near ship C, outcrop is 0.8 m high. **e** Imbricated disposal of ships F, E and B within units 1 and 2 looking to the SE: the long axes of these ships are transversal to the main paleoflow direction to NNW, and their keels dip at the same angle of the cross-bedding of the encasing sands. **f** Sandy silt, clay and vegetal debris (darker beds) draping sand in unit 1, trench close to ship C, the rod in the foreground is 0.6 m long

Cross-bedded sands in the different units indicate sediment accretion on the fronts of composite bedforms by paleoflows directed mainly towards NNW and subordinately towards WSW. These bedforms migrated within channelized troughs, mostly elongated in a NNW-SSE direction, cross-cutting older units (Fig. 12.4; Benvenuti et al. 2006).

Typically, the sands are draped by silty, clayey beds rich in vegetal debris and subordinate archaeological material (Fig. 12.3f). Archaeological and radiometric (C^{14}) dating of the materials collected in these units and performed with conventional technique indicates a maximum time interval comprised between the eighth century BC and the fifth century

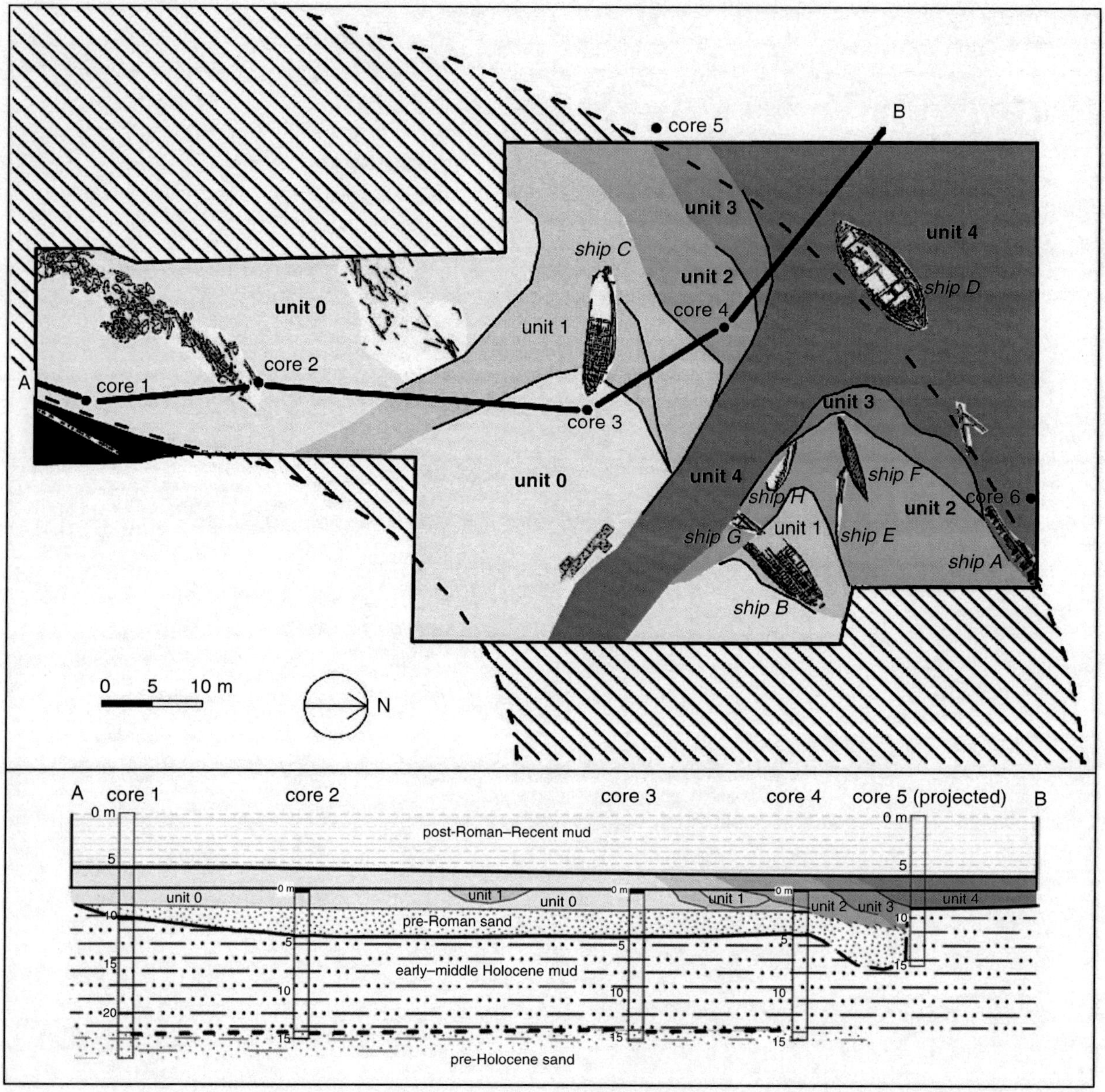

Fig. 12.4 Plan view of the distribution of major archaeological findings and encasing deposits, subdivided in units described in the text, in the Pisa-S. Rossore trench excavated in late 1998. Hatched area outlines the inferred meandering geometry of the abandoned channel where the wharf was built. Geological section A–B integrates data from the exposed deposits and from several shallow cores drilled in the site and in the surrounding area

AD (Fig. 12.5; Belluomini et al. 2002; Martinelli and Pignatelli 2008).

Each unit is interpreted as the product of catastrophic overbank floods occasionally re-entering and partially silting up an abandoned channel. This conclusion is supported by the integration of stratigraphic and sedimentological data collected in the site (cross section A–B on Fig. 12.4) and specifically by: a) the stratigraphic correlation of shallow cores drilled in the area showing an abrupt transition between fluvial sand devoid of archaeological remains (pre-Roman sand in cross-section A–B of Fig. 12.4), considered to be the active channel fill, and the overlying sand, bearing Etruscan and Roman materials, subdivided in units 1–4; b) the geometry and paleocurrent of sandy bodies stacked in these units, that point to repeated incision of

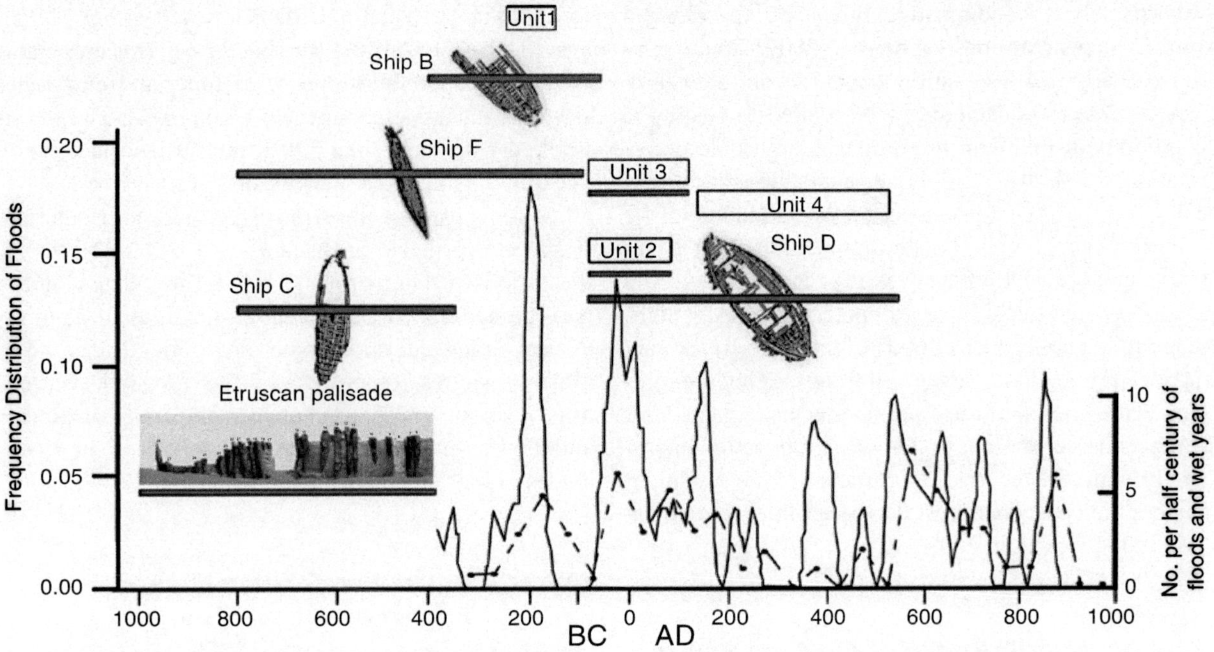

Fig. 12.5 Chronology of sedimentary units and selected archaeological materials from the Pisa-S. Rossore site compared with the Italian paleohydrological record inferred from historical sources and gauge data. Left vertical axis, thick curve: flood frequency of the Tiber River (After Camuffo and Enzi 1996); right vertical axis, hatched curve: flood and wet years recorded in Roman Italy (After Lamb 1995). Bars indicate the standard deviation (2σ) of calibrated radiocarbon ages from the different dated materials

small channels filled by the NNW-migrating composite bedforms. On the whole these lines of evidence suggest that a pre-existing river channel, reconstructed by the correlation of cores as a grossly WNW-ESE trending trough (Fig. 12.4), may have been a former reach of the Arno or of the Ausur (Auser) rivers (Bruni and Cosci 2003) affected during Roman times by recurrent catastrophic floods. These floods were strong enough to sweep away the ships from their mooring areas. The channel, evidently, provided a safe site for the establishment of a riverine wharf area exploited firstly by the Etruscans and then by the Romans (Benvenuti et al. 2006). This channel would have been hydraulically similar to the present Fiume Morto (Fig. 12.2a); that is, inactive but still connected to the sea allowing efficient navigability and suitable sea-land connection for trade exchange. Due to repeated catastrophic floods, the ships and their cargoes were deposited within the NW-dipping cross-stratified sandy bodies filling small linear troughs oriented transversally to the trend of the main fluvial channel (Benvenuti et al. 2006). The dominantly northwestward progradation of these composite macroforms suggests two things: flood waves spread out from and active channel located south of the wharf area and identified as the Arno River; and progressive migration and siltation of the channel was accommodated into a paleomeander with a maximum depth along the northern, erosional, bank (Fig. 12.4; Benvenuti et al. 2006). After each flood, the sandy deposits were draped by sandy silts and clays. Recurring coarse- and fine-grained sediments indicate recurring floods whose chronology is in good agreement with the flood frequency of the Tiber River inferred from historical sources and gauge data (Fig. 12.5; Camuffo and Enzi 1996). Physical and historical evidence from distinct and wide river catchments of central Italy indicates a dynamic late Holocene hydroclimatic regime rapidly fluctuating in terms of magnitude and frequency of flood events (Benvenuti et al. 2006; Mariotti-Lippi et al. 2007).

Pollen analyses of alluvial mudstone, included in unit 0 into which the Etruscan wharf palisade was driven, indicate a pre-Etruscan alluvial plain covered by extensive woodlands dominated by *Fagus* and *Abies* (Mariotti-Lippi et al. 2007). This vegetation, currently distributed at high elevation in the Northern Apennines (see below), is indicative of moist-cool climate for a site location near sea level. With the onset of the Roman Age, the pollen data indicated that the plain was sparsely wooded mostly with deciduous oaks and similar plants (Mariotti Lippi et al. 2007). The depletion of arboreal vegetation cover during the early Roman Age is considered to be related to hydraulic conditions of the plain more than to anthropic deforestation. As shown in Fig. 12.5, available chronological data suggest that floods in the Pisa plain may have occurred synchronously with those of the Tiber River that increased in frequency with a maximum in the second century BC. In this hypothesis, coeval high-magnitude and frequent floods of the Arno River may have determined increasing soil waterlogging that precluded the growth of trees and favored an expansion of herbaceous vegetation. Pollen data from sediments of late Roman Age record a reforestation of the Pisa plain as possible consequence of decreased flood frequency (Mariotti-Lippi et al. 2007).

12.4.2 Middle Reach of the Arno River: The Etruscan Town of Gonfienti, Near Prato

12.4.2.1 General Setting

The second study site is located in the Plio-Pleistocene Florence basin, on the Arno River floodplain at the foot of the steep, faulted flank of the northern hills, specifically the Calvana Mt (Fig. 12.6). This mountain is composed of limestone, marlstone, and claystone (Ligurid succession; Capecchi et al. 1975). The basin is filled with more than 500 m thick fluvio-lacustrine and alluvial mudstones, sands, and gravels overlain by alluvial deposits of the Arno River and its tributaries (Capecchi et al. 1975; Briganti et al. 2003). North of the Arno River, numerous wetlands, now for the most part drained out, were present on the alluvial plain in ancient times and wide coalescing alluvial fans, now mostly inactive, characterized the footslope of the hills. The site is located between the Bisenzio River and the Marina Creek at about 1 km south of the steep slopes of the Calvana Mt (Fig. 12.7b).

12.4.2.2 Description, Stratigraphy and Paleoenvironment of the Site

Despite its shallow burial the Etruscan town of Gonfienti was discovered only in the late 1990s. Although the Etruscan presence near Gonfienti was previously documented by few votive statuettes and tombs, the existence of a wide urban center was unknown. The first recovery of stones and bricks during the early excavations in 1997 stimulated an extensive Ground Penetrating Radar (GPR) survey that revealed several

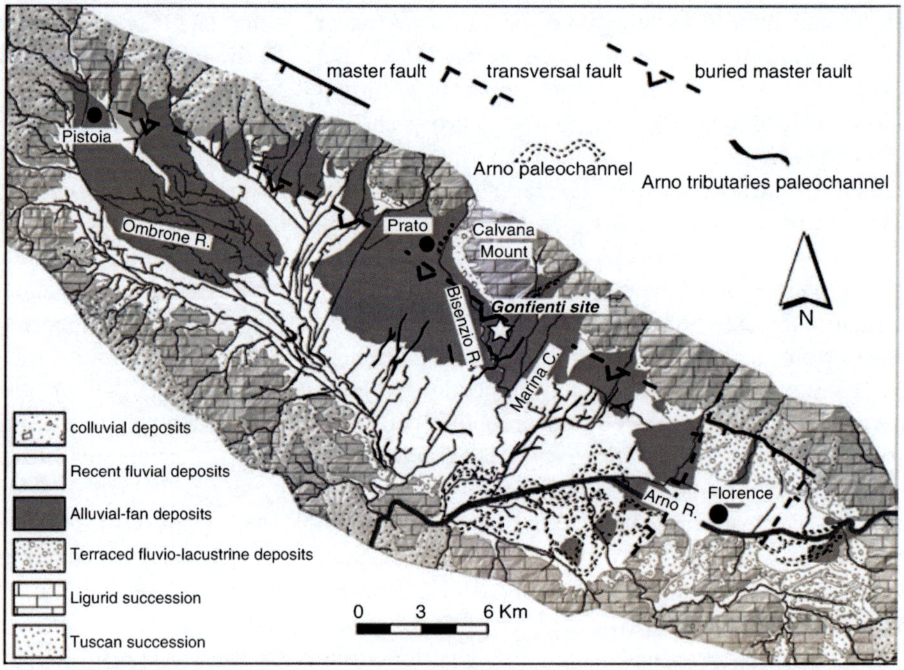

Fig. 12.6 Schematic geology superimposed on a shaded relief (*gray tone*) map of the Florence basin and location of the Gonfienti site. (Paleochannel distribution after Conedera and Ercoli 1973)

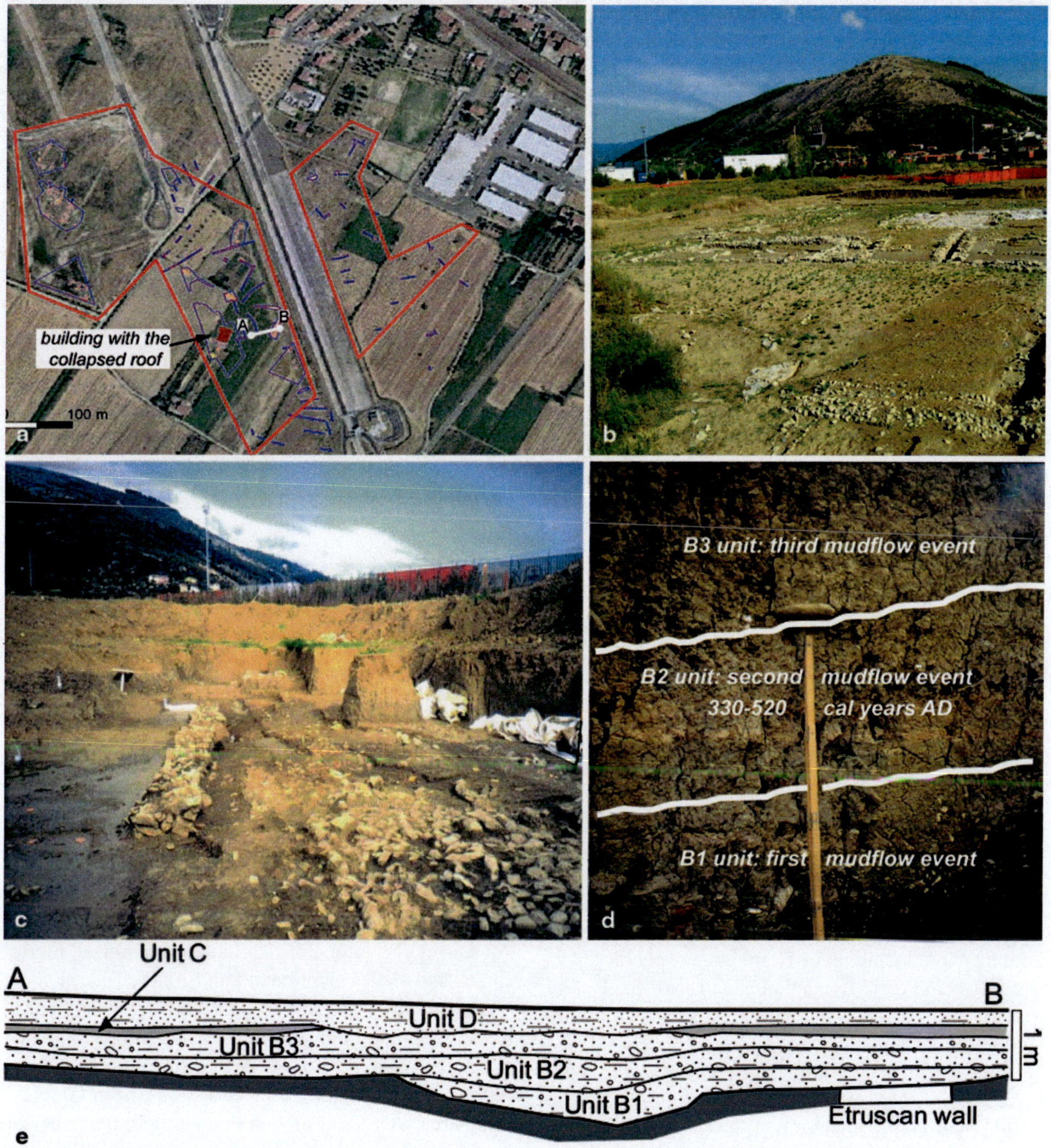

Fig. 12.7 **a** Aerial view of the Gonfienti site with annotated distribution of major findings and location of section A–B (shown in 12.7e). **b** Stony walls of houses, in the background the steep slopes of the Calvana Mt. **c** The stony walls are buried by 1.5 m thick deposit mostly of pebbly sand and silt. **d** Detailed stratigraphy of the deposits burying the Gonfienti town subdivided in units discussed in the text (scale bar is 1 m long). **e** Cross-section showing the distribution of the units described in the text

stony structures buried at a maximum depth of 1.5 m over an area of about 2 km^2 (Poggesi et al. 2005). These structures delimit the foundations of several houses (Fig. 12.7a–c) and outline a street network following an urban style common to other Etruscan towns (Poggesi et al. 2005).

To date, significant portions of the town have been excavated and among the more significant findings a large building has been found with a collapsed brick roof and many artefacts (stones, pottery and bronze utensils) preserved inside (Fig. 12.7a). This building, possibly representing a handicraft shop in the town, indicates that the site was likely hurriedly abandoned and it was never re-occupied. Its collapse may have been related to a catastrophic event. Indeed the archaeological evidence indicates that the town of Gonfienti developed and subsisted for a relative short time in the VI and the V centuries BC (Poggesi et al. 2005).

The stratigraphy of the site has been exposed in several large trenches. The substratum on which humans settled since the Bronze Age is represented by mottled sandy-clayey silts. The surface of this substrate shows shallow, wide troughs, evidenced by the distribution of Bronze Age–Middle Age archaeological materials found buried at different depths below the modern surface. This morphology, outlining a gullied surface, is particularly evident in the Etruscan town where, as example, the walls of the large building with the fallen roof are about 1 m higher than the house walls found 100 m to the ENE (Fig. 12.7a, e). Together with these gentle undulations the mudstone bedrock is locally crossed by gravely and sandy channel-like bodies indicating repeated fluvial incision of this surface.

Above the Etruscan deposits, composed of dark, charcoal and organic debris-rich mud with dispersed brick and stones clasts (unit A), a sediment cover up to 1.50 m thick can be subdivided into three main units (B–D) (Fig. 12.7d, e). Unit B has been subsequently subdivided into three sub-units (B1–B3), 0.2–0.4 cm thick, each characterized by massive clayey silts with floating clasts including brick fragments, pebbles, particularly common in B1, and rare land snail shells. The base of each sub-unit is almost planar and normally marked by an alignment of pebbles. Charcoal in the uppermost part of sub-unit B2 has been dated to 330–520 cal year AD. Unit C (0.4–0.1 m thick) is similar in lithology and texture to the previous ones. Where unit C was eroded out unit D rests directly on unit B. Unit D is up to 0.5 m thick and is composed of massive sandy silt and is affected by intense reworking due to agriculture and excavations.

The sedimentary succession burying the Etruscan town of Gonfienti is interpreted as the product of mudflows (units B, C). The older mudflows (sub-unit B1) were funnelled through shallow linear troughs incised into substrate unit A (Fig. 12.7e). Successive mudflows (B2, B3, C) overfilled the troughs and covered and smoothed out the original undulated morphology. Dating of sub-unit B2 suggests that the burying of the Etruscan remains occurred in the late Roman Age and that buildings located in relatively higher positions may have remained partly exposed for several centuries. Unit D represents a modern, reworked, alluvial deposit due to overbank flooding of the nearby Bisenzio River and Marina Creek.

Preliminary palynological analyses of unit A evidenced that in Etruscan times the plain surrounding the town was prevalently covered by herbaceous vegetation. Among the herbs, plants typical of damp environments are always well represented. Scanty pollen data don't allow any paleoclimatic conclusion simply indicating that the arboreal component was constituted by hygrophilous trees and shrubs, such as *Alnus* or *Salix*, and by elements of the mixed oak forest.

12.4.3 Upstream Reach of the Arno River: The Etruscan Sacred Site of Lago Degli Idoli, Falterona Mt

12.4.3.1 General Setting

This site is located a few hundred meters south of the source of the Arno River on the south-western flank of the Falterona Mt in the central part of the Northern Apennines (Fig. 12.8a). The mountain is composed mostly of Miocene turbiditic deposits characterized by a quasi-regular interstratification of sandstones, marlstones, and shales (Falterona Sandstones, Tuscan succession; Cibin et al. 2004). This makes the steep flanks of the mountain rather unstable and prone to mass wasting, particularly those in the down-dip direction of the strata. Landslides created various morphologies in the area, damming of small creeks, and the generation of small depressions like the Lago degli Idoli.

12.4.3.2 Description, Stratigraphy and Paleoenvironment of the Site

The Lago degli Idoli is a small sub-circular depression encircled by a thick beech forest mantling the southern slopes of the Falterona Mt (Fig. 12.8b). The site was

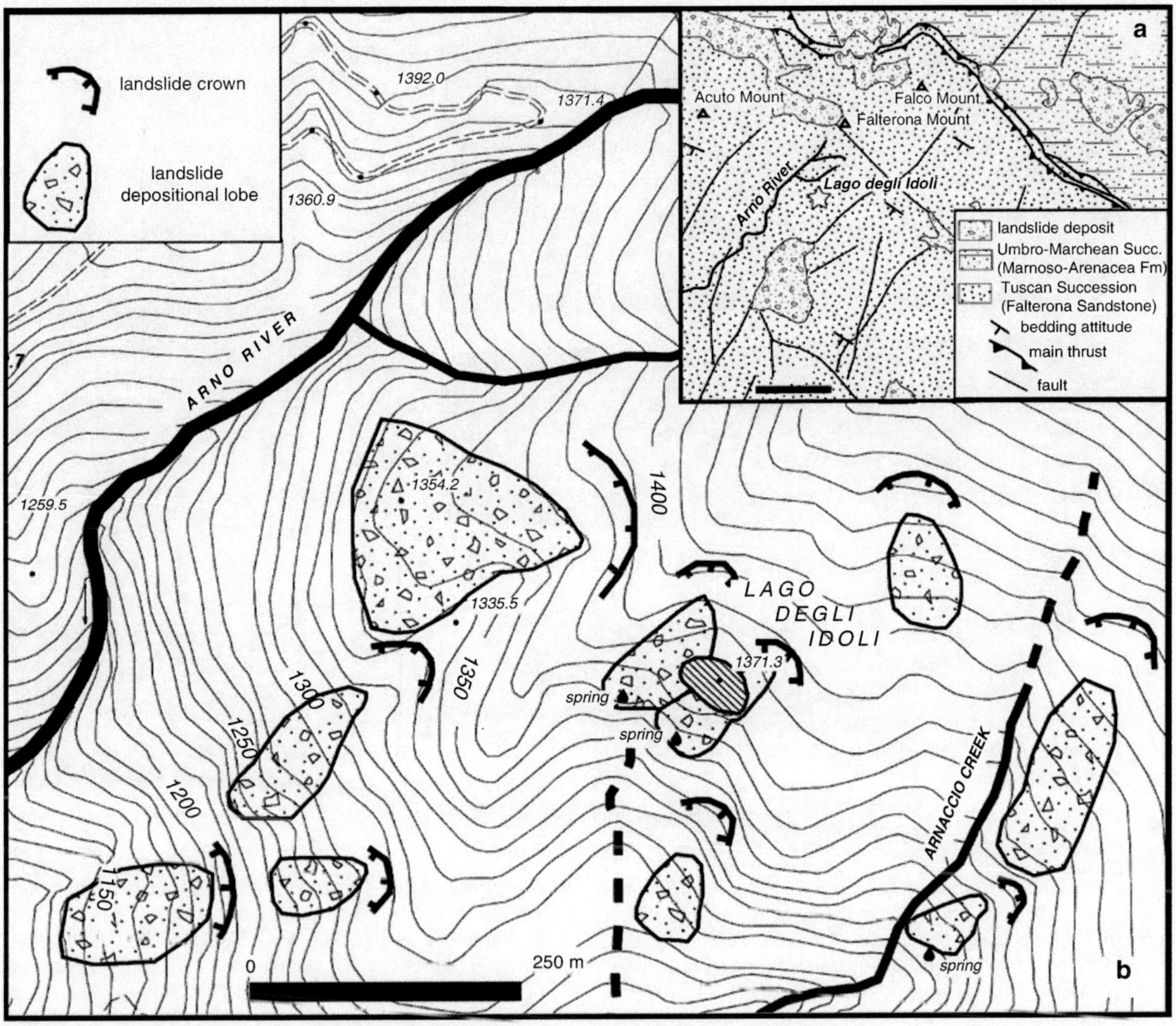

Fig. 12.8 **a** Schematic geological map of the Falterona Mt area and location of the site (see Fig. 12.1 for a general location). **b** High resolution topography and main geomorphologic features in the area of the archaeological site

occupied by a small mire up to the early nineteenth century.

In the summer of 1838 a bronze statuette found on the margins of the mire stimulated extensive but poorly supervised excavations (Fig. 12.9a). More than 600 bronze statuettes were recovered. Further studies occurred in the 1970s and between 2003 and 2006 including more detailed geomorphologic and stratigraphic analyses. Few new statuettes (Fig. 12.9b, c) were found (Ducci 2004). The archaeological material retrieved indicates that the Etruscans frequented the site during the VI and III centuries BC mostly for religious purposes (Fedeli 2005). The majority of the bronze figurines represent martial subjects (Fig. 12.9c) suggesting that this was a votive site for soldiers invoking protection and fortune in battle.

The steep slope where the site is located and the bedding attitude of the bedrock have favored the occurrence of several small and shallow planar slides (Fig. 12.8b). The geomorphic and seismic surveys indicate that the Lago degli Idoli depression may have been generated by the coalescence of two small landslide lobes (Fig. 12.8b, 12.9d, e).

A section about 3.50 m thick was logged and sampled for radiocarbon dating and pollen analysis (Fig. 12.9f). The section is composed of a recurring alternation of poorly sorted muddy strata with sandstone clasts and a variable amount of plant debris mostly deposited

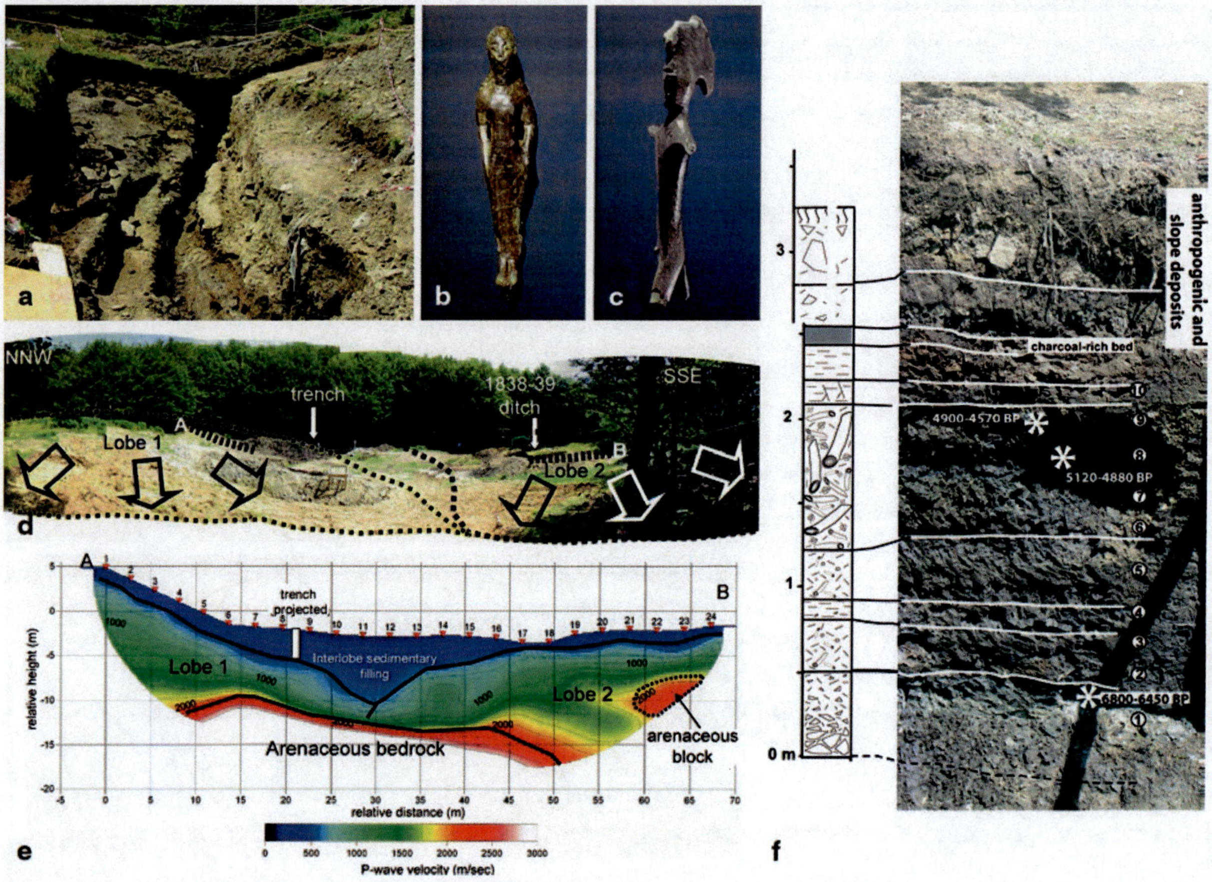

Fig. 12.9 **a** Ditch for draining the Lago degli Idoli mire done between the 1838 and the 1839 in the southern border of the depression. The channel, about 1.5 m deep and excavated in the deposits of the southern landslide lobe, was deepened in successive stages, as indicated by the original terraced section, to favor the complete outflow of the mire. **b, c** Two anthropomorphic bronze figurines found between 2003 and 2005 from sieving the debris discarded in older excavations of the site (from Ducci 2004): **b** (about 9 cm high) Greek divinity Kore, **c** (7 cm high) 'Marte gradiente' (the God Mars). **d** General view of the Lago degli Idoli depression with annotated landslide deposits, trace of the refraction seismic line A–B, and the 2005 excavation (trench). **e** Interpreted WET tomography along the A–B section showing two well defined landslide lobes. **f** Stratigraphic log of the deposits filling the interlobe depression exposed in the trench excavated in 2005, samples for pollen analysis are indicated by numbered circles, and those with dated organic matter by asterisks

by mudflows. The lower part of the section (pollen samples 1–3 in Fig. 12.9f) is characterized by a pollen assemblage dominated by *Fagus* (beech) indicating at this elevation vegetation in equilibrium with an overall warm climate. Pollen sample 4 (Fig. 12.9f; Ricciardi and Calò 2007) yielded an abundance of *Pinus sylvestris* group pollen indicating a cold-dry climatic pulse that, given the available chronological constraint (Mayewski et al. 2004), can be correlated to a global cooling event (Fig. 12.10).

A thick layer in the middle of the sections is composed almost exclusively of transported logs and plant macroremains of *Fagus, Abies* (fir), and *Acer* (maple) (Giachi and Capretti 2007), attesting to the recurrence of a thick forest cover, similar in composition to the present one, protecting the slopes from intense erosion. The analysis of macrofossils and pollen assemblages of the various layers, thus, indicate conditions for the periods antecedent to the human occupation of the area (~from 6400 to 4790 BP, Fig. 12.9f) that fluctuated from warm to colder conditions similar to the present ones (Ricciardi and Calò 2007; Cattani personal communication;). Similar environmental conditions and climatic oscillations may have continued during the period the Etruscans used the site, but the uppermost deposits have been too greatly reworked to provide specific information.

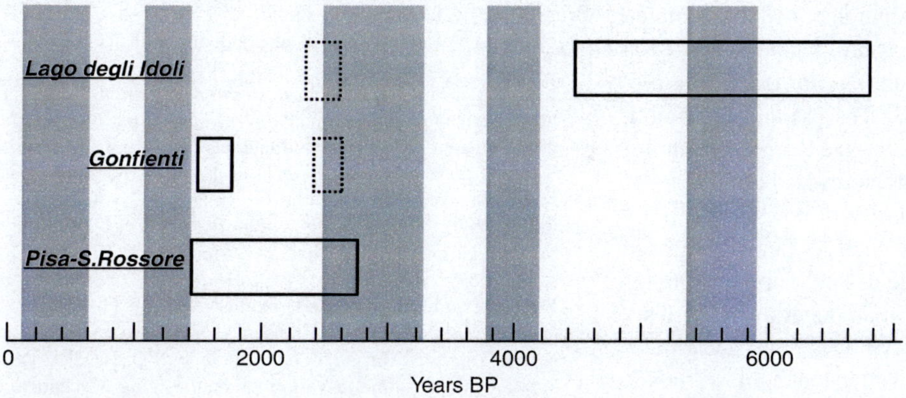

Fig. 12.10 Chronological data from the discussed sites are compared with the major climatic transitions globally recorded during the mid-late Holocene (After Mayewski et al. 2004), gray bars are cold periods. Thick chronological intervals derive from the calibrated radiocarbon ages available for the different sites. Hatched chronological intervals refer to Etruscan frequentation of Gonfienti and Lago degli Idoli sites testified by archaeological evidence

12.5 Community Adaptability

The three sites presented here are good archives of environmental changes that occurred during the middle to late Holocene, and reveal how these affected differently the Etruscan and Roman communities that occupied or used them. Differences in landscape presented different advantages and challenges and those societies adapted differently to the various sites.

The Pisa site indicates the adaptation and persistence of successful shipping activities while battling recurring disastrous flood events. That site was eventually abandoned in more recent times when siltation favored the use of more suitable localities along the coast and the main channel of the Arno River.

The Gonfienti site may have offered good advantages both for agriculture and connection across that part of central Italy known as Etruria (Poggesi et al. 2005), but was eventually doomed by recurring mudflows from steep flanks of increasingly deforested hills.

The Lake of the Idoli sites offered the peaceful, restful forest amenity of sacral remote water springs that continued to be visited over a long period of time.

The events that affected the various sites are very different but they may likely be indirectly and approximately related to regional or even global climatic conditions (Fig. 12.10; Mayewski et al. 2004). This may be in part justified by the fact that the various sites are located in the same drainage basin.

Human settlement documented in the three sites span the 3200–2300 years BP cold period (Iron–Etruscan ages) and the 2300–1400 years BP warm period (Roman Age) (Lamb 1995). The Roman exploitation of the abandoned channel for the Pisa-S. Rossore harbor was marked by repeated catastrophic overbank floods from the adjacent Arno River. The widely accepted idea that the Roman Age was warm (Lamb 1995), confirmed by pollen data, implies for this part of central Italy a fluctuating precipitation regime generating frequent and catastrophic floods (Fig. 12.5). Nevertheless, the inactive, silting up channel of Pisa-San Rossore continued to be used as a mooring site for relatively small ships. This is demonstrated by the youngest ship (D in Fig. 12.4) dated at the end of the Roman Age (Belluomini et al. 2002; Martinelli and Pignatelli 2008) and marooned by floods against the northern bank of the abandoned channel. Similarly to other societies living in flood-prone alluvial plains, the Romans seem to have considered that the economical benefit of that mooring outweighed any risk that it may have presented.

The geoarchaeological record of Gonfienti shows how hydroclimatic events may have differently affected an alluvial plain close to a steep hillslope. Though this plain was relatively protected against river floods, the vicinity of the slopes made the mass flow hazard high for the settled communities. Repeated mudflow events possibly triggered by intense rainfall and affecting slopes made unstable by forest clearance, were the likely cause of the rapid abandonment and burying of the Etruscan town. The site was abandoned by the Etruscans and it was never extensively populated again in Roman times as demonstrated by isolated farms related to the local system of land subdivision and distribution for agriculture ('centuria', Castagnoli 1948). By contrast to Pisa-S. Rossore and Gonfienti, Etruscans were attracted to the Lago degli Idoli site for the amenity of the forest and the revered remote water springs. This is a typical example of how a potentially hazardous geologic process such as landsliding, led to the formation of a beneficial landscape to humans although for spiritual, rather than material, activities.

The drainage basin of the Arno River was originally forested and with locally well developed wetlands. Along came 'man', settled the river mouth and riparian zones, drained swamps and cut down trees. As a consequence, flooding, erosion and deposition were intensified, and the human communities learned to deal with them in various ways. In the Pisa area fluvial harbors were very important and the facilities were rebuilt after each catastrophic flood by Etruscan and Romans. It was only during medieval times that riverine harbors of Pisa lost their importance when coastal ones were built more suitable for larger ships. The Gonfienti site was affected by recurring, disruptive landslides and floods and was soon abandoned during Etruscan times. The sacred mountainous springs remained pristine sites, but their use was abandoned following the demise of the Etruscan culture.

References

Aguzzi M, Amorosi A, Sarti G (2005) Stratigraphic architecture of Late Quaternary deposits in the Lower Arno plain (Tuscany, Italy). Geologica Romana 38:1–10

Aguzzi M, Amorosi A, Colalongo MC, et al. (2007) Late Quaternary climatic evolution of the Arno coastal plain (Western Tuscany, Italy) from subsurface data. Sedimentary Geology 202:211–229

Belluomini G, Mandra L, Tomassi AV, Vesica P (2002) L'eta` dell'antico porto di Pisa: datazioni con il Radiocarbonio. Science and Technology for Cultural Heritage 11:7–12

Bellwood P (2004) The first farmers: origins of agricultural societies. Blackwell, Oxford, 360 pp

Benvenuti M, Mariotti-Lippi M, Pallecchi P, Sagri M (2006) Late-Holocene catastrophic floods in the terminal Arno River (Pisa, Central Italy) from the story of a Roman riverine harbor. The Holocene 16:863–876

Briganti R, Ciufegni S, Coli M, et al. (2003) Underground Florence: Plio-Quaternary geological evolution in the Florence area. Bollettino della Società Geologica Italiana 122:435–445

Bruni S (ed) (2003) Il porto urbano di Pisa: la fase etrusca il contesto e il relitto ellenistico. Silvana Editrice, Milano, 221 pp

Bruni S, Cosci M (2003) 'Alpheae veterem contemptlor originis urbem, quam cingunt gemini Arnus et Ausur aquis.' Il paesaggio di Pisa etrusca e romana: materiali e problemi. In: Bruni, S (ed) Il porto urbano di Pisa: la fase etrusca il contesto e il relitto ellenistico. Silvana Editrice, Milano, pp 29–43

Camilli A (2002) The site of the Urban Port of Pisae. The situation. In: Colombini MP, Giachi G, Modugno F, Ribechini E (eds) The Ancient Ships of Pisa. A European Laboratory for Research and Preservation, Bruxelles 2002, pp 11–16

Camuffo D, Enzi S (1995) The analysis of two bimillenary series: Tiber & Po river floods. In: Jones PD, Bradley RS, Jouzel J. (eds) Climatic Variations and Forcing Mechanism of the Last 2000 years: Springer-Verlag, NATO – ANSI Series, 41:433–452

Capecchi F, Guazzone G, Pranzini G (1975) Il bacino lacustre di Firenze-Prato-Pistoia. Geologia del sottosuolo e ricostruzione evolutiva. Bollettino della Società Geologica Italiana, 94:637–660

Carmignani L, Lazzarotto A (eds) (2004) Geological Map of Tuscany, scale 1:250000. Regione Toscana, Direzione Generale delle Politiche Territoriali e Ambientali, Servizio Geologico. Litografia Artistica Cartografica S.r.l., Firenze, 2004

Carratori L, Ceccarelli Lemut ML, Frattarelli Fischer L, et al. (1994) Carta degli elementi natualistici e storici e dei rilievi contermini, scala 1:50.000. In: Mazzanti R (ed) La pianura di Pisa e i rilievi contermini – la natura e la storia. Memorie della Società Geografica Italiana, L, Roma (insert)

Castagnoli F (1948) La centuriazione di Florentia. L'Universo 28:361–368

Cibin U, Di Giulio A, Martelli L, et al. (2004) Factors controlling foredeep turbidite deposition: the case of northern Apennines (Oligo-Miocene, Italy). In: Lomas S (ed) Confined turbidite systems. Geological Society of London Special Publication 222, pp 115–134

Conedera C, Ercoli A (1973) Elementi geo-morfologici della piana di Firenze dedotti dalla foto interpretazione. L'Universo 53:255–262

Della Rocca B, Mazzanti R, Pranzini E (1987) Studio geomorfologico della Pianura di Pisa. Geografia Fisica e Dinamica Quaternaria 10:56–84

Ducci, M (ed) (2004) Santuari etruschi in Casentino. Comunità Montana del Casentino, Stia (Ar), 48 pp

Faegri K, Iversen J (1989) Textbook of pollen analysis. 4th Edition by Faegri K, Kaland P, Krzywinski KJ. Wiley and Sons, New York, 230 pp

Fedeli L (2005) Stia (AR). Lago degli Idoli: campagna di scavo 2005. Notiziario Soprintendenza per i Beni Archeologici della Toscana 1/2005, All'Insegna del Giglio, Firenze, pp 164–167

Giachi G, Capretti C (2007) Indagini diagnostiche sui resti arborei del Lago degli Idoli. Indicazioni paleo ambientali. In: Borchi S (ed) Gli scavi e le indagini ambientali nel sito archeologico del Lago degli Idoli. Comunità Montana del Casentino, Stia (Ar), pp 119–128

Lamb HH (1995) Climate, History and the modern world. Routledge, London, New York, 433 pp

Mariotti Lippi M, Bellini C, Trinci C, et al. (2007) Pollen analysis of the ship site of Pisa San Rossore, Tuscany, Italy: the implications for catastrophic hydrological events and climatic change during the late Holocene. Vegetation History and Archaeobotany 16:453–465

Martinelli N, Pignatelli O (2008) Datazione assoluta di alcuni relitti dal contesto delle navi di Pisa. Risultati preliminari delle indagini dendrocronologiche e radiometriche col 14C. Gradus (online journal, http://www.cantierenavipisa.it/Pubblicazioni_Gradus4.html) 3:69–78

Martini F, Poggesi G, Sarti L (eds) (1999) Lunga memoria della Piana. L'area fiorentina dalla preistoria alla romanizzazione. Guida alla mostra, Museo di Doccia, Comune di Sesto Fiorentino, Ministero per i Beni e le Attività culturali, Università degli Studi di Firenze, Università degli Studi di Siena, 105 pp

Martini IP, Sagri M (1993) Tectono-sedimentary characteristics of Late Miocene–Quaternary extensional basins of the Northern Apennines, Italy. Earth Science Reviews, 34:197–233

Martini IP, Sagri M, Colella A (2001) Neogene–Quaternary basins of the inner Apennines and Calabrian Arc, Italy. In: Vai GB, Martini IP (eds) Anatomy of an orogen: Northern Apennines and Adjacent Mediterranean Basins. Kluwer Academic Publication, Dordrecht, pp 375–400

Mayewski PA, Rohling EE, Curt Stager J, et al. (2004) Holocene climate variability. Quaternary Research 62:243–255

Poggesi G, Donati L, Bocci E, et al. (2005) Prato-Gonfienti: un nuovo centro etrusco sulla via per Marzabotto. In: Sassatelli G, Govi E (eds) Studi e Scavi numero speciale 11, Ante Quem, Bologna, pp 267–300

Pranzini E (2001) Updrift river mouth migration on cuspate deltas: two examples from the coast of Tuscany Italy. Geomorphology 38:125–132

Rau A, Tongiorgi M (1974) La geologia dei Monti Pisani a Sud-Est della Valle del Guappero. Memorie della Società Geologica d' Italia 13:227–408

Redman CE (1978) Rise of Civilization: From Early Hunters to Urban Society in the Ancient Near East. Freeman, New York, 367 pp

Ricciardi S, Calò C (2008) Le analisi palinologiche dei sedimenti del Lago degli Idoli. In: Borchi S (ed) Gli scavi e le indagini ambientali nel sito archeologico del Lago degli Idoli. Comunità Montana del Casentino, Stia (Ar), pp 109–118

Sani F, Bonini M, Piccardi L, et al. (2009) Late Pliocene-Quaternary evolution of outermost hinterland basins of the Northern Apennines (Italy), and their relevance to active tectonics. Tectonophysics 476:336–356

Vai GB (2001) Structure and stratigraphy: an overview. In: Vai GB, Martini IP (eds) Anatomy of an orogen: Northern Apennines and Adjacent Mediterranean Basins. Kluwer Academic Publ., Dordrecht, pp 15–32

Chapter 13
Landscape Influences on the Development of the Medieval–Early Renaissance City-states of Pisa, Florence, and Siena, Italy

I. Peter Martini, Giovanni Sarti, Pasquino Pallecchi, and Armando Costantini

The objective of this chapter is to analyze how the local landscape and environmental conditions influenced the evolution of the city-states of Pisa, Florence (Firenze), and Siena, primarily from the High (~1000–1300 AD) to Late Medieval Age and early Renaissance (~1300–1500 AD). The three cities are located in the province of Tuscany in the central north-western part of Italy (Fig. 13.1).

13.1 The Physical Setting

13.1.1 Landform

Tuscany can be subdivided into three roughly NW–SE oriented zones: an eastern mountainous zone that includes the North Apennines divide, a hilly zone (where Siena is located) with relatively narrow fluvial valleys and wide intramontane basins (where Florence is located), and an irregular, relatively narrow alluvial- and coastal-plain zone to the west along the Tyrrhenian Sea coast where Pisa is located. During the Medieval Age the population was concentrated in major urban centers among which Florence, Pisa, and Siena and their territories ('contadi') (Fig. 13.1; Table 13.1; Day 2002; Pinto 2002).

13.1.2 Significant Geological Features

Tuscany is located in the inner (west) part of the Northern Apennines mountain complex that formed from about 30 million years ago through the collision of the European plate with the African plate (Vai 2001). The Northern Apennines are characterized by a series of imbricate thrust-fold belts that have accreted eastward in response to a westward-dipping subduction zone (Malinverno and Ryan 1986). The migration was not continuous nor homogeneous, rather differential movements occurred along several major SW–NE oriented transverse tectonic lineaments (~transfer faults) (Pascucci et al. 2007). As the main compression deformation zone migrated eastward, extension occurred in the western, inner part of the orogen superimposing an overall extensional (rift-like) basins on the pre-existent dormant imbricate thrust zones (Fig. 13.2; Martini and Sagri 1993; Martini et al. 2001; Bonini and Sani 2002; Benvenuti et al., this volume). The formation of these basins had a complex, debated history of reactivations but essentially they developed sequentially from the Tyrrhenian Sea eastward, a first group (central basins: west of the structural/geomorphologic divide of Albano Mt. and Chianti–Cetona Ridge) starting during late Miocene, and a second group (peripheral basins: east from that structural divide) forming mainly during the Pleistocene–Pliocene. The former contain marine and continental deposits, the latter only continental alluvial and lacustrine deposits. Pisa (Pi) and Siena (Si) are located in the central basins of Viareggio and Siena respectively, and Florence (Fi) in its homonymous peripheral basin (Fig. 13.2).

I. P. Martini (✉)
School of Environmental Sciences, University of Guelph
Guelph, ON N1G 2W1, Canada
e-mail: pmartini@uoguelph.ca

See Plates 11, 13 in the Color Plate Section; also available at: extras.springer.com

I. P. Martini, W. Chesworth (eds.), *Landscapes and Societies*,
DOI 10.1007/978-90-481-9413-1_13, © Springer Science+Business Media B.V. 2010

Fig. 13.1 Medieval map of Tuscany (white: low elevation tendentially planar areas; increasing darker gray indicated increased elevation areas (*Fi* Florence, *Pi* Pisa; *Si* Siena). (After Pinto 2002)

13.1.3 Climate

Tuscany is part of the Mediterranean biome. Consequently it is characterized by hot (average high/low of about 28–32/17°C) dry summers and cool (average high/low of about 12/3°C) wet winters. Current average annual rainfall ranges from approximately 780 mm in Siena to about 900 mm in Pisa and Florence. Peak, at times torrential, precipitation may occur during Spring and Fall and severe local floods and land-

Table 13.1 Various estimates for the populations of Firenze (Florence), Pisa, and Siena from the 1000s to the early 1300s before the Black Plague of 1348. (After Day 2002) (Dates in bold, number of individuals in italics)

Cities	Bairoch et al. (1988)	Ginatempo and Sandri (1990)	Pirillo (1994)	Russell (1972)	Other authors
Pisa	**1000**–*9000* **1200**–*17,000* **1300**–*38,000*	**1300/1338**– *40,000/50,000*	**1200/1249**–*20,000/25,000* **1250/1299**–*38,000*	**1200/1230**–*15,000/20,000* **1290/1300**–*38,000*	**1228**/*25,000/28,000* **1300**–*30,000*
Florence	**1300**–*95,000*	**1300/1338**– *100,000/120,000*	**1200/1249**–*15,000/20,000* **1250/1299**–*100,000* **1300/1325**–*110,000* **1326/1345**–*110,000*	**1172**–*10,000* **1200/1230**–*15,000/20,000* **1290/1300**–*96,000*	**1175**–*10,000* **1200**–*15,000* **1252**–*60,000* **1280**–*85,000* **1300**–*105,000* **1325/1328**–*120,000*
Siena	**1200**–*12,000* **1300**–*50,000*	**1300/1338**– *40,000/50,000*	**1200/1249**–(*10,000/15,000*) **1300/1325**–(*50,000*)	**1200/1230**–*10,000/15,000* **1290/1300**–*52,000*	

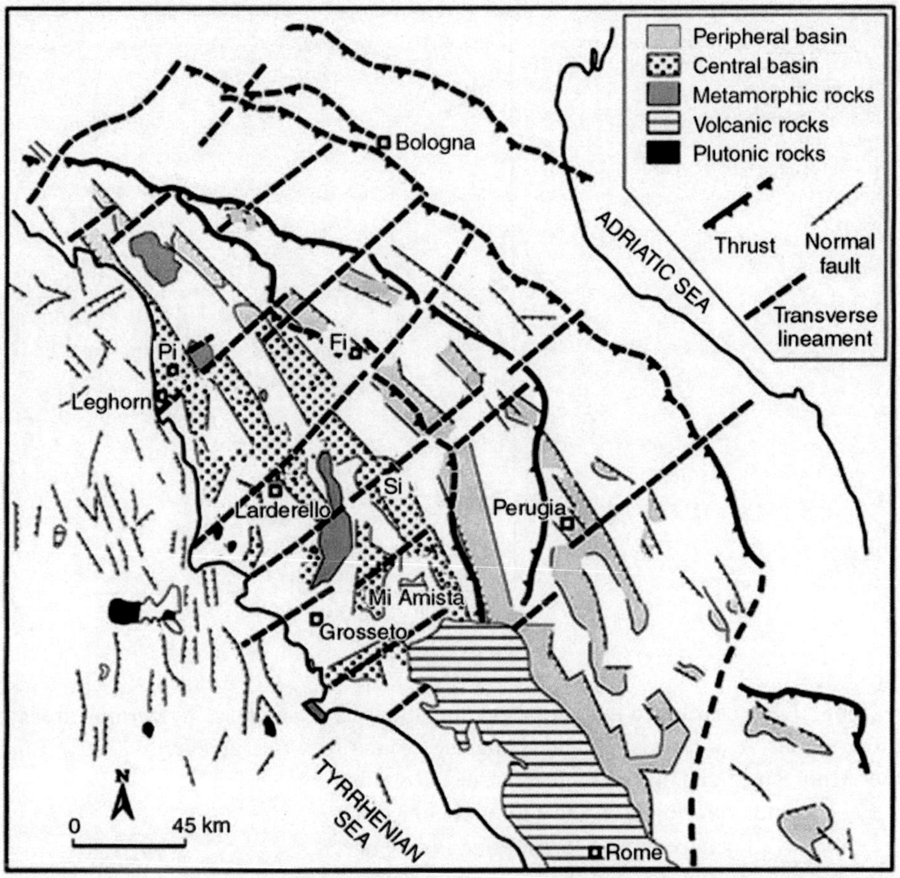

Fig. 13.2 Schematic tectonic map of the Northern Apennines (Peripheral basins: terrestrial deposits mainly Pleistocene; Central basins: marine and terrestrial deposits mainly Miocene to Pliocene with local thin Pleistocene cover; *Fi* Florence, *Pi* Pisa, *Si* Siena, *MTR* Middle Tuscany Ridge). (After Martini et al. 2001)

slides may develop. During medieval times Tuscany was affected by the so called global Medieval Warm Period (~800–1300 AD) (also called Medieval Climatic Anomaly) that produced prolonged droughts in some areas worldwide, for example of North America, and persistent rains in others (Fig. 13.3; Stine 1994; Mann et al. 1998; Bradley 2003).

13.1.4 Hydrology

A fundamental geological/geomorphologic event that determined the evolutionary trends of great parts of central Tuscany was the deviation during the Pleistocene of the Arno River from the Val di Chiana, where it ran as a tributary to the Tiber River, into the Valdarno, through the Florence basins and to the Pisa plain area (Fig. 13.1; Bartolini and Pranzini 1981). This mostly occurred naturally through differential tectonic moments of the various crustal blocks of the area, and it was finalized in the sixteenth century by efforts to drain waters from the Val di Chiana wetlands into the Arno River. The Arno River is the vital artery that has most benefited Florence, Pisa, and, indirectly, Siena. This river and its tributaries, however, have been responsible for extreme damaging floods as well (Table 13.2)

Table 13.2 Recorded floods in Florence from late 1100s to the present. (From Caporali et al. 2005)

Medium	12161; 1303; 1305; 1362; 1368; 1378; 1406; 1434; 1490; 1491; 1520; 1438; 1550; 1621; 1641; 1651; 1660; 1674; 1683; 1695; 1698; 1715; 1745; 1761
Large	1117; 1262; 1282; 1284; 1288; 1334; 1345; 1389; 1456; 1465; 1515; 1532; 1543; 1544; 1646; 1676; 1677; 1680; 1687; 1688; 1705; 1709; 1714; 1719
Exceptional	1333; 1547; 1557; 1589; 1740; 1758; 1844; 1966

Fig. 13.3 Reconstructed climatic conditions of the last two millennia. (Observations and measurements by various researchers; After http://commons.wikimedia.org/wiki/Image:2000_Year_Temperature_Comparison.png)

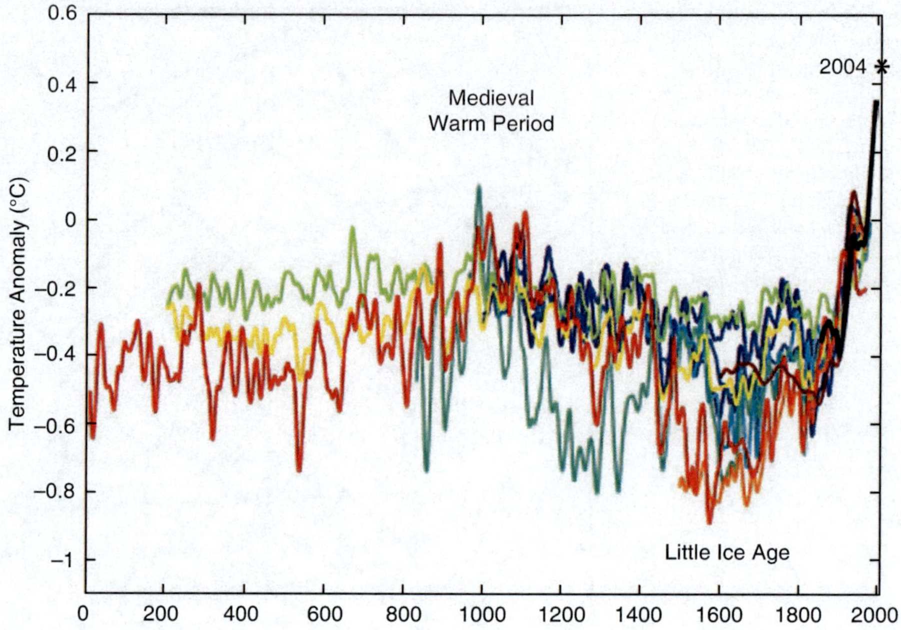

Small lakes developed in intermountain valleys and more extensive wetlands formed along the course of the Arno River and its major tributaries. Relatively large wetlands developed in the Florence area, along secondary valley in between Florence and Pisa (Fucecchio and nearby Bientina), and much more extensively in the Pisa alluvial and coastal plain (Fig. 13.1). In certain areas and times, the wetlands were obstacles to overland travel and made certain types of agriculture difficult. In other areas and times, the opposite was the case, in terms of waterway communication, food resources (mainly fishing and hunting), and materials such as straw for fodder and constructions.

13.2 The Overall Human Dimension

Medieval to early Renaissance times were exciting, innovative, turbulent, difficult, and dangerous. Societies throughout Europe, Asia, and elsewhere strived for achievements in science, art, politics, commerce, and management of the environment and natural resources. The progress continued through the centuries in spurt punctuated by short-lived retreats. Difficult periods were imposed by environmental conditions and severe sicknesses in part due to poor hygiene and inadequate knowledge. These included two diseases: the endemic malaria in certain areas and the horrifying bubonic plague everywhere in Europe.

13.2.1 *Malaria*

Malaria has been recorded throughout Italy since Roman times, and is still present in some parts of the country. Differently from ancient Rome and its surrounding territory, however, there is no direct evidence of great impact of the deadly strain of malaria associated with wetlands in the three Tuscan cities, just a few cases having been recorded in Pisa (Torelli 1882; Sallares 2002). Nevertheless some territories, such as the Val di Chiana and particularly the Sienese-dominated Grosseto area (Maremma) suffered greatly (Fig. 13.1). The population of the town of Grosseto, for instance, was at times reduced to a few hundred people and the population of the area was sustained by continuing immigration. Indeed in 1333 Siena ordered the officials of Grosseto (under Siena at that time) to leave the city during July and August every year and to move to residences in higher hills to avoid contact with the malaria-carrying *Anopheles* mosquitoes (Sallares 2002). That area was important for Siena because it was agriculturally highly productive and offered accessibility to a small harbor (Talamone located about 23 km south of Grosseto).

13.2.2 Bubonic Plague

By contrast, the bubonic plague (called also the Black Death) had dramatic effects everywhere. In 1347–1348 the first of several occurrences of the bubonic plague swept throughout Europe devastating populations and drastically changing economic and power distribution. This plague was caused by the *Yersinian pestis* bacterium, transmitted by fleas carried by rats. It probably originated in Asia and was transported into Italy and Europe following trading routes. In Italy it entered through the main harbor cities, mainly Genoa, and spread rapidly because of people fleeing infected cities into the country and other towns. Because of poor sanitation, ignorance of the causes of the sickness, and the rapidly induced death, nobody was safe anywhere. Pisa, Florence, Siena and their territories, like most other European sites, lost up to 50–75% of their populations (Pinto 2002). The Black Death had a negative influence on all aspects of the society and on the surrounding environments. Cities contracted, planned buildings were abandoned, such as the large new cathedral in Siena, most lands in the countryside were abandoned everywhere because of lower demand of foodstuff and the very expensive remaining manpower. Similarly, work to reclaim wetlands or maintain already existing drainage canals, such as in the Pisa plain, was discontinued and the land returned to its original, locally unhealthy state. Valuable skills disappeared and the economy suffered, for instance many mills closed because of diminished demand of their services and products (Cartwright and Biddiss 2004; http://www.insecta-inspecta.com/fleas/bdeath/bdeath.html). At the end of it all, larger societies, such as that of Florence, rebounded relatively rapidly after each bout of the plague, others could not.

13.3 The Three Cities: Pisa, Florence, Siena

The development of the medieval societies of Pisa, Florence, and Siena was mostly dictated by changing socio-economic, socio-political conditions, including changes in leadership. The local landscape (settings and environments) did however influence somewhat the fortune of the three cities providing both benefits and risks. The following is a brief analysis of the advantages and limitations offered by the landscape at different times.

13.3.1 Pisa

Pisa (coordinates: ~43°43′N, 10°24′E) is located on a very flat plain (Pisa plain), 4 m asl, on the banks of the Arno River, about 10 km from the sea (Fig. 13.4a). It has now about 85,000 inhabitants.

The Pisa plain is bounded to the north and south by hills with mostly Tertiary sedimentary to slightly metamorphic rocks that constitute good building material and provide for good freshwater springs. Medieval Pisa obtained poor quality, unhealthy waters from relatively shallow wells and cisterns until 1600 AD. After-

Fig. 13.4 **a** General view of modern Pisa with the "Piazza dei Miracoli" in the foreground and defaced (quarried) flanks of the Monte Pisano in the background. **b** Aqueduct (Aquedotto Mediceo) from Monte Pisano (in background) to Pisa

wards a major aqueduct was built by the Florentine Medici family from a locality (Asciano) on the flank of the Monte Pisano (Fig. 13.1). Remains of this aqueduct still run across the plain (Fig. 13.4b).

The basin fill consists of 2000 m thick, latest Miocene–early Pliocene to Holocene marine, lacustrine, and fluvial sediments (Federici and Mazzanti 1995; Pascucci et al. 2007). The town of Pisa rests directly on about 50 m of Holocene coastal to alluvial clays and sands parts of which constitute the fill of a paleovalley cut during the last glacial period (Aguzzi et al. 2007) (Fig. 13.5). The interlayered clay-sand substrate has created foundation problems for some bell towers with shallow foundations. The most famous one is the Leaning Tower (55.86 m high on one side and 56.70 m on the other) of the cathedral that started to tilt during construction (from 1173 to 1372 AD) (http://en.wikipedia.org/wiki/Leaning_Tower_of_Pisa). To compensate for that, builders chose to erect a curved, banana-like building such that its center of gravity remained within the footprint of foundations and the structure could withstand gravitational stresses (Fig. 13.6).

Medieval Pisa was surrounded by wetlands, up to the city walls on its south-western side, but some ancient representations also show some well-groomed, surrounding countryside (Fig. 13.7).

Waterworks have been major preoccupation in the Pisa plain since antiquity. Works were built for defense against floods from the larger rivers (Arno, Serchio) and the no-longer existing Auser River that flowed from the northwest to very close to town, at times acting as a defensive moat and as a navigable route of communication (Fig. 13.8a, b). Levees were built and canals were dug upstream from Pisa to deviate extremely-high floodwaters of the Arno River elsewhere (Cavazza 1994; Ceccarelli Lemut et al. 1994). To improve navigation, the course of the Arno River was straightened cutting off meander loops, and in 1606 the mouth of the Arno River was shifted a few kilometers northward to shelter it better from the direct impact of strong SW wind-driven storms that generated silting and obstructed water efflux causing floods upstream in Pisa (Fig. 13.8c, d; Pranzini 2001). Other works consisted in building canals both for reclamation of the numerous wetlands of the poorly drained plain caused by rainfalls and stream floods, and for navigability. Also of interest is the attempt by Florence to utilize river waters as a weapon against enemy cities such as Lucca, a city-state north of Pisa, and against Pisa itself (Fig. 13.1). In the first case the waters of the Serchio River were deviated to try to inundate the town. The plan failed because the town people erected

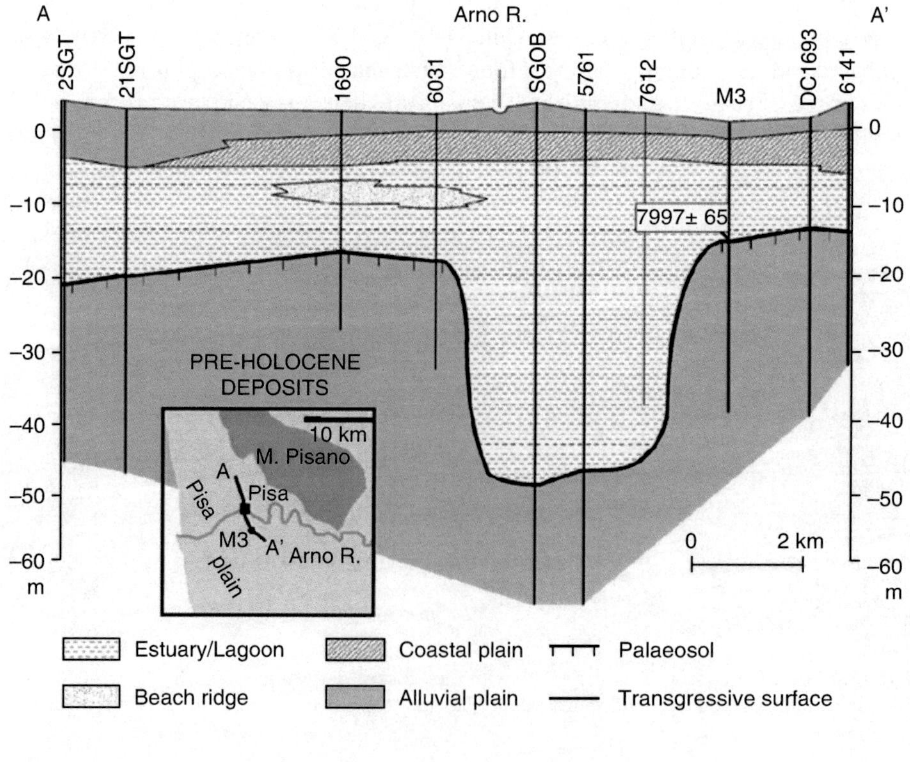

Fig. 13.5 Pisa plain (insert) and cross section showing Holocene deposits and the paleo-Arno River valley fill at Pisa (*Vertical lines* indicate core-logs; a C^{14} AMS uncalibrate date is reported in well M3). (After Aguzzi et al. 2007)

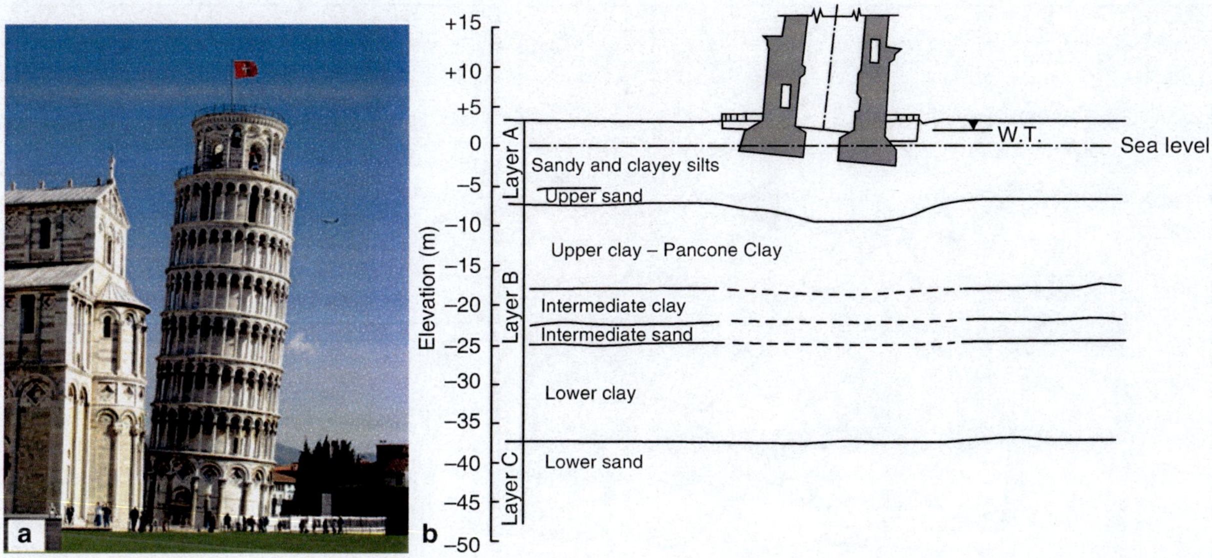

Fig. 13.6 Leaning Tower of Pisa. **a** The building shows strong inclination to the right and a slight left bend created by constructing the right flank slightly higher than the left one at various levels. **b** The tilting of the tower is caused by differential subsidence into the weak substrate, exacerbated during periods of water pumping from wells. (After Burland et al. 1998)

a higher levee on their bank of the river and eventually broke the opposite levee flooding the Florentine military camp. In the second case an ambitious plans was designed by Leonardo da Vinci at the beginning of 1500s to change the course of the Arno River upstream from Pisa (from point B on Fig. 13.8b) with the double purpose of cutting off water supply to the besieged town as well as opening an independent navigable canal to the sea for Florence (Masters 1998). The task was a difficult one, and in any case modified and botched by an incompetent, executive chief engineer. How difficult it was is demonstrated by subsequent unsuccessful attempts to construct canals to deviate floodwaters across the very flat Pisa plain. A partial success was not achieved until the late twentieth century.

Sufficient waterpower exited in the territory of Pisa for numerous mills. The principal mill complexes were built along the Arno River within and near the city itself, as well as along the Serchio River and other secondary streams discharging from the nearby hills. These mills became particularly important after the Livorno (Leghorn) harbor became established in the early Renaissance when much wheat flower was required to bake the *biscotto* (hard bread) extensively used by sailors (Pult Quaglia 1994) (Fig. 13.8).

During medieval times the wetlands of the Pisa plain were used primarily for animal husbandry (mainly cattle and sheep in the 1400s) (Ceccarelli Lemut et al. 1994; Pult Quaglia 1994). It was only later in the 1500s that numerous canals were constructed to

Fig. 13.7 Medieval–early Renaissance Pisa. (From http://www.stilepisano.it/immagini/Pisa_foto_stampe_antiche.htm)

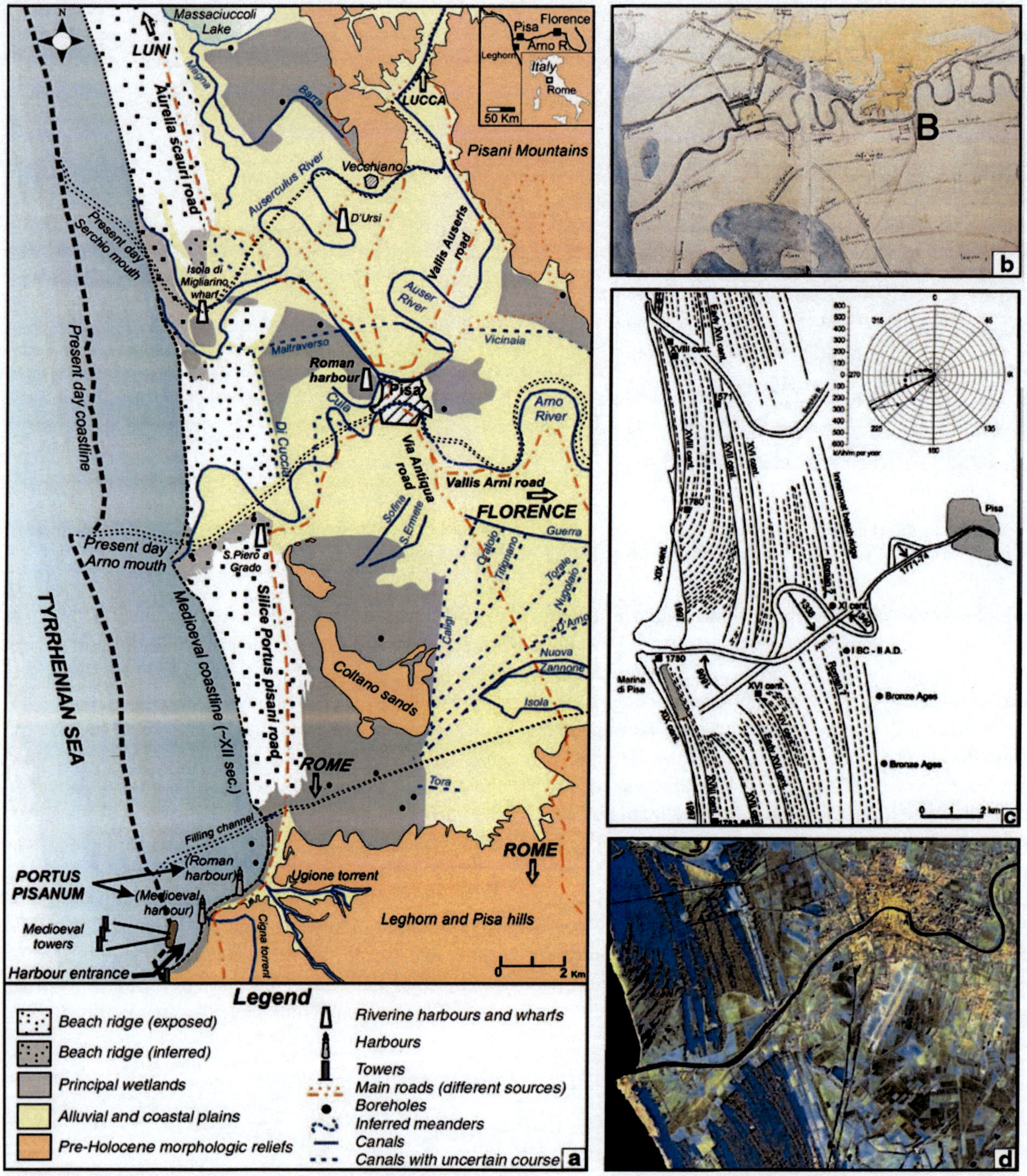

Fig. 13.8 Pisa plain area. **a** Map of the area for the tenth to twelfth centuries (After Carratori et al. 1994) (Plate 11). **b** Leonardo daVinci map of Arno River in the Pisa plain (After Starnazzi 2003) with indication of site (*point B*) where a unsuccessful major river diversion was planned in early 1500 (After da Vinci's Codice Madrid). **c** Modifications made to the lower course and river mouth of the Arno River (From Pranzini 2001). **d** Satellite image showing modern mouth area of Arno River traces of the ancient meanders and distributary channel downstream from Pisa. The small coastal city of Marina di Pisa between the ancient and recent mouth of the river is protected from erosion by artificial groins

partially drain the wetlands. Extensive grain cultivation was then implemented, provoked by the increasing demand of the growing populations of Florence and Pisa. In late 1500s the cultivation of the mulberry trees (for the silk industry), olive, and wine were introduced in the hills surrounding the Pisa plain (Pult Quaglia 1994).

All things considered, easy access to the sea was the prime asset and drawback of the town. As an asset, it allowed Pisa to become one of the first societies to flourish in Tuscany, reaching the apex of its development in the eleventh–twelfth centuries. This is when it became the dominant marine power of the Mediterranean area (http://en.wikipedia.org/wiki/Pisa) and also provided transport for the first Crusades. From those enterprises Pisa acquired great wealth and could afford to build its major monuments, such as the Piazza dei Miracoli (Fig. 13.4), and the numerous mansions that still ornament both banks of the Arno River. The drawback was the rivalry with other maritime powers such as Genoa. Genoa badly defeated Pisa in 1248 and impeded its development for one or two decades removing and imprisoning most of the male population and thus preventing the fathering of a new generation, and partially destroyed the main harbor ('Portus Pisanum', Fig. 13.8a). Furthermore Pisa harbors were necessary facilities to furnish industrial and sustenance materials for the landlocked Florence. So numerous battles were fought and eventually Pisa was subjugated first in 1406, then in 1494 after a brief revolt, and permanently in 1509.

13.3.2 Florence (Firenze)

Florence (coordinates: ~43°49′N, 11°19′E) is located on the banks of the Arno River at the southwest corner of a NW–SE elongated basin 10 km wide and 40 km long, ~50 m asl, ~100 km from the sea, and surrounded by hills (Fig. 13.9). It has now about 370,000 inhabitants.

The surrounding hills are composed of Tertiary sedimentary rocks, predominantly limestones and sandstones that are good construction materials as can be observed in almost every large palace in town. Marble for the large monuments derived mostly from the Apuane Mountains along the Tyrrhenian Sea coast (Fig. 13.1). The city is provided with a reliable supply of freshwater obtained from springs along the flanks of the hills or filtered from the river itself.

The city substrate consists of fluvial and alluvial-fan gravels and sands and some lacustrine clays and sands dating from the latest Pliocene to Holocene. Under the city itself the sediments generally coarsen upward from basal clays, through sand and gravel at the top marking ancient river beds (Fig. 13.10).

Fig. 13.9 Panoramic view of modern Florence with the Arno River in the foreground, ancient monuments in the middle ground, and forested hills in the background

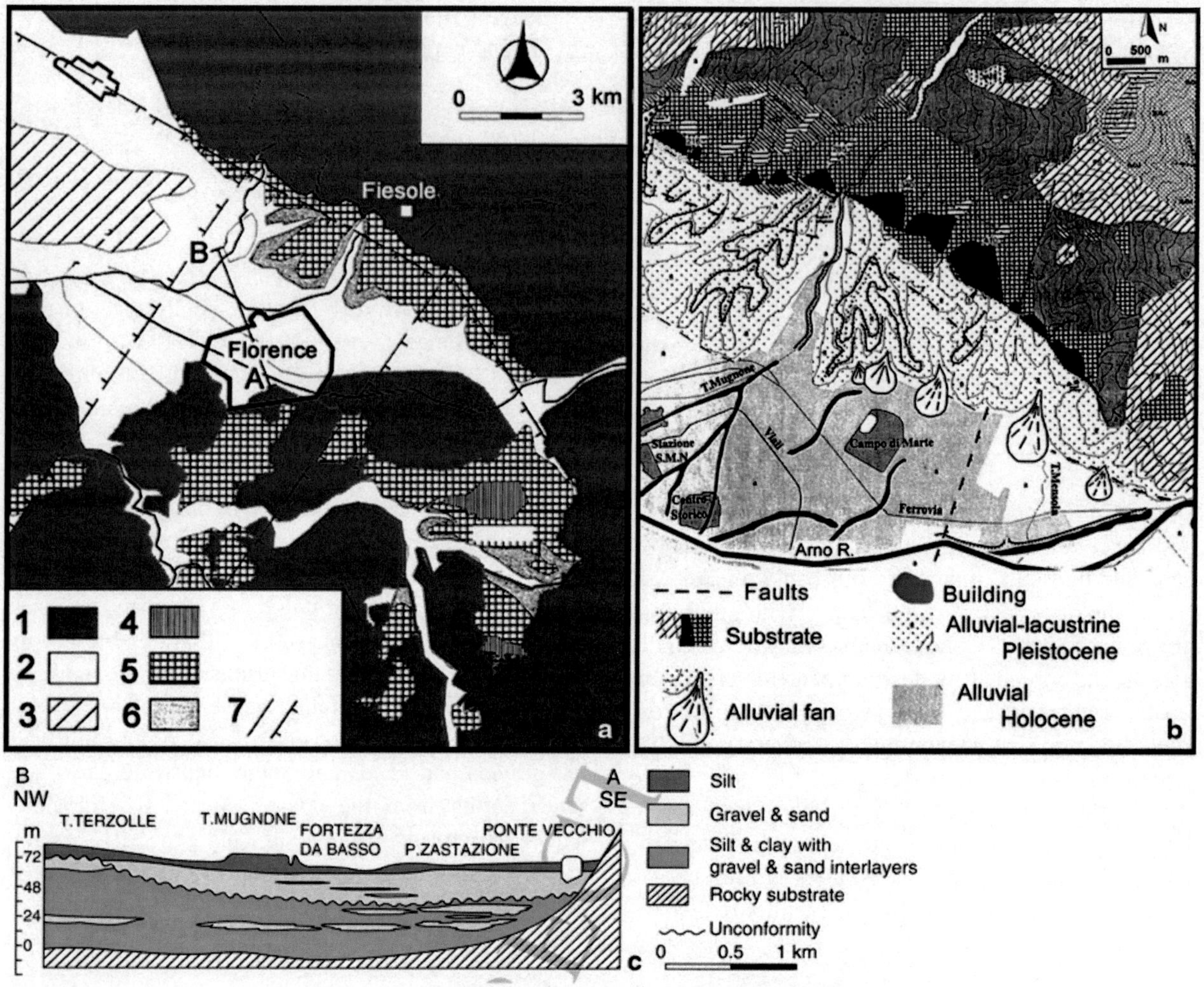

Fig. 13.10 Geology of the Florence area (After Boccaletti et al. 2001). **a** Schematic geological map (*1* Bedrock; **Holocene:** *2* fluvial gravel and sand, *3* fluvial silty clay; **Pleistocene:** *4* sandy clay with disseminated pebbles, *5* interbedded gravel, sand, and silty sand, *6* gravel with silty sand lenses; 7 faults, barbed ones indicate normal faults; A–B: trace of cross section). **b** Geomorphology. Note the different course of the Mugnone T. (creek) that was progressively trained toward the west at the margins of the growing medieval city. **c** Cross section showing Quaternary deposits under the city

Since Roman times when it was first founded at the junction of the Arno River and one of its tributaries, Florence has manipulated the local water courses to its advantage. As the city enlarged, the northern tributaries to the main river were diverted progressively westwards and in part utilized as defensives moats along the western town walls. The main streams were also canalized to prevent floods. Numerous water-powered establishments, such as gristmills (for grains) and fulling mills ('gualcherie', to make woollen cloth) were built along the Arno River, other minor streams, and along artificial canals deviating flows from these watercourses (Muendel 1984). The mills consisted either of large edifices on low banks of the river where water could be canalized off the main course, or of flat boats anchored in the middle of the river. Low dams ('pescaie') were built diagonally across the Arno River to divert part of the water into canals toward the wooden paddles of the mills (Fig. 13.11). Those early dams consisted of wooden piles driven into the river bottom, lateral twig brances, and sand and gravel fills.

The industry of making woollen cloth was one of the activities that made Florence rich, particularly from the early 1300s to mid 1400s. This city-state benefited from this industry over others because of its great availability of hydraulic energy. Several steps were required to fabricate good cloth, besides starting with good wool. Some of the first activities could be

Fig. 13.11 Ancient Florence. **a** Chain Map showing the walled town and various activities in the Arno River in the 1470s. **b** Details showing defensive constructions and pescaie (shallow dams) for mills (large structure on foreground bank) in Florence (After reproduction of lost work by Rosselli F. (1471–1482) located in the Florence, Historical and Topographic Museum "Firenze com'era", Florence, Italy) (Plate 13a)

performed directly by hand, such as cleaning, washing and selecting the wool, spinning the yarn and weaving the cloth. The next step of cleansing the cloth of oils, dirt, and other impurities, and softening and thickening it (fulling, 'gualcatura'), however, could only be efficiently done by heavy machines of the fulling mills driven by river currents (Salvini 1987). The cloth was impregnated with chemicals (usually animal and human urine as sources of ammonia), and was rhythmically beaten by wooden hammers driven by paddles of watermills. Originally several 'gualcherie' were located within Florence itself, but the continuous, loud beating noise and the stench of the ammonia (urine) forced them to be moved both upstream and downstream from the city.

The river was also a convenient waterway for small, flat-bottom crafts. However, the river constituted a risk as well, particularly as it was in part mismanaged. Numerous large floods occurred, inevitably because of the torrential regime of the rivers of Tuscany (Table 13.2). They became more damaging, though, because of canalization (restriction) of the river inside the growing city and the construction of the pescaie (low dams) for mills. These pescaie induced sedimentation and when the sand was not routinely dragged, the bed of the river rose and floods occurred (Table 13.2). Notably devastating medieval floods were those of 1269 and in particular the disastrous one of 1333 that lead to severe famine (Salvestrini 2005; Schenk 2007; http://www.liceoquadri.it/cittamed/citta_med.htm).

The fertile countryside around Florence and other parts of its territory were used for varied cultivations, cereals in particular (Pinto 2002). The cultivation of plants such vines, olive, and of mulberry in the later medieval–early Renaissance times were important as well. The local agricultural production was generally barely sufficient for the growing town population and it was subject to sudden crises due to climatic adverse conditions and floods. Crop failures occurred and famines developed, those at the beginning of the 1300s being particularly severe. Food from neighboring as well as oversea states was imported.

There was also a great demand for cattle for milk, meat and leather, and of sheep for the same things and mainly for wool. The Florentine cloth manufacturers utilized local low-quality wool supplemented by better material imported from other central Apennine areas and high prized quality wools from as far as England. Freshwater fish, partly dictated by religious practices, was a valuable staple and it was obtained from very productive rivers, ponds, and shallow lakes. The Chain Map image showing net fishing in the Arno River downstream from the city is intriguing and indicative of the poor health consciousness of those times (Fig. 13.11); on the river right bank of the fishing area shown on the map was the dump of garbage of any sort including leftover from butchering. The Arno River was indeed much polluted due to injections from the industries and of raw sewage, a situation that continued until quite recently.

13.3.3 Siena

Siena (coordinates: ~43°20′N, 11°20′E) is located on a hill at the western corner of an uplifted elongated Pliocene basin. It is at 322 m asl and about 65 km from the sea (Fig. 13.12). It has now about 56,000 city dwellers (http://en.wikipedia.org/wiki/Siena).

The Siena basin is surrounded primarily by some Mesozoic metamorphic rocks and mostly by Tertiary sedimentary and local volcanic rocks. Some of these rocks were used in the construction of older palaces, towers, and major monumental churches. Brick, though, has been the most common building material since medieval times. Marine clay from the Pliocene basin provided the raw material, and wood was obtained from the surrounding hills. Numerous furnaces originally located in town were banded to the outskirts because of fire hazards. Of the three cities discussed here, Siena had the richest territory in minerals (such as iron, copper, and silver) that were mined in the hills (Colline Metallifere) separating the Siena basin from the Tyrrhenian Sea (Mare Tirreno) area (Fig. 13.1). The city itself is underlain by Pliocene poorly cemented marine sandstones (locally called 'tufo') and conglomerates overlaying marine clays.

Siena is bounded on three sides by steep, high hillflanks, and has a flattish, elongated hilltop to the northwest (Figs. 13.13, 13.14). This gives a characteristic inverted Y plan to the town map.

The location greatly benefited the city during its early developmental stages because it offered three readily defensible sides, and was crossed by the main route (Via Francigena) followed by the pilgrims to Rome during the thirteenth and fourteenth centuries (Fig. 13.15). The pilgrims transiting through Tuscany preferred hill routes to those of the lowlands that were covered by potentially unhealthy wetlands, some being infested by malaria. The Senesi took full advantage of this opportunity catering well to the travellers constructing hospitals, providing money changing facilities and banking, and assuring security in their territory against robbers that infested the lands. This, combined with international banking and commerce primarily of wool cloth across Italy and Europe, made Siena prosperous. It reached maximum development during the last part of the thirteenth century and the first few decades of the fourteenth century. Siena reached it maximum population of about 60,000 peo-

Fig. 13.12 Siena. **a** General view of the Town Hall (Palazzo Pubblico) and the surrounding territory. The Palazzo Pubblico and its characteristically shell-shaped antecedent square were built between 1298 and 1348 AD near the apex of the Montone valley, one of those that indent the hill of Siena. **b** S. Ansano valley, another one that indents the hill of Siena. Note ancient town wall in the middle ground also enclosing cultivated fields called "orti". (See location of valleys in Fig. 13.13)

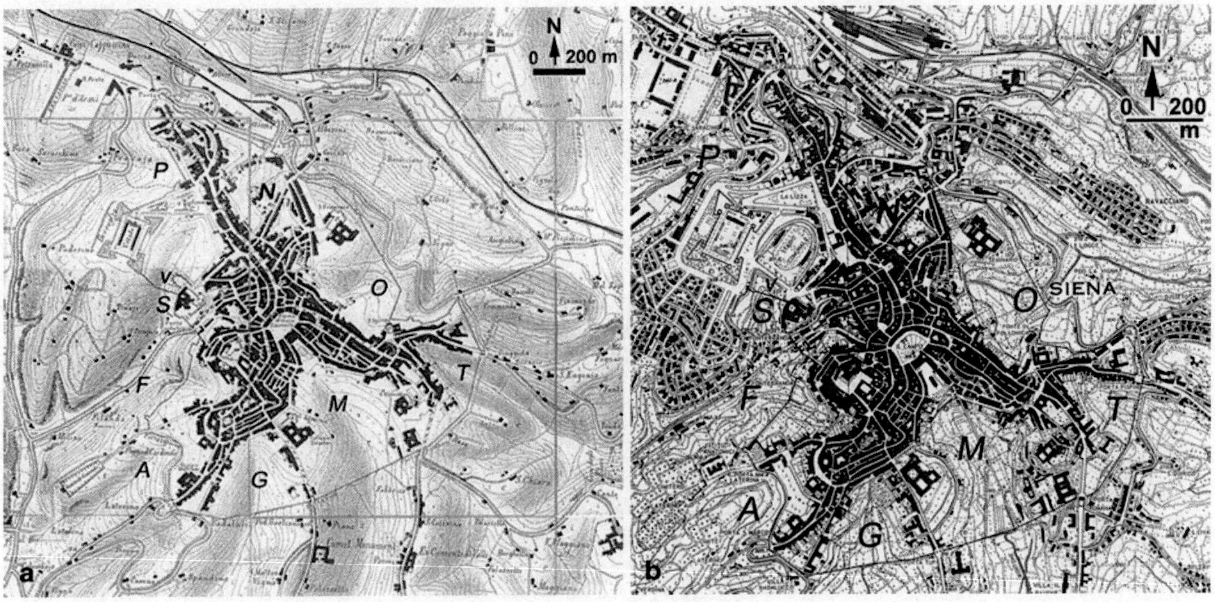

Fig. 13.13 Maps of Siena with flanks deeply cut by narrow steep valleys (After images reported in Bortolotti 1988). **a** 1869 AD map (Accademia dei Fisiocritici, Siena) (Note the San Domenico–Fortezza viaduct (*V*) just started across valley *S* (Rastrello); Palazzo Public and adjacent squares already completed (1298 AD) at the head of Montone (*M*) valley). **b** 1965 AD map (Distance between *M* and *T* is ~500 m; note San Domenico–Fortezza viaduct and Stadium completed at the head of valley *S*). (Valleys: *M* Montone; *G* Gavina; *A* St Ansano; *F* Fontebranda; *S* Rastrello; *P* Pescaia; *N* Fontenova; *O* Follonica; *T* Pispini)

ple in early 1300 living in the confined, walled space of approximately 2 km^2.

The location of Siena had disadvantages as well, primarily (a) the lack of sufficient space for expansion and (b) waterpower for its industry.

(a) Siena developed through the amalgamation of initial centers – 'castellari': castle-like communities – that settled three nearby hilltops. As the city grew several progressively wider town-walls were built to protect dwellings but also cultivable land (the 'orti' (fields) that still exist) that could produce food during also during sieges. The cathedral was built on one of the hilltops after it was partially levelled off. Flattening and carving spaces for building, roads, squares from the easy to excavate sandstone substrate (locally called 'tufo') has been the continuous practice of Siena to our present times. Other building spaces were obtained by reclaiming the upper part of the steep valleys indenting the main hilltops. Retention walls were built across the valleys and the upper parts were partially filled with

Fig. 13.14 Mid-1900s view of Siena showing still existing bounding, steep, northeastern scarp; on the lower right corner of the image is the Camollia entrance to the city from the only flatter northern side of town vulnerable to invasions. (After Etiennez (1852) reported in Bortolotti 1988)

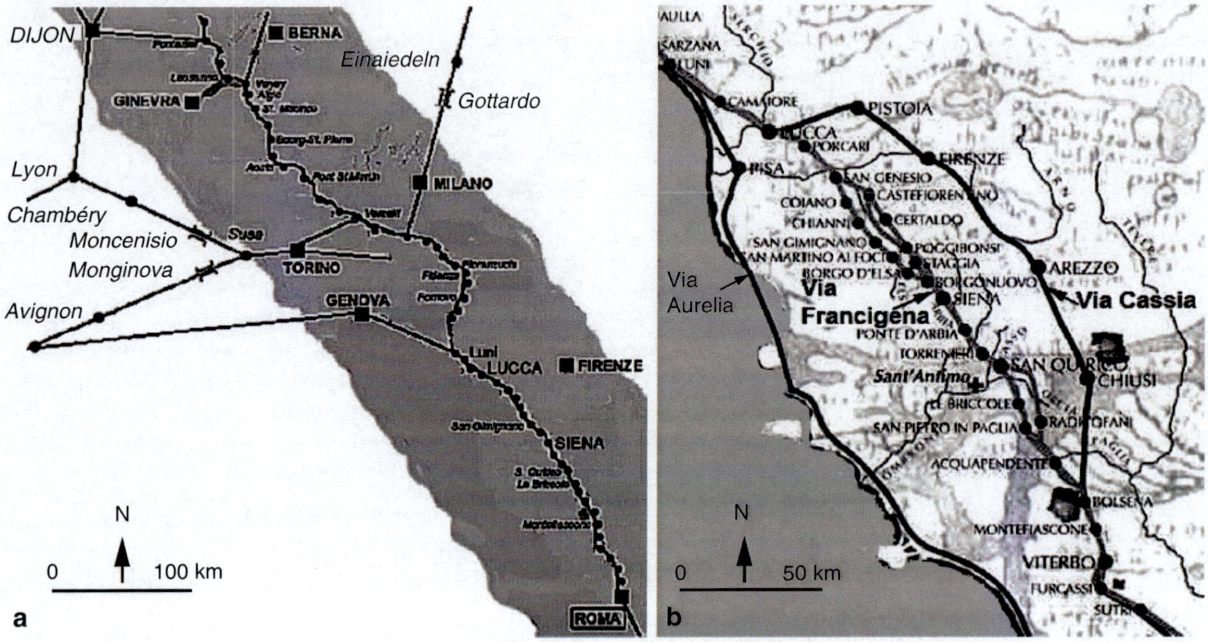

Fig. 13.15 Pre-modern principal routes in central north Italy. **a** Main pilgrim route (Via Francigena) in Italy and south-central Europe (After http://www.francigena.ch/). **b** Detailed map of the routes in central-north Italy with slight variations in the Via Francigena toward Siena, and other routes (Via Aurelia and Via Cassia) built and used by Romans and again in late medieval and subsequent times. (After Adrian Fletcher 2000–2008 http://www.paradoxplace.com/Perspectives/Maps/Via%20Francigena.htm)

sediment removed from other places and/or dump rubble and other debris generated by earthquakes, such as the one in 1319. The famous Piazza del Campo and the Palazzo Pubblico (City Hall) was built in the 1100s at the head of one of these valleys (Figs. 13.12, 13.16); the viaduct between the of church of S. Domenico and the Medicea Fortress was partly built using rubble from the 1798 earthquake across the upper part of another deep, steep valley to the NW of the city. In recent times a soccer field was constructed at its head

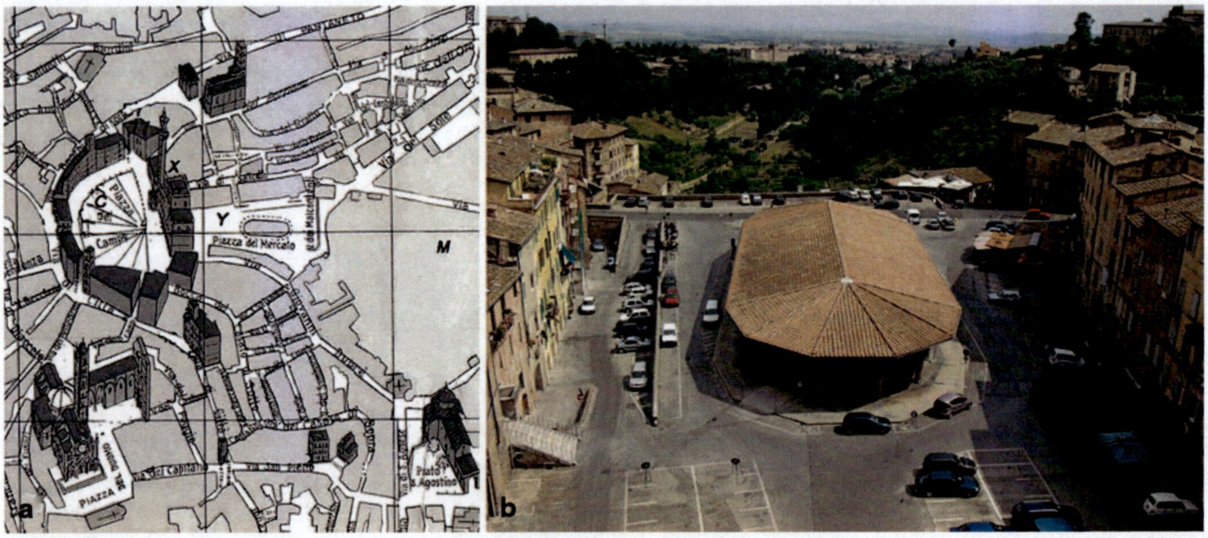

Fig. 13.16 **a** Map of Siena showing: Piazza del Campo (*C*), Palazzo Pubblico (*X*), Piazza del Mercato Vecchio (*Y*), and Montone valley (*M*). **b** Piazza del Mercato Vecchio looking eastward toward the lower part of the Montone valley (see also Fig. 13.13). The first head-valley retaining wall was built at the outer eastern edge of this square

(compare the S valley on Fig. 13.13a without viaduct V, and on Fig. 13.13b after the viaduct and the soccer field had been built).

(b) No large springs, productive water wells, or permanently flowing streams exist in the city area, and only few ephemeral torrents occur in the nearby valleys. At times cisterns collecting rainwater were used but not extensively, and their water was unhealthy and insufficient for the population. Small springs exist at the bottom of the hill, and some of these were eventually included within the town walls. Siena, the town without water, developed a water culture through extreme efforts to find it, bring it to town, implement multiple uses of it, and conserve as much as possible. The Senesi enhanced output of valley bottom springs by excavating tunnels along the boundary between the underlying clay (aquifuge) and the overlying porous sandstone (aquitard), and constructing long (approximately 25 km) underground aqueducts (tunnels, locally called 'bottini', similar to the 'qanats' of ancient Middle East and north Africa people, and to Roman underground aqueducts) to tap small springs on the hills northwest of the city (Figs. 13.17, 13.18; Kucher 2005). One of two major aqueducts carried waters to the base of the hills (Fontebranda, Fig. 13.19), the other directly inside the town (Fonte Gaia, in the main city square in front to the Palazzo Pubblico). These aqueducts still function mainly for watering gardens and other non-potable uses. However, water was never enough for medieval Siena and this eventually crippled its industries.

Although much of the countryside of the Siena basin has clay soils that desiccate quickly in summer, it could produce wheat. Wine, olive oil, and other produces were obtained from the fertile hills around. More agricultural and husbandry supply were also obtained from its distant, fertile, although unhealthy, Maremma territory. Nevertheless famines recurred and other difficult periods were caused by both natural causes as well as by wars and by disruptions caused by companies of jobless, marauding, mercenary soldiers during times of peace between major cities (Bortolotti 1988).

Notwithstanding the poor hygiene common to most medieval towns, and although it had less water than

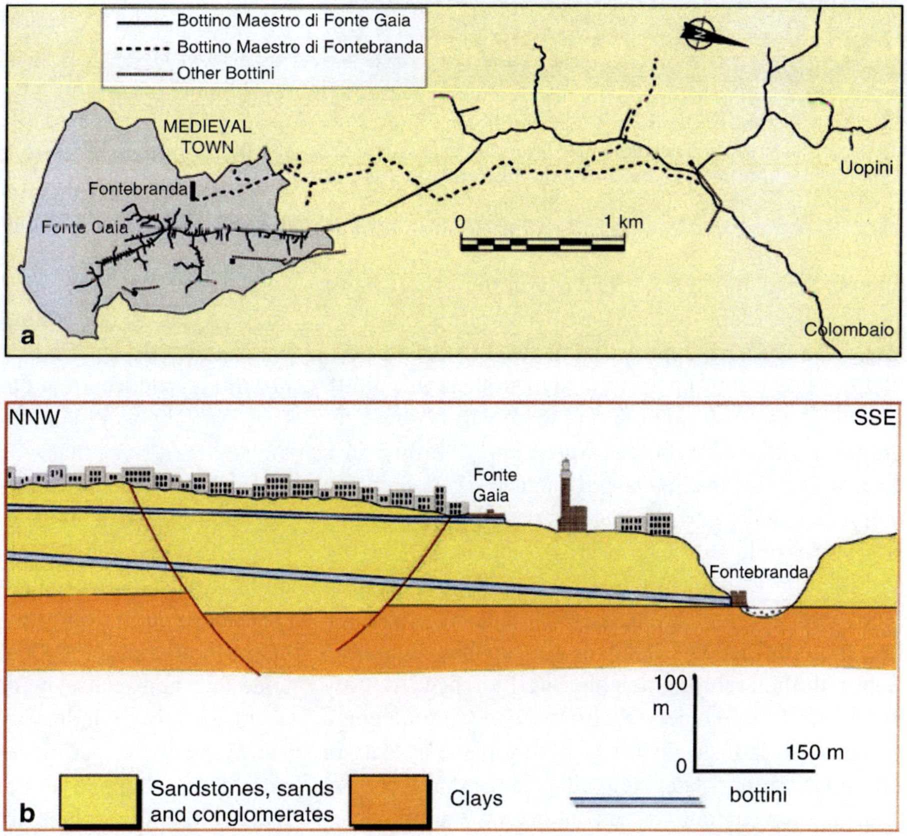

Fig. 13.17 Bottini (tunnels of underground aqueduct) of Siena. **a** Map of the underground aqueducts showing main (maestro) tunnels (Fontebranda and Fonte Gaia) and secondary (other) ones excavated to improve local water outputs or distribute waters from the Bottino Maestro di Fonte Gaia (After Marchi 1869). **b** Schematic geological cross section under the town showing the relative position of the two main aqueducts (Curved, sub-vertical lines are traces of major normal faults; scales are approximate)

Fig. 13.18 Portion of the Bottino Maestro di Fonte Gaia. Fossiliferous sandstone and conglomerate layers are exposed on the ~3 m high walls. The small channel (gorello) in the floor is carrying the waters of the underground aqueduct

others, Siena did not particularly suffer from major sicknesses, except for the Black Death. In about three months in 1348 AD approximately two third of the population died (Bowsky 1964; Mucciarelli et al. 2000). The death rate was so high that there was no time for proper burial. Many bodies were simply thrown in a deep well inside the Spedale di Santa Maria della Scala (City Hospital), and sparingly covered with lime. It is still possible to see in a cross-section the horrific stratification of bones and sediments. Siena never fully recovered in the post-plague world and its landscape had some influence in making it incapable of competion with the archrival neighboring Florence. As Italy and Europe were emerging from the terrible plague a tendency to industrialism was taking place at least in the old wool and later new silk cloth manufacturing and commerce. Being forced out of step with the times in part because of landscape limitations, many affluent families retrenched their interest and wealth in country properties. Siena progressively declined to eventually surrender after an 18 month horrendous siege in 1555 to the Spaniards who sold it to their ally Florence in 1557 (http://en.wikipedia.org/wiki/Siena) when only about 8000 people were left in Siena.

13.4 Synthesis

The development of the three city-states of Pisa, Florence, and Siena in the Middle Ages–early Renaissance is closely interwoven, complex, and primarily determined by socio-political and socio-economical conditions. Those were difficult times and each society strived to emerge or to survive, as the case might have been, with variable fortune associated to factors such enterprising citizens, the selection and loss of powerful allies, the continuous strife of internal discord and external wars, and the scourge of soldiers of fortune that from time to time pillaged the territories. Beyond this socio-political morass there were also environmental and sanitary factors that greatly contributed to the diverse success of the three societies. Some factors, such as the great scourge of the Black Plague starting in 1347–1348, affected all equally, albeit with different results commensurate to the size of the populations. Florence had the largest population at that time and although drastically reduced, it could rebound more readily.

Other factors were tied to the settings (landscape) of the three cities; one of these related to the viability by water and overland. Pisa was never an industrial power, nor did it produce great quantities of foodstuff from its vast, partially wet plain. Pisa, though, was located near the sea on a river that was navigable for much of the early part of medieval times, and could utilize good sea harbors, such as the 'Portus Pisanus', for large ships. Pisa catered to sea transport and travels thorough the Mediterranean including servicing the first Crusades and in doing so its society thrived. Pisa was also a bottleneck for landlocked towns such as Florence that had to use it for its commerce, for materials needed for its industry, and for importing foodstuff during recurring periods of famine. It was therefore a prize to be obtained and eventually it fell definitively under Florentine rule in 1509.

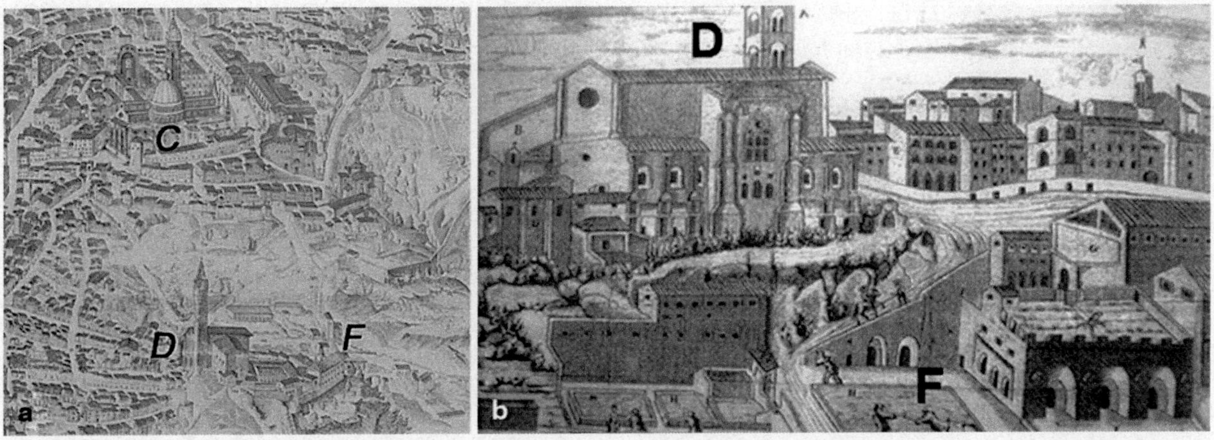

Fig. 13.19 Fontebranda area: the main industrial area of medieval Siena. **a** Detail of Vanni map (Vanni early 1595). San Domenico (*D*) with its original bell tower still standing: it was toppled by an earthquake in 1798; the Cathedral (*C*) was built on a flattened hilltop and the Baptistery in a quarry excavation just to the left (*F:* Fontebranda bottom valley with various water edifices). **b** Complex of Fontebranda showing arched building containing the main pool for potable water, and, sequentially to the left: pool for animals, pool for washing clothes, pool for washing animals, and, farther to the left and outside the walls (see Fig. 13.18a), water was used for running mills, and finally to water the fields. Parallel to this, cisterns were used for washing wools, washing and tanning pelts, for butchering and so on. Note gardens on top of the main arched-pool building. In previous times this top was used as 'tiratorio'; that is, an area where to stretch and dry colored wool cloth

Florence in the early part of the thirteenth century was surrounded by powerful towns among which were Pisa and Siena. The city was excluded from the benefits derived from major transport routes and apparently had limited prospects of success. It did not have access to the sea, the river was difficult to navigate (and only with small crafts), it was far from the principal pilgrimage routes, and it did not have primary resources. Even some of the wool and dies for the cloth industry had to be imported through the harbor of the often antagonistic Pisa. In spite of all, within a few decades it succeeded in reversing the situation. A major communication route shifted from the old Via Francigena to one crossing Florence (the ancient Roman via Cassia, Fig. 13.15b). Powerful new alliances were struck, particularly with the Vatican and other states. The city started to use profitably the great hydraulic resources of the Arno River, increasing greatly the productivity of its cloth and later its silk industries. In 1252 it coined the 'Fiorino', golden coin that overtook the coinage of Siena slowly depreciating as less and less silver was used to mint it. Florence grew to reach about 100,000 inhabitants by the end of the thirteenth century. Nevertheless the subsequent development of Florence to major power status was not a linear one. It suffered strong setbacks such as during the 1333 large flood and crop failure, and its weakening continued reaching bottom during the plague of 1347–1348 and subsequent Black Death bouts. Florence, though, weathered the difficulties better than others in part because of the benefits it derived from the fertile lands of its territory and hydropower it had. As industry evolved and grew, the commerce expanded and competitive practices were implemented such as flooding the markets with less expensive goods. Siena and other societies of Tuscany could not keep pace.

Siena is located on a hilly country away from water courses. Albeit distant, it had an opening to the sea with a small harbor (Talamone) in central south Tuscany, the Maremma. More importantly, during the religious fervour that brought thousands of pilgrims to Rome during the early High Middle Ages, Siena found itself located on one of the preferred pilgrimage route (Via Francigena) because safer and healthier than those of the adjacent bottom valleys locally inundated by wetlands infested by biting insects. The city catered well to the pilgrims offering some degree of security, hospitality, health support, and, most of all, money exchange and banking facilities. This eventually led to the temporary dominance of the Sienese silver coin and to expansion of its commerce throughout Europe. However Siena had strong landscape limitations that impeded its competiveness in expanded markets and industrial development of wool-cloth manufacturing. It had enough space and water in its major, bottom-valley, industrial area of Fontebranda to remain competitive with others during the early part of the High Middle Ages, but it simply did not have enough natural resources to compete later on. Siena like most of Tuscany became subservient to Florence by the mid sixteenth century.

As a whole, the three cities and their territories provide abundant examples of the ways in which humanity is an integral component of landscape, with two-way traffic between the two. The constraints laid upon human development by the landscape, and our attempts to minimize those constraints by the invention of appropriate technologies produce possibilities of complex feedback at all stages. Generally, our victories are short lived, and we are forced to devise new strategies on a continuing basis. We call this progress, for want of a more honest label.

Acknowledgements Research like these protracted over many years has benefited from the help of colleagues, students and enthusiastic local peoples like some witty Florentines and 'bottinieri' of Siena. Sarti likes to single out for a special thank M. Bini, S. Giacomelli, G. Gattiglia and M. Baldassarri of Pisa.

References

Aguzzi M, Amorosi A, Colalongo MC, et al. (2007) Late Quaternary climatic evolution of the Arno coastal plain (Western Tuscany, Italy) from subsurface data. Sedimentary Geology 202:211–229

Bairoch P, Batou J, Chèvre P (1988) La population des villes européennes de 800 à 1850: banque de données et analyse sommaire des résultats. Droz, Genève, 336 pp

Bartolini C, Pranzini G (1981) Plio-Quaternary evolution of the Arno basin drainage. Zeitschrift für Geomorphologie N.F., Supplementband 40:77–91

Benvenuti M, Bellini C, Censini G, et al. Floods, mudflows, landslides: adaptation of Etruscan–Roman communities to hydrogeological hazards in the Arno River catchment (Tuscany, Central Italy). This volume

Boccaletti M, Corti C, Gasperini P, et al. (2001) Active tectonics and seismic zonation of the urban area of Florence, Italy. Pure Applied Geophysics 158:2313–2332

Bonini M, Sani F (2002) Extension and compression in the Northern Apennines (Italy) hinterland: Evidence from the Late Miocene-Pliocene Siena-Radicofani Basin and relations with basement structures. Tectonics 21:1–28

Bortolotti L (1988) Siena. Editori Laterza, Roma, 238 pp

Bowsky WM (1964) The impact of the Black Death upon Sienese government and society, Speculum, 39:1–34

Bradley RS, Hughes MK, Diaz HF (2003). Climate in Medieval Time. Science 302 (5644):404–405

Burland JB, Jamiolkoswski M, Veggiani C (1998) Stabilizing the leaning tower of Pisa. Bulletin of Engineering Geology and the Environment 57:91–99

Caporali E, Rinaldi M, Casagli N (2005) The Arno River floods. Giornale di Geologia Applicata 1:177–192

Cartwright FF, Biddiss M (2004) Disease and history. Stroud, Sutton, 258 pp

Cavazza S (1994) L'idrologia attuale. La natura e i molteplici interventi umnau. In: Mazzanti R (ed) La pianura di Pisa e i rilievi contermini – la natura e la storia. Memorie della Società Geografica Italiana L:421–463

Carratori L, Ceccarelli Lemut ML, Frattarelli Fischer L, et al. (1994) Carta deli elementi natualistici e sorici e dei rilievi conermini, scala 1:50.000. In: Mazzanti R (ed) La pianura di Pisa e i rilievi contermini – la natura e la storia,. Memorie della Società Geografica Italiana, L, Roma (insert)

Ceccarelli Lemut ML, Mazzanti R, Morelli P (1994) Il contributo delle fonti storiche alla conoscenza della geomorfologia. In: Mazzanti R (ed) La pianura di Pisa e i rilievi contermini – la natura e la storia. Memorie della Società Geografica Italiana, L, Roma, pp 401–429

Day WR (2002) The population of Florence before the Black Death; survey and synthesis. Journal of Medieval History 28:93–129

Etiennez H (1852) L'Italie à vol d'oiseau. Painting of the northern side of Siena reported in Bortolotti L (1988) Siena. Editori Laterza, Roma

Federici PR, Mazzanti R (1995) Note sulle pianure costiere della Toscana. Memorie della Società Geografica Italiana 53:165–270

Ginatempo M, Sandri L (1990) L'Italia delle città: il popolamento urbano tra Medioevo e Rinascimento (secoli XIII-XVI). Le Lettere: Florence, Italy

Kucher MP (2005) The water supply system of Siena, Italy. Routledge, New York & London, 115 pp

Leonardo da Vinci. Codice Madrid, Biblioteca Nacional, ms. 8937 (II), cc. 52v–53r

Malinverno A, Ryan WBF (1986) Extension in the Tyrrhenian Sea and shortening in the Apennines as a result of arc migration driven by sinking of the lithosphere. Tectonics 5:227–245

Mann ME, Bradley RS, Hughes MK (1998) Global-scale temperature patterns and climate forcing over the past six centuries. Nature 392:779–787

Marchi C (1869) Pianta geometrica dei Bottini e loro adiacenze (redrawn form map originally designed by Pianigiani G, 1836), reported In: Baletracci D, Vigni L, Costantini A (2006) La memoria dell'acqua – I Bottini di Siena. Comune di Siena, Siena)

Martini IP, Sagri M (1993) Tectono-sedimentary characteristic of late Miocene-Quaternary extensional basin of Northern Appennines. Earth Science Reviews 34:197–233

Martini IP, Sagri M, Colella A (2001) Neogene-Quaternary basins of the inner Apennines and Calabrian arc, Italy. In: Vai GB, Martini IP (eds) Anatomy of an orogen: Northern Apennines and Adjacent Mediterranean Basins. Kluwer Academic Publications, Dordrecht, pp 375–400

Masters RD (1998) Fortune is a river: Leonardo da Vinci and Niccolò Machiavelli's magnificent dream to change the course of Florentine history. The Free Press, New York, 278 pp

Mucciarelli R., Vigni L, Fabbri D (2000) Vergognosa immunditia. Siena ambiente, Siena, 224 pp

Muendel J (1984) The 'French' mill in medieval Tuscany. Journal of Medieval History 10:215–247

Pascucci V, Martini IP, Sagri M, Sandrelli F (2007) Effects of the transverse structural lineaments on Neogene–Quaternary basins of Tuscany (inner Northern Apennines, Italy). In: Nichols G, Paola C, Williams EA (eds) Sedimentary processes, environments and basins – A tribute to Peter Friend. International Association of Sedimentologists, Special Publication 37, pp 155–183

Pinto G (2002) Campagne e paesaggi toscani del Medioevo. Nardini, Firenze, 237 pp

Pirillo P (1994) Demografia, città e territori: alcuni esempi toscani ed umbri tra la fine del XII secolo ed i primi del XIV. In: Comba R, Naso I (eds) Demografia e società nell'Italia medievale, Società per gli studi storici, archeologici ed artistici della Provincia di Cuneo, Cuneo, pp 293–311

Pult Quaglia A (1994) Gli aspetti economici e sociali in età moderna. In: Mazzanti R (ed) La pianura di Pisa e i rilievi contermini – la natura e la storia. Memorie della Società Geografica Italiana, L:359–367

Pranzini E (2001) Updrift river mouth migration on cuspate deltas: two examples from the coast of Tuscany Italy. Geomorphology 38:125–132

Russell JC (1972) Medieval regions and their cities. Studies in Historical Geography. David and Charles, Newton Abbot, Devon, 286 pp

Sallares R (2002) Malaria and Rome: A history of malaria in ancient Italy. Oxford University Press, Oxford, 341 pp

Salvestrini F (2005) Libera città sul fiume regale: Firenze e l'Arno dall'Antichità al Quattrocento. Nardini, Firenze, 155 pp

Salvini E (1987) Gualcherie e tiratori a Firenze nel Medioevo. L'Universo, LXVII:397–459

Schenk GJ (2007) '...prima ci fu la cagione de la mala provedenza de' Fiorentini...' Disaster and 'Life World'—Reactions in the Commune of Florence to the Flood of November 1333. The Medieval History Journal 10:355–386

Starnazzi (2003) Leonardo cartografo. Istituto Geografico Militare, Firenze, 159 pp

Stine S (1994) Extreme and persistent drought in California and Patagonia during medieval times. Nature 369 (6481):546–549

Torelli (1882) Route map of malaria spreading over Italy during medieval times. In: Sallares R (2002) Malaria and Rome: A history of malaria in ancient Italy. Oxford University Press, Oxford,

Vai GB (2001) Structure and stratigraphy: an overview. In: Vai GB, Martini IP (eds) Anatomy of an orogen: Northern Apennines and adjacent Mediterranean basins. Kluwer Academic Publications, Dordrecht, pp 15–32

Vanni F (1595) Sena vetus civitas Virginis. (Map of Siena, reproduced under the supervision of Bonaiuti L, 1873)

Chapter 14
Paleo-Hazards in the Coastal Mediterranean: A Geoarchaeological Approach

Christophe Morhange and Nick Marriner

Human societies in coastal zones are arguably the populations most prone to the danger of geological hazards and the need to devise strategies to live with them. Not only do settlers in coastal zones confront, the major geological problems of earthquake and volcanic eruption as do inland societies, but any such hazards are compounded by the situation of life at the interface between land and sea. Tsunamis are an obvious link between classical geological hazards and the ocean, but slower connections are also encountered, for example sea-level rise associated with the wasting away of the Pleistocene ice sheets. Slow, neotectonic changes along coasts are also significant, and starting in the Neolithic, human activities become a notable forcing factor in this zone.

In fact, the human dimension is a two way street. Pioneer settlements from the Neolithic onwards are clearly constrained by their environments. After initial colonization of the habitat, the environment is in turn manipulated by the human inhabitants, who are now recognized as a geological force in their own right. Seldom are the human manipulations without significant problems, so that humanity itself has become a geological hazard.

Geoarchaeology has long focused on paleoenvironmental reconstructions and landscape evolution (Rapp and Hill 1998; Goldberg and Macphail 2005). Recent research progress in the Mediterranean has furthered the understanding of paleohazards in the coastal areas (Marriner and Morhange 2007). In this chapter, we draw on current topical examples to focus on four types of coastal hazard: slow postglacial sea-level rise, rapid sea-level rise, coastal deformation linked to base-level sediment inputs, and human impacts.

14.1 Slow Postglacial Sea-Level Rise in the Coastal Mediterranean

Since 18,000 year BP a sea-level rise of about 120 m has drowned significant areas of Paleolithic archaeology beneath the sea (Fig. 14.1; Masters and Flemming 1983). Until recent times, human societies in coastal regions were totally at the mercy of sea-level rise. Only late in history, essentially beginning with the Roman era, did people acquire the engineering sophistication to do something about it.

Southern France provides good evidence of the effect of sea-level rise on human settlement in late prehistory. Cosquer, for example, is a partially drowned Paleolithic cave near Marseille (Fig. 14.2). The cave has an entrance 37 m below present sea level, and was partially submerged around 7000 year BP during the marine transgression of the continental shelf (Fig. 14.3; Sartoretto et al. 1995). The preserved horse paintings in it demonstrate that the present sea level is at its highest point since the postglacial period in a so-called tectonically stable setting. Many coastal Paleolithic sites may therefore have been drowned offshore, waiting the investigations of underwater archaeologists. The sea-level change was too slow to constitute a hazard in the true meaning of the word, but, in any case, no technology was yet available to protect against the inexorable rise of the sea.

C. Morhange (✉)
CEREGE Europôle méditerranéen de l'arbois
Aix Marseille University
F 13555 Aix-en-Provence, France
e-mail: morhange@cerege.fr

See Plate 14a in the Color Plate Section; also available at: extras.springer.com

I. P. Martini, W. Chesworth (eds.), *Landscapes and Societies*,
DOI 10.1007/978-90-481-9413-1_14, © Springer Science+Business Media B.V. 2010

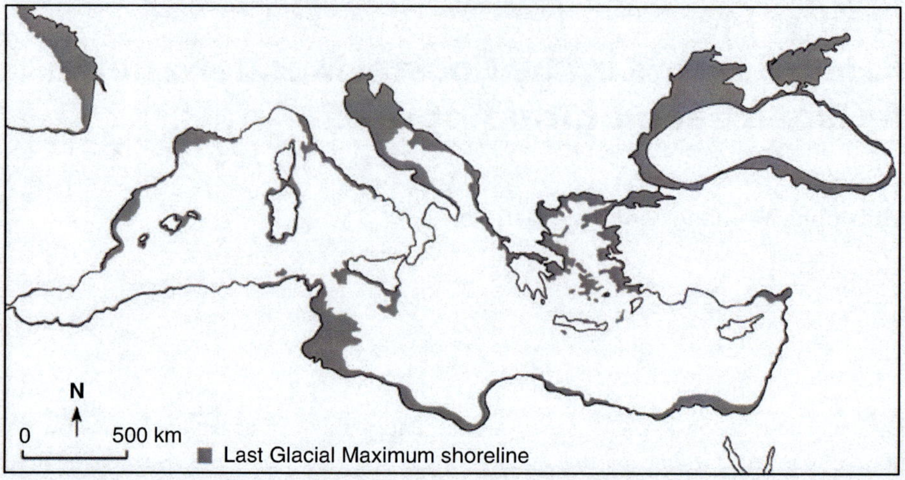

Fig. 14.1 Trangression of the Mediterranean coastal shelf since the Last Glacial Maximum

Since ca. 6000 year BP, sea-level changes have been characterized by a pronounced deceleration linked to the end of glacio-eustatic forcing. After this period, local adjustments are for the most part attributable to glacio-isostatic factors, and in the case of the Mediterranean coast, relative sea-level changes of less than 10 m are observed (Fairbanks 1989; Bard et al. 1996). Within this context, Mediterranean environments provide excellent paleobathymetric archives due to a precise biological zonation of marine species living just above or below mean sea level, and given the density of archaeological coastal remains such as harbors and drowned urban areas (Blackman 1982a, b; Franco 1996). A methodology refined by Laborel and Laborel-Deguen (1994) has been successfully applied to numerous excavations including the ancient harbor of Marseille (Pirazzoli and Thommeret 1973; Morhange et al. 2001) and Pozzuoli (Morhange et al. 2006a). Such data, fundamental to understanding the vertical distribution of coastal remains, were traditionally derived from the geological record (note the exception of Lyell's (1830) observations on the bored columns of the Roman market of Puteoli (Pozzuoli), and the intensive fieldwork of Negris (1904) in coastal

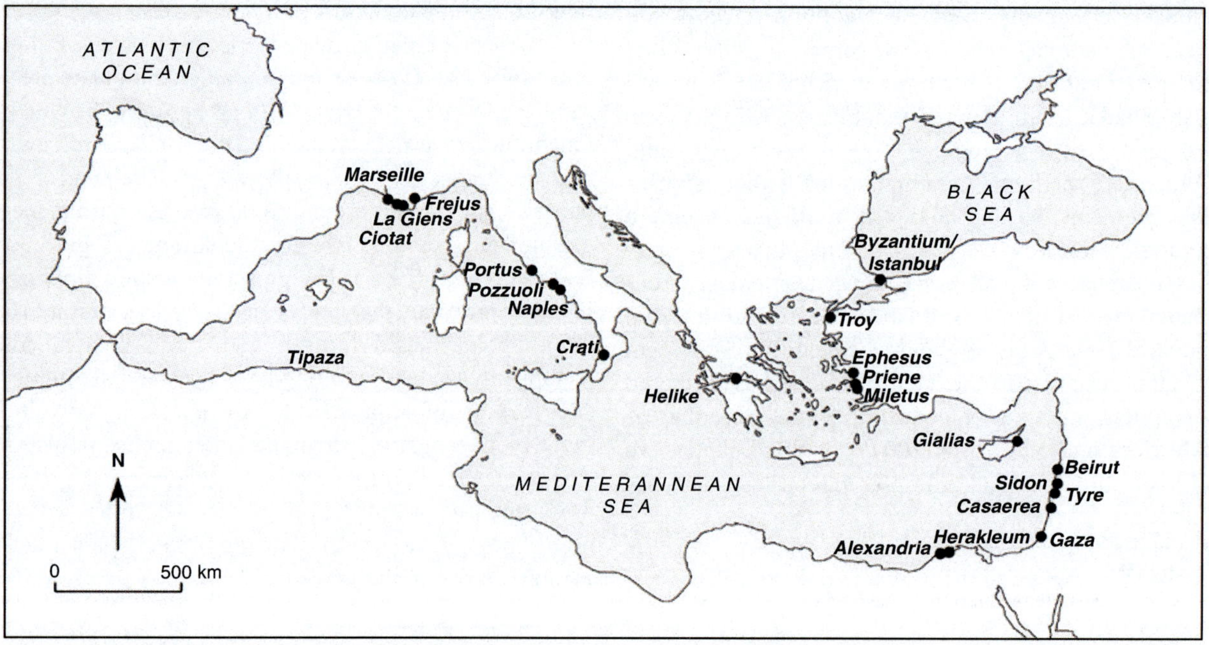

Fig. 14.2 Location of sites discussed in the text

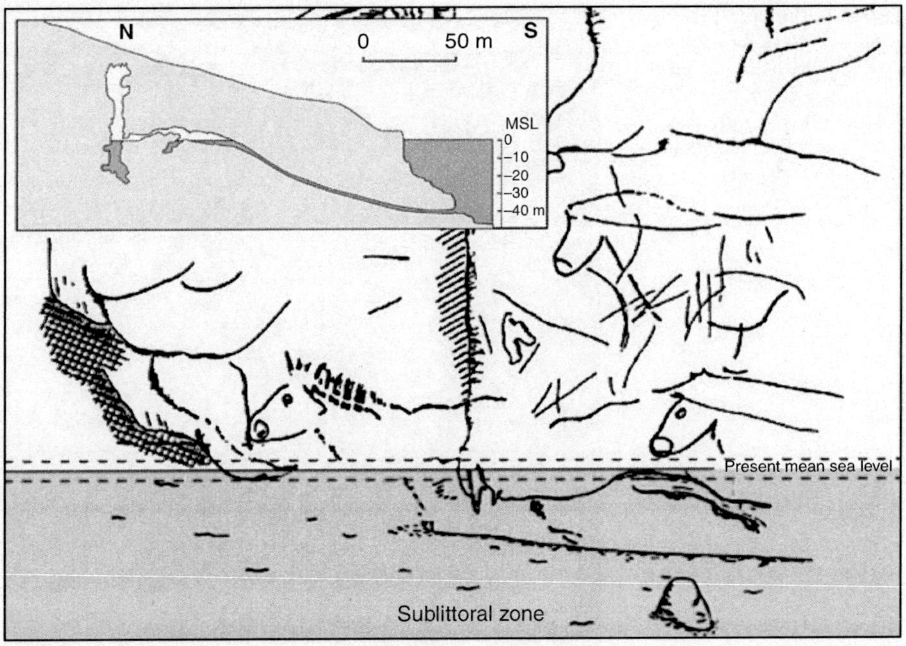

Fig. 14.3 Partial submersion of Paleolithic rock paintings in Cosquer cave, southern France. This example demonstrates that, in tectonically stable areas, no sea level higher than present is attested since the Last Glacial Maximum 18,000 years ago. (From Morhange et al. 2001)

Greece. Where precise vertical relationships can be established between archaeological structures and biological indices it has been possible to accurately reconstruct relative sea-level trends since antiquity at a number of Mediterranean sites (Pirazzoli 1976, 1979–1980, 1980, 1987a, b, 1988). Three groups of structures have traditionally been used: emerged vestiges (dwellings, stock houses, walls, mooring-stones), partially emerged structures (quays, slipways, channels), and submerged structures (shipwrecks) (Blackman 1973a, b; Flemming 1978; Flemming 1979–1980; Flemming and Webb 1986; van Andel 1989; Stanley 1999; Blackman 2003). Unfortunately, the bathymetric imprecision linked to these data is often significant, around 50 cm in most cases. Indeed, the envelope of imprecision can frequently be as important as the absolute sea-level change since antiquity.

Since the 1970s, shortfalls have been overcome using biological fossil remains attached on interface harbor structures (quays and jetties). By transposing the techniques developed on rocky coasts (Pirazzoli 1988; Stiros et al. 1992; Laborel and Laborel-Deguen 1994; Stiros and Pirazzoli 2008) to the context of ancient harbors, precise sea-level datasets have become a good source of primary data (Devillers et al. 2007). The strength of such results lies in the bathymetric precision of biological zonation with the chronological accuracy of well-dated archaeological remains. For example, the biological zoning of certain species (such as the upper limit of *Balanus* spp., *Lithophaga lithophaga*, *Vermetus triqueter*, *Chama griphoides* populations) is linked to mean biological sea level (Péres 1982). By measuring the upper altimetric difference between fossil and contemporary populations low vertical error margins of ±5 cm can be obtained (Laborel and Laborel-Deguen 1994).

Recent geoarchaeological research undertaken in the Roman harbor of Forum Julii (lower Argens valley, Frejus, southern France), demonstrates that sea-level rise of less than 50 cm has occurred during the past 2000 years. Devillers et al. (2007) have dated the upper limit of fixed *Vermetus triqueter* populations at −33 cm under the 0 N.G.F. ('Nivellement Général de la France', French 0 datum; Fig. 14.4). Two different samples yielded respective ages of 2420 ± 30 year BP (300 BC–10 AD) and 2345 ± 30 year BP (160 BC–80 AD). These radiometric datings are supported by ceramics attributed to 30–20 BC and 20–30 AD. The findings fit well with other sites from the region including Marseille (Morhange et al. 2001) and La Ciotat and Giens (Fig. 14.5; Laborel et al. 1994) characterized by a relative sea-level change of ~50 cm during the past 2000 years. It is regrettable that such a multidisciplinary approach is not more widely applied to harbor contexts. In other words, over the last 2000 years, sea-level rise has averaged less than 1 mm/year—hardly a hazard to the human population. Of course it could be said that higher sea levels meant that

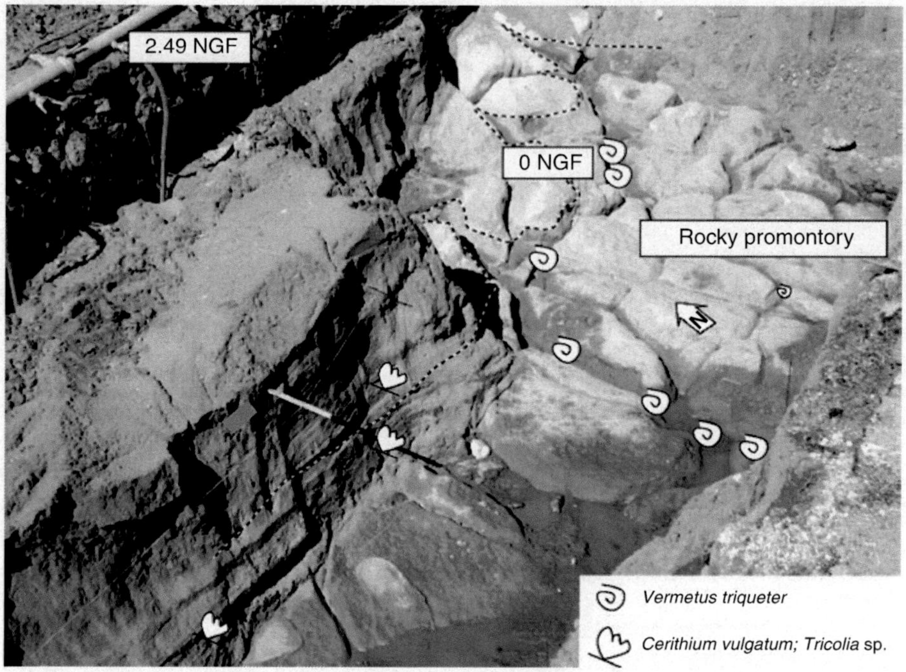

Fig. 14.4 Biological sea-level indicators showing the position of Roman sea level in Frejus' ancient harbor, southern France (after Devillers et al. 2007). The dotted line denotes the NGF ('Nivellement Général de la France' or French national 0 datum)

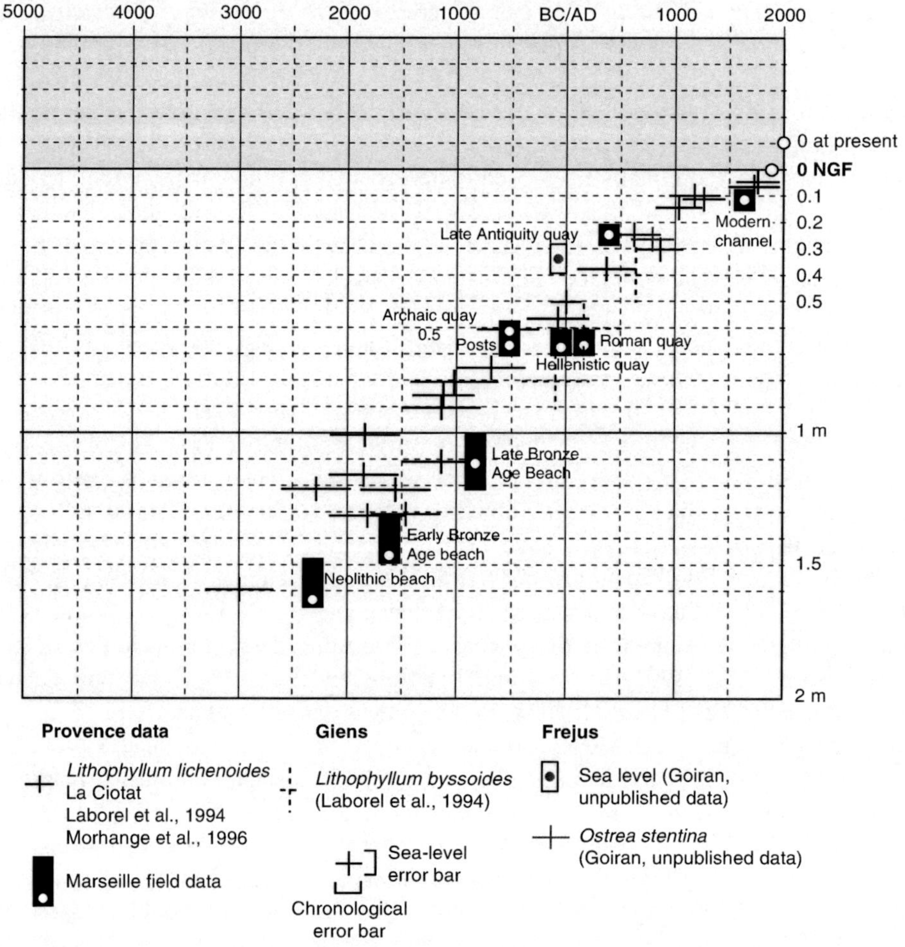

Fig. 14.5 Sea-level changes along the Provence coast since 3000 BC (modified from Morhange et al. 2001; Devillers et al. 2007)

the risk of inundation during storms, or from tsunami in tectonically unstable areas, would be increased, but the short term hazard of rapid sea-level rise in such cases is hardly to be laid at the door of deglaciation.

A consequence of moderate sea-level rise after 6000 years BP was the gradual infilling of base-level depocenters such as lagoons, river mouths, and marshlands. During the Bronze Age, for example, the Levantine coastline was characterized by an indented morphology, where lagoons and estuaries were exploited as natural harbors. Limited accommodation space and high clastic inputs from local sediment sources and the Nile River gradually infilled this indented morphology to yield a linear coastline. Bronze Age sites gradually became isolated from the sea and human populations, unable to offset the rapid rates of sedimentation, were displaced to new locations on the rapidly prograding coasts.

14.2 Rapid Sea-Level Rise and Paleohazards

Effects of rapid sea-level rise may be illustrated by two well-dated examples from Helike in Greece, and Alexandria in Egypt.

The southwestern coast of the Gulf of Corinth, Greece, lies in a region of rapid tectonic uplift and extension. In 373 BC, the city of Helike and its harbor, built on a Gilbert-type fan delta, were destroyed by an earthquake and submerged (Kiskyras 1988; Soter and Katsonopoulou 1998). Using borehole datings, Soter (1998) estimates that the Helike delta subsided by at least 3 m during the event. The opposition between gradual regional uplift and local co-seismic subsidence apparently resulted in a relatively small absolute displacement of the delta during the Holocene.

In a similar vein, the late Roman harbor of Alexandria is submerged about 6 m below present sea level (Goiran 2001; Stanley and Bernasconi 2006). To the west of the city, at ancient Menouthis and Herakleum, this offset is even more pronounced at ~8 m relative to present (Stanley et al. 2001, 2004). The mechanisms responsible for the collapse of the western margin of the Nile delta are at present unclear; scholars have attributed sediment failure to different factors including, fault tectonics, sediment compaction, offshore diapirism and slope instability due to Messinian salt outcrops.

Research has also highlighted the role of instantaneous relative sea-level changes causing harbor and settlement damage during severe storm and tsunami events. For example, major excavation works in the Byzantine port of Theodosius (Yenikapi, Istanbul) has elucidated a scenario of catastrophic seaport destruction during the sixth century AD (Periniçek, personal communication). The sedimentary sequence studied at Yenikapı represents a high-energy sequence attributed to the earthquake of 553 AD and its associated tsunami (Fig. 14.6). Harbor destruction is related to a rapid sea-level oscillation linked to exogenous forcing agents. Other well-dated tsunami sequences are known from the Levantine coasts (Morhange et al. 2006b). For example, Reinhardt et al. (2006) have analyzed high-energy facies in the offshore zone of Caesarea Maritima. They ascribe coarse biofacies to the destruction of Caesarea seawall during the fifth century AD.

In the western Mediterranean, recent work has also focused on catastrophic mega-block deposition on the Algerian coast of Tipaza, a region prone to large earthquakes. Several former tsunamis are inferred to have detached large boulders from the nearshore zone and deposited them inland (Maouche et al. 2009). The boulders, which weigh up to 200 tons, are scattered along some 150 km of coastline, isolated or in clusters, from the sub-littoral to supra-littoral zones. Radiocarbon datings of attached bio-indicators have been used to constrain two tsunamis events on the Algerian coastline between 400 and 600 AD and approximately 1700 AD.

A review of the literature written during the past 30 years shows a shift away from the drowning of

Fig. 14.6 Tsunami depositional layer at Yenikapi (Istanbul) dated to the sixth century AD. (Photo: D. Perincek)

ancient cities (Frost 1963; Flemming 1971) to a more modern paradigm of rapid sediment accretion driving coastal progradation and the landlocking of ancient coastal cities and their infrastructures (such as harbors). In the case of rapid coastal progradation, sites were invariably dislocated seawards. This is particularly true of settlements located in rias, the best examples deriving from the Ionian coast of Turkey (Brückner 1997; Brückner et al. 2002). These examples will be addressed in more detail below in Sect. 14.3. Geographical inertia means that earthquake and tsunami impacted settlements were, in most cases, rebuilt (for example Beirut). The discovery of hydraulic concrete during the early Roman period marked a watershed in coastal engineering. Natural roadsteads were no longer a prerequisite for seaport construction and completely artificial harbor basins could be built on high-energy coastlines, an enterprise which was difficult during the Bronze and Iron ages.

14.3 Hypersedimentation and Coastal Deformation

14.3.1 Delta Scale

Since 6000 year BP, Mediterranean coasts attest to exceptional coastal progradation linked to a deceleration of global glacio-eustasy at all spatial scales (Stanley and Warne 1994). This phenomenon is the rule and not the exception, and explains significant coastal changes to which ancient societies had to constantly adapt. The Bronze Age harbor of Gaza, for example, is currently landlocked due to sediment inputs from the Nile that have been reworked by the eastern Mediterranean gyre. This sweeps westward across the prodelta area before being deviated north towards the Levantine coast (Morhange et al. 2005). In a wave-dominated situation, sedimentary infilling has led to a change in the littoral geomorphology from an indented rocky coastline to a rectilinear coast comprising clastic sediments of predominantly fluvial origin. The effect on the pattern of human settlement has been a gradual dislocation of ancient settlements to keep pace with coastal progradation.

Recent research in the lower Argens (Frejus) has elucidated a coastal progradation of the shoreline by about 10 km during the last 6000 years (Dubar 2003, 2004; Excoffon and Devillers 2006; Devillers et al. 2007). In a similar vein, the Pedheios-Gialias ria (Cyprus) has undergone some 20 km of coastal progradation since the Neolithic. Ancient harbor paleogeography in this vast paleobay attests to the gradual seaward displacement of settlements in order to keep pace with the rapid sedimentation and dislocation of the shoreline (Devillers 2008). Hypersedimentation of coastal areas, therefore, clearly engendered problems of access to the sea and hence the long-term viability of settlements.

All coastal valley centers of deposition have been affected by this dynamic. Many good examples are known from the Ionian coast of Turkey, an area where human–environment interactions have a long history of research (Kraft et al. 1977, 1980; Brückner 1997; Brückner et al. 2002, 2005; Kraft et al. 2003, 2007). The watersheds of Miletus, Troy, Priene, and Ephesus correspond to narrow paleorias, or transgressed grabens, with very limited accommodation space. Recent research at Ephesus provides a good illustration of harbor displacement, or 'race to the sea', linked to rapid shoreline progradation. The ancient first artificial harbor, near Artemision, silted up as early as the sixth century BC, during a period of rapid deltaic growth. A second harbor was subsequently built to the west in the fifth century BC, before relocation of the landlocked city at the end of the third century BC.

Work by Stanley and Bernasconi (2009) the Crati River delta in Italy has focused on coastal progradation and the evolution of three ancient Greco-Roman sites. Sybaris, Thuri, and Copia were successively built up on the delta coast, between the early eighth and first centuries BC. Stanley used sediment cores to reconstruct the gradual seaward growth of the delta front and the respective isolation of each of the sites from the sea.

14.3.2 Harbor Basin Scale

In recent years, a number of studies have shown ancient harbors to be rich time-series of human-environment interactions since the Bronze Age (Reinhardt et al. 1998; Reinhardt and Raban 1999; Morhange 2000; Goiran and Morhange 2003; Kraft et al. 2003; Marriner et al. 2008). Sediment base-level accumulation in ports is the terminal transport pathway for fine-grained sediments in the coastal zone. The main problem of harbor maintenance was rapid silting up. To maintain a sufficient draught depth, ancient societies adapted techniques to evacuate sediment tracts deposited inside these artificial traps (Marriner and Morhange 2006a). Understanding how sediment accumulation rates have varied in space and time has helped to shed light on regional sediment transport conveyors, depocenters and anthropogenic impacts. Societies have had a significant role to play in coastal sedimentation, where ports act like artificial sinks accumulating thick sequences of fine-grained sediments over many millennia.

A common speculation is that primitive harbor dredging began during the Bronze Age along the Nile, Euphrates, Tigris, and Indus rivers (Fabre 2004/2005). For the Roman period, Vitruvius gives a few brief accounts of dredging, although direct archaeological evidence has traditionally remained elusive (Hesnard 2004a, b). Recent examples from Marseille (Morhange et al. 2003), Naples (Giampaola et al. 2004), Sidon (Marriner et al. 2006), and Tyre (Marriner et al. 2008) show evidence for extensive coastal dredging from the late fourth century BC onwards.

These recent case studies allow three questions to be resolved.

14.3.2.1 Why Dredge?

Two variables can be used to explain the long-term viability of ancient harbors: sea-level changes, and sediment supply and its role in modifying the draught depth. Since relative sea-level changes have been quite modest on stable Mediterranean coasts during the past 6000 years (within 2–3 m of present) this variable is of minor importance in explaining coastal deformation (Laborel et al. 1994; Lambeck and Purcell 2005). On centennial timescales, continued silting induced a concomitant thinning of the water column. On short timescales de-silting infrastructure, such as sluice gates, vaulted moles, and channels partially attenuated the problem but in the medium term these measures appear to have been relatively ineffective (Blackman 1982a, b). In light of this, repeated dredging was the only means of maintaining a viable draught depth and ensuring long-term harbor viability.

14.3.2.2 Where and When?

Marseille Archaeological excavations at Marseille have uncovered around 8000 m^2 of the buried port. Litho- and bio-stratigraphic studies elucidate a long history of human impacts stretching back to the late Neolithic period (Morhange et al. 2003). Rapid shoreline progradation is recorded following the foundation of the colony in 600 BC. During the first century BC, after over 500 years of Phocean rule, the demise and fall of the Greek city is translated by wide-reaching changes in the spatial organization of the harbor area. Although dredging phases are recorded from the third century BC onwards, the most extensive enterprises were undertaken during the first century AD, at which time huge tracts of Greek sediment were extracted down to a hard oyster-shell midden layer (Fig. 14.7). Notwithstanding the creation of artificial accommodation space the

Fig. 14.7 Example of a cut and fill talus at Marseille, as depicted by the dotted line, resulting from Roman dredging activity. The cohesive nature of the harbor sediments (>90% silts) has allowed these feature to be well-preserved in the stratigraphic record

seaport rapidly infilled and necessitated regular intervention. Repeated dredging phases are evidenced up to late Roman times, after which time the basin margins were completely silted up.

Naples In Naples, recent excavations at the Piazza Municipo show the absence of pre-fourth century BC layers due to extensive dredging between the fourth and second centuries BC (Giampaola et al. 2004). Unprecedented traces 165–180 cm wide and 30–50 cm deep attest to powerful dredging technology that scoured the volcanic tufa substratum, completely reshaping the harbor bottom.

Dateable archaeological artefacts contained within the deposits allow the decipherment of a very detailed time series of sediment fluxes with much greater temporal resolution than traditional radiometric methods. Investigated stratigraphic sections were dated to the third century BC and the beginning of the sixth century AD. Calculated fluxes are concurrent with intercentennial variability throughout this period. Rapid settling velocities of 17–20 mm/year are recorded during the second century BC and the first and fifth centuries AD. Low sedimentation fluxes of 0–5 mm/year are evidenced during the first century BC, and the late second and early fifth centuries AD. The most rapid rates are consistent with data from Archaic Marseille (20 mm/year; Morhange 1994), Roman Alexandria (15 mm/year; Goiran 2001) and Roman and Byzantine Tyre (10 mm/year; Marriner et al. 2008).

Phoenicia At Sidon and Tyre, unique chronostratigraphic patterns from over 40 radiocarbon dates have yielded strong evidence in support of the dredging findings from other sites (Marriner and Morhange 2006a). Naturally accreting marine bottoms are observed between approximately 6000 BC and 1500 BC, with a pronounced sediment hiatus spanning the Middle Bronze and Iron ages. Rapid rates of sediment accretion and persistent age-depth inversions are evidenced from the third century BC onwards, inconsistent with a natural sedimentary system. Chronostratigraphic patterns from the natural coastlines of the cities do not show similar patterns, discarding the hypothesis of radiocarbon discrepancies at the two sites.

The Romans and Byzantines significantly refashioned their seaports, notably removing great tracts of Bronze Age and Iron Age sediments. This has created a stratigraphic paradox of archive-less Phoenician harbors.

14.3.2.3 How?

The discussed data assert that Roman and Byzantine dredging was a well-organized management technique, not as crude as previously speculated. Bed shear stress in cohesive harbor clays is considerable, and powerful vessels are inferred from the depth of scour marks and the volume of sediment removed. Dredging boats, dating from the first and second centuries AD, have been unearthed and studied at Marseille (Pomey 1995; Pomey and Rieth 2005). The vessels are characterized by an open central well that is inferred to have accommodated the dredging arm. Jules Verne 3's reconstructed vessel length is about 16 m and the central well measures 255 cm long by 50 cm wide. Although the exact nature and mechanics of the dredging arms are not known, dredging taluses some 30–50 cm deep have been fossilised in the stratigraphic record.

It is only during the Romano–Byzantine period that deltaic areas could be transformed into artificial harbor environments. The basin of Portus, on the Tiber delta, is the archetype of such coastal management (Keay et al. 2005). Ancient harbors on rocky coasts were generally not subject to such intense rates of sedimentation. For example, both Marseille and Istanbul are not located in proximity to large fluvial systems; this explains why the ancient port basins are still in use today, more than 2500 years after their foundation.

14.4 Human Impacts

Relationships between human societies and environments have long been considered in quasi-independence of each other rather than as a co-evolution where both are complimentary. Recent work demonstrates that coastal sediments can be used to reconstruct the history of humans and their interactions with the environment since prehistory. The presence of human societies is manifested by a number of proxies.

a. Granulometric impacts: the construction of harborworks is recorded in the stratigraphic record by a unique fine-grained sedimentary facies. This lithoclastic signature facilitates a delimitation of the ancient basin topography. For example, Alexandria's

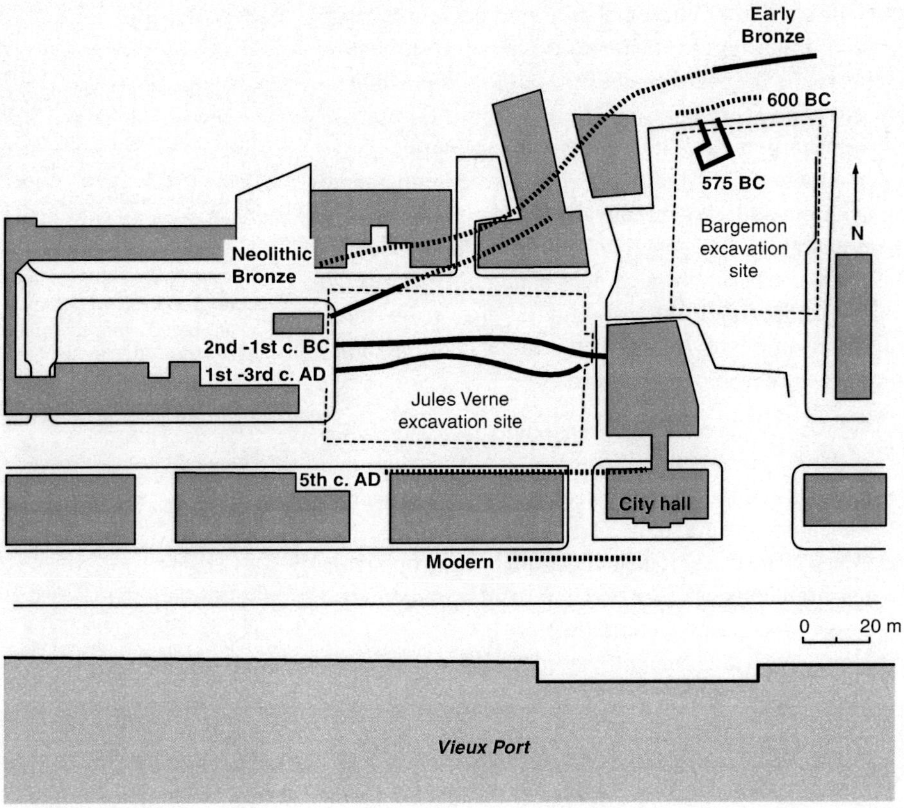

Fig. 14.8 Coastal changes in the ancient harbor of Marseille since the Neolithic period (from Morhange et al. 2001). The full and dashed black lines denote the various shoreline positions for the prehistoric and historic periods. A gradual straightening of the coastline is noted as the harbor basin infilled with fine-grained sediments

eastern harbor is characterized by very fine-grained particles, mainly silt. This harbor facies contrasts with the pre-harbor sedimentary environment, which includes coarse sand and gravels in association with open sea marine assemblages (Goiran 2001). After the collapse of the eastern bay by 6 m during late Antiquity, a transition to open sea facies is observed (post-harbor facies; Marriner and Morhange 2006b).

b. Morphological impacts: the rapid aggradation of harbor bottoms leads to accelerated coastline progradation. For example, progradation of Marseille's northern harbor coastline since the Neolithic is characterized by a progressive regularisation of the littoral geomorphology (Fig. 14.8).
c. Biological pollution: modification in faunal assemblages reworks local anthropogenic inputs such as increases in turbidity and use of the basin as a waste depocenter over many thousands of years.
d. Geochemical impacts: lead has proved to be a powerful tool in recognizing ancient industrial activities (Hong et al. 1994; Renberg et al. 1994; Nriagu 1998; Shotyk et al. 1998; Grattan et al. 2007). Within this context, ancient harbors have been demonstrated to be particularly rich archives of paleopollution. At Alexandria in Egypt, for example, lead isotope analyses have been used to elucidate the pre-Hellenistic occupation of the site (Véron et al. 2006), calling into question the Alexandria 'ex nihilo' hypothesis. The Greco-Roman apogee of the city is attested by lead pollution levels twice as high as those measured in contemporary ports and estuaries. Similar patterns have also been reconstructed in harbor sediments from Marseille (Le Roux et al. 2005), Sidon (Le Roux et al. 2002, 2003) and Tyre.

14.5 Conclusion

Coastal archaeological contexts in the Mediterranean comprise excellent sedimentary archives, yielding insights into the magnitude and direction of anthropogenically forced coastal changes during the Holocene (Marriner and Morhange 2007). In addition to reconstructing the paleoenvironmental evolution of ancient sites, it is important to move beyond the site scale of investigation to compare and contrast the now rich geoarchaeological data from around the Mediterranean and to formulate a working type stratigraphy of ancient harbors. Traditional disciplinary studies have been shown to be largely inadequate when considered in isolation and, through the above examples, we have demonstrated that a geoarchaeological approach is particularly useful in areas of data paucity. An informed earth-science approach can aid in answering three questions imperative to the better understanding of the maritime archaeological record.

a. Where? We have demonstrated that diagnostic litho- and bio-stratigraphies, consistent with geological hazards and human-modified coastal environments, are clearly recorded in the geological record.
b. When? The transition from natural to anthropogenic environments can be dated using either radiometric or ceramic dating techniques.
c. How? How did ancient hazards and local populations impact upon coastal zones.

Recent examples have demonstrated that coastal sites, and particularly ancient harbors, are also appropriate for the analysis of archaeological data at three scales.

a. Basin scale: An informed geoarchaeological approach can yield insights into the harbor basin topography, its functioning, spatial organization, and coeval infrastructure through time.
b. Urban scale: Information pertaining to the site occupation history, notably using geochemistry and geophysics, is made possible due to high rates of sedimentation through time.
c. Regional scale: Typological data can be derived on how these individual maritime sites evolved on a regional scale. It has also been demonstrated that harbor basins are important in better understanding the source to sink sedimentary conveyor and the impact of natural hazards on coastal populations.

Nowadays, most large-scale coastal archaeological projects seek to apply a multi-disciplinary approach at different temporal and spatial scales. Since 1985, harbor archaeology and geoscience workshops have furnished important scientific arenas for multidisciplinary discussion and debate, and attest to a clear growth in this domain as a focal point of research interest.

References

Bard E, Hamelin B, Arnold M, et al. (1996) Deglacial sea level record from Tahiti corals and the timing of global meltwater discharge. Nature 382:241–244

Blackman DJ (1973a) The harbours of Phaselis. International Journal of Nautical Archaeology 2:355–364

Blackman DJ (1973b) Evidence of sea level change in ancient harbours and coastal installations. In: Blackman DJ (ed) Marine Archaeology, Butterworths, London, pp 115–139

Blackman DJ (1982a) Ancient harbours in the Mediterranean, part 1. International Journal of Nautical Archaeology 11:9–104

Blackman DJ (1982b) Ancient harbours in the Mediterranean, part 2. International Journal of Nautical Archaeology 11:185–211

Blackman DJ (2003) Progress in the study of ancient shipsheds: a review. In: Beltrame C (ed) Boats, ships and shipyards: Proceedings of the Ninth International Symposium on Boat and Ship Archaeology, Venice 2000. Oxbow Books, Oxford, pp 81–90

Brückner H (1997) Coastal changes in western Turkey; rapid progradation in historical times. In: Briand F, Maldonado A (eds) Transformations and evolution of the Mediterranean coastline. Bulletin de l'Institut Océanographique, Monaco 18:63–74

Brückner H, Müllenhoff M, Handl M, van der Borg K (2002) Holocene landscape evolution of the Büyük Menderes alluvial plain in the environs of Myous and Priene (Western Anatolia, Turkey). Zeitschrift für Geomorphologie 127:47–65

Brückner H, Vött A, Schriever M, Handl M (2005) Holocene delta progradation in the eastern Mediterranean—case studies in their historical context. Méditerranée 104:95–106

Devillers B (2008) Holocene morphogenesis and anthropisation of a semi-arid watershed, Gialias River, Cyprus. Archaeopress, BAR, International series, Oxford, 197 pp

Devillers B, Excoffon P, Morhange C, et al. (2007) Relative sea-level changes and coastal evolution at Forum Julii (Fréjus, Provence). Comptes Rendus Geosciences 339:329–336

Dubar M (2003) The Holocene deltas of Eastern Provence and the French Riviera: geomorphological inheritance, genesis and vulnerability. Géomorphologie: relief, processus, environnement 4/2003:263–270

Dubar M (2004) L'édification de la plaine deltaïque du Bas-Argens (Var, France) durant la Protohistoire et l'antiquité. Application d'un modèle numérique 2D à l'archéologie. Méditerranée 102:47–54

Excoffon P, Devillers B (2006) Nouvelles données sur la position du littoral antique de Fréjus. Le diagnostic archéologique du « théatre de l'agglomération » (Fréjus, Var). Archéosciences, revue d'Archéométrie 30:205–221

Fabre D (2004/2005) Seafaring in ancient Egypt. Periplus Publishing, London, 278 pp

Fairbanks RG (1989) A 17,000-year glacio-eustatic sea level record – Influence of glacial melting rates on the Younger Dryas event and deep-ocean circulation. Nature 342:637–642

Flemming NC (1971) Cities in the sea. Doubleday, New York, 222 pp

Flemming NC (1978) Holocene eustatic changes and coastal tectonics in the northeast Mediterranean: implications for models of crustal consumption. Philosophical Transactions of the Royal Society of London A 289:405–458

Flemming NC (1979–1980) Archaeological indicators of sea level. Oceanis 5, Hors-Série:149–166

Flemming NC, Webb CO (1986) Tectonic and eustatic coastal changes during the last 10000 years derived from archeological data. Zeitschrift für Geomorphologie 62:1–29

Franco L (1996) Ancient Mediterranean harbours: a heritage to preserve. Ocean and Coastal Management 30:115–151

Frost H (1963) Under the Mediterranean. Routledge, London, 278 pp

Giampaola D, Carsana V, Boetto G (2004) Il mare torna a bagnare Neapolis. Parte II: dalla scoperta del porto al recupero dei relitti. L'archeologo subacqueo 10:15–19

Goiran J-P (2001) Recherches géomorphologiques dans la région littorale d'Alexandrie, Egypte, PhD thesis. Université de Provence, Aix-en-Provence, 264 pp

Goiran J-P, Morhange C (2003) Géoarchéologie des ports antiques de méditerranée. Topoi 11:645–667

Goldberg P, Macphail R (2005) Practical and theoretical geoarchaeology. Blackwell, Oxford, 454 pp

Grattan JP, Gilbertson DD, Hunt CO (2007) The local and global dimensions of metalliferous pollution derived from a reconstruction of an eight thousand year record of copper smelting and mining at a desert-mountain frontier in southern Jordan. Journal of Archaeological Science 34:83–110

Hesnard A (2004a) Terre submergée, mer enterrée: une "géoarchéologie" du port antique de Marseille. In: De Maria L, Turchetti R (eds) Evolución paleoambiental de los puertos y fondeaderos antiguos en el Mediterráneo occidental. Rubbettino Editore, Soveria Mannelli, pp 3–29

Hesnard A (2004b) Vitruve, De architectura, V, 12 et le port romain de Marseille. In: Gallina Zevi A, Turchetti R (eds) Le strutture dei porte e degli approdi antichi. Rubbettino Editore, Soveria Mannelli, pp 175–203

Hong S, Candelone J-P, Patterson CC, Boutron CF (1994) Greenland ice evidence of Hemispheric lead pollution two millennia ago by Greek and Roman civilizations. Science 265:1841–1843

Keay S, Millett M, Paroli L, Strutt K (eds) (2005) Portus. Archaeological Monographs of the British School at Rome 15:1–360

Kiskyras DA (1988) The reasons for the disappearance of the ancient Greek town Helice (Eliki): geological contribution to the search for it. In: Marinos PG, Koukis GC (eds) Engineering geology of ancient works, monuments and historical sites. Balkema, Rotterdam, pp 1301–1306

Kraft JC, Aschenbrenner SE, Rapp G (1977) Paleogeographic reconstructions of coastal Aegean archaeological sites. Science 195:941–947

Kraft JC, Brückner H, Kayan I, Engelmann H (2007) The geographies of ancient Ephesus and the Artemision in Anatolia. Geoarchaeology 22:121–149

Kraft JC, Kayan I, Erol O (1980) Geomorphic reconstructions in the environs of ancient Troy. Science 209:1191–1208

Kraft JC, Rapp GR, Kayan I, Luce JV (2003) Harbor areas at ancient Troy: sedimentology and geomorphology complement Homer's Iliad. Geology 31:163–166

Laborel J, Laborel-Deguen F (1994) Biological indicators of relative sea-level variations and co-seismic displacements in the Mediterranean region. Journal of Coastal Research 10:395–415

Laborel J, Morhange C, Lafont R, et al. (1994) Biological evidence of sea-level rise during the last 4500 years on the rocky coasts of continental southwestern France and Corsica. Marine Geology 120:203–223

Lambeck K, Purcell A (2005) Sea-level change in the Mediterranean Sea since the LGM: model predictions for tectonically stable areas. Quaternary Science Reviews 24:1969–1988

Le Roux G, Véron A, Morhange C (2002) Caractérisation géochimique de l'anthropisation dans le port antique de Sidon. Archaeology and History in Lebanon 15:37–41

Le Roux G, Véron A, Morhange C (2003) Geochemical evidences of early anthropogenic activity in harbour sediments from Sidon. Archaeology and History in Lebanon 18:115–119

Le Roux G, Véron A, Morhange C (2005) Lead pollution in the ancient harbours of Marseille. Méditerranée 104:31–35

Lyell C (1830) Principles of Geology, Volume 1. John Murray, London 511 pp

Maouche S, Morhange C, Meghraoui M (2009) Catastrophic boulder accumulation along the Algerian coast. Marine Geology 262:96–104

Marriner N, Morhange C (2006a) Geoarchaeological evidence for dredging in Tyre's ancient harbour, Levant. Quaternary Research 65:64–171

Marriner N, Morhange C (2006b) The 'Ancient Harbour Parasequence': anthropogenic forcing of the stratigraphic highstand record. Sedimentary Geology 186:13–17

Marriner N, Morhange C (2007) Geoscience of ancient Mediterranean harbours. Earth-Science Reviews 80:137–194

Marriner N, Morhange C, Carayon N (2008) Ancient Tyre and its harbours: 5000 years of human-environment interactions. Journal of Archaeological Science 35:1281–1310

Marriner N, Morhange C, Doumet-Serhal C (2006) Geoarchaeology of Sidon's ancient harbours, Phoenicia. Journal of Archaeological Science 33:1514–1535

Masters PM, Flemming NC (eds) (1983) Quaternary Coastlines and Maritime Archaeology. Academic Press, London, 641 pp

Morhange C (1994) La mobilité des littoraux provençaux: Eléments d'analyse géomorphologiques. PhD thesis, Université de Provence, Aix-en-Provence

Morhange C (ed) (2000) Ports antiques et paléoenvironnements littoraux. Méditerranée 94:1–112

Morhange C, Blanc F, Bourcier M, et al. (2003) Bio-sedimentology of the late Holocene deposits of the ancient harbor of Marseille (Southern France, Mediterranean Sea). The Holocene 13:593–604

Morhange C, Hamdan Taha M, Humbert JB, Marriner N (2005) Human settlement and coastal change in Gaza since the Bronze Age. Méditerranée 104:75–78

Morhange C, Laborel J, Hesnard A (2001) Changes of relative sea level during the past 5000 years in the ancient harbor of Marseille, Southern France. Palaeogeography, Palaeoclimatology, Palaeoecology 166:319–329

Morhange C, Marriner N, Laborel J, et al. (2006a) Rapid sea-level movements and noneruptive crustal deformations in the Phlegrean Fields caldera, Italy. Geology 34:93–96

Morhange C, Marriner N, Pirazzoli P (2006b) Evidence of late Holocene tsunami events in Lebanon. Zeitschrift für Geomorphologie 146:81–95

Negris P (1904) Vestiges antiques submergés. Mitteilungen des Deutschen Archeologischen Instituts 29:340–363

Nriagu JO (1998) Tales told in Lead. Science 281:1622–1623

Péres J-M (1982) Major benthic assemblages. In: Kinne O (ed) Marine ecology 5, part 1. Wiley, Chichester, pp 373–522

Pirazzoli PA (1976) Sea level variations in the northwest Mediterranean during Roman times. Science 194:519–521

Pirazzoli PA (1979–1980) Les viviers à poissons romains en Méditerranée. Oceanis 5:191–201

Pirazzoli PA (1980) Formes de corrosion marine et vestiges archéologiques submergés: interprétation néotectonique de quelques exemples en Grèce et en Yougoslavie. Annales de l' Institut de Oceanographique 56:101–111

Pirazzoli PA (1987a) Submerged remains of ancient Megisti in Castellorizo Island (Greece): a preliminary survey. International Journal of Nautical Archaeology 16:57–66

Pirazzoli PA (1987b) Sea-level changes in the Mediterranean. In: Tooley MJ, Shennan I (eds), Sea-level changes. Blackwell Publishers, Oxford, pp 152–181

Pirazzoli PA (1988) Sea-level changes and crustal movements in the Hellenic arc (Greece), the contribution of archaeological and historical data. In: Raban A (ed) Archaeology of coastal changes. Proceedings of the first international symposium "Cities on the sea – past and present". BAR International Series 404, Haifa, pp 157–184

Pirazzoli P, Thommeret J (1973) Une donnée nouvelle sur le niveau marin à Marseille à l'époque romaine. Comptes Rendus de l'Académie des Sciences 277 D:2125–2128

Pomey P (1995) Les épaves grecques et romaines de la place Jules Verne à Marseille. Compte-Rendus Académie Inscriptions et Belles Lettres, avril-juin, pp 459–484

Pomey P, Rieth E (2005). L'archéologie navale. Errance, Paris, 215 pp

Rapp G, Hill CL (1998) Geoarchaeology: The Earth-Science Approach to Archaeological Interpretation. Yale University Press, New Haven, 274 pp

Reinhardt EG, Raban A (1999) Destruction of Herod the Great's harbor Caesarea Maritima, Israel, geoarchaeological evidence. Geology 27:811–814

Reinhardt EG, Goodman BE, Boyce JI, et al. (2006) The Tsunami of December 13, 115 A.D. and the destruction of Herod the Great's Harbor at Caesarea Maritima, Israel. Geology 34:1061–1064

Reinhardt EG, Patterson RT, Blenkinsop J, Raban A (1998) Paleoenvironmental evolution of the inner basin of the ancient harbor at Caesarea Maritima, Israel; foraminiferal and Sr isotopic evidence. Revue de Paleobiologie 17:1–21

Renberg I, Persson MW, Emteryd O (1994) Pre-industrial atmospheric lead contamination detected in Swedish lake sediments. Nature 368:323–326

Sartoretto S, Collina-Girard J, Laborel J, Morhange C (1995) Quand la grotte Cosquer a-t-elle été fermée par la montée des eaux? Méditerranée 82:21–24

Shotyk W, Weiss D, Appleby PG, et al. (1998) History of atmospheric lead deposition since 12,370 14C yr BP from a Peat Bog, Jura Mountains, Switzerland. Science 281:1635–1640

Soter S (1998) Holocene uplift and subsidence of the Helike Delta, Gulf of Corinth, Greece. In: Stewart I, Vita-Finzi C (eds) Coastal Tectonics. Geological Society, London, Special Publications 146, pp 41–56

Soter S, Katsonopoulou D (1998) The search for ancient Helike, 1988–1995. Geological, sonar and bore hole studies. In: Katsonopoulou D, Soter S, Schilardi D (eds) Ancient Helike and Aigialeia. Athens, pp 68–114

Stanley DJ (1999) Evaluating use of Rock-Hewn features for sea level measurement, Israeli Coast. Journal of Coastal Research 15:326–331

Stanley DJ, Bernasconi MP (2006) Holocene depositional patterns and evolution in Alexandria's eastern harbor, Egypt. Journal of Coastal Research 22:283–297

Stanley DJ, Bernasconi MP (2009) Sybaris-Thuri-Copia Trilogy: Three delta coastal sites become land-locked. Méditerranée 112:75–86

Stanley DJ, Warne AG (1994) Worldwide Initiation of Holocene marine deltas by deceleration of sea-level rise. Science 265:228–231

Stanley DJ, Goddio F, Jorstad TF, Schnepp G (2004) Submergence of ancient Greek cities off Egypt's Nile Delta – A Cautionary Tale. GSA Today 14:4–10

Stanley DJ, Goddio F, Schnepp G (2001) Nile flooding sank two ancient cities. Nature 412:293–294

Stiros SC, Pirazzoli PA (2008) Direct determination of tidal levels for engineering applications based on biological observations. Coastal Engineering 55:459–467

Stiros S, Arnold M, Pirazzoli PA, et al. (1992) Historical coseismic uplift on Euboea Island, Greece. Earth and Planetary Science Letters 108:109–117

van Andel T (1989) Late Quaternary sea-level and archaeology. Antiquity 63:733–745

Véron A, Goiran JP, Morhange C, et al. (2006) Pollutant lead reveals the pre-Hellenistic occupation and ancient growth of Alexandria, Egypt. Geophysical Research Letters 33: L06409

Chapter 15
Mount Etna, Sicily: Landscape Evolution and Hazard Responses in the Pre-industrial Era

David K. Chester, Angus M. Duncan, and Peter A. James

Mount Etna dominates eastern Sicily, being over 3000 m in height and covering an area of some 1750 km^2. Etna has instilled a sense of awe in men and women for thousands of years (Fig. 15.1); to voyagers in the Classical Age it was considered the highest point on Earth (King 1973a) and, even before the colonization of the island by the Greeks ca. 740 BC, members of Sicel culture were practicing cults that associated volcanism with subterranean processes (Chester et al. 2000).

Volcanologically Etna is not only large, but also one of the few continental volcanoes that is continually active. Etna's volcanic activity was initiated about 300–400 ka BP in what was then a marine gulf on the east coast of Sicily (Bonaccorso et al. 2004). Etna's last catastrophic eruption occurred around 15,000 years ago, led to caldera collapse and the eruption of hot and potentially highly destructive pyroclastic flows, which swept down the southwestern flank of the volcano (Guest et al. 2003). Since then Etna has been characterized by basaltic activity, with the principal hazard being posed by lava flows. The vast majority of lava flows on Etna show *aa* morphology with a rough surface made up of irregular lava fragments. The name 'aa' is derived from the onomatopoeic Hawaiian word *a'ā*, which means hard to walk on, but a small proportion of flows show 'pahoehoe' morphology, again an Hawaiian word and typically describing a flow with smooth, lobate and undulating surfaces. There have been very infrequent explosive eruptions, which are unusual on a basaltic volcano, and these have deposited a few centimeters of tephra (volcanic ash) beyond the margin of the volcano. One example of such an event was the eruption of 122 BC when tephra fall brought about extensive damage to Catania during the Roman era and, more recently, tephra from the 2001 eruption caused considerable disruption to communications.

In terms of the relationships between people, land, and the creation of distinctive landscapes, Etna is fascinating because, in spite of the ever present threat of volcanic eruptions and earthquakes, since ancient times the region has attracted settlers in large numbers. Catania, the principal city of the Sicilian east coast, has been badly affected by both these phenomena, and during the past 2000 years has been destroyed in part by earthquakes in 1169 and 1693, and by lava flows in 1669 and probably also in 1371 or 1381, the dating being uncertain (Chester et al. 1985, 2005). This paradox between the hazardous character of the environment and the region's attractiveness for settlement may be accounted for by a number of physical and human factors which operate at two distinct scales. Etna occupies an east coast location within Sicily (Fig. 15.1) and at an all-island scale part of the reason for its development is related to deep-seated and long-recognized contrasts between the coastlands and the interior, which Italian geographers commonly refer to as the "ugly picture in a frame of gold: the dry poverty-stricken core of the island contrasting vividly with the intensively-cultivated, irrigated coastal periphery" (Fig. 15.2; Milone 1960; King 1973a, p. 112). At the more detailed scale Etna has developed in a distinctive way, not just because people have had to respond to hazards, but also through a combination of factors that include not only particularities and variations in climate, vegetation and soils, but also distinctions caused

D. K. Chester (✉)
Department of Geography
University of Liverpool
Liverpool, L69 3BX, UK
e-mail: jg54@liverpool.ac.uk

See Plate 14b in the Color Plate Section; also available at: extras.springer.com

I. P. Martini, W. Chesworth (eds.), *Landscapes and Societies*,
DOI 10.1007/978-90-481-9413-1_15, © Springer Science+Business Media B.V. 2010

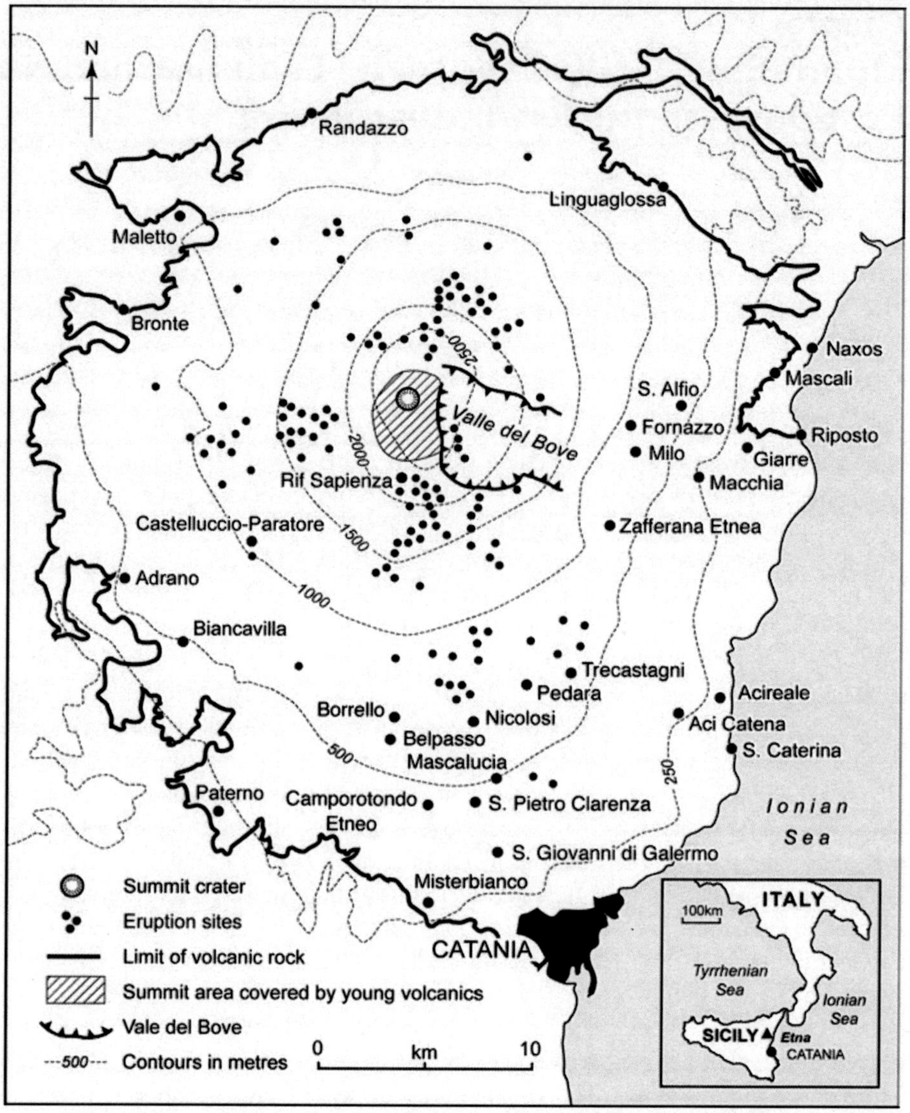

Fig. 15.1 Mount Etna region and Sicily: general location map. (From Chester et al. 1985)

by history, economy and culture. This makes the Etna region in some respects similar to, yet distinct from, other portions of the "frame of gold" (King 1973a, p. 112) and its unique character has made it of interest not just to volcanologists and other earth scientists, but also to classicists, historians and prehistorians, geographers and rural-sociologists (Leighton 1996, 1999; Chester et al. 2000; Malone and Stoddart 2000; Duncan et al. 2005; Smolenaars 2005).

Although important social and cultural particularities remain, today Sicily is closely integrated into the economies of Italy and the wider European Union, but even as late as the late 1960s and early 1970s major elements of the pre-industrial society still persisted and reflected an interplay of people and environment that had produced distinctive agricultural landscapes and which had developed over hundreds, in some cases thousands, of years of settlement. For instance, the 1971 census recorded a Sicilian population of around 5 million, of which the percentage employed in agriculture varied from 20% to 40% depending on province, with an estimated 50% of the island population still dependent on agriculture to some degree (King 1971). Administratively most of the land area of Etna is contained within the Province of Catania and, reflecting its "frame of gold" location and other more localized features of attraction, accounted for 20% of the population of Sicily and some of the highest popu-

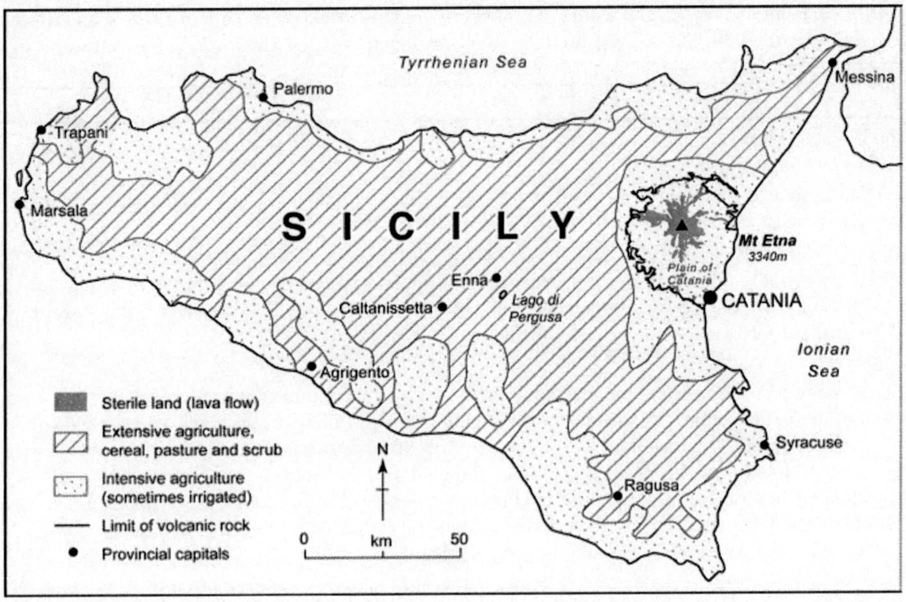

Fig. 15.2 The contrast between the intensive and extensive agricultural areas in Sicily. It should be noted that the Plain of Catania only became an area of intensive cultivation late in pre-industrial times. In the 1940s this area had a low population and little commercial agriculture due to its malaria character. From the 1950s new irrigation techniques were introduced and malarial was eradicated. (From King 1971)

lation densities found on the island. In 1971 figures of up to 800 persons/km^2 occurred in local authority areas ('comuni') located close to Catania, whilst elsewhere densities of 500 persons/km^2 were far from uncommon, densities declining with both increasing height on the volcano and with increasing distance from Catania. In contrast, for the island as a whole, an average population density in 1971 was 181 persons/km^2.

15.1 The Ugly Picture in the "Frame of Gold"

The reasons for the contrasts between central Sicily and its coastal margins, including Etna (Fig. 15.2), are due to both environmental differences and contrasting historical, social, and economic circumstances. With regard to environmental factors these include: higher annual rainfall on the north coast (up to 800 mm) and on Etna (700 mm to over 1250 mm), in comparison with totals of only 700 mm and usually far less in the interior (Fig. 15.1); the fact that flat land is rare in Sicily, plains only accounting for just 7% of the land area and with a strong concentration near to the coast; and the greater soil and irrigation potential of the island peripheral margins (Pecora 1968; Chester et al. 1985).

There is a lack of certainty about when humans first became established in Sicily, though stone implements on the plain of Catania and in the Province of Agrigento (Fig. 15.2) point to Lower Paleolithic times, but the evidence is ambiguous. By the Upper Paleolithic (ca. 18,000 ^{14}C year BP; Table 15.1) human presence is more certain and Sicily was widely inhabited (Leighton 1999), though recent palynological studies based on evidence from lacustrine sediments from the Lago di Pergusa in Central Sicily (Fig. 15.2) suggest that the first major impact of people on the vegetation of the island occurred much later, by 2800 ^{14}C year BP if not slightly earlier (Sadori and Narcisi 2001; Sadori et al. 2008). This date coincides with the early Bronze Age, a time of population growth in the lowlands and some expansion into the uplands including Etna (Table 15.1; Malone and Stoddart 2000). Indeed Leighton comments, "Etna, Europe's largest volcano, (is) characterized by an extraordinary ecosystem. Here one may detect human adaptations peculiar to the environment of its lower slopes: well-watered fertile, with a rich and varied fauna and flora, and numerous caves. If it were better documented the archaeology of Etna would constitute a remarkable case study in European prehistory" (Leighton 1999, p. 4). Lava-flow caves provide convenient sites for habitation and several on Etna have yielded Copper Age material. Lava-flow caves are lava tubes and these develop on lava flows when an active feeding channel roofs over. When lava supply ceases the active lava drains and produces a lava tube (Table 15.1).

The contrasts between the center and the coastal periphery are largely the result of both farming techniques and social and economic systems introduced by

Table 15.1 Sicily: Phases and dating of prehistoric and historical development. (Based on information in Finley 1968; King 1973a; Leighton 1996, 1999; Malone and Stoddart 2000; Ayala and French 2003, 2005). The locations of places mentioned in the Table are shown on Fig. 15.1

Phase	Date
Prehistoric Times	To ca. 740 BC
Lower Paleolithic	From ca. 500,000 BP
Evidence about population densities is unreliable for the lowlands of Sicily. Densities were low in the uplands[a]	
Upper Paleolithic	Late stage ca. 18,000 ^{14}C BP
Sicily was widely inhabited. In the lowlands high population densities occurred in river valleys and low densities were maintained in the uplands. The lowland margins of Etna became important foci for settlement from this time	
Neolithic	
Some renewed colonization took place, but there was also a continuity of settlement. Neolithic farming communities became established: in the Simeto Valley on the southwest margins of Etna at Adrano, Biancavilla, Paterno; to the north of Etna (on the Bolzano Soprane lava flow near to Maletto); and at other sites. Some of the sites in the Adrano area yield pottery with stamped and incised motifs, which is known as 'impressed ware', and dates from ca. 6000 to 5000 BC. Typical settlement pattern throughout Italy is based on floodplain sites, near to a water course where early agriculture was practiced	7th millennium BC, a jar from near Paterno yielded a date of 5000–6000 ^{14}C year BP
The Copper Age and the Bronze Age (including the Thasos and Pantalica cultures)	Copper Age—second half of 4th millennium. Early Bronze Age—late 3rd millennium BC Thapos—Mid 2nd millennium BC Pantalica—Late 2nd millennium BC
Copper Age—circular and oval huts are found near to Adrano. Lava caves in lava flows were convenient sites for habitation and are commonly developed on pahoehoe flows, but are also sometimes found on *aa* flows. Some evidence of soil erosion in the northern mountains of Sicily	
Metal items were rare until *Middle Bronze Age*. Dense Bronze Age settlement was found in the Bronte, Adrano, Biancavilla and Paterno areas on the south west margins of Etna, where closely-spaced settlements occurred and a mixture of arable farming and pastoralism was practiced. Sites have also been found at Catania and Naxos. A very important site occurs at the Salinelle di San Marco at the Paterno mud volcano, showing settlement from the Neolithic onwards. Signs of 'demographic saturation' within the limits of the agricultural economy of the time in the lowlands and expansion in upland area have been suggested during the Bronze Age	
Iron Age	ca. 1000 BC
Phoenicians	
Objects from throughout the circum-Mediterranean were traded with Sicily. Iron implements may have been introduced for the first time	
Immediately before the Greek settlement, the island was dominated by three tribal groups: Sicals (Sikels) in the east; Sicans (Sikans) in the west and the Elymians in the extreme north west. Many hill top towns were originally Sical strongholds. There are no written records of Sicily before the Greeks, but Greek myths about earlier times (Daedalus and the travels of Heracles and Odysseus) refer to the Sicans and the Sicals. These accounts should be treated with caution because they are not always substantiated by archaeological evidence. It is possible that some agro-towns date from Sical times	
Historical Times[b]	
Greek	ca. 740 BC–264 BC
The first Greek settlement occurred at Naxos in 734 BC and later Catania was occupied. The Greeks settled the eastern coastal margin of Etna, with the Sicals continuing to occupy inland sites. Within less than 140 years most of coastal and eastern Sicily, including Etna, had been settled. A further century was required before the interior was Hellenized. Lasting memorials include the Greek temples at Agrigento (Fig. 15.2). The Greeks introduced the vine, olive and fig and left a lasting imprint on the culture and language	
Roman	264 BC–827 AD
Although conquest began in 264 BC the whole island was not occupied until 210 BC. The Romans farmed large estates (latifundia) in the interior, the principal crops being wheat, olives and barley. Wine was exported and cattle, sheep and pigs were plentiful. Roman rule also strongly influenced the language and introduced Catholicism. Towards the end of Roman times, when the island was part of the Byzantine Empire, Vandals and Ostragoths raided Sicily	

Table 15.1 (continued)

Phase	Date
Arab	827–1091
Although the Islamic invaders are often termed Arabs, invaders included Berbers and Spanish Muslims. In 827 the invading army numbered over 10,000	
Norman	1091–1194
Swabian	1194–1268
Angevin	1268–1282
Spanish	1282–1713
Austrian	1720–1734
Bourbon (between 1806 and 1815 Sicily was controlled by the British)	1734–1860
United Italy	From 1860

[a] Upland is defined as land over 750 m

[b] For historical times the influence of different rulers on agricultural landscapes is discussed in the main text

later invaders, particularly the Arab (827–1091) and later Norman, Spanish, and Bourbon rulers (Table 15.1), who in combination produced forms of land use and a social system that remained virtually unchanged for centuries. Small-scale irrigation was introduced by the Romans, but it was only after the Arab conquest from the ninth century AD that these techniques became widespread. Diffusion of irrigation depended chiefly on the ready availability of ground water and here the northern coastal margin and the Etna region were particularly favored, with crops such as bananas, bitter oranges, lemons, melons, mulberries, possibly cotton, sugar cane, the date palm, and pistachio nuts being introduced. Sumac was brought into the island for tanning, hemp for caulking, and livestock rearing was intensified (Mack-Smith 1968; Benjamin 2006). By the Early Middle Ages most of crops found in Sicily today were being cultivated, though maize, the prickly pear cactus, the potato, tomatoes, and tobacco were only introduced from the New World after the close of the fifteenth century (King 1973a).

Decline in the fortunes of the interior lands may be placed firmly at the door of Arab and later rulers. The Arabs cleared vast tracts of land for settlement, later timber was exported and/or used for shipbuilding and large areas of the interior were given over to extensive grazing, an unsuitable land-use for the easily eroded soils of the rolling hills found in the interior. Grazing animals provided hides, wool, meat, and cheese for export. In studies of soils in southern Italy, a distinction is often drawn between those found on Pliocene and other clay outcrops and those which occur on other substrates (King 1973b). The former are characteristic of vast areas of central Sicily, where the clays are described as *scagliose*, or scaly. Here soils are baked during the long summer drought, but become water saturated, viscous, and impermeable in winter. Shallow ploughing in spring and autumn using traditional techniques of peasant farming, has progressively caused the upper levels of soil profiles to become unstable and rendered them highly susceptible to erosion. The soil map of Sicily (Fierotti et al. 1988) records erosion in 24 of the mapped 33 soil associations found on the island, while the CORINE organization (European Environment Agency's Co-ordination of Information on the Environment) also highlights a high risk of soil erosion (CORINE 1994). The worst-case scenario, which assumes no vegetation cover, shows not only high potential rates for the western half to two-thirds of Sicily including the interior, but also for Etna and the highlands of the north east. Actual soil-erosion risk under current land-use and vegetation cover is, in contrast, reduced for the whole island, remains significant for the center and west, particularly for the interior lands, but is greatly reduced for Etna and much of the north east (CORINE 1994). Late in pre-industrial times in the 1920s and 1930s the situation was further exacerbated and the campaign by the fascist government of Mussolini to achieve national self-sufficiency in grains, saw many of the more marginal interior lands being brought under intensive wheat cultivation using fertilizers. The land could not long withstand being sown with wheat year after year, though whether this was due to erosion or because of a decline in fertility is not made clear (Grove and Rackham 2001).

Socially as well as economically, core and periphery were very different, the latter having greater prosperity, being more cosmopolitan and with a more innovative population than was to be found in the interior. *Latifundia* estates were established in central Sicily by the Romans (Table 15.1), but their diffusion over the interior lands was largely the result of later rule, gifts of estates under a system of patronage often being granted as rewards to nobles who had rendered service to the ruler (King 1973a, b). Estates were almost invariably subdivided and under the management of agents ('gabelloti', in Italian) who controlled the renting of small plots to peasant farmers. Landowners ('baroni', in Italian) were usually absentee and the gabelloti had untrammeled power to raise rents against a background of severe population pressure that produced widespread land hunger. In order to survive, farmers had to maximize their incomes by adopting a virtual wheat monoculture, by over-cropping without rotation or fallow and by cultivating steep slopes. Widespread lawlessness made the situation worse. The gabelloti class was strongly associated with the mafia and, in addition, banditry and crime became part of the way of life and death throughout much of interior Sicily. In contrast eastern Sicily, including most of the Etna region, was relatively free of organized crime (Rochefort 1961). In the late nineteenth century especially after 1890, conditions did improve somewhat due to an opening of the safety valve of emigration especially to the USA (United States of America), but poverty remained endemic for many more decades. Demographic change introduced a further element of contrast with the coastal periphery, which did not lose population at anything like the same rate. The population of the Etna region in fact increased. This phase of migration peaked in 1913 at a figure of over 148,000 persons/annum and thereafter fell rapidly especially following the introduction of immigration controls by the USA in the 1920s (Benjamin 2006). The 30 years from 1947 saw a new phase of emigration which involved young skilled workers, rather than landless peasants. Emigration was primarily to northern Italy and other countries of Western Europe and this time the Etna region and the rest of the "frame of gold" were not immune from its effects. The impact in the interior was far more significant, however, and throughout all phases of migration the Province of Catania increased its share of the island population, due to a combination of natural increase and internal migration from other areas of Sicily. For instance, in 1861 the Province of Catania accounted for 15% of the population of Sicily, by 1901 this had risen to 16% and stood at 20% of the island's total in 1971 (Chester et al. 1985).

In Sicily agricultural settlement was not diffuse and farmers usually lived in large agro-towns or peasant cities, commuting to their agricultural holdings on a daily and sometimes on a twice daily basis. Agro-towns had populations of between 3000 and 15,000 people or even more, and in pre-industrial times had at least 50%, sometimes 90%, of their inhabitants engaged in agriculture (King and Strachan 1978). A few settlements are known to have been continuously occupied since pre-Hellenic Sical times (Table 15.1), but many have no certain date of foundation. Alfred Demangeon (1927) has argued that agro-towns were founded as a result of several interlocking factors, which included natural conditions, plus ethnic and social linkages between the first settlers, but his principal insight was that these peasant cities continued to exist because of social and economic inertia, since they served to alienate and separate farmers from their land and so reduced agricultural efficiency—not least because of lengthy journeys to work.

In the coastal margin, agro-towns are also a feature of the landscape, though on Etna they are not found on hilltops so shortening journeys to work and reducing the negative effects on agricultural efficiency. In pre-industrial times the residents of these villages were more open to outside influences (Chester et al. 1985).

15.2 The Uniqueness of the Etna Region

In contrast with the interior of Sicily, during pre-industrial times the Etna region was marked by greater affluence, which was based on more productive agriculture. Etna not only showed a unity in terms of its relative prosperity, but also contrasted with the interior in other respects. Despite often continuous cropping of some areas for over two thousand years, production on Etna was sustained without either reduction in yield or significant soil erosion, and this is further discussed in Sect. 15.2.1. At a more detailed scale and within the overall unity of relative wealth and sustainability, there was a diversity in the distribution of land uses on the volcano, which demonstrated a fine adjustment to both: the physical environment and the eco-

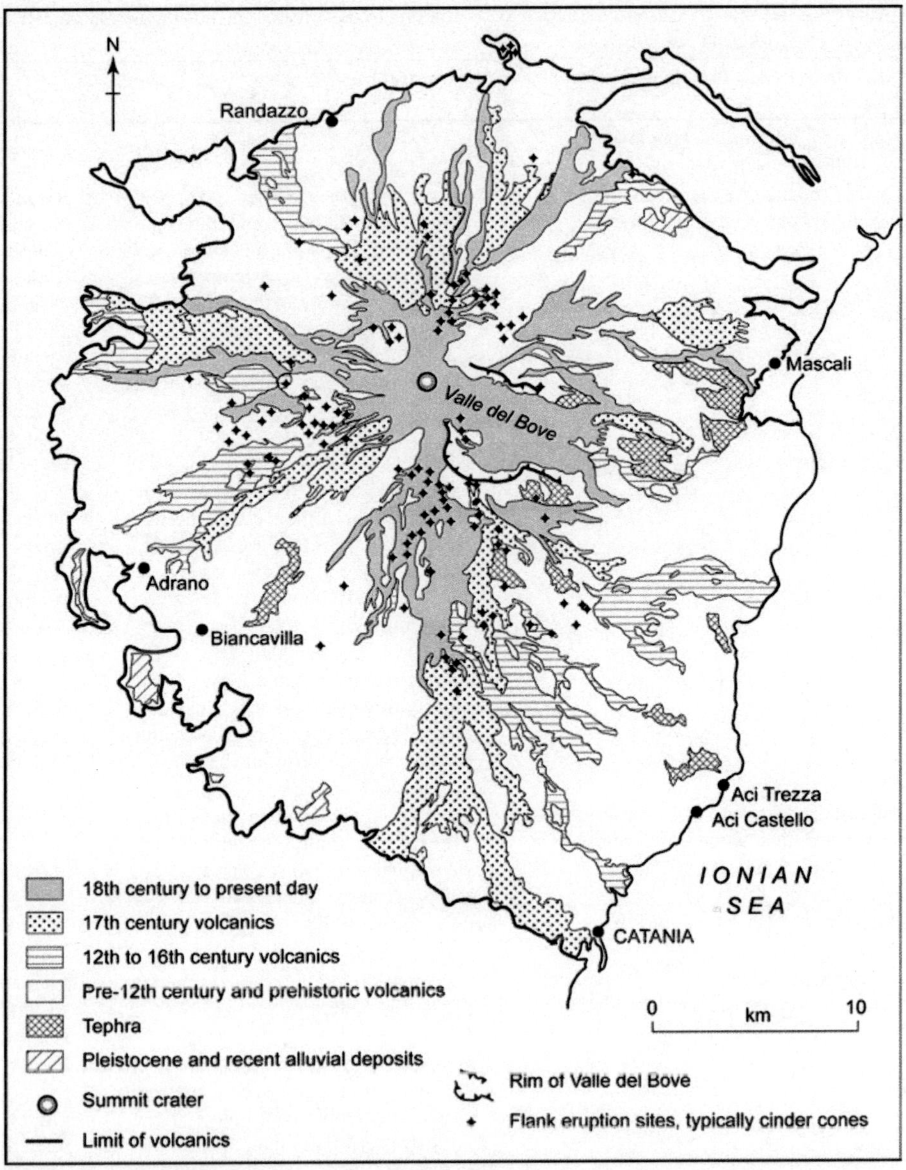

Fig. 15.3 Geological map of Mont Etna. (From Guest et al. 2003)

nomic potential of different sectors and height bands on the volcano, and the ever present danger of earthquakes and volcanic eruptions, particularly the threat of lava inundation. The proportion of the surface area of Etna covered by lava since 1600 AD is significant (Fig. 15.3; Table 15.2) and includes flow fields that have extended to the margin of the volcano. The hazard can be assessed and, as demonstrated by Guest and Murray (1979), many of the towns and villages on the flanks of the volcano are located in low hazard zones, through a process of natural adaption which is further discussed in Sect. 15.2.2. Although lava flows have caused considerable damage to agriculture, communications and occasionally settlements, on Etna they do not pose a significant risk to human life.

15.2.1 Adjustments to Environmental and Economic Factors

The agricultural landscapes of Etna have fascinated scholars for many centuries and the following account is based largely on the works of these writers, plus field observations and other referenced sources where appropriate (Rodwell 1878; Basile 1941; Milone 1960;

Table 15.2 Details of principal crops and agricultural activities during the pre-industrial era. (Based on information in Basile 1941; Chester et al. 1985; Dazzi 2007)

Altitudinal Zones, Principal Crops and Pastoralism	Location on Etna	Notes
Regione piedmontese (sea level to 1000 m)		
Citrus Fruits (mostly lemons, oranges, but also tangerines, and limes)	From sea level to ca. 550 m, the upper limit is defined by temperature. The upper limit is lower on the eastern flank because of lower temperatures. Citrus needs more irrigation in the south-western sector because of lower rainfall and is absent from the northern, north-western and western flanks because of low temperatures	a. An ideal crop for peasant cultivation because it produces a high cash return from a small area b. Intolerant of frost c. A labor intensive crop on the terraced lands of the eastern flank d. Lemons were a major export crop of Sicily. In the 1940s, 90% of Italian production was from Sicily, with 2/3 being from the Provinces of Catania and Messina
Vines	Located above the citrus belt and, with the exception of the western, north-western and northern flanks where conditions are too arid, vines are found on all sectors of the volcano. Altitudinal limits 300–900 m	a. Not drought resistant. b. Needs a chilling period before the growing season c. Very labor intensive, requiring head pruning and the use of the hoe on the terraced lands of the eastern flank
Olives	Similar altitudinal range to the vine, but more common on the drier western and south-western sectors. Specialized olive groves are found on south-western and western flanks	a. Requires low humidity for proper fruiting and is more drought resistant than the vine. Long, hot summers are essential b. Will not tolerate frost for very long
Other crops Apples, almond, apricots, avocados, cherries, figs, loquats, peach, pears, pistachio, pomegranate, quince and strawberry	Grown throughout the regione piedmontese, but areas of specialization, for example: apples—south-western flanks; cherries—east coast and pistachio—western slopes	
Hazelnut, carob, cotton, mulberry, sugar cane	Diffuse distribution	
Aubergine, broccoli, cauliflower, fennel, lettuce, pepper, spinach and tomato	Diffuse distribution, but concentrated on the south-western sector	Often cultivated in winter throughout the eastern; south-eastern; southern and south western sectors
Cereals	Diffuse distribution throughout the region and often grown beneath tree crops. Much more common on the north-western flank than in other parts of the volcano	a. Far less important than in the rest of Sicily. b. On the north-western flank, rye is rotated with wheat and beans
Animals	Mostly confined to historic lava flows, which have not weathered sufficiently to produce soils able to support cropping	Throughout the region many farmers integrate cultivation with small number of animals to produce manure for the land and milk and other products for family use
Regione boscara (1000–2000 m)		
Forest Main species—oak, (various species), beech, chestnut, pine and poplar	Sweet chestnut (*Castanea sativa*) is common on the southern and eastern flanks, where it is often managed (coppiced), with its nuts being harvested as a tree crop. The Laricio pine (*Pinus nigra ssp. Calabrica*) has a pioneer role on lava flows	The altitudinal limits and sectoral distribution of different trees were related to: environmental conditions (especially climate); and practices of felling and re-planting

Table 15.2 (continued)

Altitudinal Zones, Principal Crops and Pastoralism	Location on Etna	Notes
Grazing	Major pastoral zone, with sheep and goats grazing both on open pasture and within woodlands	a. Often large flocks (500–2000) where involved, whose ownership was split between members of the same family and/or extended family b. Frequently combined with other farming activities c. Often under the control of large land-owners (including the Church) and leased for the growing season
Regione deserta (above 2000 m approximate tree-line to the summit)	Not well named because Alpine vegetation occurs. Some low intensity grazing towards the lower altitundinal limit of the zone	

Rochefort 1961; Pecora 1968; Clapperton 1972; King 1973a; Duncan et al. 1981; Chester et al. 1985; Dazzi 2007).

A full discussion of climate and hydrology of Etna is beyond the scope of this chapter (Chester et al. 1985), but its essential features are: an increase in precipitation with height, to over 1200 mm in the summit zone; a steep environmental lapse rate of temperature, so that heavy snowfall is a feature of winter months; a temperature contrast between the warmer northern, north-western, western and other sectors of the volcano; the occurrence of a volcanic plume driven by the prevailing westerly winds, with a concomitant rainfall enhancement and slight temperature depression on the eastern and south-eastern sectors of the volcano, in comparison with the more arid northern and north-western flanks; and high irrigation potential. With regards to the latter, aquifers are recharged by winter rainfall and snow melt, and since Roman times have been exploited for irrigation (Table 15.1).

On a volcano with such strong altitudinal and sectoral variations in climate, it is hardly surprising that at first sight the link between climate/hydrology and land use appears to be a deterministic one, with much more intensive cultivation, usually under irrigation, occurring in the eastern, south-eastern, southern and south-western sectors, within the so-called 'regione piedmontese' (mountain foot (piedmont) region), which stretches from sea level to 1000 m (Fig. 15.4; Table 15.2). In reality the relationship was more complex because the northern and north-western sectors of the volcano were isolated from Catania, the major city and market center of the region, and from the principal export ports of the east coast (Fig. 15.1). Land-use intensity declined with increasing distance along all principal routes linking Catania to other parts of the Etna region. Variations in land-use intensity were striking and, based on the land below 2000 m and the percentages utilized for the cultivation of vines, orchards and other plantation crops, in the final years of the pre-industrial era intensity varied from 3% in the northwest, through 23%, 27% and 41%, respectively, on the western, northern and north-eastern flanks, to 52% in the south; 63% in the east, 67% in the south-west and 69% in the southeast (Chester et al. 1985).

The principal pre-industrial land uses of Etna are summarized in Fig. 15.4, 15.5, Table 15.2, but there are a number of additional points that are not fully captured by these illustrations. For instance, yields were maintained in areas of intensive cultivation by a combination of (a) inter-cropping of cereals and vegetables, usually sown beneath tree crops; (b) the creation and maintenance of elaborate lava-block terraces on steep slopes, that were well established in Sicily by the thirteenth century and which created flat land and also acted as an effective means of erosion control; (c) wood-mulch was widely used both to increase organic matter and reduce evaporation; (d) great care was taken over grafting, so as to maximize yields and reduce labor requirements; and (e) animals were closely integrated into the system of cropping. Animals not only produced meat and other products for both on-farm and local consumption, but also provided a vital source of manure.

Cultivars were carefully chosen for the particular conditions encountered on Etna and some, including the Nerello Mascalese and Carricante grapes, were used for many centuries, although additional varieties

Fig. 15.4 View of the Etna from Catania in the 1870s. The contrast between the intensively cultivated regione piedmontese and the barren upper slopes of Etna should be noted. (From Rodwell 1878)

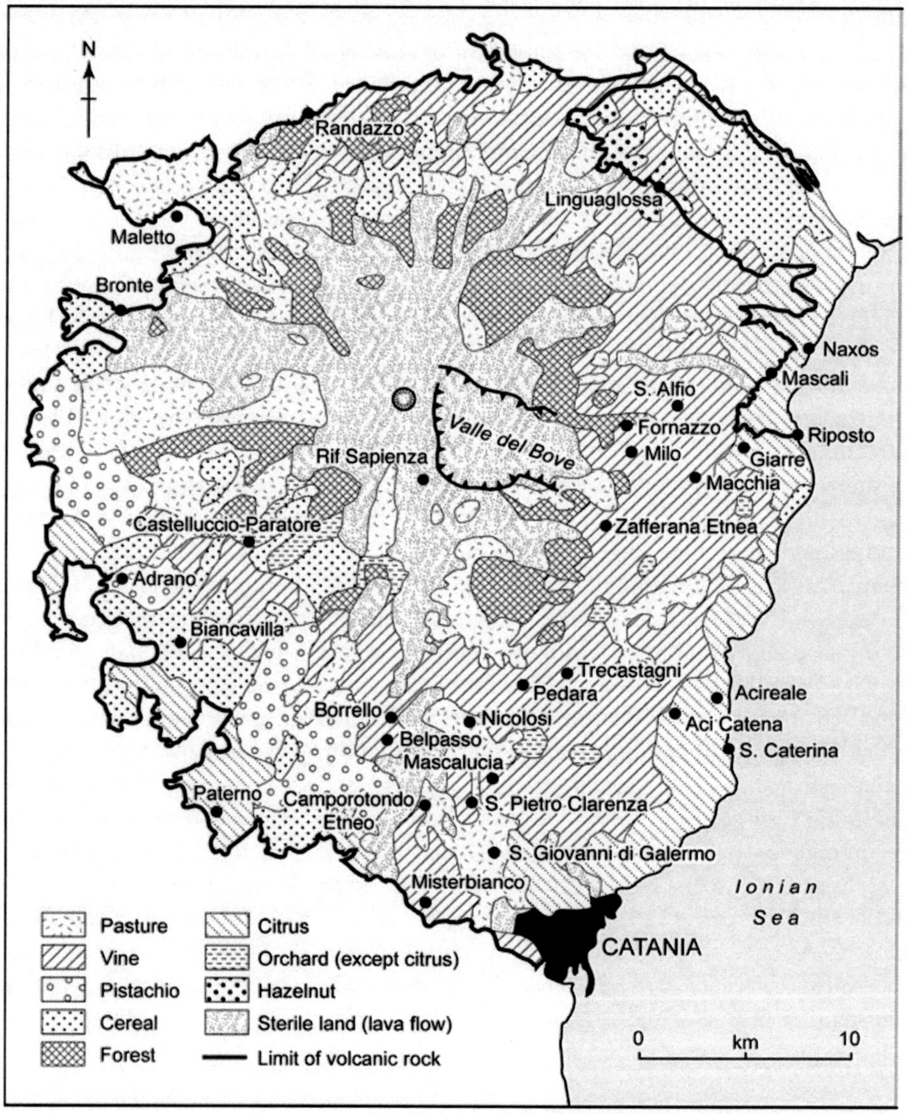

Fig. 15.5 The agricultural land-use of Etna. (After Rochefort 1961)

were introduced late in the pre-industrial era (Dazzi 2007).

Maintaining this system of agriculture demanded large drafts of family and extended family labor and by the close of pre-industrial times in the 1970s terraces abandoned from cultivation, but still in use for grazing, were a feature of the landscape. These bore witness to a reduction in the area being intensively worked especially towards the upper altitudinal limit of the regione piedmontese. This was due to a combination of: investment by the Italian State and the European Economic Communities in larger units, mechanization and more modern techniques of irrigation—so producing scale economies and spatial concentration; and to a reduction in the agricultural labor force, particularly sons and daughters of farmers who were increasingly unprepared to be employed tilling family-owned plots (King 1971; Chester et al. 1985).

15.2.2 Adjustments to Hazards

The terms pre-industrial and industrial are not only used to portray societies at different levels of economic development, but also to describe responses to disasters caused by extreme natural events (White 1973). The nature of these two characteristic forms of response are summarized in Table 15.3, and for most of historical time societies throughout the world have had to respond "pre-industrially" when having to cope with disasters. Many inhabitants of economically less developed countries continue to do so even in the early years of the twenty-first century. In pre-industrial societies losses are accepted on an involuntary basis by individuals, families and isolated regions, but as the process of economic development takes place the burden shifts to the State and in large catastrophes to the international community, involving loss-sharing by means of aid transfers, insurance and the State assuming a central role, both as an aid provider, and as an agency seeking to reduce the impact of subsequent events (White 1973). This is known as an industrial or modern technological response. In Sicily there were few industrial features of responses evident until a hundred years ago, when in 1909 the State played a leading role in providing aid following the Messina earthquake. In the twentieth and twenty-first centuries major eruptions of Etna have occurred in 1910, 1911, 1923, 1928, 1950–1951, 1971, 1979, 1981, 1983, 1991–1993, 2001 and 2002–2003 and, albeit with some exceptions, this sequence has coincided with a progressively greater State involvement in managing the impact of volcano-related losses. Although from the Classical era until ca. 1900 there are a few historical examples of State involvement during and following eruptions of Etna (Chester et al. 2005), the best recorded examples were after 122 BC when inhabitants of Catania were granted a ten year tax moratorium by the Roman authorities (Rodwell 1878), and in 1669 AD when the Spanish Viceroy deployed troops and sent monetary aid to the region (Mack-Smith 1968). The dominant coping strategy until 1900 was typically pre-industrial, it included the features listed in Table 15.3 and reflected a fine adjustment of people to the Etnean environment. Although it is reported that troops were deployed during eruptions in 1874, 1879, 1883, and 1892, this was principally for the mainte-

Table 15.3 Responses to hazards in contrasting pre-industrial and industrial societies. (Based on White (1973) and Chester et al. (2005))

Pre-Industrial Society	Industrial Society
A wide range of adjustment	A restricted range of adjustment
Action by individuals or small groups	Action requires co-ordination by the authorities
Emphasis on harmonization with, rather than technological control over, nature	Emphasis is placed on technological control over nature, rather than harmonization with nature
Low capital requirements	High capital requirements
Responses vary over short distances	Responses mostly uniform
Responses are flexible and were easily abandoned if unsuccessful	Responses are inflexible and are difficult to change
Losses are perceived as inevitable. The 'mindset' of many inhabitants is strongly influenced by notions of supernatural punishment, vengeance and the need to appease divine wrath	Losses may be managed by government action, technology, economic development and science
Responses continue over time scales ranging from hundreds to thousands of years	Not commonly observed until the mid-nineteenth century and not widespread until the mid-twentieth century

nance of law and order. In 1883 troops played a limited role and were deployed to prevent the population seeking shelter in churches because of the risk of a major loss of life caused by volcanic earthquakes, whilst in 1892 soldiers distributed bread and the very limited financial aid provided by the Italian State.

In addition to the altitudinally fluctuating frontier of cultivation produced by temporal changes in economic and demographic pressures (see Sect. 15.2.1), throughout history and prehistory the cultivated area has not only been reduced by lava invasions, but also increased by the re-colonization of lava flows and associated volcanoclastics (Figs. 15.3, 15.5). In the case of incursions the distress caused to the population has been documented by many writers (Rodwell 1878; Hyde 1916; Chester et al. 1985). Dealing with just those eruptions that have occurred since the fifteenth century, when the records became reasonably complete: the villages of Trecastagni and Pedara were destroyed in 1404; in 1537 Nicolosi was devastated; in 1646 several small villages on the north flank of the volcano were reported to have been obliterated; the 1669 eruption, the largest historical event, wiped out Belpasso, S. Pietro Clarenza, Mascalucia, Comporotondo, S. Giovanni de Gelermo, Misterbianco, fourteen smaller settlements, and most of Catania; and a number of small villages were razed in the vicinity of Macchia in 1689 (Fig. 15.1). In pre-industrial times, the outskirts of towns and villages often comprised a corona, a roughly circular rim of particularly productive agriculture, and losses within such areas were recorded following eruptions in: 1371 (possibly ~1160) and 1444, in Catania; 1536, in Randazzo; 1566, in Linguaglossa; 1595 (possibly 1062); 1607 (possibly 1610), in Adrano; 1651–53, in Bronte; 1792, in Zafferana; 1810, in Milo; 1832 and 1843, in Bronte; 1852/3, in Zafferana; 1879, in Passopisciaro (located between Randazzo and Linguaglossa); and 1883, in Nicolosi. It has been calculated that between 1500 and 1900, some 8% of the total land area of the regione piedmontese was sterilized by lava flows (Chester et al. 2005).

It is often assumed that people panic when disaster strikes, but research across a range of disasters throughout the world shows that this rarely occurs. It was certainly not the case on Etna during the pre-industrial era because, although fearfulness and apprehension are noted in some accounts, the vast majority of reports bear witness to the fact that people remained calm, normal day-to-day activities continued unabated, people still farmed their land and continued with their trades. A strong sense of solidarity is evident as family and community coped with and recovered from lava incursions in a variety of ways. On Etna, eruptions of lava affecting settlements and agricultural land often occurred several times a century, inhabitants were familiar with the risks they faced and were able to adapt to them. People whose homes were under threat frequently made use of family and extended family relationships, leaving their villages to live with relatives. For instance, in the 1883 eruption it is reported that some of the threatened population in the villages of Belpasso, Biancavilla, Borello, and Nicolosi were forced to live in makeshift accommodation, in tents or in the fields, but in fact most people merely left their homes on a temporary basis to live with their families in other villages. Almost everybody returned once the emergency ended because only the village of Nicolosi was badly affected. Spontaneous evacuations also occurred during eruptions in 1843 and 1886. Even the tent and field dwellers were not invariably the marginalized poor as many nineteenth century press accounts imply, because many farmers owned permanent shelters on family plots that were normally used for a daily siesta in the hot summer months or for storage of tools and fodder, but could be easily converted into temporary family accommodation. In 1892 relief committees are also known to have been formed in several villages.

A remarkable feature of loss-bearing on Etna was that, although cities, towns and villages could be badly affected and even destroyed, they normally recovered rapidly. Nicolosi was, for instance, destroyed by lava in 1537, but was a prosperous settlement when an earthquake again badly affected the settlement in 1633 and was probably fully re-built by the time of the 1669 eruption (Fig. 15.1; Chester et al. 1985). This was not an isolated example and more than thirty nine towns and villages suffered losses between 1400 and 1900, some on more than one occasion, yet an area was never abandoned a settlement being either rebuilt on the same site or as close as possible to it (Table 15.4; Chester et al. 2005). Close to the vent of a high-effusion rate eruption, lava may travel at velocities of several kilometers per hour, but typically by the time flows reach inhabited areas they advance more slowly, normally at no more than tens of meters per hour. Residents knew in advance if and when their village and home was about to be destroyed and there is

Table 15.4 Urban responses to major flank eruptions 1400–1900. Ticks indicate that relocation or rebuilding on the same site featured in the response. The impact of eruptions on Catania are discussed in the text and locations are shown in Fig. 15.1 (After Chester et al. 2005). The dates of historic lavas are from Tanguy et al. (2007)

Eruption	Towns/Villages effected	Relocation	Rebuilding on same site	Subsequent expansion of the settlement over the lava flow
1408	Trecastagni and Pedara were destroyed	✓		
1536	Lava approached close to Randazzo			A small settlement was built on the flow
1537	Nicolosi was destroyed		✓	
1566	Lava approached close to Linguaglossa.			Village expanded
1595 (or 1062?)	Lava approached close to Adrano			Little later urban expansion on the flow
1607 (or 1610)	Lava approached close to Adrano			Little later urban expansion on the flow
1646	Affected several villages on the northern flank			The village of Passopisciaro was built on the flow
1651–53	Lava approached close to Bronte			Village expanded
1669	Nicolosi was destroyed by a volcanic earthquake		✓	
	Belpasso was destroyed by lava	Rebuilt on a new site and called Mezzo-campo, but the air is reported to have been unhealthy and the present site, 1 km from the flow was selected in 1695		
	S. Pietro Clarenza was destroyed by lava	✓		
	Mascalucia was destroyed by lava	✓		
	Camporotondo was destroyed by lava	✓		
	S. Giovanni di Gelermo was destroyed by lava	✓		
	Misterbianco was destroyed by lava	✓		
	Fourteen small villages on the southern and south-eastern flanks were destroyed	✓		
1689	Macchia area. Several small housing clusters were destroyed on the outskirts of the village			Village expanded and the lava flow is now densely settled
1792	The outskirts of Zafferana were destroyed			Little urban expansion on the flow
1811	Lava approached close to Milo			Little urban expansion on the flow
1832	Lava approached close to Bronte			Little urban expansion on the flow
1843	Lava approached close to Bronte			Little urban expansion on the flow
1852/3	Lava approached close to Zafferana			Village expanded
1879	Lava approached close to Passopisciaro and adjacent smaller villages			Village expanded
1886	Lava approached close to Nicolosi			Little urban expansion on the flow
1892	Lava approached close to Nicolosi			Little urban expansion on the flow

good historical evidence that people salvaged all they could (Chester et al. 2005). Salvage included furniture and personal possessions, but newsreel films of the slightly later 1928 eruption show that the removal of tiles, windows and doors was a well established practice (Duncan et al. 1996; Chester et al. 1999).

Despite being devastated so many times by earthquakes and eruptions the city of Catania was never fully destroyed. It always remained functional both as a city and port, was still the focus of inland transport and, even in its ruined state, held far more locational advantages than any possible rival (Chester et al. 1985). Lava flows in contrast to earthquakes, sterilized the land over which they moved for hundreds of years and it could take several months before a flow had cooled sufficiently to allow site clearance to begin. Lavas from the 1669 eruption took over eight months to cool (Rodwell 1878). In the decades following each lava inundation, the urban area of Catania expanded over sterilized land, a policy of hazard adjustment that had the twin advantage that building sites were cheap and reconstruction did not have to compete with high-value agriculture. A similar expansion, but on a smaller scale, frequently occurred as villages spead over sterile land following episodes of lava inundation (Table 15.4).

All peasant agriculture involves maximizing family security over profit and one feature throughout Sicily is that cultivation plots are often owned in different localities. On Etna, parcels of land may be as small as 2 ha, but a family normally either rents or owns plots in several localities so that a single eruption is unlikely to wipe out all a farmer's property (Clapperton 1972). Pastoralism also provided a means of increasing security and, although high altitude pastoralism has declined in recent decades, in pre-industrial times this was far more important and provided a valuable additional source of income especially in times of distress (Chester et al. 2005). High level pastures could be leased on a seasonal basis from large landowners, including the church, by members of an extended family.

Field work and the analysis of aerial photographs have shown that even on the relatively well-watered southern, eastern and south-eastern flanks of the volcano centuries may pass before cropping can recommence (Chester et al. 1985; James et al. 2000). When studied in the early 1990s, the 1928 lava flow, which destroyed the town of Mascali, only supported low-intensity rough grazing and in the few places where vegetation succession had not been influenced by humans, the lava was only covered with a ground flora comprising lichens, mosses, and some small trees where soil and moisture have been concentrated in depressions (Duncan et al. 1996). Left on their own lavas will be re-colonized by natural processes and soils will develop. It is not surprising that, given the complexity in the interplay between geology, relief and climate, and in terms of the mix of its soils and their spatial extent, Etna is striking as the most distinctive region of Sicily (Fierotti et al. 1988). Five soil associations are mapped for the volcano the pattern shows a broad relationship to elevation (Table 15.5). In terms of agricultural quality, apart from lava rock outcrop, the poorest soils of Mt Etna are Lithosols (Leptosols according to the WRB Classification; Deckers et al. 2002), young or eroded, shallow soils of minimal development in weathering lava. Eutric refers to the relative richness in basic elements, but the quality of Lithosols is limited by shallowness and the generally rugged terrain. Regosols are weakly developed soils in loose substrates, particularly tephra, but may be cultivated. Of greater pedogenetic development and agricultural value are the eutric Cambisols and Luvisols, well developed brown soils of good depth. On generally well drained and commonly glassy volcanic ashes and pyroclastics, Fierotti et al. (1988) describe these as having the distinctive andic properties of volcanic soils, namely low bulk density, high water retention and porosity, and relatively high organic matter and nutrient content. These soils are of greatest extent on the eastern, south-eastern, southern, and south-western lower flanks of the volcano.

The broad distribution of soils on Etna is controlled primarily by climatic variation with sector and elevation, yet at many localities soil quality is related to the nature, depth, and age of soil-forming materials and varies spatially across distances as small as a few meters. Thus a local pattern of soils is likely to reflect age and morphology of lava, and the age and depth of tephra deposited across lava surfaces. Soil depth is critical in a region where it is highly variable and often limited. In agricultural terraces, soil depth is built up artificially. The authors find that terrace-soils, which were formerly cultivated but are now grazed, comprise tephra little altered by pedogenesis: the black tephra is coarse, loose and rich in little-altered volcanic glass (a soil property described as vitric), the only visible pedogenetic horizon being an organic-rich and shal-

Table 15.5 Soil associations mapped and described for Mt Etna by Fierotti et al. (1988). Soils in associations 5, 15 and 10 may have andic properties that confer low bulk density, high porosity, high water retention and good drainage. They are relatively rich in organic matter and nutrients

Soil Association[a]	Distribution on Mt Etna (see Fig. 15.1)	Land morphology	Soil character	Agricultural value
5. Lithosols; Rock outcrop (lava flows); Eutric Cambisols. Andosols as inclusions.	Restricted to lower slopes on the northern and western flanks	Sloping and moderately steep	Shallow, but Andosols may be of medium depth	Low. Some soils are capable of being cultivated
15. Eutric Regosols; eutric Cambisols; Orthic Luvisols	Extensive; particularly on eastern, south-eastern, southern and south-western flanks	Gently sloping to very steep	Shallow to moderately deep. Cambisols and Luvisols are well developed	Modest. Vineyards, orchards, citrus groves, woodland and grassland. These are the principal agricultural soils of Mt Etna.
10. Eutric Regosols; Lithosols; Eutric Cambisols. Andosols as inclusions	Between sea level and 1750 m	Fairly steep.	Shallow to moderately deep. Shallow Regosols are developed in tephra	From mediocre to good. Woodland and grassland
1. Rock outcrop (lava flows); Lithosols		On younger lava flows, mostly between ca. 500 m and the summit	Very shallow	Almost none
24. Eutric Cambisols; Eutric Fluvisols	Restricted to the base of the Eastern flanks	Level to sloping; on alluvium and the volcaniclastic fan to the south of Riposto (Fig. 15.1)	Moderately deep to deep	Good. Vineyards, orchards, citrus groves and other crops

[a] Numbers refer to the associations defined by Fierotti et al. (1988). The soil classification follows that of FAO-UNESCO (1974)

low topsoil. The importance of soil depth in relation to other factors in soil fertility may apply to relatively young soils, but the soils of the oldest flows on the lower slopes of Mt Etna are markedly more developed, being finer-textured, brown Cambisols and sometimes Luvisols. These soils, associated with mature landscapes, are also terraced and carry a rich variety of more demanding crops. The patchwork of Etna's geomorphology and soils, which occurs between very local and regional/sector scales, accounts in part for the great ecological and agricultural diversity of the volcano.

Tephra is the major constituent of many of Mt Etna's soils, and its periodic deposition rejuvenates them. During the 2002–2003 eruption, 10 cm of tephra accumulated 5 km from the source. Citrus fruits and vegetables were badly affected and grapes had to be washed to remove ash; washing machinery could not be used because of damage by the abrasive ash, and oranges could not be processed economically for juice (Dazzi 2007).

Although in the absence of human interference lava will be re-colonized by natural processes, there is evidence that farmers both understood weathering and re-colonization and assisted it to a limited degree. By comparing geological maps and aerial photographs it is possible to see that farmers had an accurate perception of re-colonization potential. Etna generates relatively little pyroclastic ash, but where it occurs mature soils are produced far more quickly than on lava flows. Around the village of Nicolosi (Fig. 15.1), located on the south-eastern flank at a height of 700 m, vines have been established for a considerable time on Andosols developed from the pyroclastic ashes of the 1669 eruption, whereas adjacent lavas of the same age only support the low intensity grazing of livestock. King (1973a) also makes the point that farmers often encouraged natural weathering and processes of soil formation, with the prickly pear cactus (*Opuntia ficus-indica*) being introduced from South America so that the plant's powerful roots could assist in the breakup of lava flows.

Although the perception that disasters are caused by supernatural forces in order to punish sinful humanity is not confined to pre-industrial societies and, indeed, transcends culture, time and religious tradition (Chester and Duncan 2010), generally in southern Italy and particularly on Etna (Chester et al. 2008) this attitude played a particularly prominent role in the ways in which people reacted to catastrophes caused by volcanic and tectonic processes. The tradition that Etna has a divine status is long-established and in fact may be traced back to before the period of Hellenic influence, with Maniscalco (2005) noting that the Sikel god, Hybla, was associated with a mud volcano at Paterno on the south-western flank of the volcano (Fig. 15.1). Later in the Classical Era the view emerged that the only course of action that people could take when threatened by the effects of an eruption was divine appeasement, with Lucilius Junior (first century AD) recording that the people offered incense to the gods who were thought to control the mountain and its eruptions (Hyde 1916). In the Christian tradition, theodicy is defined as any attempt to reconcile the notion of a loving God with the existence of human suffering, and southern Italian popular Catholicism modified the pre-Christian tradition to maintain that losses had to be accepted because they were a legitimate expression of a vengeful God's punishment of a sinful people (Chester et al. 2008). All the inhabitants of Etna could do in response was to appeal to God in prayer and supplication, to try and propitiate wrath by liturgical actions and, in planning recovery from the disaster, resolve to live better lives. In the eighteenth century and strongly influenced by enlightenment thinking, alternative models of theodicy were given prominence especially by the philosopher Gottfreid Wilhelm Leibniz, but rural Sicilian Catholicism remained wedded to notions of punishment, with the only positive feature of a disaster—what theologians term the greater good—being found in the virtues of self-sacrifice, public service and social cohesion (Chester et al. 2005).

Given the manner in which the greater good was defined in Sicily, it is hardly surprising that from the earliest times heroic acts became the stuff of legend. Seneca, for instance, records the story of Anapias and Amphinomus—the so-called 'Fratelli pii' or pious brothers—who rescued their parents from a major conflagration in Catania, lava opening in front of them like the Biblical Red Sea and so allowing escape, whereas much later in 1669 Diego Pappalardo a local leader and some fellow citizens of Catania attempted to divert a flow which threatened their city, a course of action which caused a riot since the diverted lava now put Paterno at risk. Following this and for the rest of the pre-industrial era, attempted diversion of lava was made illegal (Chester et al. 2005).

A widely-held theodicy of punishment meant that eruptions became associated with liturgies of divine appeasement. Although intercession through the agency of the Madonna and the saints was an established part of mainstream catholic teaching, in the form of popular Catholicism practiced on Etna this was expressed in a distinctive and extreme manner, with saintly relics, statues and other votive objects supposedly having the ability to prevent disasters (Chester et al. 2008). As early as 252 AD when lava approached Catania, the inhabitants processed the veil of the recently martyred St. Agatha at the flow front which it was claimed was immediately halted. Following this early success the veil and objects associated with saints were used on numerous occasions: in 1669 it was asserted that Agatha's veil prevented all Catania being destroyed; and as late as 1886 the veil was claimed to have been responsible for lava passing close to, but not destroying, the village of Nicolosi (Chester et al. 2005). The rural landscape of the Etna region was dotted with roadside shrines to saints and martyrs and an important part of the traditional way of life of each village was a festival (*festa*) dedicated to the local patron saint.

In pre-industrial times on Etna there was a marked psychological tension caused by an inconsistency between beliefs and actions. Sometimes called cognitive dissonance or parallel practice (Dibben 1999) the people of Etna, while on the one hand accepting that disasters were manifestations of divine wrath, on the other had no difficulties in reducing their risk exposure through, for example, exploiting extended family networks, having cultivation plots in different areas, and forming self-help committees. What is more from the eighteenth century some of the early geologists who studied Etna from a scientific perspective were priests, most notably the German Jesuit Athanasius Kircher and the Sicilian Canon Giuseppe Recupero who assisted the pioneer volcanologist Sir William Hamilton (Chester et al. 1985). This influence, however, did not displace or modify the indigenous popular Catholicism of the Etna region, with its emphasis on the supposed power of relics and votive objects and the efficacy of village patron saints and the Madonna to intercede for divine protection (Chester et al. 2008).

15.3 Conclusion: Landscape and Hazard Response Today

Although most characteristics of pre-industrial loss bearing on Etna have not been in evidence for a hundred years, some elements have survived (Table 15.3). Most notable are liturgies of divine appeasement which are still carried out each time there is a major eruption. Since 1970 major flank eruptions have affected or threatened settlement on the volcano in 1971, 1974, 1981, 1983, 1991–1993, 2001 (Guest et al. 2003) and in 2002–2003, and during most of these events religious rites have been performed, showing a continuity with the situation before 1900. These liturgical actions still enjoy overwhelming popular support and in July 2001, Luigi Bommarito Archbishop of Catania, celebrated mass in the village of Belpasso (Fig. 15.1; Chester et al. 2008), hopefully to halt the lava that was threatening the village. It is estimated that between 7000 and 10,000 people attended this open-air mass, representing a wide cross-section of the inhabitants of the village including many well-educated professional people. A teacher is quoted as remarking local people still believe in miracles. "If human technology can't keep the lava back, the eternal father is our only salvation" (Kennedy 2001, p. 10). Furthermore, in detailed interviews carried out in the village of Trecastagni in the 1990s (Fig. 15.1), Dibben found that "for many (members of the public) religious beliefs play a significant role in their representation of the volcano" (Dibben 1999, p. 196). The parallel practice noted in pre-industrial times, of people accepting that disasters are manifestations of godly wrath which are only capable of being moderated through mechanisms of divine propitiation, yet at the same time fully embracing attempts to reduce risk and aid recovery through State supported actions, is still part of the psychological makeup of a large number of the inhabitants of Etna. Salvaging all that may be easily removed from a building before it is engulfed by lava has also remained an important means of coping. This well established tradition is probably not unrelated to the low take up of domestic property insurance which is both costly and of only limited availability (Chester et al. 1985).

In spite of these exceptions, present day responses to periodic flank eruptions are industrial in character (Table 15.3), with emergencies being closely managed: by government, through the Ministry of Civil Protection (Dipartimento della Protezione Civile), and other scientific agencies, including the National Institute of Geophysics and Vulcanology (Istituto Nazionale di Geofisica e Vulcanologica INGV) in Catania; and by the use of technology and planning expertise, particularly hazard mapping, land-use zoning and geophysical monitoring. In this respect Etna differs little from volcanoes located in other parts of the economically more developed world (Chester et al. 2005) in terms of its hazard preparedness.

In introducing this chapter it was noted that many features of pre-industrial agriculture and the distinctive lifestyle of Sicilian agro-towns were still in evidence in the late 1960s and early 1970s. Over the past forty years poverty in Sicily has been reduced through aid and subsidies from the Italian State and the European Union, with the interior lands in particular benefiting from the programmes of land-reform which were introduced from the 1950s (King 1973a, 1973b). On Etna the regione piedmontese retains its agricultural distinctiveness and, though farmers now have access to machinery and subsidies, much of the traditional system of intensive irrigated agriculture described in Sect. 15.2.1 may still be recognized (Dazzi 2007). Since the 1970s, the policy towards concentrating investment in more environmentally favored and larger more easily worked and more accessible holdings within the regione piedmontese and on the plain of Catania has continued, and further abandonment from cultivation of more marginal land has taken place. The numbers employed in agriculture have fallen, those in service occupations have increased and many comuni, particularly those within easy driving distance of Catania, now serve as commuter settlements and contain many inhabitants who are far less aware of risks they face from the volcano than were their forebears four decades ago (Dibben 2008). The region is now a major tourist destination and since the early 1970s has attracted many second home owners. Especially in the settlements near Catania and adjacent to the coast, much of the traditional Sicilian way of life and the character of the agro-town has disappeared, though elements are still to be found in the more remote villages of the northern and north-western and western sectors of the volcano.

References

Ayala G, French C (2003) Holocene landscape dynamics in a Sicilian upland river valley. In: Howard Macklin Passmore (eds) Alluvial Archaeology in Europe. Swets and Zeitlinger, Rotterdam, pp 229–235

Ayala G, French C (2005) Erosion modeling of past land-use practices in the Fiume di Sotto di Troina River Valley, North Central Sicily. Geoarchaeology 20:149–167

Basile DG (1941) Agricultural Sicily. Economic Geography 17:109–120

Benjamin S (2006) Sicily: Three thousand years of human history. Steerforth Press, Hanover New Hampshire, 498 pp

Chester DK, Duncan AM (2010) Coping with disasters within the Christian tradition, with reference to volcanic eruptions and earthquakes. Religion 40 (2):85–95

Chester DK, Duncan AM, Dibben C, et al. (1999) Mascali, Mount Etna region Sicily: An example of Fascist planning during the 1928 eruption and its continuing legacy. Natural Hazards 19:29–46

Chester DK, Duncan AM, Dibben CJL (2008) The importance of religion in shaping volcanic perception in Italy, with special reference to Vesuvius and Etna. Journal of Volcanology and Geothermal Research 172:216–228

Chester DK., Duncan AM, Guest JE (2005) Responses to eruptions of Etna from the Classical Period to 1900. In: Balmuth MS, Chester DK, Johnston PA (eds) Cultural responses to volcanic landscape. Colloquia and Conference Papers 9, Archaeological Institute of America, Massachusetts, pp 93–107

Chester DK, Duncan AM, Guest JE, Kilburn CRJ (1985) Mount Etna: The anatomy of a volcano. Chapman and Hall, London, 404 pp

Chester DK, Duncan AM, Guest JE, et al. (2000) Human response to Etna during the classical period. In: McGuire WJ, Griffiths DR, Hancock PL, Stewart IS (eds) The archaeology of geological catastrophes. Geological Society Special Publication 171, London, pp 179–188

Clapperton CM (1972) Patterns of physical and human activity on Mount Etna. Scottish Geographical Magazine 88:160–167

CORINE (1994) Soil erosion risk and important land resources – in the southern regions of the European Community. European Environment Agency, Copenhagen. http://www.eea.europa.eu/publications/COR0-soil (accessed 24 March 2009)

Dazzi C (2007) Environmental features and land use of Etna. In: Arnalds O, Bartoli F, Buurman P, Oskarsson H, et al. (eds) Soils of Volcanic Regions in Europe. Springer, Berlin, pp 629–644

Demangeon A (1927) La géographie de l'habitat rurale. Annales de Géographie 36:1–23

Deckers J, Driessen P, Nachtergaele F, Spaargaren O (2002) World reference base for soil resources – in a nutshell. In: Micheli, E., et al. (eds) Soil Classification 2001. (EUR 20398 EN). Luxembourg: Office for Official Publications of the European Communities. European Soil Bureau, pp 173–181

Dibben CJL (1999) Looking beyond eruption for an explanation of volcanic disasters: vulnerability in volcanic environments. Unpublished Ph.D. thesis, University of Luton, Luton.

Dibben CJL (2008) Leaving the city for the suburbs – The dominance of 'ordinary' decision making over volcanic risk perception in the production of volcanic risk on Mt. Etna, Sicily. Journal of Volcanology and Geothermal Research 172:288–299

Duncan AM, Chester DK, Guest JE (1981) Mount Etna volcano: Environmental impact and problems of eruption prediction. Geographical Journal 147:164–179

Duncan AM, Chester DK, Guest JE (1996) The 1928 eruption of Mount Etna Volcano Sicily, and the destruction of the town of Mascali. Disasters 20:1–20

Duncan AM, Chester DK, Guest JE (2005) Eruptive activity of Etna before A.D. 1600 with particular reference to the classical period. In: Balmuth MS, Chester DK, Johnston PA (eds) Cultural responses to volcanic landscape. Colloquia and Conference Papers 9, Archaeological Institute of America, Massachusetts, pp 57–69

FAO-UNESCO (1974) Soil map of the world, Vol. 1 Legend. United Nations Food and Agriculture Organization (FAO) and United Nations Educational, Scientific and Cultural Organization (UNESCO), Rome.

Fierotti, GF, Dazzi C, Raimondi S (1988) Carta dei suoli della Sicilia, 1: 250,000 (with report). Regione Siciliana Assessorato Territorio ed Ambiente and Univesita degli Studi di Palermo, Facoltà di Agraria, Istituto di Agronomia Generale, Cattedra di Pedologia, Palermo

Finley MI (1968) A history of Sicily: Ancient Sicily to the Arab conquest. Chatto and Windus, London, 222 pp

Grove AT, Rackham O (2001) The nature of Mediterranean Europe. An Ecological History. Yale University Press, New Haven, 524 pp

Guest JE, Murray JB (1979) An analysis of hazard from Mount Etna Volcano. Journal of the Geological Society of London 136:347–354

Guest JE, Cole PD, Duncan AM, Chester DK (2003) Volcanoes of southern Italy. Geological Society, London, 284 pp

Hyde, WW (1916) The volcanic history of Etna. Geographical Review 1:401–418

James P, Chester D, Duncan A (2000) Volcanic soils: their nature and significance for archaeology. In: McGuire WJ, Griffiths DR, Hancock PL, Stewart IS (eds) The archaeology of geological catastrophes. Geological Society Special Publication 171, London, pp 317–338

Kennedy F (2001) Sicilians pray as technology fails to stop lava. The Independent (London), 10 July, p 30

King R (1971) Mediterranean island in torment. Geographical Magazine 44:178–185

King R (1973a) Sicily. David and Charles, Newton Abbot, 208 pp

King R (1973b) Land Reform: The Italian experience. Butterworth, London 257 pp

King R, Strachan A (1978) Sicilian agro-towns. Erdkunde 32:110–123

Leighton R (1996) Early societies in Sicily. Accordia Research Centre University of London, London, 211 pp

Leighton R (1999) Sicily before history: an archaeological survey from the Palaeolithic to the Iron Age. Duckworth, London, 312 pp

Mack-Smith D (1968) A history of Sicily: Vol. 1 Medieval Sicily 800–1713; Vol. 2 Modern Sicily after 1713. Chatto and Windus, London, 559 pp

Maniscalco L (2005) The prehistoric settlement near Salinelle di San Marco (Paterno, Catania Province, Sicily). In: Balmuth MS, Chester DK, Johnston PA (eds), Cultural responses to volcanic landscape. Colloquia and Conference Papers 9, Archaeological Institute of America, Massachusetts, pp 127–142

Malone C, Stoddart S (2000) The historical geography of ancient Sicily: implications for genetics. Journal of Cultural Heritage 1: sup. 29–sup. 31

Milone F (1960) Sicilia: La natura e l'uomo. Paulo Boringhieri, Torino, 462 pp

Pecora A (1968) Sicilia: Le regione D'Italia, Vol. 17. Unione Tipografico Editice, Torinese, 644 pp

Rochefort R (1961) Le travail en Sicile. Étude de Géographie Sociale. Presses Universitaires de France, 363 pp

Rodwell GF (1878) Etna: A history of the mountain and its eruptions. C. Kegan Paul, London, 146 pp

Sadori L, Narcisi B (2001) The postglacial record of environmental history from Lago di Pergusa, Sicily. The Holocene 11:655–670

Sadori L, Zanchetta G, Giardini M (2008) Late glacial to Holocene palaeoenvironmental evolution at Lago di Pergusa (Sicily, Southern Italy) as inferred by pollen, microcharcoal, and stable isotopes. Quaternary International 181:4–14

Smolenaars JJL (2005) Earthquakes and volcanic rruptions in Latin literature: Reflections and emotional responses. In: Balmuth MS, Chester DK, Johnson PA (eds) Cultural responses to volcanic landscape. Colloquia and Conference Papers 9, Archaeological Institute of America, Massachusetts, pp 311–329

Tanguy JC, Condomines M, Le Goff M, et al. (2007) Mount Etna eruptions of the last 2759 years: revised chronology and location through archeomagnetic and ^{226}Ra-^{230}Th dating. Bulletin of Volcanology 70:55–83

White GF (1973) Natural hazards research. In: Chorley RJ (eds) Directions in Geography. Methuen, London, pp 193–212

Part IV
The Mediterranean and European World–Cool-Temperate European Lands

Chapter 16
Romanian Carpathian Landscapes and Cultures

Adrian Cioacă and Mihaela S. Dinu

16.1 Introduction

16.1.1 The Physical Environment

The objective of this chapter is to examine human–landscape interactions in the mountainous region of the Romanian Carpathians. The most diversified range of landscapes in Romania are found there, comprising steep alpine ridges, rounded forested summits, planation surfaces, deep valleys, defiles and gorges, intramontane and submontane depressions. The Carpathians stretch across the central-eastern part of Europe (Mihăilescu 1963) as the northeastern part of the Alpine-Carpathian chain of mountains, stretching from the Vienna Basin (Bratislava, Slovakia) to the Timok's Valley (Niš, Serbia). At 1600 km length and an area of 170,000 km², the Carpathians are the largest chain of mountains in Europe. Specifically, Carpathians in Romania are 910 km long and cover 66,303 km². This is essentially 28% of the national territory (Niculescu 1987).

The Carpathians enclose the vast, hilly Transylvanian Depression as a 'corona montium' (Iordanes, sec. IV, cited by Cocean 2004–2005) (Fig. 16.1). The Subcarpathian Hills mark the border of the mountainous region to the east and south, and the Western Hills on the west. Numerous mountain passes and defiles allow ready intra-Carpathian and trans-Carpathian travel. The most important modern road along the Prahova Valley, partially correspond to the much older 'road of the table land' ('drum de plai', in Romanian), which is marked by a number of monasteries (Fig. 16.2; Ciobanu 1979).

Rugged peaks up to 2544 m high (Moldoveanu Peak) occur and are surrounded by rounded highlands and forested planation surfaces. Valleys, narrow gorges and defiles cut through the highlands, and intramontane hilly depressions or lowlands are more or less extensive and offer a variety of reliefs and ecosystems. They offered and still offer activities and resources of great diversity such as hunting, fishing, gathering, forest products, alpine pasture, good arable terrains, and a variety of minerals.

16.1.2 The Historical Contest

The pre-historic and historic periods recognized in Romania are shown in Table 16.1. Evidence of Paleolithic and Neolithic habitations are found throughout the country. In the Maramureş Depression, archaeological researches in the period 1982–1990 indicate a human presence from the Upper Paleolithic (35,000–10,000 BC). Evidence comes from the territory of the settlements Sighetu-Marmaţiei (Valea Cufundoasă, Valea Blidarului, Grădina Morii, Vadu Izei, and Câmpulung on Tisa). The earliest archaeological evidence is from Năneşti (Iza Valley), where a deposit of 13 stone artefacts was found. Sedentary life is attested since the Neolithic (6000–3000 BC) by the archaeological discoveries at localities such as Coştiui, Onceşti, Corneşti, Giuleşti, Sighetul Marmaţiei, Săpânţa, Crăciuneşti. Ciobanu and Cioacă (1975) record painted Neolithic pottery indicating sedentary settlement from about 5000 years ago.

A. Cioacă (✉)
Faculty of Geography and Geography of Tourism
Universitatea "Spiru Haret"
Timisoara Bd. 58, Sector 6
061333 Bucharest, Romania
e-mail: acioaca@gmail.com

See Plate 15 in the Color Plate Section; also available at: extras.springer.com

I. P. Martini, W. Chesworth (eds.), *Landscapes and Societies*,
DOI 10.1007/978-90-481-9413-1_16, © Springer Science+Business Media B.V. 2010

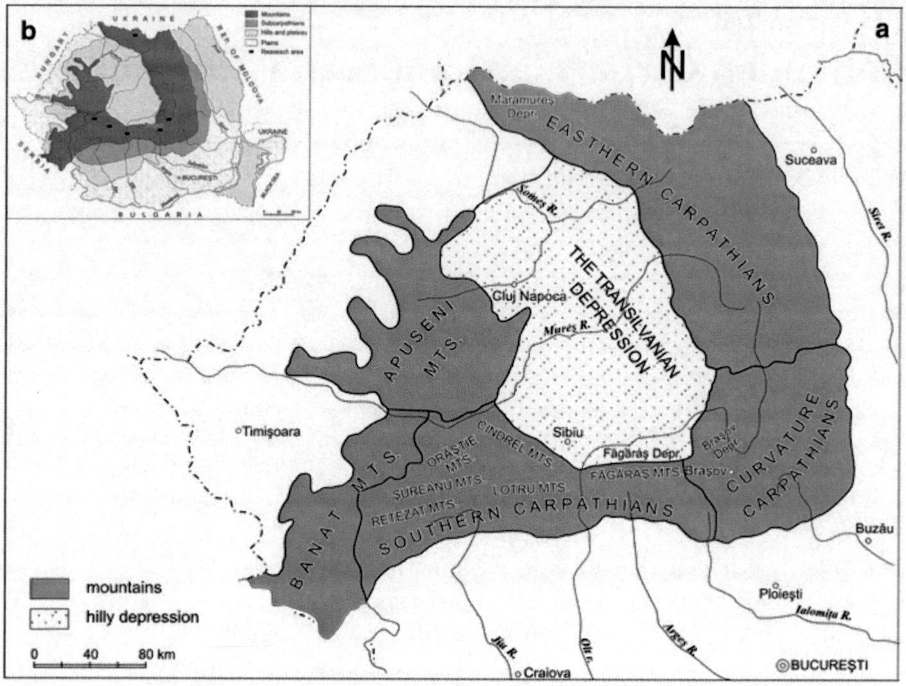

Fig. 16.1 Simplified morphological maps of Romanian Carpathians. **a** Major component units of the Carpathian Mountains and the Transilvania Depression. **b** Simplified morphological units of Romania, and surrounding states

A second important cultural stage existed during the period that began with the Geto-Dacian state and ended with the Roman retreat from Dacia. Geto-Dacians are documented in the Carpathians and the Balkans from the first century BC, and in the Carpathian arc they formed the nucleus of the Dacian state that reached its maximum extension under King Burebista (82–44 BC). After more than a century, another king of the Geto-Dacians (Decebal 87–106 AD) heroically opposed the Roman legions, but ultimately was overwhelmed by them.

The Roman conquest of Dacia also led to the interbreeding of Romans with indigenous tribes. In effect, the result was the ethnogenesis that defines the Romanian people. Indeed, the name Carpathians derives from the Carpi tribes who lived on the inner lower slopes around Transylvania.

During the long process of disintegration of the Roman Empire in both western and the eastern versions, the province of Dacia also disintegrated. Later, during the Middle Ages, the territory came under submission to the Kingdom of Hungary, and immigration of various ethnic groups occurred. This included the Saxons who cohabited with Romanians in Transylvania. Separate Romanian entities developed later and constituted the nuclei around which the three historically important provinces of Wallachia, Moldavia, and Transylvania have developed.

16.2 From Populated Lower Lands to the Conquest of the Mountains

Most evidence of early settlements occur in mountain depressions and cross-valleys, where the Carpathian cultures came into being. Later, settlements expanded to higher grounds where environmental factors were favorable or strategic advantages were to be had.

Evidence of Paleolithic and Neolithic habitations are found in all physiographic environments: depressions, alluvial plains, piedmonts, and mountains. Traces of Neolithic cultures occur both in the internal depressions of the Carpathians and peri-Carpathian areas (Dinu and Cioacă 1998).

The following are a few examples of areas where people expanded from lowlands (depressions) into the surrounding mountains.

16.2.1 Braşov Depression

This is the largest intramontane depression of the Curvature Carpathians (Figs. 16.1, 16.3). Its branched shape is dictated by a network of conjugated faults that separates horst-like massifs. The area includes the floodplain of the Olt River and its tributaries (Posea

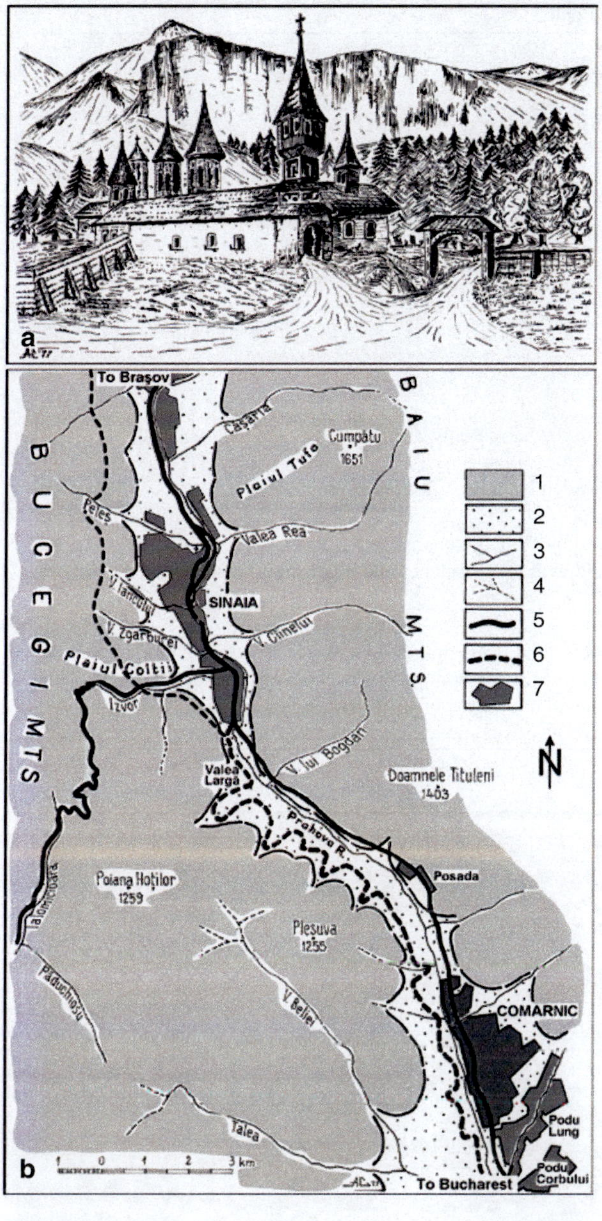

Fig. 16.2 Sinaia Monastery founded in 1695 (**a**), located on the transcarpathian road partially corresponding to the 'road of the table land' (**b**)

1981). Within the valley, a series of stepped landscapes grade from the mountains toward the foot of the floodplain. The higher lands are occupied by piedmonts and glacises from 850 m to approximately 550 m asl at the contact with the alluvial plain. The latter contact is marked by lines of springs followed closely by old as well as recent settlements (Cioacă and Dinu 2002). The best developed piedmonts are on the southern and southwestern margins, including the one in the Braşov area (Fig. 16.3). Several floodplains have been for the most part avoided by human settlements because of flooding and swampy terrains.

When the early Neolithic settlements of the lowlands were abandoned, people resettled on higher grounds. The terraces along the Olt River north of Feldioara were settled in this way, and they continue to support agriculture and large communities (Cioacă and Dinu 1997). Most communities were located either at the foot or top of the piedmonts near the contact with the mountain. The most representative sites at the end of the Neolithic occur at Noua, near Braşov, at the foot of the Postăvarul Mountains, at Ariuşd, and even on top of the peripheral Baraolt Mountains (Binder 1998).

16.2.2 Făgăraş Depression

This is located 30 km northwest of the Braşov depression at the northern margin of the Central Carpathian Mountains. It consists primarily of an alluvial plain, large, and asymmetric with terraces, and piedmont glacis developed on the left of the Olt River up to the Făgăraş Mountains (Fig. 16.1). The passage proceeds from marginal mountain summits, corresponding to the Carpathian border, to the lower relief of the alluvial plain through two levels of discontinuity, commonly at altitudes of 750–850 m and 650–700 m. Wide depositional piedmonts were formed by sediment-charged rivers sourced from the glacial circs of the Făgăraş Mountains. Many archaeological remains indicate continuity of settlements in this area. Local Neolithic pottery shows that the first settlements were situated at the edge of the Olt River floodplain and followed the pattern of relocating to higher elevations. The Dacian period is well represented by the settlements of Cumidava, Cuciulata, located on glacis, and at Breaza (see below: 16.3.1) the remains of a Dacian fortress has been located on the mountain slope which just above the edge of the piedmont.

Beginning in the fifteenth century, many of the Făgăraş Depression villages were built on the glacis at the foot of the Făgăraş and Perşani mountains (Ucea, Veneţia) and in the Braşov Depression (on the eastern flank of Perşani Mountains: Rotbav and Codlea). The earliest evidence comes from documents describing the

Table 16.1 Periods of the Romanian history (1,500,000 BC – nineteenth Century)

Pre-history (1,500,000 BC – fifth Century BC)

- Stone Age (1,500,000–3000 BC):
  - *Paleolithic* (1,500,000–12/10,000 BC)
  - *Mesolithic* (12,000–6000 BC)
  - *Neolithic* (6000–3000 BC)
- Bronze Age (approx. 3000–1150 BC):
  - *Early Bronze Age* (3000–2300/2200 BC)
  - *Middle Bronze Age* (2300/2200–1500 BC)
  - *Late Bronze Age* (1500–1150 BC),
- Iron Age (1150 BC–100 AD):
  - *Hallstat* (1150–450 BC)

Antiquity (fifth century BC)

- *La Téne* (450 BC–100 AD)
- *Roman occupation (second to third centuries AD)*
- *Post Roman/migrating peoples period (third to fourth centuries AD):* Sarmatians, Goths, Huns, Gepids, Avars, Slavic)
- *Middle Ages (fifth to ninth centuries AD)*
- End of Romanian people ethno genesis. The first political entities led by Gelu, Glad and Menumorut.

Medieval Ages

- Ninth to thirteenth centuries. Small principalities in the area lived by the Romanians.
- The fourteenth to sixteenth centuries. The constitution of the medieval independent Romanian states: Wallachia, Moldavia and the Autonomous Principality of Transylvania.

Modern Age (Romanian subdivision)

- The sixteenth to eighteenth centuries. The Ottoman domination over the Romanian principalities. The first Union accomplished by Michael the Brave (1600). The establishing of the Austrian domination over Transylvania and Banat.
- The eighteenth to nineteenth centuries. The transition from medieval to modern history. The beginning of the movement for national emancipation: Şcoala ardeleană (Transylvanian School).

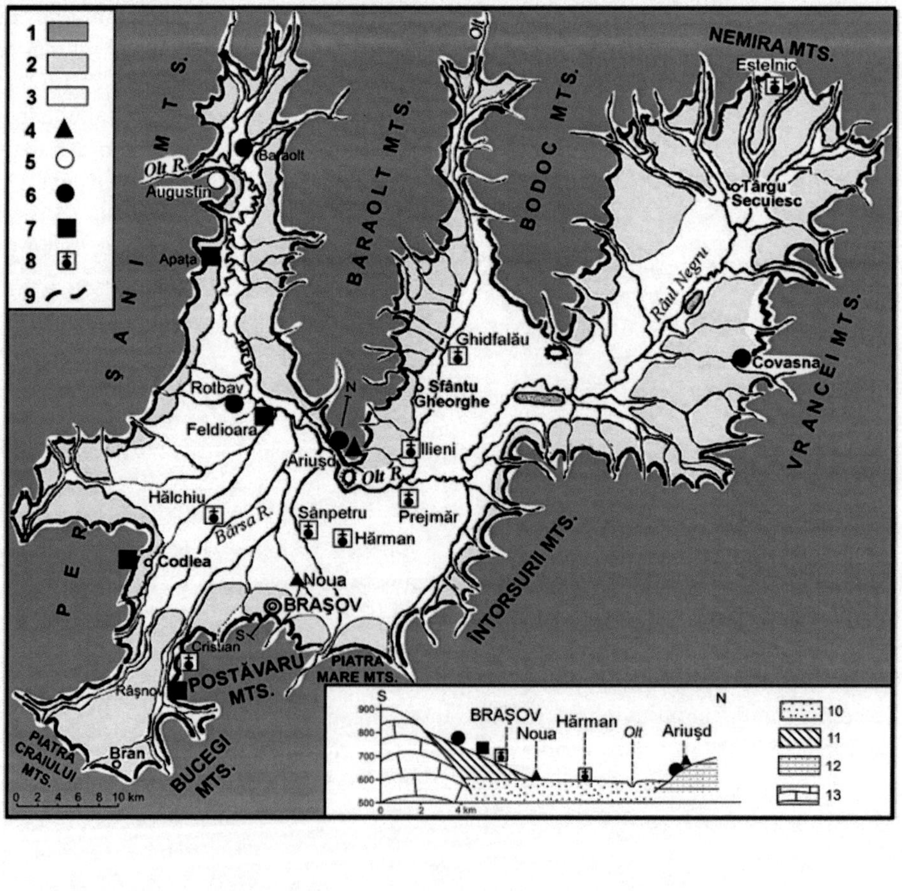

Fig. 16.3 Map and geological profile of Braşov Depression: *1* mountains; *2* piedmonts; *3* alluvial and terraced plain; *4* Neolithic vestiges; *5* Dacian vestiges; *6* Roman vestiges; *7* peasant fortresses; *8* fortified medieval churches; *9* limits of depression; *10* bedded sands and silt; *11* piedmont deposits; *12* marlstone; *13* limestone

acts of the Wallachia rulers, in the fourteenth–fifteenth centuries. Much later (eighteenth and nineteenth centuries), the settlers migrated from the lower villages to higher locations: Ucea de Sus, Veneția de Sus (where 'de sus' means 'upper').

16.2.3 Maramureş Depression

This is the most extensive intramontane depression of the Eastern Carpathians (Fig. 16.4), but being parted by the Tisza River between Romania and Ukraine, the Romanian part is smaller than the Braşov Depression (Fig. 16.3). The Maramureş Depression has a mostly hilly relief, the alluvial plains being reduced to narrow strips along the main rivers. It is surrounded by hills and high mountains, with a narrow main gate along the Tisza River valley to the north (Vasilescu 2008).

This depression was part of the ancient hilly Transylvania Plateau that contains the oldest and longest Mesozoic volcanic chain in Europe. A series of lakes formed in the area from Mesozoic times. Gradually the lakes silted up and were drained by entrenching rivers such as the Tisza River. The result was the formation of fertile intramontane depressions favorable for the establishment of first Paleolithic communities (Fig. 16.4). Communities continued to develop in relative isolation in this environment throughout Roman times and the Middle Ages, progressively increasing contact and trade with the inhabitants of the surrounding lands.

Again, people eventually moved from lower lands onto the surrounding piedmonts and mountains where communication with the Transylvanian Plateau to the south, or with Moldavia to the east was facilitated by the many valleys and gorges. An important communication route established at this periods was the 'salt road' that crosses the Oaş Depression. The latter became the Oaş Country in the Middle Ages (Cioacă and Mihai 1980).

16.2.4 Haţeg Depression

This depression lies between the Southern Carpathians and the Apuseni and Poiana Ruscăi mountains (Figs. 16.5, 16.6; Grumăzescu 1975). It is an area occupied by humans since Neolithic times It is bounded to the south by the steep slopes of the high massifs of the Retezat and Ţarcu mountains (almost 2000 m asl) (Fig. 16.6). In this, it resembles the Făgăraş Depression, but unlike the latter, the other sides of the Hate depression are closed by mountains of intermediate altitudes (900–1400 m), the Poiana Ruscăi and

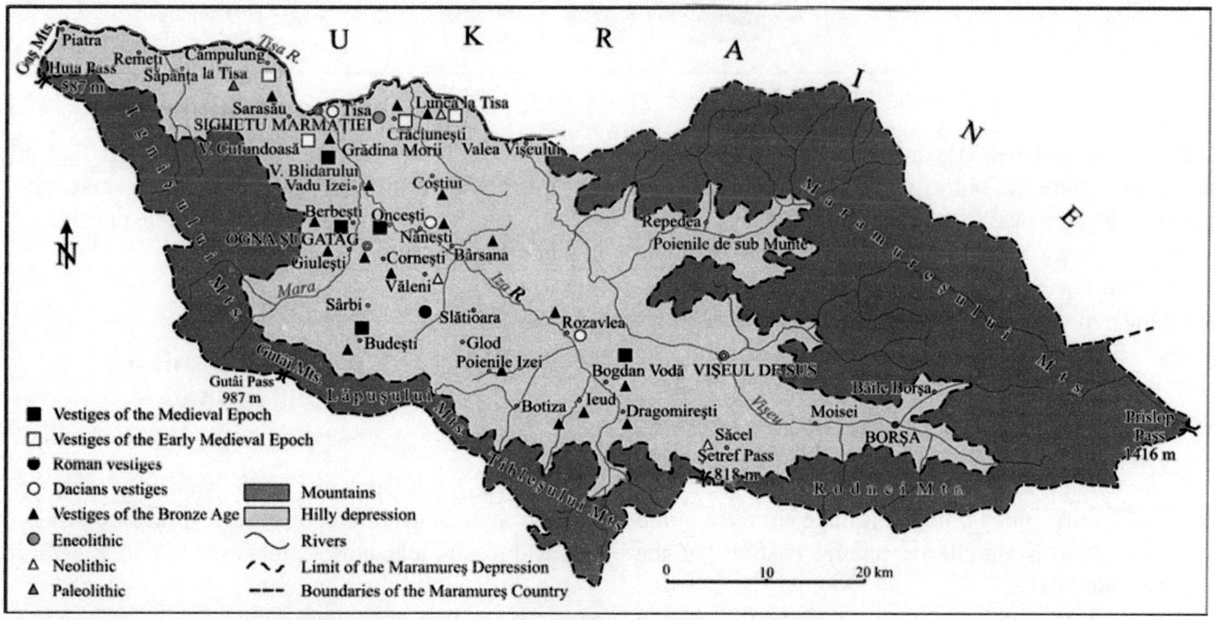

Fig. 16.4 Archaeological vestiges in Maramureş Depression. (After Vasilescu 2008)

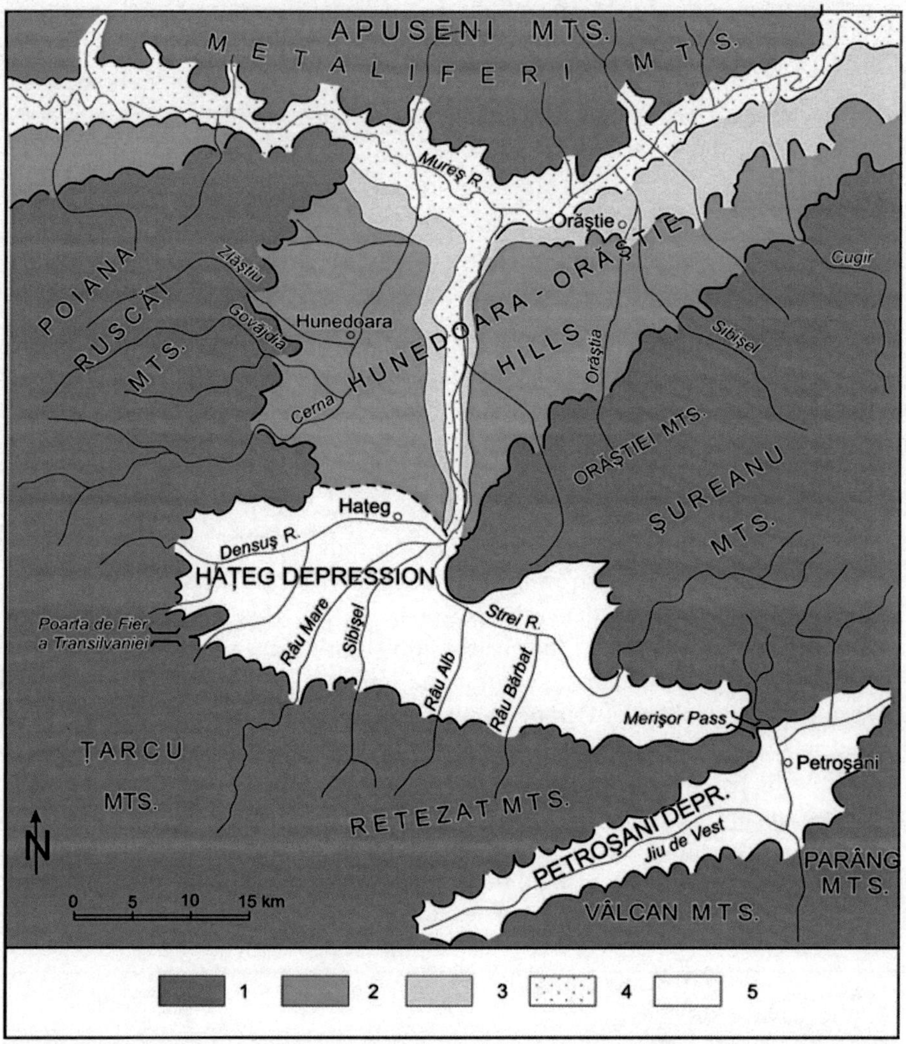

Fig. 16.5 The geographic location of the Haţeg Depression: *1* mountains; *2* hills; *3* terraces; *4* floodplains; *5* flatten-depression. (After Grumăzescu 1975)

Metaliferi mountains for example. The only links with the outside were the Mureş-Orăştie, the Iron Gates of the Transylvania pass to the west, and the Merişor Pass to the southeast (Fig. 16.5; Grumăzescu 1975). The depression resembles a fortress, and like other intramontane depressions it has provided shelter since prehistoric times. Expansion from the lower lands to the surrounding uplands began early. Archaeological evidence of Neolithic settlements were discovered along the Cugir and Sibişel rivers, followed by Bronze, Iron Age, and Geto-Dacian centers (Fig. 16.6). Much later traces of fortifications from the Roman era were found at nearly 2000 m altitude on natural pastures of the Şureanu Mountains.

Here again, archaeological investigations and modified landscapes reveal the common trend of communities to find new spaces for their activities first in the lower lands of the depressions and transverse valleys of the Carpathians, followed by later expansion to higher altitudes (Cioacă and Dinu 2003).

16.3 Short History of Romanian Carpathian Civilization

The various landscapes of the Carpathian Mountains and the abundant natural resources (pastures, wood, ores and fossil fuels, pure waters, rich in fish, together with a diverse hunting fauna) that can be found there, have the creation of various cultures since early Neolithic times. Paraphrasing the title of the book "Dacian

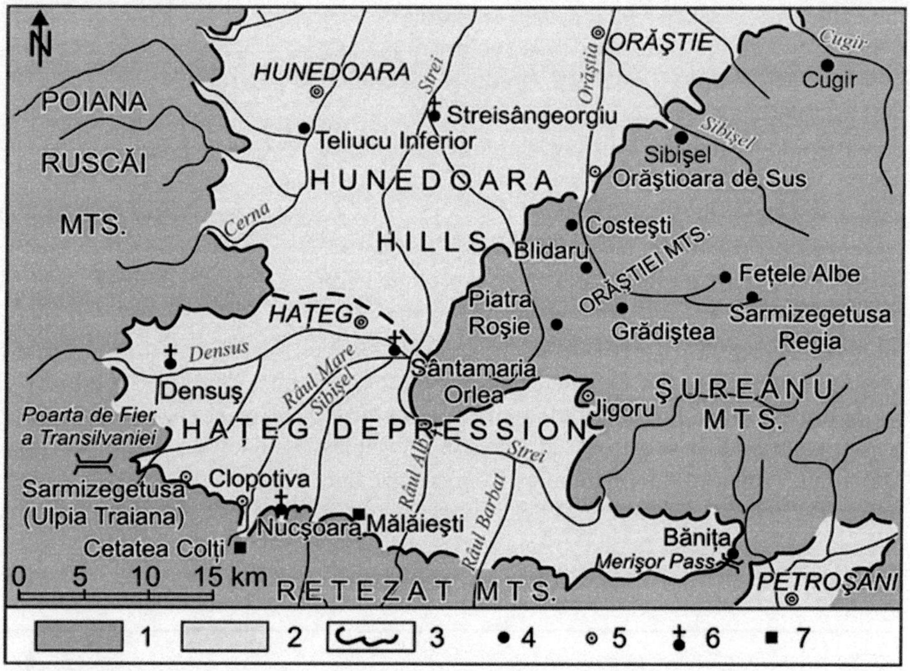

Fig. 16.6 Haţeg Depression: *1* mountains; *2* piedmonts; **3 limit of the Haţeg Depression**; *4* Geto-Dacian vestiges; *5* Roman vestiges; *6* early medieval orthodox churches; *7* early medieval fortifications

stones speak" by Mac Kendrick (1978), the Carpathian landscape has witnessed the emergence, evolution, and decline of civilizations at different stages of history. The chronicle of cultural succession shown by case studies, including those of some standing monuments such as the Dacian Sarmizegetusa Regia **and the Roman Ulpia Traiana in the Haţeg Depression** (Fig. 16.7), reveals a distinctly Carpathian narrative along the following lines.

The first inhabitants of the Carpathians clearly benefited from the protection offered by the mountains. In addition they also enjoyed an amenable, hot-humid climate during the climate optimum of the Neolithic. They progressively developed and improved various activities including hunting and fishing, artisanship in metals and wood, and manufacture of ceramics, all of which have left its mark in archaeological record. In addition, the communities established here in late Paleolithic and Neolithic have modified the mountain slopes to create conditions suitable for agriculture. In the low regions of floodplain and terrace only the front of the terraces and the sand banks were cultivated to begin with. The increasing frequency of floods during the Neolithic climatic optimum (about 4500 year BP) propelled agriculture onto the terraces and even onto the piedmonts. Practically, climatic change required a revolution in the agriculture of the Neolithic (Botzan 1996). The Holocene Climate Optimum period was favorable for the cultivation of cereals and the raising of livestock. However drought during the Bronze Age forced communities to expand into the highlands where rainfall was more abundant. Use of

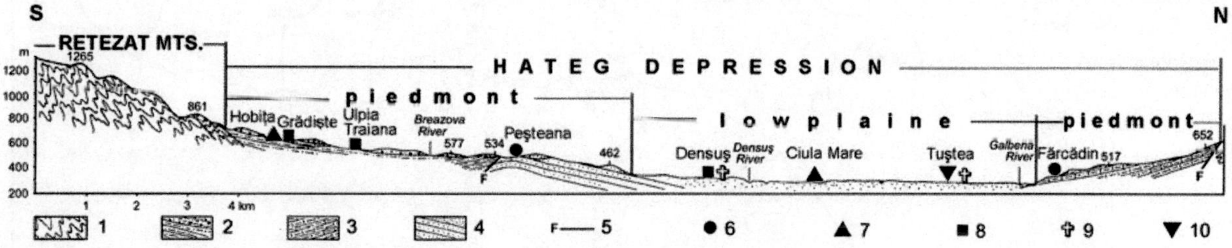

Fig. 16.7 Geological cross-section of the Haţeg Depression with location of archaeological sites: *1* crystalline schist; *2* piedmont accumulations; *3* monocline marls; *4* stratified sands and alluviums; *5* fault line; *6* Neolithic ceramics; *7* Dacian fortifications; *8* Roman castra; *9* early medieval orthodox churches; *10* medieval fortifications

the hill-slopes required the creation of terraces. The wooden plough drawn by cattle also came into use at this time, so that agriculture in the mountains brought together the culture of cereals with the breeding of domesticated animals. This agricultural revolution was accompanied by novel ways of utilizing and processing new resources.

Since the establishment of the early, sedentary Neolithic cultures in the Carpathians, there has been the evolution of various cultures including the Neolithic Boian culture (revealed in the archaeological sites of Ariuşd) and the Precucuteni (at Covasna) in the Braşov Depression, and the Bronze Age cultures at Botiza and Ieud (Fig. 16.4) at the foot of Ţibleş Mountains (south of the Maramureş Depression), and at Noua (Fig. 16.3), at the foot of Postăvarul Mountain in Braşov Depression.

16.3.1 Geto-Dacian Culture

The Geto-Dacian tribes originally differentiated from the Thracian tribes north of the Danube. The culture became distinctive in late Bronze Age and Hallstat periods (twelfth–fifth centuries BC) when systems of terraces on the mountain slopes were constructed. The terraces were protected against erosion and slumping by artificial water catchments that are still visible in many parts of the Carpathians. This period has been named in many scientific treaties "the Thracian-Geto-Dacian world", although it is known in history as the Geto-Dacian civilization (Mac Kendrick 1978). The state of the Geto-Dacians reached its maximum extension in Burebista time (82–44 BC), but conflicts with the Romans splintered their forces and their territorial extent diminished.

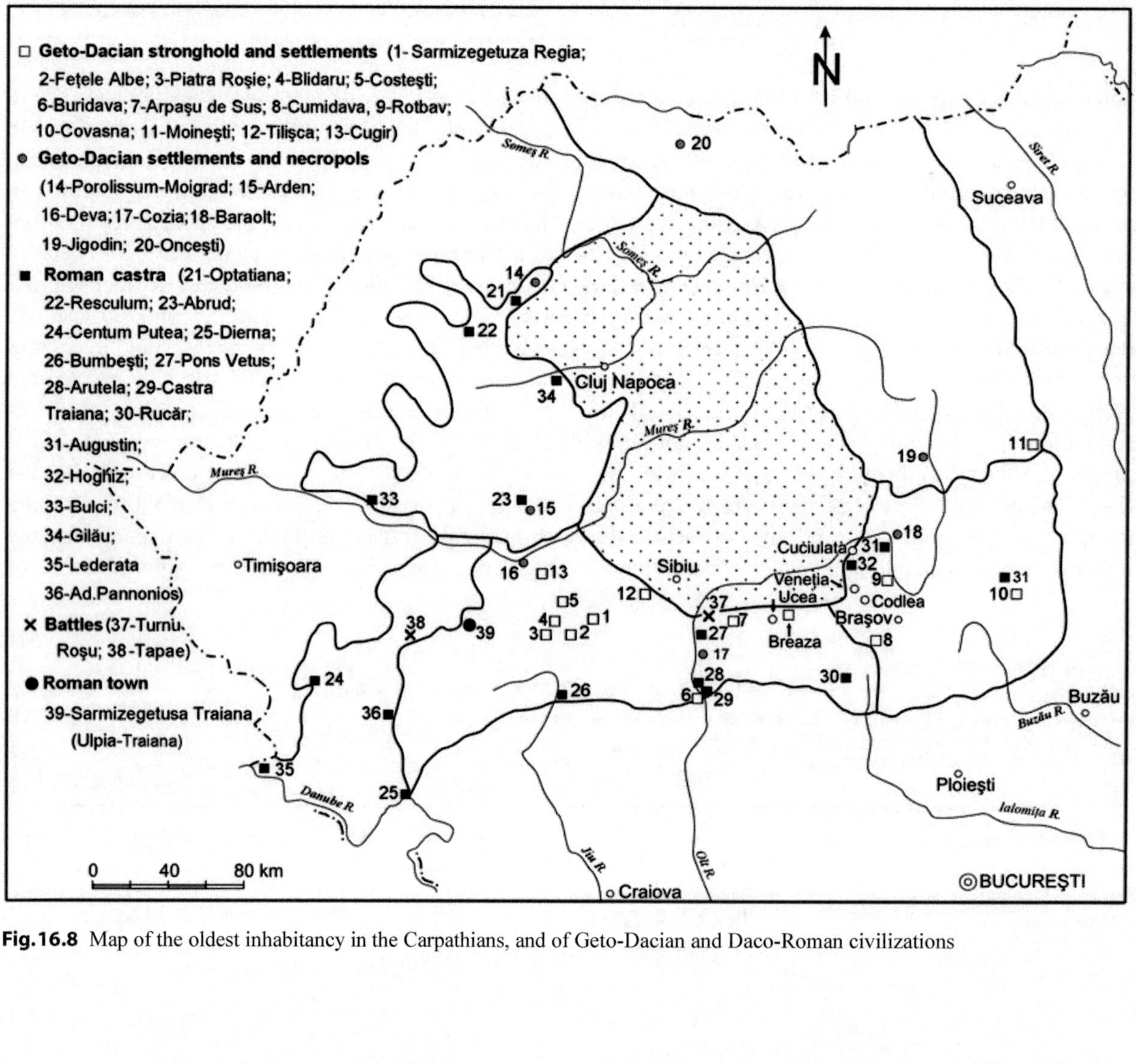

Fig. 16.8 Map of the oldest inhabitancy in the Carpathians, and of Geto-Dacian and Daco-Roman civilizations

This period was marked by a general cooling of the climate, which had severe consequences for agriculture in the mountains, for example in the reduced productivity of crops, and degradation of the increasingly abandoned lands. Like many other mountain regions in the world, these changes in climate caused mass migrations that had repercussions on the stability and continuity of mountain settlements.

Traces of the Geto-Dacian civilization are scattered throughout the piedmont and mountain areas of the Carpathians and the Transylvania Plateau, but there is a higher density in the Haţeg Depression and in the Orăştiei and the Apuseni mountains. The Apuseni Mountains became the core of the Geto-Dacian state.

Fortifications and roads exploiting the natural relief date from these times. Moreover, King Burebista (82–44 BC) founded his capital, Sarmizegetusa, in the mountains some 20 km south of Orăştie. It remained the capital under King Decebal (87–107 AD). Notable among the tools used by contemporary farmers was the Celtic plough of the La Téne period (450 BC–100 AD) to which the Geto-Dacians of the mountains brought some improvements (the bent tip of the iron ploughshare; Haywood 2001).

Other sites of note occur in the Curvature Carpathians (Hărman) and in the center of the Eastern Carpathians. Of special interest are local centers of bronze metallurgy in the north Eastern Carpathians at Coştiui, Vadu Izei, Năneşti, Bârsana, Rozavlea, Bogdan Vodă, Dragomireşti, Ieud, Brebeşti, Ocna Şugatag, Giuleşti. This area is emphasized because turned gray pottery and Dacian cups found at Onceşti, Sighetu Marmaţiei and Rozavlea are evidence that the potter's wheel was in use during the climax of this culture (Fig. 16.4).

Written information is also available. Classical Greek texts from this time describe not only the region, but also the origin and dialects spoken by the tribes north of the Danube. Both Hecateus of Millet (530–470 BC) and Herodotus (484–425 BC) give the location of the Geto-Dacian civilization in a geographical area protected by mountains and hills near forests, while Demosthenes (384–322 BC) later extolled the Getic farmers who provided cereals for Greece. Greek colonies were located on the sites of Getic settlements on the coast of the Black Sea (Pontus Euxinus), and the Greeks referred to the Gets as the inhabitants of the lower basin of the Danube, and Dacians as the Thracian-Getic tribes of the north-west. (Dragu and Tetea 1984).

16.3.2 Daco-Roman Civilization

The next step in the evolution of civilization in the Romanian Carpathians is the Daco-Roman, which began with the conquest of Dacia by the Roman legions (101–102 AD and 105–106 AD). Stylized images of these battles are carved on the metopes of the triumphal monument Tropaeum Traiani (109 AD) at Adamclisi (Dobrogea) and Trajan's Column (109 AD) in Rome. Some provide evidence of the Dacian environment with images of forested mountains as background, and of mining activities related to gold extraction. Of course the monuments commemorate moments of Roman glory, while written documents describe the many expeditions of the Roman legions needed to punish the Geto-Dacian incursions into the Roman Empire (Pârvan 1957). In fact, resistance to the Roman conquest was strong.

Romanization of the Dacian population occurred where the Geto-Dacian civilization already existed in the Carpathians, thus forming the Roman province of Dacia. It also occurred where the rest of the free Dacians lived outside the Carpathians. The hard work of local inhabitants, of colonists brought in by the Romans, as well as of Roman veterans who settled there, led to a high level of material and spiritual culture in Dacia. In essence this was what Romanization was all about. However, evidence of Dacian spirituality remains in numerous places of worship (stone sanctuaries). Moreover, the Roman historian Annaneus Florus, impressed by the large number of Dacian sanctuaries, fortifications and settlements that existed up to 1200 m altitude in Orăştiei Mountains (Piatra Roşie, Blidaru, Costeşti, Sarmizegetusa), highlighted the link between the Dacians and the mountain with the sentence "Daci montibus inhaerent", which could be translated "the Dacians cleave to their mountains".

The artistic processing of metals, stone, and painted ceramics gave names to the autochthonous cultures, including those in the mountains. The Romanian culture developed through a melding of Roman culture with Dacian traditions. Vulgar Latin became the common language for the people living in the former Roman province of Dacia.

Daco-Roman cohabitation was maintained after the withdrawal of the Roman government and armies. Even under difficult conditions during the period between 275 and 1241 nearly 400 rural settlements of

Roman Dacia survived for generations (Giurăscu and Giurăscu 1974).

16.3.3 Civilization of the Middle Ages

Succinctly stated, the societies that succeeded in Romania were involved in a continuous struggle to occupy the Carpathian–Danubian space during post-Roman and medieval times. Early civilizations of the Middle Ages continued the Daco-Roman heritage, but also produced a major new change: the Romanianization of the native population. This process was synchronous with the penetration of Christianity to the north of the Danube and in older history texts it is referred as the oriental Romanity. Dating of this time, the fourth century AD, is based on coins with the effigy of Constantine the Great that are preserved in the Christian graves of the largest necropolis of Romania at Alba Iulia, at the fringe of the Apuseni Mountains. Another achievement was the Romanian language that crystallized at that time, which preserved Latin phrases used by the locals for their agricultural and grazing activities, especially in the mountain areas, for example: to plough ('a ara' Romanian= 'arare' Latin), field ('câmp' Ro.= 'campu' Lat.), heath ('ogo'r Ro.= 'ager' Lat.), crops ('culturi agricole' Ro.= 'culture agricole' Lat.), to sow ('a semăna' Ro.= 'seminare' Lat.), cattle ('vaci' Ro.= 'vacae' Lat.), pigs ('porci' Ro.= 'porci' Lat.), sheep ('oi' Ro.='ovis' Lat.), wool ('lână' Ro.= 'lana' Lat.), potters ('olari' Ro.= 'ollari' Lat.).

Until the end of the first millennium AD, the Daco-Roman population and then the Romanians have been linked to the high Carpathians that they organized for an efficient agriculture. This does not mean these communities had abandoned the valleys, depressions, and piedmonts. Ancient documents of the time of the Hungarian King Arpad use the term "terrae valahice" for lands inhabited by the Vlachos provide some detail. At the shelter of mountains and forests, these lands were organized in the so called "countries": Maramureşului Country (Fig. 16.4), Oaşului Country, Dornelor Country in Eastern Carpathians; Bârsa Country, Vrancea Country, in Curvature Carpathians; Făgăraş/Olt Country, Loviştei Country, Haţeg Country and Severin Country in Southern Carpathians. These 'countries' later became the nuclei of duchies and principalities.

The medieval rural and urban civilization that developed both in the Carpathians and environs left numerous testimonies during the thirteenth–seventeenth centuries. A few examples of Carpathian "countries" can indicate the role they played in the populations living on both sides of the mountains. After the Daco-Roman period, the Haţeg Country was already a territory in the heart of Dacia, with numerous *villa rustica* but also with Roman cities. Situated in the northern extremity of Litovoi principality (included the old Severin Country in the south and Haţeg Country in the north), which stretched from the Danube River to Strei River, this land is the best example of cohabitation on people of both sides of the Carpathians. The medieval "Diploma of Knights of St. John" noted in 1247 that the Haţeg Country had privileges, including the right to raise princiary castles (evidences by the ruins of the Mălăieşti castle at the feet of Retezat Mountains). Subsequently, it was integrated in the Hungarian Kingdom under the name of Hátszegvidék, and 11 large Haţeg principalities were identified during the thirteenth and fourteenth centuries. There also is the Colţi Fortress built by Prince Cândea at the beginning of the fourteenth century in the Râuşor valley, 3 km from the *Râu de Mori* village, which is the source of inspiration for the novel "The Castle of the Carpathians" by Jules Verne.

Considered to be the oldest preserved Romanian medieval structures are the many monuments of stone that were raised for worship (Densuş, Nucşoara, Colţi Fortress, Mălăieşti, Sântămăria-Orlea, Streisângeorgiu) both in the depression and on the slopes of the mountains around (Figs. 16.6, 16.7). Remarkable is the Byzantine church of Densuş, built in the thirteenth century, with walls that include blocks with Latin inscriptions, taken from the nearby Roman sanctuaries at Ulpia Traiana (Fig. 16.7).

Another example, on a smaller scale, is the Loviştei Country, referred to in 1233 as "terram ... Loystha vocatam, ab aqua Lothur vocata, que fluit ad aquam Olth" (Endlicher 1849). This country is situated in the Loviştea depression in the middle of the Olt defile in the Southern Carpathians (Badea 2007).

16.4 Civilizations and Modifications to the Carpathian Landscapes

The Carpathian forests affected greatly the daily life of the inhabitants of the mountains from prehistoric times onward. Covering the mountains almost to their tops, the forests have always had a protective function, but they were also a resource of raw materials and food. Forest landscapes of the Carpathians were more extensive in the past than today as indicated by place names in the north—for example Forested Carpathians (Cernagora in Ukraine). Even the toponymy of Transylvania suggests the presence of huge forests interposed between the vast Panonian and Danubian steppe and open fields. The scholar and naturalist Emil Pop (1945) estimated that, at national level, the forests have covered about 60–70% of the territory since the dawn of the formation of the Romanian people. A similar estimate is provided by soil geography. Forest soils (reddish-brown, brown argillo-luvic, brown luvic and albic Luvisols in plains and hills; Cambisols and Spodosols in the mountains) indicate that over 70% of the country area was formerly covered by forests. Alpine meadows of the Romanian Carpathians were also important and were among the most extensive in Europe. They offered local grazing and transhumance along the Carpathian chain of mountains to the Daco-Roman civilization. The close contact with the meadows and forests of the Romanian Carpathians led to the emergence of a "wood culture" that persisted throughout the history of civilization in that area. These activities led to a major human impact on the natural mountain environment: deforestation. Among the consequences of this impact is the extension of pastures, first as gaps at the upper limit of the forests, then as clearings in the forest body where erosion developed rapidly.

Some Latin phrases from Daco-Roman and medieval times that define the effects of deforestation have been preserved until today, for example 'runc' (Latin 'runcus', deforested Engl.), 'curătură' (Latin 'curatore', clean cut), 'secătură' (to shed the bark), 'jarişti' and 'arşiţă' (setting the forest on fire). In fact, the people in the mountains occupied and utilized these places by Wallachian Law—'Jus Wallachicum'—even though they were not formally the legal owners of the forest—'Sylva Blacorum' (Dragu and Tetea 1984).

During the Daco-Roman and medieval periods, farmers created agricultural terraces on the lands obtained by deforestation, in order to ensure land cultivation and to cultivate subsistence crops. Ethnographic research and documentation by historical geographers, supplemented in recent decades by aerial photography, has led to the identification of twelve locations of intense agricultural terracing in the Carpathians (Botzan 1996). Most of these are located on the mountain sides, but some are on the rounded and deforested summits, above 1400 m. Such locations have been identified in the Apuseni Mountains and, to the south, in the Poiana Ruscă Mountains (Fig. 16.9). Other locations are on planation surfaces on top of mountains of the Southern Carpathians, on the planation surface on top of Sebeş and Şureanu mountains, on the northern slopes of Făgăraş and Retezat mountains, the southern slopes of Iezer Mountains, and also in the Loviştei Country. Agricultural terraces are found also in certain areas of the Eastern Carpathians. Agriculture on these terraces was not necessarily permanent, since there are areas where the forest has regenerated, such as those situated above 1400 m in part of Curvature Carpathians and Southern Carpathians (Dinu and Cioacă 1998).

16.4.1 Wood Culture in the Romanian Carpathians

The natural landscapes of the Carpathians, dominated by spontaneous vegetation (forests, mountain meadows), are associated today with cultural landscapes developed on land cleared of forest, drained of waters, and fallowed. Changing from picking fruits in the wild, to fishing and hunting, to cultivating the cleared land, also meant passing from traditional forestry to modern methods of wood processing. However, the development of such activities based on processing the wood from the forests around the mountain communities has generated different ways of living, synthesized by the old Romanian saying: "the forest is the Romanian's brother". Forest offered to the people, directly or indirectly, shelter, food (fruits, mushrooms, edible plants, venison, honey, amongst others), fuel, and especially raw materials for manufacturing household tools. Up to the beginning of the twentieth century, houses were built without iron nails. Even now, despite the invasion of industrially produced objects and machinery, the cradles and coffins continue to be made only of wood according to ancient traditions.

The Romanian wood civilization was born in the mountains. The culture underwent several manifestations during its history with regards to habitat (village, household, house, cemetery, places of worship), the means of making a living (agriculture, grazing, beekeeping, hunting, fishing, crafts, and popular technical installations), artistic events (folk, folk art), and

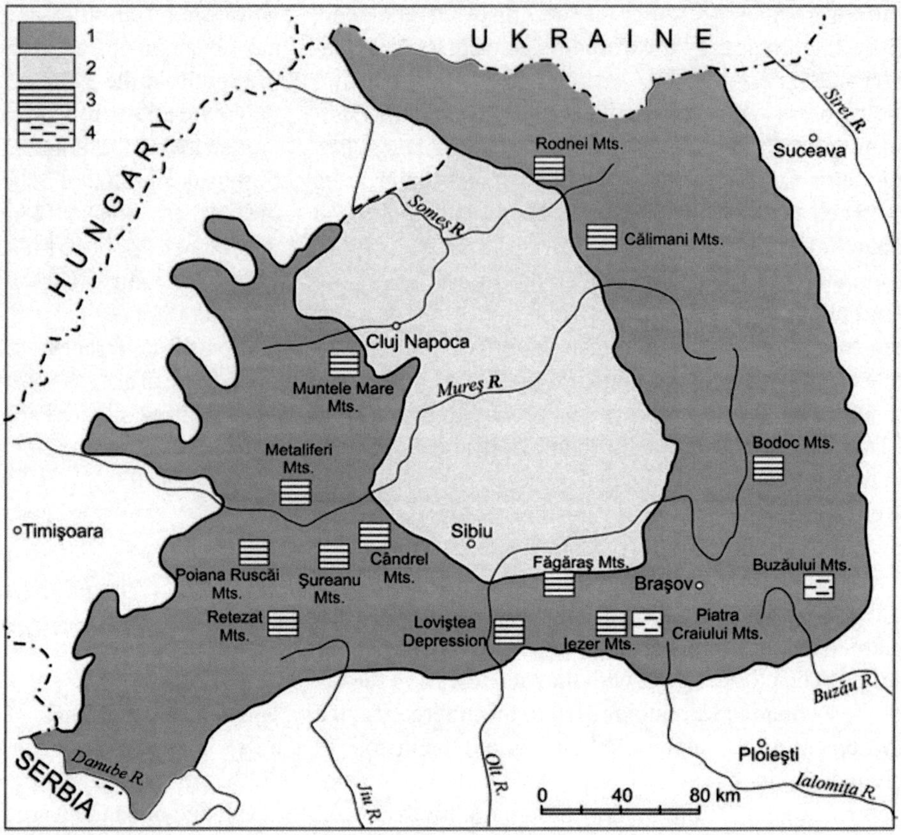

Fig. 16.9 Map of geoarchaeological attestations of the agro-terraces in the Romanian Carpathians: *1* mountains; *2* hilly-depression; *3* documented agroterraces; *4* undocumented agroterraces. (After Botzan 1996)

spiritual manifestations (holidays and customs, habits, birth, marriage and burial, mythology). As a result landscapes or ethnographic regions acquired specific cultural characteristics in the overall European context. Wooden churches are still widespread in Europe, from Scandinavia to Calabria and from Iberia to the Balkans, but the forms of artistic and spiritual expression found in the Romanian Carpathians are not found elsewhere. Out of the eight wooden churches that were included in the list of monuments protected by UNESCO, five are in the Maramureş Depression (Fig. 16.4).

16.5 Conclusions

This brief review of the archaeological vestiges of the evolution of civilization in the landscape of the Romanian Carpathians allows us to appreciate the socio-historical and geoecological conditions under which they evolved. People utilized the abundant resources the land offered according to their abilities at different times. They occupied different parts of the landscape, such as the mountainous areas, to take full advantage of changing climatic and other environmental conditions. One of the results of this adaptation to the landscape is the development of stylistic and architectural features and specific ethno-geographical traits that define the uniqueness of the people living in the Carpathians area

The other side of the coin is that the succeeding populations of the area had a considerable influence in modifying the Carpathian landscapes that had modified their behaviour. As in many other parts of the world, deforestation had a major impact on the landscape both beneficial to humans, such as opening more land for cultivation and grazing, as well as being disruptive for the environment, especially because of enhanced erosion. The forested area has been reduced to about 26% of Romanian land, but the Carpathians still retain the most representative forests and natural environments in comparison with the rest of the national territory.

Acknowledgements We wish to thank Mario Panizza for the invitation to be a contributor to this volume, and reviewers and editors for their professional help in refining the ideas and the written text.

References

Badea L (2007) Depresiunea Loviştei. Studiu de Geografie, Editura Universitaria, Craiova, 206 pp

Binder P (1998) Pământul României în literatura geografică săsească (Izvoare geografice în documente săseşti). Editura Aldus, Brasov, 414 pp

Botzan M (1996) Mediu şi vieţuire în spaţiul carpato-dunăreano-pontic. Editura Academiei Române, Bucureşti, 148 pp

Cioacă A, Dinu M (1997) Modificarea arealelor de risc geomorfologic şi hidrologic ca efect al lucrărilor de corectare a albiei Oltului în Depresiunea Braşovului între Ariuşd şi Feldioara, Analele Universităţii din Oradea, seria geografie VI:47–52

Cioacă A, Dinu M (2002) Pliocene-Quaternary Evolution in the contact area between Brasov Depression and the surrounding mountains,(Romania). Geografia Fisica e Dinamica Quaternaria 24:3–14

Cioacă A, Dinu M (2003) Aria de interferenţă carpato-subcarpatică dintre Trotuş şi Slănicul Buzăului. Influenţa reliefului asupra dezvoltării aşezărilor, Analele Universităţii din Oradea XIII:23–34

Cioacă A, Mihai E (1980) Condiţiile morfohidrografice şi procese geomorfologice actuale, din Depresiunea Ocna Şugatag, Studii şi Ccercetări de Geologie, Geofizică şi Geografie-Geografie XXVII:251–260

Ciobanu Şt R, Cioacă A (1975) Vechimea oraşului Baia - studiu de geografie istorică. Revista Muzeului Naţional de Istorie, Bucureşti II:462–466

Ciobanu Şt R (1979) The population of the Upper Prahova Valley with special consideration on the town of Sinaia. Revue Roumaine de Géologie, Géophysique et Géographie, serie de Géographie 23:131–142

Cocean P (2004-2005) Carpaţii-spaţiu mental şi arhetipal al poporului român. Studii şi Cercetări de Geografie LI+LII:79–89

Dinu M, Cioacă A (1998) The impact of relief and present-day geomorphological processes on the development of rural Carpathian settlements. In: Surd V (ed) Rural Space and Regional Development, Studia Ed., Cluj, pp 72–80

Dinu M, Cioacă A (2000) The Carpathian-Subcarpathian contact between the Dâmboviţa and the Motru valleys. Influences of landforms and settlement spread. In: Editors Bălteanu D, Ielenicz M (eds) Geomorphology of the Carpatho-Balcan Region. Proceedings of the Carpatho-Balcan Conference, Băile Herculane, Orşova, Drobeta-Turnu Severin, Romania, Oct. 11–17, 1998, pp 162–174

Dragu Gh, Tetea Al (1984) Autohtonie, permanenţa şi continuitatea populaţiei daco-romane şi româneşti. Geografia României, vol. 2, Editura Academiei Române, 634 pp

Endlicher SFL (1849) Rerum Hungaricum Monumenta Arpadiana, Scheitlin and Zollikofer, Budapest (Reprinted in 1931 by Hiersemann, Leipzig)

Giurăscu CC, Giurăscu D (1974) Istoria Românilor din cele mai vechi timpuri până astăzi. Editura Ştiinţifică, BucureştiM, pp 21–34

Grumăzescu C (1975) Depresiunea Haţegului, studiu geomorfologic. Editura Academiei, Bucureşti, pp 11–24

Haywood J (2001) Atlas of the Celtic World, Thames & Hudson, London, pp 30–37

Mac Kendrick P (1978) Pietrele dacilor vorbesc, Editura Ştiinţifică şi Enciclopedică

Mihăilescu V (1963) Carpaţii Sud-estici. Editura Ştiinţifică, Bucureşti, 373 pp

Niculescu Gh (1987) Caractere fizico-geografice: relieful. Geografia României, vol. 3, Carpaţii Româneşti şi Depresiunea Transilvaniei. Editura Academiei, Bucureşti, 655 pp

Pârvan V (1957) Dacia. Civilizaţiile străvechi din regiunile Carpato-Danubiene. (traduction by Radu Vulpe, after original inedited manuscript, in French). Editura Ştiinţifică, Bucureşti, 121 pp

Pop E (1945) Cercetări privitoare la pădurile diluviale din Transilvania, Buletinul Grădinii Botanice din Cluj 25:1–2

Posea Gr (1981) Depresiunea Braşovului (Caracterizare geomorfologică). Analele Universităţii Bucureşti, seria Geografie XXX:3–12

Vasilescu I (2008) Depresiunea Maramureşului. Studiu de geografie integrată (manuscriptum, in the Library of the Babes-Bolyai University, Cluj Napoca, Romania), 404 pp

Chapter 17
Sea-Level Rise and the Response of the Dutch People: Adaptive Strategies Based on Geomorphologic Principles Give Sustainable Solutions

Pieter D. Jungerius

17.1 Introducing the Challenge

17.1.1 The Roots of Our Conflict with the Sea

Sea-level rise due to global warming will hit the Netherlands harder than any other country in Europe. It comes on top of the sea-level rise that began at the end of the Pleistocene with 70 cm/century and gradually slowed down to the present 15–20 cm/century. This sea-level rise has a number of natural and, more recently, man-induced causes (Van Koningsveld et al. 2008). Some of these causes resulted in an increase in absolute sea-level height, others in relative height due to subsidence of the land. It means that the North Sea has inundated, or threatened to inundate, the Netherlands for more than 10,000 years. For a long time it was a barren, uninhabited coast, bordering a land of infertile sands such as we still find in the eastern part of the country. Settlers came when sandy beach ridges emerged about 5000 years ago, offering firm ground for housing and living. A wide lagoon with shallow water gradually changing into peat bogs separated the beach ridges from the mainland in the east.

It was a mystery to the Romans why people choose to live in such a hostile environment. Pliny the Elder (23–79 AD) visited the area and described the miserable living conditions as follows: "The land, invaded twice each day and night by the overflowing waves of the ocean, opens the question whether these regions belong to the land or to the sea. Here a wretched race is found, inhabiting either the more elevated spots of land, or else eminences artificially constructed, and of a height to which they know by experience that the highest tides will never reach. Here they pitch their cabins and when the waves cover the surrounding country far and wide, they are like so many mariners on board ship. When the tide recedes, their condition is that of shipwrecked men, and around their cottages they pursue the fishes as they make their escape with the receding tide. They fashion mud (Pliny means peat) with their hands, and drying it by the help of the winds more than of the sun, cook their food by its aid, and so warm their entrails, frozen as they are by the northern blasts".

Pliny indicates where the answers to his questions might be found: the rich fishing grounds. Already in Mesolithic times groups of fishermen roamed the coast, long before the Romans arrived in the country. Later on, the extensive peats and alluvial sediments filling the lagoon were converted into fertile soils supporting a prosperous agriculture. Reason enough to stay!

At first there was not much that the inhabitants of the Low Countries could do about the inundations, apart from fleeing to higher grounds. Already in the Iron Age people made 'terps', artificial mounds, in areas that were flooded during storm surges. This submissive attitude changed after 1000 AD when the people developed the techniques to defend themselves against the sea or even to fight back, instead of fleeing to higher grounds (Dietz et al. 2004). This happened after a series of storm events culminating in the twelfth century with huge loss of land and the splitting up of the northern and southern sections of the Dutch coast in a number of islands. Building of dikes coincided with the construction of dams across river outlets to

P. D. Jungerius (✉)
Stichting Geomorfologie en Landschap
Oude Bennekomseweg 31
6717 LM Ede, The Netherlands
e-mail: p.d.jungerius@uva.nl

See Plate 16 in the Color Plate Section; also available at: extras.springer.com

I. P. Martini, W. Chesworth (eds.), *Landscapes and Societies*,
DOI 10.1007/978-90-481-9413-1_17, © Springer Science+Business Media B.V. 2010

prevent the sea from entering the river mouth during floods. These dams gave Amsterdam, Rotterdam and many other towns the characteristic 'dam' ending to their name.

The techniques have evolved throughout the ages, but so has the need to keep the water out. Some human malpractices helped nature in her efforts to drown the land and land has been gained and lost ever since, often with dramatic consequences and sometimes with unexpectedly positive side effects. An example is the reclamation of large peat areas, which lowered the hinterland by several meters and caused the enormous inroad of the sea known as Zuiderzee. Much land was lost to the sea, but the harbor facilities offered by the sheltered inland sea also opened the way for many small towns to participate in the European-wide trading network of the Hanseatic League and were the basis of our Golden Age in the seventeenth century. Another example: at one time the western part of the country was pockmarked by lakes left behind by peat digging. Some of these lakes were drained and transformed to fertile fields; others were abandoned but are now popular recreational lakes. By and large, the Netherlands survived, even though millions of people live permanently as much as 6 m or more below sea level (Visser and Misdorp 1998).

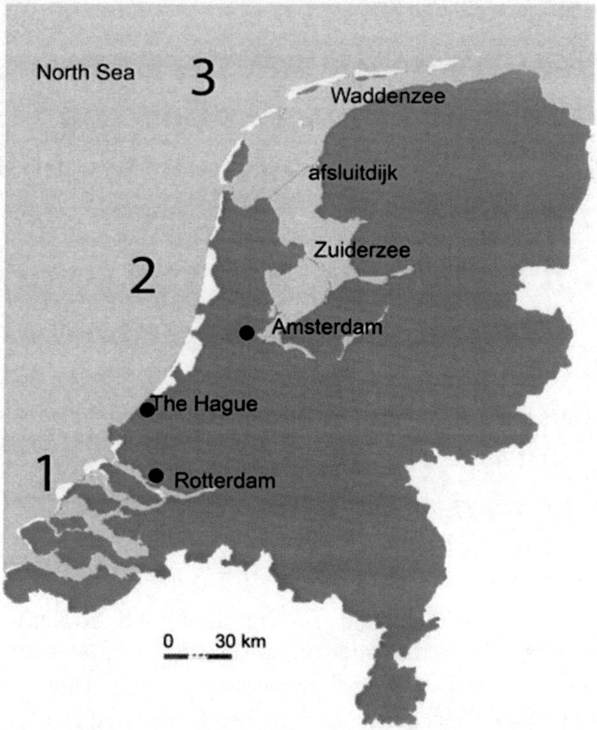

Fig. 17.1 The three physiographic units of the Dutch coast. (*1* the estuaries in the southwest; *2* the closed barrier coast of the central parts; *3* the open barrier coast in the north. Dark gray: land; light gray: sea; white belt along the coast: dunes)

17.1.2 The Coast Line in Geomorphologic Perspective

Google Earth shows for everyone to see how the North Sea coast curves in a wide sweep from Calais to Denmark (coordinates 50°50′–53°50′N, 1°20′–8°00′E). The Netherlands occupy the pivotal position halfway along, where the curve bends eastward around the Pleistocene nucleus of Texel Island. The smooth curve exemplifies the way marine and aeolian processes tend to work: protruding parts are abraded, bays are filled in. The longitudinal drift along the coast, the so-called 'sand river' (Wong et al. 2007) also has a straightening effect. Remarkably, even the Rhine, Meuse, and Scheldt with their supply of sediments from the hinterland, have not been able to build a delta. Their estuaries cause no more than irregularities in the smooth curve.

The length of the Dutch coast is about 350 km of which 254 km consist of dunes and 34 km of dikes. Customarily the Dutch coast is divided into three physiographic units: the estuaries in the southwest, the closed barrier coast of the central parts, and the open barrier coast in the north (Fig. 17.1).

The estuaries of the coast in the SW are the embouchures of the Rhine, Meuse, and Scheldt, widened by strong tidal scouring (Fig. 17.1: 1). The islands in between the estuaries inherited their west–east orientation perpendicular to the coast from the flow-direction of the rivers. Safety measures taken in the past which took the natural orientation into account were generally sustainable. This applies also for future coastal defense measures. The units of the coastal landscape are dunes, mud flats and tidal marshes. The latter are under threat because the recent closure of estuaries by the Deltaworks has lowered the hydraulic transport capacity and reduced the supply of the fresh sediment needed for sustaining the ecosystem.

The closed coastline of the central mainland of the Netherlands is fringed by dune ridges (Fig. 17.1: 2). The so-called Old dunes, formed about 5000 years ago, consist of a 10 km wide system of parallel sandy beach ridges separated by marshy depressions.

The dry sandy ridges were soon to be inhabited. Even now the pattern of long and straight streets of The Hague reflect the underlying ridges, and the name of one of the main shopping streets, the Venestraat, meaning peat street, is a reminder of the peaty marsh that had to be traversed when crossing from one ridge to the other. The barrier ridges protected the coast right from the start and continued to do so for thousands of years, allowing the growth of peat in the quiet of the wetlands which separated them from the high Pleistocene eastern part of the Netherlands. As the area filled up, the invasions of the sea became less frequent. Finally, the peat blanket stuck out more than 4 m above sea level, effectively preventing inundation of the wetlands behind the ridges.

Formation of the much higher young dunes along the coast started around the tenth century AD. The source of the sudden enormous supply of sand is still somewhat mysterious. Most interesting is the hypothesis that it is caused by soil erosion following deforestation in the European hinterland in the ages preceding the thirteenth century. Forest areas were converted to arable land on a large scale to provide food for the rapidly growing population of the many new cities (Poeteray et al. 1984). Moreover, a period of disastrous soil erosion occurred in Germany in the early fourteenth century. The rivers transported the eroded material to the sea, longshore drift spread it out along the coast, surf deposited it on the beach, and the wind blew it inward. The resulting dunes reach heights of up to 50 m.

The width of the strip of dunes varies along the length of the coast. A number of wider dune wedges, some 5 km wide, are arranged like beads on a string, beginning at The Hague in the south and ending at the tip of the mainland. If they ever existed south of The Hague, they have since been destroyed by the North Sea. Some, and perhaps all narrow dune strips in between, represent the outlets of former streams draining the hinterland. These were the weak spots where the sea could force its way in. The northern dune wedges became islands as a result of storm surges in the twelfth century.

The open barrier coast in the north comprises the Wadden Sea fringed by a string of islands (Fig. 17.1: 3). The geomorphology of the islands is basically similar (Oost 1995). Each consists of one to three sandy core areas where the settlements are located. These are the islands left by the storm surges of the twelfth century and in origin comparable to the dune wedges along the barrier coast further south. The core areas of the islands were separated either by large inlets held open by tidal currents, or by tidal marshes. Dunes were formed along the North Sea edge of the tidal marshes, but there was apparently not enough sand to form a continuous foredune ridge to connect the core areas. The inhabitants helped nature by planting marram grass or constructing fences of reed or brushwood to trap sand. As a result, the island cores are now interconnected across the tidal marshes which were turned into grasslands, with sand dikes mimicking foredune ridges. Ecologists would like to remove these sand dikes and restore the tidal marshes, but that is presumably easier said than done. The sand dikes are fairly stable, because they have been constructed according to the geomorphological principles underlying foredune formation.

17.2 The Problems

Subsidence of land and sea-level rise have always been the main driving forces behind the problems that the Dutch have with the sea. There are many natural causes of the relative sea level changes, the most important being structural activity and isostatic rebound since the last glaciation. The actual risks came from unpredictable storm events such as the extensive inundation some 800 years ago.

Reclamation of the coastal peat areas provides a good example of the environmental problems the Dutch created for themselves. As was explained earlier, the Holocene peat bogs had grown up to 4 m above sea level, effectively keeping the sea from entering the coastal lowlands. Draining the coastal peat by digging ditches to make the land suitable for agriculture started as far back as Roman times and has gained in importance since 900 AD. Due to oxidation and compaction, peat shrinks when drained, leading inevitably to lowering of the land. This started an endless feedback-loop of stepping up drainage followed by further subsidence. In this way, land subsidence has become a self-perpetuating process, with rates up to 5 cm/a. Peat cutting for fuel aggravated the problems. Without these practices much of the Netherlands might still be above sea level. As it is, much of the surface of the peat lands in the coastal zone had sunk to 1–2 m below sea level by about 1000 AD. The surplus water could no longer be removed by natural drainage and other solutions had to be found. Perhaps even more seriously, subsidence of the peat areas weakened the coastal defense and allowed the sea to penetrate, erode large tracks of peat land and inundate the land behind the dunes during storm floods.

The inundations were not always seen as hazardous. By some, it was considered a blessing in disguise when sea water inundated peat. They dug up the water-soaked peat and burned it after drying. The residue was then washed to remove the salt. This practice, called 'moernering' provided the salt for conserving the fish and was a substantial source of income in former times when salt was very valuable. It was so profitable that inundations were in fact encouraged in some parts of the country. Nevertheless, the situation became increasingly more perilous. The combination of NW storms raising the sea water in the funnel-shaped area between Great Britain and the Netherlands, spring tide, and high discharge of the rivers made them particularly dangerous.

The storm flood of 838 AD is one of the first recorded disasters, reported by a French bishop and a Carolingian source. Many villages were lost. The storm surge of 1014 was allegedly the first to breach the coastline. The flood of 1–2 November 1170 and subsequent floods changed the map of the Netherlands drastically. The North Sea penetrated the coast of the northern part of the Netherlands, splitting up the dune coast in a number of islands and giving the sea access to the inland lake then known as Almere. Later storm floods continued to erode the peat lands, enlarging Almere until it became the Wadden Sea (Waddenzee) and the Zuiderzee.

Table 17.1 shows two peaks in the frequency of storm floods. The peaks appear to be associated with climate change. The well-documented increase of storminess and storm flood frequency culminated in the thirteenth century and coincides with a major transgression phase. Currently, we seem to live in a period of increasing storm frequency (Van Baars 2007).

Table 17.1 Storm floods through the ages (from Wikipedia)

Year	Number of storm floods
1000–1100	1
1100–1200	5
1200–1300	9
1300–1400	5
1400–1500	5
1500–1600	6
1600–1700	3
1700–1800	2
1800–1900	2
1900–2000	4

Floods were commonly named after the saint of the current day. Some saints must have been particularly fear-inspiring. There have been three enormous storm surges round the 19th November, the saint's day of St. Elisabeth. The St. Elisabeth flood of 1421 broke many records. Perhaps the level of the high tide was not exceptionally high, but the dikes failed because they were neglected due to political quarrels in the country. All-Saints day, 1st November, was the scene of disaster at least five times. Even in 1953, many people in the SW part of the deeply religious Dutch Bible belt saw the disastrous storm surge of that year as a divine act of retribution. This storm which affected also parts of Great Britain, Germany, and Belgium claimed 1835 dead in the Netherlands alone. Dikes were damaged over a length of about 500 km, 2000 km^2 of land was flooded, and over 72,000 people were evacuated.

17.3 The Answers

In the course of their defense against the sea, the Dutch developed five strategies. Each time a strategy was no longer sufficiently effective a new strategy had to be applied. All five are still in use. The first strategy was learning to cooperate and coordinate the activities against the common foe. The next and simplest strategy was preventing the floods from inundating the land by building dikes around polders (Van de Ven 2004). Sluices in the dike were opened during low water outside to dispose of excess water in the polder. When subsidence and sea-level rise caused the level of the polder to become too low to dispose of the water in that way, windmills had evolved sufficiently to be used to pump out the water in the third strategy. Technological progress enabled the Dutch to drain increasingly large water volumes: steam engines replaced wind power, to be substituted in their turn by the present electrical- and oil-driven pumps. But dikes were not the only weak spots in the defense. In the nineteenth century a new strategy was needed to strengthen the natural dune belt. Finally, at the end of the last century, the Dutch decided to push the sea back and build new land in the North Sea. This is the fifth strategy.

17.3.1 First Strategy: Organizing the Defense

Various principles have been applied to solve the problems created by rising sea levels but what was needed first of all was organizing a common defense. A flood left the land in great disorder: people had to be moved, the dead had to be buried, houses had to be rebuilt, in short, a multitude of activities was necessary at a time that the roads were washed away and all other communication lines were destroyed. The ever-present danger of flooding required short lines of communication between the seat of authority and the place of action. This calls for a high level of decentralization and collective cooperation. In the thirteenth century the Dutch took a wise decision: they decided to form water boards. Widespread experience with decentralized government and communal cooperation appears to have been an important factor in the formation of the Dutch Republic in the sixteenth century.

Water boards are regional organizations in charge of all aspects of water management: they keep the sea outside, our feet dry, and the water clean. Over time, they developed into prestigious and well-managed bodies that were highly respected throughout. This is reflected in the splendid Polder Houses, the seats of their administration (Fig. 17.2). The water boards are called 'waterschappen' or 'hoogheemraadschappen'. The latter term is used for water boards in charge of sea dikes or those that drained their surplus of water off to the sea.

Water boards are the oldest democratic institutions in the country. The board is chosen and consists of people representing the various categories of stakeholders: landholders, leaseholders, owners of build-up areas, companies and, more recently, the residents as well. The government appoints the chairperson, the 'dijkgraaf', literally: "dike count", emphasizing the noble office he held.

Water boards have always been an enigmatic element in the administrative structure, mainly because they overlap with municipalities and provinces and have the power to overrule the decisions of these authorities for the sake of safety. The organizational structure of the water boards is currently changed and brought in line with other government institutions, with more influence of the political parties. What remains from the old days of poor communication when people could not leave their farms for travelling, is that voting can be done by mail.

A good example of a multifunctional, modern water board is the "Hoogheemraadschap Amstel, Gooi en Vecht", one of the 26 water boards left in the Netherlands, out of the 2600 that once existed. Territorially, the municipality of Amsterdam is part of this water board. On the other hand, Amsterdam and the water board co-operate in Waternet an executive body integrating the duties of the water board with the supply

Fig. 17.2 The "Gemeenlandshuis", ancient seat of the administration of the "Hoogheemraadschap Amstel, Gooi en Vecht"

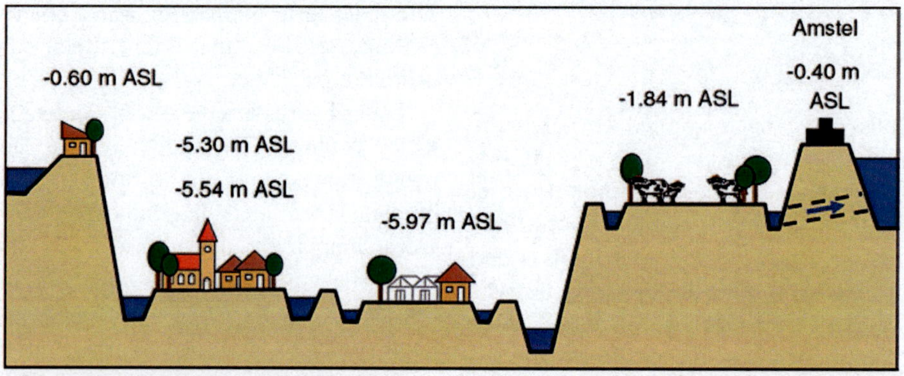

Fig. 17.3 A schematic cross section of three polders of different elevation south of Amsterdam (not to scale). ASL = standardized Amsterdam sea level in meters, a fixed datum to which all elevations are referred to. The minus signs mean that the whole system is below sea level. Gray: land, black: water in the canals. Excess water of the polders (*the built-up areas*) is pumped out into the 'boezem' canal that surrounds the polder, and from there up to the river Amstel (see *arrow on the right*) which brings the water to the sea

of public water. To give an impression of its size and complexity, this organization:

- has a surface of 700 km^2 of which water occupies 95 km^2,
- includes 25 municipalities with a total of 1.2 million inhabitants,
- provides them with 26,000 m^3 drinking water daily,
- comprises 111 polders (Fig. 17.3),
- maintains 1700 km length of dikes and 1300 km length of canals,
- operates 221 pumping stations and 108 locks and sluices,
- works 13 sanitation plants processing 130 million m^3 of waste water, and employs 1700 staff members.

In the eighteenth century with the scaling up and increase of the size and complexity of the drainage systems, there grew the need for a country-wide integrated approach which exceeded the territory of one or even a cluster of water boards. A dangerously high discharge of the Rhine, for instance, needs a coherent and coordinated set of measures from the point where the river water enters the country down to the place where it reaches the sea. From this need arose the foundation of the Rijkswaterstaat as a department of the Ministry of Public Works and Water Management. Rijkswaterstaat also has the main responsibility for the safety of the coast. Its task is to maintain the strength of the three Ds of coastal defense: dunes, dikes and dams.

17.3.2 Second Strategy: Building Dikes to Keep the Water Out

When the first farmers started developing the peat areas, the peat was dry enough to allow the growth of cereals. The peat was so high above sea level that any excess of water could drain off naturally to the sea through ditches, creeks and rivers. When the peat shrank, and natural drainage was insufficient, the farmers shifted to grassland and dairy farming. This lies at the roots of the well-known "green heart" region in the western part of the country, now a major tourist attraction, under continuous threat of urbanization from the existing settlements. The unrelenting subsidence made it imperative to protect the land against flooding. This was generally accomplished by building dikes to keep the water out. The areas they protected could only be drained by opening culverts at times of low water in the surrounding streams and canals.

Interestingly, the traditional practice of increasing or raising the height of land was only practiced in Amsterdam. Early in the seventeenth century the municipality prudently decided that new town developments were to be realized on a 6-m thick layer of sand covering the peat and alluvial sediments. The sand was brought in from the coastal dunes and from a Pleistocene moraine east of the city. This is why many parts of Amsterdam will not be faced with insurmountable problems when the dikes should no longer hold. The rule was abandoned after the completion of the

Afsluitdijk in 1932 when the water of the North Sea could no longer reach Amsterdam and the town was considered safe. Urban developments since are as vulnerable as the rest of the lowlands.

The first dikes to keep the sea water out were constructed round 1100 AD, around small pieces of land between creeks. Soon dikes were constructed to connect these small 'polders', even across the creeks which always remained weak spots where dikes could relatively easily be breached. The process ended with dike rings; that is, large areas enclosed by a common dike. That this was considered to be an important accomplishment, is shown by a quotation of the Rustinger Law of the twelfth century: "That we, Frisians, have to found a stronghold against the sea, a golden hoop which lies around all of Friesland, of which every dike-measure shall be equal to the other …" (Forbes 1993, p. 55).

The first dikes were often built with peat or soft clay. From the fifteenth century onwards special methods were developed to protect the outside slope of the dikes. In the Northern provinces seaweed or sea grass were heaped up against the dike to protect the outside slopes, preferentially at places where reed marshes were absent and the sea had immediate access to the foot of the dike. Wooden pile constructions were erected to prevent parts of the seaweed from slipping away into the sea. In the eighteenth century the pile worm (*Teredo navalis*) caused much damage to the protective wooden constructions of the dikes, particularly during its large infestation of 1731–1732. Within a short time the lower parts of the country lost all their protection against floods, and new dikes had to be constructed rapidly. Rather than trying to directly protect the foot of the dikes with seaweeds and piles, easily displaced by plunging storm waves, dykes were built with seaside slopes over which the incoming waves could lose their energy. Since then, columnar basalt mainly from the Eifel region in Germany has often been used for the outer mantle because the polygonal pillars fit tightly together and are not easily pried loose by the waves of the sea. The newest trend is to make the outer lining ecologically friendly to encourage the colonization of crustaceans which are an important part of the food chain.

Geomorphologically interesting is the fact that river dikes and sea dikes were given different profiles, because they are subject to different forces. A river dike has to resist the pressure of high water and therefore has a gently sloping inner bank which is not easily pushed aside. A sea dike, in contrast, has a gently sloping outer bank to allow surf to slacken off gradually. The inside slopes of dikes should not be too steep, because steep slopes suffer much damage by turbulence of water pouring over the crown of the dike. Slumping of the inside slopes hollow them out and weaken the dike until it collapses.

Dikes are not necessarily beneficial. By keeping the water away from the empoldered surfaces they reduce the area of flooding, and this pushes up the water level in the remaining open part, the retention area, of the drainage basin. As long as polders were small, this gave no problems, but the success of the first dike builders soon invited others to follow. The areas left to accommodate the floods became small and the water level accordingly so high that a storm tide could easily lead to critical situations. Increasing the height of the dikes was only a temporary solution. In the end, some polders had to be given up to find a stable balance between safety and area of reclaimed land. Surprisingly, this practice is currently reintroduced: dikes are removed to allow the sea to recapture its lost territory and reduce flood height. Understandingly, these measures meet much opposition by the local population because it is seen as an act of treason to the forefathers, who did so much to win land from water.

Along the larger rivers the problem could also be solved by shifting the dike landward to increase the storage capacity of the river bed. Alternatively the slope of the river bed could be levelled by removing obstructions that slow down the rate of flow. This is a controversial issue: those in favor point to the advantages of increased safety for the people living near the river, those opposing the measures regret the loss of the 'uiterwaarden', the semi-natural landscape of the winter bed with its characteristic oxbow lakes, cut-off meanders and abandoned levees.

Another negative side effect of dikes is the fact that they interrupt the sedimentation on the land they protect. This land has therefore been sediment-starved for up to 1000 years. In addition, peat decomposition and soil compaction caused by drainage have resulted in considerable land subsidence. The amount of sediment required to keep up with sea-level rise in the diked-in lowlands of the Netherlands has been calculated by multiplying the surface of the deprived area with the expected rise of sea level. Depending on the chosen scenario of climate change and sea-level rise, Van der

Meulen et al. (2007) estimate the sediment deficit at 136 ± 67 million m^3/a. About 85% of this volume is required to keep up with sea-level rise, 15% to balance the effect of land drainage (peat decomposition and compaction). The deficit is only partly compensated by the input of the 23 million m^3/a of filling sand from outside the polders for building purposes and creating higher areas.

It is remarkable that water changed its role and became an ally in times of war. When the Netherlands were in danger of being occupied by foreign armies the dikes were breached to allow water to inundate the land and prevent the enemy advance. Even in Roman times Tacitus reported that rebellious Dutch tribes breached quays to flood their land. Incurring the help of the water for military activities, both offensive and defensive, played an important role in the period of the struggle for freedom from Spanish rule in the sixteenth century. Sometimes the Spanish troops had to abandon the siege of a town because the Geuzes, the military arm of the Dutch rebels, inundated the surroundings and approached in boats with shallow draft. In the course of time several plans were devised to create systematic defense strategies by means of inundation. Basically the design of these 'stellingen', as they are called, was simple. The sluices of many low-lying polders were opened and water was let in: not too little, for in that case the enemy troops could wade through the water; on the other hand, not so much water that the enemy could use ships. A few decimeters of water was ideal, because it made the many ditches invisible and therefore very treacherous.

During the nineteenth and twentieth century large parts of the countryside were set aside for the defense of the Netherlands. Remains of the constructions made to inundate the country can still be found in many places. One of the most important was the Stelling van Amsterdam designed by Cornelis Krayenhoff, inspector-general of fortifications. The design of this defense work is based on knowledge of the geomorphology of the terrain and the position of the stelling in the landscape. For this reason it has been called a geo-military complex (Van den Ancker 2005). It is a 135 km long water defense circle around Amsterdam, consisting of water bodies and lowlands that could be flooded with a complicated system of dikes, dams and locks. No less than 42 forts protected the weak points in the armour such as roads on dikes which the enemy could use to cross the water bodies. The fortresses are monuments of great cultural-historical value. Indeed, the Stelling van Amsterdam has been included in the world heritage list of UNESCO. However, the real, creative, military genius lies in the way the designers made use of the possibilities offered by the terrain. The Stelling van Amsterdam could never have been realized without the geomorphological characteristics of the landscape around the capital. This landscape of peat, lakes, rivers, canals, polders, and sea possessed all the ingredients for the necessary strategies of inundations.

The main resources used by the military planners for this water defense line were: peat areas, which become soggy and lose their carrying capacity when inundated; the many rivers draining the peat; the natural lakes and 'plassen' (lakes left after peat-cutting); the polders that could be inundated; the Zuiderzee. These obstacles were used in combinations. Preferably the enemy had to pass at least three of these hurdles to reach the inside of the Stelling.

The Stelling van Amsterdam was kept in service until decommissioned in 1963. The use of aircraft rendered it obsolete after World War I. Still, it cannot be denied that the German troops in 1940 as well as the Allied forces in 1945 avoided the direct assault on the capital of the country when conquering the Netherlands. Without doubt the Stelling had fulfilled the real function of a fortress: not to withstand a siege, but to discourage the enemy from attacking.

17.3.3 Third Strategy: Pumping to Get Rid of the Excess of Water

Compaction of peat can be minimized by maintaining a high water level in the ditches between the parcels. The water boards issued strict regulations to the farmers for the authorized lowering of the water level in the polders. But sometimes subsidence had lowered the agricultural lands so much that the water level outside the polders was permanently too high for natural drainage. Fortunately the windmills were already well developed as flourmills when they were introduced in water management. They were adapted to drain off water by means of a paddle-wheel. Two types of windmills for drainage were developed almost simultaneously from basic types of mills for grinding cereals

Fig. 17.4 An octagonal mill (*left*) and a post mill, with a modern pumping station in between

that existed in different parts of the country (Fig. 17.4). The octagonal mill, with a thatched cap which turned from the inside, was developed from the tower mills found on the ramparts of towns where they could catch most wind. The post mill rests some meters above the ground on a colossal oak tree trunk, the post, which can be turned by an outside beam construction, the 'tail'.

The first windmill for drainage, recorded in historical sources, was built in 1408. Soon hundreds of the wind mills followed throughout the countryside in spite of the high costs involved. It has been calculated that the construction of a windmill would have amounted to €600,000 in the current value (Van de Ven 2004). In addition, maintenance was very expensive because of the complex machinery. Sometimes during foul weather conditions the windmill remained in operation for too long, as it was essential to keep the fields dry. This often caused vital parts to wear out quickly and occasionally even to break off. From the beginning these technical works were therefore carried out by groups of specially trained hydraulic engineers, who acquired international fame and were often consulted for reclamation projects in other European countries. Even before 1250 AD Dutchmen were involved in the reclamation of marshy land in North Germany. Recently, Dutch expertise has been called in to prevent repetition of the damage done by Hurricane Katrina in New Orleans.

Large-scale digging of peat for burning as fuel in households and industries began in the sixteenth century. This practice lowered the land too much for water to be pumped out and left many lakes. Windmills with paddle wheels could lift the water 1.5 m at the maximum. An important innovation was that a new lifting device, the open Archimedes screw, was introduced in the windmill. It could lift water up to 4 m. Leeghwater, a famous engineer of the seventeenth century, connected the wind mills in series which made it possible to drain big lakes such as the Beemster, another recognized UNESCO World heritage site.

The structure of water management in the Netherlands had reached full growth by that time. Great bodies of water had been turned into productive agricultural areas. Men had strongly taken the initiative. But by the end of eighteenth century, the windmill was at the end of its possibilities, though large parts of low-lying land remained flooded during the winter months. The Dutch had to wait for the invention of the steam engine that was used in combination with huge water pumps in the English mines in the mid-nineteenth century, to drain these areas. These pumps could not be used without modification. The water of mines was pumped up in small quantities from a depth of up to hundreds of meters, whereas in the Dutch polders large quantities of water had to be pumped up for no more than a few meters. The first experiments with steam engines in 1769 were a disaster because the engineer in charge, a clockmaker by profession, had underestimated the enormous forces within the large constructions. But new experiments with an improved pump and steam engine, the James Watt engine, were a resounding success. In 1787 four polders could be kept dry by one pump, in spite of heavy rainfall.

The most spectacular lake reclamation project was that of the Haarlemmermeer in 1852. This lake had over the years increased so much in size that it threatened to engulf the city of Amsterdam. The main reason for the reclamation was therefore not the gain of new land but to protect Amsterdam. Two centuries before, Leeghwater had made a plan for the reclamation of the Haarlemmermeer with 160 windmills. The work was now done by no more than three huge steam-powered pumping stations.

The early colonization of the former lake bottom was a social tragedy. It is a sad story of a community that faced conditions far beyond its control, with poverty, disease, alcoholism, criminality, and internal strife. The main problem was that the mud left on the former lake bottom took many years of ripening before it turned into a soil that could carry crops. Moreover, the settlers of the slightly higher grounds regulated the groundwater level and that meant that the fields of the less privileged settlers in the lower areas were inundated almost the year round. The dikes built around the land by the owners of the lower grounds were destroyed by those of the higher grounds. All this is long forgotten. The Haarlemmermeer is now one of the most productive agricultural areas of the Netherlands.

In the first part of the twentieth century, by far the largest project was the reclamation of the Zuiderzee. By then, electric and diesel pumps had replaced steam pumping. The decision for this reclamation was taken shortly after 1916 when a storm tide breached the dikes along this tidal inlet, and sea water inundated large areas of agricultural land, making them unsuitable for agriculture for a long period. This happened during the First World War when the country was cut off from the supply of grain from other parts of the world and it was realized by the government how this endangered the nation's food supply. The construction of the Afsluitdijk separating the Zuiderzee from the Wadden Sea and the North Sea was completed in 1932. With 32 km it was allegedly the largest dam ever built. It turned this inland sea into a large freshwater basin now known as IJsselmeer. The original plan aimed at constructing five polders within this lake, of which four polders have now been realized. The first two polders were attached to the mainland, but the difference in surface height created drainage problems. Later polders are separated from the mainland by a wide canal that acts as a buffer. As the need for agricultural production decreased in the second part of the twentieth century, much of the new land in the fourth polder was reserved for nature development and urbanization. The fifth polder was never realized and remains a large lake reserved for the storage of fresh water, presumably until food shortage revives the old plans.

17.3.4 Fourth Strategy: Strengthening the Dunes and Sand Nourishment

From around the tenth century AD, the coastal dunes have been, and remain the Netherlands natural defense against the sea, but it is only from the middle of the nineteenth century that they have been managed for that purpose. It started with the stabilization of the foredune. The choice of the measures applied reflects the lack of understanding of the coastal managers of that time in the geomorphologic processes involved. They saw the foredune as a pitiful erosional remnant of the past which would disappear without their help. It was not realized that a foredune is a contemporary construction of the combined actions of wind and sea.

What foredune formation needs is a sand-catching vegetation, preferably marram grass (*Ammophila arenaria*) that can withstand the arid conditions of sandy beaches and binds the sand with its fibrous rhizome system. It thrives on burial and can survive yearly burials of more than 1 m of sand. Where burial ceases, plant cover is reduced and the plants eventually die, because their root system is easily affected by nematodes and moulds (Van der Putten 1989). What marram grass therefore needs is a constant supply of fresh sand to outgrow the infested roots. This is the reason that naturally growing vital marram grass shoots stand far enough apart to let the wind blow in between, take up sand and deposit it again around the shoots after rubbing harmful substances from the grains. The coastal managers were not aware of this and were worried about the openness of the vegetation structure. They increased the density of the cover by planting extra marram grass shoots in between. The marram grass, old and new, received insufficient clean sand and lost its vitality and protective effect. This was solved by planting more marram grass, which repeated the process. From that time on, planting marram grass was a never-ending necessity. In fact, it was forbidden by law to allow any form of wind erosion such as blowouts in the dunes, for fear of weakening the strength of the coast.

Around 1975, one of the water boards in charge of coastal defense tacitly consented to an illegal experiment to see what suspension of stabilization measures meant for the development of blowouts. The experiment was carried out in the inner dunes where the safety of the country was not directly at stake, and with the instruction to keep it out of sight of the public. After 2 years of measurements it was clear that it was best to let nature have its way: the blowouts stabilized spontaneously (Jungerius et al. 1981). Repetition of the experiment in other dune terrains confirmed these results. Dune managers and dune ecologists welcomed the outcome, but for different reasons: not having to plant marram grass saved the dune managers thousands of guilders a year, and ecologists were happy because it stimulated the growth of pioneer vegetation which rejuvenated the dune landscape and increased biodiversity. When it was clear that the safety of the country was not at stake, similar experiments were hesitatingly allowed even in the foredunes along the beach. They had the expected result: natural geomorphologic processes led to a more natural and stable foredune. Society could have saved millions of guilders if the natural system had been left alone right from the middle of the nineteenth century.

Even if the foredune is not blown away by the wind, the sea keeps on eroding the coast along much of the Dutch coastline. Past coastal engineers often relied on hard artificial structures to protect the dunes, although these techniques defy the geomorphologic rules of nature. Many groynes to protect beach erosion were built in the nineteenth century, but they are effective for a limited length of coast only. They disrupt the longshore drift of sediment; therefore beaches downstream are deprived of sand supply and become eroded. This necessitates the construction of more and more groynes in the down drift direction. This has been called the domino effect. In 1990 it was decided by the Dutch government to preserve the position of the coastline of that year by sand nourishment. This is much cheaper: hard coastal defense measures cost yearly €180,000/km, whereas sand nourishment costs yearly €130,000/km. Sand is extracted with suction dredgers from the bottom of the North Sea where the water is more than 20 m deep, and spouted via pressure pipes onto the coast. The yearly supply of sand amounts to 12 million m^3.

Sand nourishment mimics the natural geomorphologic process of sediment brought to the sea by the rivers and spread out along the coast by longshore drift. When first applied in the seventies of last century, the sand was deposited on the beach of endangered sections of the coast. It was left to the wind to blow the sand inward to restore and strengthen an eroded foredune ridge. It is important that this sand is similar to that of the foredune. Gravel which is often present in the glacial bottom sediments of the North Sea leaves a lag deposit on the beach, which is unacceptable for holiday-makers and upsets the ecology of beach and foredune. The next step is foreshore nourishment which is the cheapest and most natural variant. This means that sand dredged from the bottom of the North Sea is deposited on the foreshore where the surf will pick it up and bring it to the beach (Fig. 17.5). From there the wind will blow it to the foredune ridge. A problem with sand nourishment is finding sufficient suitable sand on the bottom of the North Sea. Even where the sand meets the requirements, it may not be available due to ecological restrictions and the dense network of cables, conducts and pipes for communication and transport of various forms of energy.

Fig. 17.5 Foreshore nourishment. Sand dredges from the bottom of the North Sea beyond the 20 m depth contour is deposited on the foreshore slope

17.3.5 *Fifth Strategy: Gaining New Land in the North Sea, the Seaward Option*

A major effort to develop the bottom of the North Sea was made by the city of Rotterdam to acquire much needed land to expand their harbor facilities. Google Earth (52°00′N, 4°00′E) clearly shows that the resulting protuberance in the smooth line of the North Sea coast defies all geomorphological rules of coastal development. Ecological and geomorphological damage to the adjoining parts of the coast is appreciable. This is compensated by allotting space for dune formation and nature building between the new harbors.

Sustainable building in the North Sea requires building with nature. It means making use of materials, forces and interactions present in nature. Attention should be given to the type and shape of the coast under consideration, the present eco-geomorphology and the historic development of the coast. The use of solid seawall elements should be minimized. These principles are realized in the Plan Waterman, which involves a progradation of the coastline by massive deposition of sand between the Hague and the Nieuwe Waterweg, the gate of the harbor of Rotterdam (Fig. 17.6). The new land will have a multifunctional purpose: housing and industry, overflow of the greenhouse horticulture of the adjacent mainland, recreational facilities and new nature.

The design of the Plan Waterman is based on geomorphologic rules. The concave shape of the new coastline minimizes coastal erosion. It replaces a problematic stretch of coastline where continuous erosion necessitated the construction of nearly a hundred groyns. The new coastline merges harmoniously with the stable coastline north of The Hague where no groynes were ever needed. A subtle touch: the ancient beach ridges which are such a characteristic feature of the coast of this part of the Netherlands and had disappeared in the sea south of The Hague, now find their continuation in the morphology of the new land.

Fig. 17.6 An artificial seaward shift of the coastline south of The Hague

Building new islands is another option. The oil industry has constructed a number of drilling platforms in the North Sea. Some of these islands have served other purposes, such as housing commercial radio transmitters. Planning islands in the North Sea for other applications has become a hype for a number of reasons. With a population density of 450 inhabitants/km^2, space for new developments within the national frontiers is considered insufficient. There is also a fear that large parts of the Netherlands will eventually be submerged due to worldwide climate warming and accelerated sea-level rise. Then there is the urge to show the world what the Dutch can do when it comes to creating new land. A plan that comes up now and again is to transfer the international airport of Schiphol to an artificial island in the North Sea. Another plan is to make islands for offshore wind farming. Geographical Information Systems (GIS) are used to find the best locations for wind parks or wind farms in environmental sensitive or congested coastal areas. Most of these plans are ill-considered and may inflict serious damage to the coastal system if executed.

17.4 The Future

The prestigious Deltaworks, which were triggered by the great flood of 1953, connected the islands in the southwestern part of the country with dams and storm surge barriers. This project accomplished what nature had been unable to do: closing the coastline across the estuaries. But the sea is unrelenting. Based on climate change scenarios, the Royal Netherlands Meteorological Institute (KNMI) predicts a sea-level rise by 2100 AD ranging from 35 to 85 cm. Rijkswaterstaat, the government department in charge of the safety of the coast, takes into account a rise of 60 cm until the end of this century. This is far more than the present sea-level rise of 15–20 cm/century. It has been calculated that we can cope with a sea-level rise of less than 1 m/century with the present techniques. A solution for protecting the country against the ever-rising sea level is to continue raising the dikes. The dikes must satisfy delta safety standard, which means that they must be able to resist a storm surge with a recurrence interval of once in 1250 years to once in 10,000 years depending on the degree of exposure to storm surges. Raising the dikes makes them more vulnerable. It increases the danger that the heavy load of the dike causes the weak subsoil to extrude. Other solutions have been discussed in the previous paragraphs, such as improving the discharge of the rivers, sand nourishment, and building new land in the North Sea.

Recently an old idea has emerged to prevent these hazards: make superdikes several hundreds of meters wide. The loss of land can be compensated by using the dikes for other purposes, such as recreation, horticulture and building houses. Another option, covering all of the western part of the Netherlands with 6 m of imported sand is technically and economically not feasible, and raises the question where all the sand must come from. Another scenario, resettling the population to the higher grounds in the eastern part of the Netherlands, is not likely to receive universal support.

In 2007 a new Delta Commission was installed, to advise the Dutch government on the steps that have to be taken to meet the problems of the predicted sea-level rise. The members were recruited mainly from politicians, representatives of the media, managers of large enterprises and contractors. Nation-wide, meetings were organized to hear the opinion of everybody who wanted to have a say in the matter and in the summer of 2008 the commission issued its advice. Their Delta programme is based on the premise of the worst case scenario: a sea-level rise of 65–130 cm by 2100 AD and even 2–4 m in 2200 AD, far exceeding the most pessimistic forecasts of KNMI and Rijkswaterstaat. Although there is as yet little proof of accelerating sea-level rise, the public is advised to expect a disastrous future if it keeps on feeling safe with the current measures being taken by the water boards and Rijkswaterstaat. The Delta commission has opted mainly for increasing the measures already taken. The estimated amount for sand nourishment is 85 million m^3/a, which will cause a 2 km seaward shift of the coast line by the year 2100. The Delta programme will cost 1.2–1.8 € annually until 2100 AD. It will contribute to the sustainability of these measures that generally the geomorphologically most acceptable solutions are chosen.

References

Dietz T, Hoekstra P, Thissen F (eds) (2004) The Netherlands and the North Sea: Dutch Geography 2000–2004. Nederlandse Geografische Studies 325. Koninklijk Nederlands Aardrijkskundig Genootschap/IGU Section The Netherlands, 156 pp

Forbes RJ (1993) Studies in ancient technology, irrigation and drainage. Brill, Leiden, pp 49–64

Jungerius PD, Verheggen AJT, Wiggers AJ (1981) The development of blowouts in 'De Blink', a coastal dune area near Noordwijkerhout, The Netherlands. Earth Surface Processes and Landforms 6:375–396

Oost AP (1995) Dynamics and sedimentary development of the Dutch Wadden Sea with emphasis on the Frisian inlet; a study of the barrier islands, ebb-tidal deltas, inlets and drainage basins. Utrecht University, Utrecht, 455 pp

Pliny the Elder (77) The natural history, Book 16 (Edition and translation Bostock J, Riley HT) Britannica Online Encyclopedia

Poeteray FA, Riezebos PA, Slotboom R (1984) Rates of subatlantic surface lowering calculated from mardel-trapped material (Gutland, Luxembourg). Zeitschrift für Geomorphologie Neue Folge 28:467–481

Van Baars S (2007) The causes and mechanisms of historical dike failures in the Netherlands. Report Delft University of Technology, 6 pp; http://www.geo.citg.tudelft.nl/vanbaars/default.htm (accessed 20 January 2009)

Van den Ancker JAM (2005) The Stelling van Amsterdam als aardkundig militair complex [The water defence line of Amsterdam, a military geographic complex]. Provincie Noord-Holland, Haarlem, 40 pp

Van der Meulen MJ, Van der Spek AJF, De Lange G, et al. (2007) Regional sediment deficits in the Dutch lowlands: implications for long-term land-use options. Journal of Soils and Sediments 7:9–16

Van der Putten WH (1989) Establishment, growth and degradation of Ammophila arenaria in coastal sand dunes. PhD thesis Agricultural University, Wageningen, The Netherlands, 154 pp

Van de Ven, GP (ed) (2004) Man-made lowlands. History of water management and reclamation in the Netherlands. Uitgeverij Matrijs, Utrecht, 432 pp

Van Koningsveld M, Mulder JPM, Stive MJF, et al. (2008) Living with sea-level rise and climate change: a case study of the Netherlands. Journal of Coastal Research 24:367–379

Visser J, Misdorp R (eds) (1998) Coastal dynamic lowlands – the role of water in the development of The Netherlands: past, present, future. Journal of Coastal Conservation 4:105–168

Wong T, Batjes DAJ, De Jager (eds) (2007) Geology of the Netherlands. Royal Netherlands Academy of Arts and Sciences, Amsterdam, 354 pp

Wikipedia, http://nl.wikipedia.org/wiki/Stormvloed (accessed 3 December 2008)

Chapter 18
Perception of Volcanic Eruptions in Iceland

Thorvaldur Thordarson

Iceland is one of the most active volcanic regions on Earth. On average a volcanic event occurs every 4 or 5 years. Every generation of Icelanders has been exposed to volcanic eruptions and their consequences over the 1140 years of habitation. Therefore, volcanism has been a significant force in shaping Icelandic society, to such an extent that volcanic eruptions are imprinted into the cultural landscape. The social as well as volcanic history is well documented in numerous contemporary accounts, which provide a platform for evaluating social aspects of volcanism. In this short communication the perception of Icelanders to volcanic eruptions through time and its communal significance is assessed and discussed.

18.1 Introduction

The island of Iceland is a hot spot situated in the middle of the North Atlantic. It owes its existence to rifting and volcanism along the divergent plate boundary that transects the country. In Iceland the plate boundary sits on top of a deeper-rooted mantle plume, which has elevated this particular segment of the Mid-Atlantic Ridge above sea level (Fig. 18.1). The volcanic zones that delineate the southwest–northeast trending plate boundary, and to lesser extent the flanking intraplate volcanic belts (Fig. 18.2), have constructed the island through semi-steady volcanism from its emergence ~25 Ma to the present day (Guðmundsson 2000; Saunders et al. 1997). This display of nature has only been observed by man for 11 centuries because Iceland was not settled until the late ninth Century AD (Landnámabók; Thorarinsson and Sæmundsson 1979; Thorsteinsson and Jonsson 1991; Smith 1995).

The settlers did not have to wait long to be introduced to the ferocity of Icelandic volcanism. Within the first 80 years of their arrival at least six eruptions had taken place (Thordarson and Larsen 2007). This eruption sequence included the Great 934–940 Eldgjá flood lava eruption, which produced ~20 km^3 of lava and tephra and represents the largest happening of its type and one of the greatest volcanic pollutant-event on Earth over the last several thousand years (Larsen 2000; Thordarson et al. 2001). Since then many eruptions brought desolation and hardship, but some produced positive changes that benefited society in the long term (Finnsson 1796; Guðmundsson et al. 2008). A prime example of a positive change is the lava flow produced by the 1973 Eldfell eruption on Heimaey (Vestmannaeyjar archipelago), which now provides a perfect natural barrier for the towns harbor against the formidable North Atlantic swells.

Consequently, volcanism has been a strong force in shaping Icelandic society. The relatively short and well documented history of Iceland provides the ideal platform for assessing societal perception towards volcanic eruptions and changes therein with time. Nonetheless, this particular perspective as revealed by the historical archives has not yet been subjected to detailed and systematic examination. The principal purpose of this short chapter is to introduce the concept in hope that it will provoke further interest in the subject. I begin with a brief overview of volcanism in history in order to set the scene. I then describe and discuss the perception towards volcanic eruptions as revealed by written records.

T. Thordarson (✉)
Earth and Planetary Sciences, School of GeoSciences
University of Edinburgh, Grant Institute
The Kings Buildings, West Mains Road
EH9 3JW, Edinburgh, UK
e-mail: thor.thordarson@ed.ac.uk

See Plate 17 in the Color Plate Section; also available at: extras.springer.com

I. P. Martini, W. Chesworth (eds.), *Landscapes and Societies*,
DOI 10.1007/978-90-481-9413-1_18, © Springer Science+Business Media B.V. 2010

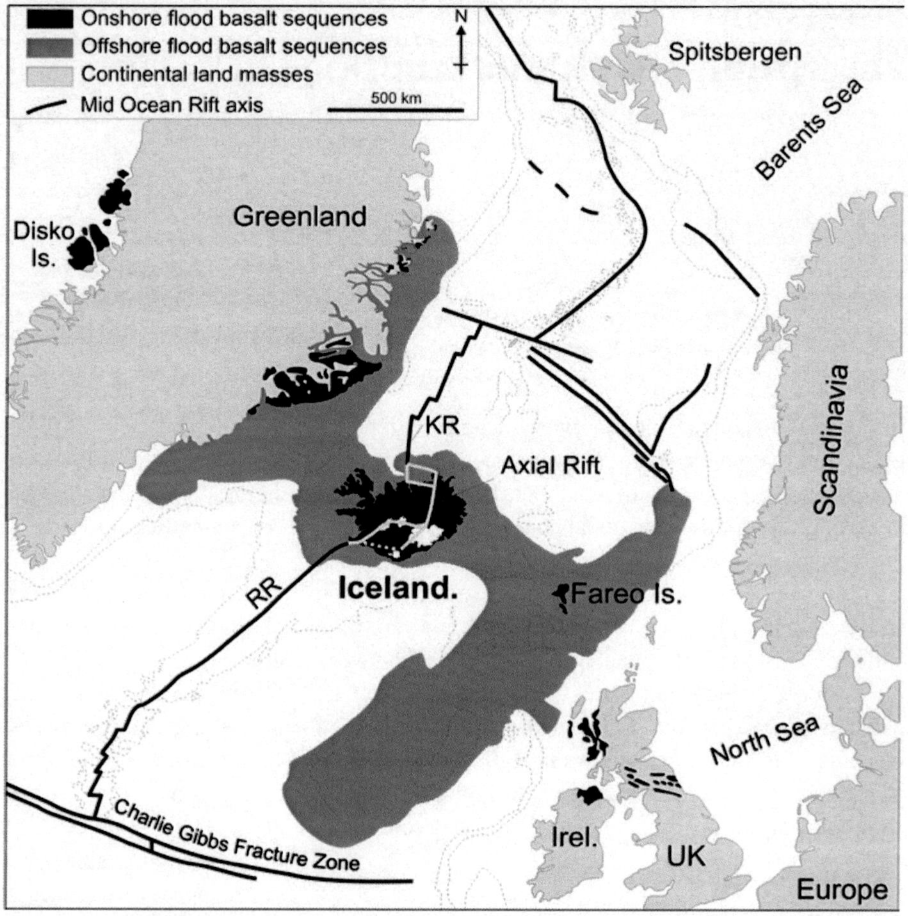

Fig. 18.1 Iceland is a basalt plateau in the middle of the North Atlantic, rising more than 3000 m above its surroundings and situated at the junction between the Reykjanes (RR) and Kolbeinsey (KR) Ridge segments. The construction of the Iceland Basalt Plateau is the product of an interaction that begun about 24 million years ago between the spreading plate boundary and a deeper rooted mantle plume. The geological architecture of Iceland is a representation of this interaction as manifested by the elevation of the plateau and the rather unusual configuration of the axial rift (volcanic zones) across Iceland. Also shown: the axis of the Mid-Atlantic Ridge (heavy solid line), the North Atlantic basalt plateau (black) and their submarine equivalents (dark gray). (After Thordarson and Larsen 2007)

18.2 Volcanism in Historical Time

The volcanic zones in Iceland, which collectively cover about 30,000 km^3, are divided into volcanic systems, each characterized by conspicuous volcano-tectonic architecture that features a fissure (dyke) swarm or a central volcano or both (Fig. 18.2). A total of thirty volcanic systems are currently considered active and eruptions have occurred on 16 of them in historical time. The most active systems, Grímsvötn, Bárðarbunga-Veiðivötn, Katla, and Hekla-Vatnafjöll (Fig. 18.2), are all situated within the East Volcanic Zone, which has been the most prolific producer among the neo-volcanic zones in historical time and responsible for >80% of the eruptions and ~60% of the erupted magma volume (Thordarson and Larsen 2007).

Despite the dominance of mafic volcanism in Iceland, the style and type of volcanic activity is very diverse and not surpassed by any other volcanic region on Earth. Emission of basalt magma has produced effusive eruptions of Hawaiian and flood lava magnitudes. Expulsion of basaltic, intermediate and silicic magmas have produced explosive phreatomagmatic and magmatic eruptions spanning almost the entire style range; from Surtseyan to phreato-Plinian in the case of wet eruptions and Strombolian to Plinian in terms of dry eruptions. Consequently, the eruptions in Iceland can be purely explosive or effusive or a combination of both (Thordarson and Höskuldsson 2008).

Iceland is also one of the most active terrestrial volcanic regions on Earth and features an eruption every four–five years on average. In total, 205 eruptions have been identified in historical time (Fig. 18.3) and >65% are recorded in written accounts. Of the 205 events, 157 are explosive (including 33 events where eyewitness accounts are the only evidence and explosive eruption is inferred), 14 are effusive and 21 are mixed eruptions. The remaining 13 events are classified as 'Fires', which include two or more eruptions that define an episode of volcanic activity that lasts for months to years.

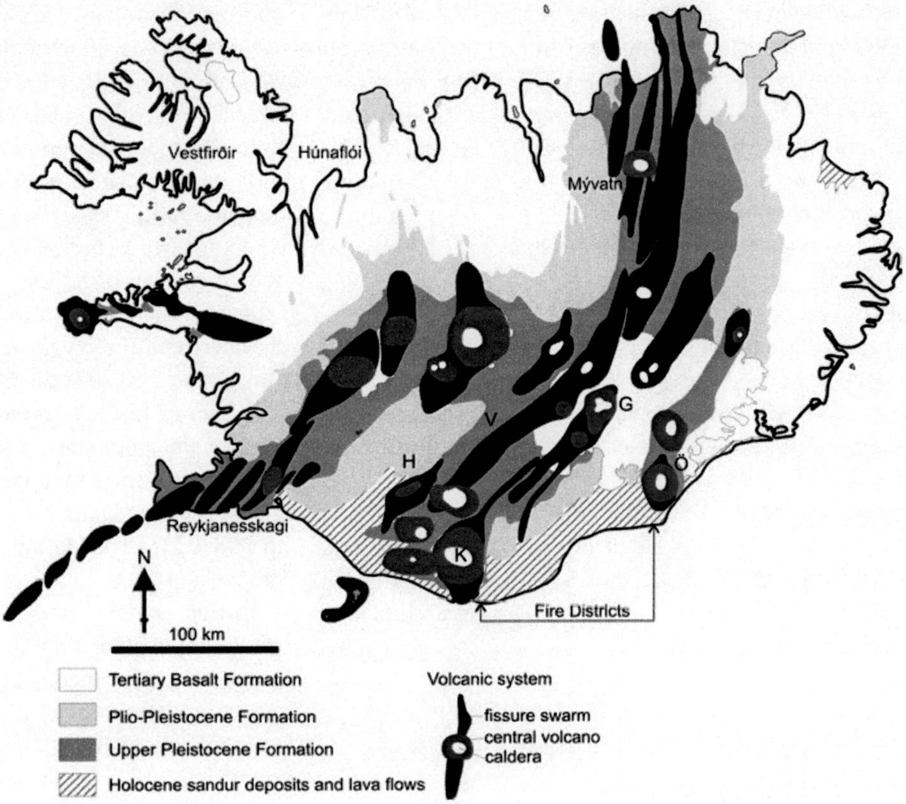

Fig. 18.2 Distribution of active volcanic systems in the Icelandic *rift system* and intraplate volcanic belts. Volcanic systems, volcanoes and place names mentioned in the text are also indicated. Abbreviations are: G, Grímsvötn; V, Bárðarbunga-Veiðivötn; K, Katla and H, Hekla-Vatnafjöll. (After Thordarson and Larsen 2007)

However, this eruption list is incomplete because in the period 870–1200 AD the total number of small subglacial explosive eruptions from volcanoes within Vatnajökull are unknown (Fig. 18.3). The magma output by these eruptions in historical time amounts to ≥87 km^3 when calculated as lava or about 2.2×10^{11} tons. This is equivalent to loading about 200 million tons onto the land annually. The volume of magma produced by

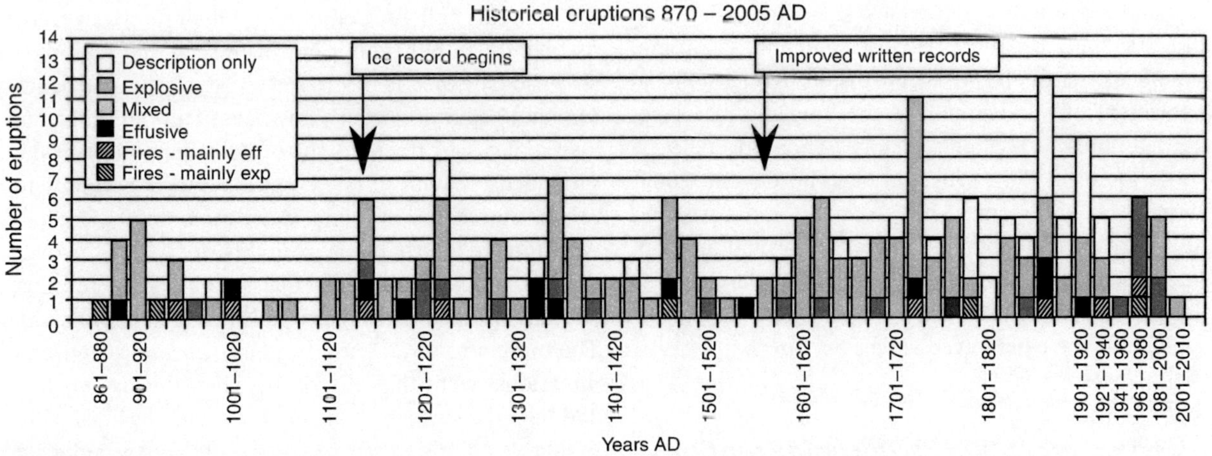

Fig. 18.3 Frequency of volcanic eruptions in Iceland since the Norse settlement around 870, where the data is binned per 20 years. As the exact timing of one pre-thirteenth century eruption is not known the total number of events used in the plot are 204, where 191 are counted as individual eruptions and 13 as "Fires", which designates year to decade long volcano-tectonic episodes with 2–11 eruptions. The category "description only" are volcanic events described in historical accounts but have not been verified by identification of their volcanic products; explosive = tephra-, mixed = tephra- and lava-, effusive = lava-producing events; ice-record = tephra layers preserved in glacier ice. (After Thordarson and Larsen 2007)

individual events spans the range from 1 m^3 to ~20 km^3 DRE and in historical time four flood lava (large-scale effusive) events, such as Eldgjá 934–940 AD and Laki 1783–1784 AD, account for >50% of the total erupted volume (Thordarson and Larsen 2007). In addition the volcanism has produced numerous explosive eruptions, including five explosive events of magnitudes greater than VEI 5 (Guðmundsson et al. 2008). Off those, the silicic Hekla 1104 AD and Öræfajökull 1362 AD events as well as the 1477 AD basaltic Veiðivötn eruption, stand out in terms of their magnitude and coverage (Thorarinsson 1958, 1967, 1970; Larsen 1984). More recent, yet smaller events, worth mentioning because of their impact on local communities, are the 1724–1729 Mývatn and 1975–1984 Krafla Fires in North Iceland (Sæmundsson 1729; Sæmundsson 1991), the 1947–1948 Hekla eruption in South Iceland (Thorarinsson 1976) and the 1973 Eldfell eruption on the island of Heimaey, Vestmannaeyjar archipelago (Thorarinsson et al. 1973).

18.3 Imprint of Volcanism on Culture and Landscape

We are fortunate that the volcanic history in Iceland since settlement is comparatively well documented both in the historical and the geological records. The principal written sources documenting volcanic eruptions are chronicles of various ages, including historical as well as church registries and annals, official reports on particular volcanic events along with an assortment of publications on the volcanic history at various times in Iceland's history (Thoroddsen 1925; Thorarinsson 1958, 1967, 1974; Thordarson 2003; Thordarson and Larsen 2007; Larsen and Eiríksson 2008). Besides containing valuable information about the volcanic events and their effects, these descriptions also provide a glimpse of how volcanic and other natural hazards were perceived at various times.

18.3.1 Perception of Volcanic Eruptions as Revealed by Historical Accounts

Landnámabók (Book of Settlement, compiled in the twelfth Century) is the oldest document containing information on historical eruptions, reporting events during the time of settlement (up to ~950 AD). There we find a brief account of the very large 934–940 AD Eldgjá event. Considering the fact that the eruption caused major environmental and social disturbance well beyond the shores of Iceland (Stothers 1998; Larsen 2000; Thordarson et al. 2001), the description is remarkably to the point and strictly a matter-of-fact documentation of its local consequences: "The settler Molda-Gnúpur took possession of land in Álftaver district between the rivers Kúðafjót and Eyjará. At that time a large lake was there and good swan hunting. He sold part of his settlement to many newcomers. The area became populated before it was overrun by jarðeldur (an earth fire), then they fled west to Höfðabrekka and set up a camp at Tjaldavellir" (Landnámabók pp. 328–331; Translation in Pálsson and Edwards 1972, Chap. 86).

In this context, it is informative to compare this description of the Eldgjá eruption to that presented in a near contemporaneous European compilation, '*Liber miraculum*', written in the years 1178–1180 by Herbert the monk, chaplain in the monastery of Clairvaux, later archbishop of Torres in Sardinia (Thorarinsson 1967): "In the North there is known to be a great island called Iceland (Hyslandia), belonging to the Christian faith. There is found a certain steep and enormous mountain, taking up a large part of that country, beneath and in which the inhabitants believe the greatest inferno exists .... Nor do I think it proper to pass over in silence the fact that this hell-fire sometimes, albeit rarely, exceeds its bounds. In our time it has been seen that it erupted so furiously that it destroyed most of the surrounding land. It burnt not only towns and all buildings but also plants and trees down to the roots and even the ground itself with its very bones. And, marvelous to relate, mountains of stone, even mountains of metal, were melted utterly like wax in this fire—they ran over the earth and covered it, so that the valleys were filled with the mire and hill-land was levelled with the plain. The molten rocks, having run out over the earth in all directions, were thus dispersed when the fire abated, and then the surface of the earth appeared as if it were of marble and like a street paved with stones, and what was previously land inhabitable and fertile was reduced to desert. When in its insatiable greed this most cruel fire had destroyed this land and all that was in it, then it yielded still more terrible marvel for it invaded the sea on the shore. And when it reached the deep sea, it began to burn and consumes the water with unheard-of fury right down to the depths. In addition to this, it carried with it in its course huge fragments of mountains and hills, overthrown by other devouring fire, so that as the waters withdrew, their place was filled with land, and mountains were carried into the heart of the sea. And when they had completely filled the sea both far and wide and made the depths of ocean level with the shore, the sea became dry land, so that where there had previously been water, there was now firm ground for twelve miles, and perhaps it is there still. Moreover in this conflagration was destroyed a famous and populous city, and an excellent port, which it had on an arm of the sea, was also destroyed, as it is established, when the sea dried up .... What is more marvelous than these marvels, or what can be conceived more terrible? Who is now so perverse and incredulous that he will not believe that eternal fire exists to make souls suffer, when with his own eyes he sees that fire of which we now speak burn in such fearful fashion not only the ground and marbly rocks, but also the invincible water which is accustomed to extinguish other fire with such ease? But those who will not believe in the punishments of eternal fire prepared for the devil and his angels, nor hear of them, they will afterwards be themselves cast among those torments which they disdain to avoid while is yet time" (Thorarinsson 1967, pp. 26–28).

Thorarinsson considered this to be a collective account of several eruptions, including the first two historical eruptions at the Hekla volcano which already was infamous outside of Iceland the time of Herbert's writing. However, he did not consider Eldgjá to be one of these events. Larsen (1979) was the first to link the description in this account to the Eldgjá eruption. The extravagance of the account is consuming, but as pointed out by Thorarinsson (1967), if the irrationalities and the exaggerations are removed the kernel that is left contains such realistic details that we have to acknowledge it is describing Icelandic eruptions. The extract from Herbert's account presented above appears to be describing the damage caused by the tephra fall, lava and jökulhlaups from the great 934–940 Eldgjá eruption (Larsen 1979, 2000). The difference between Herbert's description and the corresponding Icelandic account is astounding; Herbert's account is overpowered by the religious overtone and insinuations, whereas the Icelandic one is matter of fact.

Another and perhaps more insightful account on the perception towards volcanic eruptions in the early days of settlement is the one mentioned in association with the debate on Christianization of Iceland at the Alþing (parliament) in the year 1000 AD. The debate was nearing its end and apparently without consensus being reached between the two factions, the pagan and Christian followers. Then news was brought of an eruption in the Ölfus district, a fissure eruption that formed Kristintökuhraun (The Christianity lava) in southwest Iceland (Jónsson 1983). It was apparent that the lava would overrun the farm of the pagan priest Þóroddur and the pagan followers spoke: "We are not surprised that the heathen gods are enraged at such a debate" (Kristni Saga, p. 33). Then the heathen priest Snorri replied: "At what were the gods enraged when the lava which we are now standing on formed ..." (Kristni Saga, p. 33). Apparently, this reply was the turning point in the debate and ended the protest by the pagan followers.

The Icelandic Annals are the prime source on eruptions in the period 1100–1600 AD. The descriptions are brief and often only state that an eruption occurred. For example, the 1104 AD eruption at the Hekla volcano, the second largest silicic explosive eruption in historical time in Iceland, is simply documented as "The first coming up of fire in Mount Hekla" (Thorarinsson 1967, p. 24). As time passed, the descriptions of the eruptions and their effects became more elaborate as demonstrated by the example below from the Hekla eruption in 1341 AD: "Coming up of fire in Mount Hekla with great sandfall and with such great crashes when rocks struck together in the fire that they were heard over nearly the whole country. It was also so dark while the sandfall was at its height that it was not light enough to read in those churches situated nearest the source of the fire. Great famine. Great loss of stock, both sheep and cattle, so that between the flitting days to the feast of St. Peter (that is, early May to 1 August) Skálholt alone lost 80 cattle"(Thorarinsson 1967, p. 53).

The report on the 1625 Katla eruption written by Þorsteinn Magnússon (Magnússon 1625) is the first day-by-day documentation of a volcanic eruption in Iceland and marks a change in the manner by which volcanic events were recorded (Thordarson 1990). Since that time, most of the eruptions that were observed directly have been documented in this mode, in reports that are commonly referred to as 'Eldrit' (Books of Fire). All of the post-1625 reports are methodical documentation of the events focused on describing the nature of the activity as well as the damage inflicted by the eruption onto the land and affected communities.

They contain remarkably realistic and detailed descriptions of the activity, such as explosive events, dispersion of eruption plumes, emplacement modes of lava flows and frequency and intensity of the eruption-associated earthquakes. The authors of these reports were not only keen observers of nature but made ample efforts to describe the events objectively (Thordarson 2003).

18.3.2 Impact of Historical Eruptions

As evident from the outline given above, no generation of Icelanders has escaped the experience of volcanic eruption and their occurrence inevitably became part of the cultural landscape. Although the experience of the phenomenon is a common theme it is unlikely that the perception obtained is the same between and even within generations. The perspective is variable and primarily a function of the volcanology (nature, frequency and magnitude of events) along with the geography (the site location relative to the volcanic source). Several inhabited regions in Iceland have felt the effects of volcanic eruptions more than others because of their geographic location and the eruption frequency at the volcanic system of prime influence. The regions most affected are those situated closest to the most active volcanic systems as well as those positioned downwind from the most profuse producers of explosive eruptions. The perspective is even more skewed when event sizes and magnitudes are taken into account, because almost all large-scale historical eruptions have been sourced from volcanoes or vent systems situated in southcentral Iceland (Fig. 18.2). Another variable to consider is eruption style. The perception obtained from exposure to an effusive eruption will differ significantly from that of a purely explosive event.

The inhabited regions north of the Hekla volcano from Húnaflói to Mývatn (Fig. 18.2) have repeatedly been exposed to the tephra fall from its explosive eruptions (8–9 out of 18 events; Larsen et al. 1999). These regions only experienced fallout of ash and its impact on vegetation and grazing animals, due to their distance relative to the volcano (Thorarinsson 1967, 1979). Other regions, such as the Reykjanesskagi and Mývatn area, have experienced frequent recurrence of relatively small and mostly effusive eruptions, which on certain occasions have become threat to communities (Sæmundsson 1729). There are also regions in Iceland that have had minimal exposure to eruptions and their effects. In this regards the NW Peninsula (Vestfirðir; Fig. 18.2) stands out, having only been affected by distal ash fall from a few events (Thordarson et al. 2005).

The Fire Districts in south-central Iceland (Figs. 18.2, 18.4) is the region that has been most often exposed to volcanic eruptions and their consequences. The district is guarded by the subglacial Katla volcano in the west, and hidden under Vatnajökull glacier in the northwest is Grímsvötn, the most active volcano on Iceland (Fig. 18.4). When these volcanoes erupt they do so explosively with accompanying tephra fall and jökulhlaups (Thorarinsson 1974; Larsen 2000). Consequently, the region has on average been affected by tephra fall two times a century and endured several jökulhlaups across the adjacent outwash plain (Larsen 2000). The Fire Districts were also the center stage for the two largest eruptions since the settlement of Iceland: the basaltic flood lava eruptions of Eldgjá in 934–940 and Laki in 1783–1784 (Fig. 18.4), which collectively cover about 1250 km^2 and caused significant modification to the landscape in the region, including loss of valuable farm and pasture land (Steingrímsson 1788; Guðbergsson and Theodórsson 1984; Larsen 2000; Thordarson and Self 1993). The emplacement the Laki lava flow field in 1783–1784 demonstrates well the extent and magnitude of these changes.

The lava that issued from the Laki fissures between 8 June and 29 July formed two branches; one that advanced down the upper reaches of the Skaftá River gorge and another that flowed through the valley of Varmárdalur and then across the highland pasture to Mt. Leiðólfsfell. These two branches are joined just west of Mt Leiðólfsfell and emerged as such onto the lowlands in front of the gorge for the next 45 days and formed the western sector of the Laki lava flow field (Fig. 18.5). By the time the lava of the western sector stopped, it had covered ~350 km^2, obliterated the glacial river Skaftá and a few of its tributaries. On 3 August the channel of the glacial river Hverfisfljót dried up and on 7 August the lava surged out from the Hverfisfljót River gorge (Fig. 18.4). The lava flowed steadily from the fissures until October 1783, both to the north over the flood plain in front of the outlets of the rivers Skaftá and Hverfisfljót and down the Hverfisfljót River gorge adding 250 km^2 to the Laki lava flow field. Thereafter the lateral growth of the

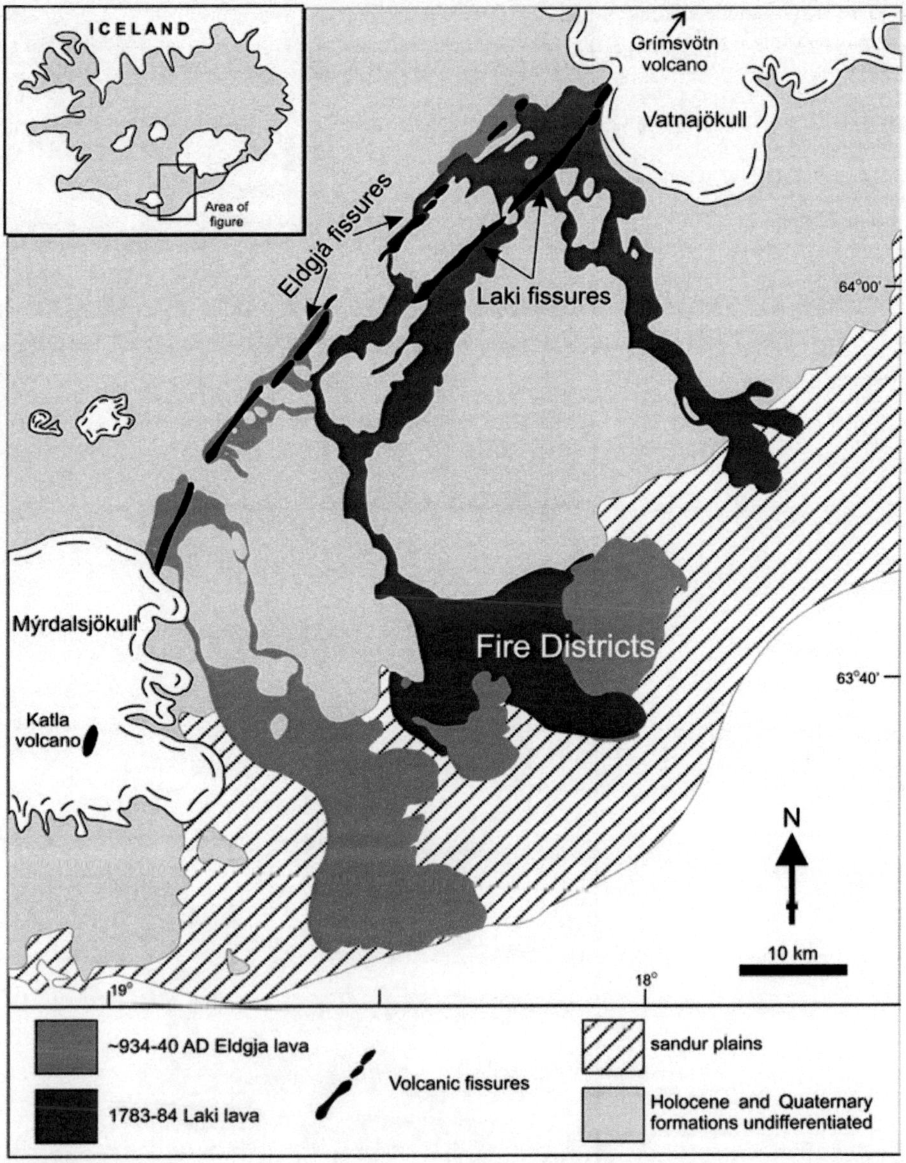

Fig. 18.4 Map of the Fire Districts showing its position to the volcanoes Katla and Grímsvötn (just off the map) and the fissures and lava flow fields of the 934–40 AD Eldgjá and 1783–1784 AD Laki flood lava eruptions. (After Thordarson et al. 2001) (Plate 18a)

flow field came to halt, however the Laki fissures continued to produce lava until 7 February 1984, when the fires at the fissures finally extinguished (Thordarson and Self 1993).

By the end of the Laki eruption, the flow field had inundated about 600 km² of the land, or approximately 30% of the total area in the districts directly affected by the lava, and raised the land surface by about 25 m on average (Fig. 18.4). The lava also completely fills the gorges of the rivers Skaftá and Hverfisfljót raising their baseline by several tens of meters and possibly over hundred meters in places (Thordarson and Self 1993; Thordarson et al. 2003). The lava eradicated everything in its path, including the rivers Skaftá and Hverfisfljót, which upon return established new channels within and along the margins of the new lava field (Fig. 18.4), resulting in significant and ongoing environmental changes yet to be quantified. It also destroyed 19 farms along with their cultivated fields (Fig. 18.4). By 1785 the population in the Fire districts had dropped to approximately 37% of what it was prior to the eruption. About 22% had died and the remainder (41%) had fled the region.

In about 15 years, most of those who fled during the eruption had returned, the population had risen to >70% of what it was in 1783 and fourteen of the farms (about

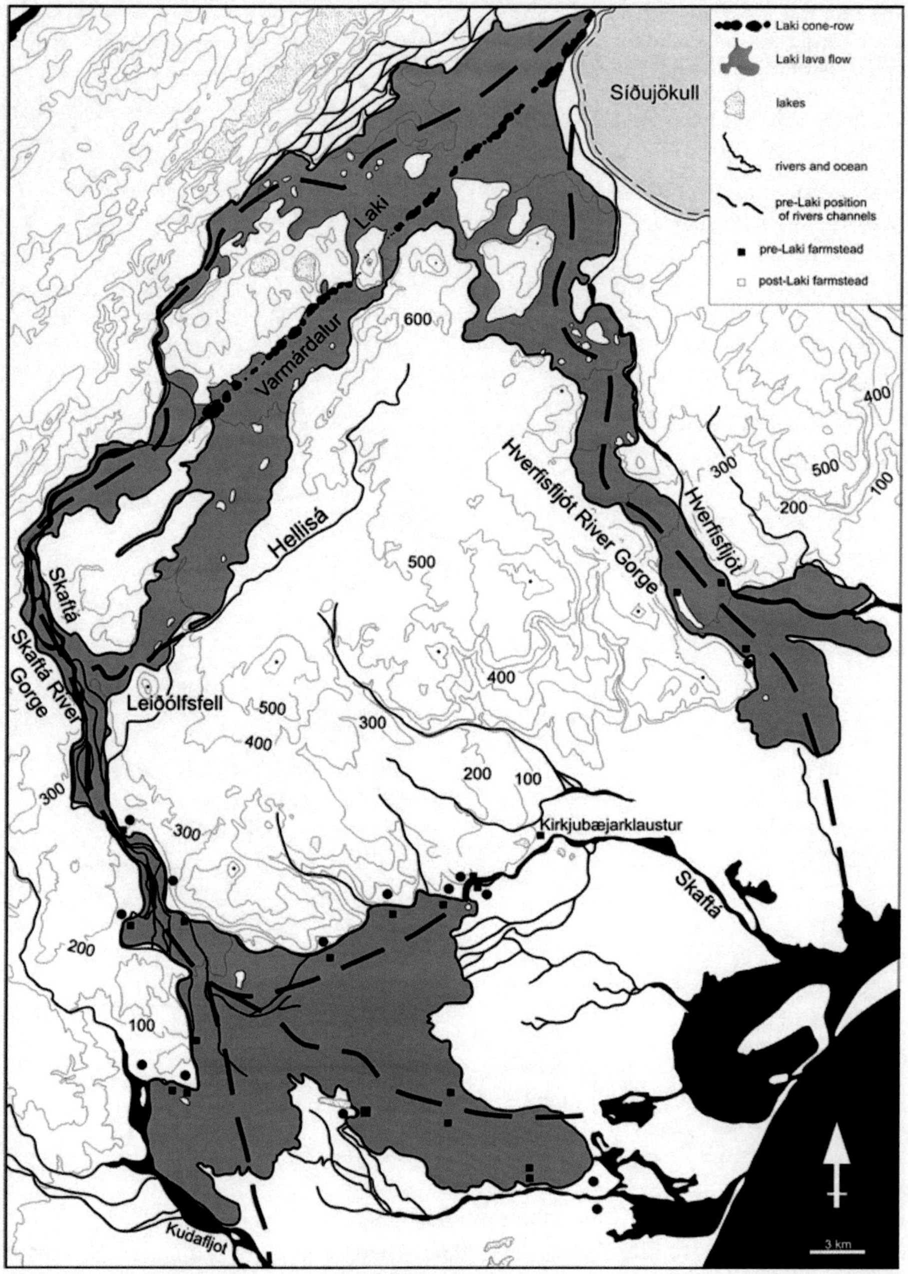

Fig. 18.5 Map of the Laki region, showing locations of vents, lava flows as well as place names and rivers mentioned in the text. The heavy broken lines show approximate position of main river channels prior to the Laki eruption. Solid squares show the pre-eruption position of farms destroyed by the lava; solid circles show farms that were relocated and rebuilt. (After Thordarson and Self 1993) (Plate 18a)

75%) had been rebuilt after relocation (Guðbergsson and Theodórsson 1984; Gunnlaugsson 1984).

18.4 Discussion

The examples presented above in Sect. 18.3.1 on the view of Icelanders towards volcanic eruptions are a small extract from the vast archives that contain information on historical eruptions in Iceland. These archives have underpinned studies on the volcanic history of Iceland since time of settlement (Thoroddsen 1925; Thorarinsson 1958, 1965, 1967, 1970, 1974, 1979, 1981, 1984; Thorarinsson and Sæmundsson 1979; Larsen 1979, 1982, 2000, 2002; Larsen et al. 1998; Thordarson and Self 1993; Thordarson et al. 2003). The archives contain information on more than 130 volcanic events or >65% of the historical eruptions known to date (Thordarson and Larsen 2007). The records are usually written and compiled by people of some societal stature and authority (usually government or church representatives). Although the records vary in quality and commonly are biased towards the authoritative viewpoint, they do provide sensible indication of the perception of Icelanders to volcanic eruptions and other natural hazards.

18.4.1 Perception of Volcanic Eruptions: Origin

Historical records give the impression that Icelanders looked upon volcanic eruptions as a natural phenomenon and had acquired basic understanding of such events very early on, as can be inferred by Snorri's reply during the parliamentary debate on Christianization of Iceland in the year 1000 AD. Irrespective of the accuracy of this account, the simple fact that it links older lava flows with eruptions in the past shows that a general understanding of such events existed among the inhabitants at the time when 'Kristni Saga' was written in the mid thirteenth Century. The account on the Christianization debate also reveals other aspects of the cultural landscape in the tenth–thirteenth Century: the authoritarian figures in society appear to hold and promote pragmatic views towards natural events, and a portion of the population was disposed to connect such events to the supernatural. Although these two opposing perspectives have co-existed throughout Iceland's history, the pragmatic perception recurs consistently and prevails (Thorarinsson 1967). Even though the matter-of-fact perception revealed by written sources may be biased towards the official or authoritarian view, the connotation is that it is representative of the view held by the majority in Icelandic society.

So, why did this pragmatic view develop and persist in Iceland? It is unlikely that the collective intellect of Icelandic society was any different from the rest of Europe, so it is improbable that this development can be attributed to enhanced cultural maturity. However, although the original settlers in Iceland were unfamiliar with volcanic eruptions, they were accustomed to the wrath of Nature in the form of extreme weather, severe winters and harsh seas. It is possible that these experiences, along with a strong stance of the settler's on the importance of independence, provided the appropriate cultural backdrop for the development of pragmatic views towards volcanic eruptions as well as other hazardous natural events. Nonetheless, it is more likely that this view originated and endured out of social and cultural necessity. From the very start Iceland was a marginal society from geographical and climatic point of view and was largely established by individuals faced with no point of return (Thorsteinsson and Jonsson 1991). In other words, they had little option but to make the settlement work. Furthermore, no part of Iceland was exempt from the effects of volcanic eruptions, although the nature and magnitude of the impacts was variable between regions. Also, the impact of these eruptions was non-discriminatory. It affected people from all levels of the society and brought hardship and devastation to estates of prominence (such as Episcopal sees, churches, and manors) as well as farms and crofts. Attributing volcanic eruptions and their effects to punishment by supernatural forces was self-defeating; it would undermine integrity of the civil and spiritual authority and in doing so would destabilize the establishment.

18.4.2 Perception of Volcanic Eruptions: Communal Importance

The foundation and perseverance of pragmatic views towards volcanic eruptions is not purely of academic interest because at the communal level such a perception does contain some elements of practical significance. Firstly, it promotes a new understanding of the phenomenon among the populace, which in turn provides the rational means of responding fittingly and positively to adverse situations. The historical records contain information that appears to substantiate such a conclusion, although it remains to be demonstrated by systematic analysis. Throughout their history, Icelanders have shown resilience when faced with the consequences of volcanic eruptions (Thorarinsson 1967, 1979). Although many eruptions, in particular the biggest explosive and flood lava events (Thoroddsen 1925; Thorarinsson 1958, 1970; Larsen 2000; Thordarson et al. 2003), have caused destruction and temporary abandonment of households, no district has yet been deserted completely. The reports indicate that tephra was cleared from cultivated fields in order to restore grazing and hay production. They also recognized the volcanic origin of fluorine poisoning in the grazing livestock and developed the means to manage the problem. The changes and consequences imposed by newly emplaced lava flows prompted improvisation. The response of the inhabitants in the Fire Districts to the emplacement of the 1783–1784 Laki lava flow field is an ample example, which includes relocation of 14 farms as well as construction of tracks across the new flow field in order to have access to pastures completely enclosed by the lava (Pálsson 1794).

18.4.3 Books of Fire

As noted in Sect. 18.3.1, it is interesting that Icelandic reports progressed to become more factual and instructive documentation of the volcanic events, while the European descriptions of Icelandic volcanism strayed to become more exaggerated and inflated on the supernatural front (Saxo 1514; Magnus 1539; Munster 1544; Krantz 1546; Apianus and Frisius 1584; Peerse 1561). Such was the extreme of these exaggerations that in the late sixteenth and early seventeenth centuries, Icelandic scholars, urged on by Guðbrandur Þorláksson Bishop at Hólar in North Iceland, found it necessary to correct these misguided claims by their European colleagues in writing (Einarsson 1590; Jónsson 1593, 1609). At the same time the accounts on volcanic eruptions written in Iceland become markedly more detailed and complete. Why did this change occur at this particular time?

The social and political landscape in Iceland changed significantly in the fifteenth–seventeenth centuries. The establishment of the Kalmar Union (1397–1523), uniting under a single monarch the kingdoms of Denmark, Norway, and Sweden, resulted in Iceland coming under Danish rule although it maintained autonomy. The abolishment of Catholicism and institutionalization of Lutheranism in 1550 caused a further shift of political power and assets from Iceland to Denmark. The transformation from independence to full Danish rule was completed by the introduction of Danish trade monopoly in 1602 and loss of autonomy in 1660 (Karlsson 2000). It is possible that the observed change in style of reporting volcanic events stems from the need to obtain external support in time of crisis. In other words, it did not become important to produce detailed reports of volcanic eruptions until Iceland had become totally dependent on the Danish economical and political system and to ease affairs when dealing with the Danish authorities in wake of hazardous natural events. The most straightforward way to accomplish this was to keep thorough and accurate records of the events and their effects. The course of this transition was undoubtedly easier because a matter-of-fact view of volcanic eruptions was the norm within the community.

18.5 Conclusions

Assessment of historical documents dating back to the twelfth Century demonstrate that from early days of settlement Icelanders viewed volcanic eruptions as an act of nature and generally did not link them to the supernatural. It is suggested that this pragmatic perception developed and prevailed for practical reasons. Accrediting volcanic eruptions and their effects to chastisement by God or other supernatural beings may have been viewed as to undermine the integrity of the authority. Such rationalization would challenge the established social structure because within any one region all individuals, irrespective of their social status, would have been equally vulnerable to the effects from an eruption. By the same token, the more enhanced descriptions of volcanic eruptions that emerged after acceptance of Danish rule and Lutheranism are more likely to have surfaced from need rather than newly gained enlightenment due to increased exposure to European culture. It is also suggested that the matter-of-fact view towards volcanic eruptions, which appears to have filtered through in society from the onset of settlement, upheld a better public understanding of these events and promoted practical solutions in dealing with their consequences. However, these topics have only been dealt with here in a qualitative manner and thus the conclusions should be regarded as provocative proposals, nothing more, put forth as a challenge to encourage more systematic and qualitative analysis on these matters. The historical archives that exist in Iceland provide the ideal platform to conduct such an analysis.

Acknowledgments Thanks to Guðrún Larsen and anonymous reviewer for constructive comments and useful suggestions that helped with improving the final version of this paper. This study was supported by Icelandic Research Fund Grant for Exellence No. 070272011.

References

Apianus P, Frisius G (1584) Cosmographia, siue Descriptio vniuersi Orbis, Petri, Apiani and Gemmæ Frisij, Mathematiorum insignium, iam demum integritati, suæ restituta. Adiecti sunt alij, tum Gemmæ Frisij, tum aliorum Auctorum eius argumenti Tractatus ac libelli varij, quorum seriem versa pagina demonstrate. Printed by Joannes Withagius, Antwerpen, 478 pp

Einarsson O (1590) Qualiscunque description Islandiae. Bókaútgáfa Menningarsjóðs, Reykjavík 1971, 159 pp (Icelandic translation)

Finnsson H (1796) Decimation of the population in Iceland due to famines. Rit þess konunglega íslenska lærdóms-lista félags XIV, Copenhagen 1796, pp 30–226

Guðbergsson GM, Theodórsson, T (1984) Effect of the Laki eruption on the community and population in the counties of Leiðavallahreppur and Kleifahreppur. In: Einarsson T, Guðbergsson GM, Gunnlaugsson GA, et al. (eds) Skaftáreldar 1783–1784. Ritgerðir og Heimildir. Mál og Menning, Reykjavík, pp 99–117

Guðmundsson A (2000) Dynamics of volcanic systems in Iceland: Example of tectonism and volcanism at juxtaposed hot spot and mid-ocean ridge systems. Annual Review of Earth and Planetary Sciences 28:107–140

Guðmundsson MT, Larsen G, Höskuldsson Á, Gylfason GÁ (2008) Volcanic hazards in Iceland. Jökull 58:251–268

Gunnlaugsson GÁ (1984) Population of Vestur-Skaftafellssýsla and the Laki eruption. In: Einarsson T, Guðbergsson GM, Gunnlaugsson GA, et al. (eds), Skaftáreldar 1783–1784; Ritgerðir og Heimildir. Mál og Menning, Reykjavík, 1984, pp 118–128 (in Icelandic)

Jónsson A (1593) Brevis commentarius de Islandica. First printed in Copenhagen 1593; reprinted in Latin by Endurprent, sf., Reykjavík, 1968, 102 pp

Jónsson A (1609) Crymogæan sive Rerum Islandicarum Libri III. Printed by Hamburgi: Typis Henrici Carstens, 1610 (privilege dated 1609), 172 pp

Jónsson J (1983) Volcanic eruptions in historical time on the Reykjanes peninsula, SW Iceland. Náttúrufræðingurinn 52:127–139 (in Icelandic)

Karlsson G (2000) Iceland's 1100 years: the history of a marginal society. Mál og Menning, Reykjavík, 418 pp

Krantz A (1546) Chronica regnorum aquilonarium Daniæ, Svetiæ, Norvagiæ. Published by Johannes Schott, Strassburg, 1548 (privilege dated 1546). Folio 8, 755(11) pp

Kristni Saga, manuscript AM 371 4to, Hauksbók. In Biskupasögur I, part 2. Hið Íslenska Fornritafélag, Reykjavík, 2003, pp 3–48

Landnámabók, manuscript AM 107 fol. Sturlubók, manuscripts AM 371 4to and AM 105 fol, Hauksbók. In Íslenzk fornrit I. Hið Íslenska Fornritafélag, Reykjavík, 1968, pp 1–397

Larsen G (1979) About the age of the Eldgjá Lavas. Náttúrufræðingurinn 49:1–26 (in Icelandic)

Larsen G (1982) Tephrochronology of Jökuldalur and the surrounding areas. In: Eldur er í Norðri, Thorarinsdottir H, Oskarsson OH, et al. (eds). Sögufélag Reykjavíkur, Reykjavík, pp 51–66 (in Icelandic)

Larsen G (1984) Recent volcanic history of the Veidivötn fissure swarm, southern Iceland – an approach to volcanic risk assessment. Journal of Volcanology and Geothermal Research 22:33–58

Larsen G (2000) Holocene eruptions within the Katla volcanic system, south Iceland: Characteristics and environmental impact. Jökull 49:1–28

Larsen G (2002) A brief overview of eruptions from ice-covered and ice-capped volcanic systems in Iceland during the past 11 centuries: frequency, periodicity and implications. In: Smellie JL, Chapman MG (eds) Volcano-ice interaction on Earth and Mars. Geological Society London Special Publication 202: 81–90

Larsen G, Eiríksson J (2008) Holocene tephra archives and tephrochronology in Iceland – a brief overview. Jökull 58:229–250

Larsen G, Dugmore AJ, Newton AJ (1999) Geochemistry of historical-age silicic tephras in Iceland. Holocene 9:463–471

Larsen G, Gudmundsson MT, Björnsson H (1998) Eight centuries of periodic volcanism at the center of the Icelandic hotspot revealed by glacier tephrostratigraphy. Geology 26:943–946

Magnus O (1539) Carta Marina. Venice 1539. A map reprinted by Gamby E, Bogillet, Uppsala, 1964.

Magnússon Þ (1625) Description of the Katla eruption in 1625. Safn til sögu Íslands 4:200–215 (in Icelandic)

Munster S (1544) Cosmographia universalis. 2nd editon printed by Henry Pierre, Basel 1552, 500 pp

Palsson H, Edwards P [translators] (1972) The Book of Settlements – Landnamabok., University of Manitoba Press, Winnipeg, 159 pp

Pálsson S (1794) Sveinn Pálsson's Journals – notebooks and essays 1791–1797. Örn og Örlygur, Reykjavík 1983, 813 pp (in Icelandic)

Peerse G (1561) Van Ysslandt. Hamburg 1561. Reprinted by Seelmann W 1883 in Jahrbuch des Vereins fur niederdeutsche Sprachforschung IX:110–125

Saunders AD, Fitton JG, Kerr AC, et al. (1997) The North Atlantic igneous province. In: Mahoney JJ, Coffin MF (eds) Large igneous provinces: Continental, oceanic, and planetary flood volcanism, Geophysical Monographs 100, American Geophysical Union, Washington DC, pp 45–93

Saxo G (1514) Danorum Regum heroumque Historiae. Printed 15th March 1514 by Jodocus Badius Paris, France

Smith K (1995) Landnam: The settlement of Iceland in archaeological and historical perspective. World Archaeology 26:319–347

Steingrímsson J (1788) A complete description on the Síða volcanic fire. Safn til Sögu Íslands, Copenhagen, 1907–1915, pp 1–57 (in Icelandic)

Stothers RB (1998) Far reach of the tenth century Eldgjá eruption, Iceland. Climatic Change 39:715–726

Sæmundsson J (1729) A new description on the sulphur mountains at Lake Mývatn in Iceland. *Safn til Sögu Íslands*. Copenhagen 1907–1915, IV, pp 389–398 (in Icelandic)

Saemundsson K (1991). Geology of the Krafla volcanic system. *In: Náttúra Mývatns*. Garðarson A, Einarsson Á (eds). Reykjavík, Hið íslenska náttúrufræðifélag, pp 25–95 (in Icelandic)

Thorarinsson S (1958) The Öræfajökull eruption of 1362. Acta Naturalia Islandica II 2:1–100

Thorarinsson S (1965) Submarine eruptions around Iceland, Nátturufræðingurinn 35:49–74 (in Icelandic)

Thorarinsson S (1967) The eruptions of Hekla in historical times. In: Einarsson T, Kjartansson G, Thorarinsson S (eds) The eruption of Hekla 1947–48 I. Societas Scientiarum Islandica, Reykjavík, pp 1–177

Thorarinsson, S (1970) Hekla. Almenna Bókafélagið, Reykjavík (in Icelandic)

Thorarinsson S (1974) The history of Grímsvötn eruptions and jökulhlaups. Bókaútgáfa Menningarsjóðs, Reykjavík (in Icelandic)

Thorarinsson S (1976) Eruption of Hekla 1947–48: Course of events. In: Einarsson T, Kjartansson G, Thorarinsson S (eds) The eruption of Hekla 1947–48 IV. Societas Scientiarum Islandica, Reykjavík, pp 5–32

Thorarinsson S (1979) On the damage caused by volcanic eruptions with special reference to tephra and gases. In: Sheets PD, Grayson DK (eds) Volcanic activity and human geology. Academic, New York, pp 125–159

Thorarinsson S (1981) Greetings from Iceland: Ash-falls and volcanic aerosols in Scandinavia. Geografiska Annaler 63:109–118

Thorarinsson S (1984) The annals of the Skaftá Fires. In: Einarsson T, Guðbergsson GM, Gunnlaugsson GA, et al. (eds) Skaftáreldar 1783–1784; Ritgerðir og Heimildir. Mál og Menning, Reykjavík 1984, pp 11–36

Thorarinsson S, Sæmundsson K (1979) Volcanic activity in historical time. Jökull 29:29–32

Thorarinsson S, Steinþórsson S, Einarsson T, et al. (1973). The eruption on Heimaey, Iceland. Nature, London 241:372–375.

Thordarson T (1990) The eruption sequence and eruption behavior of the Skaftá Fires, 1783–1785, Iceland: Characteristics and distribution of eruption products. M.Sc. Thesis, The University of Texas at Arlington, Arlington

Thordarson T (2003) 1783-85 Laki-Grímsvötn eruptions I: A critical look at the contemporary chronicles. Jökull 51:1–10

Thordarson T, Höskuldsson Á (2008) Postglacial volcanism in Iceland. Jökull 58:197–228

Thordarson T, Larsen G (2007) Volcanism in Iceland in historical time: Volcano types, eruption styles and eruptive history. Journal of Geodynamics 43:118–152

Thordarson T, Self S (1993) The Laki (Skaftár Fires) and Grímsvötn eruptions in 1783–1785. Bulletin of Volcanology 55:233–263

Thordarson T, Johannsdottir GE, Geirsdottir A (2005) Tephra layers in lake sediments from South and West Iceland, Geophysical Research Abstracts. European Geoscience Union General Assembly, Vienna; 1607-7962/gra/EGU05-A-10766

Thordarson T, Larsen G, Steinthorsson S, Self S (2003) 1783-85AD Laki-Grímsvötn eruptions II: Appraisal based on contemporary accounts. Jökull 51:11–48

Thordarson T, Miller DJ, Larsen G, et al. (2001) New estimates of sulfur degassing and atmospheric mass-loading by the 934 AD Eldgjá eruption, Iceland. Journal of Volcanology and Geothermal Research 108:33–54

Thoroddsen T (1925) Die Geschichte der Islandischen Vulkane, D. KGL. Danske Vidensk. Selsk. Skrifter, Naturvidensk. og Matem. Afd., 8. Række, IX., AF Høst and Søn, Konglige Hof-Boghandel, Copenhagen

Thorsteinsson B and Jonsson B (1991) Iceland's Saga to the present. Sögufélagið, Reykjavík, 539 pp

Part V
South and East Asia

The evolutionary pattern of Asian societies is strongly similar to that of the Mediterranean and European world, particularly in eastern China. Similarities include the fact that cultures appeared there during the earliest Holocene, and some became the cradle of eastern civilizations. Furthermore, through the Holocene the land and their societies responded to climate change in similar fashion to the west. Desertification of some areas was confronted in the north, while temperate to hot monsoonal conditions persisted, despite recurring droughts, in the center and south. Other similarities consist of (a) the comprehensive anthropic modification of the landscape in the temperate zone, including a striking parallel with the Netherlands: the winning and defending against a rise in seal level, of large areas of land from the sea along the vast eastern coastal plains, and (b) in the impact of diseases such as malaria between the Pacific islands and some Mediterranean countries. A major difference is presented by China and other countries of east Asia, namely the cultivation of rice as the staple crop. Sri Lanka, with its unique "hydraulic" civilization based on tanks, is also noteworthy. Finally Japan, as quintessentially eastern as its great neighbor China, yet now so thoroughly westernized as to be unrecognizable in terms of its social structure, from its pre-nineteenth century precursor.

Chapter 19
Holocene Environmental Changes and the Evolution of the Neolithic Cultures in China

Duowen Mo, Zhijun Zhao, Junjie Xu, and Minglin Li

19.1 Introduction

China has three great topographic zones (Fig. 19.1). The first, in the eastern part of the mainland, contains plains ranging from a few tens of meters to ca. 100 m in elevation and hills and low mountains ranging from hundreds of meters to ca. 1000 m asl. The first zone passes gradually westwards into the second, which is characterized primarily by hills and mountains over 1000 m high and by deep river valleys. The third topographic zone is the Qinghai-Tibetan Plateaus with average altitude over 4000 m asl.

China also has three climate zones: the monsoon controlled zone in the east, a semi-arid to arid zone in the northwest, and the Tibetan Plateau zone in the southwest. The monsoon zone covers the regions of the first topographic zone and the eastern part of the second zone.

The north-western part of China, with its inland location, is usually dry, and desertification is common. The Tibetan Plateau is cold and dry by virtue of its inland position and high altitude. The high latitude of the Northeastern China ensures cold conditions compared with the country to the south.

The sub-tropical monsoon region (the middle and lower Yangtze areas) and temperate monsoon region (the Yellow River areas and the Western Liao River areas) are commonly viewed as the main regions where Neolithic cultures developed early. Neolithic cultures emerged later in areas of extreme climatic conditions in comparison. In South China, because of the easy accessibility of wild plant and animal food resources in the tropical forests, domestication and cultivation developed slowly.

As we show in this chapter, archaeological data and environmental reconstructions indicate that regional environmental conditions and their changes affected the ways in which local cultures evolved.

19.2 Environmental Changes During the Holocene in China

No complete scenario of the climatic history of China is yet possible, in spite of the accumulation of work on the subject since the 1960s (Shi et al. 1992; Lu and Zhang 2008). Partly this is due to the limitations of the methods applied in paleoclimatic reconstruction and the accuracy of dating techniques.

Based on the studies of the past 50 years, however, it is possible to reconstruct general trends of climatic change during the Holocene in different regions. The three main climatic scenarios are:

(a) a cool, gradually warming period in the early Holocene,
(b) a megathermal middle Holocene, and
(c) a katathermal period in the late Holocene (Fig. 19.2).

There were also several millennial-scale fluctuations that differ in different regions, while sharing the same pattern of change. From the Last Glacial Maximum to early Holocene there was a general though fluctuating warming trend (Shi et al. 1992; Morris et al. 2003; Wang et al. 2008; Lu and Zhang 2008). Around 9000 year BP, the average temperature was similar or higher than now. From 9000 to 8000 year BP, after an initial cooling, the temperature remained relatively

D. Mo (✉)
College of Environmental and Urban Sciences
Peking University, Beijing, 100871, China
e-mail: dmo@urban.pku.edu.cn

See Plate 18 in the Color Plate Section; also available at: extras.springer.com

I. P. Martini, W. Chesworth (eds.), *Landscapes and Societies*,
DOI 10.1007/978-90-481-9413-1_19,

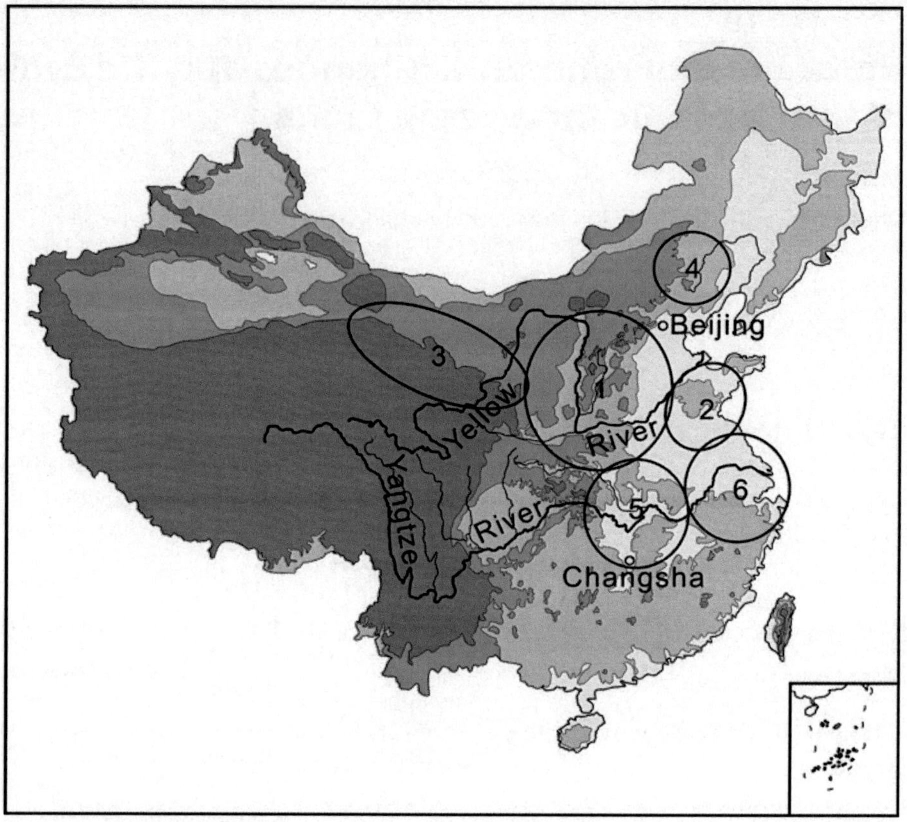

Fig. 19.1 Map showing the six Neolithic culture areas of central-eastern China. (*1* Zhongyuan region; *2* Haidai region; *3* Gan-Qing region; *4* The western Liao River region; *5* The middle Yangze River region; *6* The lower Yangze River region)

high. The middle Holocene megathermal period from 8000 to 5000 year BP was warmer than the present in spite of short-lasting slight cooling around 7000 year BP (An et al. 2005; Chen et al. 2005; Lu and Zhang 2008; Li et al. 2009). After 5000 year BP the average temperature showed a fluctuating declining trend (Shi et al. 1992; Mo et al. 1996, 2002; Morris et al. 2003; Feng et al. 2006) and some research results indicate one or several cooling events at 5000; 4000 and 3000 year BP (Wang et al. 1990, 2008; Kong et al. 1990; An et al. 2005; Chen et al. 2005, 2008; Lu and Zhang 2008).

The East Asian monsoon was a second major influence, resulting in seasonally contemporaneous high temperature and high humidity. Rain is carried by the humid Pacific airflow, and precipitation decreases from southeast to northwest. The southeastern part is humid and warm throughout the year, while the northwestern part is much drier and colder by comparison. Temperature and precipitation also vary greatly within the eastern part of China in response to the various landscapes. These diverse conditions constitute the climatic background for the evolution of the Chinese Neolithic cultures in the eastern part of China (Fig. 19.3).

19.3 Relationship between Regional Environmental Changes and Evolution of the Neolithic Cultures in the Eastern and Central Parts of China

With a warming climate at the end of the Pleistocene, a gradual transition began from the hunter gatherer to the settled life of the farmer. After developing in the early Holocene, six regions of Neolithic culture arose in the east of China (Fig. 19.1; Table 19.1). The middle Holocene (ca. 6000 year BP, Table 19.1) saw great prosperity of every culture in each region and extensive communication between them. The culture of the Zhongyuan region (mainly the middle Yellow River area), as reflected in archaeological and environmental evidence, became the first major center of

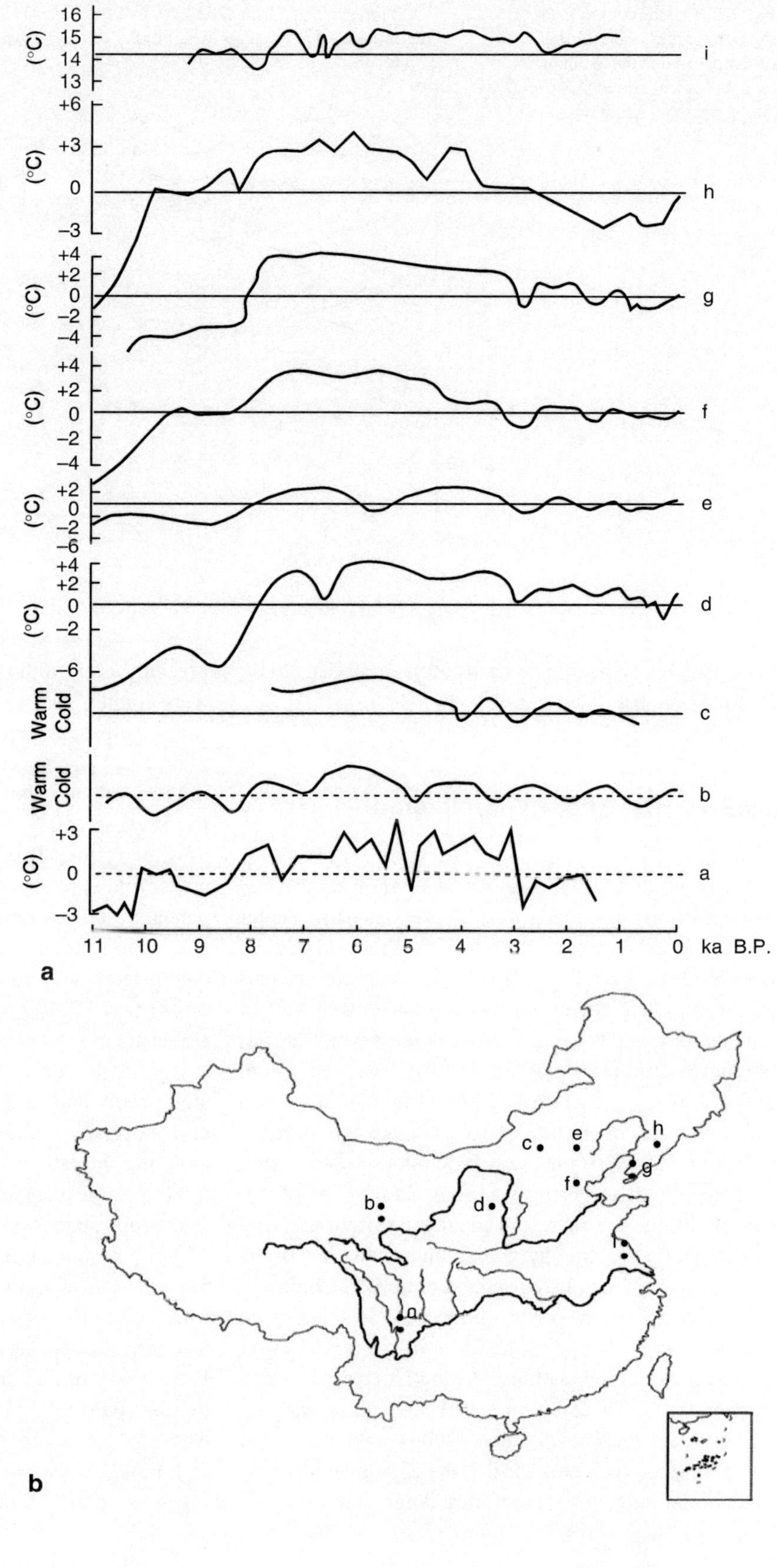

Fig. 19.2 Diagrams (**a**) and location map (**b**) of temperature estimated in various areas of China from pollen spectrums of sedimentary sections of loess deposit (After Shi et al. 1992) (*a* Luoji Mts. (27°35′N, 102°15′E) (After Wang et al.1990); *b* Qinghai Lake (30°54′N, 100°11′E) (After Kong et al. 1990); *c* Inner Mongolia (41°20′N, 113°E) (After Cui and Kong 1992); *d* The Loess Plateau (After Sun et al. 1991); *e* Beijing (40°N, 110°E) (After Zhang et al. 1981); *f* Eastern Hebei Province (After Tong et al. 1991); *g* Southern Liaoning Province (39°30′N, 122°E) (After Guiyang Institute of Geo-Chemistry 1977); *h* Western Changbai Mts. (42°30′N, 126°20′E) (After Wang et al.1990); *i* Qingfeng of Jiangsu Province (After Tang and Shen 1992))

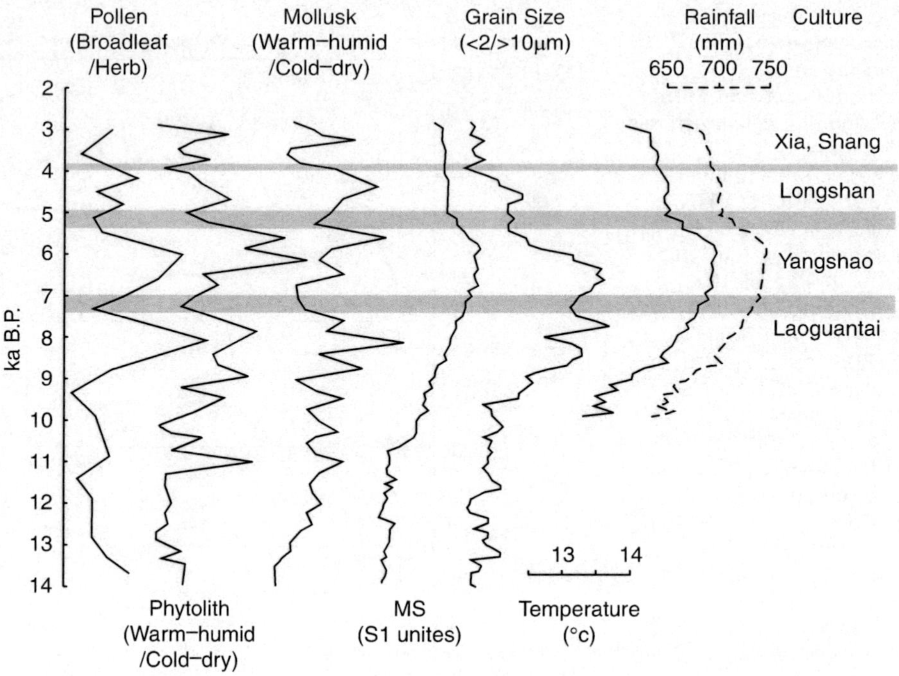

Fig. 19.3 Different environmental indexes and cultures of the profile of the Weinan archaeological site in Guanzhong region. (After Lu and Zhang 2008)

the Chinese Civilization at ca. 4000 year BP (Liu 2004; Zhao 2006a, b).

19.3.1 The Zhongyuan Region

The Zhongyuan region is part of the warm-temperate, deciduous broad-leaved forest biome. It is in the center of the monsoon area, traversing the first and second topographic zones (Figs. 19.1, 19.4). The principal part of the region includes the middle and lower Yellow River valley, and is covered by thick loess which acts as parent material for highly fertile soils. The area is cool and dry in winter, warm and humid in summer, has a mean annual temperature of 10–14 °C, annual precipitation of 550–800 mm, with precipitation decreasing from southeast to northwest (Song 2002). The physical conditions are therefore ideal for agriculture, and consequently the Zhongyuan region became one of the most important centers of ancient culture in China.

In this region, the earliest pottery vessels date from 12,000 year BP (Baoding Institute of Cultural Relics, Department of Archaeology, Peking University 1992). Domestication of common millet may be as early as 10,000 years ago here (Lu et al. 2009).

From 9000 to 7000 year BP, the Zhongyuan region had precipitation and temperature ranges similar to the present before 8000 year BP. After 8000 year BP they were higher. Advanced Neolithic cultures evolved in several areas, such as the Peiligang culture in hill and plain areas of the center and south, the Cishan culture in the northeast plain area, and the Laoguantai culture in the west valley of Wei River (the principal branch of the Yellow River, Fig. 19.4). All three cultures are similarly characterized by round or oval-shaped, partially sunken houses, and large numbers of stone and bone artefacts. The bone flute, the earliest musical instrument known, was discovered at the Jiahu site (Fig. 19.5a) (Zhang et al. 1999). Painted pottery appeared in sites of the late Laoguantai period. Funerary goods in burials are usually rare. Most are ground stone tools and pottery vessels (Fig. 19.5c) that indicate a division of labor between male and female. For example, millstones (Fig. 19.5b) are usually found in the graves of women whereas males were often buried with stone spades and sickles.

Crop remains have been found at some sites of this period. At the Cishan site (Fig. 19.4), foxtail millet ashes were discovered in 88 ash pits, mostly at 1 m in depth, though at over 2 m in 10 pits (Tong 1984). Rice was found at sites in the south plain and valleys (Shaanxi Provincial Institute of Archaeology 1999). The earliest domesticated dog was identified at the Jiahu site (Fig. 19.4), whilst the earliest domesticated pig was found at the Cishan site (Fig. 19.4). It has

Table 19.1 Neolithic Cultures of six regions of central-eastern China

The middle Yangzi River region	The lower Yangzi River region	The Gan-Qing region	The Zhongyuan region	The Haidai region	The Western Liao River region
Pengtoushan Culture (9-7.9)	Shang-shan Culture (10-8.5)	Laoguantai Culture (8-7.3)	Peiligang, Cishan & Laoguantai Culture (9-7)	Houli Culture (8.5-7.5)	Xinglongwa Culture (8-7.3)
Zaoshixiaceng Culture (7.9-6.8)	Kuahu-qiao Culture (8-7)	Yangshao Culture (7-5.3)	Yangshao Culture (7-5)	Beixin Culture (7.4-6.3)	Zhaobaogou Culture (7-6.4)
Tangjiagang Culture (6.8-6.3)	Majia-bang Culture (7-6)	Majiayao Culture (5.3-4.8)	Miaodigou II Culture (5-4.6)	Dawenkou Culture (6.3-4.6)	Hongshan Culture (6.6-4.8)
Daxi Culture (6.3-5.5)	Hemudu Culture (7-5)	Banshan-Machang Culture (4.8-4)	Zhongyuan-longshan Culture (4.6-4)	Longshan Culture (4.6-4)	Xiaoheyan Culture (4.8-4.3)
Qujialing Culture (5.3-4.6)	Songze Culture (6-5.3)	Qijia Culture (4.2-3.8)	Erlitou Culture (3.9-3.5)	Yueshi Culture(3.9-3.6)	Xiajiadian-xiaceng Culture (4.2-3.5)
Shijiahe Culture (4.5-4)	Liangzhu Culture (5.3-4)				
	Maqiao Culture (3.9-3.1)				

been argued that these are the earliest domesticated animals known in China (Yuan 2001). Though cultivation began at this period, the major subsistence practices were still hunting, fishing, and gathering (Zhao 2005).

Between 7000 and 5000 year BP, the Neolithic Yangshao culture developed. Temperature was generally 2–3 °C higher and precipitation was 150 mm greater than now in the Zhongyuan region. There was a slight cooling and drying toward the end of the period (Shi et al. 1992). Typical sites in the early and middle Yangshao periods covered areas up to about 2–3 ha. Partially sunken houses each about 10 m^2 in area were common. In some settlements there is usually one large public meeting house at the center of a site. Red pottery shards with geometric patterns of plants and animal dominate the findings (Fig. 19.5d, e). No evidence of social stratification has ever been found. Trade between settlements was common; household lithic workshops have been discovered that utilized the materials derived from outside territories. Animal breeding and farming with crops such as foxtail millet and broomcorn millet were the major subsistence practices. Rice remains increase throughout this period. The settlements of the early Yangshao period are distributed along the valleys, while in the middle Yangshao period (Miaodigou Phase), the culture expanded to the surrounding area and occupied most of the Zhongyuan region as well as affecting all the other five regions.

During the late Yangshao period (5500–5000 year BP), temperature and precipitation remained relatively high, though a tendency to decrease in both parameters began. Agriculture and husbandry developed further. Many new settlements occur indicating a rapid growth of population at this time. Houses were all built above the ground. Differentiation between settlements

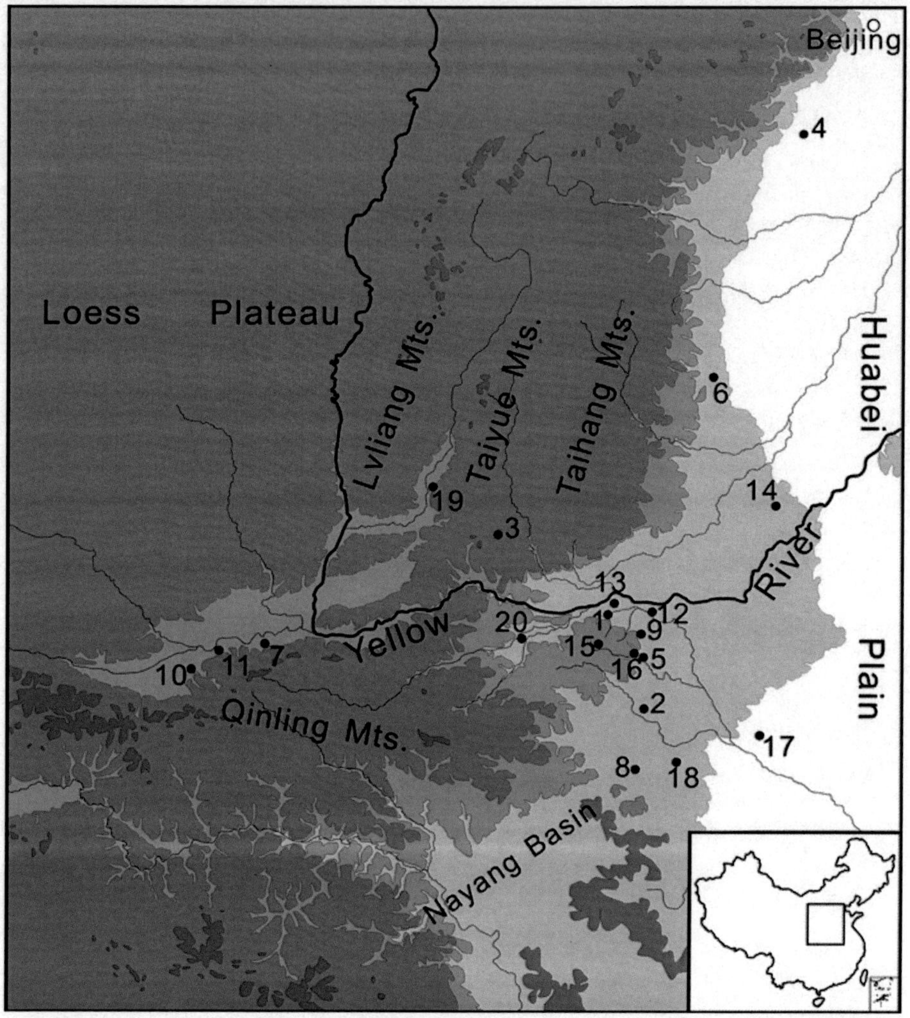

Fig. 19.4 Map of the Zhongyuan region, showing major geographic features and the Neolithic sites. (Sites: *1* Zhijidong; *2* Lingjing; *3* Xiachuan; *4* Nanzhuangtou; *5* Peiligang; *6* Cishan; *7* Laoguantai; *8* Jiahu; *9* Shawoli; *10* Banpo; *11* Jiangzhai; *12* Dahecun; *13* Xishan; *14* Xishuipo; *15* Wangchenggang; *16* Guchengzhai; *17* Pingliangtai; *18* Haojiatai; *19* Taosi; *20* Erlitou; *21* Xiaonanhai)

appeared and their sizes varied greatly. Enclosures for defensive use suggest tensions between different tribes or communities. Xishan is the largest city site of this period. It is walled, surrounded by a moat and covers an area of 34,500 m^2 (Participants of the Training Course for Excavation Managers 1999). Difference in grave goods indicates the emergence of social stratification. Skeletons indicate that the frequency of violent death had increased and it is possible that some of the victims were sacrificial. The occupant of a grave at Xishuipo was buried with dragon and tiger images made of shells, and is identified as a shaman by some scholars (Fig. 19.6a; Puyang Wenguan Hui 1988). The differentiation of the settlements and social stratification can be taken to mark the inception of civilization in that region.

From 5000 to 4000 year BP, temperature and precipitation decreased although remaining slightly higher than present (Shi et al. 1992; Lu and Zhang 2008). The Miaodigou II (5000~4700 year BP) and Longshan (4600~4000 year BP) cultures appeared and started another period of rapid cultural development. The frequency of foxtail and broomcorn millets increased, though rice is not commonly seen in the assemblages. The differences among local cultures increased, and by the end of the period a population peak was reached (Li et al. 2009). Numerous settlements and cities of various size appeared and wars between them occurred frequently (Henan Institute of Cultural Relics and Archaeology 1983, 1992, 2002; Zhoukou Relics Office 1983). The small settlements covered about 10,000 m^2 whereas the large ones reached millions of square meters. The Taosi site in the southern Shanxi Province provides a good example. Remains of palaces, roads, wells, and observatories (Fig. 19.6c; Wu et al.

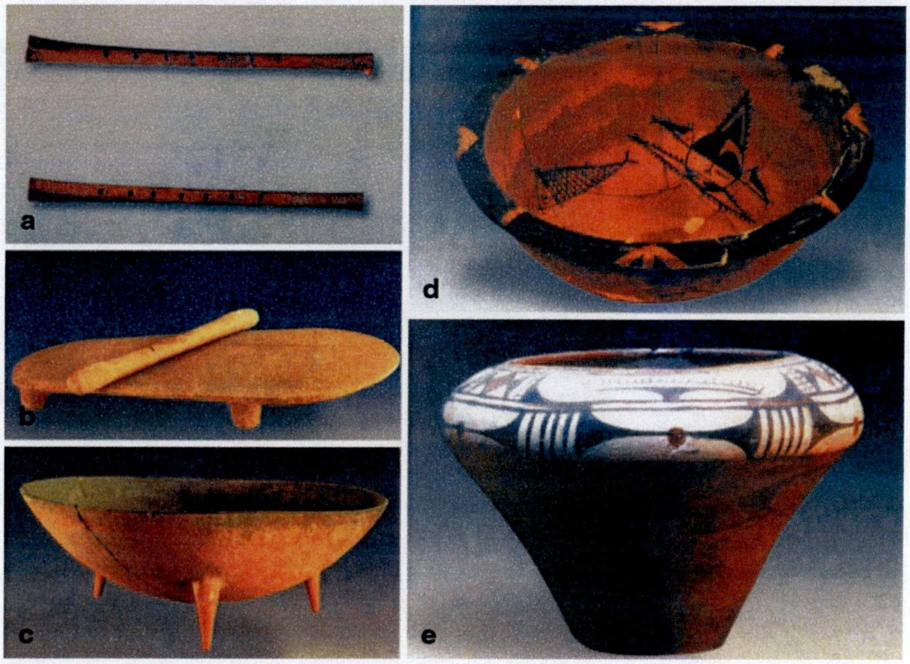

Fig. 19.5 Early Neolithic relics in the Zhongyuan region. **a** bone flute (Jiahu site; after Li 2003, Fig. 6). **b** Millstone (Peiligang site; after Li 2003, Fig. 2). **c** Earthen bowl with three legs (Peiligang site; after Li 2003, Fig. 4). **d** Pottery basin (Banpo site; after Gong 2002, Fig. 1). **e** Earthen Bowl (Dahecun site; after Gong 2002, Fig. 5)

2009) have been discovered there over an area of some 3 million m^2. The residences for nobles and for ordinary civilians were built in different neighborhoods. Eighty seven percent of burials were small and lacking burial goods. Nine large excavated burials of high-ranking people, however, contain wooden coffins and one to two hundred grave goods, including normal and painted pottery, musical instruments, painted wooden objects, jade and stone ritual artefacts (Fig. 19.6f–i), various ornaments, and entire pig skeletons.

After 4000 year BP temperature and precipitation decreased further, but on average remaining slightly higher than the present ones. The Erlitou culture dating to 3900–3500 year BP succeeded the Longshan culture. The core site at Erlitou was 2.5 km wide from east to west and 1.5 km from south to north (Erlitou Working Team 2003). Many features were found, including three palace foundations, and workshops for bronze, pottery, stone tools, and turquoise respectively. Ritual places were maintained, and more than 400 burials have been found. Bronze ritual instruments and elaborate jade objects were excavated from this site (Fig. 19.6b, d, e).

Most of the Erlitou sites are found in the central and southern Zhongyuan regions. Some scholars believe that the city site at Erlitou was the capital of the Xia dynasty, the first Chinese dynasty in antiquity (Liu 2004; Xu and Du 2005). In any case, starting at ca. 4000 year BP the Zhongyuan region became a political and cultural center from where the Civilization of China developed (Liu 2004; Zhao 2006a, b).

19.3.2 The Haidai Region

The Haidai region includes the lower reaches of the Yellow River and extends to Shandong Peninsula (Fig. 19.7). It is approximately at the same latitude as the Zhongyuan region, and is flanked by the Yellow Sea to the east and the Bohai Sea to the north. The average altitude is low, though the region does contain uplands and the Taiyi Mountain ranges are located in its middle part. Influenced by monsoon winds the mean annual temperature fluctuates between 11 and 14 °C, and the mean annual precipitation usually varies between 500 and 950 mm. The major surface water system is composed of the branches of the Yellow River supplemented by other rivers and mountain creeks. Climate changes in this region throughout the Holocene were similar to those of the Zhongyuan region (Lu 1989).

The oldest Neolithic culture in the Haidai region is the Houli culture, which is radiocarbon dated to 8500 year BP. It was followed by the Beixin culture after 7500 year BP (Table 19.1; The radiocarbon laboratory of Peking University 1996). The Houli culture

Fig. 19.6 Late Neolithic relics in the Zhongyuan region. **a** Tomb with skeleton and design made with mussel shells shaped to be dragon and tiger (Xishuipo site) (After Gong 2002, Fig. 8). **b** Bronze jue (cup with two ears and three legs) (Erlitou site) (After Xu and Du 2005). **c** The basement of an observatory (Taosi site). **d** Jade ge (dagger like artefact used in ceremony (Erlitou site) (After Xu and Du 2005). **e** Jade zhang (tablet like artefact used in ceremony) (Erlitou site) (After Xu and Du 2005). **f** Jade huang (pendant of semi-circular shape) (Taosi site) (After Xie 2007). **g** Jade cong (an octagonal jade piece with a round hole in the center) (Taosi site) (After Xie 2007). **h** Jade shaped to be a monster (Taosi site) (After Xie 2007). **i** Pottery drum (Taosi site) (After Xie 2007)

sites are widely distributed on the plains to the north and west of the central mountain ranges and also in the valleys and on the terraces between the uplands. The tools used were primarily made of stone and bone. The pottery vessels were tempered with sands, coarse, unadorned, and mostly have rounded bottoms. House remains are mostly square or rectangular in shape and have semi-underground structures. Agricultural activities are indicated by the presence of foxtail millet and broomcorn millet. Principle subsistence practices included fishing, hunting, and gathering.

Beyond the areas traditionally occupied by the Houli culture, the Beixin sites are also widely identified on the plains and uplands to the south and southwest of the central mountain ranges. In the early Beixin period, tools are primarily of chopped stone, whilst during its late period polished tools are dominant. Compared with the pottery ware of the Houli culture, vessel types and surface treatment methods of the Beixin culture are more complex and varied. Most Beixin houses are round or oval-shaped with semi-underground structures. The farming of foxtail millet of this period developed further, and rice cultivation began to appear. Domesticated animals included pigs, dogs, chicken, cattle, and buffalos. However, fishing, hunting, and gathering were still important activities

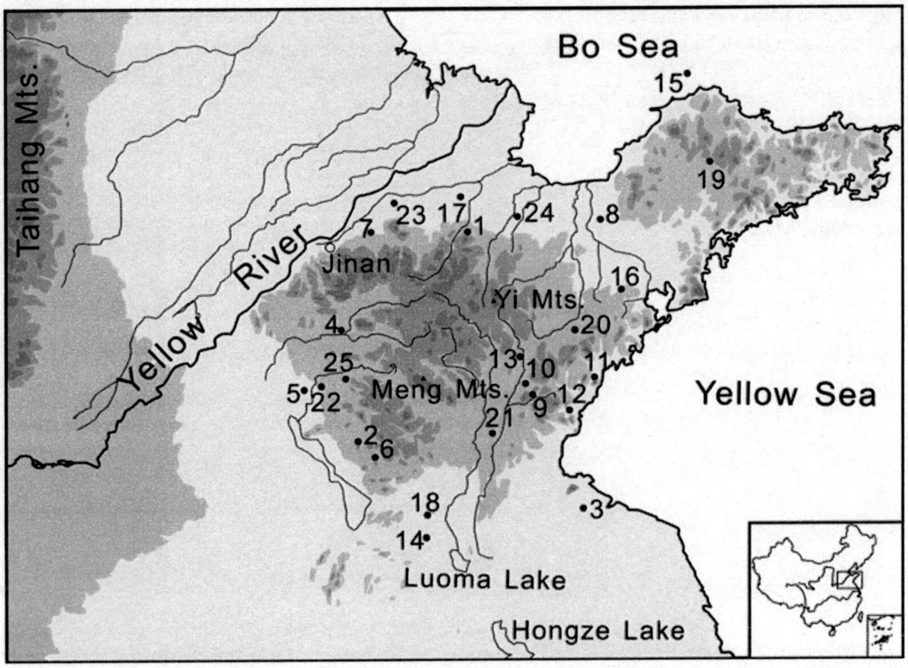

Fig. 19.7 Map of Haidai region, showing major geographic features and the Neolithic sites. (Sites: *1* Houli; *2* Beixin; *3* Erjiancun; *4* Dawenkou; *5* Wangyin; *6* Jianxin; *7* Chengziya; *8* Yueshi; *9* Lingyanghe; *10* Tangzi; *11* Liangcheng; *12* Yaowangcheng; *13* Shagnyu; *14* Liulin; *15* Daheishan; *16* Sanlihe; *17* Fujia; *18* Dadunzi; *19* Yangjiajuan; *20* Chegnzi; *21* Dafanzhuang; *22* Xixiahou; *23* Dinggong; *24* Bianxianwang; *25* Yinjiacheng)

(Shandong Fieldwork Team, Institute of Archaeology of the Chinese Academy of Social Sciences, the Museum of Teng County, Shandong 1984).

The Dawenkou culture dating to 6300 year BP is the third dominant archaeological culture in this region (Table 19.1; Gao and Luan 2004). It is distributed on the uplands and plains around the central mountains and is characterized by the rapid development of farming (foxtail millet, broomcorn millet, and rice) and animal domestication (mainly pigs and dogs). Yet, hunting and fishing continued to play an important role in the Dawenkou subsistence economy. Polished stone tools are dominant in the archaeological assemblages of this period. Most of the tools are perforated for attachment of handles. Painted pottery wares of various shape and surface ornamentation are numerous. The degree of craft specialization is relatively high. Houses of this period are usually round or square semi-underground structures. A strong indication of social stratification is seen in the settlement pattern, with the central areas occupied by several large houses. The difference of the funerary goods among the burials also suggests an orderly ranking of social status (Yan 1986).

From 5500 year BP onwards, climate became cooler and drier, though the average temperature and precipitation were above present values. The late Dawenkou culture continued to develop and expand (Gao and Luan 2004). It spread to the Zhongyuan region in the west and the Yangtze River valley areas in the south. Agriculture developed to a new stage. Hoes were used to cultivate foxtail millet and broomcorn millet, and rice and beans became important domesticated plants (Institute of Archaeology, CASS 1988; Wang 1995). Domesticated livestock (mainly pigs) developed further as well. The increased number of stem cups (Fig. 19.8a) found in the burial context indicates that large amounts of beverages might have been produced and consumed by the Dawenkou people. Fast-wheel methods might have been used in pottery making. Stone tools are usually delicately refined and increase in variety. New types primarily include stone axes, stone arrowheads, and other weapons. A large number of jade and refined pottery burial goods have been discovered in the high ranking burials (Shandong Administration of Relics 1974). The discovery of jade and stone axes in some tombs suggests that the occupants may have had military power (Yan 2000). By contrast the ordinary graves were small and usually without burial objects. This too indicates a social stratification and a complex society.

The Longshan culture dating to around 4600 year BP inherited the farming tradition of the Dawenkou culture and developed further. Foxtail millet and broomcorn millet were the main cultivated plants, but rice reached nearly 50% of the cultivated plants in the assemblages at some Longshan sites in the

Fig. 19.8 **a** Stem cup in the Longshan culture of Haidai region (After Gao and Shao 2005). **b** Pottery bottle of early Yangshao culture in Gan-Qing region (After Gansu Provincial Institute of Cultural Relics and Archeology 2006)

south-eastern part of the region. The range of species of domesticated animals increased. Fast wheels were widely used in pottery making. Pottery vessels of this period have thin walls, such as the well-known egg-shell stem cup. Parent materials of such pottery vessels had to be meticulously prepared to achieve such fine quality. Large numbers of jade ritual artefacts and ornaments were excavated from the burials of the Longshan culture (Shandong Provincial Institute of Cultural Relics and Archaeology 1989). Tombs of the nobles and those of the low-ranking people were usually separately arranged. Large differences in quality and quantity of burial goods are evident in the tombs of this period. Early bronze production at some sites is indicated by finds of small pieces of bronze. Size and number of settlements at this period increased rapidly. Settlement enclosures built with rammed soil have been frequently found (Gao and Luan 2004). Large city sites have been discovered in the south-eastern part of the Haidai region. Social stratification intensified at this time and some scholars believe that early city-states emerged (Zhang 1993; Gao 2000).

After 4000 year BP, temperature and precipitation decreased to the present level. In the late Longshan period as well as the subsequent Yueshi culture period, the intensive use of bronze indicates the beginning of the Bronze Age in this region. However, the number of sites, which reflect population and prosperity, decreased significantly (Fang 2003). Pottery manufacturing techniques degraded considerably (Luan 1997). On the whole the level of economy and social development in the Haidai region was significantly lower than for the Zhongyuan region.

The decline of the Neolithic culture since the late Longshan period in the Hadai region may have had several causes. First, since the late Dawenkou period, people had migrated westwards and southwards to the flood-prone areas of the Yellow River. However, sea level rose at around 6000 year BP, which resulted in the rise of river water level and the silting of the lower-lying plains, bottom valleys, and river channels, the latter favoring an increase in frequency and intensity of floods. The intensification of floods severely disrupted human activities in the lower Yellow River plain. Second, in the middle Longshan period, rice cultivation reached a higher proportion in agriculture in south-eastern area that was affected by the climatic fluctuations characterized by a drying and cooling climate. Third, the conflicts between the competing city–states within the Haidai region and with the neighboring city–states of the Zhongyuan and lower Yangtze River regions also occurred (Yan 1986).

19.3.3 The Gansu-Qinghai Region

In this large region, the Neolithic cultures are mainly distributed across the loess hills in east Gansu, the Hexi Corridor and the Yellow River–Huang River

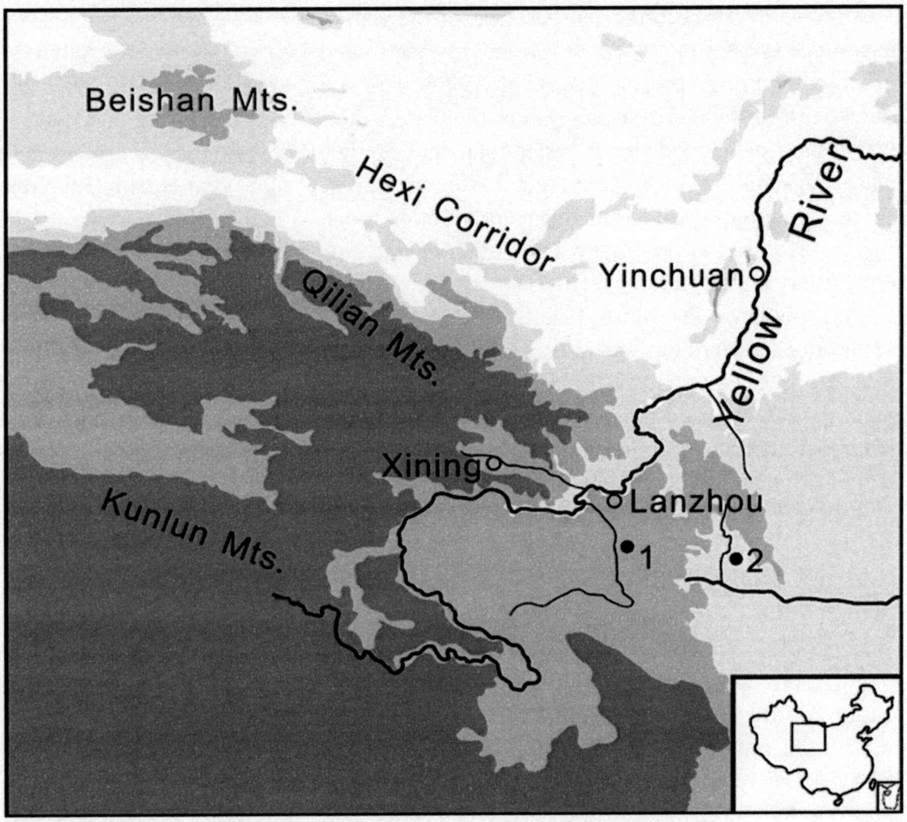

Fig. 19.9 Topographic map of Gansu and Qinghai region, showing Neolithic sites. (Sites: *1* Majiayao; *2* Dadiwan)

region in the north-eastern part of Qinghai (Fig. 19.9). There are various landscapes in this wide region. (a) The main landforms in eastern Gansu are hills of loess and river valleys. The mean annual precipitation in this region is 350–650 mm and decreases from east to west. The vegetation changes from broad-leafed forest to grasslands. (b) In the Hexi Corridor, the mean annual precipitation is 160 mm in the east, and gradually decreases to less than 50 mm in the west. This area is known as the warm temperate desert zone. The main landforms consist of plains and deserts.

The Qilian Mountains on the north-eastern edge of the Qinghai-Tibet Plateau reach an average elevation of more than 4000 m asl. Numerous oases are formed along the rivers that originate from Qilian Mountains. So the Hexi area has been a transport and migration corridor from the Zhongyuan region to the west for thousands of years. (c) Broad river valleys, mesas, and hills covered with thick loess occur in the Yellow River-Huang River areas in north-eastern Qinghai. The mean annual precipitation is 400–500 mm. The temperature and the precipitation increased in the early Holocene (ca. 8500 year BP) and were higher than present. Between 8000 and 6000 year BP, the temperature was 2–4 °C and the precipitation was 100–300 higher than today. Between 6000 and 4000 year BP, it became cooler and drier (Mo et al. 1996; Tang and An 2007).

The earliest Neolithic culture in the Gansu-Qinghai region is the Laoguantai culture (8000–70,000 year BP). Its Dadiwan site is regarded as the most important, and was contemporaneous with cultures in the western part of the Zhongyuan region. Broomcorn and foxtail millets have been discovered at many sites. Chopped stone shovels, mussel shovels, and knives were used in agriculture. Pigs were domesticated. The houses are round in shape and have semi-underground structures, and a single room usually covers an area of around 6–7 m^2 (Gansu Provincial Museum, Qin'an Cultural Centers, Dadiwan Excavation Team 1981). After 7000 year BP, the culture in this region merged with the Yangshao in the Zhongyuan region. Agriculture based on millets and livestock breeding further developed. The pottery of this period became more refined. Remains of square semi-underground houses are commonly seen at the sites (Gansu Provincial Institute of Cultural Relics and Archeology 2006).

After 5500 year BP, the temperature of the region began to decrease, but it was still relatively warm and humid compared to present (Mo et al. 1996). The late Yangshao culture in the eastern part of the Gansu-Qinghai region developed much faster than its previous phase. The number of archaeological sites increased, and millet agriculture expanded. The production tools, such as blade and knives, were made of stone and pottery. Pottery manufacturing techniques were improved. Painted pottery was hard, beautiful and colorful. Square houses were built on the ground. Burned mud (mixed with grass or calcareous matter) floorings were found in many house relics. Remains of a large public building covering an area of 290 m^2 was discovered at the Dadiwan site, which suggest a complex society at the site (Archaeological Team of Gansu Province 1986).

Around 5000 year BP, social complexity in the eastern part of the Gansu-Qinghai region was reduced. Some people moved westwards to the Yellow River and Huang River valleys and the middle Hexi Corridor, and south to northern Sichuan. Animal husbandry became much more important. The Majiayao culture (5300–4800 year BP) emerged and millet, hemp and other dry crops were cultivated, while pigs, goats, and cows were domesticated. Stone knives, pestles and mortars were the primary agricultural production implement. The painted pottery of the Majiayao culture was made exquisitely and decorated colorfully (Fig. 19.8b). A bronze knife was discovered at the Majiayao site (Xie 2002), which indicates the emergence of bronze smelting techniques. The Majiayao culture gave way to a so called Banshan-Machang Phase (4800–4000 year BP), which inherited the Majiayao's painted pottery tradition, but this waned. Husbandry was developed further and the number and the distribution of archaeological sites decreased.

The Qijia culture in the east of the Gansu-Qinghai region emerged from 4200 to 3800 year BP. It expanded toward the eastern Hexi Corridor and replaced the Machang culture and the number of Qijia sites increased. Bronze smelting technique became evident. Millet agriculture and husbandry still played an essential role in subsistence economy. After the Qijia a number of different husbandry-based cultures occupied the Gansu-Qinghai region (Xie 2002). The difference represented in the archaeological remains suggests the isolation of local people between the sub-areas.

The Gansu-Qinghai region is located at the western extremity of East Asian monsoon affected areas, and is characterized by drier conditions. The Neolithic cultures emerged later here than those in the Zhongyuan region. In the warm and humid period between 8000 and 5000 year BP, the development of Neolithic cultures in the region was synchronized and related to those of the Zhongyuan region. The cooling event of ca. 5000 year BP might have influenced the process of their early social stratification. In the following 1000 years, from the Majiayao culture through the Qijia culture, the process of social stratification declined, despite the emergence of relatively well-developed bronze smelting techniques. Cultural stagnation after 4000 year BP in the Gansu-Qinghai region was due not only to the drier and cooler conditions but also to the isolated geographical conditions between the sub-areas.

19.3.4 The Western Liao River Region

The western Liao River region lies in the southern part of Northeast China, on the eastern edge of the Inner Mongolia Plateau (Fig. 19.10). This region has a high elevation in the west and is low in the east. Mountain, upland and terrace form the major landforms of the region. The Horqin sandy land covers the central and eastern portions of this region. The western Liao River and its branches comprise the dominant water network. Controlled by the temperate continental monsoon, the mean annual temperature in the region is 4–6 °C and the mean annual precipitation is 300–620 mm. Temperate deciduous broad leafed forests are primarily found in the south-eastern part of the region, and temperate forest-grasslands are concentrated in the north-west. The major areas in the Horqin sandy land are steppes with sparse forests. In the early Holocene, the temperature and the precipitation increased gradually to the value similar to present. It was warm and humid between 8500 and 6000 year BP with temperatures 3–4 °C and the annual precipitation was 200–300 mm higher than today. The Horqin sandy land gradually became smaller and some dunes stabilized during this period. However, there was a cooling event just before 7000 year BP. After 6000 year BP though the climate became cooler and drier, the temperature and the precipitation were all higher than today, until around 4000 year BP (Kong et al. 1991; Mo et al. 2002; Tarasov et al. 2006).

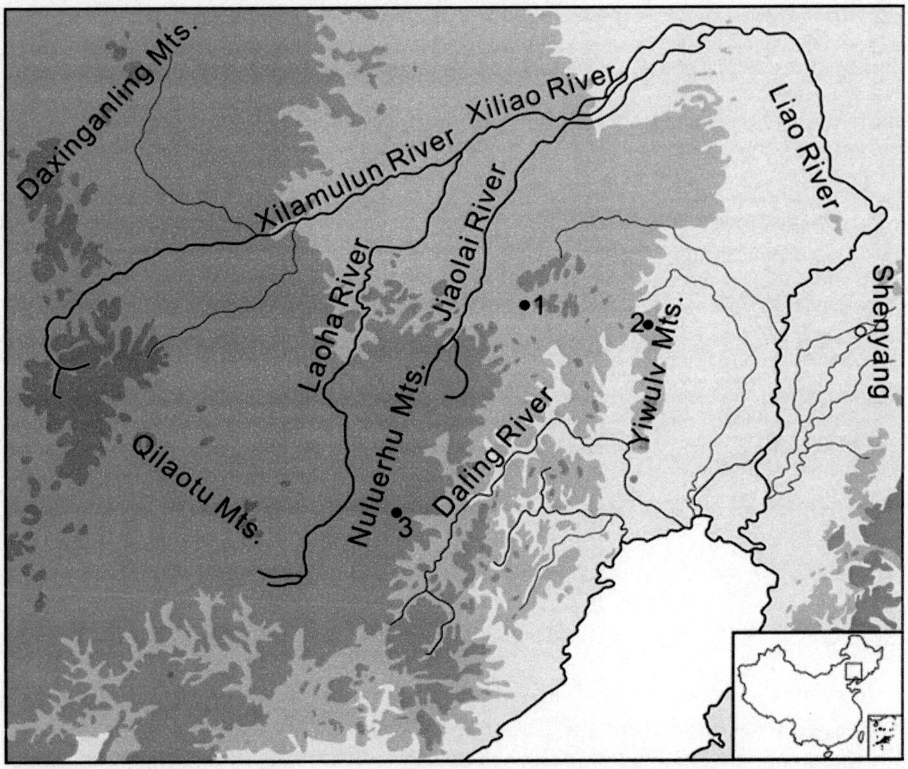

Fig. 19.10 Topographic map of the western Liao River region, showing Neolithic sites. (Sites: *1* Xinglonggou; *2* Chahai; *3* Niuheliang)

The Xinglongwa culture dating to 8000–7300 year BP is the earliest Neolithic culture in this region (Table 19.1). People began to cultivate broomcorn millet, though the major food resources were still provided by fishing, hunting, and gathering. The discovery of agricultural tools, such as stone hoes, and stone shovels, suggests the beginning of early agriculture (The Inner Mongolia Work Team of Institute of Archaeology, Chinese Academy of Social Sciences 2004). The dragon patterns found on pottery vessels are the earliest dragon totemic images in the Neolithic of China. Jade cups ('jue'), jade spoons ('bi'), and jade axes are the earliest jade artefacts so far unearthed in China (Xue 2008). The semi-underground houses at the Xinglongwa site were arranged on an E–W, N–S grid, which indicates a clear planning for these densely distributed houses. During 7000 to 6400 year BP, the Zhaobaogou culture emerged (Table 19.1) and expanded toward the east, west, and south (Tian 2004). The number of ground stone tools increased. A new type of stone plough ('si') began to be used at this period (Xia et al. 2000).

From 6600 to 4800 year BP, the Hongshan culture appeared and around 5500 year BP developed rapidly in terms of subsistence economy and religion. The various types of agricultural implements indicated that agriculture had developed to a high level. However, fishing and hunting still existed (Tian 2004). The number of archaeological sites increased and became densely distributed. An important high social-ranking cemetery and a large ritual site were found at Niuheliang. Thirty burials are covered with stone slabs. A temple to the female deity, a set of altars for making sacrifices to spirits, and the remains of many other large building have been unearthed. Mud female and animal figurines in various sizes were found in the remains of the temple. Various types of exquisite jade artefacts were unearthed in burials, showing that the degree of jade craft specialization was relatively high (Fig. 19.11). The remains of the Niuheliang site indicate a highly complex society (Su 1988).

After 5000 year BP, the western Liao River region was primarily occupied by the Xiaoheyan culture (Table 19.1; 4800–4300 year BP) but the number of archaeological sites sharply decreased and the distribution was relatively localised. The common stone digging tools of the Hongshan culture almost disappeared, indicating the lower level of agricultural production at this period. The cooling and drying event around 5000 year BP may be a possible explanation of this decline.

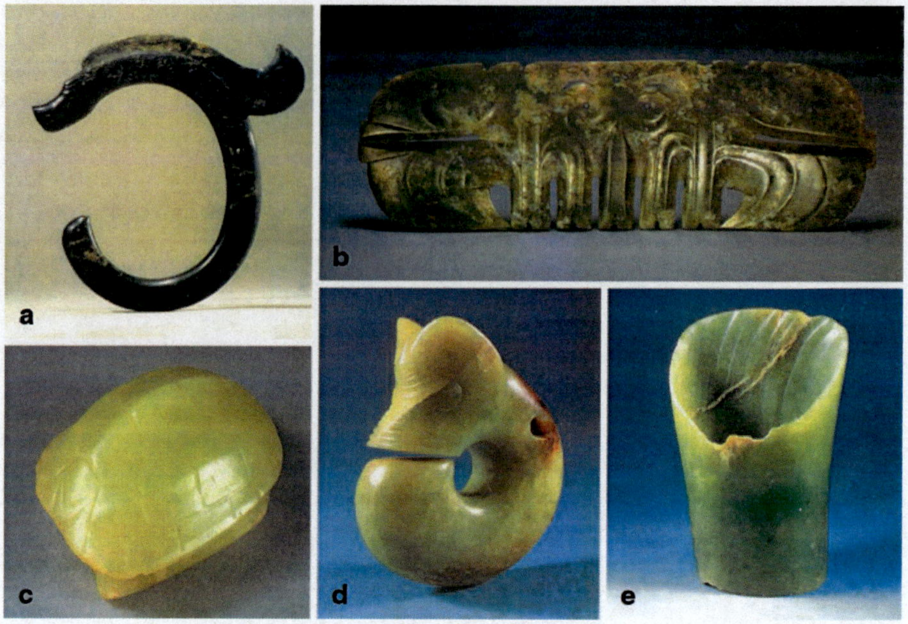

Fig. 19.11 Jade of Hongshan Culture in the Western Liao River region. **a** Jade dragon (Sanxingtala Site); after Liaoning Institute of Archaeology 1997, Fig. 4. **b** Jade pendant (Niuheliang Site); after Liaoning Institute of Archaeology 1997, Fig. 46. **c** Jade turtle (Niuheliang Site); after Liaoning Institute of Archaeology 1997, Fig. 42. **d** Jade dragon (Niuheliang Site); after Liaoning Institute of Archaeology 1997, Fig. 2. **e** Jade hoop (Niuheliang Site); after Liaoning Institute of Archaeology 1997, Fig. 6

In summary, under the warm and humid conditions from 8500 to 5000 year BP, the Neolithic cultures in this region underwent a rapid and continuous development, especially in agriculture. However, fishing and hunting were still important in the subsistence economy. A cooling and drying event around 5000 year BP resulted in a decline in agriculture. This decline might have brought about the loss of social wealth and lower population growth. Consequently, the size and the number of the settlements decreased sharply. The Horqin sandy land area was reduced in the warm and humid period of the Holocene until the cooling and dry event around 5000 year BP when it enlarged again. The expansion of the sandy land may have been another cause for the socio-cultural recession (Song 2002).

19.3.5 The Middle Yangtze River Region

The Neolithic cultures in the middle Yangtze River region are mostly concentrated in the Jianghan-Dongting basins (Fig. 19.12). The basin is a low-lying plain surrounded by uplands and mountain ranges. This region was formed primarily by neotectonic subsidence. The Yangtze River and its branches constitute the major river network of the region. During flood seasons the water level of the Yangtze can rise to more than 30 m asl, so the farmlands and towns in the basin must rely on levee bank protection, but disastrous floods still occur. Controlled by a subtropical monsoon climate the mean annual temperature of the area is 14–18 °C and the mean annual precipitation is 1000–1400 mm, which is mainly concentrated in the spring and summer. Droughts also occur occasionally because the precipitation may vary widely. It is an area where evergreen and deciduous broad-leafed forests co-existed. The wetlands in valleys and basins are appropriate for rice cultivation, which played a major role in the local subsistence pattern in both the prehistoric and early historic periods.

Between 13,000 and 10,000 year BP, stone tools used by the local inhabitants were small and primarily made of flint. People also began to make pottery and domesticated and cultivated rice (Zhang and Yuan 1998; Zhao and Piperno 2000).

After 9000 year BP, the Pengtoushan culture emerged (Table 19.1; Hunan Provincial Institute of Archaeology and Cultural Relics 2006) with sites mainly distributed in the plains to the south-western part of the Dongtinghu Basin. Both wild and domesticated rice species were important food resources. However, hunting, fishing and gathering still played an important role in the subsistence economy. Agricultural tools were primarily chipped stone tools, polished tools are very rare. Pottery making was often characterized by a low-fired and irregularly shaped paste, typical of a rather early pottery tradition. The settlements

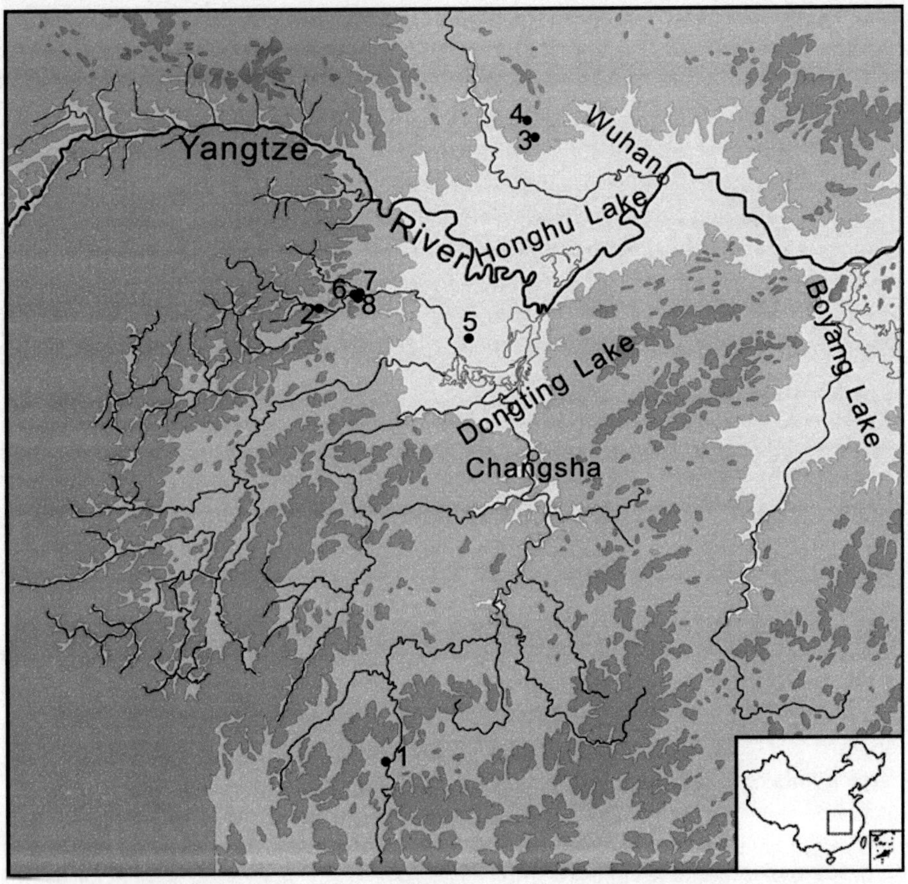

Fig. 19.12 Map of the middle Yangze River region and Neolithic sites. (Sites: *1* Yuchanyan; *2* Zaoshi; *3* Qujialing; *4* Shijiahe; *5* Tangjiagang; *6* Pengtoushan; *7* Jijiaocheng; *8* Chengtoushan)

of this period are frequently enclosed by trenches used both for defense and for flood control. During the middle Pengtoushan period, the culture expanded to the northwest of the Dongtinghu Plain and along the middle section of the Yangtze River. This expansion and subsequent localization gave rise to new types of Neolithic cultures such as the Chengbeixi culture.

The Pengtoushan culture was followed by the Lower Zaoshi culture (7900–6800 year BP) and the Tangjiagang culture (6800–6300 year BP). Rice agriculture expanded further and the frequency of polished stone tools increased. Pottery manufacturing techniques developed rapidly. Pastes were usually regular in shape and surface treatment methods were various. Painted pottery began to appear and the size and number of the settlements increased. Archaeological evidence suggests that the Tangjiagang culture expanded to the center of the basin and spread to reached other river valleys. Settlements in the northern part of the Jianghan Plain probably had close contacts with the contemporaneous cultures of the neighboring Zhongyuan region.

The Daxi culture period (6300–5500 year BP, Table 19.1) in the middle Yangtze region was marked by rice agriculture (Guo 1985). The pottery manufacturing techniques of this period were fairly advanced, jade artefacts started to emerge and settlements were larger and more than five times the number of the previous period (Guo 1992). Social stratification is indicated by better housing at the center and lower ranking ones at the margin of the settlements. Around 6000 year BP, the first city in China with large walls and wide trench enclosures was built in the western part of the Dongtinghu Plain (Hunan Provincial Institute of Archaeology and Cultural Relics 2007). Residential areas, pottery workshops, cemeteries were arranged separately inside the city. Wealth polarization and human sacrifices indicate that there were conflicts and tension between different social groupings of the Daxi society.

The Qujialing culture period (5300–4600 year BP) saw an increase of number and density of settlements. The culture at this period expanded to the upper Yangtze River and also to the Zhongyuan region. Pottery manufacturing techniques were relatively advanced and textile technique developed (Zhang 1992). However, social conflicts increased and wars were frequent (Guo 2003). Intensified social stratification is reflected in differentiation of the settlements. More than ten central city sites with defensive structures have been found. The most important discovery is the Shijiahe city site, a square city in the northern part of the Jianghan Plain, which is considered the capital city of the Qujialing culture. It covers an area of over 1 million m^2, with each wall built with rammed soil, 1100–1200 m in length, 50 m in width and an average height of 5 m remaining. The city is enclosed by a deep moat approximately 100 m wide (Zhao 2006b). The Shijiahe culture period (4500–4000 year BP) saw an eastward expansion and withdrawal from the Zhongyuan region in the north. The number of settlements increased. Shijiahe city was possibly still the political and economic center in this region. Houses, palaces, cemeteries for the ordinary people and the nobles were all orderly planned and separately arranged in different zones. One religious site was found in the city. Over 1 million red pottery cups for ritual use were discovered at the site (Zhao 2006b). Large numbers of jade items were also unearthed. Evidence suggests that manufacture and use of bronze artefacts might have emerged at this period (Zhang 1992).

In the late Shijiahe period (ca. 4000 year BP), civilization in this region declined and all the cities were abandoned. The social development of this region did not resume until the southward expansion of the Bronze Age Shang and Zhou cultures from the Zhongyuan region occurred.

After developing over several thousand years, the civilization of the middle Yangtze region waned quickly around 4000 year BP. Though the exact reasons for these declines are not clear, three important things might be implicated. First, while human settlements spread into the low-lying plains during 5000–4000 year BP, these same plains may have subsided, with attendant rise in sea level, the water level rose, and silts accumulated in rivers and lakes. These processes caused intensive flood disasters before 4000 year BP, which is shown by the fact that some archaeological sites were buried by several meters of fluvial silts. Second, a cooling-and-drying climate may have reduced local crop production. Third, conflicts within the region, brought on by the environmental factors, may have been intensified by invasions from the Zhongyuan areas.

19.3.6 The Lower Yangtze River Region

This region encompasses the plains and the hills along the lower section of the Yangtze River and areas near the Hangzhou Bay (Fig 19.13). The plain's elevation

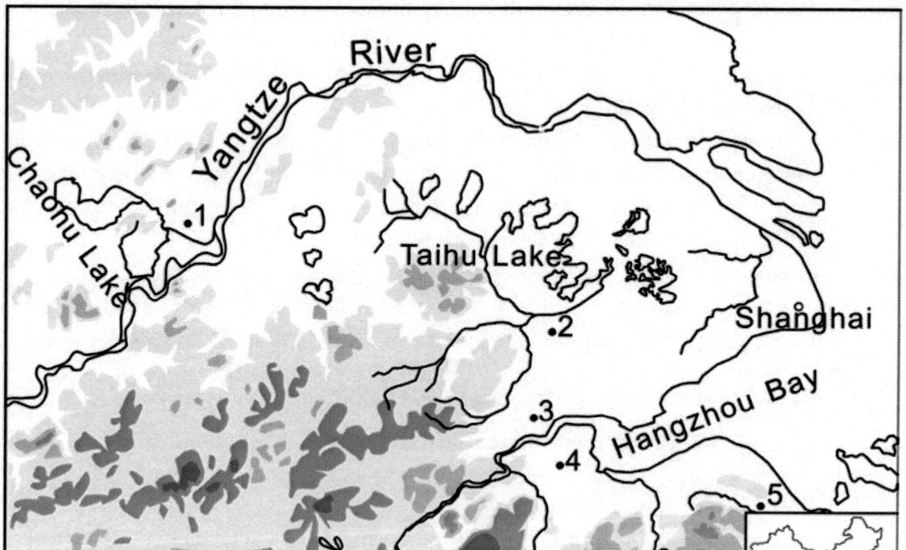

Fig. 19.13 Map of the lower Yangze River and Neolithic sites. (Sites: *1* Lingjiatan; *2* Qianshanyang; *3* Liangzhu; *4* Kuahuqiao; *5* Hemudu; *6* Xiaohuangshan; *7* Shangshan)

is generally lower than 5 m asl. Numerous rivers and lakes are distributed in the lowland of this region. The mean annual temperature is 15–16 °C and the mean annual precipitation is 900–1500 mm. It is an area where evergreen and deciduous broad-leafed forests co-existed. The Holocene climate changes of this region were probably similar to those of the middle Yangtze River region (Zhao et al. 2007).

The Shangshan culture (10,000–8500 year BP) in the river valley south of the Hangzhou Bay is known as the earliest Neolithic culture in this region (Table 19.1; Jiang and Sheng 2007). Though plant evidence of rice cultivation has been found, hunting and gathering may still have played an essential role in the subsistence economy. Production tools were principally made with rough stones. Common pottery vessels were dishes with inverted rims and flat bases.

The Kuahuqiao culture (8000–7000 year BP) was discovered on the south bank of Hangzhou Bay. Large amounts of charred rice grains from the site are often viewed as evidence of rice agriculture. Domesticated animals included dogs, pigs and buffalos (Zhejiang Provincial Institute of Cultural Relics and Archaeology, The Xiaoshan Museum 2004; Zong et al. 2007). Hunting and gathering still provided the major food resources for the local people. Cultural assemblages of the Kuahuqiao culture are dominated by polished stone tools and refined pottery vessels. Well-preserved wooden boats and paddles and basketry dustpans have been discovered at the Kuahuqiao site.

Later, the plains south to the Hangzhou Bay were occupied by the Hemudu culture (7000–5000 year BP). The Majiabang culture (7000–6000 year BP) and the Songze culture (6000–5300 year BP) successively occupied the plains between the north Hangzhou Bay area and the Yangtze River.

At the Hemudu site, large amounts of charred rice and a bone plough suggest an advanced agricultural economy. Livestock breeding appeared, though hunting and gathering were still important. Artefacts made of stone, wood, pottery, bone, tooth, lacquer, jade, and fabrics show advanced craft techniques were practiced. The pottery vessels were often tempered with charcoal. Development in construction techniques is reflected in wood building remains (Liu 2006; Jiao 2007).

The Majiabang culture shares many similarities with the contemporaneous Hemudu culture. Yet, red pottery shards are more commonly seen in the assemblages of the former. Both the Majiabang and Songze are well-known for polished stone tools and refined jade artefacts. Settlement differentiation occurred during the late period of the Songze culture, which indicates a complex society (Wang 2006).

The Songze culture was replaced by the Liangzhu culture (5300–4000 year BP), that has become known as one of the most advanced Neolithic cultures of this period in China. The culture is characterized by the intensive use of stone ploughs, animal domestication, outstanding jade ritual artefacts with animal and human images (Fig. 19.14), silk fabricating techniques, and black pottery production (Zhang 2004). Differentiation of settlements occurred, including hundreds small ordinary ones, tens of local and bigger regional central settlements, and a large capital city at Mojiaoshan. The city site extends over an area of over 3 million m^2 with a central square platform nearly 1 km^2 in size of rammed earth several to over 10 m thick, probably the basement of some important construction. Large wooden beams

Fig. 19.14 Jade in Liangzhu Culture sites. **a** Jade hoop (Liangzhu Culture site) (After Jiang 2007). **b** Jade Comb (Liangzhu Culture site) (After Jiang 2007)

were unearthed near this platform (Yan 2000). The city is enclosed by walls 40–60 m wide with rounded corners, built with rammed soil, and erected directly on top of the foundation stones. The surrounding trench outside the city is over 100 m wide (Liu 2008). The construction of the city, large cemeteries, and numerous jade ritual artefacts indicate that the Liangzhu complex society was highly stratified and ruled by a strong power base.

The late Liangzhu period (ca. 4200 year BP) saw a rapid decline of the Neolithic culture and local economy. Hydrological problems might have driven this. Due to the increase of the population the human settlements of the Liangzhu culture widely expanded and became densely distributed in the low-lying plains within the river and lake network. Neotectonic subsidence, raising sea level since the middle Holocene (Zong 2004), and gradual siltation of the Yangtze River delta added a few additional meters rise to river water levels. Some archaeological sites of the Liangzhu culture once located at or near sea level are now found at the bottom of Taihu Lake or are buried in some 3 m of fluvial silts. The climatic change to cooler and dryer conditions and the conflicts between local settlements and with those of the lower Yellow River region may also have been other reasons for the decline of the Liangzhu culture.

19.4 Summary

Climatic and environmental changes in the central-eastern part of China were conductive to the early emergence of Neolithic cultures and to their continued development. By virtue of warm temperatures, copious precipitation, and a receptive landscape, the vast low-lying plains of the middle and lower Yangtze River region were ideal for the development of rice agriculture. Thus the Neolithic culture that is characteristic of rice farming and livestock breeding can be traced back there to 10,000 year BP. In the Zhongyuan region, mainly in the middle reaches of the Yellow River, not only climate and landform, but also the high fertility of the loess soils has contributed to the development of dryland agriculture based on foxtail and broomcorn millets. These crops, together with livestock breeding are characteristic of the Neolithic cultures there, which also date back to 10,000 year BP. By about 8000 year BP Neolithic cultures had begun in many regions west of the Liao River, in the upper, middle, and lower Yellow River, and the Yangtze River, and continued to develop and flourish for the next 2000 years under stable warm and humid climatic conditions.

In spite of similarities these Neolithic cultures show differing characteristics and follow different evolutionary paths. More research is needed to fully explain the differing developmental paths, though three hypotheses may be put forward at this time. First, it is a normal consequence of human development that cultures that are isolated from each other will develop different social structures as a response to local conditions. Second, the same isolation, as well as variations in the availability of material resources, will lead to technical differences, for example in methods and intensity of resource exploitation and production. Third, cultural and material (trade) exchanges between the cultures could account for similarities that may develop.

From 6000 to 5000 year BP climatic conditions were most favorable particularly in the middle and lower Yellow River and Yangtze River. A big leap in the development of production techniques and many new cultural changes occurred, in particular the Neolithic societies became stratified but not sufficiently to reach the level of the subsequent 'Chinese Civilization' stage. The climate quality declined after the Holocene Optimum, between 5000 and 4000 year BP and the Neolithic cultures in different regions presented different evolutionary histories due to the different regional environment conditions. The Gansu-Qinghai and western Liao River regions had the lowest precipitation of the six regions. The cooling and drying climate around 5000 year BP resulted in difficult conditions for the foxtail millet and broomcorn millet agriculture, the development of culture and the process of social stratification of the two regions slowed. In the middle and lower Yangtze River regions, the numerous floodplains with alluvial soils, the temperature still high, and abundant rainfall, were good for the cultivation of rice. The Neolithic cultures developed and the population grew rapidly between 5000 and 4000 year BP (Li et al. 2009), based on rice cultivation and rich plant and animal resources. However, before 4000 year BP, hydrological changes led to frequent floods and cooler and drier climate rendering conditions difficult. The situation was further exacerbated by densely distributed human settlements in low-lying plains as a response to population increase. This and the increasing conflicts between settlements in the region and with populations of other areas, led to a collapse of the social structures of the area. The regression is also clearly seen in the Haidai region. Early civilizations developed a complex social organization there with the building of cities. However, around 4000 year BP, the number of settlements decreased, pottery technique regressed, and the overall cultural development slowed because of hydrological and climate problems and the conflicts within the region. By contrast, in the Zhongyuan region the indigenous culture continued to develop. Around 4000 year BP, with the formation of the first dynasty, Xia Dynasty, though there are still some contrary arguments, increasing evidence indicates that the society of this region evolved into the early 'Chinese Civilization' stage (Liu 2004; Xu and Du 2005). Subsequently, the Zhongyuan region became the center of Chinese Civilization gradually influencing the culture and politics of the other regions. This was the process that has been called the 'unification of multiple cultures into one civilization: the Chinese Civilization' (Liu 2004; Zhao 2006a, b).

References

An, CB, Tang, LY, Loukas, B, Chen, FH (2005) Climate change and cultural response around 4000 cal yr BP in the western part of Chinese Loess Plateau. Quaternary Research 63:347–352

Archaeological Team of Gansu Province (1986) Excavation brief report of the Dadiwan No. 901 Room site, Qin'an, Gansu. Cultural Relics 2:1–12

Baoding Institute of Cultural Relics, Department of Archaeology, Peking University (1992) Brief report of excavation on the Nanzhuangtou site in Xushui Country, Hebei Province. Archaeology 11:961–970

Chen Z, Wang Z, Schneiderman J, et al. (2005) Holocene climate fluctuations in the Yangtze delta of eastern China and the Neolithic response. The Holocene 15:915–924

Cui HT, Kong ZC (1992) Primary climate change analysis of Megathermal Maximum in central and eastern region of Inner-Mongolia. In: Shi YF, Kong ZC (eds) Climate and environment of Holocene in China. Ocean Press, Beijing, pp 72–79

Erlitou Working Team (2003) Achievement in palace excavation of the Erlitou site. Communication of Ancient Civilization of China 5:50–53

Fang H (2003) The decline reason of Yueshi Culture. Literature History and Philosophy 3:139–144

Feng ZD, An CB, Wang HB (2006) Holocene climatic and environmental changes in the arid and semi-arid areas of China: a review. The Holocene 16:119–130

Gansu Provincial Museum, Qin'an Cultural Centers, Dadiwan Excavation Team (1981) Early Neolithic remains of the Dadiwan site, Qin'an, Gansu. Cultural Relics 4:1–8

Gansu Provincial Institute of Cultural Relics and Archeology (2006) Excavation Brief Report of the Dadiwan Neolithic site, Qin'an. Cultural Relics Press, Beijing, 945 pp

Gao GR (2000) The Pre-Qin Archaeological collection of Haidai Region. Science Press, Beijing, 341 pp

Gao GR, Luan FS (2004) Dawenkou Culture. Cultural Relics Press, Beijing, 205 pp

Gao GR, Shao WP (2005) Haidai Culture and Qilu Civilization. Jiangsu Education Press, Nanjing, 744 pp

Gong QM (2002), Yangshao Culture, Cultural Relics Press. Beijing, 239 pp

Guiyang Institute of Geo-Chemistry (1977) Natural environment change in the southern part of Liaoning Province. Science in China 6:605–614

Guo F (1992) Prediction of settlements size and population growth trend: study on population at different stages of neolithic cultures in the Middle Yangtze River Region. Relics From South 1:55–60

Guo L (2003) Conflict and war: from Daxi to Shijiahe. In: Committee of State Council, State Cultural Relic Bureau (eds) Three Gorges Project Construction. Academic Papers on Heritage and Archaeological Research in Three Gorges. Science Press, Beijing, pp 137–156

Guo SB (1985) The original agriculture remains of the Chegushan site. Hunan. Agricultural Archaeology 2:86–89

Henan Institute of Cultural Relics and Archaeology (1983) Excavation on the Wangchenggang site, Dengfeng. Cultural Relics 8:8–20

Henan Institute of Cultural Relics and Archaeology (1992) Excavation on the Haojiatai site, Yanshi. Huaxia Archaeology 3:71–100

Henan Institute of Cultural Relics and Archaeology (2002) Brief report of excavation on the Guchengzhai site of Longshan Culture, Xinmi, Henan. Huaxia Archaeology 2:53–82

Hunan Provincial Institute of Archaeology and Cultural Relics (2006) Pengtoushan and Bashidang. Science Press, Beijing, 734 pp

Hunan Provincial Institute of Archaeology and Cultural Relics (2007) Chengtoushan in Lixian – Excavation report of a Neolithic Site. Cultural Relics Press, Beijing, 809 pp

Institute of Archaeology, CASS (1988) The Sanlihe site of Jiao County. Cultural Relics Press, Beijing, 201 pp

Jiang LP, Sheng DP (2007) The Shangshan site and Shangshan Culture. In: Mo DW, Cao JY, Zheng WH, et al. (eds) Research of Environmental Archaeology (Volume 4). Peking University Press, Beijing, pp 25–42

Jiang WD (2007), Sanctity and Refinement, Zhejiang Photography Press, Hangzhou, 293 pp

Jiao J (2007) A 5600-year-old wooden well in Zhejiang Province. Journal of Hydrogeology of China 15:1021–1029

Kong ZC, Du NQ, Liu GM (1991) The Preliminary Study of Environmental Archaeology 8000a-2400a BP in Chifeng City, Inner-Mongolia. In: Zhou KS (ed) Researches of Environmental Archaeology (Volume 1). Science Press, Beijing, pp 112–119

Kong ZC, Du NQ, Shan FS, et al. (1990) Holocene vegetation change and climate analysis around Qinghai Lake: the palynological data analysis in QH85-14C Bore. Marine Geology & Quaternary Geology (China) 10 (3):79–96

Li XQ, Dodson J, Zhou J, Zhou XY (2009) Increases of population and expansion of rice agriculture in Asia, and

anthropogenic methane emissions since 5000 BP. Quaternary International 202:41–50
Li YM (2003), Peiligang culture, Cultural Relics Press, Beijing, 211 pp
Liaoning Institute of Archaeology (1997), Jade of Hongshan Culture. Cultural Relic Press, Beijing, pp a-52, b-71, c-69, d-51, e-53, f-50
Liu HW (2008) The germination of jade carving remains of Liangzhu Culture. Archaeology and Cultural Relics 1:29–34
Liu J (2006) Hemudu Culture. Cultural Relics Press, Beijing, 192 pp
Liu L (2004) The Chinese Neolithic –Trajectories to Early States. Cambridge University Press, Cambridge, 310 pp
Lu HY (1989) The spore-pollen assemblages in the sediments in the southern Bohai sea since the late Pleistocene and its palaeoenvironment analysis. Journal of Oceanography of Huanghai and Bohai Seas 7:11–26
Lu HY, Zhang JP (2008) Neolithic culture evolution and Holocene climate change in the Guanzhong Basin, Shaanxi, China. Quaternary Sciences 28:1050–1060
Lu HY, Zhang JP, Liu KB, et al. (2009) Earliest domestication of common millet (Panicum miliaceum) in East Asia extended to 10,000 years ago, Proceedings of the National Academy of Sciences 106:7367–7372
Luan FS (1997) The Archaeological Research on Haidai Region. In: Luan FS (ed) The Archaeological Research on Haidai Region. Shandong University Press, Jinan, pp 318–347
Mo DW, Li F, Li SC, Kong ZC (1996) A preliminary study on the paleoenvironment of the Middle Holocene in the Hulu River region in Gansu Province and its effects on human activity. Acta Geographica Sinica 51:59–69
Mo Duowen, Yang Xiaoyan, Wang Hui (2002) Study on the environmental background of the Niuheliang site, Hongshan Culture, and the relationship between ancient man and environment. Quaternary Sciences 22:174–181
Morris C, Overpeck JT, Cole JE (2003) A synthesis of abrupt changes in the Asian summer monsoon since the last deglaciation. The Holocene 13:465–476
Participants of the Training Course for Excavation Managers (1999) Excavation of a city in Yangshao Period in the Xishan site, Zhengzhou City. Cultural Relics 7:4–15
Puyang Wenguan Hui (1988) Brief report of excavation on the Xishuipo site, Puyang, Henan Province, Cultural Relics 3:1–6
Shandong Administration of Relics (1974) The Dawenkou site. Cultural Relics Press, Beijing, 164 pp
Shandong Fieldwork Team, Institute of Archaeology of the Chinese Academy of Social Sciences, the Museum of Teng County, Shandong (1984) The excavation report in the Bexin site in Teng County, Shandong. Journal of Archaeology 2:159–191
Shandong Provincial Institute of Cultural Relics and Archaeology (1989) The clearing of burials of Longshan Culture in Xizhufeng, Linqu. In: Zhang XH (ed) Archaeology in Haidai Region(Volume 1). Shandong University Press, Jinan, pp 219–224
Shaanxi Provincial Institute of Archaeology (1999) 50th Anniversary of Archaeology Research in Shaanxi Province. Cultural Relics Press, Beijing, 1264 pp
Shi YF, Kong ZC, Wang SM, et al. (1992) The basic features of climate and environment in China during Holocene Megathermal period. In: Shi YF, Kong ZC (eds) The climate and the environment during Holocene Megathermal period in China. Ocean Press, Beijing, pp 1–18
Song YQ (2002) The Simple Comment on the Man-land Relationship of the Origin of Chinese Civilization. Science Press, Beijing, 227 pp
Su BQ (1988) The New Dawn of Chinese Civilization. Southeast Culture 5:1–7
Sun JZ, Ke MH, Sun XR, et al. (1991) Holocene paleoenvironment in Loess Plateau. In: Sun JZ, Zhao JB (eds) Quaternary in Loess Plateau. Science Press, Beijing, pp 186–205
Tang LY, An CB (2007) Sporopollen record of vegetation change and drought events in Longzhong Loess Plateau Holocene. Progress in Natural Sciences 17:1371–1382
Tang LY, Shen CM (1992) The vegetation and climate during Holocene Megathermal period in northern Jiangsu. In: Shi YF, Kong ZC (eds) The climate and environment during Holocene Megathermal period in China. Ocean Press, Beijing, pp 80–93
Tarasov P, Jin GY, Wagner M (2006) Mid-Holocene environmental and human dynamics in northeastern China reconstructed from pollen and archaeological data. Palaeogeography, Palaeoclimatology, Palaeoecology 241:284–300
The Inner Mongolia Work Team of Institute of Archaeology CASS (2004) The excavation on the Xinglonggou site in the year of 2002-2003, Chifeng City, Inner-Mongolia. Archaeology 7:3–8
The Radiocarbon Laboratory of Peking University (1996) The Carbon 14 dating report. Cultural Relics 6:91–95
Tian GL (2004) The Origin of Civilization of Western Liao River Region in Northeastern China. Chinese Publishing House, Beijing, 290 pp
Tong GB, Zhang JP, Yan FH, Mai XX (1991) Sporo-pollen sequence and division of climatic period in the eastern North China plain since late Pleistocene. Seismology and Geology, 13:259–265
Tong WH (1984) Agricultural remains and its correlative problems in the Cishan site. Archaeology 1:194–207
Wang J, Lu HY, Shen CM (1990) Factor interpretation method and quantitative analysis of climatic changes over the last 12000 years in China. In: Liu C, Zhao SQ, Zhang PY, Shi PJ (eds) Global Changes and Environmental Evolution in China. Proceedings of Section III, Regional Conference on Asian Pacific Countries of IGU, The Inner Mongolia Press of Science and Technology, Hohhot, CN, pp 62–68
Wang SM (2006) Preliminary studies on prehistorical social complex evolvement in Taihu Lake Area. Southeast Culture 3:10–16
Wang YJ, Cheng H, Edwards RL, et al. (2008) Millennial- and orbital-scale changes in the East Asian monsoon over the past 224,000 years. Nature, 451:1090–1093
Wang ZL (1995) Application of opal phytolith analysis in the Yuchisi site, Mengcheng, Anhui. Archaeology 1:88–90
Wu J, Chen M, Liu C (2009) Astronomical function and date of the Taosi observatory. Science in China Series G: Physics, Mechanics & Astronomy 52(1):151–158
Xia ZK, Deng H, Wu HL (2000) Geomorphologic Background of the Prehistoric Cultural Evolution in the Xilamulun River Basin, Inner-Mongolia. Acta Geographica Sinica 55:329–336
Xie DJ (2002) Prehistoric archaeology of Gansu-Qinghai region. Cultural Relics Press, Beijing, 258 pp

Xie XG (2007) Research on Xiangfen Taosi site, Science Press, Beijing, 724 pp

Xu H, Du JP (2005), Research on Yanshi Erlitou site, Science Press, Beijing, 852 pp

Xue ZQ (2008) Discussion on the position and function of the Xinglongwa culture in the origin of the Chinese culture. Journal of Dalian University 29(5):11–15

Yan WM (1986) The Lepan site in Zhangqiu County from Dawenkou Culture to Shang Dynasty. In: Chinese Society of Archaeology (ed) The Archaeological Almanac of China. Cultural Relics Press, Beijing, CN, pp 136

Yan WM (2000) Agricultural Occurrence and Origin of Civilization. Science Press, Beijing, 338 pp

Yuan J (2001) Origin of domestic animals during neolithic period in China. Cultural Relics 5:51–58

Zhang J, Harbottle G, Wang C, Kong Z (1999) Oldest playable musical instruments found at Jiahu early Neolithic site in China. Nature 401(6751):366–368

Zhang WX, Yuan JR (1998) A Preliminary study on the ancient rice excavated from Yuchanyan, Daoxian, Hunan Province. Acta Agronomic Sinica 24:416–420

Zhang XH (1993) The Chengziya site and Chinese Civilization. In: Zhang XH (ed) Academic Papers on Archaeological Research for 60th Anniversary of Excavation on the Chengziya site. Qilu Book House Press, Jinan, pp 13–25

Zhang XQ (1992) Outline of Neolithic Cultures in the Middle Yangtze River. Hubei Science and Technology Press, Wuhan, 326 pp

Zhang ZB, Wang D, Ding JX (1981) Natural environment change around Beijing City 13000 a BP. Scientia Geological Sinica 3:259–268

Zhang ZH (2004) Neolithic Culture in Lower Yangtze River. Hubei Education Press, Wuhan, 268 pp

Zhao BC, Wang ZH, Chen ZY, WuG (2007) Climate, vegetation and geomorphology evolution since 8 ka BP recorded by sediments from dish-like depression of Taihu Lake Plain. Journal of Palaeogeography 3:321–330

Zhao H (2006a) The prehistoric basis of China: the central region of China(Zhongyuan) as the center of the historical trend. Cultural Relics 8:50–54

Zhao H (2006b) The dawn of Chinese civilization. In: Yan WM, Li L (ed) The history of Chinese civilization (1). Peking University Press, pp 31–94

Zhao ZJ (2005) Botanical archaeology and recent developments. Archaeology 7:42-50

Zhao ZJ, Piperno D (2000) Late Pleistocene/Holocene environments in the middle reaches of Yangtze River Valley, China and rice (*Oryza sativa* L.) domestication: the phytolith evidence. Geoarchaeology: An International Journal 15: 203–222

Zhejiang Provincial Institute of Cultural Relics and Archaeology, The Xiaoshan Museum (2004) Kuahuqiao. Cultural Relics Press, Beijing, 379 pp

Zhoukou Relics Office (1983) Brief report of excavation on Pingliangtai City of Longshan Culture, Huaiyang, Henan Province. Cultural Relics 3:21–36

Zong Y (2004) Mid-Holocene sea-level highstand along the Southeast Coast of China. Quaternary International 117:55–67

Zong, Y, Chen, Z, Innes, JB, et al. (2007) Fire and flood management of coastal swamp enabled first rice paddy cultivation in east China. Nature 449:459–463

Chapter 20
Landscape and Subsistence in Japanese History

Gina L. Barnes

20.1 Introduction

Japan consists of four main islands plus the Ryukyu archipelago south of Kyushu Island (Fig. 20.1). It stretches from 20°24′N to 45°30′N, thus crossing cold to warm temperate and subtropical climatic zones. Because of its backbone mountain ranges, the climate of Honshu Island is radically different on the Japan Sea side and Pacific Seaboard. Seasonal monsoonal winds bring heavy snowfall to the western side in winter, and a June rainy season and typhoons from the south/southeast in summer. Average annual precipitation ranges from 944 to 4060 mm but mostly exceeds 1020 mm/year. The forest types vary by latitude and altitude, with broad-leaf evergreens at low elevations in the southwest, deciduous forest at higher altitudes in the southwest and through the northeast into Hokkaido Island, while high northeastern mountains host conifers (Statistics from www.britannica.com and Kojima 2004).

Deep green forests on steep mountainsides dominate the Japanese Islands. This inviting yet forbidding topography is a product of rapid and continuing uplift due to the subduction of the Pacific and Philippine plates under the Eurasian plate at the Japanese archipelago's eastern edge (Barnes 2003). The mountains themselves are mostly folded and faulted due to the pressures of the opposing plates, but subduction also gives rise to volcanic and earthquake activity (Barnes 2008). In northeastern and southern Japan volcanoes dominate the landscape, while most areas nationwide are at continuous earthquake risk.

The hazards of living in a tectonically active area might seem a deterrent to human occupation. Yet Japan is the fifth most densely populated country in the world, with an average of 343 persons/km^2. However, the population is unevenly spread over Japan's total land area: seemingly insignificant for island composition are the flattish lands of fringing coastal plains and mountain basins, yet this is where Japanese civilization resides. Residential land accounted for only 3.0% of Japan's land area and is concentrated in such lowlands (2005 calculation: MLIT 2007). In 2005, Tokyo's population density was 17 times higher than the national average, at 5751 persons/km^2 (MIAC 2008).

In line with the aims of this volume, the question arises: how did Japanese society adapt so successfully to this archipelago where both population and agricultural production are squeezed onto the plains, which account for only 35% of Japan's total land area. I define three major adaptations within Japan's historical sweep that address changes in landscape use (Table 20.1): the prehistoric hunting-gathering-fishing-horticultural subsistence systems of the Jomon peoples, the historic system of wet-rice agriculture, and the modern opening of Hokkaido and importation of new foods and food technologies. Between Phases 1 and 2, occupation shifted from the uplands to the coastal plains, and between Phases 2 and 3, agriculture in Japan itself ceased to be the major food supplier for the Japanese people. This scheme is outlined in the first section below.

The dichotomy between upland and lowland is not a new phenomenon in Japanese history, nor is it unusual amongst other nations. Yet Japan did not develop ethnically separate hill tribes who exploited the uplands while

G. L. Barnes (✉)
Department of Art & Archaeology at SOAS (School of Oriental & African Studies), University of London, Thornhaugh Street London WC1H 0XG, UK
e-mail: gb11@soas.ac.uk

See Plate 19 in the Color Plate Section; also available at: extras.springer.com

I. P. Martini, W. Chesworth (eds.), *Landscapes and Societies*,
DOI 10.1007/978-90-481-9413-1_20, © Springer Science+Business Media B.V. 2010

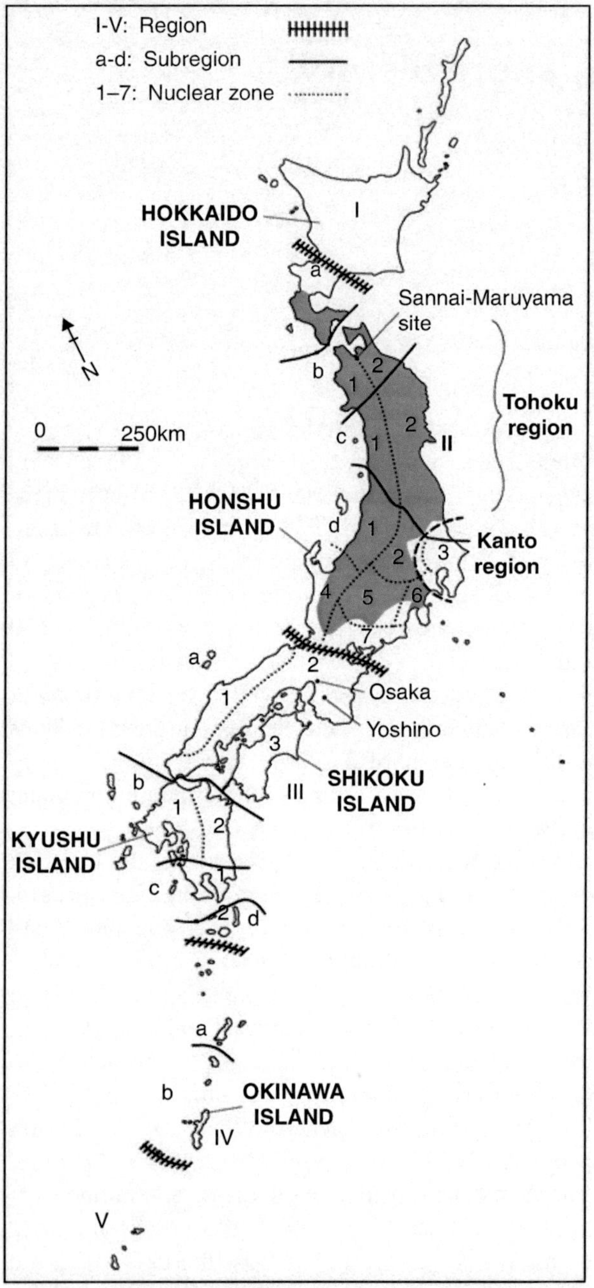

Fig. 20.1 Zones of Jomon pottery styles and regional divisions (After Kobayashi 2004). (*I–V* regional divisions; *a–d* subregional divisions; *1–7* nuclear zones of Jomon pottery distributions; *gray area* northeastern deciduous; broad-leaf forest)

agriculturalists occupied the lowlands. One focus below is how the Jomon exploited hills and coastline. A second focus is the integration of the Jomon into agricultural Yayoi society based on the plains; another focus is the development of the plains themselves, which are of extremely recent formation by geological standards. There is a tendency to think that once rice agriculture was introduced into the lowlands, other forms of food production became unimportant. A fourth focus will illustrate how the development of dry field agriculture has been vital to the continuing success of the historical economy. In particular, I follow up Ohnuki-Tierney's work (1993) demonstrating that no matter how ritually and symbolically important rice was and is to the historic Japanese, they did not and do not live by rice alone.

20.2 Historical Context of Landscape Exploitation

With each major subsistence transition, the carrying capacity of the Japanese archipelago grew until it reached modern proportions. The first phase witnessed a temporary concentration of communities in the central mountains and Kanto shorelines in the Middle Jomon period, but these were not sustainable (Kobayashi 2004). During that time, a symbiosis developed between Jomon peoples on the coasts, where fishing and shellfish collecting were important subsistence strategies, and those living in the mountains. The complementary concepts of *yama-no-sachi* (mountain treasures) and *umi-no-sachi* (sea treasures) in Japanese cuisine reflect this Jomon orientation towards the landscape. But the Jomon were more than hunters, fishers and gatherers: they were also horticulturalists, tending nut trees and cultivating several species of grains and pulses. Imamura (1996, p. 90) boldly states that "larger populations are linked to the prominence of plant foods, and smaller populations are linked to the relative importance of hunting".

The coastal plains were the focus of the second (historic) and third (modern) phases where a reliance on rice agriculture developed. Around the world, agriculture of some sort is anthropologically recognized to be essential for state formation. Whereas the Jomon existed for thousands of years without hierarchical social forms, it only took between 600–1000 years after rice came under cultivation that the first complex societies and the state arose in the Kofun period.

Rice can be grown in either wet or dry conditions, but it is not native to Japan—the seeds and technology are continental imports. Dry rice was possibly grown first in the Final Jomon along with other dry-field crops. Wet-rice technology, ultimately from south-

Table 20.1 The three subsistence phases of Japanese society and historical periodization, with emphasis on Phase 2—mainly compiled from Farris (2006, 2009) and Smith (1959). (c=century (e.g. 18c); ? = estimates by Farris; mil = million; dev = development; eml = early, middle, late; BCE = BC, CE = AD)

PERIOD	FOOD & POPULATION	TECHNOLOGIES	LANDSCAPE & ENVIRONMENT
Phase 1	PREHISTORIC	Hunter-Gatherer-Fisher (HGF)-Horticulturalists	
JOMON	14,000–300BCE [Calibrated dates BCE, Pearson, 2006: Incipient Jomon 13,680–9250; Initial (Earliest) Jomon 9250–5300; Early Jomon 5300–3360/3500; Middle Jomon 3360/3350–2580/2510; Late Jomon 2580/2510–1260/1230/1220; Final (Latest) Jomon 1260/1230/1220–(410)BCE]		
	HGF, fishing: riverine, coastal, shellfish collection, detachable harpoon, deep sea; horticulture	Chipped & polished stone tools, postulated but unproven swidden, low-fired earthenware, sedentary settlements	Clearance around settlements, possible nut tree curation, rise & fall of postglacial sea level
Phase 2	HISTORIC	Rice Agriculturalists	
YAYOI	800BCE–250CE [900/800–500BCE Initial Yayoi, 500–200BCE Early Yayoi (EY), 200BCE–0CE Middle Yayoi (MY), 0–200 Late Yayoi, 200–250 Terminal Yayoi]		
	HGF, wet-rice agriculture in southwest, dry-field agriculture in northeast; green manuring; iron shoes on agricultural tools; dramatic population increase;	Chipped & polished stone tools, from MY bronze casting & iron forging, low-fired earthenware, continental trade from MY (mirrors, weaponry), moated villages	Paddy clusters on coasts and river bottomlands, dry fields on levees and terraces, cold spell & lowered sea level in EY, flooding in M/LY
KOFUN	250–710 [250–400 Early Kofun, 400–500 Middle Kofun (MK), 500–710 Late Kofun]*		
	Reduced reliance on hunting, still gathering & fishing, continued population increase; emergence of chieftains, then state by 6c	New continental iron agricultural tools, prestige goods, iron armour & stoneware production; agricultural hamlets, elite housesteads (dispersed settlement pattern)	increasingly geometric paddy fields, large canals for transport, deforestation for kiln fuel
NARA-HEIAN	710–1185 [Nara period 710–794: Nara capital 710–784, Nagaoka capital 784–794. Heian period 794–1185: Heian capital 794–1185; Civil War 1180-85]		
	8c, 20-25% dry farming; 730 population ?5.8-6.4 mil; 698–1061 smallpox epidemic every 30 years; depopulation then recovery to ?5.5-6.3 mil by 1150; 20-30% arable abandoned by mid-Heian; reclaiming of arable in 12c	Urbanization but dispersed rural settlement; after 950 adoption of ox-drawn mouldboard plough; seedling transplanting	Gridded paddy field reorganizations in 690–900, 900–1050, 1050–1200; 670–1100 worldwide hot dry spring-summers; 125 famine years, more by drought than deforestation for kilns and monumental architecture
MEDIEVAL	1185–1603 [Kamakura 1185–1282; Nanbokucho 1382–1393; Muromachi 1392–1568; Momoyama 1568–1603; Wars 1467-1568, 1568-1600]		
	Subsistence agriculture, heavy rice taxes; 15c beginning cotton production; population decline but rebound after 1280 to ?15-17mil in 1600; floating popu. then stem family & corporate village dev.; 5% urban by 1600; warfare, servitude, piracy, bandits, riots from 1428	Great improvements after 1280: double-cropping, Champa rice, waterwheel, more livestock & fertilizer; iron smelting advances, better tools; cities increased (ports, provincial capitals); commercial economy, coinage, commercial fishing; public baths, castle building	Expansion of paddy on hills & moors; irrigation networks & flood prevention on plains; 12c early Little Ice Age?; Yowa, Kangi & Shoga famines & plagues; syphilis introduction in e16c; endemic smallpox, climate flux
EDO	1603–1867		
	Shift from cultivation by pseudo-extended families to that by nuclear families or tenants; 1697 first scientific agricultural treatise; population increase to ?31.3 mil in 1721 then 250 year stasis. Pax Tokugawa; increasing literacy; castle towns	Growth of rural trade & industry from 18c; development of market economy, cash cropping, ecological specialization; urban expansion; treadwheel irrigation, new rice harvesting tools & machines; night soil use & commercial fertilizers; sericulture intensification; advent of wage labour	Expansion of irrigation networks, reclamation of marshy lowlands for paddy; cotton-growing overtakes rice agriculture; Kyoho, Tenmei & Tempo famines; deforestation;
Phase 3	MODERN	Industrial Society	
MODERN	1868–present [Meiji 1868–1912; Taisho 1912-1925; Showa 1926–1989; Heisei 1989–present]		Colonization of Hokkaido, dairy & wheat farming; afforestation; urbanization of coastal lowlands; aggregation of lowland paddies; trainline & motorway construction
	60% dry field cultivation in late 19c; population increase from 35 to 128 mil§; from 20% urban::80% rural in 1920	Dev cold-resistant rice strains, wheeled vehicles; industrialization, militarization; agricultural	

* These are the standard Kofun dates based on material culture, but I would like to revise them to 250-350 Early Kofun, 350-475 Middle Kofun, 475-645 Late Kofun, 645-710 Terminal Kofun or preferably Asuka, based on political organization.

§ http://www.stat.go.jp/english/ data/chouki/02.htm, rounded up ern China, was imported from the Korean Peninsula early in the first millennium BC and formed the basis of Yayoi agricultural society. Wet rice was initially grown on the lowland plains, but by the end of the Kofun period the lower reaches of small river valleys had been brought into cultivation.

Wet-rice technology was imported, rather than developed indigenously, but this does not diminish its importance. What was crucial in its adoption and exploitation was that it was supremely suited to the geography of the archipelago in its capacity to support large populations on relatively little land through intensive cultivation and high yields. Nevertheless, to obtain such yields required the development of drainage, irrigation, harvesting and fertilization technologies (in that order) that allowed the achievement of those high yields; even so, these were not all assembled until almost two millennia after the introduction of wet-rice agriculture, while upland areas remained exploited by other strategies. In the transition from Jomon to Yayoi, we see a reversal in the importance of regions for the carrying capacity of the archipelago. As the agricultural population of southwestern Japan grew and exceeded the northeastern Jomon, the southwest became the vibrant center of civilization while the northeast became an agricultural backwater.

The historic Japanese culture is founded on wet-rice cultivation, to the extent that 'rice' is equated to 'self' (Ohnuki-Tierney 1993). The high yields and landboundedness that were required of such cultivators gave rise to chiefly societies of the Yayoi period, succeeding the Jomon. Rulers supported by agricultural products and peasant labor eventually formed political hierarchies, and the Yamato state emerged in the late fifth century during the Kofun period. This state, in various forms, has survived to the present day—one of the longest-lived monarchies in the world—and historically, it has been underwritten by rice.

Therein, however, lay a contradiction: though lowland wet rice produced high yields able to sustain high population densities, rice eventually became a commodity and currency of the historic elite, leaving the plebeian majority to subsist on other grains and natural products. Rice was so heavily taxed towards the end of the early historic Nara and Heian periods that cultivators fled their lands to take refuge under powerful tax-free elite estates. Double-cropping of rice and wheat began in the Heian period, as did the practice of using human waste to fertilize fields (MOE 2008), and by 1550 the forests of west-central Honshu were heavily exploited for green fertilizer (Totman 1989). Hunting had been forbidden under Buddhism, introduced in the mid-sixth century, but the discovery of butchered dog bones in medieval sites and records of specialist hunters known as 'matagi', who exploited mountain game throughout the historic periods, raise questions about the impact of these prohibitions.

The medieval period was characterized not only by war but by peasant immobility, heavy rice taxes and frequent peasant rice rebellions—these latter conditions continuing into the Edo period. The nature of rice as commodity and currency rather than food dominated its existence in the Edo period, with daimyo domain size measured in the number of 'koku' (1 koku = 180 L) of rice produced in their lands and with many samurai becoming indebted to rice merchants in support of their entertainment allowances. In the era of high taxes, even households which grew rice needed extensive swidden areas to grow subsistence crops, since the rice was either paid as taxes or sold to provide cash income (Miyamoto 1959). Peasants were thus forced to rely on dry-field grains and vegetables, and the area of dry-field cultivation was greatly expanded. Population remained stable at 26–27 million between 1720 and 1845 (Hayami and Kito 2004), interpreted by Koyama as having reached maximum carrying capacity and exploitation of the flatlands (Koyama 1983). Within ten years thereafter, Japan was opened to the West and a new era of food importation began.

Japan today prides itself on its ability to feed its own people on home-grown rice, but it has only been in the postwar period that milled white rice, the favored food, has been available to everyone. The best varieties developed in the early Modern period are now grown in the cool temperate climates of northeastern Japan, a previous agricultural backwater. Ironically, postwar farmers have received double subsidies: one to let land lie fallow because of overproduction, the other to sell their rice to the government at subsidized prices; much of that rice has been stored uneaten in the face of other cuisines that have captured the Japanese imagination (Ohnuki-Tierney 1993). The colonization of Hokkaido in the late nineteenth century, where the main products are now wheat and dairy goods, contributed greatly to a westernized diet.

However, the modern urban economy is supported by trade, not agriculture, and the coastal plains not only give crucial access to ports but facilitate increasingly extensive rail and motorway connections. The high population densities of the lowlands create economies of scale for transport and interaction that are unrivalled in the modern world. Thus, Japan in its third adaptational phase has moved beyond exploitation of its own landscape; its geography, with long coastlines and many harbors, has facilitated food importation to support a carrying capacity that would be unsustainable in isolation.

20.3 Prehistoric Hunting and Gathering

The Jomon is probably the world's longest archaeological period, lasting over 14,000 years. It was based mainly on the exploitation of the natural environment but included horticultural manipulation of indigenous and maybe some imported plant communities (Imamura 1996; Habu 2004; Kobayashi 2004). As hunter-gatherer-horticulturalists who created and used ceramics, Jomon people came to live in sedentary communities for much of the year (Pearson 2006), facilitated by rich exploitable environments within limited catchment areas and food storage technologies. The very late adoption of intensive agriculture in Final Jomon, in the form of wet rice technology as discussed below, was due not only to Japan's isolated position but also to the fact that the Jomon belonged to an East Asian littoral horizon of ceramic hunter-gatherers that included Fujian, Taiwan, the Ryukyus and the southern Korean Peninsula (Barnes 1999).

Jomon pottery styles suggest basic social divisions that correlate with topography and climate (Fig. 20.1): the central backbone mountain ranges divided communities into east slope and west slope groups; and then from north to south, major breaks occurred every several hundred kilometers (subregions). The significant division between northeast and southwest (between Regions II and III) at the waist of Honshu towards the end of Jomon has lasted throughout Japanese history, as manifested in myriad ways from politics and dialect distributions to subsistence practices and food preferences (Amino 1998; Obayashi 1990).

The Jomon population was concentrated in Region II, in northeastern Japan (Koyama 1983). This area had a markedly different post-glacial climatic orientation from southwestern Japan, as reflected in climax forest types. The major division between pottery Regions II and III does not exactly match that of the transition between the two forest types—temperate deciduous broad-leaf in the northeast (Fig. 20.1, gray area) and subtropical broad-leaf evergreen in the southwest—but it is notable that most of the northeastern populace had access to temperate forest products in their hinterlands, perhaps accounting for their greater population densities. The subtropical broad-leaf evergreen forest of southwestern Japan was far less densely occupied, judging by the fewer, smaller known settlements (Koyama 1979). The differential edibility of sweet chestnuts and walnuts in the northeast versus southwestern evergreen oak acorns is often cited as one factor in the varying population densities (Fujiyama 2004). The ecozonal strategy suggested by the ceramic zonations, running from shore to mountain watershed, is similar to that developed in Hawaii and Bali, where the round volcanic islands were traditionally divided into pie-slice divisions which integrated mountain and coasts into single social communities.

The coasts became important resource locations during the Jomon Transgression, which peaked up to +3 m above present-day levels between 5300–5000 cal years BP (Yonekura et al. 2001). As the post-glacial sea level rose and drowned the valleys that had been downcut during the glacial age, newly created shallow-water lagoons and estuaries provided habitats for shellfish and inner coastal fisheries. Shellfish collection, bay fishing and open sea fishing are attested in some of the earliest shellmound sites (Imamura 1996). The range and variety of sea protein boosted the terrestrial diet. Large shell middens developed at coastal sites where shellfish were processed; trade to the interior might account for the fact that 60% of shellmound sites are located in the Kanto (split between nuclear zones 3 and 6) (Habu 2004). The upland nuclear zone 5 in Fig. 20.1 was land-locked and did not extend to the coast, yet these central mountains were where the highest population densities developed in Middle Jomon. Instead of having direct coastal access, a trade network is postulated between coastal zones 1 and 4 in the west and zones 3 and 6 in the east, providing an ecozone symbiosis in both foodstuffs and other important resources such as obsidian and jade (Barnes 1999; Kobayashi 2004). However, Chisholm's work (2004) on C_3 isotope analysis of Honshu Jomon dietary patterns shows a difference between coastal and inland sites that indicates a regionality of diet, perhaps not fully eliminated by trade.

The Jomon diet was thus based on a variety of mountain and sea products according to seasonal, regional, and temporal availability—as illustrated in the Jomon calendar (Kobayashi 2004). Among these resources, deer and wild boar supplied terrestrial protein, but it is still unclear to what extent the salmon and trout of cold water streams were utilized (Habu 2004). Nut meats, particularly of the northern forests (chestnuts, horse chestnuts, walnuts and deciduous acorns), were important starchy staples with considerable protein content. Sweet-chestnut tree pollen accounts for 90% of arboreal pollen around the Sannai-Maruyama site in the Middle-Late Jomon periods (Takahashi 2001; Matsui and Kanehara 2006), suggesting horticultural care of this important, immediately useable resource. Ground into powder and mixed with binders such as deer or boar blood, nutmeats were sometimes made into Jomon cookies or hamburgers (Imamura 1996).

Cultivars are known in all phases of the Jomon period, from before 9000 years ago; but even with the presence of major cereal crops such as barley and dry rice in Late Jomon, these comprised less than 1% of the food remains (Matsui and Kanehara 2006). Although the Jomon were skilled horticulturalists, it is clear that grain crops utilized by the Jomon did not lead to an agricultural society. Nevertheless, there are strong suggestions that they might have cultivated their grains in swidden fields, a topic dealt with below.

Beyond hillside cultivation and the exploitation of natural habitats, there were two specific uses of the landscape in procuring Jomon foods. Pit-traps have been discovered covering hillsides or arranged in linear arrays on sloping ground (Fig. 20.2); these pits narrowed at the bottom to longitudinal slits so that an

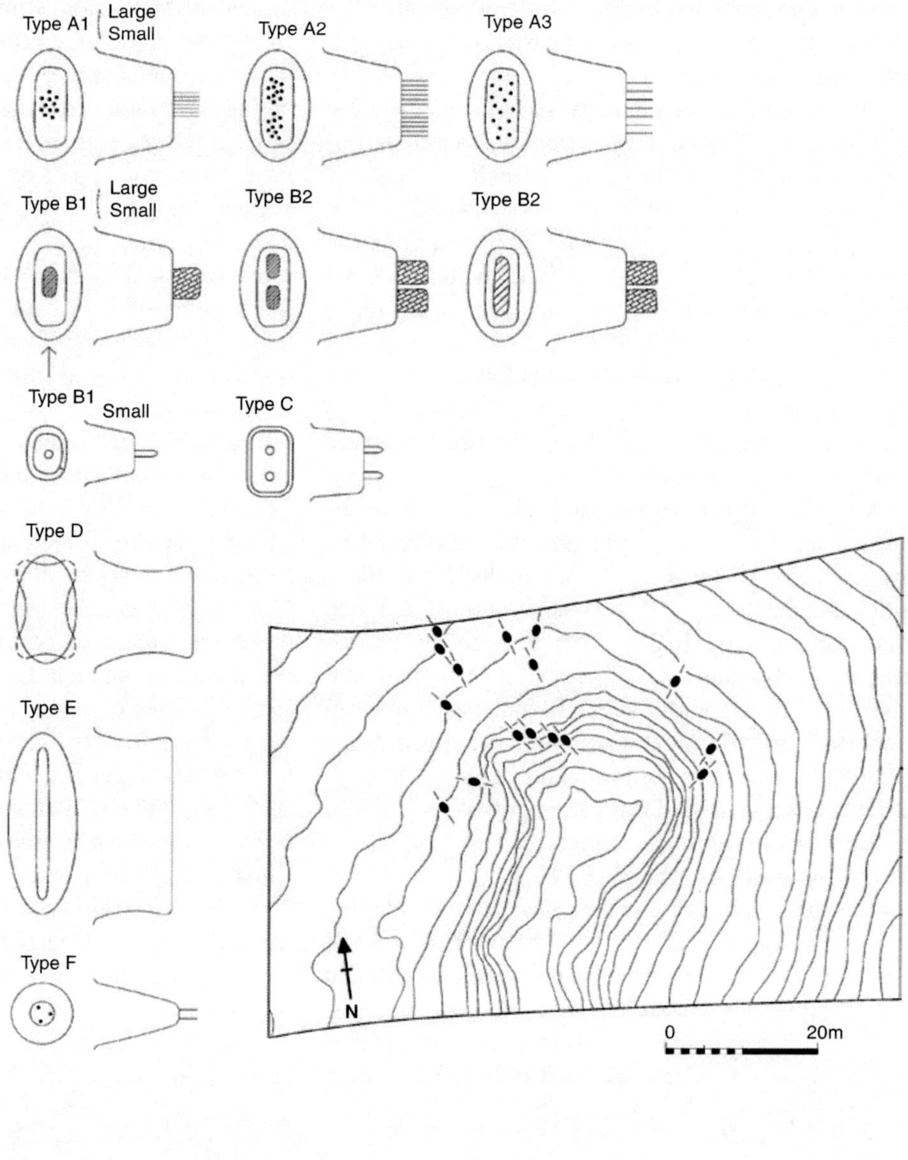

Fig. 20.2 Jomon pit-traps arrayed on a hillside, with a sampling of pit-trap types. (After Imamura 1996)

animal which fell in found their legs unable to make purchase for jumping out (Imamura 1996). The second use of pits gives insight into the unusual use of alluvial flats in the Jomon period: in western Japan storage pits were often dug in areas of high water table so that the nuts within were constantly immersed, either to prevent sprouting or to leach out tannic acid to make them edible (Habu 2004). The latter is one of the few preferential uses of alluvial plains by the Jomon, who carried out most of their subsistence activities in the uplands or at the shoreline. These plains were extremely limited in distribution, particularly along the coasts due to the high sea levels of the Jomon Transgression. The plains, growing in extent as the sea receded as discussed below, became the focus of the second, historic phase of Japanese subsistence when a reliance on rice agriculture developed.

20.4 Identity of the Agriculturalists

Wet-rice technology, ultimately from southern China, was imported from the Korean Peninsula to north Kyushu early in the first millennium BC. It has been thought that the social disturbance of the Warring States period in China (ca. 475–221 BC) caused this out-migration, but the new early dates of rice in both Korean and Japan push back the social context to the Zhou take-over of Shang (at 1027 BC) and its ramifications in the Western Zhou period (1027–771 BC). No one yet has postulated an emigration scenario for this time period.

Wet rice was preceded by dry rice and millet, but these were little exploited by the Jomon; wheat and barley followed rice and were cultivated in northern Japan by 700 AD (Crawford 2006). Formal rice paddy fields have been recovered from Final Jomon sites in north Kyushu, now reclassified as Yayoi precisely because of the presence of agriculture. It is assumed that migrants from Bronze Age Korea initially brought this technology to the islands, and skeletal remains of migrant types have been excavated from Yayoi cemeteries. Physical anthropologist Hanihara (1991) identifies this continental migration as the first of two waves which changed the physical character of the occupants of the Japanese archipelago, the second having occurred in the fifth century also from the Korean Peninsula.

Significantly, Yayoi agriculture took hold in southwestern Japan where Jomon populations were relatively scarce and where flatlands were little used. The increased productivity of rice yields enabled north Kyushu Yayoi population levels to swell, and by 300 BC further migrations through the Inland Sea area of Japan resulted in agricultural settlements along those coasts. It is still a mystery what happened to the native Jomon peoples in this region, but there is no evidence of warfare between the two peoples. A brief overlap between Jomon and Yayoi sites around Kawachi lagoon, now filled in and occupied by the city of Osaka, indicates a short period of co-existence. If disease did not decimate the Jomon population, then intermarriage and apprenticeship to field tasks would have served to absorb them into Yayoi villages, a possibility suggested also by the survival of some Jomon ceramic decorative patterns in early Yayoi products.

It is interesting that there did not develop a clear hill peoples/plains peoples dichotomy in western Japan as attested in areas of Southeast Asia—the Kachin and Shan of Myanmar (Burma) being the classic example (Leach 1970). Nevertheless, the hills were not abandoned, as hints in the eighth century 'Nihon Shoki' and 'Kojiki' chronicles attest. Peoples living in upstream areas were socially and politically distant from the developing chiefdoms of the Yayoi peoples. Some of these were groups referred to as 'kawakami' (up-river), and they were often portrayed as brigands (Aston 1972). By the late fifth century AD, a mountain people of Yoshino in Nara, called the Kuzu, were sending such produce as trout, mushrooms, and mountain vegetables such as bracken and fern fronds to the Yamato Court as tribute (Aston 1972). Takatori (1993) portrays the early Kuzu and their later historic occupational specialists, the Kuzu-be (Barnes 1987), as hunters; moreover, he emphasizes that they were valued as culturally distinct from the plains court culture and in the early tenth century were even brought to the court to play music and sing as well as present their tribute. These glimpses suggest that the hills and mountains were indeed occupied by peoples with their own subsistence, traditions and rituals different from the plains agriculturalists. So far however, they are invisible archaeologically and did not form a coherent hills culture across broad areas of western Japan.

The spread of rice technology into northeastern Japan encountered greater resistance from the settled

Jomon peoples exploiting their rich coastal fishing grounds. Akazawa (1986) postulated that the new techniques spread first to groups in mountain basins and gradually to populations on the coast, farming partly being inhibited by the rocky nature of the coast. Kobayashi identified the increasing ritualization in the Final Jomon period as an effort to preserve their own way of life in the face of the threat of the newcomers. Although there is evidence of spot landings of Early Yayoi agriculturalists in the Tohoku region of northern Honshu Island, their settlements did not survive probably because rice was not yet cold-adapted to the far north. Recent finds in northern Nagano prefecture (Fig. 20.1, zone IId5) reveal that some migration occurred from the west during the Middle Yayoi period (Chunichi Newspaper 11 Nov 2007 evening edition), but basically the northeastern Jomon people gradually took up Yayoi culture themselves. Rice-growing was instituted where possible, but alternative grain products such as millet mark the northeastern Yayoi.

During the early historic period, the peoples of the northeast were referred to as 'emishi'. Although there is a tendency to identify these people as the ancestors of the Ainu, recent scholarship asserts that the word emishi was an administrative term of the state, used to refer to people outside their jurisdiction in the north. The emishi led a fully agricultural life consistent with northeastern climate (they relied mainly on millet and non-rice grains, sea and mountain products). Even the later Ainu, who were unattested until the seventeenth century, were not exclusively hunter-gatherers as they are often portrayed but practiced simple dry-field agriculture using the foot plow.

Three named groups of forest exploiters survive from the medieval period into the twentieth century: 'kobiki' (lumberjacks), 'kijiya' (wood-workers), and 'matagi' (hunters). Matagi hunters are known from their collected rituals and legends to date back to the early Historic period (Table 20.1), when they ranged across the deep forests; by the beginning of the Edo period, they had been turned into farmers through pressure from the local daimyo and decimations of the Medieval wars (Takatori 1993). Nagamatsu (2005) is careful to emphasize that these later matagi were neither a separate ethnic group nor full-time hunters: they were northeastern farmers of status, most active in the mid-eighteenth century, who lived in villages and practiced slash-and-burn agriculture as well as rice cultivation. Nevertheless, they had feudal permits to range throughout domain forests in order to seasonally kill a variety of animals, birds and fish.

The forests provided not only game but timber. The term 'soma' was applied to loggers that worked the forests and maintained hereditary rights from the eighth–nineth centuries through the medieval period. In the mid-fourteenth century, another term, 'kobiki', came into use, applying to men of the timber trade who felled trees and reduced them to lumber in the mountains or fashioned logs into lumber in villages (Setagaya Kyoi 2005). Like the matagi, kobiki became settled villagers in the Edo period, timbering in the agricultural off-season (Setagaya Kyoi 2005). In the Yoshino mountains, after great cedars were felled for timber and the area fired, millets, buckwheat and beans were planted alongside new cedars in the swidden fields and were cultivated for four to five years until the trees grew tall again (Miyamoto 1959). Such symbiosis illustrates a degree of cooperation between forestry and farming that is seldom seen elsewhere in the world.

'Kijiya' woodworkers, who formed specialist groups from the medieval period, were known particularly for their lathe-turned wood products (Takatori 1993). Since lathe-turned wooden pedestaled bowls are known from the Yayoi period, these skills were deep-rooted. Kijiya were originally itinerant specialists but later settled into villages. Unlike the matagi or kobiki, their craft was based on partnership with their wives, who chiselled out bowls and pulled the ropes to turn the lathes.

Thus, throughout the first millennium of historical society in Japan, the hills and mountains were exploited by specialists who had liminal status vis-à-vis the central government. In the Nara period, many villages around the country supplied specialist food and forestry products to the court, with most tax rice grown in the nearby home provinces (Kidder 1972). But by the late Edo period, almost all populations in Japan, regardless of where they resided (including fishing villagers), had been settled as farmers first and foremost to grow rice for tax purposes.

20.5 Plains Development and Paddy Field Landscapes

By the time wet-rice technology was introduced to Japan, the coastal plains had been developing for two millennia and were available for farming. However, in order to understand the patterns of lowland exploitation, the term plains as it is used in Japanese deserves cultural interpretation. The land unit commonly called 'heiya' (translated as plains) comprises 35% of Japan's total land area (Fig. 20.3), but it is internally very diverse, consisting of 11% rolling hills (often dissected former terraces), 11% terraces (both marine and riverine, and 13%

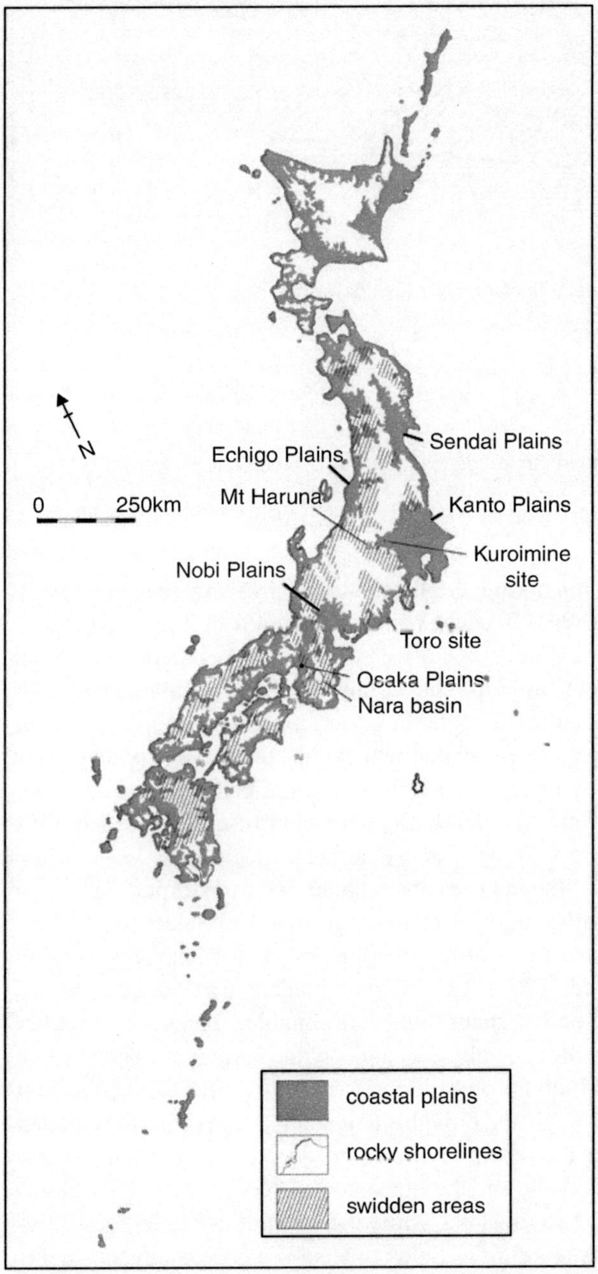

Fig. 20.3 Gross distribution of plains (*heiya*) in Japan, with rocky coastlines and areas of traditional swidden agriculture indicated. (After Yonekura et al. 2001; Chiba 1988)

Table 20.2 Composition of some major plains in Japan. (After Yonekura et al. 2001)

	Kanto Plains (%)	Echigo Plains	Nobi Plains (%)	Osaka Plains (%)
hills	<30	<30%	<30	<30
terraces	>30	ca. nil	<30	<30
alluvium	>30	>30%	>30	>30

former alluvial fans), and recent alluvium (Table 20.2). Of 22 major heiya, five have negligible recent alluvium; the Kanto Plains, where Tokyo is now located, is fairly evenly divided between hills, terraces, and alluvium.

Numerous kinds of plains have been defined by Japanese geomorphologists (Fujii 1972; Yonekura et al. 2001) by taking into account the type of landforms exhibited and the agent or environment of sedimentation. Takahashi (2001) identified eleven stages of Holocene micro-geomorphologic changes in plains development, resulting in three types of current alluvial plains. His F-type plains (the majority) exist in places where the Jomon Transgression did not extend far inland. In these plains, alluvial fans cover much of the Pleistocene terrace surfaces, while Holocene I and II terraces and current floodplains are clearly distinguishable. His L-type plains developed sandbar barriers during sea level regression, enclosing lagoons, while landward areas were occupied by two types of deltas. D-type plains developed after the formation of Holocene II terraces and are characterized by a particular delta formation, usually backed by relatively large inland basins where alluvial fans are particularly visible on Pleistocene terraces I and II.

All these plains have undergone constant modification in the last 5000 years, obscuring to a great extent previous landforms such as alluvial fans, floodplains, natural levees, back marshes, lakes, sand dunes, lagoons, bays, tidal flats and deltas, each of which had different potential for human activities. When archaeological sites are found within those plains deposits, they cannot be interpreted vis-à-vis the surface topography but instead must be related to the microtopography of a previous age. Reconstruction of the Echigo Plains topographic succession (Fig. 20.4), for example, has confirmed a paleosurface at 3000 BC where an entire ancient barrier sand body, lagoon, back marsh and delta have been buried between −12 m and −19 m depth. Middle-to-Late Jomon pottery recovered from within the lagoonal back marsh suggests fishing activities (Urabe and Takahama 2003).

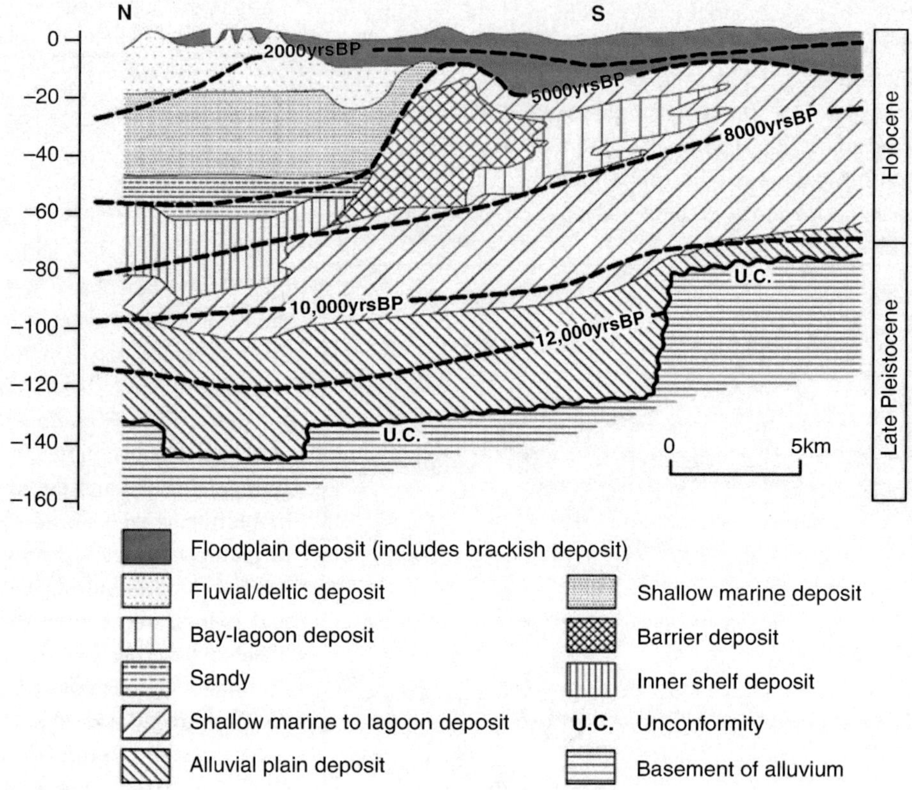

Fig. 20.4 Stratigraphy of the former lagoon and reconstructed landscape at 3000 BC under the Echigo Plains. (After Urabe et al. 2004; See Fig. 20.3 for location)

Wet rice does not necessarily need a formalized field system for growing: several cultures around the world broadcast their rice seed in floodwaters and wait for the rice to grow naturally in the wet alluvium. However, it is clear from the archaeological record that wet rice production in Japan was initiated in a fully formed field system introduced from the continent that included bunds separating fields, irrigation and drainage canals. Takahashi (2001) notes that the earliest paddies were constructed on the floodplains of small streams cutting through F-type plains; these early fields were then isolated on Holocene terraces by river down-cutting between Early and Middle Yayoi. L-type plains, consisting of lagoonal feeder deltas, became prime spots for Middle Yayoi agriculture, where settlements were positioned on higher delta ground and paddy fields located in the adjacent marshes—previously fished by Jomon peoples.

Since rice growing began in marshy gley soils, fields and ditches could be dug with simple wooden tools; but once higher ground was brought into cultivation, field levelling and irrigation became important. Sandy gravelly soils necessitated installing an impermeable paddy base to keep water from draining through the soil during the three-month growing season (Barnes 1986). Because paddy soils build up through time by siltation, the cultivated soils are regenerated without salt buildup, and in any case, rice is tolerant of salt. (Barnes 1990) However, a different environmental impact associated with paddy fields is the potential for methane production during the growing season. Continuously flooded fields can emit up to 14.8 $g\,m^2$ during a single season (Yagi et al. 1996).

Tsude (1989) postulated two transformations in the imported tool technology that facilitated paddy construction and cultivation. In the Middle Yayoi period (ca. 200–1 BC), longer bladed, iron-edged types of wooden spades and hoes enabled breaking of harder more gravelly fan and deltaic alluvium. Then in the Middle Kofun period (ca. fifth century AD), the foot plough and plough and u-shaped iron shoes for wooden spades and hoes were introduced. The first transformation allowed the expansion of rice-growing from gley to the sandy soils, while the second increased productivity within the fields themselves. These advancements underwrote the hierarchization and centralization of political society, leading to state formation in the latter fifth century. Essentially these types of digging tools were used throughout the entire historic period until field mechanization in the twentieth century.

Ancient paddy fields have been excavated from under the plains surface at numerous sites in Japan. In some areas, temporal successions of several layers of paddies are known as flooding, tephra fall, loess sedimentation from China, or general silting through irrigation has raised the land surface. Those paddies in lowland marshy gley soils tend to be larger (30×50 m at the Late Yayoi Toro site) with bunds reinforced with stake and board. Smaller fields, from 2×2 m but averaging 40–50 m^2, are known from sloping ground on higher alluvium or terraces; they are separated by low mounded-earth bunds but often laid out within a larger bunded area (Tsude 1989). Early small fields tend to be of irregular size and oriented with natural stream directions, often tucked in between different strands of riverbeds (Fig. 20.5).

In the Late Yayoi and Kofun periods, regularized fields became common, raising the question whether centralized or directed field construction was coming into play. In Osaka, evidence of regular field layout on a N10°W axis matches the directionality of mounded tomb orientation and ancient roadways (Fig. 20.6). Tsude (1989) concludes that not only did the new fifth century tool technology enable the opening up of Pleistocene terraces for paddy field agriculture, but the scale of the canals and field systems indicate considerable political control over labor and resources. Noto (2000) emphasizes the political importance of rice by noting that sixth-century paddies covered with volcanic ash eruptions in Gunma prefecture (Fig. 20.1, zone IId2) were quickly brought back into service with new irrigation facilities.

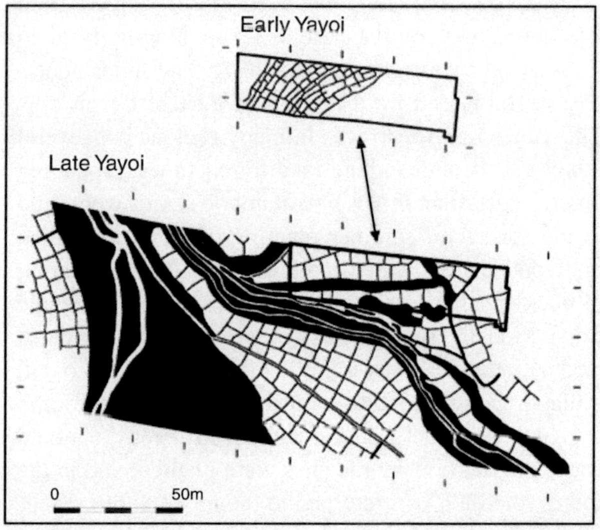

Fig. 20.5 Changing orientation of Yayoi paddy fields in plan through time; surrounding black shows irrigation canals and river. (After JAA 1988) (*Top* Lower Early Yayoi, of same scale as inset box. *Bottom* Late Yayoi)

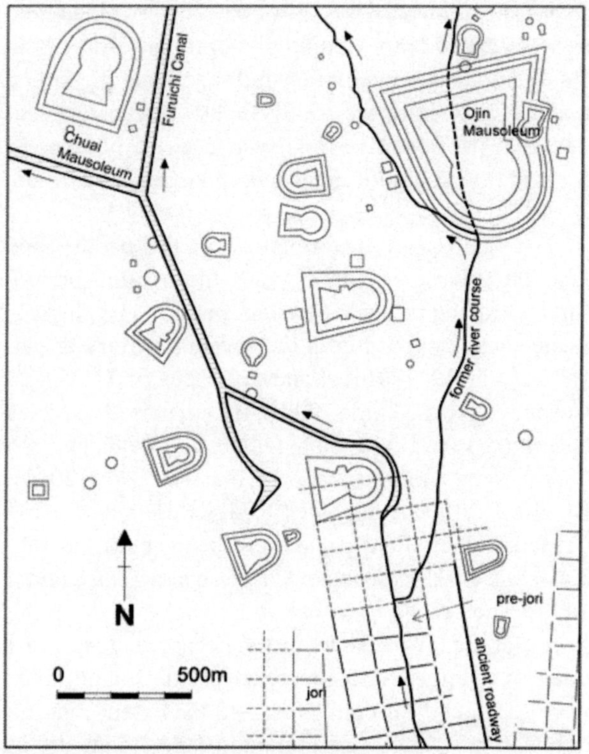

Fig. 20.6 The Furuichi Tomb Cluster in southeastern Osaka, located atop a Pleistocene terrace; diagonally oriented fields at lower center are orientated with several of the tombs and an ancient roadway on their eastern edge, while the newer jori system of field layout oriented more northerly surrounds them on both sides. Large canals 5–8 m wide and 4–5 m deep run through the tomb group from the natural river course at lower center. (After Tsude 1989)

The rapid expansion of paddy in Middle Yayoi across alluvial surfaces and again in Middle Kofun onto upper terraces represents a thorough reworking of Japan's lowland landscape during a millennium's time. In the seventh century, the landscape was transformed yet again near the capital areas for another millennium's duration, as a gridded field system was instituted using Chinese administrative techniques. From 645, the Yamato Court adopted a gridded city plan and extended the grid across the lowlands to divide paddy land into approximately 1-hectare squares in order to facilitate the allocation of plots to individual peasants for rice taxation purposes (Fig. 20.6). This gridded field system, called 'jori', required enormous resources in surveying the land, rerouting and canalizing the rivers, and reconstructing the paddy fields. The jori system has been the operative field system in many areas of the Kinai region until urbanization encroached on the flatlands.

The final two transformations of the paddy field landscape have occurred since the Kofun period. First, terracing of the lower slopes and river valleys is assumed to have begun in the seventh century at elevations of 130–170 m asl and up slopes of 1/15–1/20 steepness (Nakashima 1999). Heian period policies encouraged field expansion, with paddy fields built up valley floors and into surviving marshes (Noto 2000). By the twelfth century, terraced fields had a name ('tanada': shelf fields) and were being built at 300–750 m asl on slopes up to 1/6 steepness (Nakashima 1999). The second transformation was wrought in the lowest of the lowlands: in the nineteenth century, most tidal flats, estuaries, lagoons, and some inland lakes were reclaimed for cultivation—to the extent that the only natural estuaries remaining in Japan today are in distant Hokkaido (Yonekura et al. 2001).

The landscape affected by the expansion of rice paddy took in the lowland alluvial plains, upland terraces, and lower mountain valleys and basin flanks—two of the three geomorphologic divisions of the plains. The third division, of rolling hills, came under exploitation by dry field agriculture, which also expanded up the lower slopes of mountains near inhabitable regions.

20.6 Dry-Field Cultivation

It is easy to overlook the presence of other field types in the Japanese landscape because of the overwhelming visual impression of the paddy fields. Yet in the late nineteenth century, rice paddy accounted for only 58.5% of arable, while dry fields amounted to 41.5%; and in the Kanto region, dry fields accounted for 1.6 times as much arable as paddy, as it has done since the Kofun period (Kimura 1996). By the 1950s, Miyamoto asserts that there were numerically more dry fields than wet fields (1959).

All traditional farmhouses had a kitchen garden, and most had additional dry fields for growing grains, pulses and vegetables—the types of fields being distinguished by location, cultivation techniques and crops grown. In short, permanent dry fields were generally flat (including terraced) while swidden fields were located on steep hillsides; fields for growing dry rice were separately identified. However, many swidden fields, once opened, were turned into permanent fields, and rice paddies could be used to grow winter vegetables or winter cash crops such as rush. Thus, outlining the history of dry-field cropping is a larger task than can be attempted here, but two or three problems need exploration.

One of the main issues is the depth of history for swidden cultivation. A limited amount of horticultural activity is acknowledged for the Jomon period, and Sasaki Komei has argued since 1971 that Jomon cultivars (Table 20.3) were grown in swidden. Matsui and Kanehara (2006) propose instead that cultivation began on trash middens, not swidden, and then expanded into kitchen gardens. There is no definitive archaeological evidence for burning fields—as opposed to copious ethnographic and textual evidence from later times. Several poems in the eighth century poem anthology, 'Manyoshu' suggest the firing of fields; but the specific Chinese character used to identify swidden vis-à-vis normal dry fields only appears in the medieval period (Kimura 1996). Chiba (1993) cautions against interpreting swidden as the original or prehistorical mode of agriculture, since a majority of swidden fields were opened under pressure from an exacting taxation system in order for farmers to feed themselves.

The size of historic swidden fields ranged from less than a tenth of a hectare to 1.5 ha, judging from an Edo-period cadastral map (Fig. 20.7). The fields generally were divided from the top of a ridge downslope. The mountain growth was in many regions cut several months in advance and fired at different times, depending on the vegetation: in Yoshino, bamboo stands were fired in the spring for summer planting of barnyard millet, but wooded areas were fired in summer for fall planting of buckwheat and vegetables (Miyamoto 1959). By the Edo period, the succession of crops was fairly standardized nationwide: barnyard millet in the first year, foxtail millet in the second; by the third year, taro was common or replaced by buckwheat, adzuki or soy beans; if a field was exhausted, white radishes were planted; and in the fourth, the field was returned to the mountain to regenerate forest for fifteen years or so. Swidden cultivation survived into the late twelfth century, particularly in the Inland Sea area where summer drought is common. However, between 1935 and 1950, the area in swidden had declined from 76,795 ha to 9457 ha (Obayashi

Table 20.3 Traditional foodstuffs grown in Japanese dry fields (% indicates the proportion of 72 villages in Sasaki's ethnographic sample growing that crop in swidden fields (Sasaki Komei's chart in Sakamoto 1983); *species included in the six traditional miscellaneous grains ('zak-koku') (Sakamoto 1983). J = present in Jomon period (Matsui and Kanehara 2006); i = Initial, e = early, m = middle, l = late)

Traditional dry-field crops	Japanese name	% of 72 villages	J	Scientific name
Grains/Seeds				
Foxtail millet	*Awa*	83.3	lJ	**Setaria italica* (L.) P. Beauv.
Buckwheat	*Soba*	80.5	emJ	*Fagopyrum esculentum*
Barnyard millet	*Hie*	56.9	lJ	**Echinochloa utilis* *Echinochloa esculenta*
Wheat/Barley	*Mugi*	20.8		
Barley	*Omugi*		lJ	
Common millet	*Kibi*		lJ	**Panicum miliaceum* L.
Finger millet	*Shikokubie*			**Eleusine coracana* Gaertn.
Sorghum	*Morokoshi*			**Sorghum bicolor* Moench
Job's tears	*Hatomugi*			**Coix lacryma-jobi* L. var. *mayuen* (Roman.) Stapf
Perilla	*Aburae/Egoma*	6.9	iJ	*Perilla ocimoides*
Dry rice	*Okabo*	4.2	lJ	*Oryza sativa*
Pulses				
Adzuki beans	*Azuki*	76.4		
Soybeans	*Daizu*	76.4		
Beans	*Rokuto*		iJ	*Vigna* spp.
Vegetables				
Taro	*Sato imo*	19.4		*Colocasia antiquorum*
Sweet potato	*Satsuma imo*	15.3		
Turnips	*Kabu*	6.9		
White radish	*Daikon*			
Bottle gourd	*Hyotan*		iJ	*Lagenaria siceraria var. gourda*
Burdock	*Gobo*		iJ	*(Arctium) Lappa major*
Beefsteak plant	*Shiso*		iJ	*Perilla frutescens var. crispa*
Rapes			emJ	Cruciferae (*Brassicaceae*)
Goosefoot			iJ	*Chenopodium* spp.
Yam			emJ	*Dioscorea japonica*
Non-edibles				
Paper mulberry	*Kozo*		J	*Broussonetia papyrifera*
Flax/Hemp	*Asa*		emJ	*Cannabis sativa*

1990); by the 1970s, it had become such a rare activity that ethnographic and botanical rescue projects were mounted to record the last of its practices and preserve the species cultivated (Sakamoto 1983).

Whereas field burning has not been recovered archaeologically, dry fields are known from Late Yayoi onwards. These are identified by cultivation features rather than field boundaries. The evidence is of two types, which are difficult to distinguish and are often mistaken for each other (Satou 2000): a type of ridge-and-furrow ('une': ridge) or plough marks, both of these usually fossilized by a covering of flood silts, tephra, landslides or even sand dunes. Plough marks excavated from at least 30 sites on the Sendai Plains, dating mostly from the Kofun through Heian periods, are almost all located on natural levees in coastal marshes (Fig. 20.8). This land-use pattern of dry-cropping on lowland levees continues today, as can be seen in the Nara Basin where as little as 50–70 cm height differential will distinguish between natural levee fields and rice paddies.

Une ridges have been recovered from only three of those sites on the Sendai Plains, giving an impression that while turning the soil with a plow was widespread, the labor invested in creating ridges of substantial soil depth was reserved for certain crops. At the Kofun-period Kuroimine site in Gunma prefecture, une were discovered immediately adjacent to houses

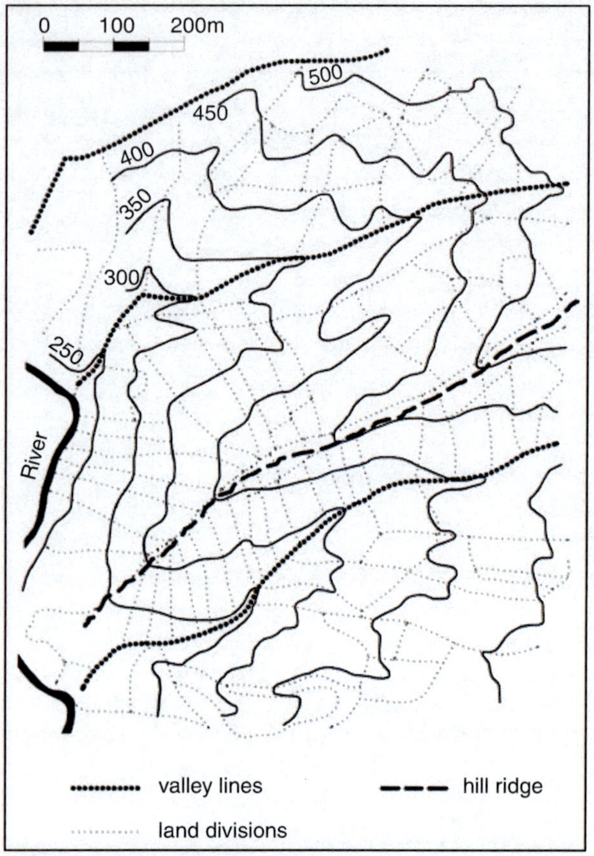

Fig. 20.7 Divisioning of modern swidden fields in an upper mountain valley; note field alignment at ridgetop. (After Chiba 1993)

(Fig. 20.9)—all having been buried under two meters of pumice from the sixth century eruption of nearby Mt. Haruna. The ridges ran downslope and were of two shapes: narrow and high, or broader and flat. Plant macro-remains recovered there included rice, adzuki beans, bottle gourds and job's tears (Kimura 1996).

Historically, the hills and lower slopes accessible to plains villages were heavily farmed with dry fields. In the late Edo period, cash crops became important so that a wide array of food and non-food plants were cultivated: orchards, paper mulberry, and even cotton. A form of terraced dry field, *dandan-batake* (stepped fields), was used on steeper surfaces, usually faced with stone walls. In narrow upper mountain valleys all three field types were used where possible (Fig. 20.10): paddy and permanent fields on the lower slopes and at terrace edge, swidden on the mountain slope. The crops in the permanent lower fields at this site in Niigata prefecture (Fig. 20.1, zone IId1) were mostly vegetables, but cedar seedlings were nurtured for replanting, chestnut trees were tended, and mulberry was grown for making tools and other objects—often with the lathe.

Since the 1970s, the rolling hills and lower mountain slopes have become targets for residential expansion, with large housing developments encroaching on land previously used for dry fields and orchards. The last of the plains are now under urban development, but due to the lack of zoning laws, houses are cheek to jowl with surviving fields and paddies.

20.7 Agriculture Past, Present and Future

The sections above have traced the beginnings of Japan's agricultural systems and, briefly, their manifestations in the historic period. In the twentieth century, population concentrations reversed from 80% rural/20% urban in the first half century to 80% urban/20% rural in the second half. The agricultural population correspondingly declined from 17 million in 1950 to 3.15 million in 2000 (Okutsu et al. 2004). The maximum extent of rice paddy was reached in the early 1930s (at ca. 3,200,000 ha, Ohura 2001) and has been declining since then, down from 15.3% total land area in 1975 to 12.5% in 2007 (MLIT 2008). From 50 to 70% of former farmland is converted to residential or commercial property, but about 10% goes to road, rail and water infrastructure.

A substantial amount of arable is also being abandoned due to population aging: the amount of disused arable has risen from 6% in 1995 to 10% in 2005 (MAFF 2008). The government is devising policies to keep this land open for agriculture or, in valleys with low productivity, letting it revert to meadow and forest (MAFF 2008). Abandoned and winter paddies are also being proposed to grow cattle feed, since currently 51% of animals raised in Japan are fed imported grain (MAFF 2008), while some paddies are being flooded to reintroduce wetland plants and animals and aid flood control. The agricultural landscape is, therefore, becoming more diverse and past natural habitats are being re-established.

In western countries, food miles are a great concern, but Japan is worried about food self-sufficiency, now at an all-time low of 39% (Table 20.4). The opening of Hokkaido, with sponsored emigration in the

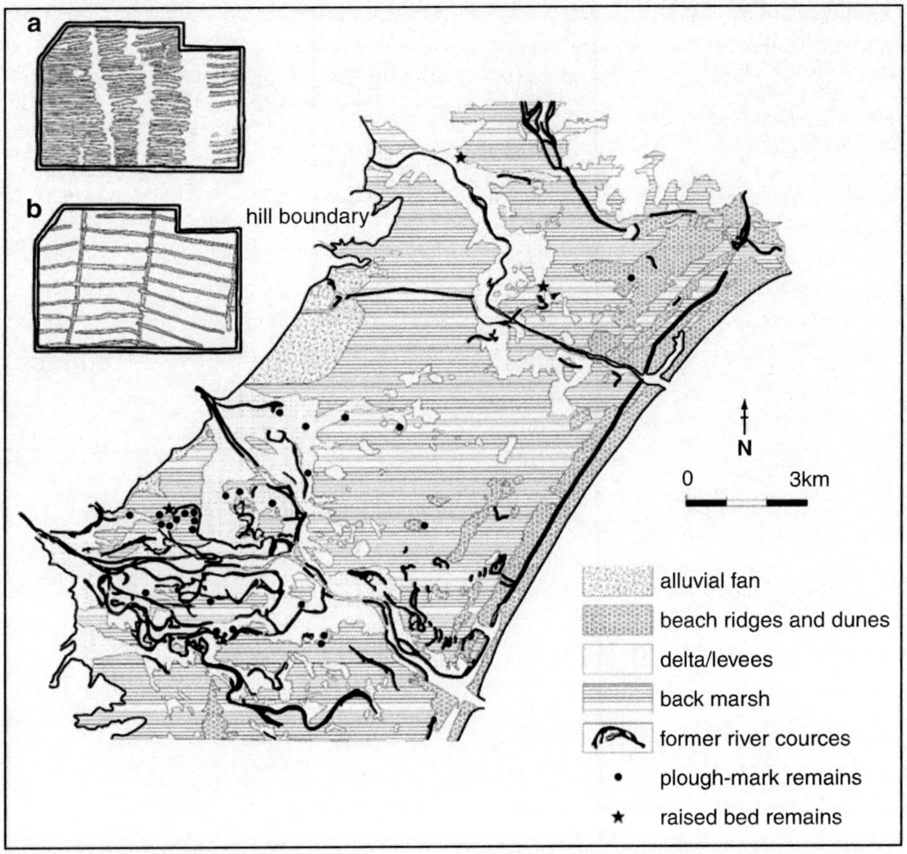

Fig. 20.8 Reconstructed medieval-period Sendai Plains with ploughed sites (*large dots* and inset **b**) and ridge-and-furrow remains (asterisks and inset **a**) on former river levees. (After Satou 2000)

late nineteenth century, eventually added 241,000 ha of rice paddy accounting for 8.6% of Japan's total paddy area as of 1994 (Fujiki 1998); this extra capacity, however, has begun to erode and is matched by a decline in individual rice consumption from a high in 1963 of 113.3 kg/year–61 kg/year in 2007 (MAFF 2008). Wheat grown in Hokkaido increased from less than 250 kg/10a in 1961 to almost 550 kg/10a in 2004 (MAFF 2008), but still 90% is imported at high prices. The government is now promoting the use of rice flour as a wheat substitute, hoping to put fallow rice paddy back to work. In 2007, the amount of land devoted to paddy and dry fields was almost equal: 2.53 million ha in wet rice contributing 22% of agricultural output, and 2.12 million ha in dry fields (grains, vegetables, fruit) contributing 25% output (MAFF 2008; MLIT 2008). Since rice, vegetables and fish have relatively high self-sufficiency ratings, the government is encouraging a return to the traditional diet based on these foods, particularly for breakfast.

The urban landscape now hosts 24% of farm households (MAFF 2008), and urban adults and children alike are being encouraged to become more participatory and knowledgeable about local production through pick-your-own, experiential farm days for both adults and school groups, and even multi-day lodgings for working holidays and instruction in farming. These activities were boosted by the 'satoyama' (village mountain) movement in the 1980s (Takeuchi et al. 2003), and by laws passed in 1989 and 1990 allowing the conversion of farmland to rentable allotment gardens (Azuma and Wiltshire 2000). An Edo-period saying that 'eating produce from "within 3-*ri* squared" (12 km radius of home) is best for the health' is reviving (MOE 2008), and a new trend of roof-top vegetable gardens is appearing in Tokyo. The extraordinary responsiveness of Japanese people to these initiatives may signal a reversal of the westernized diet, helped along by high prices and oil worries, to more local and traditional fare. This will be the stimulus for bringing back into production much fallow land subsidized for non-production, and perhaps it will even involve returning to the farming lifestyle, as many urban retirees are

Fig. 20.9 Sixth century houses at the Kuroimine site surrounded by ridged fields; all preserved by tephra fall. (After Tsude 1989; See Fig. 20.3 for location)

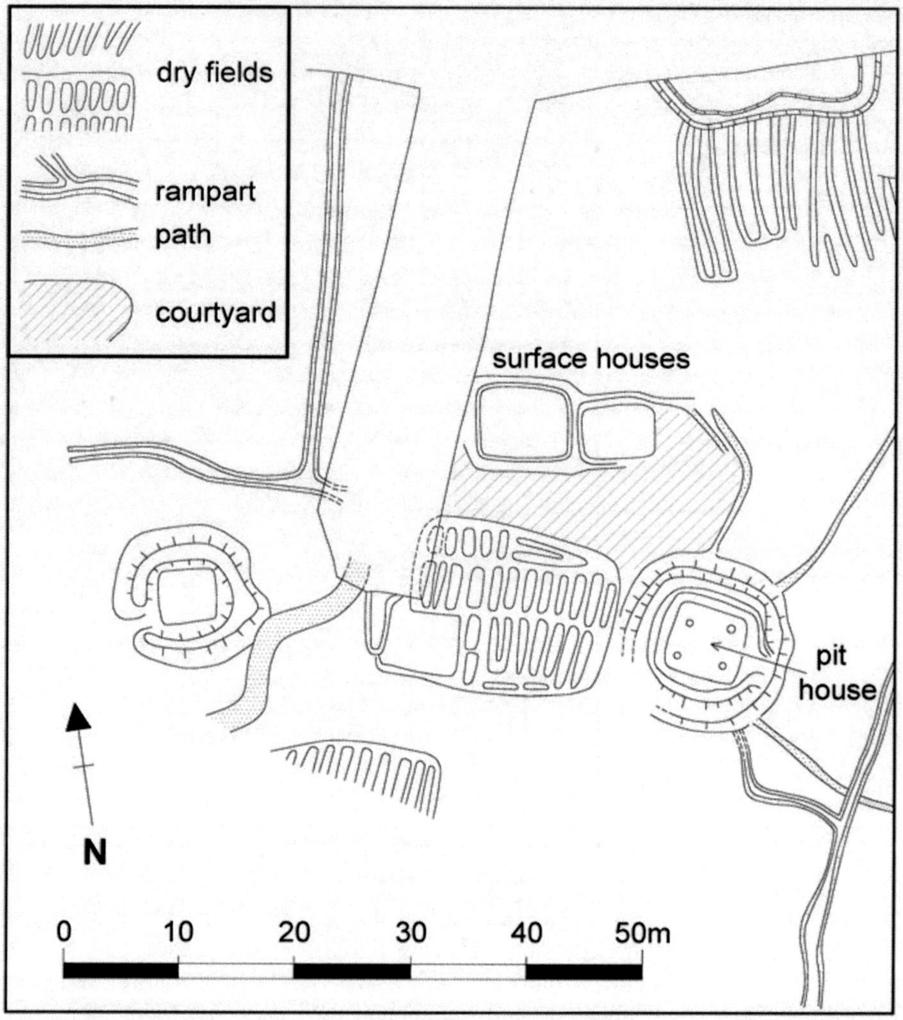

now doing. Thus, the current complicated economics between urban expansion, farmland contraction, population decline, and aging farmers may yet have a happy ending for Japan.

However, sea-level rise is one problem looming on the horizon that is not being well addressed in Japan. Kojima (2004) writes that very few preparations are being made for ocean transgression. He notes especially that relocation policies are not being considered; instead, measures against the erosion of sandy beaches and a one-meter sea-level rise are currently being implemented. For the twenty-first century, one study (Suzuki et al. 2005) comparing two models of sea-level rise calculates a global average of 20–37 cm rise; these figures are corroborated by a recent Australian study (Church and White 2006) but narrowing the range to 28–34 cm. However, sea-level rise will be higher in southwestern Japan, due to activities of the Kuroshio current washing the seaboard there (Suzuki et al. 2005). A rise of 30 cm has already been documented in the last half-century by the Okayama University of Science, with a predicted 80 cm rise by the end of the century (Tsukamoto 2007).

Calculations in Table 20.5 show the land area, population numbers and asset levels that would be affected by differing projections of sea-level rise.

These figures give projections for different conditions under the stated sea-level rise, so that even at present, over 17 times more area is affected under storm conditions than normally. In the worst scenario, the area affected would comprise about 2.5% of Japan's land area, which is considerable given that it is the coastal plains where the current population, economy and agricultural system is focussed.

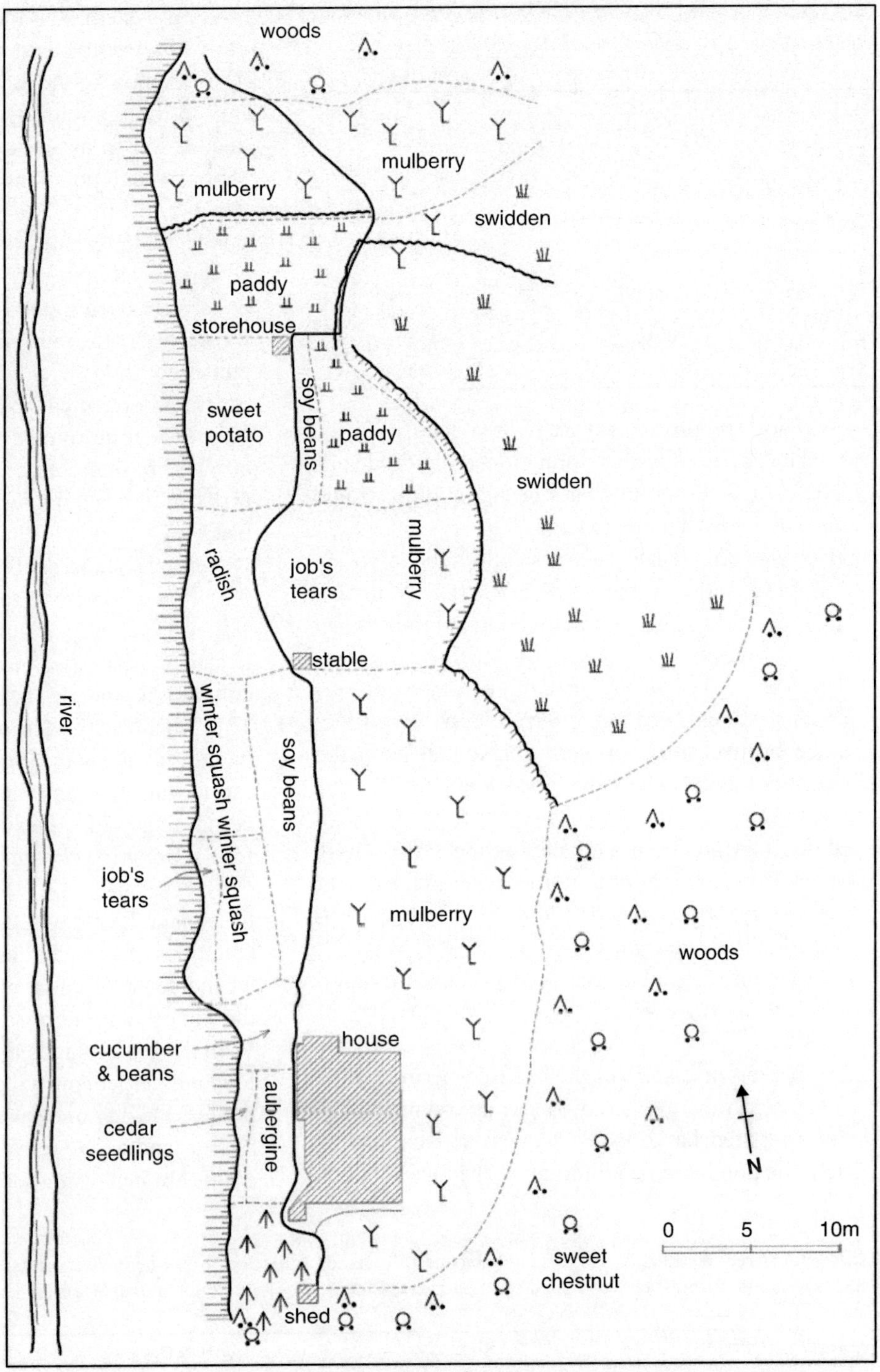

Fig. 20.10 Field systems in an upper mountain valley. (After Sasaki 1971) All three types of fields—paddy and permanent dry fields near river's edge, and swidden on the lower hillslopes—are represented here

20.8 Conclusions

a. The Jomon are one of two unusual cases (the other being Danish Neolithic) of temperate hunter-gatherers with settled lifestyles in rich environments. The Jomon have been referred to as 'ceramic hunter-gatherers' or 'affluent foragers'; the term Neolithic is sometimes applied because of their polished stone tools and ceramics, but they were not reliant on agriculture, a hallmark of other Neolithic societies.

Table 20.4 Declining levels of self-sufficiency in Japan's food supply. (After MAFF 2008)

	1965 (%)	2006 (%)
Overall	73	39
Rice	100	94
Meat	47	16
Cooking oils	33	4
Wheat	28	13
Sugar	31	32
Fish	110	59
Vegetables	100	76
Soybeans	41	25
Fruit	86	35
Other	68	23

b. Had not the postglacial rising sea levels caused lowland alluvial plains formation around Japan's fringe, it is doubtful that rice agriculture would have taken hold to the extent it did.
c. Had western Japan been as fully occupied by Jomon groups as the northeast, a hills people antagonistic to the rice-cultivating migrants may have developed.
d. The Yayoi, though the first intensive agriculturalists in Japan, are not referred to as Neolithic, since bronze and iron were introduced from the continent soon after the adoption of wet rice technology.
e. If rice had not been available as the state staple, population densities and political hierarchies from the Kofun period onwards could not have been supported to the same extent.
f. Peasant abilities to fall back on dry-field cultivation allowed their historic survival and the survival of the state.
g. The precarious nature of the late Edo-period political system was revealed by rice riots and rebellions due to failed harvests, epidemics, and over-taxation; the population equilibrium of the historic agricultural system was thus under great threat in the early nineteenth century.
h. The opening of Japan in the late nineteenth century to foreign trade allowed a new avenue of food acquisition from abroad. Importation of food now supports high population densities.
i. The type of plough was relatively unimportant in paddy field agriculture because the field extent is so small, labor being more important; draft animals were used in the medieval period, as attested by scroll paintings and footprints preserved in stratified paddy fields. The post-war innovations of the mechanical plough and seedling transplanter revolutionized the labor requirements and allowed field enlargement so that the smaller historic fields are being consolidated into larger field parcels.
j. Nevertheless, the decline of farming together with the reliance on imported foods threatens Japanese self-sufficiency, particularly if the lowlands are sacrificed to urban sprawl and individual knowledge of cultivation practices is also lost.
k. The Japanese government and agricultural cooperatives are working hard to re-establish interest in farming among the Japanese population, coinciding with grass-roots movements for slow food, organic production, local produce, and conservation.
l. Rising sea levels will encroach on the densest distributions of rice paddies, housing and urban sprawl on the coastal plains, endangering large sectors of the Japanese economy.
m. With the bottomlands at risk, the historic pattern of lowland settlement will see an intensification of a process already underway: that of a movement onto higher land as the terraces and hilly regions of the plains are increasingly exploited.

Table 20.5 Area, population and assets on Japanese lowlands subjected to sea-level rise. Units: area (km^2); Population (10,000 persons); Assets (trillion yen) (compiled by Impact Assessment Working Group, Global Warming Committee, Environment Agency; After APEC 2007)

Sea level	Present			30 cm rise			50 cm rise			100 cm rise		
	Area	Popul.	Assets	Area	Popul.	Assets	Area	Popul.	Assets	Area	Popul.	Assets
Average sea level	364	102	34	411	114	37	521	140	44	679	178	53
Full tide	861	200	54	1192	252	68	1412	286	77	2339	410	109
Typhoon or tsunami	6268	1174	288	6662	1230	302	7538	1358	333	8898	1542	378

Acknowledgements This research has been supported by SOAS, University of London; International Research Institute for Japanese Culture, Kyoto; Durham Archaeological Services.

References

Akazawa T (1986) Hunter-gatherer adaptations and the transition to food production in Japan. In: Zvelebil M (ed) Hunters in transition. Cambridge University Press, pp 151–65

Amino Y (1998) Japanese history as told by east and west. Kodansha Gakujutsu Bunko, Tokyo (in Japanese)

APEC (2007) Global warming. APEC Virtual Center for Environment Technology Exchange website. http://www.epcc.pref.osaka.jp/apec/eng/earth/global_warming/dounaru.html (accessed 13 October 2008)

Aston WG (1972) Nihongi: chronicles of Japan from the earliest times to A.D. 697. Tuttle, Rutland VT, 850 pp

Azuma R, Wiltshire R (2000) Rural Allotments and Sustainable Development: A Japanese Perspective. Paper Presented at the Conference People, Land and Sustainability held at the University of Nottingham, September 13-16, 2000. http://www.btinternet.com/~richard.wiltshire/pls.htm (accessed 11 October 2008)

Barnes GL (1986) Paddy field archaeology. Journal of Field Archaeology 13.4:371–379

Barnes GL (1987) The role of the be in state formation. In: Brumfiel E, Earle T (eds) Production, exchange and complex societies. Cambridge University Press, pp 86–101

Barnes GL (1990) Paddy soils now and then. World Archaeology 22:1–17

Barnes GL (1999) The rise of civilization in East Asia: archaeology of China, Korea and Japan. Thames & Hudson, London, 288 pp

Barnes GL (2003) Origins of the Japanese Islands: the new "big picture". Nichibunken Japan Review 15: 3–50

Barnes GL (2008) The making of the Japan Sea and the Japanese mountains. Nichibunken Japan Review 20:3–52

Chiba T (1988) Hata and hatake: regional differences and rituals. In: Sasaki K, Matsuyama T (eds) Hatasaku bunka no tanjo: Jomon kosakuron no apuroochi. Nippon Hoso, Tokyo, pp 307–24 (in Japanese)

Chiba T (1993) Society and culture of mountain peoples, fishing peoples. In: Amino Y, Obayashi T (eds), Nihon minzoku bunka taikei, vol. 5: Mountain peoples and sea peoples—the life and customs of non-plains peoples. Shogakkan, Tokyo, pp 65–98 (in Japanese)

Chisholm B (2004) Paleodiet studies in Japan using stable isotope analysis. Bulletin of the International Jomon Culture Conference 1:25–33

Church JA, White NJ (2006) A 20th century acceleration in global sea-level rise (abstract). Geophysical Research Letters 33.1:602

Crawford GW (2006) East Asian plant domestication. In: Stark M (ed), Archaeology of Asia. Blackwell, Malden MA, pp 77–95

Farris WW (2006) Japan's medieval population: famine, fertility and warfare in a transformative age. University of Hawai'i Press, Honolulu, 372 pp

Farris WW (2009) Daily life and demographics in ancient Japan. Center for Japanese Studies, University of Michigan, Ann Arbor, 137 pp

Fujii S (1972) Some problems of the coastal plains in Japan. Chishitsugaku Ronshu 7:189–196

Fujiki H (1998) Japanese rice market liberalization: a competitive equilibrium approach. www.kier.kyoto-u.ac.jp/DP/DP479.pdf (accessed 11 October 2008)

Fujiyama R (2004) Living behaviours at the end of the Ice Age. Kokogaku Kenkyu 20:1–23 (in Japanese with English abstract)

Habu J (2004) Ancient Jomon of Japan. Cambridge University Press, 332 pp

Hanihara K (1991) Dual structure model for the population history of the Japanese Islands. Japan Review 2:1–33

Hayami A, Kito H (2004) Demography and living standards. In: Hayami A, Saito O, Toby RP (eds) The Economic History of Japan, v. 1., Oxford University Press, pp 213–246

Imamura, K (1996) Prehistoric Japan: new perspectives on insular East Asia. University College Londn Press, London, 246 pp

JAA (Japanese Archaeological Association) (1988) The initiation and development of rice agriculture in Japan. Nihon Kokogaku Kyokai Shizuoka Takai Jikko Iinkai, Shizuoka-ken Kokogakkai, Numazu, 493 pp

Kidder JE Jr (1972) Japan before Buddhism. Thames & Hudson, London, 212 pp

Kimura S (1996) Dry fields and the Japanese: one other agricultural culture. Chuko Shinshu 1338. Chuo Koronsha, Tokyo, 220 pp (in Japanese)

Kobayashi T (2004) Jomon reflections: forager life and culture in the prehistoric Japanese archipelago. Oxbow Books, Oxford, 240 pp

Kojima H (2004) Vulnerability and adaptation to sea-level rise in Japan. Proceedings of APN/SURVAS/LOICZ Joint Conference on Coastal Impacts of climate Change and Adaptation in the Asia-Pacific Region, pp 81–88 www.survas.mdx.ac.uk/pdfs/3kojima.pdf (accessed 11 October 2008)

Koyama S (1979) Jomon subsistence and population. Senri Ethnological Studies 2:1–65

Koyama S (1983) The establishment of rice agriculture and changes in population. In: Sasaki K (ed) Nihon noko bunka no kigen. Tokyo: Nippon Hoso, Tokyo, pp 347–73 (in Japanese)

Leach E (1970) The political system of highland Burma: a study of Kachin social structure. Athlone Press, London, 324 pp

MAFF (Ministry of Agriculture, Forestry and Fisheries) (2008) Shokuryo nogyo noson hakusho [White paper on foodstuffs, agricultural industry, and farming villages]. Norin Suisansho, Tokyo, 269 pp (in Japanese)

Matsui A, Kanehara M (2006) The question of prehistoric plant husbandry during the Jomon period in Japan. World Archaeology 38.2:259–273

MIAC (Ministry of Internal Affairs and Communications) (2008) Statistical handbook of Japan. http://www.stat.go.jp/English/data/handbook/c02cont.htm#cha2_6 (accessed 30 August 2008)

Miyamoto, T (1959) Dry field cultivation. In: Omachi T (ed) Nihon minzokugaku taikei, vol. 5, Seigyo to minzoku. Heibonsha, Toko, pp 49–74 (in Japanese)

MLIT (Ministry of Land, Infrastructure and Transport) (2007) Summary of White Paper on Land and Real Property. http://tochi.mlit.go.jp/h19hakusho/chapter3_sec1/setsu_3-1_eng.html (accessed 10 Sept 08)

MLIT (Ministry of Land, Infrastructure and Transport) (2008) Tochi hakusho [Land white paper]. Kokudo Kotsusho, Tokyo, 214 pp (in Japanese)

MOE (Ministry of Environment) (2008) Kankyo junkan-gata shakai hakusho [White paper on environment and recycling society]. Kankyosho, Tokyo, 416 pp (in Japanese)

Nagamatsu A (2005) A study on hunting folklore: real and folkloric images of hunters in the Early Modern Age. Hozokan, Kyoto, 362 pp (in Japanese with English title and abstract)

Nakashima M (1999) Terraced fields in Japan. Kokin Shoin, Tokyo, 240 pp (in Japanese)

Noto T (2000) Dry cultivated fields in the Kanto region: excavations of the dry cultivated fields covered by the ash fall. In: Ikehata K, Nakamura K, Dogome H, et al. (eds) The archaeology of field agriculture. Nihon Kokogaku Kyokai, Kagoshima, pp 26–29 (in Japanese with English title and abstract)

Obayashi T (1990) East to west, sea to mountain: Japanese cultural spheres. Shogakkan, Tokyo, 263 pp (in Japanese)

Ohnuki-Tierney E (1993) Rice as self: Japanese identities through time. Princeton University Press, 184 pp

Ohura Y (2001) 5th Presentation. Seminar, International Environmental Economics, University of Tokyo. park.itc.u-tokyo.ac.jp/iee/seminar/ohura0613.pdf (accessed 26 August 2008)

Okutsu M, Tomosue T, Kataoka T, Sawada J (2004) Agriculture and employed labour force: the current situation and future direction. JILPT (Japan Institute for Labour Policy and Training) website. www.jil.go.jp/english/documents/JILPT-NoL2.pdf (accessed 28 August 2008)

Pearson R (2006) Jomon hot spot. World Archaeology 38.2:239–58

Sakamoto S (1983) The miscellaneous grains of Japan and surrounding areas. In: Sasaki K (ed) Nihon noko bunka no kigen. Nippon Hoso, Tokyo, pp 59–106 (in Japanese)

Sasaki K (1971) Before rice cultivation. NHK Books 147. Nippon Hoso, Tokyo, 316 pp (in Japanese)

Satou K (2000) Issues of identification of cultivated traces in fields: cases in Sendai city (Tohoku region). In: Ikehata K, Nakamura K, Dogome H, et al. (eds) The archaeology of field agriculture. Nihon Kokogaku Kyokai, Kagoshima, pp 16–25 (in Japanese with English title and abstract)

Setagaya Kyoi (2005) Lumbermen and the culture of trees. Setagaya Ward Board of Education, Tokyo, 28 pp (in Japanese)

Smith TC (1959) The agrarian origins of modern Japan. Stanford University Press, Stanford, 246 pp

Suzuki T, Hasumi H, Sakamoto TT, et al. (2005) Projection of future sea level and its variability in a high resolution climate model: Ocean processes and Greenland and Antarctic ice melt contributions. www.earthsimulator.org.uk/meetings/Nov_2005/WS_11_2005_ppt/Tatsuo_Suzuki.ppt (accessed 12 October 2008)

Takahashi M (2001) Environmental archaeology for disaster risk management. www.ritsumei.ac.jp/acd/cg/lt/rb/593PDF/takahasi.pdf (accessed 11 October 2008) (in Japanese)

Takatori M (1993) Ancient mountain peoples. In: Amino Y, Kinoshita T, Kamino Y (eds) Umi, kawa, yama no seisan to shinkyo. Yoshikawa Kobunkan, Tokyo, pp. 80–103 (in Japanese)

Takeuchi K, Brown RD, Washitani I, et al. (eds) (2003) Satoyama: the traditional rural landscape of Japan. Springer, Tokyo, 229 pp

Totman C (1989) The green archipelago: forestry in preindustrial Japan. University of California Press, Berkeley, 297 pp

Tsude H (1989) Trends in the establishment of Japanese agricultural society. Iwanami, Tokyo, 497 pp (in Japanese)

Tsukamoto O (2007) What can we do to prevent the Setonaikai sea level rise: special interview. Save the Earth in Kurashiki website. http://www.krashjapan.com/v5/eco/index02_e.html (accessed 13 October 2008)

Urabe A, Takahama N (2003) Echigo Plains: reconstruction of the Middle Jomon palaeotopography in Nishi-Kanbara area. In: Kobayashi S (ed) Zen-kindai no Niigata kasen kotsu to iseki ritchi no chiikishi-teki kenkyu. Niigata University, Niigata, pp 39–53 (in Japanese)

Urabe A, Takahama N, Yabe H (2004) Identification and characterization of a subsided barrier island in the Holocene alluvial plain, Niigata, central Japan. Quaternary International 155–166:93–104

Yagi K, Tsuruta H, Kanda K, Minami K (1996) Effect of water management on methane emission from a Japanese rice paddy field: automated methane monitoring. Global Biogeochemical Cycles 10.2:255–267

Yonekura N, Kaizuka S, Nogami M, Chinzei K (2001) Regional geomorphology of the Japanese Islands, Vol. 1: Introduction to Japanese geomorphology. University of Tokyo Press, 349 pp (in Japanese)

Chapter 21
Evolution of Hydraulic Societies in the Ancient Anuradhapura Kingdom of Sri Lanka

P. B. Dharmasena

21.1 Introduction

The objective of this chapter is to describe an ancient community of Sri Lanka, which developed a sophisticated hydraulic civilization based on a dry-zone agricultural system utilizing a cascade of tanks. The word tank was derived from the Portuguese word 'tanque' used for reservoir. The western invasion of the country began with the Portuguese in 1505 followed by the Dutch in 1656 and subsequently by the British in 1798. The local term 'wewa' is still in use for the reservoir. The hydraulic civilization evolved as a network of streams and tanks in varying sizes spread over the dry zone of Sri Lanka forming tank clusters referred to as tank cascades. A cascade is a connected series of tanks organized within micro-catchments of the dry zone landscape, storing, conveying and utilizing water from ephemeral rivulets (Madduma Bandara 1985).

These small tank cascade systems, constructed in ancient times, could be considered well designed irrigation systems with a distinctive assembly of land uses. The land and water management practices that were refined and perfected over several centuries in order to match the capricious nature of rainfall, along with the special attributes of the landscape led to the system of irrigation development referred to as cascading system (Panabokke et al. 2002). The small tank builders must have had a profound understanding of landscapes and hydrology of the region (Panabokke 1999).

Medieval villages developed in association with small tank cascade systems in the dry zone of Sri Lanka. The center of the village is the tank, and houses are grouped on one or both sides of the tank on the relatively higher elevation around the tank bund (embankment). The wells for domestic water tap the shallow phreatic groundwater table. The irrigable area below the tank bund is located along the main axis of the inland valley. Rainfed upland crops are grown on the unirrigable slopes of the upland under a system of shifting or 'chena' cultivation. Tank water is conserved with tree belts ('gas gommana' and 'kattakaduwa') around the tank, upstream bunds ('iswetiya') and water holes ('godawala') above the tank bed as described below in Sect. 21.5.3. This is known as the traditional tank-village system typical of Sri Lanka.

Over the course of history, a large number of communities developed their own culture and lifestyle that are intricately tied to nature and the local landscapes. The ancient civilization of Sri Lanka is not an exception, but its social, political and economic interactions with neighboring countries have interfered during the entire process of the evolution. The known history of Sri Lankan civilization records two main discontinuities due to external invasions taken place in 3000 BC and 1200 AD. This chapter attempts to describe the evolutionary process of the ancient hydraulic civilization of Sri Lanka, leading to the formation of the ancient Anuradhapura city that was the capital of Sri Lanka for about 1500 years. Available evidence from such a long period is also reviewed in this chapter to explore the evolutionary process of the Anuradhapura-based dry zone civilization influenced by landscape and other factors.

P. B. Dharmasena (✉)
Field Crops Research and Development Institute
Mahailluppallama, Sri Lanka
e-mail: dharmasenapb@ymail.com

See Plate 20 in the Color Plate Section; also available at: extras.springer.com

I. P. Martini, W. Chesworth (eds.), *Landscapes and Societies*,
DOI 10.1007/978-90-481-9413-1_21, © Springer Science+Business Media B.V. 2010

21.2 Natural Setting

The island of Sri Lanka (65,625 km^2) lies off the southern tip of India between latitudes of 5°55′–9°51′N and longitudes of 79°41′–81°53′E. Topographically the island consists of a south central mountainous region that rises to an elevation of 2500 m and is surrounded by broad lowland plains at an elevation of 0–75 m above sea level. From the mountainous regions nine major rivers and 94 other streams flow across the lowlands into the Indian Ocean. Anuradhapura city is located at 8°20′N and 80°25′E. It is situated 205 km north of the current capital city of Sri Jayawardenapura (Colombo) in the north central province on the banks of the river Malvathu Oya.

21.2.1 Geology and Geomorphology

The greater part of island is made up of crystalline rocks of Precambrian age belonging to one of the most ancient and stable parts of the Earth's crust. On the basis of lithology, structure and age, the Precambrian rocks are subdivided into three major groups as Western Vijayan Complex, Highland complex (Highland series and the Southern group), and Eastern Vijayan Complex (Fig. 21.1). The Anuradhapura district includes both Western Vijayan Complex and Highland series. Main rock types include a variety of gneisses, migmatites and granitoid gneisses including charnokite gneisses.

Three broad geomorphologic divisions can be recognized in Sri Lanka: the coastal plain, three peneplains (planation surfaces): the lowest from 0–125 m, middle from 125–750 m, and the highest from 750–2500 m, and dissected transitional surfaces in between the planation surfaces. Flood and coastal plains and dunes characterize the coastal plain; the lowest peneplains can be subdivided into mantled, rocknob, and remnant plains; and the middle and highest peneplains have a variety of mountains, hills, and escarpments. This variability of landforms results in a great diversity of ecological and hydrological conditions.

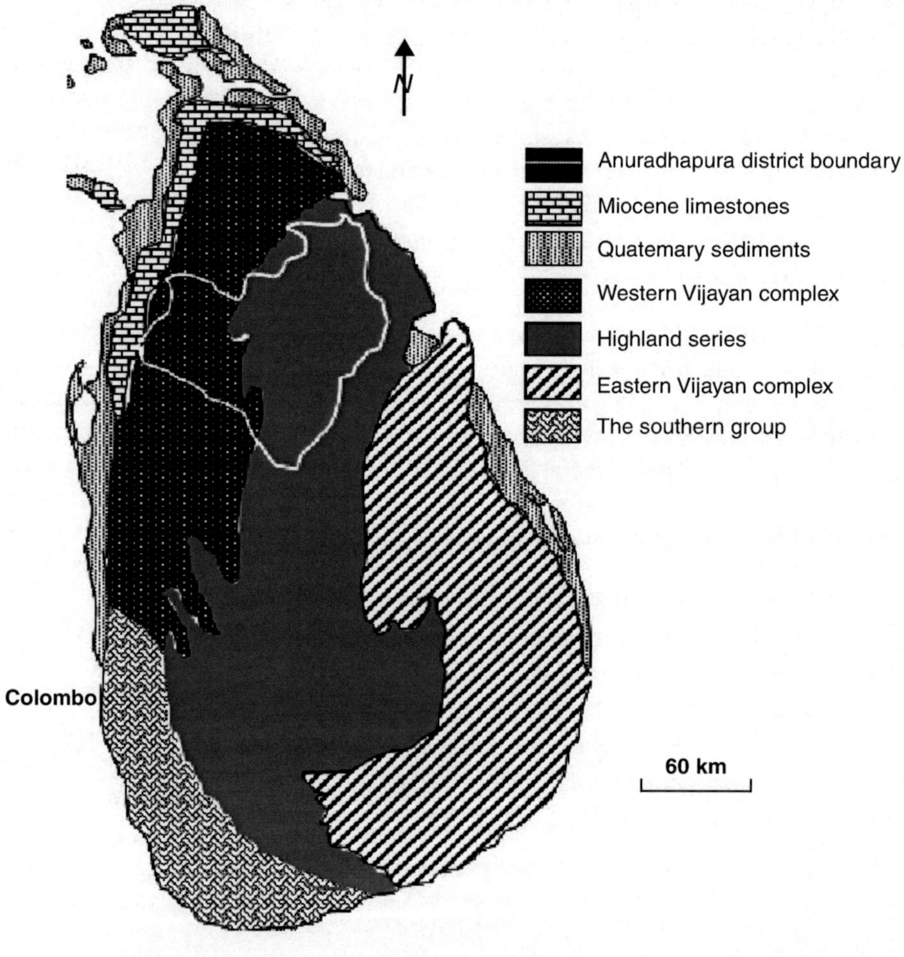

Fig. 21.1 Main geological domains in Sri Lanka

21.2.2 Soils

Diversity of soil in Sri Lanka is very high. Most abundant soils are: Reddish Brown Earths (RBE) (Rhodustalf) in 1.61 million ha, Red Yellow Podzolic soils (RYP) (Tropudalts) in 1.44 million ha, and Low Humic Gley soils (LHG) (Tropaqualfs) in 0.95 million ha. Great soil groups recognized in the Anuradhapura city area are RBE, LHG and alluvial soils (Tropaquents) of variable texture (Fig. 21.2). The general soil landscape has been recognized as RBE and LHG undulating terrain. The RBE occupies the crest and the upper and mid-slopes of the landscape. The LHG occupies the lower parts of the slope and upper parts of the valley bottom. A narrow strip of alluvial soils occurs along the natural drainage system. In the modal form, the RBE occupies around 60% of the land surface, the LHG around 30%, and the remaining 10% is made up of alluvial soils and rock knob plains.

The RBEs (Rhodustalfs) are well drained and do not retain much water in the soil and associated bedrock. Specific yield of this soil is as low as 0.015–0.065 and hydraulic conductivity is also in a low range (0.056–1.2 m/day). Nevertheless absence of mottles and stains indicates no fluctuation of groundwater table in the RBE profile. The texture varies from sandy loam to sandy clay. A subsoil horizon with a high proportion of quartz gravel is present, and the depth to the gravel layer is variable. The gravel layer impedes water movement and small showers can moist the upper layer of soil. This is favorable for chena (shifting) cultivation. Burning the vegetation after clearing the chena lands creates an ash layer over the soil and discourages the erosion. They are suitable for a variety of crops including perennials, but land characteristics promote soil erosion when exposed to rains.

LHG soils (Tropaqualfs) have developed from colluvial deposits on the foot-slopes of undulating land-

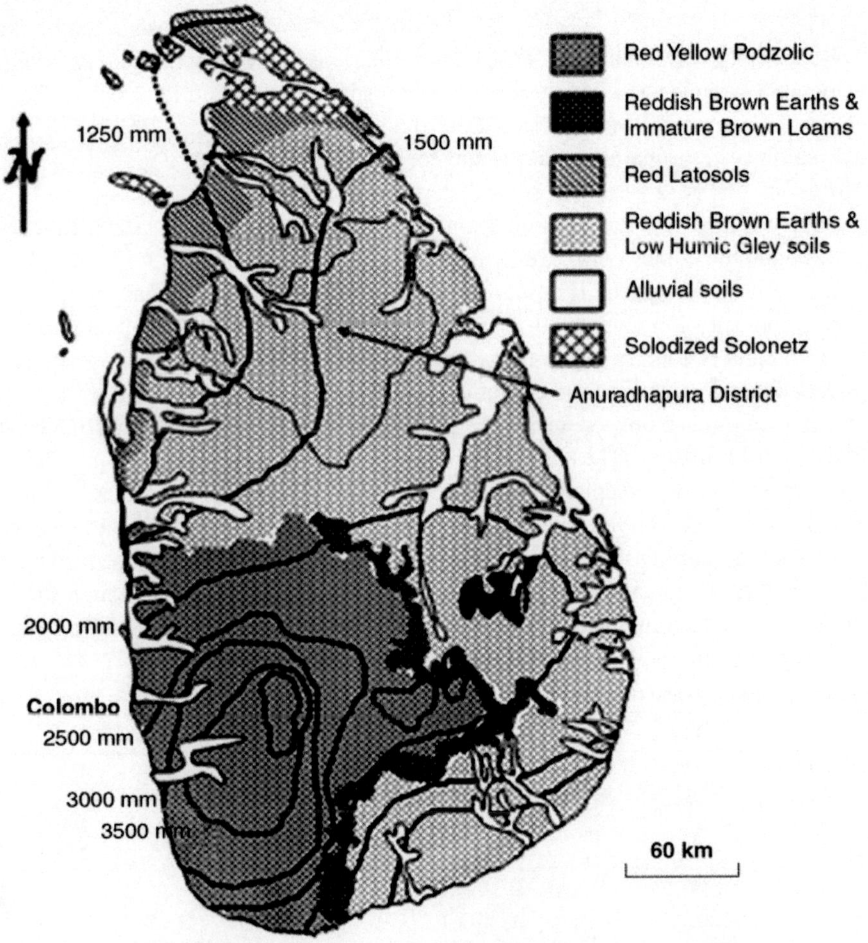

Fig. 21.2 Soil type distribution and rainfall isohyets of Sri Lanka

scapes. The color of the surface soil is dark grayish brown to dark brown, and in the sub-soil it is grayish to yellowish brown with distinct mottles and gleying. This indicates fluctuation of the groundwater table within the soil profile throughout the year. These soils are extremely hard when dry and sticky when wet and drainage is poor. The soils have a high CEC (cation exchange capacity) and are more suitable for rice, though other crops can also be grown with proper drainage. The flat landscape with poor drainage is ideal for water storage systems.

Alluvial soils (Tropaquents) are found on flat floodplains. The poorly drained alluvium shows a grayish color; better-drained soils are brownish to yellowish brown. Poorly drained soils are suitable for rice. In well drained soils many arable crops can be grown. Shifts of the river and stream routes can be generally recognized from the distribution of these alluvial soils.

21.2.3 Climate

Despite the relatively small size of the country, there is considerable variation in climate over time and space. The mean annual rainfall varies from 1250 to 5000 mm in the southwestern quarter of the island to less than 1250 mm in the northwest and southeastern parts of the country (Fig. 21.2). Based on the rainfall patterns, three geographic zones, dry, wet, and intermediate zones have been identified. They develop respectively in topographically low country 0–300 m, mid country 300–900 m, and upcountry higher than 900 m.

Sri Lanka experiences four climatic seasons in a year. First the inter-monsoon season begins in March and lasts until mid-May. The southwest monsoon prevails from mid-May to September. The second inter-monsoon season commences in October and ends in November. The northeast monsoon is responsible for the larger part of annual showers from November to February. Thus, both monsoons occupy a greater part of the year, namely 7.5 months or 62.5% of the length of the year. The inter-monsoon would therefore be 4.5 months or 37.5% of the year. Rains during southwest monsoon are mainly concentrated in the southwestern part of the island.

Annual average rainfall in Anuradhapura area is 1450 mm. Some climate data for Anuradhapura is shown in Table 21.1. A distinct bimodal rainfall distribution is shown, dividing the year into two main cultivation seasons. 'Yala' (the minor) season receives highly variable rainfall from first inter-monsoon. 'Maha' (the main) season receives the largest part of annual rainfall from both the second inter-monsoon and the northeast monsoon. This seasonal rainfall not only contributes to rainfed agricultural production but also to replenish the reservoirs in the dry zone.

There is little seasonal variation of temperature, but diurnal variation is wide for the island. Diurnal range increases from the coastal areas towards the interior. In general, the diurnal range of temperature is about 8°C in the coastal areas and increases towards inner portion of the country recording 14°C in the central highlands. The decrease in temperature with altitude is greater on the western than eastern slopes of the highlands. Relative humidity varies from 70% during the day to 90% at night.

21.2.4 Flora and Fauna

Sri Lanka was once a land rich in tropical forests nurtured by abundant radiation, high temperatures and rainfall, and long growth periods. Over the last century much of this resource has been destroyed, along with many of its material benefits. The island's flora represents six floristic elements namely Sri Lankan, Indo-Sri Lankan, Himalayan, Malaysian, African, and pantropic and cosmopolitan. The fauna represents Indo-Sri Lanka, Malaysian, African, and Sri Lankan elements. Although the predominant named forms of Sri Lankan plants and animals are present in peninsu-

Table 21.1 Rainfall and temperature at Anuradhapura (8.33°N 80.40°E, at 89 m asl)

Data	Jan.	Feb.	Mar.	Apr.	May	Jun.	Jul.	Aug.	Sep.	Oct.	Nov.	Dec.
Rainfall (mm)	102	47	79	175	91	22	32	41	74	258	253	231
Temp. Max. °C	30	32	35	33	32	32	32	33	33	32	31	30
Temp. Min. °C	21	22	23	24	25	25	25	24	24	23	22	22
Temp. Avg. °C	25	26	28	27	28	27	27	27	27	27	26	25

lar India, there is evidence of a high concentration of diversity and endemicity. Around 30% of the angiosperm flora, 18% of the ferns, and 16% of the terrestrial vertebrates are endemic to the island. In addition, thousands of years of settled agriculture have produced a storehouse of cultivated biological resources of crop plants, fruit trees, spices and livestock, and a century ago, in the plantation sector, numerous local new cultivars developed.

Sri Lanka has a rich treasure of rice genes. Several thousand years of selection and cultivation, augmented by traditional farming practices and the country's ecoedaphic variability have produced a wide variety of rice crop cultivars. Of the 120,000 rice cultivars in the entire world, this small island has recorded 2800 varieties, vastly disproportionate to its size.

21.3 Historical Synopsis

The dry zone civilization might have first emerged around natural lakes ('vil') and streams ('oya') in varying sizes. Subsequently, human beings developed rainwater-harvesting methods to make living possible in other areas of the dry zone. Intrinsic networks of minor tanks and streams evolved with the formation of organized communities. More logical water-use systems were used when farming societies developed. Culture, customs, agriculture, traditions, and all sorts of life styles have evolved on this landscape with influence of migration and invasion from northern side of the island. These external factors might have accelerated or sometimes disrupted this evolutionary process of civilization. Two main types of supportive factors of the evolutionary process of this ancient civilization can be identified: need based and problem based respectively. Basic needs were water, food, fibre, shelter, and protection. Problem-based driving forces of the evolution of our human community were strategies to address natural disasters such as droughts, epidemics, floods, cyclones, and external invasions. Strategies adopted to face these challenges were infiltrated to the social systems in form of their lifestyle and as accepted facts in their traditions. The pre-historic evolution that took place in Sri Lanka and India shows great similarities due to continuous interaction between the two countries.

21.3.1 Paleolithic (Old Stone Age)

During the last one million years, when human population is known to have existed in various parts of India, Sri Lanka was connected to the sub-continent on numerous occasions. The rise and fall of sea level (due to temperature fluctuations in the global climate) determined the periodicities of these connections, the last separation having occurred at 7000 year BP (Deraniyagala 1992). Hence it is impossible to view Sri Lankan prehistory in isolation from India. By about 125,000 year BP it is certain that there were prehistoric settlements in Sri Lanka. It is estimated that during certain subsequent pluvial episodes the population density in the dry zone of northern, eastern, and southern Sri Lanka could have ranged between 1.5 and 0.8 individuals/km^2, whereas the wet zone in the west would have had densities of 0.1 or less.

21.3.2 Mesolithic (Middle Stone Age)

From about 34,000 year BP onwards the prehistoric record is much more complete. It appears to have been a remarkably static situation over so long a period, relatively undisturbed by the arrival of new populations with diverse physical traits. Their life style could not have been too different from that described for the Veddas of Sri Lanka, the Kadar, Malapantaram, and Chenchus of India, the Andaman Islanders, and the Semang of Malaysia. They would have been moving from place to place on an annual cycle of foraging for food.

21.3.3 Neolithic (New Stone Age)

The transition from the Mesolithic Culture to the proto-historic Early Iron Age has been inadequately documented in Sri Lanka as the relevant transitional deposits have been disturbed by the extraction of fertilizer (guano) from prehistoric cave habitations. However, at a site there are indications of pottery for the cultivation of a cereal. The discovery of a few pieces of copper-working slag from the Mesolithic context could signify the first identification of a Chalcolithic horizon in Sri Lanka.

21.3.4 Early Iron Age

The Early Iron Age settlement at Anuradhapura does not have a Megalithic cemetery to which it can even remotely be linked. The Megalithic mortuary complex could possibly have been associated with just a special group of people, such as pastoralists, on the periphery of those who occupied Anuradhapura. Migration from the North to the island could have occurred primarily before 500 BC. What attracted these people who intruded on the scene at this early date? It is probable that the agricultural potential of Sri Lanka, notably its abundant supplies of water, with iron technology capable of subjugating the dense equatorial rainforest and heavy soils, was a major factor. Thereafter, Sri Lanka attraction for settlers from further afield than South India appears to have gained rapidly. This swell coincided with the so-called Second Urbanization of the Indo-Gangetic Plain. The settlement at Anuradhapura spread over more than 10 ha by 800 BC. By then prehistoric stone tool technology had been completely superseded by that of iron at this site. Other advanced traits were the manufacture of copper-alloy artefacts, high-quality pottery (notably Black and Red Ware), the breeding of cattle and horses, and the cultivation of rice. By 700–500 BC Anuradhapura exceeded over 50 ha.

21.3.5 Historic Period

Evolution of a great civilization in the north central part of Sri Lanka was principally driven by the process of collecting water from rainfall for people and agriculture. It is somewhat a sacrifice made to form societies around small sources of water in a water scarce area and not moving to water rich areas found in the central highlands of the island. This is debatable and difficult to understand unless one deeply probes the interactive process of historic evolution. One might conclude that although the apparent problem in this area was water, the historic fate of the country nonetheless had enabled the erection of the civilization through skilful management of the available water.

With the progressive acceleration of cognitive abilities and technology in the later pre-historic period (30000–1000 BC), the stage was set for a radical transformation in the interactive balance between man and environment. People steadily proceeded to dominate nature and bend it to the collective will. The advent of iron technology in Sri Lanka around 1000 BC is witness to this ascendance of man. Excavations in the citadel of Anuradhapura have produced important evidence of iron technology, breeding of horses and cattle, and paddy cultivation, from cultural horizons nearly ten meters below the present ground surface. There is incidental evidence (faunal, sedimentological) for water management associated with paddy cultivation. Agriculture would undoubtedly have been dominated by paddy, which can only be intensified in the Sri Lanka dry zone, where Anuradhapura is situated, by the adoption of water management measures to control supplies from seasonal rainfall, streams, and perennial rivers (Deraniyagala 2002).

The history and cultural heritage of Sri Lanka extends over a period of more than 5000 years. The great King Ravana ruled the land of Lanka, which covered also part of India, around 3000 BC as mentioned in the popular Indian epic 'Ramayana' (The Epic of Lord Rama) thought to have been written in 500 BC. The capital of the Asura King Ravana is said in the epic Ramayana to be situated in Anuradhapura. The invasion of Lanka by Vijaya through the Princess Kuweni (Princess of land) occurred in 600 BC. The Princess Kuweni ruled the area 'Chalaka' (western region), which was one of the ten divisions at that time. Cities of these local kings were said to be on mountains and four of them were around the Anuradhapura city, which was built 250 years later.

Although according to historical records the city was founded in the fifth century BC, the archaeological data put the date as far back as the tenth century BC (Proto-historic period). Excavations in Anuradhapura have uncovered information about the existence of a Proto-historic habitation of humans in the citadel. The Lower Early Historic period spanning from 500–250 BC is studied on the lines of the chronicles. During this time, King Pandukabhaya formally planned the city, with gates, quarters for traders, and so on. The city would have covered an area of 1 km^2, which makes it one of the largest in the continent at the time.

21.4 The Water Resource

Sri Lanka receives about 12 million ha m of water annually from rainfall, of which more than 50% is lost directly through evapotranspiration. Another 20% seeps down to replenish groundwater whereas only 30% or about 3.5 million ha m is available for irrigation through streams, rivers, and lakes. Surface water from the high watersheds is transported by 103 distinct natural river basins, of which 9 cover 90% of the island; the remaining 94 small coastal basins contribute only marginally to water resources. River basins originating in the wetter parts of the hill country are perennial while the majority of those in the dry zone are seasonal. Annual surface water estimates vary from 4.0 to 5.13 million ha m. A part of this is used for irrigation and hydropower projects, and the balance discharges to the sea. Rivers flowing for long distance through flat to undulating landscape especially Malwathu Oya, Kala Oya, Yan Oya, and Deduru Oya replenish groundwater creating more favorable ecology for human living. These four rivers collectively contributed to evolve the spatial base for the ancient hydraulic civilization of Sri Lanka.

Inadequacy of rainfall especially the poor distribution, higher evaporation, and transpiration rates with respect to precipitation, occurrence of droughts, and resulting agricultural failures led the ancient rulers to construct reservoirs of varying sizes to create a favorable environment for human life. Landscape hydrology prevailing in the above mentioned river basins characterized by flat to undulating landscape with catenary formation, bi-modal rainfall distribution, two soil types with distinctly different soil drainage conditions, and highly impervious underlying hard rock basement favored the development of small tank cascade systems in the area.

There are six main type of groundwater aquifers demarcated and identified in Sri Lanka. They are shallow karstic aquifers, coastal sand aquifers, deep confined aquifers, lateritic aquifers, alluvial aquifers, and shallow regolith aquifers in the dry zone hard rock region. The regolith aquifer found in Anuradhapura area is made up of three main subdivisions namely a collapsed zone extending up to 3 m in thickness and lying above the water table, a saprolite zone that is commonly highly weathered rock, and a saprock that is highly weathered rock where the fractures are likely to be more open than in the fresh bedrock (Panabokke 2003).

21.5 Irrigation and Agriculture

The legendary story of civilization in Sri Lanka begins when the island was inhabited by people of the original tribes Yakka, Raksa, and Naga. These early people gradually developed systems of sedentary agriculture based on irrigation, and folklore maintains that the Yakka built some ancient irrigation tanks. By the arrival of Vijaya around 500 BC, small scale irrigation systems were already operational.

The first extensive Sinhalese settlements were along rivers in the dry northern zone of the island. Because of early agricultural activity, primarily the cultivation of wet rice was dependent on unreliable monsoon rains, the Sinhalese constructed canals, channels, water-storage tanks, and reservoirs to provide an elaborate irrigation system to counter the risks posed by periodic droughts. Such early engineering attempts reveal the brilliant understanding these ancient people had of hydraulic principles and trigonometry. The discovery of the principle of the valve tower, or valve pit (*bisokotuwa*), for regulating the escape of water is credited to Sinhalese ingenuity more than 2000 years ago. By the first century AD, several large-scale irrigation works had been completed.

The mastery of hydraulic engineering and irrigated agriculture facilitated the concentration of large numbers of people in the northern dry zone, where early settlements appeared to be under the control of semi-independent rulers. In time, the mechanisms for political control became more refined, and the city-state of Anuradhapura emerged and attempted to gain sovereignty over the entire island. The state-sponsored flowering of Buddhist art and architecture and the construction of complex and extensive hydraulic works exemplify the Sri Lanka's classical age, which roughly parallels the period between the rise and fall of Anuradhapura (from 437 BC to 1040 AD).

After King Pandukabhaya (437–367 BC) founder of the Anuradhapura city, 122 kings reigned for 1477 years up to 1040 AD, until the kingdom was moved to Polonnaruwa (80 km to the west). The establishment of forests and construction of ponds, reservoirs and irrigation systems were considered great meritorious acts in accordance with popular Buddhism, the faith of the leaders and of the large majority of the people. Sri Lanka history is full of achievements of kings who contributed to the development of water resources. Since the first century AD, kings such as Vasabha (67–111 AD), Mahasena (276–303 AD), Dhatusena (455–473 AD), Agbo II (575–608 AD), and Parakrambahu (1153–1186 AD) built numerous reservoirs and irrigation systems that fed vast expanses of paddy field in the dry zone. Construction and upkeep of these irrigation systems became massive undertakings. An indigenous expertise developed over the centuries, which appears to have been called upon by other countries of South Asia.

This ancient hydraulic civilization of the dry zone disappeared after the twelfth century AD. Climatic change, malaria, depletion of soil fertility, foreign invasions, and famine are some of the reasons cited. The breakdown of the efficient irrigation management system may have resulted from annihilation of the 'kulinas' (the dry zone nobility who possessed irrigation expertise) by invading South Indian forces (Paranavithana 1960).

21.5.1 Water Management in Paddy Cultivation

The early hydraulic societies thrived on small irrigation systems with unique assemblages of land uses and agricultural attributes. Possibly these systems evolved from early rain-fed shifting agriculture into small scale irrigation that, in turn, led to major systems. The sedentary way of life facilitated by this hydraulic base led to land tenure, property inheritance, and social organizations that persisted for centuries. Community leadership patterns had to be strong and effective with increasing size and complexity of irrigation systems. These conditions eventually led to centralized authority representing a form of oriental despotism (Witfogel 1957).

Paddy cultivation under rain-fed conditions or with supplementary irrigation water from reservoirs of varying sizes evolved to become a sustainable farming practice in the dry and intermediate zones of Sri Lanka. Basic problems faced by farmers were shortage of water in less rainy seasons, development of salinity in certain part of the field, and damage from wild animals, pest, and diseases. Strategies emerged to address these problems are not specific to a certain problem but collective. The following strategies can be traced:

a. 'Bethma' practice—It is a practice that temporarily redistributes plots of land among shareholders (paddy landowners) in part of the command area (territory) of a tank (reservoir) during drought periods.
b. 'Pangu' method—The tank had to be maintained properly to avoid breach, leak, and excess seepage. Repair and desiltation of tanks and cleaning of canals during dry periods are shared tasks assigned to each farmer proportionately to land ownership.
c. 'Kekulama'—Farmers advance the cultivation time using early seasonal rains whenever they feel that tanks would not get enough water to cultivate the command area. They have the experience that if September (2nd inter-monsoonal) rains are high, the total seasonal rainfall is not adequate to fill the tank.
d. 'Karahana'—This is a water distribution device fixed across the canal made up of log with two weir-shape cuts. The size and bottom level of these cuts are made according to flow requirements of the two canals below, and the *karahana* is fixed by the village head ('Gamarala').
e. Village commons—micro-landscapes are utilized to reduce tank water losses, mitigate salinity effects, prevent tank sedimentation and so on (Fig. 21.3).

Organization of small tanks into a cascading sequence within micro-catchments allowed greater efficiencies in water use. Drainage from the paddy fields in the upper part of the cascade flowed into a downstream tank for reuse in the paddy fields below. Efficient use of available water within the system reflects the wisdom of managing droughts, flash floods, and land degradation in a tropical environment that suffers chronically from seasonal water shortages. The technology that evolved has withstood the test of time and lasted for over two millennia (Madduma Bandara 2009).

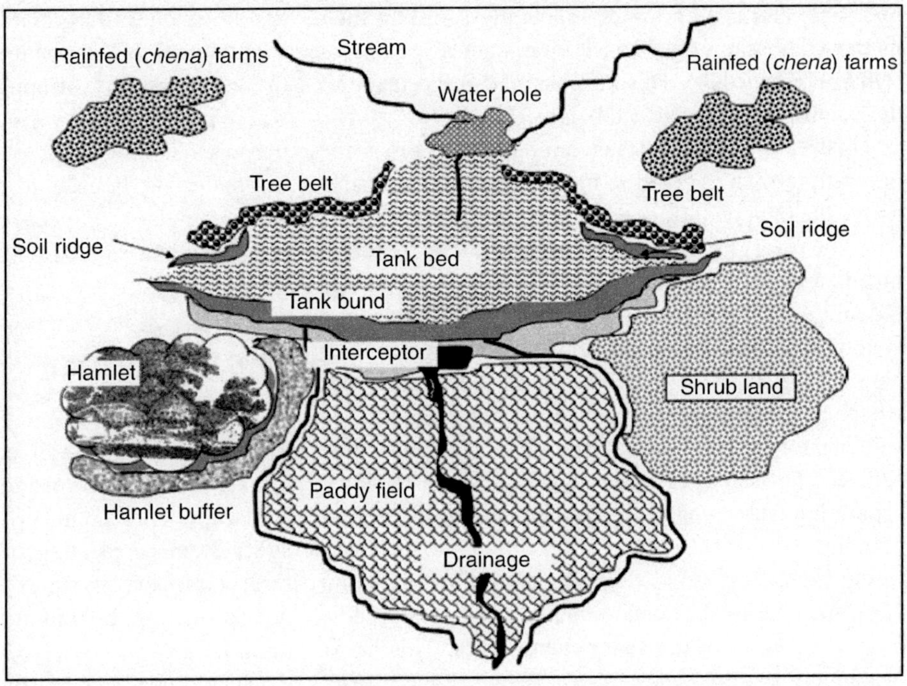

Fig. 21.3 Landscape management to mitigate natural stresses

21.5.2 Rain-Fed Cultivation

A three-fold farming system evolved in the tank-villages of the dry zone: rain-fed upland cultivation, lowland paddy cultivation ('wela'), and dwelling gardens ('gangoda').

As in many countries of the world shifting cultivation is the oldest way of farming in Sri Lanka. Many similarities are found in these countries such as clearing of forest or savannah, burning, sowing a mixture of seeds with onset of rains, cultivation under rain fed condition for few seasons, and moving to other lands once the cultivated land becomes less fertile. However, chena cultivation in Sri Lanka evolved on undulating terrains with variable soil and rainfall conditions, hence specific practices as given below were developed.

a. Land for chena cultivation was selected from middle part of the land catena with gentle slopes, where soil is relatively deep.
b. Risks of farming due to factors such as rainfall, drought, pest and diseases, and damages from wild animals were reduced by adjusting the cultivation to the best times through long experience.
c. Favorable environment for crops was maintained by adoption of various soil and moisture conservation practices and through shade management.
d. Land productivity was maintained by posing least disturbance to soil and using high amounts of burnt biomass.
e. Diverse crop combinations were adopted to cope with variation of climate, soil, and other biotic as well as abiotic stresses.
f. Simple farm implements were used with lesser energy consumption.
g. Land races were improved as family secrets to utilize most suitable crop varieties for the area.

21.5.3 Sustainability

Sustainability of the traditional tank-village system was maintained in the past not only for its structural maintenance. Each and every component of the landscape was utilized at its best to create a sustainable system (Fig. 21.3). The traditional communities addressed and continue to address all issues as a systematic whole rather than individually. In addition to rain-fed farming (chena) area, hamlet and paddy fields, they take care of the commons (their territory) surrounding them. It is vital to understand why these commons have been maintained purposely. These communities had the wisdom to mitigate the effects of droughts, floods, cyclones, and epidemics by managing properly their natural landscape and conserving natural resources (Dharmasena 2004). Descriptions and importance of the commons are discussed below.

Upstream tree belt ('Gasgommana')—This is the naturally grown vegetation in the upstream land strip above the tank bed, flooded only when spilling occurs from the reservoir. Large trees and climbers are found in this area. The gasgommana acts as a wind barrier reducing adverse effects of strong winds, minimizing evaporation from the tank, and lowering water temperature. It extends up to the bund (embankments) where roots of large trees create watery conditions suitable for breeding and living places for some fish species. This strip of trees demarcates the territory separating somewhat the human and wild animal domains.

Upstream meadow ('Perahana')—This meadow developed below the gasgommana filters out the sediment flowing from upstream chena lands. This helps in maintaining the tank capacity reducing the threats of floods and droughts.

Upstream soil ridge ('Iswetiya' or 'potawetiya')—This is an upstream bund constructed to prevent entering eroded soil from upper land slopes.

Upstream water hole ('Godawala')—A manmade water hole to trap sediment and. It also provides water to wild animals.

Hamlet buffer ('Tis-bambe')—A fertile land strip found around the settlement area (*gangoda*) that is not privately owned. Tree species such as *Maduka longifolia*, mango, coconut are grown in scattered manner. This area was mostly used for sanitary purposes and as the resting place of buffaloes. Buffaloes were used as a protection mechanism from wild animals and malaria.

Downstream drainage ('Kiul-ela')—This is the old natural stream utilized as the common drainage. Tree species, which could absorb salts, and a few rare species of small fish are also found in water holes along the kiul-ela. Most importantly it filters salts and iron polluted water, and improves the drainage condition of the paddy tract.

Interceptor ('Kattakaduwa')—This is a reserved area of land below the tank bund. Its micro-landscape consists of three micro-climatic environments: water hole, wetland, and dry upland, therefore, diverse vegetation is developed. This land prevents soluble salts and ferrous iron entering the paddy field. The water hole referred to as 'yathuruwala' minimizes bund seepage by raising the groundwater table. Villagers plant *Pandanus kaida* along the toe of the bund to strengthen the bund stability. It appears to be a village garden, where people utilize various parts of the vegetation for purposes such as fuel wood, medicine, timber, fencing materials, household and farm implements, food, fruits, vegetables. Specifically they harvest raw materials from this vegetation for cottage industries.

21.6 Discussion

The ancient water–soil conservation systems of Sri Lanka are a classic example of human adaptation to landscape. They include river diversion and water storage systems consisting of small, medium, and large reservoirs. These systems were constructed over a long period of time, beginning about the mid first millennium BC.

Hence, one may reasonably postulate that the distinct water management systems of Sri Lanka had their origins in the technological development that occurred during its proto-historic Early Iron age from 1000 to 500 BC. It constituted the springboard for what followed during the Early and Middle Historic periods up to the end of the first millennium AD: the development of a water management system that in its technology and organization is one of the most sophisticated in the world (Deraniyagala 2002).

This civilization evolved through millennia as a result of trying to find solutions to basic needs (food, water, and protection) and basic problems (drought, flood, cyclone, epidemics, and invasions). It is difficult to trace specifically the timing of the special strategies adopted to mitigate such disasters or to what degree these approaches were indigenous or influenced by invading/migrating foreigners.

Three main factors were most responsible for evolution of the water resources management system in the dry zone. They are the morphology of the landscape, the amount and distribution of the rainfall, and the nature of the substratum (in terms of pedology and geology).

Water was managed by sharing between river basins through diversion or feeder canals to avoid excess or shortage. Some examples (Fig. 21.4) are the Dambulu Oya–Malwathu Oya diversion canal (860 AD), the Malwathu Oya–Kanadara Oya diver-

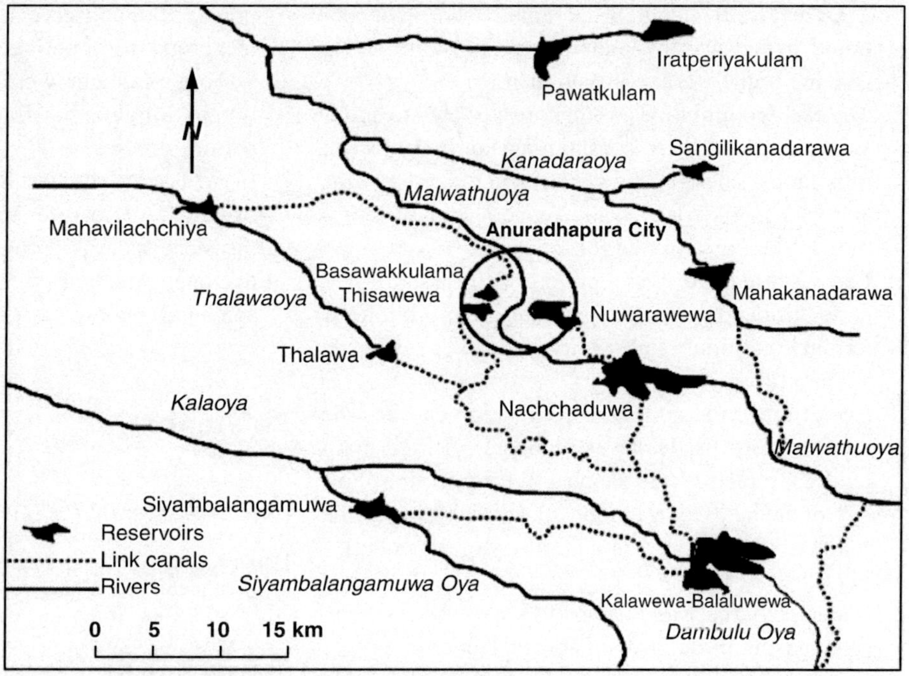

Fig. 21.4 Link canals and streams network in Anuradhapura District

sion canal (860 AD), and the Yoda Ela–Nachchduwa feeder canal (540 AD). Within some river basins, water flow had been also regulated through link canals to avoid imbalance of water. Examples are the feeder canals of Kalawewa–Thisawewa Yodha Ela (470 AD), Nachchaduwa–Nuwarawewa (290 AD), Balaluwewa–Siyambalangamuwa feeder canal (290 AD), and Basawakkulama–Maha Vilachchiya (470 AD).

Cascading form of small tanks and streams was developed to manage the water resource within river sub-basins. These tank cascades are mostly rainfed but locally they may be supplied by feeder or diversion canals. Those people developed the knowledge to construct long canals with extremely low gradient, such as the Yoda Ela, which carried water from Kalawewa to the city tanks of Anuradhapura (Thisawewa) along a canal 87 km long (Fig. 21.4). This 'Yoda Ela', which had a gradient of less than 10 cm/km within its first 27 km continued to maintain itself as a natural stream. It feeds numerous tank cascade systems found along its way, at the same time receiving excess drainage waters dispersed on the land from upstream tank cascades.

Indigenous agriculture is mostly based on inborn principles. People observed natural phenomena operating around them and studied how they could be manipulated for their needs. This included management of the forest by observing its anatomy, association of coexistent species, regeneration after fire, spatial variations and so on. The farming system, which includes chena (shifting cultivation), paddy and home garden cultivation, evolved through interaction of man with the environment and developed in harmony with the natural landscape. Their experience and observations of rainfall pattern, wind, temperature, humidity, and landscape have used to adjust their cultivation activities. To reduce the risks of natural disasters they developed the tank-village community system that spread all over the dry zone. When they found that some of the disasters they faced in farming were beyond their control, they appealed to the support of religion and spiritual and cosmic influences. The most important fact they realized was that sustainability of their food sources was not possible without giving due respect to the resources used in farming.

21.7 Conclusions

The ancient dry zone civilization evolved with two distinct discontinuities occurring in 3000 BC and 1200 AD due to external invasion from the northern side of the island.

a. The early settlements of the country were concentrated in the dry lowlands along few main rivers flowing to the western coastal belt.
b. The early communities were more attracted to undulating terrains than to coastal plains or highly steep hilly lands. These terrains are characterized by hills and valleys underlain by the Western Vijayan Complex of the Precambrian rocks.
c. Early settlements were preferentially established under moderate regime of climate, geomorphology, terrain conditions, soil quality, and water resources availability.
d. Evolution of the hydraulic civilization can best be ascribed to a particular assemblage of three main landscape factors: the shape and form or morphology of the landscape, amount and distribution of the rainfall, and soil types and lithology of the underlying substratum.
e. Water resource was managed by sharing between river basins through diversion or feeder canals to avoid excess or shortage of water. Water was also stored in upstream reservoirs, used to irrigate the land, and in part returned to the rivers at a downstream point as excess amount drained out from the fields. Retention of water was enhanced by developing intrinsic networks of canals and reservoirs within small watersheds (tank cascades). Water was shared within tank cascades to avoid water shortages to crops. Tank water was conserved and efficiently utilized through strategies such as practicing kekulama, bethma, pangu, using karahana and maintaining gasgommana, kattakaduwa, iswetiya, godawala described in Sects. 21.5.1 and 21.5.3.
f. Hard rock basement with regolith aquifers found in these areas do not show any enrichment of groundwater. The ancient city was supplied with water filtered through the substrate from the three large man-made surface reservoirs.
g. Combination of two main types of soils RBE (Reddish Brown Earths) and LHG (Low Humic Gley) frequently associated with alluvials found along valley bottoms creates some of the best environments for agricultural communities who adopted the three-fold farming system (paddy field, chena, and home garden).
h. Chena cultivation was practiced on undulating terrains with variable soil and rainfall conditions, with practices to conserve soil and moisture and to protect crops from biotic and abiotic stresses.
i. Various practices were adopted in paddy cultivation to mitigate the effect of water deficit and other natural stresses.
j. In Sri Lanka, the best example to illustrate the use of landscape features to reduce the risks of natural disasters such as droughts, floods, epidemics, and cyclone is the tank-village farming community that spread all over the dry zone.

Reference

Deraniyagala SU (1992) The prehistory of Sri Lanka, parts I, II. Archaeological Survey Department, Colombo

Deraniyagala SU (2002) A note on man and water in Sri Lanka: from prehistory to history. In: Mendis DLO Water Heritage of Sri Lanka. Sri Lanka Pugwash Group, Colombo, 258 pp; to order it: pugwash@sltnet.lk

Dharmasena PB (2004) Exploring tank village farming system in the dry zone. Journal of Soil Science Society of Sri Lanka 16:17–28

Madduma Bandara CM (1985) Catchment eco-systems and village tank cascade in the dry zone of Sri Lanka: A time tested system of land and water management. In: Lundqvist J, Lohm U, Falkenmark M (eds) Strategies for River Basin Management. Linkoping, Sweden, pp 302–306

Madduma Bandara CM (2009) Village tank cascade systems of Sri Lanka–A traditional technology of drought and water management. Paper presented at the Annual Workshop on Disaster Reduction Hiperbase - Asian Application (DRH-Asia) held from 8-9 January 2009 Tokyo, Japan. http://drh.edm.bosai.go.jp/Project/Phase2/2Events/20_WS3/5-6_DRH_WS3_Contents_TIK_(bandara).pdf

Panabokke CR (1999) The small tank cascade systems of the Rajarata: Their setting, distribution patterns, and hydrography. Mahaweli Authority of Sri Lanka, Colombo, Sri Lanka, 39 pp

Panabokke CR (2003) Nature of occurrence of the north regolith aquifer in the hard rock region of the north central dry zone, and its rational exploration for agro-well development. In: Proceedings of symposium "Use of groundwater for agriculture in Sri Lanka," Peradeniya, Sri Lanka, 30 Sept 2002. Agricultural Engineering Society of Sri Lanka, Peradeniya, pp 10–22

Panabokke CR, Sakthivadivel R, Weerasinghe AD (2002) Small tanks in Sri Lanka: Evolution, present status and issues. International Water Management Institute, Colombo, Sri Lanka, 73 pp

Paranavithana S (1960) The withdrawal of Sinhalese from the ancient capitals. History of Ceylon, vol 1, University of Ceylon, Peradeniya, PP 713–720

Witfogel KA (1957) Oriental despotism; a comparative study of total power. Yale University Press, New Haven. Connecticut

Chapter 22
Disease in History: The Case of the Austronesian Expansion in the Pacific

Robert Sallares

22.1 Introduction

The expansion of speakers of Austronesian languages during the last 6000 years, from a homeland in Taiwan or the adjacent parts of mainland China, thousands of miles across the Pacific Ocean to Hawaii, New Zealand and Easter Island, and across the Indian Ocean to Madagascar, is the most extraordinary migration in history (Fig. 22.1). This chapter differs from others in this volume in focusing on a predominantly maritime landscape, peppered with thousands of islands. The Austronesian expansion has generated a great deal of controversy over the last 30 years or so. An interdisciplinary debate is taking place in which it is necessary to evaluate and accommodate the contributions of several different subjects, especially archaeology, linguistics, and genetics. The focus of the debate has been on attempts to relate the historical expansion of the Austronesian language family, which displays its greatest diversity in Taiwan (Blust 1995), to the prehistoric people who spread the so-called Lapita cultural complex (Spriggs 1984; Bedford 2003)—named after the type site in New Caledonia—and to the population history of the region as suggested by modern genetic and molecular analyses (Cox 2003). A number of different models have been proposed to relate the various categories of evidence. The "express train" and the "slow boat" models of human population movement have been widely discussed, as well as various intermediate models which incorporate elements of both of the two extremes (Gibbons 2001). Of these models Green's "Intrusion/Innovation/Integration (Triple I) model" is probably the most popular way of explaining the origins of the Lapita cultural complex among archaeologists today (Green 1991).

It is not feasible to discuss all the various theories in detail here. Nevertheless the position adopted in this chapter is that the totality of the evidence of archaeology, of human genetics, of historical linguistics, and of the phylogenies of microorganisms that attack humans in Pacific populations, all agree in supporting an expansion out of Taiwan (or possibly the neighboring Fujian province of China) within the last 6000 years, punctuated by pauses in which Austronesian populations interbred to some extent with non-Austronesian populations, especially in the coastal areas of New Guinea and neighboring islands, before moving on again towards Polynesia (Gray et al. 2009; Moodley et al. 2009). The alternative "slow boat" theory (Oppenheimer and Richards 2001), which posits the origins of the Austronesian peoples in island south east Asia over a much longer timescale dating back to approximately 17,000 years ago, depends on a method of dating human genetic data which has recently been shown to frequently give unreliable results (Cox 2008) and is otherwise unsupported by evidence.

Many environmental factors have been introduced into the debate. For example it has been suggested that changes in sea-level in the mid-Holocene, affecting the resources available in coastal habitats, compelled people to search for new homes, leading to the spread of the Lapita cultural complex (Gibbons and Clunie 1986; Dickinson 2003; Cabioch et al. 2008; Pope and Terrell 2008). The role of the spread of farming and the subsequent transformation of landscapes by farmers has also played a major role in

R. Sallares (✉)
UMIST, 184 Albert Avenue, Prestwich,
Manchester M25 0HF, UK
e-mail: rsallares@aol.com

I. P. Martini, W. Chesworth (eds.), *Landscapes and Societies*,
DOI 10.1007/978-90-481-9413-1_22, © Springer Science+Business Media B.V. 2010

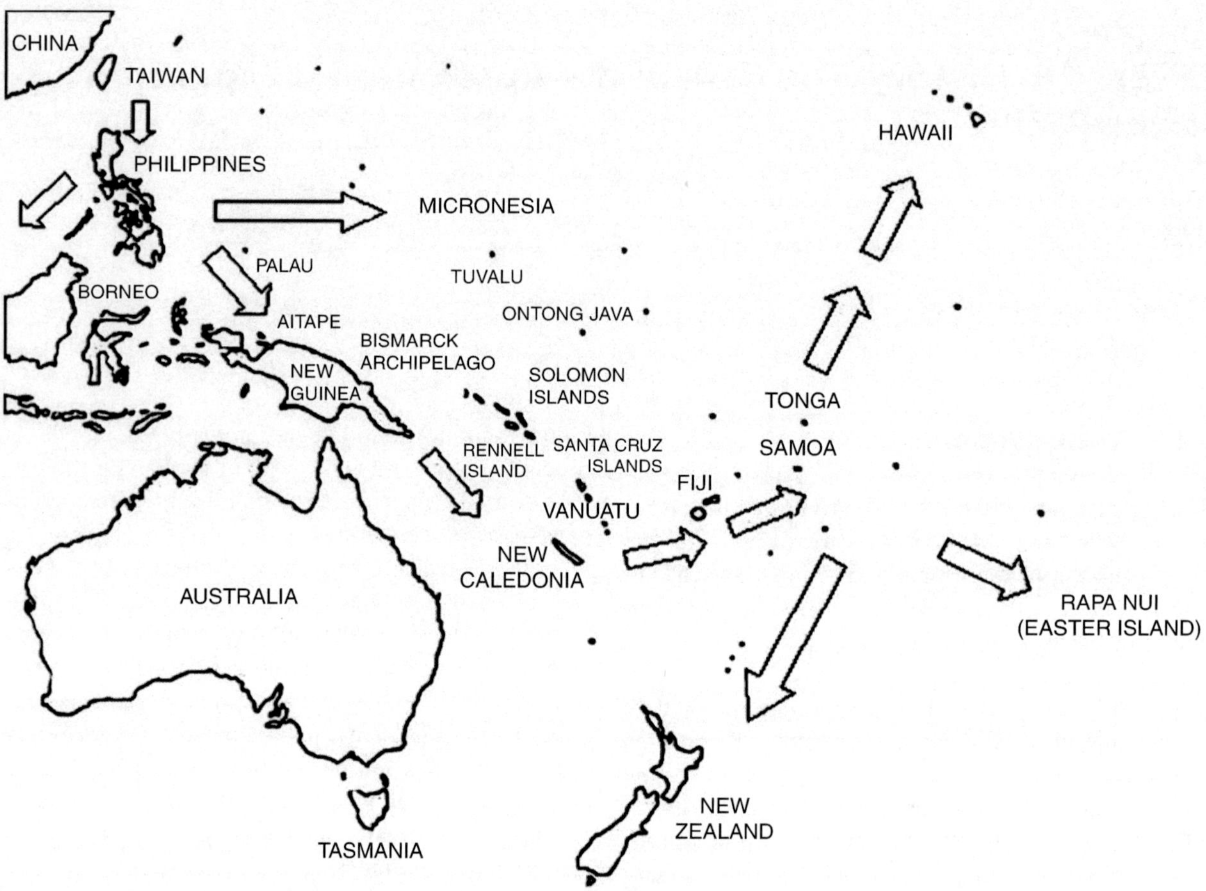

Fig. 22.1 Schematic map of the Western Pacific area with identification of the island mentioned in this chapter. (Arrows indicate the direction of migration)

the debate (Bellwood 1975). In general the various environmental factors are not necessarily mutually exclusive; the overwhelming balance of probability is that a comprehensive explanation of the Austronesian expansion would have to incorporate several different factors.

This chapter concentrates on one factor, the role of some of the smallest inhabitants of the landscape, namely pathogenic microorganisms. Several different studies have demonstrated the utility of pathogens as a proxy tool for investigating human population history in the Pacific. For example the phylogeny of Pacific populations of the JC polyomavirus suggests that the spread of this rapidly evolving DNA virus follows and parallels the spread of its human hosts, albeit on timescales that are the subject of debate (Yanagihara et al. 2002; Takasaka et al. 2006; Kitchen et al. 2008). Similarly the phylogeny of strains of *Helicobacter pylori*, implicated in stomach ulcers and gastric cancer, has proved to be a very useful tool for following the spread of human populations across the Pacific. The strain HpSahul, found in Australian aborigines and New Guinea, diverged from the Asian populations of *Helicobacter pylori* 17,000–37,000 years ago (the statistical methods used in phylogenetic analysis can only yield a range of variation, not absolute dates). However, HspMaori (a subpopulation of the HpEast-Asia strain) has its greatest diversity among Taiwan aborigines and was spread to Melanesia and Polynesia later, accompanying the Lapita migrations (Moodley et al. 2009). Such studies are potentially possible in relation to other diseases as well. For example, the balance of strains of tuberculosis that occurs among the Taiwan aborigines is different from that among Han Chinese settlers in Taiwan (Dou et al. 2008). Filariasis is another very important disease in the Pacific which has been studied in relation to prehistoric human migrations (Pichon et al. 1982). These examples show that pathogens have followed the differentiation of human populations in the Pacific into different environments and habitats. Malaria is one disease that has greatly affected Indonesia and other islands in the western Pacific area.

22.2 Malaria in the Western Pacific Islands

Many of the important diseases that affect human populations in the Pacific, such as tuberculosis, yaws and leprosy, are transmitted directly from person to person. Consequently their spread is not affected directly by the physical environment. However, there is one extremely important disease which is intimately linked to the physical environment because of its mode of transmission, namely malaria. Malaria is normally transmitted by various species of mosquitoes, each of which has its own peculiar requirements for breeding purposes. For example, the species of mosquito which transmitted malaria to humans in the past in Mediterranean countries frequently chose marshes as breeding sites, while *Anopheles gambiae*, the most important malaria vector in tropical Africa, likes to breed in small pools created by human disturbance of the ground for agriculture or other activities, and the mosquito species of Southeast Asia often favor forested environments. In the western Pacific islands there are numerous wetlands which have been divided into the following categories: coral reefs, seagrass beds, shoreline mangrove swamps, riverine wetlands (principally in New Guinea), lacustrine wetlands (either freshwater or saline), freshwater swamp forests, and freshwater marshes with grasses or reeds as the dominant vegetation type (Ellison 2009). Several of these types of wetlands have potential for supporting the mosquito vectors of malaria. A discussion of how differences in microenvironments affect the incidence of malaria in the Admiralty Islands of Papua New Guinea is given by Ataka et al. (2001). Some areas seem to have a higher prevalence of malaria than others because they have a higher proportion of flat terrain which is more likely to be suitable for mosquito breeding. Human modification of microenvironments, for example by digging open wells and reservoirs, also has a significant influence on the prevalence of malaria.

The possible role of malaria in Pacific prehistory has attracted interest because it is still today a common disease of major significance on many islands in the western Pacific. Little is said in this paper about the eastern Pacific islands, such as Hawaii, Fiji, Samoa, Tonga and Easter Island, precisely because *Anopheles* mosquitoes never reached them and malaria has never occurred there (Fig. 22.2). Nevertheless the key geographical contrast in this paper is between the western and the eastern Pacific islands. Human population densities are considerably lower in western Pacific islands than in eastern Pacific islands. Even though the western islands have been occupied by humans for much longer, their frequently still heavily forested environments are less degraded (despite the prevalence of shifting cultivation) than the more densely populated eastern Pacific islands (Rapaport 2006; Rolett 2008). The most notorious and controversial example of environmental degradation (deforestation) associated with the spread of humans across the Pacific is of course Rapa Nui, or Easter Island, the furthest outlier of the Polynesian expansion, where leprosy was the commonest disease at the beginning of the twentieth century, but there is no sign of malaria (Bahn and Flenley 1992; García-Moro et al. 2000). Human population densities and the extent of environmental degradation across the Pacific are closely correlated with the presence or absence of malaria (Rolett 2008).

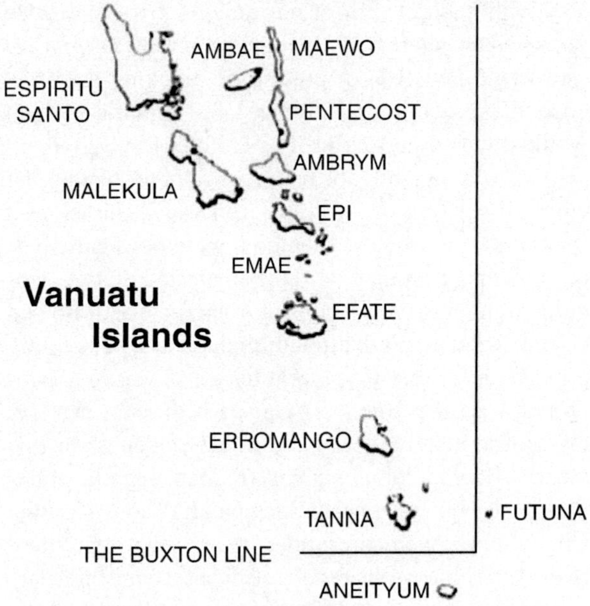

Fig. 22.2 Map of the islands of the Vanuatu group, also showing the Buxton line that marks the modern boundary of the geographical distribution of malaria in the southwest Pacific

Malaria's importance in the past has been stressed, for example, by Kirch (1997). Blust (1980) listed Austronesian words for the fevers and chills associated with malaria and for mosquitoes. Groube (1993) proposed an interesting hypothesis for an even earlier (pre-Lapita) episode of prehistory, invoking the presence of epidemiologically unstable malaria (which would attack adults as well as children) in a gradient zone in what is now northern Australia to explain the cultural isolation of Australia from New Guinea at a time of low sea levels when both were physically part of the Sahul continent. Groube suggested that the isolation of human populations in Australia was followed by the eventual elimination of malaria as a very thinly populated territory dried up after the physical separation of Australia from New Guinea with rising sea levels. Unfortunately there is no direct evidence for this theory. Haemoglobinopathies (human genetic mutations that confer resistance to malaria) are unknown among Australian aborigines today, although there is one possible fossil report of thalassaemia in a late Pleistocene human skull from Australia (Webb 1990).

Malaria is very common today in the areas that were the home of the Lapita cultural complex, the islands of the Bismarck Archipelago and the northern coast of New Guinea from Aitape eastwards. In fact the prevalence of malaria is higher in Melanesia than anywhere else in the world outside Africa today. Moreover malaria was until recently also common in the lowlands of Taiwan, the likely point of initial diversification of the Austronesian language family. There were 1.2 million cases of malaria on Taiwan as recently as 1952 and the disease was not finally eradicated until 1965, although how long it has been present there is unknown. This eradication helped to facilitate the modern economic growth of Taiwan. The prevalence of malaria on Taiwan in the past may have been a major force of natural selection on human populations on the island. The distribution of one particular immunoglobulin haplotype is significantly associated with malarial endemicity among lowland aboriginal tribes in Taiwan (Schanfield et al. 2002).

The fact that Austronesian language speakers, including the bearers of the prehistoric Lapita cultural complex, succeeded in colonizing coastal lowlands where malaria is a major problem today, raises the possibility that their spread was assisted by the possession or acquisition of genotypes conferring a selective advantage in environments where malaria is common. Attention has been drawn to the possible role of immunoglobulin haplotypes in the migrations of Austronesian peoples (Kelly 1990, 2000; Clark and Kelly 1993). This hypothesis has been criticised as unsupported by evidence and relying on an over-simplified distinction between malaria-resistant and malaria-susceptible populations (Searjeantson and Gao 1995). The Austronesians could also of course have been spreading the disease themselves at the same time, and also the mosquitoes that transmit it. In fact modern research has shown that human movement is essential for long distance malaria gene flow in the Pacific (Lum et al. 2007), because mosquitoes generally only fly short distances on their own, often no more than a few hundred yards. The enforced movement of Africans to the western hemisphere by means of the slave trade offers a good parallel for the postulated scenario in the prehistoric Pacific; Africans carried both malaria itself and the malaria-resistant sickle cell trait (haemoglobin S) genotype across the Atlantic. To give another example for comparative purposes, in the Mediterranean countries population movements of Greek and Phoenician colonists in antiquity led to the spread of both the vector mosquitoes and thalassaemia mutations giving some resistance to malaria (Sallares et al. 2004).

Human malaria itself undoubtedly has a very long history in Southeast Asia. Recent research has demonstrated that *Plasmodium knowlesi*, previously regarded as essentially a parasite of macaque monkeys, in fact infects humans in Borneo with a sufficiently high frequency that it should be regarded as a fifth species of human malaria, in addition to the four other species (*P. falciparum*, *P. vivax*, *P. malariae* and *P. ovale*) that are generally recognized as human pathogens. The role of *P. knowlesi* was not previously recognized because the early stages of its erythrocytic forms resemble those of *P. falciparum*, the most dangerous species of human malaria, while its late stages resemble those of *P. malariae* (the agent of quartan fever), leading to confusion during microscopic analysis of blood films (Singh et al. 2004; Luchavez et al. 2008). It can cause potentially fatal complications in about 10% of cases (Daneshvar et al. 2009), and it has been identified in archived blood films, proving it is not a human disease that has just started to emerge (Lee et al. 2009; Cox-Singh and Singh 2008). *P. knowlesi* occurs widely today in Malaya, Burma, Singapore and the Philippines, as well as Borneo. This geographical distribution suggests that it probably occurred right across the ancient continental shelf of Sundaland at times of low sea level during Pleistocene glacial periods, awaiting the arrival of the first hominids. As a generalist parasite which is not confined to a single host species, it is likely that it would have infected *Homo erectus* as soon as it reached Sundaland. Consequently it is a reasonable hypothesis that malaria was a problem for the earliest hominids in island Southeast Asia, not just the earliest modern humans to reach the area.

Research in molecular evolution indicates that the simian malaria parasites have diversified in south Asia within the last 5 million years, following the diversification of the monkeys themselves. *P. vivax*, the most abundant human malaria species in Asia, is an offshoot of this evolutionary process (Mu et al. 2005). Since it is now known (from the finds at Dmanisi in Georgia) that hominids were moving out of Africa into Asia by about 1.9 million year ago, it is an attractive hypothesis that the speciation of *P. vivax* soon followed the arrival of hominids in southern or Southeast Asia. There is little doubt that hominids and humans in Southeast Asia have suffered from malaria for a very long time.

However, both *P. vivax* and *P. malariae* which is also an ancient human parasite, produce diseases with fairly low direct mortality rates (at least in comparison with *P. falciparum*; see below for further comments on this in the case of *P. vivax*). Consequently their effects as agents of natural selection on human populations may be limited. It is not clear at the moment whether the frequency of infections with *P. knowlesi* and the resulting burden of morbidity and mortality were sufficiently high to lead to the evolution of human genetic mutations that would give some resistance against it. Most of the numerous mutations in the human genome that are thought to be the products of natural selection for resistance to malaria are likely to be responses to *P. falciparum*, the most dangerous species of human malaria. Unlike the other species of malaria that have been mentioned so far, *P. falciparum* undoubtedly evolved in Africa from the chimpanzee parasite *P. reichenowi*. The worst effects of malaria on human populations in inland Southeast Asia would not have been experienced until *P. falciparum* reached the area from Africa. The timing of this spread is critically important for assessing the possible effects of malaria on human population history in the western Pacific. Most probably the beginning of the dispersal of *P. falciparum* out of Africa accompanied the spread of early populations of *Homo sapiens* out of Africa (Jongwutiwes et al. 2005). The discovery of human skeletal remains with a pathology suggestive of thalassaemia in an archaeological site in Thailand dating to circa 2000 BC (Tayles 1996) could indicate that *P. falciparum* had reached Thailand during the Bronze Age, but it is now conceivable that *P. knowlesi* was involved instead. In the absence of direct evidence, it is not known for sure precisely when *P. falciparum* started to exercise a strong influence on human populations in the western Pacific, whether it antedated or post-dated the prehistoric migrations of the people responsible for the Lapita cultural complex. The spread of malaria is still an ongoing process in the western Pacific in modern times. For example it was only introduced to Ontong Java in the late nineteenth century, but had not yet reached the nearby Rennell Island 50 years later even though *Anopheles* mosquitoes are now present on that island. Malaria was mild on Ontong Java at that time, presumably *P. vivax* predominated (Hogbin 1930/31; Lambert 1931/32). Modern experience in the Pacific suggests that the spread of malaria does not necessarily immediately follow the spread of the vector mosquitoes (Black 1956).

In Vanuatu, for example, both *P. vivax* and *P. falciparum* are widely distributed today. Both species occur on all the inhabited islands, except for Futuna (Mills 1954). The Polynesian inhabitants of Futuna cannot go to work or live on other islands because they lack any resistance to malaria (Buxton 1926). On Tanna in 1974 10% of all patients admitted to hospital were found to be infected with malaria, but only 2–3% displayed febrile symptoms, indicating that a high proportion of all malaria cases are asymptomatic carriers of the disease. 55.8% of cases were infected with *P. vivax*, 42.2% with *P. falciparum*, and 2.0% with *P. malariae* (Bouree 1976). Malaria is more frequent in the wetter and hotter Solomon Islands to the north of Vanuatu, where in the 1920s 80% of children had malaria, the majority of whom (70%) were infected with *P. vivax* (Crichlow 1929). In the late 1980s across Vanuatu as a whole reported prevalences were *P. vivax* 6.7% and *P. falciparum* 5.2%. *P. malariae* is rare in Vanuatu and occurs mainly on Ambrym (Kaneko et al. 1998). All species are commoner in the larger, hotter, and more densely forested northern islands like Espiritu Santo than in the southern islands like Tanna (Cheesman 1933). *P. falciparum* is active mainly during the wet season from December to April, and is rare from August to October. *P. vivax*, in contrast, occurs all the year round, almost entirely in children under the age of 10. It tends to peak during the dry season (Buxton 1926; Cheesman 1933). Transmission patterns of malaria on Vanuatu are rather unstable, ranging from hypoendemic to mesoendemic. This is particularly true of the smaller islands, where malaria often appears cyclically in an epidemic form (Kaneko et al. 1998).

All species of human malaria in Vanuatu are transmitted by a single mosquito species, *Anopheles farauti*1, one of the twelve cryptic species of the *A. punctulatus* complex in Oceania which can only be differentiated by molecular techniques. This complex speciated during the Pleistocene. Molecular analysis suggests that *A. farauti*1 has evolved recently, even though it is now widely distributed (Beebe et al. 2000; Beebe and Cooper 2002). It is an anthropophilic species occurring principally in coastal environments, which is active at night and undertakes feeding flights of 1–2 km. Swamps and taro fields were formerly regarded as its most important breeding sites, although now abandoned wells near settlements and seashore seeping pools are assuming a greater significance as breeding sites (Maffi 1977, 1989). In many parts of the western Pacific human settlement is often concentrated on small islets near the large islands (a pattern not found in the eastern Pacific where malaria does not occur). This pattern has also been observed in island groups in other parts of the world in the past, for example in the Caribbean (Keegan et al. 2008). The question arises of whether this is an adaptation of the settlement pattern to avoid mosquitoes or alternatively simply an adaptation to a maritime culture in which the resources of the sea made a considerable contribution to human subsistence. The indigenous inhabitants of Vanuatu themselves suggest that this is to avoid being bitten by mosquitoes. Parsonson (1965) and Kirch (1997) noted this situation on Taumako Island in the Santa Cruz group, north of Vanuatu. However, malarial infection is still observed on the offshore islets, for example on Atchin off Malekula in Vanuatu where there are no mosquito breeding sites (Buxton 1926).

Research in the Santa Cruz Islands suggests that the relationship between human behaviour, malaria and mosquitoes is more complicated than previously supposed. By living on the offshore islets the inhabitants may indeed be escaping some mosquito bites, but what they are avoiding are daytime bites by other species of mosquito which indeed give nasty bites, but prefer to hide in dense vegetation on the main islands and, most importantly of all, are not vectors of human malaria. *A. farauti*1, in contrast, ventures out into the open to seek prey at night. It can fly across to the islets, but its bites only leave small marks on its victims. Consequently settlement patterns do help people to avoid certain types of mosquito, but not necessarily *A. farauti*1, the vector of malaria (Taylor 1998). Specht (2007) also observed that mosquitoes and malaria can be as troublesome on small offshore islets as they are on the main islands.

Molecular analysis of the population structure of *A. farauti* in the Solomon Islands (along the route of the Lapita expansion towards Vanuatu from the Bismarck Archipelago) suggests that this mosquito species comprises a single monophyletic clade, indicating a single founder event for the population. This population has expanded considerably since then, like the New Guinea population, but more recently than the New Guinea population. In fact the estimates for the timing of these population expansion events very approximately coincide (given statistical margins of error) with archaeological evidence for the first human occupations of these regions. Moreover the important point is made that only anthropophilic mosquito species from the *Anopheles* group have spread beyond the Bismarck Archipelago (Hasan et al. 2007). Consequently the evidence strongly suggests that humans were unwittingly responsible for the dispersal of this dangerous vector of malaria in Melanesia. This new evidence settles the question raised by Kaneko (2001) as to whether anophelism preceded or accompanied the human settlement of Vanuatu. Similar research to ascertain the population history and structure of this mosquito species on the coast of northern Australia would also be of interest in order to shed further light on the hypothesis of Groube (1993) about the role of malaria in Australian prehistory mentioned above.

In Melanesia today *P. vivax*, the commonest species of malaria, only causes clinical symptoms in children under the age of 10 (Kaneko et al. 1998). People above that age have acquired and/or inherited immunity and so do not develop clinical symptoms, although they are continuously reinfected by mosquito bites. Where malaria is endemic the burden of clinical disease falls principally on children. There is an important ongoing debate about the effects in terms of morbidity and mortality of infection with *P. vivax* in the western Pacific. Groube (1993) maintained that *P. vivax* was powerful enough to act as a mechanism of human population regulation in northern Australia in prehistory. This view is congruent with historical evidence from parish registers for severe demographic consequences of vivax malaria in some parts of Europe in the past (Sallares 2002). Nevertheless, it has been argued that malaria is a relatively mild disease in Espiritu Santo in Vanuatu with no malaria-specific mortality among children and no cerebral malaria, its most dangerous clinical syndrome in Africa (Maitland et al. 1996, 1997; Queyrel and Moranne 1998). It was postulated that this was caused by cross-species immune reactions between the relatively mild *P. vivax*, the more severe *P. falciparum*, **and the α-thalassaemia human genotypes** (Williams et al. 1996). The idea of mild malaria on Vanuatu is supported by statements of early modern observers (McNabb 1894; Baker 1929). When these observers described malaria as a mild disease it is possible that they were just thinking about adults and did not consider its effects on infants and children.

The pendulum of opinion has begun to swing again recently. New research indicates that *P. vivax* did cause severe symptoms in a significant proportion of cases in two studies, one at Timika in the southern lowlands of New Guinea (associated with severe anaemia) (Tjitra et al. 2008) and the other in the East Sepik region of northern New Guinea (associated with respiratory distress) (Genton et al. 2008). No evidence was found that mixed infections give any protection against severe clinical symptoms. It is not clear at the moment how the results of these studies can be reconciled with the apparently contradictory results obtained earlier by Maitland et al. (1996, 1997). It is possible that the effects of malaria and the way in which the different species interact with each other immunologically vary at a very local level. If vivax malaria is more dangerous in New Guinea than it is in Vanuatu, for example, it is possible that the people of the Lapita cultural complex, far from having a selective advantage against malaria (resistance may only have developed later—see below) were, at least in part, fleeing severe malaria when some of them decided to go ahead with migratory voyages away from the Bismarck Archipelago. Undoubtedly malaria has sometimes forced people to migrate to find new homes elsewhere in this part of the world. For example Salisbury (1975) describes how one of the main coastal settlements of the Tolai people in New Britain became infested with mosquitoes and unhealthy, leading to its eventual abandonment as the people moved elsewhere. In general malaria does make a major contribution to infant and child mortality and the low human population densities in western Pacific islands, but it may not be the immediate cause of death in many cases because its effects are indirect rather than direct.

Kelly (1990, 2000) has argued for the importance of immunoglobulin haplotypes in permitting speakers of Austronesian languages to settle in coastal lowlands that were infested with malaria. He suggested that it was not resistance to malaria itself that gave Austronesian speakers their selective advantage, but resistance to an immune system disease, hyperreactive malarial splenomegaly (Kelly 1996). However, since that hypothesis was first proposed, plenty of research based on analysis of DNA sequence data has been conducted in relation to other human genetic polymorphisms that are considered to be directly related to malaria resistance in the geographical areas in question. Research on molecular human population genetics provides a way of investigating whether malaria was an environmental factor that favored or hindered the Austronesian migrations in prehistory. Malaria appears to have acted as an agent of natural selection on the human genome to a greater extent than any other infectious pathogen. The glycophorin genes (Mayer et al. 2009)—among the fastest evolving human genes—were probably the first targets of selection by malaria, perhaps starting in the hominid lineage as long ago as 10 Myr. Natural selection for direct resistance to malaria has produced numerous genetic responses in the human populations of the areas under consideration here. Consequently the debate cannot be restricted any longer to the immunoglobulins, but needs to be expanded to include all the other mechanisms that are thought to be implicated in resistance to malaria. The immunoglobulins do play an important role in acquired immunity to malaria and display a significant degree of heritability (Courtin et al. 2009; Duah et al. 2009), but the question still remains why all the other polymorphisms evolved if the immunoglobulins were so important on their own. These other polymorphisms are briefly reviewed below, with an emphasis on the islands of Vanuatu as an example.

22.2.1 Thalassaemia

α-Thalassaemia is caused by a deletion of one or more of the α-globin genes, leading to an imbalance in globin chain production which confers some resistance to malaria (Flint et al. 1986). **The α-thalassaemia**

of Southeast Asia does not occur in Vanuatu, but the α^+ form is frequent, particularly on the northern islands of the group. The rightward $-\alpha^{3.7}$ III variant, which is unique to Melanesia and Polynesia (Hill et al. 1985), occurs mainly in the northern islands of Vanautu, where malaria is common. Its frequency is about three times higher than the leftward $-\alpha^{4.2}$ variant found principally on the southern islands of Emae, Tanna and Futuna where there is less (or in the case of Futuna, no) malaria. The frequency of α-thalassaemia varies not only between islands but also within islands. For example, on Espiritu Santo its frequency is higher in the lowlands, where there is more malaria, than in the highlands, giving the impression of very localised population evolution (Ganczakowski et al. 1995a). These forms of α-thalassaemia are single gene deletions, which produce a high frequency of anaemia but no disease more severe than that.

In general the α-thalassaemia forms of New Guinea and Melanesia do not occur west of the Wallace line, so important for biogeography, and are shown by their haplotypes to have evolved in their current locations. In New Guinea the impact of malaria produced independent α-globin deletions in Austronesian and non-Austronesian speaking (Papuan) populations. The rightward $-\alpha^{3.7}$ III variant is the commonest deletion in Austronesian populations on the north coast of New Guinea and in the Bismarck Archipelago, but is very rare among non-Austronesian populations. Conversely the $-\alpha^{4.2}$ variant is very common among non-Austronesian populations on the north coast of New Guinea (occurring in up to 80% of the population), but is also found today in Austronesian populations throughout island Melanesia but not further afield in Polynesia (Oppenheimer and Richards 2001). Its presence in Melanesia but not in Polynesia can be explained by gene flow mediated by male migration into matrilocal Austronesian societies after Lapita.

The impression of local evolution is reinforced when another form of thalassaemia, β-thalassaemia, is considered. β-thalassaemia takes the form of single nucleotide mutations rather than major length polymorphisms. It is much rarer in the Pacific than α-thalassaemia, but nevertheless occurs with a significant frequency (12.6%) on Maewo in Vanuatu, where homozygous β-thalassaemia is a major cause of infant mortality, associated with a high degree of inbreeding. Its absence from other islands which are only a few miles away illustrates the importance of local population isolation in evolution in Vanuatu (Bowden et al. 1985; Hill et al. 1988; Ganczakowski et al. 1995a).

22.2.2 *Haemoglobin Variants*

Not only are haemoglobins S and C, characteristic of Africa, absent from Vanuatu, but also Hb E, which is common in Southeast Asia and associated with resistance to malaria there. Nevertheless two other haemoglobin variants are known to exist on Vanuatu, namely Hb J Tongariki, which is most frequent on Efate and Espiritu Santo, and Hb G Philadelphia, which is only known from Epi (Bowden et al. 1985). Just as in the case of the thalassaemias, the picture that emerges is one of substantial local evolution within Vanuatu.

22.3 Glucose-6-Phosphate Dehydrogenase (G6PD) Deficiency

G6PD deficiency is yet another important defense mechanism against malaria, generally taking the form of single nucleotide mutations. In Vanuatu, where it causes neonatal jaundice, its prevalence varies considerably both from island to island and on different parts of the larger islands. Its frequency is positively correlated with malaria prevalence. One survey revealed three phenotypic patterns corresponding to four DNA sequence variants, three of which are not found in other parts of the world (Ganczakowski et al. 1995b). Kaneko (2001, p. 115) suggested that "a reasonable hypothesis in Vanuatu is that malaria endemicity was introduced to the islands with the first human settlement from the northwest a long time ago. A geographical pattern of malaria endemicity similar to the present situation was then probably established, the north being more malarious than the south. Transmission of malaria then possibly selected for G6PD deficiency over many generations, and the G6PD rates may now be either at equilibrium or still increasing." Of course it is still possible that malaria arrived some time after the very first immigrants, and that different species of malaria arrived at different times.

22.3.1 Tumour Necrosis Factor-Alpha (TNF-α)

Promoter polymorphisms of the gene coding for TNF-α, a cytokine that affects the pathology of infections, have recently been positively associated with malaria endemicity in Vanuatu (Ubalee et al. 2005) and elsewhere. It was found that the frequency of the TNF-D allele is inversely correlated with malaria prevalence on the various islands.

22.3.2 Southeast Asian Ovalocytosis

In relation to the question of the effects of malaria in relation to human population movements ovalocytosis is the most interesting of the various resistance mechanisms displayed by Pacific populations. Ovalocytosis is a deformation of the red blood cell. There are at least two quite different genotypes that cause ovalocytosis in Pacific populations (Patel et al. 2004). Of these, the band 3 deletion is of particular interest here. Band 3 is the most important transmembrane protein of human red blood cells. A 27 base pair deletion in the gene for this protein causes the changed morphology of the red blood cell that is characteristic of ovalocytosis (Jarolim et al. 1991; Kuma et al. 2002). This mutation is only known in the heterozygous form; presumably homozygotes are lethal 'in utero'. The resulting alteration in morphology does not necessarily prevent malaria parasites invading red blood cells, since at least one parasite strain is known to be able to overcome the deformation (Cortés et al. 2004). Moreover ovalocytosis is positively associated with increased anaemia, but nevertheless it does seem to prevent cerebral malaria, the most dangerous manifestation of malaria (Allen et al. 1999). Consequently ovalocytosis does not influence the chances of infection, even during pregnancy (O'Donnell et al. 2007), but it does substantially reduce the chances of mortality from malaria.

Despite its name, the band 3 deletion does not occur commonly on the mainland of Southeast Asia, for instance in Thailand (Kimura et al. 1998). The view has been expressed that its origins lie in the aboriginal populations of island Southeast Asia and that it spread before the Austronesian migrations (Searjeantson and Gao 1995). It has not been reported from Taiwan so far, but recent population studies indicate that it is strongly correlated with the presence of populations speaking Austronesian languages of the Western and Central Malayo-Polynesian groups (Baer 1995). It is frequent in Austronesian populations on the north coast of New Guinea, where it is strongly associated with the 9 bp east Asian mitochondrial DNA deletion, and has spread across New Guinea to the south coast, in the process spreading to some Papuan populations, presumably by the same sort of gene flow process in New Guinea already noted in relation to the α-globin variants (Kimura et al. 2003; Patel et al. 2004; Tsukahara et al. 2006). Moreover, unlike for example the α-thalassaemia mutations of Melanesia which do not occur west of the Wallace line in Indonesia, the band 3 deletion does occur at the westernmost limit of the Austronesian migrations, far away in Madagascar (Rabe et al. 2002). The linguistic and genetic affinities of the Austronesian people in Madagascar are with southeastern Borneo. The band 3 deletion causing ovalocytosis is the best candidate for a malaria resistance gene spread by the Austronesian migrations. However, it does not occur today in Vanuatu or further east, in Polynesia. The deletion's single haplotypic background (band 3 Memphis) indicates that the mutation occurred just once, probably somewhere in the vicinity of the Philippines or Borneo. It then spread widely in the western Austronesian region, onwards to Madagascar, and to New Guinea, where it flowed across the boundaries between Austronesian and non-Austronesian populations (like the $-\alpha^{4.2}$ thalassaemia variant), but only after some sections of the Proto-Oceanic community had already sailed eastwards to Vanuatu. This suggests that the pressure of natural selection by malaria that has led to the mutation's successful spread, bearing in mind that it is lethal in homozygotes, has occurred essentially within the last 3000 years, after Lapita.

22.4 Conclusions

There are three possible scenarios for the possible effects of malaria on the Austronesian migrations in the western Pacific in prehistory.

a. The proto-Austronesians already had a significant degree of resistance to malaria; this gave them a positive selective advantage when migrating to regions where malaria was already present or could flourish if introduced by the Austronesians themselves.

b. The effects of malaria were essentially negative rather than positive as far as the Austronesians were concerned. The early Austronesians struggled to resist the disease, and the human genetic resistance mechanisms that exist today only evolved and achieved high population frequencies later. Consequently the Austronesians were confined to the fringes (small islets off the coast of larger islands) and had to keep moving to try to find healthier environments in which to live, until some of them had moved beyond the Buxton Line (which marks the modern limits of the geographical distribution of malaria in the southwest Pacific) to New Caledonia and Fiji, at which point they were beyond the reach of the disease and could start to experience higher rates of population growth (Figs. 8.1, 8.2; Brewis 1995).
c. It is also quite possible that malaria was not after all an important determining factor in the voyages of the Austronesians. Their focus on living in coastal environments may have been instead primarily an adaptation to a lifestyle that was heavily geared towards exploiting the resources of the sea (although the evidence of archaeobotany and historical linguistics demonstrates that they were farmers as well as fishermen). Even if this was the case, there is no doubt that Austronesian settlements west of the Buxton Line were gradually forced to adapt to the spread of endemic malaria, although this could have happened mainly after the spread of the Lapita cultural complex.

All three scenarios have had advocates. Apparently Nieuwenhuis (1930) was the first author to try to link the obvious physical differences between Melanesians and Polynesians to malaria. Dempwolff (1937), the founder of Austronesian studies, is said to have argued that the proto-Austronesians initially had little resistance to malaria. As they migrated they encountered other peoples who already had some experience of malaria and were more resistant to it. They intermarried with these other peoples to some extent and were influenced by them. Dempwolff suggested that where malaria was prevalent the genes of the non-Austronesian peoples prevailed, but in Polynesia where there was no malaria the original Austronesian gene pool was better preserved (Dempwolff as discussed by Pawley and Ross 1993).

This is essentially the second of the three possible scenarios described above. As a refinement to it, it should be noted that the proto-Austronesians could have had experience of *P. vivax*, but that would not have given them resistance to the more dangerous *P. falciparum*. If a migrating Austronesian population with prior experience of *P. vivax* only had tried to move into an area like New Guinea, and if *P. falciparum* was already present among the non-Austronesian indigenous Papuan peoples in the coastal regions of New Guinea, then the Austronesians might well have struggled (compare the views of Bellwood 1975). The discussion above has shown that nearly all the mechanisms of human genetic resistance to malaria that occur in the area today have a very local distribution and presumably evolved where they are found today, in other words they were not brought to their current locations by long-distance migrations. The only exception is ovalocytosis which was carried as far away as Madagascar. It evolved and spread in the western Austronesian regions after the community of Proto-Oceanic speakers migrating eastwards had already diverged from populations speaking the western Austronesian languages. This would seem to confirm Dempwolff's theory, which was also supported by Searjeantson and Gao (1995). On the currently available evidence, the second scenario is more plausible than the first scenario, while the third scenario still remains a possibility. We cannot say that the proto-Austronesians, or even the later Austronesian sub-group responsible for the Lapita cultural complex, spread one particular mechanism of resistance to malaria throughout the entire, large, geographical areas which they occupied, in the same way that black Africans spread the sickle cell trait to the western hemisphere in the course of the slave trade.

The very abundance of malaria resistance mechanisms in Melanesia may well be congruent with models favored by many archaeologists, in which the Austronesian progenitors of the Lapita cultural complex spent a considerable period of time accumulating genetic variation and intermarrying with other peoples, before continuing their migrations eastwards. Such an interpretation of the evolution of patterns of human genetic resistance to malaria in Melanesia is congruent with the most recent data on human population history. Recently a new human mitochondrial DNA subclade, B4a1a, has been identified that directly connects Polynesian maternal lineages with maternal lineages of the

Taiwan aborigines. According to Trejaut et al. (2005, p. 9) their "findings provide the first direct phylogenetic evidence for the common ancestry of Austronesian and indigenous Taiwanese maternal lineages and their maturation phase in East Indonesia or Melanesia." It is a plausible hypothesis that it was during this maturation phase that the great diversity of malaria resistance mechanisms observed today in Melanesia really began to evolve.

Similarly the first direct phylogenetic evidence from the Y chromosome has recently been adduced for an ultimate Austronesian ancestry in Taiwan with links all the way through to Polynesia (Kayser et al. 2008). Just as in the case of maternally inherited mitochondrial DNA, the Y chromosome evidence implies a phase of intermarriage and genetic mixing between Austronesians and non-Austronesians in Melanesia. On the Admiralty Islands, for example, the archaeological, genetic and linguistic evidence combine to show that the Austronesian arrival on these islands was accompanied by a complete linguistic replacement, but an incomplete genetic replacement, with gene flow into the Austronesian population of non-Austronesian Y chromosomes, facilitated by the matrilocal nature of Austronesian society. In New Guinea the opposite trend led to the flow of Austronesian mitochondrial DNA haplotypes into non-Austronesian societies, made possible by the patrilocal nature of Papuan societies (Lum et al. 2002; Hage and Marck 2003; Kayser et al. 2008). This two-way process of genetic mixing facilitated the interchange, for example, of α-globin deletions conferring some resistance to malaria.

The conclusion of this chapter is that malaria was an environmental factor that probably hindered rather than assisted the Austronesian migrations in the western Pacific in prehistory. Another significant hypothesis can also be proposed here. Several authors have wondered why malaria seems to have stopped in Vanuatu and did not spread further eastwards. Recently there has been some discussion in scholarly literature of similarities between the Austronesian languages and cultures of Vanuatu and the non-Austronesian languages and cultures of New Guinea (Blust 2008; Donohue and Denham 2008), building on the concept of Melanesianization that was introduced by Spriggs (1997, p. 159). These similarities suggest that besides the Lapita migrations, there were early migrations of Papuans, probably inspired by the Lapita seafarers, which brought Papuan peoples from New Guinea or neighboring islands to Vanuatu and New Caledonia, but no further. If malaria was not spread by the Austronesians themselves, it could have been spread by Papuans instead. It is suggested here that the spread of malaria in Near Oceania may have been one aspect of Melanesianization, and the failure of Papuan populations to spread further may help to explain why malaria did not reach Polynesia. This line of argument is congruent with the idea that the spread of malaria may have helped to disrupt communications between Melanesia and Polynesia.

22.5 Future Research

The author is involved in a laboratory project with Dr H. Buckley which is attempting to find ancient DNA from malaria in human skeletal remains from a Lapita archaeological site in Vanuatu.

References

Allen, SJ, O'Donnell, A, Alexander, NDE, et al. (1999) Prevention of cerebral malaria in children in Papua New Guinea by Southeast Asian ovalocytosis band 3. American Journal of Tropical Medicine and Hygiene 60:1056–1060

Ataka Y, Ohtsuka R, Inaoka T, et al. (2001) Variation in malaria endemicity in relation to microenvironmental conditions in the Admiralty Islands, Papua New Guinea. Asia Pacific Journal of Public Health 13:85–90

Baer AS (1995) Human genes and biocultural history in Southeast Asia. Asian Perspectives 34:21–35

Bahn P, Flenley J (1992) Easter Island, Earth Island. Thames & Hudson, London, 240 pp

Baker JR (1929) The northern New Hebrides. The Geographical Journal 73:305–325

Bedford S (2003) The timing and nature of Lapita colonisation in Vanuatu: the haze begins to clear. In: Sand C (ed) Pacific archaeology: assessments and prospects. Proceedings of the International Conference for the 50th anniversary of the first Lapita excavation, Koné-Nouméa 2002, pp 147–158

Beebe NW, Cooper RD (2002) Distribution and evolution of the *Anopheles punctulatus* group (Diptera: Culicidae) in Australia and Papua New Guinea. International Journal for Parasitology 32:563–574

Beebe NW, Cooper RD, Morrison DA, Ellis JT (2000) A phylogenetic study of the *Anopheles punctulatus* group of malaria vectors comparing rDNA sequence alignments derived from the mitochondrial and nuclear small ribosomal subunits. Molecular Phylogenetics and Evolution 17:430–436

Bellwood P (1975) The prehistory of Oceania. Current Anthropology 16:9–28

Black RH (1956) The epidemiology of malaria in the south-west Pacific: changes associated with increasing European contact. Oceania 27:136–142

Blust R (1980) Austronesian etymologies. Oceanic Linguistics 19:1–181

Blust R (1995) The prehistory of the Austronesian-speaking peoples: a view from language. Journal of World Prehistory 9:453–510

Blust R (2008) Remote Melanesia: one history or two? An addendum to Donohue and Denham. Oceanic Linguistics 47:445–459

Bouree P (1976) Aspects de la pathologie aux Nouvelles-Hébrides. Médecine et Maladies Infectieuses 6:293–301

Bowden DK, Hill AVS, Higgs DR, et al. (1985) Relative roles of genetic factors, dietary deficiency, and infection in anaemia in Vanuatu, Southwest Pacific. Lancet ii:1025–1028

Brewis AA (1995) Fertility and analogy in Pacific palaeodemography. Asian Perspectives 34:1–20

Buxton PA (1926) The depopulation of the New Hebrides and other parts of Melanesia. Transactions of the Royal Society of Tropical Medicine and Hygiene 19:420–458

Cabioch G, Wirrmann D, Sémah A-M, et al. (2008) Évolution des paléoenvironnements dans le Pacifique lors de la dernière déglaciation: exemples en Nouvelle-Calédonie et au Vanuatu. Journal de la Société des Océanistes 128:25–40

Cheesman LE (1933) The island of Malekula, New Hebrides. The Geographical Journal 81:193–207

Clark JT, Kelly KM (1993) Human genetics, palaeoenvironments, and malaria: relationships and implications for the settlement of Oceania. American Anthropologist 95:613–631

Cortés A, Benet A, Cooke BM, et al. (2004) Ability of *Plasmodium falciparum* to invade South-East Asian ovalocytes varies between parasite lines. Blood 104:2961–2966

Courtin D, Oesterholt M, Huismans H, et al. (2009) The quantity and quality of African children's IgG responses to merozoite surface antigens reflect protection against *Plasmodium falciparum* malaria. PLoS ONE 4(10):e7590

Cox MP (2003) Genetic patterning at Austronesian contact zones. PhD, University of Otago, NZ

Cox MP (2008) Accuracy of molecular dating with the Rho statistic: deviations from coalescent expectations under a range of demographic models. Human Biology 80:335–357

Cox-Singh J, Singh B (2008) Knowlesi malaria: newly emergent and of public health importance? Trends in Parasitology 24:406–410

Crichlow N (1929) The prevalent diseases of the British Solomon Islands. Transactions of the Royal Society of Tropical Medicine and Hygiene 23:179–184

Daneshvar C, Davis TME, Cox-Singh J, et al. (2009) Clinical and laboratory features of human *Plasmodium knowlesi* infection. Clinical Infectious Diseases 49:852–860

Dempwolff O (1937) Vergleichende Lautlehre des austronesischen Wortschatzes. Deduktive Anwendung des Urindonesischen auf austronesische Einzelsprachen. Beiheften zur Zeitschrift für Eingeborenen-Sprachen 17. Reimer, Berlin

Dickinson WR (2003) Impact of mid-Holocene hydro-isostatic highstand in regional sea level on habitability of islands in Pacific Oceania. Journal of Coastal Research 19:489–502

Donohue M, Denham T (2008) The language of Lapita: Vanuatu and an Early Papuan presence in the Pacific. Oceanic Linguistics 47:433–444

Dou HY, Tseng FC, Lu JJ, et al. (2008) Associations of *Mycobacterium tuberculosis* genotypes with different ethnic and migratory populations in Taiwan. Infection, Genetics and Evolution 8:323–330

Duah NO, Weiss HA, Jepson A, et al. (2009) Heritability of antibody isotype and subclass responses to *Plasmodium falciparum* antigens. PLoS ONE 4(10):e7381

Ellison JC (2009) Wetlands of the Pacific Island region. Wetlands Ecology and Management 17:169–206

Flint J, Hill AVS, Bowden DK, et al. (1986) High frequencies of α-thalassaemia are the result of natural selection by malaria. Nature 321:744–750

Ganczakowski M, Bowden DK, Maitland K, et al. (1995a) Thalassaemia in Vanuatu, south-west Pacific: frequency and haematological phenotypes of young children. British Journal of Haematology 89:485–495

Ganczakowski M, Town M, Bowden DK, et al. (1995b) Multiple glucose 6-phosphate dehydrogenase-deficient variants correlate with malaria endemicity in the Vanuatu archipelago (southwestern Pacific). American Journal of Human Genetics 56:294–301

García-Moro C, Hernández M, Moral P, González-Martín A (2000) Epidemiological transition in Easter Island (1914–1996). American Journal of Human Biology 12:371–381

Genton B, D'Acremont V, Rare L, et al. (2008) *Plasmodium vivax* and mixed infections are associated with severe malaria in children: a prospective cohort study from Papua New Guinea. PLoS Medicine 5(6):e127

Gibbons A (2001) The peopling of the Pacific. Science 291:1735–1737

Gibbons JRH, Clunie FGAU (1986) Sea level changes and Pacific prehistory: new insight into early human settlement of Oceania. Journal of Pacific History 21:58–82

Green RC (1991) The Lapita cultural complex: current evidence and proposed models. Bulletin of the Indo-Pacific Prehistory Association 11:295–305

Gray RD, Drummond J, Greenhill SJ (2009). Language phylogenies reveal expansion pulses and pauses in Pacific settlement. Science 323:479–483

Groube LM (1993) Contradictions and malaria in Australian and Melanesian prehistory. In: Spriggs M, Yen DE, Ambrose W, et al., (eds) A community of culture: the people and prehistory of the Pacific. Occasional Papers no. 21, Dept. of Prehistory, Research School of Pacific Studies, Australian National University, Canberra, pp 164–186

Hage P, Marck J (2003) Matrilineality and the Melanesian origin of Polynesian Y chromosomes. Current Anthropology 44:S121–S127

Hasan A-U, Suguri S, Fujimoto C, et al. (2007) Phylogeography and dispersion pattern of *Anopheles farauti senso stricto* mosquitoes in Melanesia. Molecular Phylogenetics and Evolution 46:792–800

Hill AVS, Bowden DK, Trent RJ, et al. (1985) Melanesians and Polynesians share a unique α-thalassemia mutation. American Journal of Human Genetics 37:571–580

Hill AVS, Bowden DK, O'Shaughnessy DF, et al. (1988) β-thalassemia in Melanesia: association with malaria and characterization of a common variant (IVS-1 nt 5 G→C). Blood 72:9–14

Hogbin HI (1930/31) Spirits and the healing of the sick in Ontong Java. Oceania 1:146–166

Jarolim P, Palek J, Amato D, et al. (1991) Deletion in erythrocyte band 3 gene in malaria-resistant Southeast Asian ovalocytosis. Proceedings of the National Academy of Sciences, USA 88:11022–11026

Jongwutiwes S, Putaporntip C, Iwasaki T, et al. (2005) Mitochondrial genome sequences support ancient population expansion in *Plasmodium vivax*. Molecular Biology and Evolution 22:1733–1739

Kaneko A, Taleo G, Kalkoa M, et al. (1998) Malaria epidemiology, glucose 6-phosphate dehydrogenase deficiency and human settlement in the Vanuatu Archipelago. Acta Tropica 70:285–302

Kaneko A (2001) Comment on Terrell JE, KM Kelly and P Rainbird. Foregone conclusions? In search of "Papuans" and "Austronesians". Current Anthropology 42:115

Kayser M, Choi Y, van Oven M, et al. (2008) The impact of the Austronesian expansion: evidence from mtDNA and Y chromosome diversity in the Admiralty Islands of Melanesia. Molecular Biology and Evolution 25:1362–1374

Keegan WF, Fitzpatrick SM, Sealey KS, et al. (2008) The role of small islands in marine subsistence strategies: case studies from the Caribbean. Human Ecology 36:635–654

Kelly KM (1990) Gm polymorphisms, linguistic affinities, and natural selection in Melanesia. Current Anthropology 31:201–219

Kelly KM (1996) IGHG3 G and the pathogenesis of hyperreactive malarial splenomegaly. Medical Hypotheses 46:135–139

Kelly KM (2000) Malaria and immunoglobulins in Pacific prehistory. American Anthropologist 101:806–809

Kimura M, Shimizu Y, Settheetham-Ishida W, et al. (1998) Twenty-seven base pair deletion in erythrocyte band 3 protein gene responsible for Southeast Asian ovalocytosis is not common among Southeast Asians. Human Biology 70:993–1000

Kimura M, Tamam M, Soemantri A, et al. (2003) Distribution of a 27-bp deletion in the band 3 gene in South Pacific islanders. Journal of Human Genetics 48:642–645

Kirch PV (1997) The Lapita Peoples: Ancestors of the Oceanic World. Blackwell, Oxford, 384 pp

Kitchen A, Miyamoto MM, Mulligan CJ (2008) Utility of DNA viruses for studying human host history: Case study of JC virus. Molecular Phylogenetics and Evolution 46:673–682

Kuma H, Abe Y, Askin D, et al. (2002) Molecular basis and functional consequences of the dominant effects of the mutant band 3 on the structure of normal band 3 in Southeast Asian ovalocytosis. Biochemistry 41:3311–3320

Lambert SM (1931/32) Health survey of Rennell and Bellona islands. Oceania 2:136–173

Lee K-S, Cox-Singh J, Brooke G, et al. (2009) *Plasmodium knowlesi* from archival blood films: Further evidence that human infections are widely distributed and not newly emergent in Malaysian Borneo. International Journal for Parasitology 39:1125–1128

Luchavez J, Espino F, Curameng P, et al. (2008) Human infections with *Plasmodium knowlesi*, the Philippines. Emerging Infectious Diseases 14:811–813

Lum JK, Jorde LB, Schiefenhovel W (2002) Affinities among Melanesians, Micronesians, and Polynesians: a neutral, biparental genetic perspective. Human Biology 74:413–430

Lum JK, Kaneko A, Taleo G, et al. (2007) Genetic diversity and gene flow of humans, *Plasmodium falciparum*, and *Anopheles farauti s.s.* of Vanuatu: inferred malaria dispersal and implications for malaria control. Acta Tropica 103:102–107

Maffi M (1977) Contribution to the knowledge of the mosquito fauna of the New Hebrides island group proper (Diptera: Culicidae). Rivista di Parassitologia 38:194–215

Maffi M (1989) The mosquitoes (Diptera: Culicidae) of Vanuatu: a further contribution. Rivista di Parassitologia 50:3–11

Maitland K, Williams TN, Bennett S, et al. (1996) The interaction between *Plasmodium falciparum* and *P. vivax* in children on Espiritu Santo Island, Vanuatu. Transactions of the Royal Society of Tropical Medicine and Hygiene 90:614–620

Maitland K, Williams TN, Petal TEA, et al. (1997) Absence of malaria-specific mortality in children in an area of hyperendemic malaria. Transactions of the Royal Society of Tropical Medicine and Hygiene 91:562–566

Mayer DCG, Cofie J, Jiang L, et al. (2009) Glycophorin B is the erythrocyte receptor of *Plasmodium falciparum* erythrocyte-binding ligand, EBL-1. Proceedings of the National Academy of Sciences, USA 106:5348–5352

McNabb D (1894) Diseases in the New Hebrides. Journal of the Anthropological Institute of Great Britain and Ireland 24:393–395

Mills AR (1954) A malaria survey of Futuna in the New Hebrides. Journal of Tropical Medicine and Hygiene 57:99–107

Moodley Y, Linz B, Yamaoka Y (2009) The peopling of the Pacific from a bacterial perspective. Science 323:527–530

Mu J, Joy DA, Duan J, et al. (2005) Host switch leads to emergence of *Plasmodium vivax* malaria in humans. Molecular Biology and Evolution 22:1686–1693

Nieuwenhuis AW (1930) Die Entstehung der Polynesier und ihrer Kultur; der Einfluss der endemischer Malaria in Ozeanien. Internationales Archiv für Ethnographie 30:129–168

O'Donnell A, Raiko A, Clegg JB, et al. (2007) Southeast Asian ovalocytosis and pregnancy in a malaria-endemic region of Papua New Guinea. American Journal of Tropical Medicine and Hygiene 76:631–633

Oppenheimer S, Richards M (2001) Fast trains, slow boats, and the ancestry of the Polynesian Islanders. Science Progress 84.3:157–181

Parsonson GS (1965) Artificial islands in Melanesia: the role of malaria in the settlement of the Southwest Pacific. New Zealand Geographer 22:1–21

Patel SS, King CL, Mgone CS, et al. (2004) Glycophorin C (Gerbich antigen blood group) and band 3 polymorphisms in two malaria holoendemic regions of Papua New Guinea. American Journal of Haematology 75:1–5

Pawley A, M Ross (1993) Austronesian historical linguistics and culture history. Annual Review of Anthropology 22:425–459

Pichon G, Riviere F, Laigret J (1982) Filariose et préhistoire océanienne. Journal de la Société des Océanistes 38:283–297

Pope KO, Terrell JE (2008) Environmental setting of human migrations in the circum-Pacific region. Journal of Biogeography 35:1–21

Queyrel V, Moranne O (1998) La médecine au Vanuatu: voyage au bout du monde sur l'île d'Espiritu Santo. Revue de Médicine Interne 19:619–628

Rabe T, Jambou R, Rabarijaona L, et al. (2002) South-East Asian ovalocytosis among the population of the highlands of Madagascar: a vestige of the island's settlement. Transactions of the Royal Society of Tropical Medicine and Hygiene 96:143–144

Rapaport M (2006) Eden in peril: impact of humans on Pacific island ecosystems. Island Studies Journal 1:109–124

Rolett BV (2008) Avoiding collapse: Pre-European sustainability on Pacific Islands. Quaternary International 184:4–10

Salisbury RF (1975) Non-equilibrium models in New Guinea ecology: possibilities of cultural extrapolation. Anthropologica 17:127–147

Sallares R (2002) Malaria and Rome: a history of malaria in ancient Italy. Oxford University Press, Oxford, 341 pp

Sallares R, Bouwman A, Anderung C (2004) The spread of malaria to southern Europe in antiquity: new approaches to old problems. Medical History 48:311–328

Schanfield MS, Ohkura K, Lin M (2002) Immunoglobulin allotypes among Taiwan aborigines: evidence of malarial selection could affect studies of population affinity. Human Biology 74:363–379

Searjeantson SW, Gao X (1995) Homo sapiens is an evolving species: origins of the Austronesians. In: Bellwood P, Fox JJ, Tryon D (eds) The Austronesians: historical and comparative perspectives. Australian National University Press, Canberra, pp 177–193

Singh B, Kim Sung L, Matusop A, et al. (2004) A large focus of naturally acquired *Plasmodium knowlesi* infections in human beings. Lancet 363:1017–1024

Specht J (2007) Small islands in the big picture: the formative period of Lapita in the Bismarck Archipelago. In: Bedford S, Sand C, Connaughton SP (eds) Oceanic explorations: Lapita and western Pacific settlement. Australian National University Press, Canberra, pp 51–70

Spriggs M (1984) The Lapita culture complex: origins, distribution, contemporaries and successors. Journal of Pacific History 19:202–223

Spriggs M (1997) The island Melanesians. Blackwell, Oxford, 326 pp

Takasaka T, Ohta N, Zheng H-Y, et al. (2006) JC polyomavirus lineages common among Kiribati Islanders: implications for human dispersal in the Pacific. Anthropological Science 114:133–140

Tayles N (1996) Anemia, genetic diseases, and malaria in prehistoric mainland Southeast Asia. American Journal of Physical Anthropology 101:11–27

Taylor B (1998) Mosquitoes and man in the Pacific: more than just words. Antenna 22:201–204

Tjitra E, Anstey NM, Sugiarto P, et al. (2008) Multidrug-resistant *Plasmodium vivax* associated with severe and fatal malaria: a prospective study in Papua, Indonesia. PLoS Medicine 5(6): e128

Trejaut JA, Kivisild T, Loo JH, et al. (2005) Traces of archaic mitochondrial lineages persist in Austronesian-speaking Formosan populations. PLoS Biology 3(8):e247

Tsukahara T, Hombhanje FW, Lum JK, et al. (2006) Austronesian origin of the 27-bp deletion of the erythrocyte band 3 gene in East Sepik, Papua New Guinea inferred from mtDNA analysis. Journal of Human Genetics 51:244–248

Ubalee R, Tsukahara T, Kikuchi M, et al. (2005) Associations between frequencies of a susceptible TNF-α promoter allele and protective α-thalassaemias and malaria parasite incidence in Vanuatu. Tropical Medicine and International Health 10:544–549

Webb S (1990) Cranial thickening in an Australian hominid as a possible palaeoepidemiological indicator. American Journal of Physical Anthropology 82:403–411

Williams T, Maitland K, Bennett S, et al. (1996) High incidence of malaria in α-thalassaemic children. Nature 383: 522–525

Yanagihara R, Nerurkar VR, Scheirich I, et al. (2002) JC virus genotypes in the western Pacific suggest Asian mainland relationships and virus association with early population movements. Human Biology 74:473–488

Part VI
Central and North America

Three cases from the Americas illustrate landscape/society adaptations at the extremes. The first comes from hot, sub-tropical, Central America, where lowlands are rich in forests and wetlands, nevertheless suffer from irregularities in precipitation. A second is provided by the Arctic, where icy landscapes vary greatly through the year as they have throughout geologic time. The third, from the temperate American west, shows human maladaptation to a region rendered dry by orographic climatic effects.

The fertile, luxurious Central American landscape allowed for the development of one of the great civilizations of antiquity, the Maya, but droughts, fuelled in part by El Niño, played an important role in their eventual collapse (Fagan 2009). Human activities, in particular deforestation, played their part in the downfall of this society that had become large and highly stratified. Great efforts were made by the Maya to combat the deteriorating environmental conditions, amongst them attempts to store large quantities of water to survive drought—but all in vain. The arrival of the European conquistadors was the coup de grace.

The Arctic landscape is cold, harsh, very variable, and experiences the long polar night. Its limited resources can support only a small permanent population of nomadic hunters, the Inuit, known previously as Eskimos. The Inuit adapted to their inhospitable landscape, and developed cultures characterized by simple temporary habitations and a small, hunting tool-kit, but they never reached the stage of a civilization as the word is used neutrally in this book. They did not have the wherewithal to modify the Arctic landscape notably. However, the landscape was greatly influenced by global climatic changes, currently due in part to the activities of societies far to the south. Southern influence has also induced Inuit people to abandon their ancient nomadic lifestyle and live in permanent settlements. This has brought benefits—an assured food supply, medicine—but also drawbacks such as long range pollution, the psychological problems of isolation, and dependency on supplies from the south. The emerging culture of the modern Inuit cannot be sustained by the local landscape.

The third case is from the State of Colorado where a style of agriculture developed on the European model, was imposed on this arid landscape initially in the nineteenth century. Here, the Rocky Mountains have created drylands by blocking the access of moisture from Pacific. John Wesley Powell's warning that there would never be enough water available here to sustain large populations and humid climate agriculture has always been ignored (Powell 1875). Rivers east of the Rockies have been dammed and over-exploited, aquifers have been tapped, and every drop of runoff has been utilized. By the late 1800s technology was used to divert large rivers from the wet, western side of the mountain range, create large reservoirs, and conduct water eastward into Colorado by canal. Again a society turns to powerful technologies in the perennial hope that they can sustain the unsustainable.

Chapter 23
Farms and Forests: Spatial and Temporal Perspectives on Ancient Maya Landscapes

Nicholas P. Dunning and Timothy Beach

23.1 Introduction

"Hills rose around us on every side, and, for that country, the scene was picturesque, but all waste and silent. The stillness of the grave rested upon the ruins, and the little notes of a flycatcher were the only sounds we heard" (Stephens 1843, p. 113). So wrote the American explorer John L. Stephens of his visit to the ruins of Xculoc, Campeche, Mexico in 1842. Through the works of Stephens, Waldeck, Maler, Maudsley, and other nineteenth century adventurers and scholars the ruins of ancient Maya Civilization were brought to the attention of the outside world (Wauchope 1965). It is perhaps these glimpses of forest-draped ruins that gave rise to the over-used epithet 'the mysterious Maya'. While such scenes of ancient cities moldering in the forest still exist in parts of the Maya Lowlands (Fig. 23.1), rapid repopulation and deforestation is transforming forest to fields and pastures and laying bare the relicts of the ancient Maya and a once densely populated landscape. This is not the first time that the Maya forest has been in retreat. In its multifarious forms, the forest has waxed and waned many times over the past several millennia, largely at the hands of the ancient Maya, though climate changes may also have contributed to changes in vegetative cover. In this light, the Maya are hardly mysterious. Over the centuries, like people everywhere they strove to wrest a living from their environment, adapted often with great success to the rhythms of the region's tropical wet-dry climate, and sometimes failed—episodes that were marked by human population decline and forest resurgence.

Over the past three decades, the spatially and temporally variable nature of the Maya Lowlands environment has become more widely appreciated, though it remains far from completely understood. A thorough review of the history of human-environment interactions in the Maya Lowlands is impossible in a few pages. In this chapter we highlight a few case studies that illuminate both the broader trends of that history as well as the nuances brought about in part by the varied nature of the lowland environment. Our review focuses particularly on the period from about 300 BC to around 900 AD; that is, from the Maya Late Preclassic through Terminal Classic period. It was during these 1400 years that Maya peoples accomplished their most enduring deeds in art and architecture, and during which their effects upon the landscape were most profound.

23.2 Environment and Cultural History

Much early scholarly thinking on Maya Civilization viewed the environment as static and culturally limiting, essentially a tropical forest prison that limited agriculture, population density, and urbanism. More recent scholarship has come to recognize the complexity of the Maya Lowlands environment and the varied ways in which the Maya transformed it. Increasingly, we are discerning the many and varied experiences of Maya populations in different places during the course of Maya Civilization.

N. P. Dunning (✉)
Department of Geography
University of Cincinnati
Cincinnati, OH 45221-0131, USA
e-mail: nicholas.dunning@uc.edu

See Plate 21 in the Color Plate Section; also available at: extras.springer.com

I. P. Martini, W. Chesworth (eds.), *Landscapes and Societies*,
DOI 10.1007/978-90-481-9413-1_23,

Fig. 23.1 In 2004, forest still covered the ruins of the Plan de Ayutla acropolis (Chiapas, Mexico)

23.2.1 The Maya Lowlands Environment

The cultural region of the Maya Lowlands includes the Yucatan Peninsula, other low-lying contiguous areas of Central America, and a small area of higher elevation in the Maya Mountains (Fig. 23.2). Environmental factors that create a mosaic of habitats across the lowlands include variations in intra- and inter-annual rainfall, soil and geomorphic processes, and slope gradients and drainage conditioned by structural geology (Dunning et al. 1998; Beach et al. 2008). Only a few of the many habitats found in the region will be looked at in detail here, as we discuss examples of ancient Maya adaptations to their heterogeneous environment. Embedded in the development of all of these habitats,

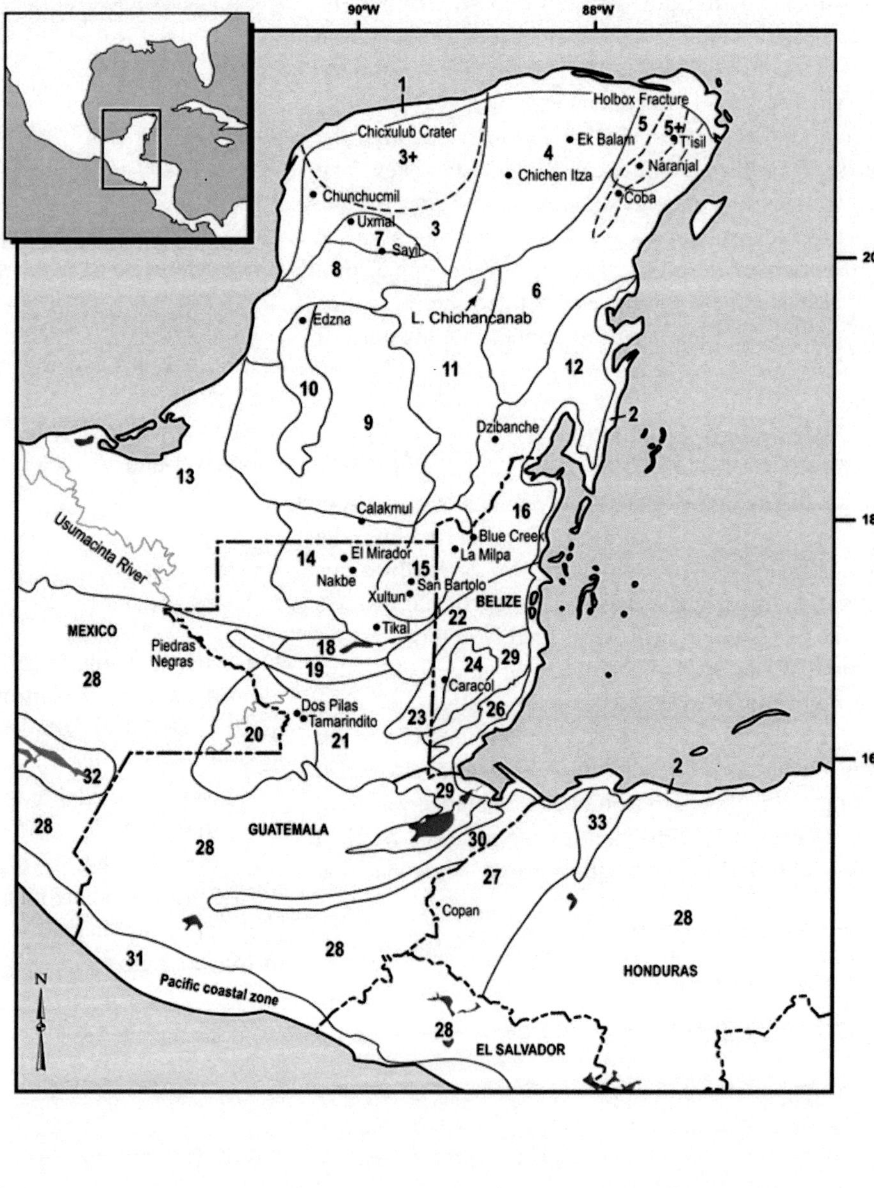

Fig. 23.2 Map of the Maya Lowlands showing physiographic sub-regions and sites mentioned in the text. (Numbers refer to sub-regions: *1* North Coast; *2* Caribbean Reef and Eastern Coastal Margin; *3* Northwest Karst Plain; *3+* Chicxulub impact feaure; *4* Northeast Karst Plain; *5* Yalahau; *5+* Holbox Fracture; *6* Coba-Okop; *7* Puuc-Santa Elena; *8* Puuc-Bolonchen Hills; *9* Central Hills; *10* Edzna-Silvituk Trough; *11* Quintana Roo Depression; *12* Uaymil; *13* Río Candelaria-Río San Pedro; *14* Peten Karst Plateau and Mirador Basin; *15* Three Rivers Horst and Graben; *16* Rio Hondo; *17* Lacandon Fold; *18* Peten Itza Fracture; *19* Libertad Anticline; *20* Río de la Pasión; *21* Dolores; *22* Belize River Valley; *23* Vaca Plateau; *24* Maya Mountains; *25* Hummingbird Karst; *26* Karstic Piedmont; *27* Motagua and Copan Valleys. (After Dunning et al. 1998; Beach et al. 2008)

of course, were humans, especially during the past three millennia of profound change.

Annual rainfall totals generally range from a low of about 500 mm in the northwest to a high near 2500 mm in the southern extremes of the lowlands. The intruding elevation of the Maya Mountains, however, induces more rainfall on their eastern flank (including the Hummingbird Karst and Karstic Piedmont) and a rain shadow to their west (Dolores region). Year-to-year variation is quite high, and Maya farmers have responded to this by developing many inventive approaches for managing risk. Furthermore, the distribution of rainfall is highly uneven throughout the year, with a pronounced and variable dry season generally from November through April.

Most of the Maya Lowlands is underlain by limestone, a simple fact with serious implications for human settlement because the regional geology exerts a strong influence on water movement. Across much of the peninsula drainage is largely subterranean with the karst interior acting as an enormous bedrock sponge for much of the region's rainfall. On the more arid northern plains freshwater is largely available sporadically in 'cenotes' (sinkholes). In the central interior of the peninsula, only a few deep caves and surface ponds ('aguadas') offer water in the dry season. In these elevated interior areas structural faulting as well as limestone and gypsum dissolution have created innumerable basins ('bajos') of varying size which drain entirely internally into the bedrock and are sometimes linked with one another by seasonal surface channels. In the wetter, more elevated south seasonal surface runoff and perennial spring discharge feed rivers including the Usumacinta and Candelaria to the west and the Hondo and Belize rivers to the east.

Broadly speaking, Maya Lowlands soils have relatively high fertility, and are well-drained, clayey, and calcareous, though they tend to be shallow. In the southern lowlands, soils are deeper, relatively poorly drained, and more leached and weathered but still clayey and calcareous. Within the lowland subregions, geologic structure is an important factor that creates variation in the spatial distribution of soils because of its influence on drainage and slope gradient as well as the age of soilscapes. Faulting and fracturing have created a landscape with enough variation in drainage to have produced a mosaic of land surfaces that vary from well drained uplands with generally shallow, but fertile Rendoll soils, to areas of seasonal inundation with deeper Vertisols (shrinking and swelling clay soils) such as in the interior bajos, and some perennially flooded wetlands with water-logged, organic Histosols (Dunning et al. 1998). Natural vegetation across the Maya Lowlands follows rainfall patterns and the seasonality of water supply, though edaphic factors, like soil drainage and chemistry, cultural factors, and lithology are also influential (Greller 2000). From the northwest coast of Yucatan southward through the peninsula, the beach ridges of the coast merge into a 10–20 km swath of swampy estuarine wetlands, and a thorn woodland-savannah that grades into a dry, deciduous, seasonal forest in the central peninsular uplands. The deciduous forest then grades into tropical, dry forest that stretches from central Yucatan into the northern Peten District of Guatemala. This dry forest breaks up into a savannah zone centered on the La Libertad Anticline, which grades southward into the species-rich, tropical, moist forest of the wetter southern half of the Peten. These forests are not truly tropical rainforests, but they have diverse species assemblages nonetheless (Greller 2000). An east to west rainfall gradient also leads to more evergreen forms of vegetation in the east and more xerophytic forms in the west particularly in the northern portion of the peninsula.

23.2.2 *Cultural History*

Pre-Hispanic human occupation of the Maya Lowlands is often divided into the following chronological periods:

- Pre-7000 BC: PaleoIndian
- 7000–2200 BC: Archaic
- 2200 BC–250 AD: Preclassic
  - 2200–1000 BC: Early Preclassic
  - 1000–400 BC: Middle Preclassic
  - 400 BC–100 AD: Late Preclassic
  - 100–250 AD: Terminal Preclassic
- 250–900 AD: Classic
  - 250–600 AD: Early Classic
  - 600–770 AD: Late Classic
  - 770–900 AD: Terminal Classic
- 900–1500 AD: Postclassic
  - 900–1250 AD: Early Postclassic
  - 1250–1500 AD: Late Postclassic.

Portions of this chronology were established on the basis of older thinking about the Maya, particularly the definition of a Classic period as the time characterized by the erection of carved stelae with dated hieroglyphic inscriptions. Stelae and inscriptions are now known from both the Preclassic and Postclassic periods. Similarly, both population numbers and cultural accomplishments were once commonly believed to have peaked throughout the Maya Lowlands in the Late Classic, followed by population collapse and cultural decline. Scholars now understand that the course of Maya Civilization was much more complex. Major population apogees and ensuing collapses occurred in both the Terminal Preclassic and Terminal Classic periods, with lesser fluctuations at the end of the Middle Preclassic and Early Classic. This population growth and decline, however, was not uniform across the entire lowlands. In each period, some sites and regions experienced greater growth or more complete abandonment than their neighbors while others experienced much less fluctuation. Some regions were abandoned for hundreds of years at a time, while others were never abandoned. In part, this spatial variability in population patterns through time reflects the underlying mosaic of habitats with the lowlands, as well as pan-regional climate pulses. Environmental variables, however, can explain only some of the patterning because human response and adaptation played a major role in the turbulent course of Maya history (see Demarest (2005) and Webster (2002) for more discussion on Maya cultural history).

23.2.3 Environmental Change

Scholarly thought on the role of environmental change over the course of Maya Civilization has evolved (Dunning and Beach 2004). For much of the twentieth century, scholars treated the tropical Maya Lowlands environment as largely unchanged over time. It is now recognized that the region's environment has been very dynamic over both long and short time scales. For instance, recent studies have brought attention to the ongoing impacts of the massive Chicxulub meteorite, which struck near the north end of the peninsula at the end of the Cretaceous, some 65 million years ago, and is implicated in the demise of the dinosaurs. Some attributes of the region owe their origin to orbitally-driven, long-term climate fluctuations that influenced global sea-levels and the emergence/submergence of the Yucatan Platform and its composition. More recently, the submergence of large portions of the platform with sea-level rise at the end of the Pleistocene (~17,000–10,000 years ago) and during the early Holocene (ca. 10,000–5000 years ago), as well as establishment of a generally warm, humid climate about 10,000 years ago, had important impacts on human settlement of the peninsula. Continued climate and environmental changes in the Preclassic and Classic compelled the Maya to adapt to continued slow sea level rise and spread of wetlands in the coastal plain. Although large volcanic eruptions devastated parts of the Maya Highlands, as yet there is little evidence for deleterious effects of volcanic activity in the limestone lowlands.

On long time scales, thousands to tens of thousands of years, climate in the Maya Lowlands has been quite variable. For instance, sediment cores from Lake Peten-Itza in northern Guatemala suggest there were dramatic changes in temperature and moisture availability during the last 85,000 years (Hodell et al. 2008). Even since the onset of the Holocene, climate conditions have also displayed considerable variability, especially with respect to effective rainfall (Brenner et al. 2002; Hodell et al. 2005). Wetter conditions in the mid-Holocene gave way to a general, progressive drying trend that began about 4000 years ago (Mueller et al. 2008). For the late Holocene, it has been proposed that short-term (centennial) fluctuations in rainfall may have been driven by the 208-year cycle of solar output (Hodell et al. 2001; Wahl et al. 2006). Such cycles may account for droughts in some areas of the lowlands, and may have had significant impacts on Maya populations (Gill 2000; Hodell et al. 2005; Haug et al. 2003). Accumulating paleoenvironmental data indicate that periods of particularly intense drought may have afflicted wide areas of the Maya Lowlands in the fourth century BC, and the second, sixth, ninth and eleventh centuries AD, as well as the Little Ice Age. These were not single megadroughts, but clusters of increased drought frequencies.

Even today, modern instrumental records from the Maya Lowlands indicate the region is characterized by large variations in rainfall from site-to-site, year-to-year, and across seasons. Inter-annual variation in rainfall can be large, on the order or 30–40%, and this spatial and temporal variability in precipitation undoubtedly played an important role in Maya agricultural strategies over time. Another potentially destabilizing and largely unpredictable aspect of Maya Lowlands climate is hurricanes. In recent times, major hurricanes have raged across the region frequently (for example, Gilbert in 1988, Mitch in 1998, and Isadora in 2002), bringing a combination of wind damage, flooding, catastrophic crop loss, and population displacement. The ancient Maya perceived this violence as akin to divine warfare.

The ancient Maya themselves were agents of both regional and local environmental change, which has been recognized for many years (Deevey et al. 1979). In particular, deforestation and soil erosion have been cited as environmental factors that might have negatively affected Maya subsistence options (Beach and Dunning 1995; Dunning and Beach 2004). The relationship between human population density and environmental disturbance has proven to be more complex than previously envisioned. The relationship between population numbers and environmental degradation is rarely linear, as we shall see in the following sections.

23.3 Preclassic Landscapes

For many years, scholars viewed the Preclassic period as one of essentially slow, steady population growth, largely pre-urban settlement organization, and relatively minor environmental impact. That perspective has changed radically over the past 20 years with more investigations targeting the earlier periods of Maya Civilization. Archaeologically, such investigations are hampered because Preclassic materials are often buried under later Classic constructions. Similarly, the environmental changes triggered by Preclassic peoples are often masked by later changes. Nevertheless, we are now beginning to understand these early times including climatic perturbation, the origins of urbanism, the diffusion and development of agriculture, and the dramatic impacts on local and regional environments brought about by forest clearance, population concentration, conservation practices, and monumental urban construction.

23.3.1 *Lake Tamarindito*

Lake Tamarindito lies in a structural depression 100 m below the Petexbatun Escarpment, and adjacent to the ancient town of Tamarindito in the southwestern part of Guatemala's Peten District. Sediment cores from this small lake provided evidence for long-term human–environmental interactions (Dunning et al. 1998). Although the intended focus of that research was the Late Classic period, the Lake Tamarindito cores contained a paleoenvironmental record reaching back over 10,000 years, to the onset of the Holocene. The lower portions of the cores record the origins of the lake from a dry, elevated depression in the late Pleistocene, to a water-filled basin that accumulated sediments as wetter conditions developed during the Holocene. Pollen grains from the lower depths of the core show the establishment of a regional, species-rich, tropical moist forest. The first evidence of human impacts appears in the Late Archaic, after about 4000 BP in the form of slightly increased levels of inorganic sediment inputs, greater charcoal concentrations, and increased relative abundance of grass pollen indicative of the first episodes of forest clearance or climate drying in the region. Definite evidence of agriculture makes its first appearance with *Zea mays* (corn) pollen about the same time. Maize production has been documented palynologically between 5000 and 3000 BP at many sites in the Maya Lowlands. Figure 23.3 represents data from the sections of the 1995B core from about 1200 BC to about 400 AD.

For much of this period, archaeological evidence indicates that permanent human settlements were largely confined to the shores of Lake Petexbatun and the Río Petexbatun, some 2–3 km east of Lake Tamarindito. Notably, human impacts detected in the Lake Tamarindito sediments are expressed as pulses of forest clearance and regrowth, though overall there is a diminution in forest plants and an increase in disturbance taxa and cultigens over time. Changes in the species-composition of the disturbance taxa (weeds) indicates that field fallow times decreased over time, probably culminating in the establishment of a largely fixed field agricultural system by ~100 AD. Around that time, the ratio of forest and field indicators in the pollen record largely stabilized. Arboreal species overall remained diminished, though far from absent, and the number of economically valuable tree species was greater than during pre-Maya times. In combination, the pollen data indicate that by the onset of the Early Classic period around 250 AD, the landscape around Lake Tamarindito consisted largely of fixed, or short-fallow agricultural fields and sizeable patches of managed forest (essentially woodlots and orchards).

Similarly, shifting rates of inorganic sediment accumulation into Lake Tamarindito indicate pulses

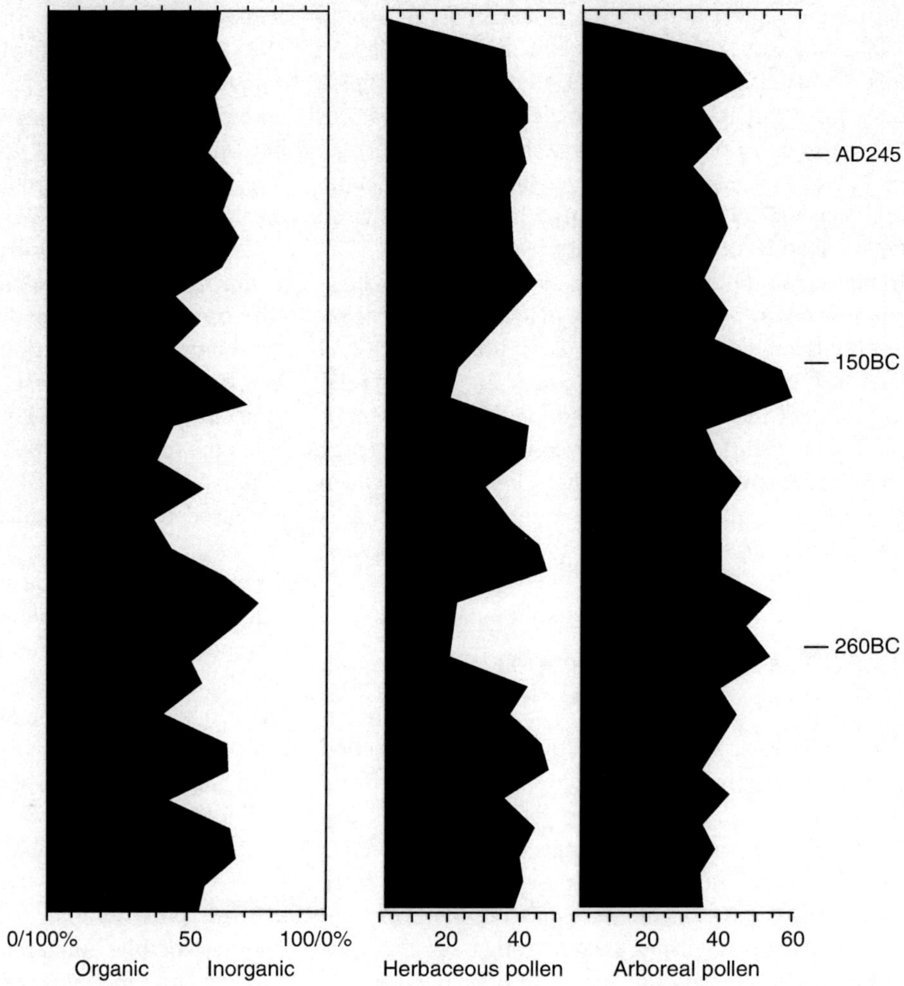

Fig. 23.3 Sediment composition and pollen frequency diagram for a sediment core extracted from Lake Tamarindito, Guatemala in 1995. Inorganic sediments are dominantly silicate clays with minor amounts of calcium carbonate and iron oxide. Dates are calibrated radiocarbon intercepts. Pollen analysis by David Rue (Archaeological and Historical Consultants, Inc., Centre Hall, PA) and John G. Jones (Department of Anthropology, Washington State University)

of forest clearance and spikes in soil erosion during much of the Preclassic. By around 100 AD, inorganic sediment input into the lake was about 25 times higher than in pre-and post-Maya times. During the ensuing Early Classic period, sediment inputs actually decreased, probably due, in part, to the development of soil conservation terracing practices on nearby sloping lands, which we discuss later. This picture of landscape stability at the end of the Late Preclassic stands in sharp contrast to environmental conditions in some other regions of the Maya Lowlands at this time.

23.3.2 Mirador Basin

About 130 km north of Tamarindito in the northernmost portion of the Peten District lies the Mirador Basin (Hansen et al. 2002; Wahl et al. 2006). It is a large, elevated subsidence basin with numerous, irregularly distributed bajos of varying size and internal, karst drainage. Archaeologically, the region is known for its numerous and very large Preclassic centers, the two most investigated being Nakbe and El Mirador. Nakbe was a large town by the Middle Preclassic (ca. 700 BC), while El Mirador rose to become the dominant center of the basin, and likely the Southern Maya Lowlands, by the Late Preclassic (ca. 300 BC). Despite its early grandeur, much of the Mirador Basin was largely abandoned by around AD 150. We still know little about the residential or sustaining areas that lay beyond the core of these sites. El Mirador, Nakbe, and other large sites in the region are perched on the edges of relatively large bajos. Paleoecological data from the bajo adjacent to Nakbe include two important findings. First, this depression once contained larger areas of biologically productive perennial wetlands than are found there today, and which may have occurred in other bajos of the Mirador Basin. Second, the degradation of this resource may have contributed to the abandonment of the region towards the end of the Late Preclassic (Hansen et al. 2002; Wahl et al. 2006).

The monumental architecture, most notably the triadic pyramid complexes of El Mirador, include the largest masonry structures ever constructed in the Maya Lowlands. The Dante Complex at El Mirador alone contains some 2.8 million m^3 of masonry, and other architectural groups at El Mirador and other sites in the region were not far behind in size. The Preclassic architecture of the Mirador Basin is notable for the enormous quantities of lime plaster that cover relatively crude stonework and form the decorative masks and other stucco ornamentation. One study by Thomas Schreiner concluded that it would have been necessary to cut down a large part of the region's forests to fuel fires needed to burn limestone and produce plaster. Indeed, large ancient quarries are still evident across the region today, and many adjacent bajos show evidence of massive influxes of eroded sediment (Hansen et al. 2002; Wahl et al. 2006). In essence, the glorification of deities and divine kings through the erection of ever-larger buildings may have spelled doom for the populace of the Mirador Basin, which was largely abandoned by 150 AD. Regional abandonment may have been linked to the destruction of forest resources, agricultural land both on the uplands and in portions of the bajos, and perhaps to the onset of more frequent droughts suggested in several paleoenvironmental records (see Sect. 23.2.3 above).

23.3.3 La Milpa

One hundred kilometers due east of El Mirador is the ancient city of La Milpa, located in the Three Rivers region of northwestern Belize. The site center is situated on a high karst ridge amidst a series of upland bajos that today contain seasonal swamp vegetation (Dunning et al. 2002). Investigations show that the small (ca. 2 km^2) Far West Bajo on the northwest side of La Milpa contained a perennial wetland and areas of permanent open water prior to about 100 AD. We attribute this transformation from a perennial to seasonal wetland largely to anthropogenic sedimentation associated with Preclassic deforestation, cultivation, and erosion in the surrounding watershed. These human-mediated impacts caused long-term hydrologic changes within the bajo itself, though natural climatic drying that peaked in the second century AD may have compounded the anthropogenic hydrologic changes. In either case, the associated erosion and sedimentation also created swaths of alluvial/colluvial soils on the margins of the bajo, which became the focus of intensive agriculture in ensuing Classic times, as is evident in extensive terracing and field wall systems situated on these slopes. The loss of the water resources in the bajo may have also spurred the development of reservoirs in the site center, a transformation that also appears to date to the Late Preclassic/Early Classic transition. Some have suggested that these reservoirs served the dual purpose of providing dry season water for domestic consumption and irrigation (Scarborough et al. 1995), but the capacity of these artificial water bodies is likely to have only barely met the consumption needs of the population in the immediate core area (Tourtellot et al. 2003). Hence, the reservoirs may have had as much a symbolic as practical function, and the ritualized control of water would have been a potent political tool for Maya rulers, especially in a region with a long dry season.

23.3.4 San Bartolo and Xultun

Between Tikal, Mirador, and La Milpa are the ancient towns of San Bartolo and Xultun, which lie 8 km from one another in the northeastern part of the Peten District (Garrison and Dunning 2009). Despite their close proximity, these neighboring communities experienced very different settlement histories. San Bartolo was an important centre in the Middle and Late Preclassic, but was abandoned at the end of the Late Preclassic around 150 AD, only to be lightly reoccupied in the Late Classic around 700 AD. Although excavations are just beginning at Xultun, architecture exposed both at the surface and in numerous looter's tunnels, and inscriptions on the site's many stelae indicate that it too was a sizeable site in the Late Preclassic, but, unlike San Bartolo, went on to grow into a large and important Early, Late, and Terminal Classic center.

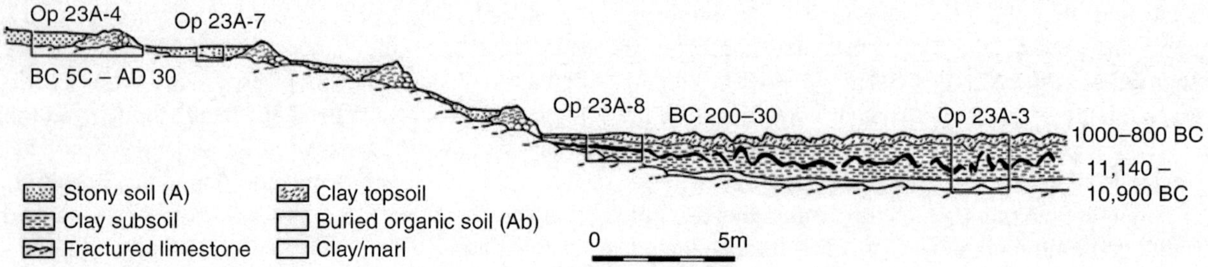

Fig. 23.4 Cross section of the southern margin of Bajo Donato near San Bartolo, Guatemala. Vertical and horizontal scales are identical. Dates are calibrated 90% probability ranges

Paleoenvironmental investigations in two bajos near San Bartolo provide some clues as to what happened in this region at the end of the Preclassic. A core taken from the Aguada Tintal, a reservoir pond in a small bajo northeast of San Bartolo, contained pollen from maize, manioc, and cotton dating to the Middle Preclassic, ca. 500 BC. Investigations in the adjacent Bajo Donato indicate increasing sedimentation throughout the Middle and Late Preclassic periods as surrounding sloping land was cleared of forest for farming. By the first century AD, Maya farmers had recognized and responded to the soil erosion problem and constructed agricultural terraces on the flanks of the bajo (Fig. 23.4).

Excavations in the small Bajo Majunche midway between San Bartolo and Xultun indicate that a shallow lake existed in this depression around 800 BC and later desiccated, probably due to a combination of anthropogenic sedimentation and regional climate drying. The divergent histories of San Bartolo and Xultun may be connected with how their inhabitants responded to the drying climate. The inhabitants of Xultun invested in the construction of numerous aguada reservoirs; their counterparts at San Bartolo, for unknown reasons, did not. Xultun prospered; San Bartolo was abandoned, only to be reoccupied many centuries later, probably by people from Xultun.

23.4 Early Classic

The Terminal Preclassic (AD 100–250) was marked by both regional-scale and individual site abandonments. The onset of the Early Classic was characterized by a return of relative stability in most parts of the Maya Lowlands. This period also saw significant increases in urban population nucleation both at small and large centers, a change brought about both by increasingly fixed and intensive forms of cultivation, including farming within urban precincts, and by the apparent development of significant trade in foodstuffs, at least in some areas. The dawn of the Early Classic was also marked by the founding of numerous royal dynasties that would hold sway over much of the Maya Lowlands for the next five to six centuries (Martin and Grube 2008).

23.4.1 Tamarindito: Landscape-Modification and Urban Agriculture

The ancient city of Tamarindito came into existence in the Early Classic period through the apparent coalescence of several Preclassic villages. Perched on a spur of the 100-meter-high Petexbatun Escarpment, Tamarindito inhabitants looked out over the nearby lake (see Sect. 23.3.1 above). During the Early Classic, Tamarindito, together with the smaller site of Arroyo de Piedra, was the twin capital of a small, regional dynasty that controlled much of the Petexbatun landscape (Demarest 2005).

Over many generations, the people of Tamarindito terraformed the hilly terrain of the growing urban center, stabilizing sloping land and making it sustainably productive by constructing many types of agricultural terraces such as contouring slope terraces, box terraces, and check dams (Fig. 23.5; Beach and Dunning 1995).

Based on their architectural elaboration and midden deposits, many of the urban farmsteads of Tamarindito centered on elite or at least middle class residen-

Fig. 23.5 Elite residential complexes and agricultural terracing in the urban zone of Tamarindito, Guatemala. Long terraces in the foreground are about 35 m in length. (After Dunning 2004)

tial groups. Typically these compounds also included lower-status residences, likely lower-status members of the same lineage (Dunning 2004).

The structure of ancient Maya society was largely lineage-based (lineages are descent or name groups in which all members claim descent from an apical ancestor). Individuals and their families were assigned by birth to a lineage and a hierarchical position both within their own lineage and a ranked system of lineages. Lineage status was closely tied to land wealth. Within garden cities such a Tamarindito there was a high correlation between residential location and either prime or well-developed land, likely reflecting the use of the 'first founder' principle. In Yukatek Mayan, for example, the expression 'yax chibal way tiluum' (first founding lineage of the land) captures this principle whereby the first settlers of land claim it as their own. Patricia McAnany (1995) has aptly called the Maya lineage system a crucible of inequality, in which first families enjoyed considerable land wealth and privileged social positions and others farmed less desirable land or labored for the elite. Even within lineages, however, were significant disparities in wealth, which likely accelerated with time as land became increasingly scarce and land inheritance less assured. A study of human skeletal remains from the Petexbatun region indicates that while the average health of the population remained remarkably stable throughout Classic times, the disparity in quality of diet between rich and poor increased markedly over several centuries (Wright 2006).

The vast majority of the ancient Maya were farmers and the vast majority of Maya settlements should be understood first as farming communities (Dunning 2004). A fundamental aspect of many Classic period Maya communities, such as Tamarindito, seems to have been agricultural self-sufficiency, through the use of intensive cultivation of infields within the towns themselves, and the control of production in lands surrounding the town (outfields and tributary settlements). This general characteristic applies to even the largest Classic Maya cities such as Tikal and Calakmul. Nevertheless, although the garden city model of Maya communities certainly applies to many, if not most, ancient Maya communities, there were exceptions in which other forms of economic, political, or other activities were paramount and urban agricultural production played a lesser role in the function of the central place.

23.4.2 Dzibanché: Urbanization in a Wetland Setting

The ancient conurbation centered on the site of Dzibanché (or Tzibanche) is situated on an isthmus of higher terrain that separates two large bajos (Bajo Morocoy and Bajo Acatuch) in southern Quinana Roo state, Mexico. Study of aerial photographs in the 1970s revealed that these bajos contain by far the most extensive areas of wetland fields and associated canals known anywhere in the Maya Lowlands, with field areas covering as much as 240 km^2 (Harrison 1977). Unfortunately, these field and canal systems have never been systematically excavated and are essentially undated. For many years, it was assumed that this agricultural system must date to the Late Classic period, an assumption based on only circumstantial evidence and the widespread belief that population and land use intensity peaked uniformly across the Maya Lowlands in the Late Classic. However, recent excavations within the Dzibanché settlement system, which is surrounded by fields and canals, indicate that this urban center reached its apogee in the Early Classic (Nalda 2005). The conurbation includes four large architectural clusters lying 2–2.5 km from one another and connected by causeways. Residential settlement spreads around and between the centers covering an area of 15–20 km^2. Late Preclassic, large monumental construction is found in both the Dzibanché (Grupo Principal) and Kinichná complexes and extensive areas of residential occupation centring on large aguadas. Growth continued at Dzibanché in the Early Classic, and towards the end of that period the site seems to have become the seat of the Kaan dynasty, which may have been centered at El Mirador in the Late Preclassic and shifted to Calakmul around AD 630 (Martin and Grube 2008). The remainder of the Late Classic is more poorly represented (Nalda 2005).

Unfortunately, the large and elaborate system of wetland agriculture surrounding Dzibanché remains poorly documented and large areas have been obliterated by modern development. The degree to which this system co-evolved with the large urban node near its center and other questions such as whether larger canals were articulated as part of the urban plan of Dzibanché remain unanswered. However, Dzibanché and its neighboring wetland field systems give the appearance of a grand experiment in a different kind of Maya garden city, wetter than others. As near Blue Creek (discussed below), these experiments may have been responses to rising water tables and regional climate drying over Maya history.

23.4.3 The Curious Case of Chunchucmil

Not all ancient Maya communities fit the garden city model as outlined above.

Chunchucmil lies some 25 km inland of the Gulf of Mexico on the karst plain of northwest Yucatan, separated from the coast by broad bands of swamp estuary and savannah. In many ways Chunchucmil is the ideal test case for the idea that Maya cities developed with agricultural self-sufficiency in mind (Fig. 23.6). Chunchucmil was among the largest and most densely

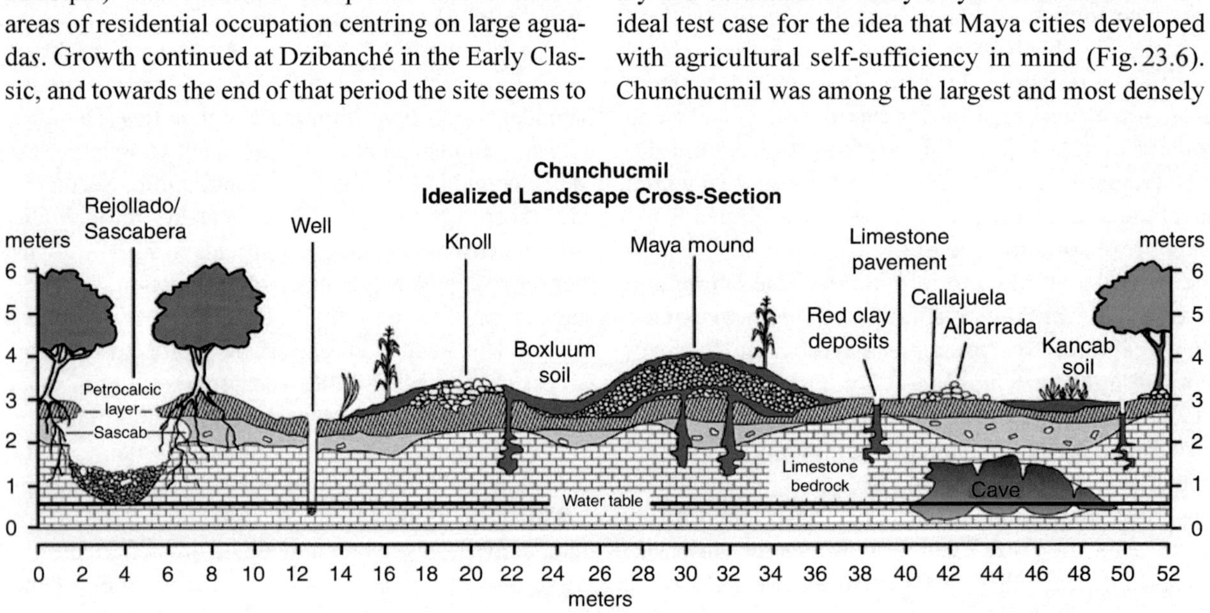

Fig. 23.6 Idealized cross section of the landscape around Chunchucmil, Yucatan

settled cities in the Maya Lowlands; the central urban area (21 km^2) population is conservatively estimated at between 42,400 and 47,600 with densities ranging variably from 1325 to 2975 per km^2 across the area, not including an additional 20–40 km^2 of less dense residential settlement (Dahlin et al. 2005, 2007). These exceptionally high population numbers stand in sharp contrast to the apparent carrying capacity of both the urban area itself and its hinterland. Chunchucmil is situated in what is widely regarded as one of the poorest areas for traditional agriculture in the Maya Lowlands with scant soil cover and relatively low rainfall (~900 mm/year). Despite these limitations, evidence suggests that the people of Chunchucmil made valiant efforts to supply themselves with food through intensive gardening, probably including tree-cropping, pot irrigation, and mining organic soil from the nearby savannah zone to build up raised beds within the city's walled 'solares' (kitchen gardens) (Beach 1998). Recent studies of underlying limestone cavities at Chunchucmil, and near Yalahau (Sedov et al. 2008), have shown significant sedimentation, which may indicate erosion of thicker and more productive soils in antiquity. Nevertheless, even models that incorporate the most creative agricultural strategies and the most conservative population estimates, cannot reconcile Chunchucmil's burgeoning population with the poor carrying capacity of its environment (Beach 1998).

Chunchucmil is not a typical Maya regal-ritual city. Monumental architecture is modest and Classic Maya dynastic attributes such as stelae are lacking. These factors, plus the relative wealth of the city's inhabitants in imported goods, the apparent existence of a central marketplace and manufacturing enclaves, and the city's location between the coast and more productive agricultural land farther inland, have led investigators to suggest that Chunchucmil functioned largely as a trade center, possibly based on its large central marketplace (Dahlin et al. 2007). The city's connection to the coast and its trade routes (as well as salt beds and other marine resources that Chunchucmil may have sought to exploit or control) was apparently enhanced by canals that traversed some intervening terrain. Although originally thought to be a Late and Terminal Classic city, recent work indicates that the city reached its zenith in the Early Classic (Hutson et al. 2008) and declined in size and importance thereafter. This chronology thus predates the major occupation of the agriculturally-rich Puuc region which lies 25–30 km southeast of Chunchucmil, suggesting that the city could have tapped that region's surplus agricultural production either through the intermediary city of Oxkintok, or perhaps directly through seasonal movement of the population into the Puuc.

23.5 Late Classic

The transition from the Early to Late Classic periods was marked by far fewer disruptions than occurred at the end of the Preclassic. Many centers simply continued to grow at the end of the Early Classic. A few centers, however, most notably the large city of Tikal and its allies, experienced a century-long downturn in their fortunes referred to by archaeologists as "the Hiatus" (Martin and Grube 2008). Such downturns seem to have been the result of escalating warfare among Classic polities, most likely related to increasing land scarcity and a growing number of royal descendents, some of whom sought to carve out a place in the political landscape (Demarest 2005). There is, however, some paleoenvironmental evidence suggesting increased drought frequency during the sixth century AD, possibly related to an equatorial volcanic eruption that disrupted climate across widespread areas in the Northern Hemisphere (Gill 2000; Larsen et al. 2008; Rosenmeier et al. 2002). If droughts led to significant agricultural failure during this time, this event may have contributed to the pattern of escalating warfare between centers. But, only one region shows a widespread pattern of site abandonment and for reasons other than drought: the Yalahau region in the northeast suffered changing groundwater levels that may have disrupted wetland agriculture (Fedick and Morrison 2004).

23.5.1 *Coba*

The largest urban center in the northeastern Maya Lowlands was Coba, which reached its zenith in the Late Classic and sprawled over some 30 km^2. Coba enjoyed a number of significant environmental advantages. Located in a comparatively elevated and well-drained southern portion of the Holbox Fracture Zone, it was blessed with several perennial lakes, and was not as adversely affected by changes in groundwater level that created havoc further to the north in Yalahau, though water levels in the Coba lakes have clearly fluctuated (Leyden et al. 1998). It also benefited from the atypically high and dependable precipitation associated with the Coba rainfall anomaly in northeastern Yucatan. Karst solution depressions within the Holbox Fracture also contain deeper, more humid, soils than those outside. In short, the residents of Coba enjoyed relative advantages in drinking water supply, soil, and abundant rainfall to grow crops. These factors aided in the development of an extensive urban zone where staple crops, tree crops, and other economic plants were grown extensively and intensively. The city had sprawling residential zones with open garden spaces carefully partitioned by networks of field walls (Folan et al. 1983), though even this fortunate place may have relied on imported food during its zenith. These favorable local environmental factors may also help explain Coba's relative longevity. The city reached its apparent zenith in the seventh and eighth centuries AD, declined in the ninth century (possibly as the result of conflict with Chichen Itza), but continued to be an important town through the Postclassic (Leyden et al. 1998).

23.5.2 Caracol: Garden City Extraordinaire

The large urban site of Caracol is situated on the Vaca Plateau on the northwest flank of the Maya Mountains of Belize. Paralleling what the Maya of Dzibanché accomplished in the bajos of southern Quintana Roo, the Maya of Caracol transformed the hillslopes of a large swath of the Vaca Plateau into a systematically agro-engineered landscape (Chase and Chase 1998). The lithology of this region (horizontally bedded and strongly outcropping limestone) may have aided the Maya in this effort, but human agency was clearly paramount in the construction of agricultural terracing on an unrivaled scale. Caracol's population may have been as high as 115,000–150,000 in ca. 700 AD. This huge and dense population may suggest that its unprecedented investment in terracing (for the Maya) was an attempt to feed itself, although production of other non-food economic crops such as cotton to support urban industry is also possible (Chase and Chase 1998). Caracol's elaborate settlement and terrace systems have been invoked as evidence of centralized organization and control of agricultural production (Chase and Chase 1998), although equally or more extensive and intricate systems of terracing in the Andes are arguably the product of collectively organized labor and planning by farmers (Erickson 2006).

Caracol was intimately involved, and indeed a major player, in the sustained conflict between Calakmul and Tikal that dominated significant aspects of Classic period geopolitics in the Southern Maya Lowlands (Martin and Grube 2008). In particular, Caracol appears to have contributed to the military success of the Calakmul alliance in the sixth and seventh centuries AD over Tikal and its allies. This was notably the period of Caracol's greatest urban growth and the period of most intensive terrace construction. It is unclear whether Caracol's success in warfare led to its growth during this period or whether its growth and productive agricultural system contributed to its success in warfare—or whether these two attributes were mutually reinforcing trends. As noted above, drought may have afflicted the central lowlands in the sixth century helping foment conflict between rival communities.

23.5.3 Tikal: Agroforestry and Temples

Situated near the southern end of the central Peten karst plateau, Tikal is often used as the archetypal large Classic Maya city. It indeed has typical elements of ancient Maya towns, with an elaborate core of elite residential and other monumental architecture and an extensive residential zone with large amounts of open space, presumably left open to incorporate intensively cultivated gardens and infields. Tikal is situated on an area of higher ground between two large bajo systems, the Bajo de Santa Fe and the Bajo Socotzal.

Early Colonial sources from several parts of the Maya Lowlands describe the existence of ancestral estates, including managed forests under the control of socially prominent lineages (McAnany 1995). These estate forests likely included a wide variety of economically useful tree species such as 'chicozapote' or 'sapodilla' (*Manilkara zapota*) and 'ramon' (*Brosimum alicastrum*). Estate forests may well have been the source of timber used in the construction of monumental architecture, such as the enormous wooden beams (lintels) used in some temples and palaces.

Many paleoenvironmental and archaeological sources indicate that the lowland forest retreated as population grew during the Late Classic. In this context of declining forest reserves, the demand for large beams used in monumental construction would have been increasingly difficult to meet. A study by David Lentz and Brian Hockaday (2009) of the pattern of timber use in Late Classic temple and palace construction at Tikal suggests that even the rulers of this large city and kingdom had trouble obtaining the wood they desired. After a period of stagnation and decline associated with several military defeats, Tikal's fortunes improved dramatically after 695 AD. Between 695 and ca. 800 AD Tikal rulers engaged in an unprecedented building campaign, including numerous expansions of the royal palace (Central Acropolis) and the erection of six towering temple pyramids (Temples 1–VI) that have become the icons of this ancient city (Martin and Grube 2008). At Tikal, as at many other Classic Maya cities, the wood preferred for lintels in otherwise masonry buildings was chicozapote (*Manilkara zapota*), both a fine lumber and fruit tree. This tree is a relatively slow-growing hardwood found on well-drained upland terrain. Lentz and Hockaday found that the monumental architecture erected between about AD 730 and AD 750 employed massive beams of chicozapote. Between about AD 750 and AD 790, builders switched to using beams of 'palo de tinto' (*Haematoxylon campechianum*), also known as logwood. Palo de Tinto is much harder to harvest and cut, and is found in the bajos located in the hinterlands of Tikal. This switch to palo de tinto suggests that available supplies of chicozapote had been exhausted. By about AD 810, however, chicozapote was again in use in monumental construction, though the beams being employed were much smaller and likely taken from immature trees. Not long after AD 810, the construction of major monumental architecture at Tikal effectively ceased and the city diminished until it was finally abandoned some 100 years later.

23.5.4 La Milpa: Late Glory, Rapid Decline

In contrast to Tikal, La Milpa was a community of relatively modest size until the eighth century AD, when it experienced explosive growth of residential population (to about 46,000 in a central area with a 5 km radius) and monumental construction (Hammond and Tourtellot 2004). This rapid rise in population has prompted speculation that the city and region experienced substantial immigration at that time, most specifically from the Tikal region, which perhaps pushed the region over its sustainable carrying capacity (Hageman and Lohse 2003). During that period, a huge investment was made in field walls and terracing in many both urban and rural contexts. Gair Tourtellot has suggested that by the early ninth century AD, the area around La Milpa may have become a landscape of desperation, with every scrap of cultivable land pressed into service and hardscrabble conditions prevailing in the region's farmsteads (Tourtellot et al. 2003). Food may also have been brought to La Milpa and other centers by developing market systems (Scarborough and Valdez 2003). Nevertheless, La Milpa was rapidly abandoned prior to about AD 830. The changing landscapes around La Milpa are reflected in the pollen recovered in a sediment core taken in 2003 from Lagunita Elusiva, an aguada in the eastern exurbs of La Milpa (Fig. 23.7).

Although few, small house mounds lie near the aguada, there are many agricultural field walls. It is not known when the aguada was either first constructed, or when it was modified from a shallower natural depression to increase its water-holding capacity. However, it appears that sediment was allowed to accumulate in at least the deepest section of the aguada after about AD 400; that is, accumulating sediment was not removed. Pollen from the seventh and eighth centuries shows a very strong human-disturbed/modified-forest signal, most notably high percentages of Poaceae (grasses), Asteraceae (Asters), and *Zea mays* (corn). *Zea* percentages range as high as 5–6%, indicating the cultivation of large quantities of maize close to the aguada. This, along with the network of field walls, suggests sustained, intensive farming. Around AD 800–900, there is a dramatic decrease in cultigens, a large spike in fern spores, followed by an increase in high forest arboreal species (Combretaceae). The fern spike may well represent an incursion of bracken fern (*Pteridium aquilinium*), an aggressive species that thrives in nutrient-poor soil and is a well known invasive in depleted fields in the Southern Maya Lowlands and is an inhibitor to succession (Pérez-Salicrup 2004). By AD 900, the area around the aguada was largely, if not completely abandoned and recolonized by upland forest.

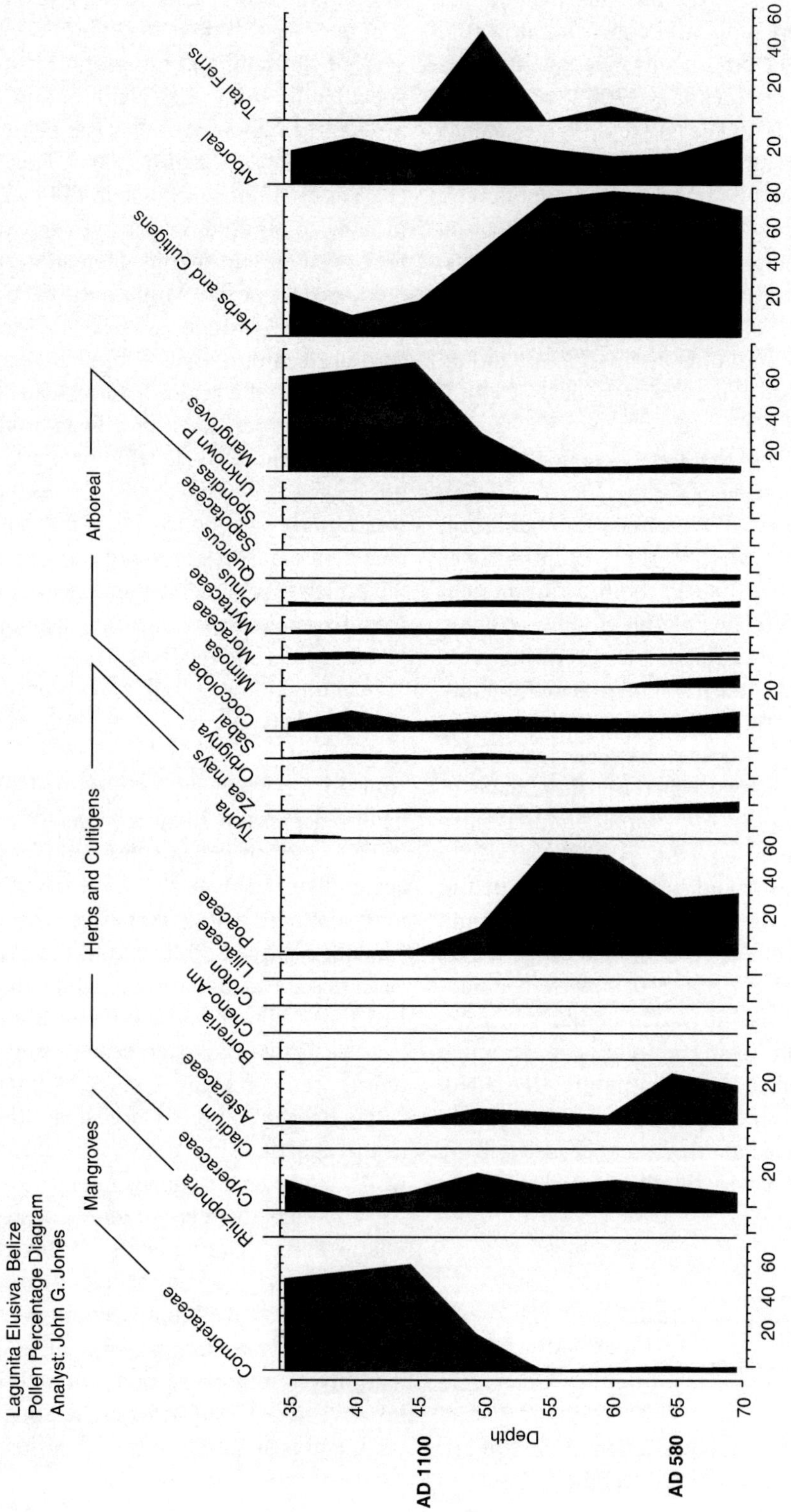

Fig. 23.7 Pollen frequency diagram for Lagunita Elusiva, an aguada on the urban periphery of La Milpa, Belize. (Pollen analysis by John G. Jones)

23.5.5 Dos Pilas: Urban Predator

In the mid seventh century AD, the Petexbatun region came under the political sway of a new center, Dos Pilas (Houston 1993; Martin and Grube 2008). Apparently established by an intrusive rogue dynasty from Tikal, Dos Pilas was located on marginal land and appears to have invested little labor in developing local agricultural potential. Rather it seized control of nearby centers such as Tamarindito that had a long history of landscape modification and intensive agricultural production. In short, Dos Pilas was a predatory capital, exacting tribute through expansionistic military domination. The fall of the Dos Pilas polity a century later in the 700 s ushered in a period of escalating regional warfare (Houston 1993; Demarest 2004). One effect of this warfare was to increasingly concentrate population into defensible positions on the uplands of the Petexbatun Escarpment. Initially walls were used to demarcate parcels of rural farm land as cultivation was intensified across the uplands. Eventually, even the rural farmsteads themselves were fortified as political authority in the region disintegrated (O'Mansky and Dunning 2004).

23.6 Terminal Classic

The Terminal Classic period in the Maya Lowlands was, overall, a time of spreading social chaos and declining populations. Embedded within this time period was the complex phenomenon often referred to as the Classic Maya Collapse. The collapse of Classic Maya Civilization is sometimes portrayed as an abrupt and nearly complete depopulation. Although some regions of the Maya Lowlands indeed experienced significant population declines, elsewhere it was less dramatic. The archaeological record remains frustratingly ambiguous concerning how rapidly population collapsed in many areas. In some places there are indications of a prolonged, multi-generational decline, while at others there seems to have been abandonment in a matter of a few years, or even days (Dahlin 2000; Demarest 2004). Thus the story must have varied from place to place according to differential resilience and the severity of both historical and environmental circumstances. In short, the collapse was a prolonged and complex phenomenon that defies simple description or explanation.

As discussed above in Sect. 23.2.3, paleoenvironmental data derived from lake sediment cores taken in several parts of the Maya Lowlands, and elsewhere in the Caribbean, indicate that the Terminal Classic included several periods of intensive drought. Given the dependence of Maya agriculture on rainfall and the difficulties of obtaining drinking water in many areas, Maya Civilization was vulnerable to drought. Hence, it is no surprise that the dry Terminal Classic period was tumultuous. Environmental problems associated with drought would have been intensified by the population growth in the preceding centuries of the Late Classic that created a need for higher levels of agricultural production. Severe or prolonged droughts would have challenged the Maya ability to feed themselves and threatened the stability of their political system. However, Normark (2006) notes that models of drought impacts on the pre-Hispanic Maya overemphasize descriptions of devastating droughts from the Spanish Colonial period, during which time much of the resilience attributable to earlier Maya adaptations may have been destroyed by the mandates of the colonial administration. In particular, the Spaniards prevented Maya population dispersal to areas with back-up water sources in times of drought and they siphoned off food supplies from many regions.

Although regional droughts may have played a major role in the depopulation of many Maya centers, droughts do not explain many aspects of the collapse. In places like the Petexbatun region, the Classic Maya political system was unravelling even before evidence for drought is found in the paleoenvironmental record. Increasingly belligerent relationships between competing royal dynasties and a political system dependent on an insatiable demand for tribute likely were major contributing factors to the collapse (Demarest 2005). The disintegration of kingdoms appears to have spread like a contagion through wide areas of the Maya Lowlands between about 750 and 950 AD (Webster 2002). By the end of that period, the Classic Maya political system with all its trappings of elaborate dynastic monuments and inscriptions had essentially disappeared. Nevertheless, in many areas population levels, while diminished, remained relatively high, and new and transformed variants of Maya culture developed in the ensuing Postclassic period.

23.6.1 The Decline of the Puuc Cities and the Rise of Uxmal

The Colonial Era town of Santa Elena sits within the middle of the broad structural valley that earth scientists have designated the Santa Elena Physiographic District (Wilson 1980). The pre-Hispanic name for Santa Elena was Nohkakab (the place of good earth), a name fitting to this valley that contains extensive tracts of fertile Mollisols and Alfisols, and the richest agricultural zone in the northern lowlands. Along with the Bolonchen Hills, a broad expanse of cone karst hills and solution valleys, archaeologists refer to this area as the Puuc Zone—culturally distinguished by a distinctive style of concrete core/decorative veneer stone architecture that appeared late in the Late Classic period.

Due to an almost complete lack of natural water sources, year-round settlement in the region was only possible through the capture and storage of rain. This feat was largely accomplished in the Puuc through the construction of thousands of underground cisterns ('chultuns') and associated collection surfaces. However, variation in the lithology of surface bedrock made the construction of chultuns much easier in some places than in others, a factor that likely shaped the distribution of settlements across the region and the residential patterns within sites

The eighth century AD Puuc community of Sayil has been described as a classic example of a garden city (Tourtellot and Sabloff 1994). The oldest and largest lineage residences at Sayil are found immediately adjacent to the most favorable tracts of agricultural land within the urban zone. Geoarchaeological studies indicate that the people of Sayil intensively farmed the land adjacent to their homes, likely growing a great variety of vegetables and fruits, including large quantities of maize and other staples. Later residential compounds filled in vacant land and peripheral areas. Settlement patterns studies indicate that Puuc cities and towns were formally bounded and their populations were perhaps reined in by strict control over the construction of chultuns needed to survive the dry season (Dunning 2004). Rural lands between towns were only seasonally occupied and likely farmed on an extensive basis.

The late ninth century rise of Uxmal as the paramount urban center in the Puuc has often been viewed as the apogee of regional Maya culture and a curious exception to the decline of Classic Maya Civilization that occurred in most other regions at that time. However, the large majority of Puuc centers, including Sayil, were themselves in steep decline by the last decades of the ninth century AD (Carmean et al. 2004). Given the complete dependence of the Puuc population on rainfall, it would have been particularly vulnerable to droughts of any significance and the depopulation of many centers in the latter half of the ninth century may have been, in part, attributable to this vulnerability (Dunning 1992). On the other hand, Uxmal underwent an unprecedented program of monumental architectural construction towards the end of the ninth century AD, including the building of the House of the Governor and Nunnery Quadrangle, two of the most stunning examples of Maya architecture ever erected. The late expansion of Uxmal may well have been at the expense of other communities in the Puuc as reflected in its bellicose and militaristic art and iconography. It is possible that Uxmal was preying on neighboring polities already under stress, perhaps from drought and crop failure in an already over-taxed regional agricultural system. Uxmal's method of population concentration may be reflected in its investment in numerous sizeable aguada reservoirs. Uxmal is situated along an unusual series of fractures that created a local abundance of trough-like sinkholes, at least partially plugged with clay sediments. Here, the Maya adapted these sinkholes into reservoirs. Uxmal thus had a natural advantage with regard to concentrating a large urban population during the troubled times of the late ninth century. In this light, Uxmal's late florescence may be less the apogee of Puuc Maya culture than its last gasp (Prem and Dunning 2004)

23.6.2 Successful Adaptations to Large Scale Environmental Changes Around Blue Creek

Unlike the other cases, the region around Blue Creek in the Three Rivers Region of northwestern Belize lasted from the early Middle Preclassic into the Terminal Classic. It even has evidence of Archaic and Postclassic occupations. This region had a fortunate, ecotonal location, situated near the perennially wet coastal plain and river valleys, the fortress-like high ground of a steep escarpment, and deep, well drained back slope,

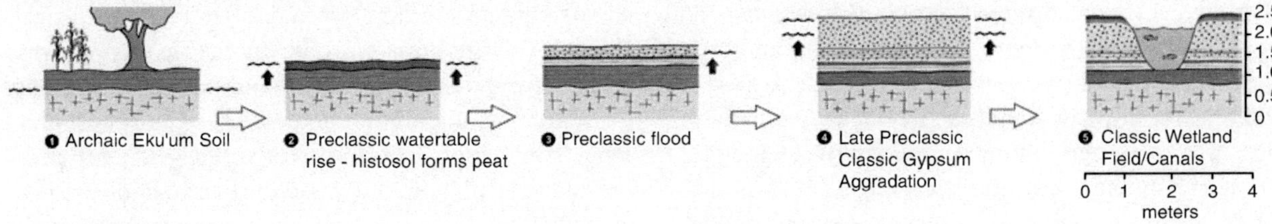

Fig. 23.8 Model of wetland evolution near Blue Creek, Belize

and bajo soils. Population estimates of the site proper are not high, but it was surrounded by numerous other smaller sites that in aggregate add up to a significant regional population of 12,500 (Guderjan 2007). These rural sites clustered around both the thick, upland soils and bajos for agriculture, clay, lime, and lithic sources and the wetlands for a range of resources that occur in these productive ecosystems. Ancient Maya farmers had many options in this landscape, including longer growing seasons in the bajo areas with high water tables, recessional farming in the floodplains, and perennial wetland systems.

The Blue Creek region provides a significant example of both ancient, intensive wetland agriculture and Maya adaptation to environmental changes. In the Archaic period, the wetlands of today were a varied combination of well drained and seasonally wet soils. Archaic to Preclassic farmers were modifying this landscape with slash-and-burn agriculture to grow maize and other crops (Fig. 23.8). But in the Preclassic, as sea levels lifted water tables, these fields became submerged, and upland erosion and large-scale flooding, some driven by a mega flood, started to bury fields. The continued rise of a water table saturated in calcium and sulfate caused precipitation of gypsum that buried the fields further by another meter. In response, from the Late Preclassic to Late Classic Maya farmers dug canals to drain and reclaim at least 7 km^2 of fields to grow corn and a variety of other crops (Beach et al. 2008, 2009; Luzzadder-Beach and Beach 2009).

Ancient Maya reclamation ranged from smaller-scale, piecemeal adaptations at the site of Chan Cahal to pre planned, larger scale systems, including the Terminal Classic and later Birds of Paradise fields. Although much of the region's upper (later) archaeological stratigraphy has been disturbed by modern farmers, there is significant evidence of Postclassic occupation around and in some of these wetland fields. Thus, the Blue Creek region experienced large-scale environmental changes but the Maya adapted through the Terminal Classic with sophisticated, intensive farming practices, likely abetted by intra-regional trade in this heterogeneous area. Moreover, this region with its wetlands and rivers, similar to the pattern at the nearby site of Lamanai, displays evidence for cultural continuity, but with diminution into the Postclassic.

23.7 The Postclassic and Beyond

Overall population numbers across the Maya Lowlands in the eleventh century AD were probably at their lowest levels since about the fifth century BC. Nevertheless, Maya Civilization persisted in many locations, most notably in the eastern-most and northern-most parts of the lowlands. By the fourteenth century AD, population levels were again increasing significantly and some areas, like the Peten Lakes District and Yalahau, which had been largely depopulated for centuries were once again thriving centers. Across the lowlands, Postclassic communities exhibited much less emphasis on monumental architecture and a greater focus on mercantilism and pan-regional commerce, especially along coastal trade routes.

Reflecting the lower population densities in most areas, relative to Classic times, swidden agriculture predominated in most regions. Spanish accounts from the sixteenth century describe a system in which fields were cultivated for 2–3 years, followed by 8–10 years of forest fallow, though intensively cultivated gardens and orchards were also commonplace. In the century following Spanish contact and the catastrophic introduction of Old World epidemic diseases, population levels in the Maya Lowlands plummeted with regional decreases as high as 90%. Spanish overlords forced remaining native people to relocate to small, more easily controlled enclaves, leading to abandonment of many areas. With population levels remaining low in the ensuing 2–3 centuries, long-fallow swidden farming became widespread across the lowlands—and became fixed in the minds of scholars as the normal system of cultivation among the Maya. This perception of the Maya Lowlands as having low population densities, low agricultural productivity, and largely vacant ceremonial centers, became the model of the Maya as an anomalous, mysterious tropical civilization. This model persisted into the 1960s until many studies pursued new questions with a host of new methods that revealed the much more complex nature of Maya Civilization (Dunning and Beach 2004).

23.8 Discussion

The examples considered herein illustrate the variability of landscapes that existed in the Preclassic and Classic Maya Lowlands. This variability was the product of spatial differences in the underlying environment, several environmental changes, and Maya adaptations over time. On a basic level, spatial variation in the adaptive strategies used by the Maya under different environmental conditions is expressed in whether Maya peoples chose to focus agriculture in wetlands or dry uplands, how they dealt with the annual problem of dry season water availability, and how diverse and intrinsically fertile soils were across the broader landscape.

Over time, the Maya devised many successful site-specific adaptations to fulfil their basic needs in the varied landscape. The environment of the Maya Lowlands was, nonetheless, far from stable and Maya communities either adapted or failed to adapt as circumstances changed. From the perspective of landscape archaeology, the overall trajectory of ancient Maya Civilization is perhaps best represented by a metastable equilibrium model; that is, periods of relative stability punctuated by episodes of significant change. This model is aptly applied to many ancient civilizations where environmental and cultural variables were often delicately balanced and even minor perturbations could have catastrophic results (Butzer 1980). To understand the history of Maya Civilization, both periods of stability and instability must be considered (Dunning and Beach 2000).

Instability was introduced by both natural and human factors. Although the Maya adapted successfully to the generally predictable annual wet-dry (seasonal) rhythm of the lowland climate, longer-term fluctuations such as episodic intensification of the frequency and intensity of drought posed a serious threat to their adaptive strategies. These episodes, plus long-term regional drying trends and sea-level rises introduced both temporary and long-term hydrological changes that sometimes led to major regional abandonment such as in the Mirador Basin, but also to successful adaptations, such as near Blue Creek.

As we study ancient human–environment interactions it is important to bear in mind that the physical evidence for environmental change generated by natural fluctuations and by human activities is often indistinguishable in the paleoenvironmental record (Rosenmeier et al. 2002; Dunning et al. 2006; Beach et al. 2008, 2009; Yeager and Hodell 2009). For example, both soil erosion rates and hydrology are affected by changes in climate or human activity and leave similar signals of disturbance. Similarly, some drought signals, such as palynological evidence of forest loss in the paleoenvironmental record, may result from climate change, from anthropogenic deforestation or a combination of these factors (Yeager and Hodell 2009).

Humans also introduced instability into the regional landscape dynamics. Most simply, population growth over time often necessitated greater amounts of agricultural land, thus depleting forest resources and disturbing the soil cover. In many parts of the world, deforestation has significantly diminished the resilience of human-environment systems (Chew 2007), a problem that was repeated at different times and places in the course of Maya Civilization. As mentioned earlier, the relationship between human population densities and soil erosion, or other forms of land degradation, is not always straightforward and linear. Indeed, there is compelling evidence from the Petexbatun region and the Central Peten Lakes that relatively small numbers of Preclassic farmers generated the greatest amount of soil loss as a consequence of initial forest clearance, when low population densities perhaps created the perception of an inexhaustible supply of land (Dunning and Beach 2000; Anselmetti et al. 2007; Beach et al. 2008). In some instances, disturbances related to Maya activities degraded, but at the same time improved local environments. For example, soil erosion generated by Preclassic agriculture may have contributed to rapid sediment accumulation in some bajos and consequent hydrologic alteration. Although this may have destroyed important water sources, the alluvial/colluvial aprons of base-rich sediments generated by the upslope erosion developed into prime farmland along the margins of these same bajos (Dunning et al. 2002). In time, the Maya also became conscious of soil erosion. They developed soil conservation technologies and created more stable agricultural landscapes in many areas during the Classic period.

Social or cultural factors could also have introduced instability. A prime example was the enormous demand for lime plaster engendered by the drive to construct ever larger and more elaborate temples and palaces, such as occurred in the Mirador Basin in the Preclassic and the consequent wholesale destruction of forest resources and land surfaces (Hansen et al. 2002; Wahl et al. 2006). In the Late Classic, the apparent proliferation of royal heirs may have exacerbated growing instability resulting from population growth as dynasties fragmented and fought with one another to control labor and land (Demarest 2005; Webster 2002). In the Petexbatun region this social problem was characterized by a century of escalating warfare, and the development of a fortified landscape of fear, preceding regional abandonment.

Today, deforestation in many parts of the Maya Lowlands is proceeding at a rapid pace. Many modern Maya agricultural colonists have moved into places like the Petexbatun region and face a daunting challenge in developing a stable and sustainable landscape -lack of clear or irrevocable land title. Because many of the colonists do not hold secure title to their land and survive largely on a harvest-to-harvest basis, their commitment to land stewardship is minimal. Likewise, in northern Belize, industrialized farming, albeit with secure land tenure, is bulldozing ancient Maya archaeological sites and leaving farmland unprotected in the face of torrential wet season rains. Soil loss in both areas is often horrendous, with whole hillsides denuded of their soil cover within a few years (Beach and Dunning 1995; Beach et al. 2006, 2008). The unfortunate irony is that as forests are cut down and slopes bulldozed across the region, relict agricultural terrace systems are revealed—terraces that had been adaptations to soil erosion and could, under more favorable circumstances, be reconstructed to serve today's farmers.

Acknowledgements We would like to thank the numerous colleagues and students who collaborated in our research and in particular Drs. Sheryl Luzzadder-Beach and John G. Jones. Funding was provided by grants from the National Science Foundation, National Geographic Society, and other agencies.

References

Anselmetti FS, Hodell DA, Aritegui D, et al. (2007) Quantification of Soil Erosion Rates related to Ancient Maya Deforestation. Geology 35:915–918

Beach T (1998) Soil constraints on northwest Yucatan: pedoarchaeology and Maya subsistence at Chunchucmil. Geoarchaeology 13:759–791

Beach T, Dunning N (1995) Ancient Maya terracing and modern conservation in the Peten rainforest of Guatemala. Journal of Soil and Water Conservation 50:138–145

Beach T, Dunning NP, Luzzadder-Beach S, et al. (2006) Ancient Maya impacts on soils and soil erosion. Catena 65:166–178

Beach T, Luzzadder-Beach S, Dunning N, Cook D (2008) Human and natural impacts on fluvial and karst systems in the Maya Lowlands. Geomorphology 101:301–331

Beach T, Luzzadder-Beach S, Dunning N, et al. (2009) A review of human and natural changes in Maya Lowlands wetlands over the Holocene. Quaternary Science Reviews 28:1710–1724

Brenner M, Rosenmeier MF, Hodell DA, Curtis J (2002) Paleolimnology of the Maya Lowlands. Ancient Mesoamerica 13:41–157

Butzer KW (1980) Civilizations: organisms or systems? American Scientist 68:517–523

Carmean K, Dunning N, Kowlaski JK (2004) High times in the hill country: a perspective from the Terminal Classic Puuc region. In: Demarest AA, Rice PM, Rice DS (eds) The Terminal Classic in the Maya Lowlands: collapse, transition, and transformation. University of Colorado Press, Boulder, pp 424–449

Chase AF, Chase DZ (1998) Scale and intensity in Classic Maya agriculture: terracing and agriculture in the "garden city" of Caracol, Belize. Culture and Agriculture 20:60–77

Chew SC (2007) The recurring dark ages: ecological stress, climate changes, and system transformation. AltaMira Press, Walnut Creek, CA, 314 pp

Dahlin BH (2000) The barricade and abandonment of Chunchucmil: implications for northern Maya warfare. Latin American Antiquity 11:283–298

Dahlin BH, Beach T, Luzzadder-Beach S, et al. (2005) Reconstructing agricultural self-sufficiency at Chunchucmil, Yucatan, Mexico. Ancient Mesoamerica 16:229–247

Dahlin BH, Jensen C, Terry R, et al. (2007) In search of an ancient Maya market. Latin American Antiquity 18:363–385

Deevey ES, Rice DS, Rice PM, et al. (1979) Mayan urbanism: impact on a tropical karst environment. Science 206:298–306

Demarest AA (2004) After the maelstrom: collapse of the Classic Maya kingdoms and the Terminal Classic western Peten. In: Demarest AA, Rice PM, Rice DS (eds) The Terminal Classic in the Maya Lowlands: collapse, transition, and transformation. University of Colorado Press, Boulder, pp 102–124

Demarest AA (2005) Ancient Maya: the rise and fall of a rainforest civilization. Cambridge University Press, Cambridge, 390 pp

Dunning NP (1992) Lords of the hills: ancient Maya settlement in the Puuc region, Yucatan, Mexico. Monographs in World Prehistory No. 15. Prehistory Press, Madison, WI, 318 pp

Dunning NP (2004) Down on the farm: Classic Maya houselots as farmsteads. In: Lohse J, Valdez F (eds) Ancient Maya Commoners. University of Texas Press, Austin, pp 96–116

Dunning NP, Beach T (2000) Stability and instability in prehispanic Maya landscapes. In: Lentz D (ed), An imperfect balance: landscape transformations in the precolumbian Americas. Columbia University Press, New York, pp 179–202

Dunning NP, Beach T (2004) Noxious or nurturing nature? Maya Civilization in environmental context. In: Golden, CW and Borgstede G. (eds) Continuities and changes in Maya archaeology: perspectives at the Millenium. Routledge, New York, pp 125–141

Dunning NP, Beach T, Luzzadder-Beach S (2006) Environmental variability among *bajos* in the Southern Maya Lowlands and its implications for ancient Maya civilization and archaeology. In: Lucero L Fash B (eds), Precolumbian Water Management. University of Arizona Press, Tempe, pp 81–99

Dunning NP, Beach T, Farrell P, Luzzadder-Beach S (1998) Prehispanic agrosystems and adaptive regions in the Maya Lowlands. Culture and Agriculture 20:87–100

Dunning NP, Luzzadder-Beach S, Beach T, et al. (2002) Arising from the bajos: the evolution of a Neotropical landscape and the rise of Maya Civilization. Annals of the Association of American Geographers 92:267–282

Dunning NP, Rue D, Beach T, et al. (1998) Human - environment interactions in a tropical watershed: the paleoecology of Laguna Tamarindito, Guatemala. Journal of Field Archaeology 25:139–151

Erickson C (2006) Intensification, political economy, and the farming community; in defense of a bottom-up perspective of the past. In: Marcus J, Stanish C (eds) Agricultural Strategies. Cotsen Institute, Los Angeles, pp 233–265

Fedick S, Morrison B (2004) Ancient use and manipulation of landscape in the Yalahau region of the northern Maya lowlands. Agriculture and Human 21:207–219

Folan WJ, Fletcher LA, Kintz, ER (1983) Coba: a Classic Maya metropolis. Academic Press, New York, 253 pp

Garrison TG, Dunning NP (2009) Settlement, environment, and politics in the San Bartolo – Xultun territory, El Peten, Guatemala. Latin American Antiquity 20:525–552

Gill RB (2000) The great Maya droughts. University of New Mexico Press, Albuquerque, 464 pp

Greller A (2000) Vegetation and floristic regions of North and Central America. In: Lentz D (ed), An imperfect balance: landscape transformations in the precolumbian Americas. Columbia University Press. New York, pp 39–88

Guderjan TH (2007) The nature of an ancient Maya city: resources, interaction and power at Blue Creek, Belize. University of Alabama Press, Tuscaloosa, 244 pp

Hageman J, Lohse J (2003) Heterarchy, corporate groups, and Late Classic resource management in northwestern Belize. In: Scarborough V, Vadez F, Dunning N (eds), Heterarchy, political economy, and the ancient Maya: the Three Rivers Region of the east-central Yucatan Peninsula. University of Arizona Press, Tucson, pp 109–121

Hammond N, Tourtellot G III (2004) Out with a whimper: La Milpa in the Terminal Classic. In: Demarest AA, Rice PM, Rice DS (eds) The Terminal Classic in the Maya Lowlands: collapse, transition, and transformation. University of Colorado Press, Boulder, pp 288–301

Hansen RD, Bosarth S, Jacob J, et al. (2002) Climatic and environmental variability in the rise of Maya Civilization: a preliminary perspective from the northern Peten. Ancient Mesoamerica 13:273–295

Harrison PD (1977) The rise of the bajos and the fall of the Maya. In: Hammond, N (ed) Social process in Maya prehistory. Academic Press, New York, pp 469–508

Haug GH, Gunther D, Peterson LC, et al. (2003) Climate and the collapse of Maya Civilization. Science 299;1731–735

Hodell DA, Anselmetti FS, Ariztegui D, et al. (2008) An 85-ka record of climate change in lowland Central America. Quaternary Science Reviews 27:1152–1165

Hodell DA, Brenner M, Curtis JH (2005) Terminal Classic drought in the Northern Maya Lowlands inferred from multiple sediment cores in Lake Chichancanab (Mexico). Quaternary Science Review 24:1413–1427

Hodell DA, Brenner M, Curtis JH, Guilderson T (2001) Solar forcing of drought frequency in the Maya Lowlands. Science 292:1367–1370

Houston SD (1993) Hieroglyphs and history at Dos Pilas: dynastic politics of the Classic Maya. University of Texas Press, Austin, 181 pp

Hutson S, Hixson D, Magnoni D, Mazeau A, Dahlin B (2008) Site and community at Chunchucmil and ancient Maya urban centers. Journal of Field Archaeology 33:19–40

Larsen LB, Vinther BM, Briffa KR, Melvin TM, et al. (2008) New ice core evidence for a volcanic cause of the A.D. 536 dust veil. Geophysical Research Letters 35: L06102

Lentz DL, Hockday B (2009) Tikal temples and timbers: ancient Maya agroforestry and the end of time. Journal of Archaeological Science 36:1342–1353

Leyden BW, Brenner M, Dahlin B (1998) Cultural and climatic history of Coba, a lowland Maya city in Quintana Roo, Mexico. Quaternary Research 49:111–122

Luzzadder-Beach S, Beach T (2009) Arising from the wetlands: mechanisms and chronology of landscape aggradation in the northern coastal plain of Belize. Annals of the Association of American Geographers 98:1–25

Martin S, Grube N (2008) Chronicle of the Maya kings and queens: deciphering the dynasties of the ancient Maya. Second edition. Thames and Hudson, London and New York, 240 pp

McAnany P (1995) Living with the ancestors: kingship and kinship in ancient Maya society. University of Texas Press, Austin

Mueller AD, Islebe G, Hillesheim MB, et al. (2008). Climatic drying and associated forest decline in the lowlands of northern Guatemala during the late Holocene. Quaternary Research 71:133–141

Nalda E (2005) Kohunlich and Dzibanché: parallel histories. In: Shaw JM, Mathews JP (eds), Quintana Roo archaeology. University of Arizona Press, Tucson, pp 228–244

Normark J (2006) The roads in-between: causeways and polyagentive networks at Ichmul and Yo'okop, Cochuah region, Mexico. Göteborg University, Göteborg, 342 pp

O'Mansky M, Dunning NP (2004) Settlement and Late Classic political disintegration in the Petexbatun Region, Guatemala. In: Demarest AA, Rice PM, Rice DS (eds) The Terminal Classic in the Maya Lowlands: collapse, transition, and transformation. University of Colorado Press, Boulder, pp 83–101

Pérez-Salicrup D (2004) Forest types and their implications. In: Turner BL II, Geoghagen J, Foster DR (eds), Integrated land-change science and tropical deforestation in the Southern Yucatan. Oxford University Press, London and New York, pp 63–80

Prem HJ, Dunning NP (2004) Investigations at Hunto Chac, Yucatan. Mexicon 26:26–36

Rosenmeier MF, Hodell DA, Brenner M, Curtis J, et al. (2002) A 4000-year lacustrine record of environmental change in the southern Maya Lowlands, Peten, Guatemala. Quaternary Research 57:183–190

Scarborough VL, Becher ME, Baker JL, et al. (1995) Water and land at the ancient Maya community of La Milpa, Belize. Latin American Antiquity 6:98–119

Scarborough VL, Valdez F (2003) The engineered environment and political economy of the Three Rivers region. In: Scarborough VL, Valdez F, Dunning N (eds) Heterarchy, political economy and the ancient Maya. University of Texas Press, Austin, pp 1–13

Sedov S, Solleiro-Rebolledo E, Fedick S, et al. (2008) Micromorphology of a soil catena in Yucatán: pedogenesis and geomorphological processes in a tropical karst landscape. In: Kapur S, Mermut A, Stoops G (eds) New trends in soil micromorphology. Springer Berlin Heidelberg, pp 19–37

Stephens JL (1843) Incidents of travel in Yucatan. 2 volumes. Harper and Rowe, New York, 308 pp

Tourtellot G, Estrada-Belli F, Rose JJ, Hammond N (2003) Late Classic heterarchy, hierarchy, and landscapes at La Milpa, Belize. In: Scarborough V, Vadez F, Dunning N (eds) Heterarchy, political economy, and the ancient Maya: the Three Rivers Region of the east-central Yucatan Peninsula. University of Arizona Press, Tucson, pp 37–51

Tourtellot G, Sabloff J (1994) Community structure on Sayil: a case study of Puuc settlement. In: Prem HJ (ed), Hidden among the hills: Maya archaeology of the northwest Yucatan Peninsula, Acta Mesoamerciana 7. Verlag von Flemming, Mockmuhl, pp 71–92

Wahl D, Byrne R, Schreiner T, Hansen R (2006) Holocene vegetation change in the northern Peten and its implications for Maya prehistory. Quaternary Research 65:380–389

Wauchope R (1965) They found the buried cities: exploration and excavation in the American Tropics. University of Chicago Press, Chicago, 382 pp

Webster DL (2002) The fall of the ancient Maya: solving the mystery of the Maya Collapse. Thames and Hudson, London and New York, 368 pp

Wilson EM (1980) The physical geography of the Yucatan Peninsula. In: Mosely E, Terry E (eds), Yucatan: a world apart. University of Alabama Press, Tuscaloosa, pp 5–40

Wright LE (2006) Diet, health, and status among the Pasion Maya: a reappraisal of the Collapse. Vanderbilt University Press, Nashville, 256 pp

Yeager J, Hodell D (2009) The collapse of Maya civilization: assessing the interaction of culture, climate, and environment. In: Sandweiss, DH, Quilter J (eds), El Niño, catastrophism, and culture change in ancient America. Harvard University Press, Cambridge MA, pp 187–242

Chapter 24
Water Follows the People: Analysis of Water Use in the Western Great Plains and Rocky Mountains of Colorado, USA

Ellen Wohl

24.1 The Drylands of Colorado: Regional Setting

The State of Colorado in the western United States has a semiarid climate (Fig. 24.1). This single fact dominates physical and ecological processes in the region, as well as the historical subsistence patterns of indigenous peoples. Rates of bedrock weathering are slow and soils are thin, with limited organic horizons. Vegetation cover is discontinuous in many parts of the state, varying from semiarid steppe characterized by bunch grasses on the eastern plains, through montane pine forests at the middle and lower elevations in the mountains and subalpine spruce and fir forests at the higher elevations, to semiarid steppe and desert shrublands on the western plateaus. Plant adaptations to aridity include reduced stomata and small leaves that may have a waxy or hairy texture to limit evaporation of moisture by the wind, a large percentage of total biomass in roots (between 60% and 80% of the plant's weight in grasses), and short stature and wide spacing (Savage 2004). Native animal species also adjust to aridity, relying on physiological adaptations or behavioural adaptations such as the migration practiced by native herbivores including bison (*Bison bison*). Indigenous people living in the region at the time of the first written historical descriptions were primarily hunter-gatherers who also practiced seasonal migration, with temporary settlements clustered along water courses (Holliday 1987), as they followed their primary food sources.

The adaptations of native plants and animals and indigenous peoples reflect the limitations imposed by lack of water. Paleoclimatic records indicate that this portion of the western United States has been arid and semiarid for approximately 60 million years, since uplift of the Rocky Mountains created a rain shadow across the western Great Plains for moisture-bearing air masses coming inland from the Pacific Ocean. Similarly, uplift of the Coast Ranges and the numerous north-south-trending ranges of the Basin and Range province fostered aridity of the lower elevation lands between the Pacific coast and the Rocky Mountains.

Precipitation across much of the lower elevation lands immediately east and west of the Rocky Mountains is sparse and highly unpredictable, with substantial inter-annual variability in yearly precipitation. Although snow falls on these regions during winter, the combination of low relative humidity, high percentage of days with sunshine, and frequent winds causes much of the snow to sublimate rather than melt and infiltrate the soil. Streams originating at these lower elevations are likely to be ephemeral, with brief intervals of surface flow following snowfall or summer convective storms that generate intense precipitation and flash floods. Perennial streams will likely have larger channels that head in the adjacent mountains.

The Rocky Mountains serve as a water tower supplying stream flow generated primarily by snowmelt runoff to the adjacent lowlands. The mountains stand 2000–2700 m above the adjacent lowlands, and are thus effective in forcing orographic precipitation from air masses carrying moisture eastward from the Pacific Ocean or northwestward from the Gulf of Mexico in the Atlantic Ocean. The headwaters of the Rio Grande, the Colorado, the South Platte, and the Arkansas rivers all rise in the Colorado Rocky Mountains, and

E. Wohl (✉)
Department of Geosciences, Colorado State University
Fort Collins, Colorado 80523-1482, USA
e-mail: ellenw@warnercnr.colostate.edu

See Plate 22a in the Color Plate Section; also available at: extras.springer.com

I. P. Martini, W. Chesworth (eds.), *Landscapes and Societies*,
DOI 10.1007/978-90-481-9413-1_24,

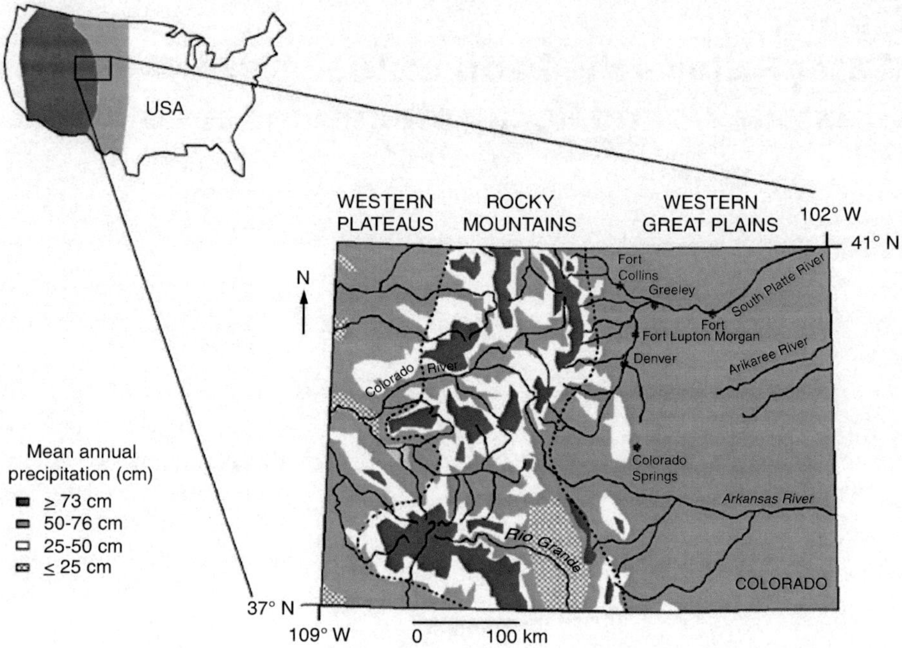

Fig. 24.1 Location map of Colorado. The three primary physiographic/ecoregions are labeled at the *top* of the figure, with *dashed lines* showing their boundaries. Mean annual precipitation is shown by *shading*. Principal rivers, and towns or cities mentioned in the text, are labelled. Inset map of continental USA shows the Rocky Mountains (*dark shading*) and western Great Plains (*light shading*)

each of these rivers provides a major water supply for irrigated agriculture, hydroelectric power generation, and municipal and industrial uses in Colorado and adjacent states. Runoff and groundwater flow from the mountains also supply the major aquifers that underlie the adjacent western and eastern lowlands, particularly the Ogallala aquifer of the western Great Plains (Fig. 24.2). These aquifers lie at varying depths beneath the surface, and may or may not be connected to surface streams.

For the most part, there is very little additional contribution of discharge once the larger rivers flow beyond the mountain front. The first written descriptions of these rivers emphasized the highly seasonal flow regime in which large annual peak flows caused by snowmelt runoff subsided to very low base flows during the remainder of the year; the magnitude ratio of peak flow to base flow commonly exceeded ten.

People of European descent began to settle across the plains and mountains of Colorado during the nineteenth century. They had a saying that 'rain follows the plow', which reflected the commonly held belief that settlement and the planting of crops would promote climate change, facilitating agriculture in the drylands of the central USA. Rain did not follow the plow, and many of the small agricultural communities on the Great Plains were abandoned during droughts such as the 1930s Dust Bowl. Farmers in some por-

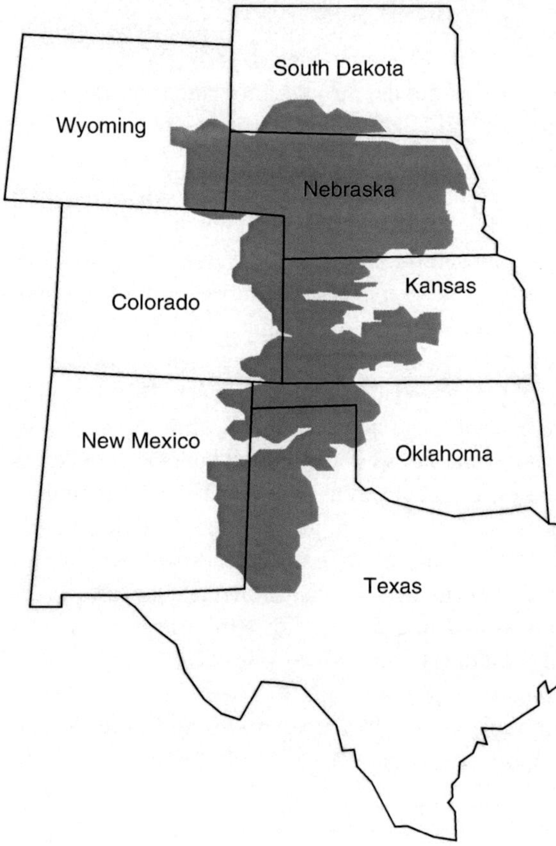

Fig. 24.2 Location of the Ogallala aquifer (*shaded*) in the central and western USA

tions of the Great Plains did create a situation where water followed people, however, by manipulating surface and ground water supplies for irrigation. Rather than adapting to regional aridity as indigenous peoples had done, nineteenth-century European settlers in Colorado immediately began to engineer water supplies in order to maintain agricultural and urban patterns developed in the wetter regions from which these people emigrated. The remainder of this chapter briefly reviews the history of these modifications to regional hydrologic conditions, as well as some of the unintended consequences of these modifications. The chapter concludes with an overview of the contemporary situation at the start of the twenty-first century, and likely future trends of resource use and human settlement.

Fig. 24.3 Late nineteenth-century view of placer mining in Illinois Gulch. Water is piped to the site and mixed with sediment along the length of the wooden rocker box in the center foreground to separate out placer metals. (Photograph courtesy of the Colorado Historical Society)

24.2 The Water Rush

Historians refer to the gold rush—actually a series of gold rushes spread in time and space across the western United States—but it is equally appropriate to describe the nineteenth century history of Colorado as a water rush. The first people of European descent to settle in Colorado were mostly fur trappers seeking new sources of beaver fur as over-trapped beaver populations in the eastern half of North America declined. Starting in the 1820s, individuals founded a few small settlements along major rivers, such as Fort Lupton, but human population densities remained very low (<30,000 people throughout the state until 1859) and water supplies were not altered to meet local demands. This situation changed when miners returning east from the California gold fields discovered placer gold along Cherry Creek, near the present location of Denver, in 1859. Subsequent discoveries of various placer metals at several locations throughout the Colorado Rocky Mountains triggered multiple gold rushes during the period of 1860–1890.

Miners and those who supplied their infrastructure—entrepreneurs, merchants, road- and railroad-builders—moved quickly to each newly found mineral deposit. Human population at such a site might increase from zero to several thousand people within a couple of months (Wohl 2001). Placer miners, in particular, relied on dependable water supplies to work their mining claims. Placer metals are disseminated within sediments deposited by streams or glaciers across the valley bottom. The most commonly used nineteenth-century techniques for concentrating the metals relied on physical separation facilitated by water such that the heavier placer metals would settle out when sediment and water were mixed in gold pans, rocker boxes, or wooden flumes (Fig. 24.3).

Miners working the Colorado deposits relied on legal frameworks developed in California, which are summarized in the doctrine of prior appropriation. Sometimes paraphrased as ‘first in time, first in right’, prior appropriation states that the first legal claimant to a water supply is guaranteed a fixed amount of water relative to subsequent claimants. The legal standing of earlier claimants was particularly critical in preventing a miner who subsequently claimed an upstream mineral deposit from diverting so much water from the river that the downstream miner became unable to work his claim. Prior appropriation also facilitates the legal transfer of water rights through buying and selling, and the physical transfer of water through diversions. Miners physically moved water from existing stream channels and lakes to placer deposits that did not have a water supply, and thereby initiated the intensive engineering of surface water supplies that now characterizes the drylands of the western USA. Nineteenth-century mining in Colorado extended nearly the length of the Rocky Mountains, from the northern headwaters of the South Platte River to the headwaters of the Rio Grande River (Fig. 24.4).

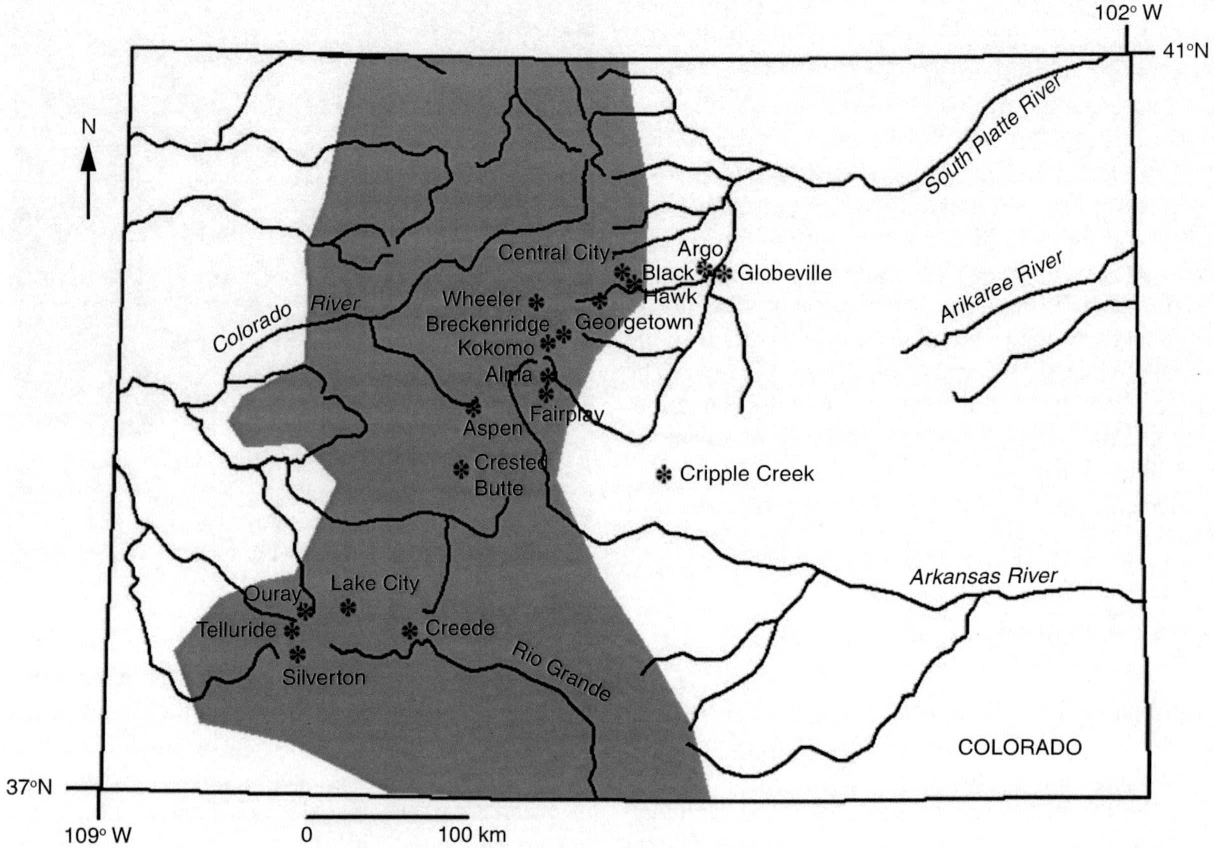

Fig. 24.4 Principal nineteenth-century mining sites in Colorado. Shaded area indicates the extent of the Rocky Mountains

Shortly after the first gold rush in 1859, farmers began to settle along the eastern base of the Rocky Mountains, as well as in the mountains (Fig. 24.5). Unable to support crop varieties imported from the wetter eastern and central United States, individual farmers dug small irrigation ditches from the nearest river. Some of the earliest ditches date to 1859, followed by many more during the next few years. As settlement density increased, particularly at the base of the mountains and across the plains, the availability of surface water quickly became a limiting factor. Farmers formed groups and irrigation districts and constructed larger, communally-owned diversion canals starting in the 1870s (Eschner et al. 1983). The establishment of the Fort Collins Agricultural Colony upstream along the Poudre River from Greeley in 1872, for example, led to the construction of two irrigation canals that had the combined capacity to divert the entire flow of the Poudre River during low-flow years. The dry summer of 1874 caused irrigators in Greeley a great deal of anxiety and ultimately led to the incorporation of prior appropriation in the 1876 Colorado Constitution as a means of preventing a junior claimant (in this case, Fort Collins) from diverting the water supply of a downstream senior claimant (Wohl 2001).

Because prior appropriation facilitates physical transfer of water, irrigators quickly began to store water and to divert it across drainage divides. Dams tended to be built in the mountains first, as illustrated by reservoirs built during the 1860s–1880s that served as the starting point for an agricultural supply ditch (Gerlek 1977). Irrigators used transbasin water diversions to supplement stored water supplies starting in the 1890s (Eschner et al. 1983). The Cameron Pass Ditch (1882), Grand Ditch (1892), and Skyline Ditch (1893) were early diversions across the Continental Divide in Colorado, taking water from the headwaters of the Colorado River on the western slope of the Rocky Mountains and diverting it to the headwaters of

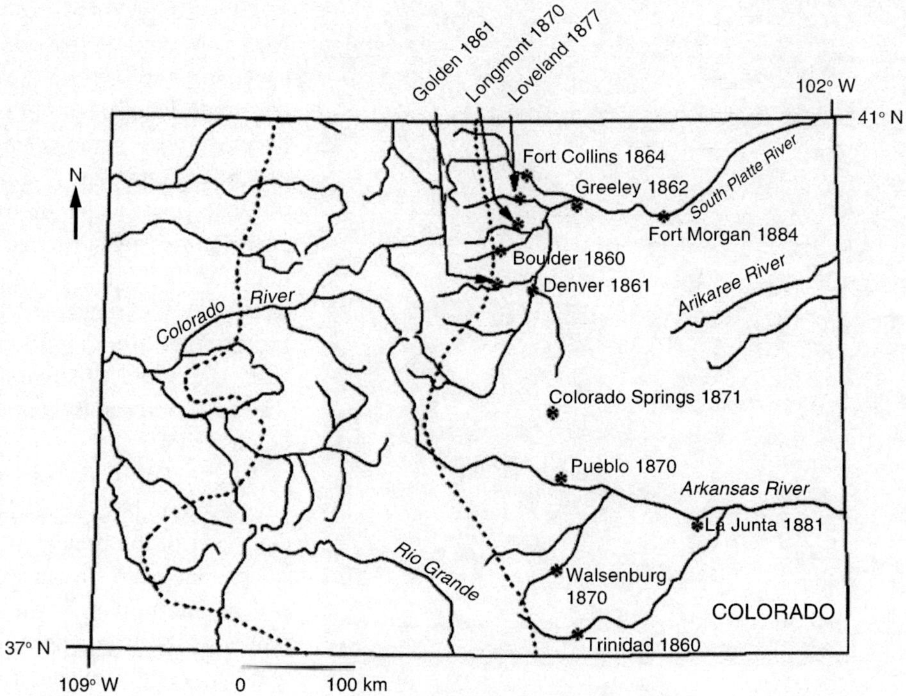

Fig. 24.5 State map of Colorado showing major rivers, boundaries between plains, mountains, and plateaus (*dashed lines*), and principal cities of the plains, with date of establishment

the South Platte River on the eastern slope. Groundwater pumping began during the 1890s as yet another supplement to water supplies on the western plains (Eschner et al. 1983).

Increased availability of water stimulated agricultural growth, which then required additional water (Wohl 2001). Approximately 180,700 ha were being farmed by 1899 in the five counties that form the northern plains portion of the South Platte River basin. This represented an increase of 85% over the previous decade. By 1909 irrigated acreage in the region increased an additional 68% to 303,300 ha. The region's population grew nearly 58% during the first decade of the twentieth century, while the population of the entire state grew from 800,000 in 1910 to 940,000 in 1920.

Twentieth-century water engineering on the Colorado portion of the western Great Plains culminated in the Colorado-Big Thompson Project of 1938–1956 (Tyler 1992). Created to supplement the water supply for 249,000 ha of irrigated land in northeastern Colorado, the project diverts water from the Colorado River basin into the South Platte basin. Through this and other projects, water users in eastern Colorado collectively remove nearly 802 million m^3 of water from the western side of the Rocky Mountains via 37 transmountain diversions. Out-of-basin imports now account for an average of one-third of the surface runoff in the upper South Platte River basin, and 94% of this imported water comes from the Colorado River (Wohl 2001).

24.3 Consequences of the Water Rush: Physical and Ecological Changes in Rivers

River ecosystems on the eastern side of the Colorado Rocky Mountains have been substantially altered by changes in flow regime, sediment supply, and channel configuration associated with nineteenth- and twentieth-century land uses. Most of the streams on the eastern side of the Rockies drain into the South Platte and Arkansas rivers. Headwater streams in the mountains flow along steep, narrow, forested valleys (Fig. 24.6). Beaver were present in large numbers prior to 1800, and probably helped to retain wood that entered the streams through mass movements or individual tree fall. The cobbles and boulders of the streambed and banks are relatively stable during annual snowmelt peak flows, but channel geometry can be substantially reworked during rare outburst floods from failure of landslide dams or rainfall-generated flash floods. Rainfall-generated floods typically occur at elevations of 2300 m

Fig. 24.6 Upstream view of a mountain stream in the Colorado Rocky Mountains

or lower, and both outburst and rainfall floods have recurrence intervals greater than a century. Suspended sediment loads are typically low. Aquatic organisms of the mountain streams are those that thrive in clear, cold, highly oxygenated water with a strongly seasonal flow regime, stable streambed, and limited access to floodplain habitats. Fish are primarily different species of trout (*Salvelinus*, *Salmo*, *Onchorhynchus* spp.; USDA 1980). Macroinvertebrates are primarily benthic species (Ward 1992). The longitudinal profiles of the mountain streams are segmented as a result of downstream changes in glacial history and geology. Lower-gradient channel segments tend to have broader valley bottoms and wider riparian zones dominated by willow (*Salix* spp.). Steeper channel segments that are more closely confined between valley side slopes have limited riparian zones, but the riparian species river birch (*Betula fontinalis*), alder (*Alnus incana*), aspen (*Populus tremuloides*), chokecherry (*Padus virginiana*), and cottonwood (*Populus angustifolia*) all grow along mountain streams. These were the rivers that early European-Americans described as very wild and beautiful (Fremont 1845).

Continuing downstream, the rivers of the eastern Rocky Mountains pass through a transition zone at approximately 1600 m elevation as they flow through the foothills. Stream gradient decreases, the rivers become more sinuous, more extensive riparian forests of cottonwood (*P. angustifolia*, *P. deltoides*) and other deciduous riparian trees are present, and aquatic communities are particularly diverse as the geographic range of the higher elevation coldwater species overlaps with the range of lower elevation warmwater species that are common on the Great Plains to the east. The bouldery streambeds of the mountains give way to cobbles and then sand and fine gravel further downstream.

Within 50 km of the mountain front, the streams have become largely sand-bed channels. Larger rivers that head in the mountains are perennial, but smaller rivers that head on the plains are typically ephemeral or intermittent if they are individually fed by a groundwater spring. Some channels are broad, shallow braided systems, others are deeply entrenched arroyos (Fig. 24.7). Some of the earliest written accounts of the mainstem South Platte and Arkansas rivers suggest that the downstream portions of these rivers were nearly or completely dry during the late summer and autumn. The large annual snowmelt peak kept the stream channels mobile and largely free of woody riparian vegetation. Native plains fishes had to cope with a highly seasonal flow regime, small size of potential habitats, and physically stressful conditions of late summer (high temperatures, low dissolved oxygen, limited lon-

Fig. 24.7 A view across the plains of eastern Colorado, showing an intermittent prairie stream (South Pawnee Creek) at the base of a bluff. Fencepost in the foreground is 1 m tall. Arrows indicate the flow direction through a wide bend of the river

gitudinal connectivity between pools). They adapted by developing strategies such as relatively small body size (<10 cm), short lifespan (<6 years, averaging 1–2 years), high mobility, adaptability to different habitats and food types in response to changes in what is available, accelerated reproduction and growth during brief periods of higher stream flow, and buoyant eggs (Fausch and Bestgen 1997). These fish survived quite well in the rivers that early European-American settlers described as 'too thick to drink but too thin to plow' and 'a mile wide and an inch deep'.

Within less than a century, the appearance of many of the mountains and plains streams had changed so dramatically that the people who penned the first written descriptions of these channels would have difficulty in recognizing them. The mountain streams were most affected by beaver trapping, placer mining, timber harvesting, tie drives, construction of transportation corridors, and flow regulation. The plains streams were most affected by flow regulation and groundwater withdrawal.

24.3.1 Historical Changes in Mountain Streams

Beaver dams help to slow the passage of flood waves, reduce bed and bank erosion and sediment movement, and increase the diversity of aquatic and riparian habitat (Olson and Hubert 1994; Butler 1995). These functions are lost when the beaver are removed and their dams fall into disrepair, and early nineteenth-century beaver trapping was the first widespread human activity that began to decrease the diversity and stability of mountain streams in the Colorado Rockies (Fremont 1845; Wohl 2001). Placer mining subsequently greatly exacerbated stream instability. Miners disrupted the coarse surface layer that limits sediment transport and channel mobility along many mountain streams. Historical photographs and written descriptions attest to the substantial increases in sediment yield from mined regions (Fig. 24.8); one observer described an unmined portion of Clear Creek as "a noble stream, rapid and clear" (Harrington 1874), whereas water discharged from mining operations created "a thick muddy flow that at times approached the consistency of thin cream" (Wood 1935). Miners also altered natural flow regimes by diverting water from, or to, individual channel segments, and introduced toxic contaminants such as mercury that were used to process placer metals (Wentz 1974).

Fig. 24.8 Aerial view of farmlands at the eastern base of the Colorado Rocky Mountains, taken from a hot-air balloon on 13 August 1894. Multiple streams drain eastward from the mountain front in this view, but only Clear Creek is readily apparent; the pale-gray zones on either side of the creek are recently deposited sediments transported downstream from placer mining districts in the mountains. The view is toward the west. (Photograph courtesy of the Colorado Historical Society)

Timber harvest further altered water and sediment yield to mountain streams. There was a tremendous demand for lumber in newly founded mining communities; wood for buildings, local railroad lines and corduroy roads (logs laid parallel to one another across the road), wood to shore up the mine shafts being blasted and dug into bedrock and to supply the steam that powered stamp mills to crush the ore, and wood for ties along the transcontinental railroad line built across the treeless plains during the 1860s (Wohl 2001). Photographs from the late nineteenth century show large swaths of mountain terrain that have been stripped of trees, and contemporary observers wrote of debris flows and landslides roaring down from the newly bare hillslopes. As farmers settling along the plains at the base of the mountains lost diversion structures buried in the excess sediment carried downstream from the mining districts, they called for limits to timber harvest, but widespread deforestation did not end until the federal government established the first national forest reserves in 1897.

Trees cut for railroad ties were stacked beside the mountain streams during winter and then floated downstream to collection booms at the base of the mountains during tie drives (Wroten 1956). Overbank areas were blocked off and natural obstructions such as wood jams or large boulders and bedrock outcrops were removed from mountain streams to facilitate the downstream passage of logs, which were often sent down in a large burst when a temporary 'splash dam' full of spring snowmelt runoff and logs was dynamited. Although tie drives usually lasted for only a few years on any given stream, hundreds of thousands of logs could be sent down the stream during this brief period (Schmal and Wesche 1989; Young et al. 1994).

Road and railroad routes in the Colorado Rocky Mountains typically trend east–west and follow the mountain valleys. Construction of these routes further constricted the already-narrow valley bottoms, limiting or removing the riparian zone and increasing sediment yield to the adjacent stream (Balog 1977; Wohl 2001).

Alteration of the natural flow regime affected the ability of mountain streams to adjust to the excess sediment loads associated with activities such as placer mining, deforestation, and road construction. Many headwater streams were partially or completely diverted. Some of this water was sent into canals or pipelines, but some was also routed downstream through larger channels. Individual streams might thus lose or gain discharge. By and large, the magnitude of annual snowmelt peak flow decreased on mountain streams. Where the stream was used to convey water downstream to irrigation diversions, base flow during the latter portion of the growing season increased. Where water was diverted from the channel, both peak and base flows decreased dramatically.

In addition to the physical effects of these human-induced alterations in water and sediment movement and channel configuration, aquatic and riparian species along mountain streams were substantially altered during the nineteenth and twentieth centuries. Habitat diversity and stability, disturbance regime, and the timing and magnitude of the natural flow regime to which many aquatic and riparian species tie their lifecycles were all altered.

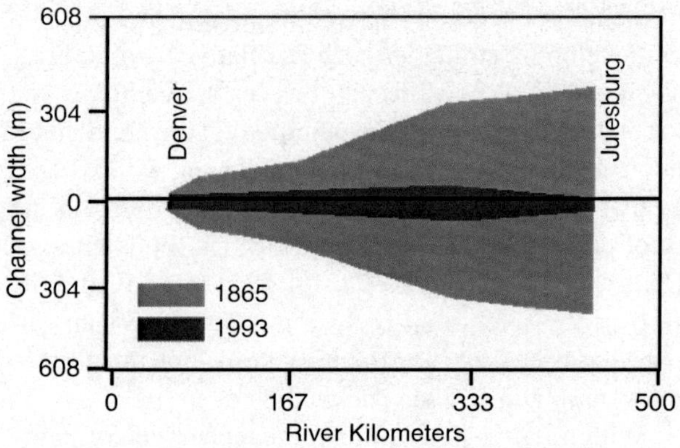

Fig. 24.9 Schematic illustration of downstream trends in active channel width along the South Platte River historically and at present. Julesburg is at the eastern border of Colorado. Most of the change documented here occurred before 1938. (After Dennehy et al. 1998)

24.3.2 Historical Changes in Plains Streams

The larger rivers of the plains, which head in the mountains, have been the most obviously altered by reduced peak flows and increased base flows associated with flow regulation. The loss of erosive annual flood peaks, the steadier supply of stream flow, and the locally increased water tables associated with flood irrigation of bottomland farm fields all helped woody riparian vegetation to establish along rivers such as the South Platte and Arkansas. Historical maps and photographs clearly document the changes from braided river segments that might be 1400 m wide in 1865 to narrow, sinuous, densely vegetated channels only 90 m wide by 1965 (Fig. 24.9; Williams 1978). As sediment transport decreased and clear, cooler water moved through the deeper, narrower, more densely shaded channels, aquatic and terrestrial species began to migrate westward along the river corridors from wetter environments to the east, increasing competition for already stressed native species.

The stresses of altered habitat and competition are reflected in the status of many native fish species. Six of 38 native plains fish species in Colorado are known to have been lost since the first fish collections were made in the late nineteenth century. An additional 13 species are listed by the State as endangered, threatened, or of special concern. Half of the native plains fish species have thus either declined or been lost (Propst and Carlson 1986; Cross and Moss 1987; Peters and Schainost 2005). Many of these are small-bodied fish inhabiting intermittent streams, although larger fish such as pallid sturgeon (*Scaphirhynchus albus*) in the perennial mainstem rivers also struggle to survive.

Migratory birds have also been affected by historical changes in the major rivers of the western Great

Plains, as exemplified by the central Platte River in Nebraska. The central Platte River lies below the Central Flyway, a primary north-south corridor for migratory birds. Islands, sandbars, and floodplain wetlands provide important habitat for migratory and resident birds, including three federally listed species; the whooping crane (*Grus americana*), the piping plover (*Charadrius melodus*), and the interior least tern (*Sterna antillarum athalassos*) (Graf 2005).

Critical whooping crane habitat along the central Platte River includes foraging areas of shallow water that are largely devoid of vegetation greater than a meter in height and have open horizons. At these sites the cranes eat the invertebrates, amphibians, reptiles, and tubers of marsh plants that provide the birds with the energy to continue their long and strenuous migration. The lack of tall vegetation and open vistas allows the cranes to avoid predators while eating, and if the cranes are too crowded, they become more vulnerable to disease and to predators. Roosting sites surrounded by deeper waters that limit predation from land are also critical to migrating cranes (Graf 2005).

Crane and other bird populations have declined from historical levels as a result of loss of open, sandy nesting habitat in and along rivers. This habitat loss reflects inundation by reservoirs, channelization, large-scale changes in flow regime, and replacement of open areas with woodlands, sand and gravel mines, housing, and roadways (Graf 2005).

Historical changes to the smaller intermittent streams that head on the plains have been less well documented, although a few recent studies that focus on native plains fish indicate how severely many species are being stressed by changes in these smaller streams. Intermittent streams depend on groundwater springs that may be fed by a large regional aquifer such as the Ogallala, or a local, shallow aquifer that largely supplies water during spring and early summer.

The heterogeneous sequence of clays, silts, sands and gravels that forms the Ogallala aquifer extends beneath 370,000 km^2 of the western Great Plains, with the top of the sediments at depths of 30–120 m below the surface (Gutentag et al. 1984). Groundwater pumping for irrigated agriculture at rates that exceed natural recharge has caused extensive declines in the aquifer, particularly in the middle and southern portions of the western Great Plains. The Ogallala watertable has dropped more than 8 m across approximately 5200 km^2 of eastern Colorado.

24.4 The Contemporary Situation in the Drylands of Colorado

Choices about resource use during the nineteenth and twentieth centuries continue to influence twenty-first century landscapes and resource use in the drylands of Colorado. Water use governed by prior appropriation, established during the nineteenth century, today facilitates transfer of agricultural lands to urban uses, as well as continued rapid population growth. Water shortages, however, trigger legal challenges between municipalities and cause court-ordered shutdowns in irrigated farming. Declining water quality results from intensive irrigated farming and densely populated urban areas, both facilitated by patterns of water use and transfer developed during the nineteenth century. Poor water quality alters aquatic communities, destroys the productivity of agricultural lands, and creates health hazards for human communities.

The Front Range urban corridor, which consists of a nearly contiguous urban area stretching north from Colorado Springs to Fort Collins, has one of the fastest rates of population growth in the United States at the start of the twenty-first century. The US population increased by 13% between 1990 and 2000, while the population of the Front Range urban corridor increased by 31%. The Colorado State Demographer's Office projects an additional 63% increase by 2030. As of 2000, approximately 10% of the state's population lived on the western Great Plains, 10% on the lowlands west of the Rockies, and about 78% in the Front Range urban corridor.

As of 1999, an estimated 36,000 ha of farm and ranch land were being converted to urban areas in Colorado each year. Patterns of water use change as formerly agricultural land is converted to urban uses along the Front Range urban corridor. In 1987, agriculture accounted for 85% of out-of-stream water use in Colorado. Because prior appropriation facilitates purchase and transfer of water rights that are separated from acquisition of riparian lands, municipalities within the Front Range urban corridor have acquired at least some of their water supplies by purchasing water rights from agricultural irrigators, as well as purchasing agricultural lands outright. Developers and residents demonstrated a willingness to pay high prices for water, and the value of water traded from agricultural to urban lands in Colorado doubled from 1989 to 1991 (Strange et al. 1999). Much of this water came from adjacent drainage basins; nearly a quarter of the South Platte's flow came from the western side of the Continental Divide. At the start of this century, 15 interbasin diversions brought water into the South Platte system, where a thousand reservoirs and 500 irrigation ditches stored and distributed the water.

Despite all of these imports, water remains scarce. During the last twenty years of the twentieth century, water levels in the major bedrock aquifer underlying the Denver Basin dropped as much as 240 m and some domestic wells in the southern part of the metropolitan area went dry (Strange et al. 1999). Public awareness of water scarcity contributes to local governmental initiatives to conserve water through mandatory (installation of household water meters, increased water costs) and voluntary (appliances that use less water, use of drought-tolerant plants in landscaping) programs.

Population in Colorado has been disproportionately concentrated along the eastern base of the mountains for more than a century, and water engineering has ensured that water follows the people. Rapidly growing small cities in the mountains and on the western slope now increasingly resent and try to prevent the diversion of water from mountain streams to the Front Range urban corridor, even as water demand continues to increase along the eastern slope because of irrigated agriculture and population growth.

In the western Great Plains as a whole, total farmland area increased rapidly from 1900 to 1945, peaked in 1959, declined slightly to 1964, and has since remained stable (Parton et al. 2007). This is not the case, however, in the Colorado Front Range urban corridor, where large population increases between 1950 and 2000 resulted in a 35% reduction in farmland (Parton et al. 2003).

In addition to loss of farmlands to urbanization along the Front Range, farming practices are threatened by water shortages. Five years after a persistent regional drought began in 2002, 4000 of 9000 irrigation wells had stopped pumping along a portion of the South Platte River near Fort Morgan, leaving thousands of hectares of maize and sugar beet fields barren (Smith 2007). The Colorado Supreme Court approved strict new rules requiring well users to return more water to the river or cease pumping, and state officials and farmers quickly began exploring options for more efficient water use and greater water storage. Although contemporary water usage in both agricultural and municipal settings is mostly so inefficient that much less water can be used to support farms and communities, implementing water conservation technologies will be expensive and will likely only occur when water prices rise so high that the improved technologies are viewed as economically beneficial by potential users.

Beyond the obvious shortages and increasing costs of water, twentieth century increases in irrigated agriculture on the western Great Plains have produced several unintended side effects. Crop production has increased 100–300% in different parts of the Great Plains since 1940, primarily as a result of increased irrigation, increased applications of pesticides and fertilizers, and improved tillage practices and plant varieties (Parton et al. 2005, 2007). Water for increased irrigation has come largely from aquifers, causing regional declines in water tables (Gutentag et al. 1984). Expanded wheat and cotton fields have increased losses of soil carbon (Metherell et al. 1995) and fluxes of nitrous oxide from the soil (Mosier et al. 1997). Greater applications of pesticides and fertilizers have caused contamination of surface waters and shallow aquifers (Matson et al. 1997; Rabalais 2002).

Patterns of water quality in the South Platte River basin are particularly well documented as a result of the US Geological Survey's National Water Quality (NAWQA) program during 1992–1995. One of the key findings of the NAWQA study is that local land use and site characteristics have more effect on stream habitat, the integrity of aquatic communities, and water quality, than do upstream or basin-wide patterns of land use (Dennehy et al. 1998). The importance of local effects partly reflects the fact that upstream influences are reduced at the numerous sites within the drainage where much or all of the stream flow is removed from the river by diversions, used for irrigating crops or municipal supplies, and then returned to the river. Many of these dry points where all stream flow is removed are associated with the more than one hundred municipal wastewater treatment plants located within the South Platte River basin. The largest of these are the 25 plants along the Front Range urban corridor from Fort Collins south to Denver, which discharge approximately 94 million L of effluent per day, or about 95% of the total daily effluent discharge in the river basin. One of the problems arising from this pattern of water use is the concentration of nutrients downstream from the wastewater treatment plants. These plants do not effectively remove

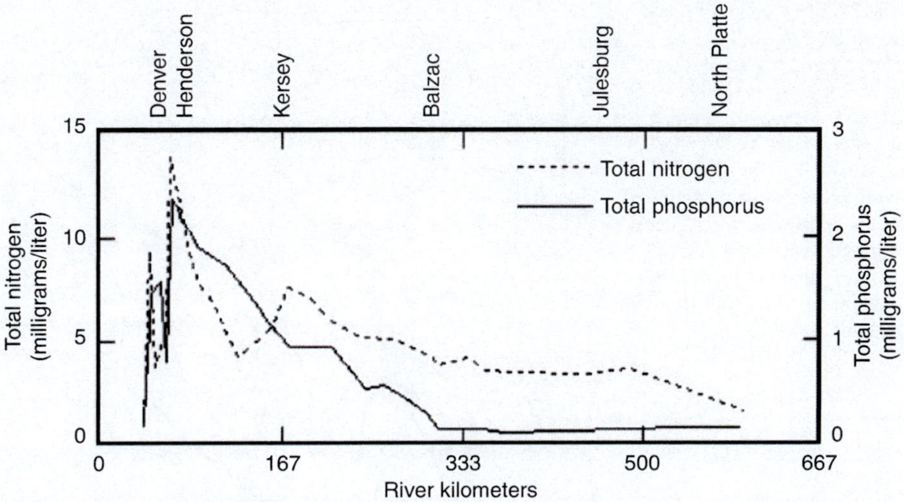

Fig. 24.10 Downstream patterns of nitrogen and phosphorus concentrations along the South Platte River during April 1994, as identified by NAWQA scientists. Concentrations were largest just downstream from the major metropolitan area of Denver and then generally decreased downstream. (After Dennehy et al. 1998)

nitrogen and phosphorus from water that is returned to streams; an estimated 6350 metric tons of nitrogen and 1100 metric tons of phosphorus now enter the South Platte River basin each year from the wastewater treatment plants (Fig. 24.10). These point sources of nitrogen and phosphorus lead to nutrient concentrations in the South Platte River that are substantially higher than elsewhere in the United States (Dennehy et al. 1998). This creates direct hazards to humans because nitrate can be converted to nitrite within the gastrointestinal tract, and nitrite reacts with haemoglobin to cause impairment of oxygen transport. In extreme cases, this leads to potentially fatal conditions such as blue-baby syndrome (methemoglobinemia), in which the lack of oxygen in the blood of a human infant creates a blue tinge in the baby's skin (Dennehy et al. 1998).

Return flows or infiltration of irrigation water can also produce high concentrations of nitrogen and phosphorus in surface water and groundwater. An estimated 36,000 metric tons of phosphorus and 180,000 metric tons of nitrogen are applied to crops in the South Platte basin each year either as chemical fertilizers or as manure from the many feedlots present in the basin. Water quality in agricultural portions of the South Platte basin is the most degraded of any type of land use, mainly as a result of nitrates and salinity in groundwater, and salinity and suspended sediment in surface water (Dennehy et al. 1998).

High nitrate concentrations have so degraded groundwater quality that wells in some cities on the eastern plains of Colorado can no longer be used to supply drinking water. Nitrate concentrations exceeded the US Environmental Protection Agency drinking-water standard of 10 mg/L in about half of the wells sampled by NAWQA scientists (Dennehy et al. 1998) (Fig. 24.11).

Groundwater in agricultural portions of the South Platte basin is derived almost entirely from irrigation return flows that infiltrate into the subsurface, and is therefore highly affected by salinity and nitrate. Salinity increases when water spread across agricultural fields partly evaporates, leaving behind dissolved salts that then increase in concentration in the remaining water. Dissolution of salts in the soil, and concentration of salts within the soil as plants take up and then transpire water from the soil to the atmosphere, also concentrates salts (Dennehy et al. 1998). Agricultural losses from soil salinization in the Arkansas River basin, immediately south of the South Platte River basin, are now estimated at $4.3 million annually (Houck et al. 2004).

Other contaminants enter surface and groundwater supplies from irrigated agricultural fields. Irrigated fields occupy only 8% of the South Platte River basin, but account for 71% of water use within the basin. More than 900,000 kg of active pesticide ingredients are applied to these crops each year. NAWQA scientists found 25 different pesticides in surface water and 15 different in groundwater of the South Platte River basin, and some type of pesticide was detected in 29 of the 30 groundwater wells sampled in the agricultural portion of the basin between Fort Lupton, Colorado and North Platte, Nebraska. Although the concentrations of pesticides present in the water and sediment of agricultural areas were generally low, NAWQA scientists caution that the long-term effects of exposure

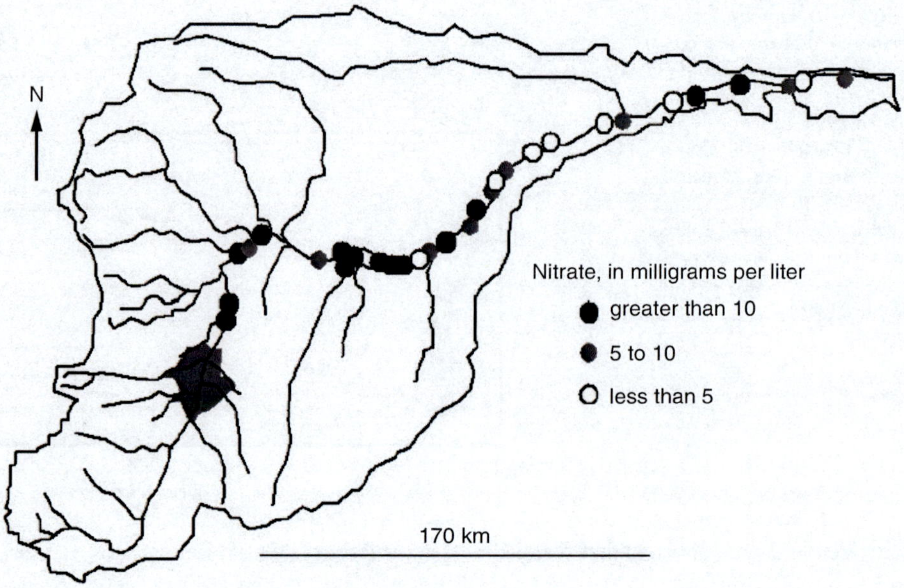

Fig. 24.11 The South Platte River basin, with Denver (*large gray area*) and nitrate concentrations in shallow groundwater samples taken along the river by NAWQA scientists. Nitrate concentrations exceed the U.S. Environmental Protection Agency drinking-water standard of 10 mg/L in about half of the wells sampled. (After Dennehy et al. 1998)

to mixed pesticides remain unknown (Dennehy et al. 1998).

The cumulative effects of all of these alterations in the quality and quantity of surface and groundwater, as well as habitat, show up in aquatic communities reduced in number and diversity of organisms, and in aquatic organisms impaired in health. The appearance of fish with both male and female reproductive parts in the South Platte River downstream from the Denver wastewater treatment plant has been tied to the presence of endocrine-disrupting chemicals in the water supply (Stein and Moffeit 2004). These chemicals also present health hazards to humans (Colborn et al. 1996; Colborn and Thayer 2000), and their appearance in concentrations substantial enough to alter aquatic communities is a warning sign to human communities consuming the same water.

24.5 The Future

Contemporary trends in population and water use show no signs of slowing or stopping in the drylands of Colorado. Consequently, the primary challenges to sustaining human and wildlife populations in the region come from changes in water supply associated with global warming and increasing population growth and associated water use, as well as from declining quality of surface and groundwater supplies. Contemporary uses of land and water resources reflect strategies designed to mitigate hazards of limited water quantity and quality.

Relatively few studies have directly estimated the potential effects of future climate warming on river systems in the Colorado Rocky Mountains and western Great Plains. Studies that have used even conservative scenarios for future climate indicate that changes in stream flow and water supply will likely severely stress natural and human communities dependent on stream flow (Rense 2000).

Much of the effect of warming climate on the larger rivers of the western plains comes from changes in the snow pack and snow melt. Rivers draining the Rocky Mountains discharge more than 70% of their annual water budget during two to three months of snowmelt and have instantaneous discharges ten to one hundred times the mean low flow. Stewart et al. (2004) observed a shift in the timing of spring snow melt toward earlier in the year along many western rivers during the period 1948–2000. Under twenty-first century warming trends predicted by the Parallel Climate Model, the timing of peak flow on rivers in the western United States will be earlier, particularly for rivers fed by snow melt from the Rocky Mountains, where many rivers will run 30–40 days earlier by 2099 as a result of changes in spring temperatures and the melting of the snow pack (Stewart et al. 2004). Precipitation will also likely decline, as indicated by computer simulations which predict that greater warming in winter at high latitudes will weaken the polar air masses, resulting in fewer winter storms passing over the portions of the Rockies that drain to the Colorado and South Platte River basins (Byrne et al. 1999). As the snow pack decreases and melts earlier each spring, runoff from rainfall may become dominant. It is more difficult to efficiently store and transfer rainfall runoff because of its unpredictability, creating challenges for managing human water supplies (Stewart et al. 2004). And aquatic and riparian organisms adapted to cold water and consistent floods in late spring-early summer may not be able to survive the changes in flow regime (Covich et al. 1997).

Climate will also grow warmer and drier across the western Great Plains (Ojima et al. 1999; NAST 2000). Average temperatures have risen more than 1°C in the past century across the northern and central Great Plains, and annual precipitation has decreased by 10%. Climate models predict continued increases in temperature, particularly in the western plains and during the winter and spring. Drying will be greatest in the rain shadow of the Rockies, where more frequent and prolonged periods of drought, together with diminished water supplies, will force changes in cropping and grazing, or make agriculture infeasible in some areas.

Under these scenarios of future changes in water supply, farmers who irrigate will have to use more efficient irrigation methods, return to dryland cropping systems, or convert agricultural lands to other uses (Parton et al. 2007). Municipal and industrial users will also need to improve water efficiency. Loomis et al. (2000) showed that residents along a 75 km section of the Platte River are willing to pay higher water bills to enhance the five ecosystem services of dilution of wastewater, natural purification of water, erosion control, habitat for fish and wildlife, and recreation. The level of funding generated in this manner along the entire river greatly exceeds estimated costs for the water leasing and conservation reserve program farmland easements necessary to produce the increase in ecosystem services.

Scenarios of climate warming and consumptive use of water, along with contemporary declines in water quality, present serious obstacles to sustaining river ecosystems and human communities on the western plains, but many efforts are underway to actively preserve and restore river ecosystems and to develop more sustainable patterns of human resource that will preserve water supply and quality. The volume and timing of water flow in rivers of the western plains is the single most fundamental issue that restoration efforts must address, and the most difficult. The form and function of a river reflect the supply of water and sediment to the river. Any sustainable attempt to restore form and function in order to provide ecosystem services such as clean water or wildlife habitat must therefore be based on restoring water and sediment fluxes. This realization has come slowly and painfully because human societies have for so long treated rivers as collections of disparate resources that can be partitioned, rather than as integrated ecosystems which can be destroyed by resource extraction. The realization that restoring water and sediment fluxes is crucial to river restoration has also been difficult because human society as presently constituted in the Colorado Rocky Mountains and western Great Plains is based on altering the movement of water and sediment across the landscape and into rivers. The irrigated agriculture that forms the economic basis for most communities in the western Great Plains absolutely depends on these alterations.

Unable at present to remove the dams and diversions that have fundamentally changed rivers of the western plains during the past century, people seek to mitigate the negative effects of these structures using techniques such as artificial recharge of groundwater. The Tamarack Project in northeastern Colorado on the border with Nebraska provides an example (Fardal 2003). The project uses groundwater recharge to augment stream flows in the South Platte River during periods of low flow in order to sustain critical habitat. During times when there is no consumptive demand for water withdrawal from the river, ten pumping wells located next to the river pump the river water into recharge basins several hundred meters away. Once in the ground, the water slowly returns to the river, reaching the channel during times when flow augmentation is needed. This experimental project is a sort of water-banking; water is stored temporarily in the subsurface in order to cushion the stream flow against extreme fluctuations that could stress aquatic organisms.

The Tamarack Project began in the autumn of 1996 and is considered successful in increasing water levels in warm-water sloughs and wetland complexes along the river. Other, similar experimental projects are currently being implemented elsewhere in Colorado along the South Platte River.

Other restoration attempts focus on specific segments of a river and attempt to restore the flux of water and sediment, and thereby form and function, to that segment. The most advanced version of this approach focuses on a segment of the Platte River in Nebraska where large concentrations of migratory birds still use the river (Smith 2006). The river banks and floodplain wetlands provide habitat to migratory birds and the channel supports endangered fish. Restoration of flows and habitat for these species is complex because the necessary water and land are already privately owned, but the first phase of the thirteen-year recovery plan began in 2006. The entire effort is valued at $300 million, including in-kind contributions of water and land by individual states and the federal government. Much of the money will be used to purchase or lease land from farmers along the river. Water from federal reservoirs will be purchased to replenish river flows during periods of the year critical for bird and fish migration and reproduction.

Preservation of migratory bird habitat is an example of the 'string of beads' restoration concept, in which river health is protected and restored through acquisition of key floodplain habitats such as areas near tributary confluences or remnant backwaters. These areas form beads along the river corridor that provide wildlife habitat and ecosystem services. The beads can be passively managed by removing existing structures and allowing naturally occurring floods to occur during periods of high flow. The beads can also be actively managed by pumping river water into distribution channels to flood wetland areas that remain disconnected from the main channel, or artificially enhancing recharge as at Tamarack. The beads not only sustain wildlife populations, but also help to recharge groundwater supplies and improve water quality.

24.6 Conclusions

In marked contrast to the indigenous peoples, the first people of European descent to settle in the mountains and plains of Colorado during the nineteenth century altered the landscape to suit their needs for reliable and abundant water by transferring water between drainage basins and pumping water from the subsurface. Continued development of water resources during the twentieth century supported extensive development of irrigated agriculture and an urban corridor along the eastern base of the Rocky Mountains. Resource patterns and societal choices at the start of the twenty-first century are increasingly governed by the need to mitigate unintended consequences of past resource use (surface and groundwater contamination, loss of ecosystem services, salinization of agricultural lands), while maintaining existing communities. Balancing these needs for mitigation and maintenance requires that people develop more efficient patterns of water use and reduce environmental contamination resulting from use of land and water.

References

Balog JD (1977) Big Thompson River tributaries: geomorphic activity and its controlling factors during the 1976 flood. Unpublished MA thesis, University of Colorado, Boulder

Butler DR (1995) Zoogeomorphology: animals as geomorphic agents. Cambridge University Press, New York, 231 pp

Byrne JM, Berg A, Townshend I (1999) Linking observed and general circulation model upper air circulation patterns to current and future snow runoff for the Rocky Mountains. Water Resources Research 35:3793–3802

Colborn T, Dumanoski D, Meyers JP (1996) Our stolen future: how we are threatening our fertility, intelligence, and survival – a scientific detective story. Dutton, New York, 306 pp

Colborn T K, Thayer K (2000) Aquatic ecosystems: harbingers of endocrine disruption. Ecological Applications 10: 949–957

Covich AP, Fritz SC, Lamb PJ, et al. (1997) Potential effects of climate change on aquatic ecosystems of the Great Plains of North America. Hydrological Processes 11:993–1021

Cross FB, Moss RE (1987) Historic changes in fish communities and aquatic habitats in plains streams of Kansas. In: Matthews WJ, Heins DC (eds) Community and evolutionary ecology of North American stream fishes. University of Oklahoma Press, Norman, pp 15–165

Dennehy KF, Litke DW, Tate CM, et al. (1998) Water quality in the South Platte River basin, Colorado, Nebraska, and Wyoming, 1992–95. U.S. Geological Survey Circular 1167, 38 pp

Eschner TR, Hadley RF, Crowley KD (1983) Hydrologic and morphologic changes in channels of the Platte River Basin in Colorado, Wyoming, and Nebraska: A historical perspective. U.S. Geological Survey Professional Paper 1277-A

Fardal LL (2003) Effects of groundwater pumping for irrigation to stream properties of the Arikaree River on the Colorado plains. Unpublished MS thesis, Colorado State University, Fort Collins, 71 pp

Fausch KD, Bestgen KR (1997) Ecology of fishes indigenous to the central and southwestern Great Plains. In: Knopf FL, Samson FB (eds) Ecology and conservation of Great Plains Vertebrates. Springer-Verlag, New York, pp 131–166

Fremont JC (1845) Report of the exploring expedition to the Rocky Mountains in the year 1842, and to Oregon and North California in the years 1843–1844. Gales and Seaton, Washington, DC

Gerlek S (1977) Water supplies of the Platte River basin. Unpublished MS thesis, Colorado State University, Fort Collins

Graf WL (ed) (2005) Endangered and threatened species of the Platte River. Committee on Endangered and Threatened Species in the Platte River Basin, National Research Council, The National Academies Press, Washington, D.C., 299 pp

Gutentag ED, Heimes FJ, Krothe NC, et al. (1984) Geohydrology of the High Plains Aquifer in parts of Colorado, Kansas, Nebraska, New Mexico, Oklahoma, South Dakota, Texas, and Wyoming. U.S. Geological Survey Professional Paper 1400-B, 63 pp

Harrington C (1874) Summering in Colorado. Richards and Company, Denver

Holliday VT (1987) Geoarchaeology and late Quaternary geomorphology of the Middle South Platte River, northeastern Colorado. Geoarchaeology 2:317–329

Houck EE, Frasier M, Schuck E (2004) The regional effects of waterlogging and soil salinization on a rural county in the Arkansas River basin of Colorado. Western Agricultural Economics Association Meeting, Honolulu, HI

Loomis J, Kent P, Strange L, et al. (2000) Measuring the total economic value of restoring ecosystem services in an impaired river basin: results from a contingent valuation survey. Ecological Economics 33:103–117

Matson PA, Parton WJ, Power AG, et al. (1997) Agricultural intensification and ecosystem properties. Science 277:504–509

Metherell AK, Cambardella CP, Parton WJ, et al. (1995) Simulation of soil organic matter dynamics in dryland wheat-fallow cropping systems. In: Lal R, Kimball J, Levine E, Stewart BA (eds) Soil Management and Greenhouse Effect. Lewis, Boca Raton, Florida, pp 259–270

Mosier AR, Parton WJ, Valentine DW, et al. (1997) CH_4 and N_2O fluxes in the Colorado shortgrass steppe: 2. Long-term impact of land use change. Global Biogeochemical Cycles 11:29–42

NAST (National Assessment Synthesis Team, U.S. Global Change Research Program) (2000) Climate change impacts on the United States: the potential consequences of climate variability and change. Overview: Great Plains, 15 pp

Ojima D, Garcia L, Elgaali E, et al. (1999) Potential climate change impacts on water resources in the Great Plains. Journal of the American Water Resources Association 35:1443–1454

Olson R, Hubert WA (1994) Beaver: water resources and riparian habitat manager. University of Wyoming, Laramie, 48 pp

Parton WJ, Gutmann MP, Travis WR (2003) Sustainability and historical land-use change in the Great Plains: the case of eastern Colorado. Great Plains Research 13:97–125

Parton WJ, Gutmann MP, Williams SA, et al. (2005) Ecological impact of historical land-use patterns in the Great Plains: a methodological assessment. Ecological Applications 15:1915–1928

Parton WJ, Gutmann MP, Ojima D (2007) Long-term trends in population, farm income, and crop production in the Great Plains. BioScience 57:737–747

Peters EJ, Schainost S (2005) Historical changes in fish distribution and abundance in the Platte River in Nebraska. American Fisheries Society Symposium 45:239–248

Propst DL, Carlson CA (1986) The distribution and status of warmwater fishes in the Platte River drainage, Colorado. The Southwestern Naturalist 31:149–167

Rabalais NN (2002) Nitrogen in aquatic ecosystems. Ambio 31:102–112

Rense WC (2000) Hydrologic impact of climate change in the Colorado Front Range. Am. Water Resources Assoc., 2000 Spring Specialty Conference, Anchorage, Alaska, 7 pp

Savage C (2004) Prairie: a natural history. Greystone Books, Vancouver, Canada, 308 pp

Schmal R, Wesche T (1989) Historical implications of the railroad crosstie industry on current riparian and stream habitat management in the Central Rocky Mountains. In: Gresswell RE, Barten BA, Kershner JL (eds) Practical approaches to riparian resource management: an educational workshop. U.S. Bureau of Land Management, Billings, Montana, 189 pp

Smith J (2006) Platte River recovery. Rocky Mountain News, 30 January 2006

Smith J (2007) Water crisis grows dire amid dry skies, wells. Rocky Mountain News, 9 July 2007, pp 6–9

Stein T, Moffeit M (2004) Mutant fish prompt concern. Denver Post, October 3, 2004, pp 1A, 18A

Stewart IT, Cayan DR, Dettinger MD (2004) Changes in snowmelt runoff timing in western North America under a 'business as usual' climate change scenario. Climatic Change 62:217–232

Strange EM, Fausch KD, Covich AP (1999) Sustaining ecosystem services in human-dominated watersheds: biohydrology and ecosystem processes in the South Platte River basin. Environmental Management 24:39–54

Tyler D (1992) The last water-hole in the West: the Colorado-Big Thompson project and the Northern Colorado Water Conservancy District. University Press of Colorado, Niwot, CO, 613 pp

USDA (U.S. Department of Agriculture Forest Service) (1980) Cache la Poudre wild and scenic river: final environmental impact statement and study report. Washington, DC, 150 pp

Ward JV (1992) A mountain river. In: Calow P, Petts GE (eds) The rivers handbook: hydrological and ecological principles. Blackwell, Oxford 1:493–510

Wentz DA (1974) Stream quality in relation to mine drainage in Colorado. In: Hadley RF, Snow DT (eds) Water resources problems related to mining. American Water Resources Association, Minneapolis, Minnesota, pp 158–173

Williams GP (1978) The case of the shrinking channels – the North Platte and Platte Rivers in Nebraska. U.S. Geological Survey Circular 781, 48 pp

Wohl EE (2001) Virtual rivers: lessons from the mountain rivers of the Colorado Front Range. Yale University Press, New Haven, Connecticut, 210 pp

Wood JP (1935) Report on mine tailings pollution of Clear Creek, Clear Creek-Gilpin Counties, Colorado. Report submitted to Davis and Wellbank, attorneys, Denver, Colorado, 35 pp
Wroten WH (1956) The railroad tie industry in the Central Rocky Mountain region: 1867–1900. Unpublished PhD dissertation, Colorado State University, Fort Collins, 287 pp
Young MK, Haire D, Bozek MA (1994) The effect and extent of railroad tie drives in streams of southeastern Wyoming. Western Journal of Applied Forestry 9:125–130

Chapter 25
Frozen Coasts and the Development of Inuit Culture in the North American Arctic

Robert W. Park

25.1 Introduction

The northernmost part of the North American continent has seen some of the most fascinating human adaptations anywhere. In the New World this huge area extends some 11,000 km from the Aleutian Islands in the west to Greenland and Labrador in the east (Fig. 25.1). Geographically and in terms of human occupations, the Arctic is perhaps best defined as the area beyond the tree line (the northern limit of continuous forest). Some other attributes that help define the Arctic include persistence of cold (long winters and short cool summers), a largely treeless environment, permafrost (year-round frozen ground), large seasonal differences in the amount of sunlight, and very few plant foods that are consumable by humans. At the time of initial European contact a majority of human groups inhabiting the North American Arctic spent at least a part of the year living near the coast and making use of both land mammals on the tundra and sea mammals in the open ocean during the summer. However, in the winter the landscape changes dramatically. In Alaska, Greenland, and Labrador a wide strip of thick landfast ice forms along the coast while in the Canadian Arctic Archipelago the channels between islands freeze completely for much of the year. In all of these regions the sea ice and its resources formed a vital part of most human adaptations. Although it may perhaps seem incongruous to discuss at length the sea ice environment in a work focusing on landscape, it was the development of a successful adaptation to a frozen coast environment encompassing both the land and the sea ice that ensured the long term success of human occupations throughout the Arctic.

An additional feature that makes the archaeological record of these Arctic cultures so interesting is the combination of the richness of the Inuit ethnographic record, the complexity of their material culture, and the often marvelous preservation of that material culture due to permafrost. All these factors mean that archaeologists have a great deal of information to work with. In terms of the specific goals of this volume, the development of the adaptation to a frozen coast environment is interesting for several reasons. First, in this part of the world there are some of the most profound seasonal landscape changes anywhere on Earth: the terrestrial landscape expands enormously with the addition of the frozen ocean, the composition and abundance of the fauna changes dramatically, and the amount of daylight also varies. Second, the human adaptation to the frozen coast and sea ice environment involved complex material culture that is recoverable archaeologically due to the preservation often afforded by permafrost. Finally, since the sea ice environment disappears every summer, all direct traces of human use of that landscape have been lost. Therefore, to a greater extent than in most archaeological situations, our understanding of the history of human use of that landscape must be inferential.

25.2 Background

Due to its high latitude, the entire region exhibits an Arctic climate, characterized by long winters and pronounced seasonal differences in the amount of solar

R. W. Park (✉)
Department of Anthropology, University of Waterloo,
200 University Ave. West, Waterloo, ON N2L 3G1, Canada
e-mail: rwpark@uwaterloo.ca

See Plate 22b in the Color Plate Section; also available at: extras.springer.com

I. P. Martini, W. Chesworth (eds.), *Landscapes and Societies*,
DOI 10.1007/978-90-481-9413-1_25, © Springer Science+Business Media B.V. 2010

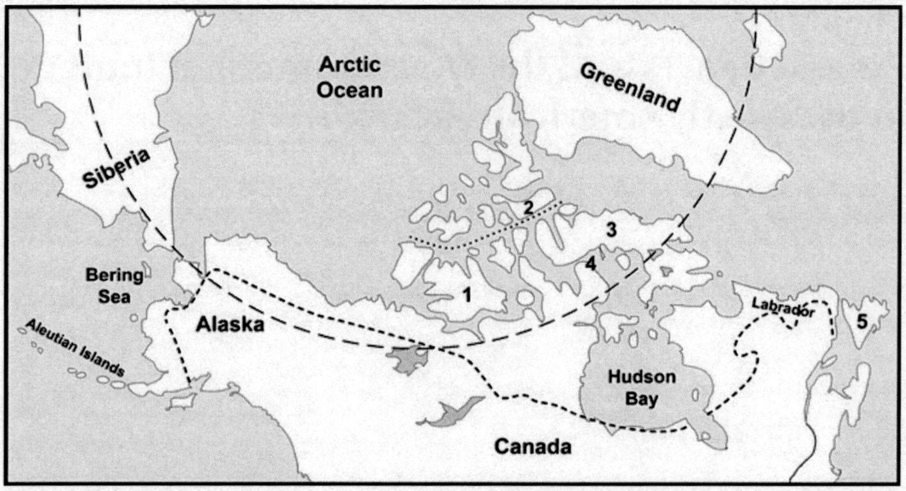

Fig. 25.1 Map showing locations and features mentioned in the text. The *long-dash line* indicates the Arctic Circle; the *narrow-dash line* marks the tree line. The Canadian Arctic Archipelago consists of all of the islands north of mainland Canada. The High Arctic is usually defined as the coasts on either side of Parry Channel (*the dotted line*) and all the islands to its north. Islands mentioned in the text include: *1* Victoria; *2* Devon; *3* Baffin; *4* Igloolik; *5* Newfoundland

radiation. The degree of winter cold varies regionally to some extent, but throughout the entire Arctic summers are relatively cool, with mean July daily temperatures of less than 10 °C in most regions. The overall climate has fluctuated somewhat over the course of human occupation. As reconstructed from ice cores in Greenland and Canada, at the time people first arrived in the Canadian Arctic some 4300 years ago the climate was warmer than today but entered into a long slow cooling trend culminating around 2000 year BP. A subsequent warming trend peaked around 1000 year BP. The climate again cooled, reaching its coolest point in the nineteenth century. Since that time the climate has steadily warmed (Jacobs 1979; Freskild 1996). The Arctic Circle is the latitude north of which there is an increasing period during the summer when the sun does not set and during the winter when the sun does not rise. For example, at the latitude of Devon Island the summer sun does not set for a period of approximately three months and the winter sun does not rise above the horizon for a similar length of time. During the winter darkness the moon may provide enough light for many kinds of activities, however.

25.2.1 The Land

Turning to the landscape and its resources, the bedrock geology of Greenland and the eastern portion of the Canadian Arctic consist of granites, gneisses and schists whereas the central and western portions of the Canadian Arctic are characterized by various sedimentary rocks, especially limestones, sandstones, and shales. Alaska is geologically more active, with volcanic and earthquake activity. The surficial geology of Alaska also differs from that of Arctic Canada and Greenland because the latter regions were fully glaciated during the Pleistocene, whereas almost all of the current tundra regions of north Alaska remained unglaciated. Glacial erosion in Arctic Canada has exposed some geological formations, which in many parts of Alaska are overlain by deep glacial outwash deposits. Stone resources that were important for human populations prior to European contact included chert, quartz, quartzite, slate, and soapstone. The first four were used for creating cutting and scraping tools; soapstone was used for containers and for lamps. These types of stone were available in most parts of the Arctic but distributed in a patchy fashion so not every population had easy access to them. Metal in the form of float copper and meteoritic iron was available in a few regions and appears to have been traded or carried long distances (McCartney and Mack 1973; McCartney 1991). These metals were cold-hammered into cutting implements.

The topography of this enormous region is extremely diverse. High mountain ranges are found in north Alaska, and along the eastern margin of the Canadian Arctic and in many parts of Greenland. However, most of the entire region, and the majority of the areas used by its human occupants, is characterized by low relief. The coastal regions, where the most intensive and extensive human occupations took place, include both rocky coasts incised by deep narrow fjords in some regions and wide relatively flat coastal plains in others. Parts of Alaska and the western portion of the Canadian

Arctic are characterized by several major rivers that carry driftwood from below the tree line downstream to the ocean where, depending on prevailing winds, currents and ice drift, the wood is deposited on some coastlines becoming an important resource for people living in that area. In all other parts of the Arctic, wood was a very scarce commodity and was conserved and recycled accordingly.

Throughout the Arctic soil development is slow and soil cover is thin or lacking although regolith is always present except in some costal rocky shores polished by waves. Almost the entire Arctic region is characterized by tundra, which is defined by the presence of permafrost—subsurface deposits that remain frozen year-round. Above the permafrost, only a thin active layer at the ground's surface thaws each summer. The depth of the active layer varies depending on location, aspect, and pedological characteristics of the deposit. Depending on the latitude and on local conditions the thickness of the active layer may vary from just a few centimeters to more than a meter. Many Arctic cultures created semi-subterranean winter houses—pit houses with low raised walls and rafter roofs that were thickly insulated with skins, rocks and turf. When their inhabitants abandoned these houses the roof and walls would collapse into the house interior, thereby burying anything left on the floor deeply enough so as to be below the active layer. The excavation of such houses from the Thule tradition, which occupied the region from Siberia through the North American Arctic to Greenland approximately 1000 years ago, has produced a wide variety of extremely fragile organic artefacts preserved through being frozen within the permafrost.

The amount and nature of tundra vegetation varies. Many tundra areas within the High Arctic are almost completely unvegetated but in most regions the vegetation found in tundra consist of mosses, lichens, and grasses. Very little tundra vegetation is edible for humans, apart from some berry species. The few trees that do exist tend to be miniature species, like dwarf willow, that grow prostrate.

The most important large land mammal species is caribou. Many caribou populations are migratory, moving hundreds of kilometers from their southern wintering territories around the tree line to their summering territories far out on the northern tundra and arctic coastal areas. Muskox and to a much lesser extent, dall sheep are other important large land mammal species. Small mammal species such as arctic fox and arctic hare are also present. Fish are of considerable importance to some human populations. Species include Arctic char, salmon, grayling, trout, lampreys, lingcod, pike, and whitefish. Almost all of the bird species that were of economic importance to human populations, like ducks and geese, were migratory and so were only available in the Arctic from spring through autumn.

25.2.2 The Coasts

Some Inuit populations had an entirely inland adaptation at the time of Euro-American contact, either focusing to a large extent on fish in a riverine environment such as along some of the major rivers in the interior of southwest Alaska, or focusing on caribou in the interior of north Alaska or in the barren grounds west of Hudson Bay. However, most Inuit populations had an adaptation that included coastal regions and their resources so it is useful to consider the distinctive characteristics of that environment.

The terrestrial plant and animal resources of coastal regions were sometimes less abundant than in the interior. However, the marine resources of the coast more than make up for this shortcoming. Sea mammals are central to most adaptations in the wood-poor regions of the Arctic because, in addition to food, they also provide heating and cooking fuel in the form of blubber. Important sea mammal species include ringed, harp, harbor and bearded seals, sea lions, walrus, and narwhal, beluga, and bowhead whales. Polar bears are probably best categorized as sea mammals since they spend most of the year on the sea ice, away from land. Ringed seals are especially important for human populations in many regions because they are non-migratory and thus are available year round; the other species are migratory to varying degrees and therefore less available or unavailable during the winter.

The availability of both terrestrial and marine resources has made coastal regions appealing places to inhabit for human populations around the world. However, polar coastal regions are distinctive. Maps of the North American Arctic often leave a misleading impression since they tend to show the land surrounded by blue water. But of course the Arctic seas are ice-free for only a relatively small portion of the year. For the rest of the year they are covered by a thick layer of ice and for human inhabitants of the Arctic the frozen sea ice forms a vital part of the landscape. Within the channels separating the islands of the Canadian Arctic Archipelago the sea freezes in almost unbroken expanses. Elsewhere in the Arctic on the coasts facing the open ocean a solid strip of landfast ice that may be many kilometers wide forms during the winter. The edge of the landfast ice, where it meets the open ocean, is known as the floe edge.

The sea ice is an extremely dynamic environment. It is recreated annually at freeze-up sometime during the early autumn and then disappears at break-up sometime during the summer. Even in midwinter it can be a varied and changing environment. Pressure ridges in the ice can be created by winds or currents crushing and pushing sea ice together. Conversely, leads in the ice—linear stretches of open water that are created when cracks in the ice are forced open by wind or currents—form and then close or freeze over hours or days later. In some locations there are local sea patches of permanently or semi-permanently open water called polynyas that are kept that way by fast-moving currents. Polynyas are important because sea mammals can remain there throughout the winter, and, early in the spring before the sea ice elsewhere breaks up, polynyas attract migratory birds and sea mammals (Dunbar 1981). Another useful resource that is unevenly distributed across the sea ice environment is multi-year ice—essentially pieces of sea ice that did not melt during the preceding summer. Multi-year ice can be useful to people living and travelling on the sea ice as sources of fresh water since its salt content will have become sufficiently depleted, unlike single-year ice which generally still retains enough salt to be undrinkable. In areas of unbroken sea ice, ringed seals will maintain breathing holes through the ice that allow them to survive throughout the winter. However, if a lead opens up nearby, the seals are likely to abandon some or all of their old breathing holes. As the open water in the lead freezes, the seals will maintain new breathing holes there. Anywhere there is an interface between sea ice and open water—either at the floe edge, at a polynya, or at a lead in the ice—the entire food chain, including the sea mammals that are important for humans, is likely to be richer than in other locations due to the upwelling that occurs at the ice edge (Dunbar 1984).

25.2.3 Ethnography

It is beyond the scope of this chapter to describe in all its richness and variety the diversity of Inuit adaptations to the Arctic coastal landscape. The following capsule ethnographic sketch most closely reflects the traditional ways of life of the so-called Central Inuit of the Canadian Arctic at the time of European contact but aspects of it apply to all the Inuit groups that lived on the coast. Starting at the time the sea ice broke up, in early or mid summer depending on the location, many groups would be camping in sites along the coast, hunting seals or fishing from the remaining ice edge or from the shore, or using small skin boats (kayaks) and large skin boats (umiaks) to hunt sea mammals. The animals, which could vary in size from ringed seals up to bowhead whales, were hunted using throwing harpoons and drag floats, and then stabbed with lances. Ducks and geese would be caught at this time of year, especially during the moult when they cannot fly, using darts and nets. Later in the summer and into the autumn some people would move inland to hunt caribou with bow and arrow; at this time of year their skins were in the best condition for manufacturing skin clothing. At that time people would also use weirs to obtain fish such as Arctic char as they returned from the ocean to overwinter in lakes. By early winter everyone would have returned to the coast to await the sea ice becoming strong enough for travel. Fishing might take place through the ice of rivers and lakes. In regions where the floe edge was not too distant, people might continue to camp at the coast throughout the winter, travelling to the floe edge for hunting. Travel, both for camp movement and for hunting, was by dog sled. Travel by sled was generally much easier on the sea ice than on land. They either lived in semi-subterranean houses constructed of wood, boulders, turf, and sometimes whale bones, or in snow houses (igloos). In regions where the ocean froze completely people had to rely on hunting ringed seals at their breathing holes. This necessitated moving out onto the sea ice and living in snow houses, moving camp every ten days to two weeks throughout the winter as the majority of the seals in the immediate vicinity of each campsite were killed. Because breathing-hole sealing was a low-return form of hunting, very complex rules governed the distribution of the meat and blubber from seals caught at this time of year to ensure that each seal was shared as widely and evenly as possible among the people. Furthermore, the largest population aggregations took place at this time of year, maximizing the number of hunters participating and ensuring the widest sharing network. By the early summer the snow on the ice was melting and the ringed seals emerged from their breathing holes to bask on the ice. They are very wary at this time, however, so approaching one close enough to hit it with a throwing harpoon required great skill. As the time of breakup approached, groups who were camped out on the sea ice would move to locations on the coast, to begin the cycle again.

Variations from the above ethnographic sketch were great but the pattern of exploiting the resources of both the land and the sea was a common theme in most societies, to such an extent that the land-sea dichotomy figured prominently in Inuit ideology. In North Alaska the groups that took advantage of the coastal area and its terrestrial and marine resources were collectively known as the Tareumiut, or people of the sea, while the groups who lived inland and focused on caribou where collectively designated Nunamiut, or people of the land (but see Burch 1976). Even more widespread and pervasive were beliefs and practices governing the relationship between things of the land and things of the sea. In many groups there were strict prohibitions against bringing things of the land and things of the sea into contact with one another. Practically, this meant that caribou meat and seal meat could never be stored or cooked together, and seal sinew thread could not be used to sew caribou skin clothing. Due to its unparalleled warmth, caribou skin clothing was used throughout the winter but it had to be manufactured on land, not out on the sea ice. This meant that women were very busy at the autumn coastal camps, having to sew all the clothing their families would need for the winter before they could move out onto the sea ice (Jenness 1922; Rasmussen 1932).

25.3 Archaeological Sequence

The primary means to understand the origins and development of this way of life, embodying as it does a distinctive relationship with the Arctic landscape, is through archaeology. Archaeological research in the Arctic presents a number of difficult challenges including immense logistical problems due to the vast size of the region and the difficulty and expense of getting there. In addition, archaeological field seasons are necessarily brief due to the shortness of the Arctic summer and the process of excavation is rendered very slow due to the difficulty of excavating into permafrost. For these reasons the prehistory of that region is probably less well understood than that of most other parts of the Americas. Nonetheless, archaeologists have been able to reconstruct the human history of the region, the broad outline of which is summarized below and in Fig. 25.2.

The earliest sites in the North American Arctic following the drowning of Beringia (the name given to the 'land bridge' that joined Siberia to Alaska during the last glaciation) are collectively assigned to the Paleo-Arctic tradition and date to approximately 11,000 BP–8500 BP. These sites are found in the unglaciated parts of Alaska and Yukon. Most interpretations emphasize the similarities between the Paleo-Arctic tradition and earlier cultural manifestations of the Upper Paleolithic, suggesting cultural continuity from them. However, on chronological and stylistic grounds it is unclear whether there is any cultural continuity from the Paleo-Arctic to later cultures in the North American Arctic. It is followed in Alaska by the Northern Archaic tradition, which is known from 6000–4000 BP. Stylistic similarities suggest that it has a very close connection to the other Archaic populations found throughout much of North America, so the Northern Archaic probably represents a northward expansion of those populations to occupy the expanding Boreal forests of the interior. Culturally and biologically the Northern Archaic peoples do not appear to be related to the later coastal populations of these regions.

The Arctic Small Tool tradition is the collective name given to a distinctive group of cultures that date from approximately 4300–2700 BP and which are found from Western Alaska all the way to Greenland. The earliest sites are found in Alaska and the later widespread distribution of the Arctic Small Tool tradition appears to have resulted from one of the most geographically dramatic population expansions in recent human history: the initial colonization of the Canadian Arctic and Greenland. The effects of the last glaciation lasted far longer in northern Canada and Greenland than elsewhere in North America so the Arctic Small

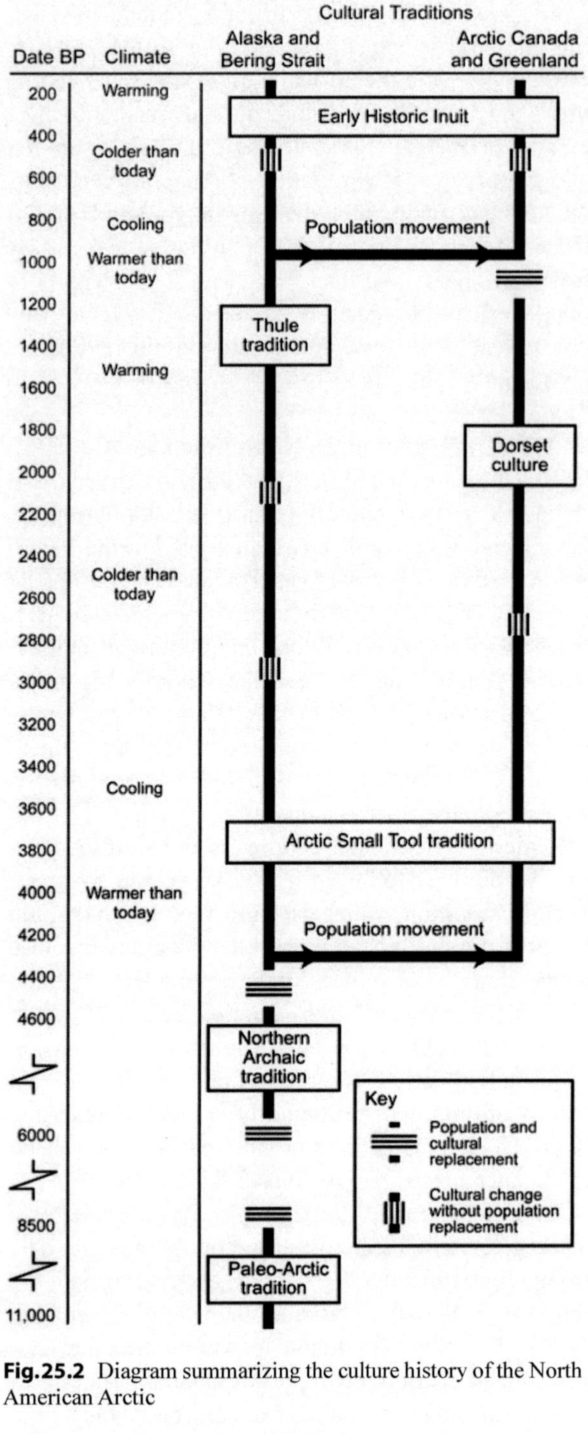

Fig. 25.2 Diagram summarizing the culture history of the North American Arctic

Tool tradition peoples appear to have migrated into the region relatively shortly after it first became suitable for human habitation. Based on the results from radiocarbon dating, their expansion from Alaska all the way to northern Greenland was extremely rapid, taking no more than a few centuries. Over subsequent centuries their descendants expanded and familiarized themselves with the different environments and resources of this vast region.

In Arctic Canada and Greenland the way of life and technology of the descendants of the Arctic Small Tool tradition had become sufficiently transformed by 2700 BP that archaeologists give them a new name: Dorset culture. Dorset sites are found from Victoria Island in the west to Greenland in the northeast and to Newfoundland in the southeast. Dorset culture persisted until at least 1200 BP but at around that time Dorset populations appear to have undergone a dramatic decline and by 1000 BP they had completely disappeared from most parts of their Arctic homeland.

During the centuries when cultures of the Arctic Small Tool tradition and then the Dorset culture flourished in Arctic Canada and Greenland, cultural developments amongst the Arctic Small Tool tradition descendants living on the Siberian and Alaskan sides of the Bering Strait eventually led to the emergence of what is known as the Thule tradition approximately 2100 BP. The cultures of this tradition developed a new form of economic and social adaptation that centered on the open-water hunting of large sea mammals from skin boats, including the largest of the Arctic whales, bowheads, which can reach 20 m in length. Sometime between 1100 and 800 BP small groups of Thule pioneers appear to have begun to move eastward from Alaska into the Canadian Arctic and Greenland and colonize that entire region (McGhee 1984; Park 1993, 2000; Morrison 1999; Friesen and Arnold 2008). The exact degree and nature of interaction between the earliest Thule immigrants and the last of the Dorset is not clear–almost all of the Dorset likely had died prior to the arrival of the Thule. At any rate, from Alaska to Greenland the diverse Inuit groups who greeted Europeans when the latter eventually entered those regions appear to have been the direct biological and cultural descendants of the Thule people.

25.4 Cultures and the Landscape

In the context of the environmental and ethnographic information presented above, three issues dealing with the relationship in prehistoric times between cultures and landscapes are explored in the following sections: the development of adaptation to a frozen coast; the cultures' ideologies of the landscape; and the effects of changes in the coastal landscape due to climate change and isostatic adjustments.

25.4.1 *The Development of a Frozen Coast Adaptation*

Neither the date nor the precise nature of the earliest adaptation to a frozen coast environment is yet completely understood, in part because the evidence for prehistoric human use of the sea ice environment must be indirect and inferential. However, it seems probable that the earliest populations inhabiting the North American Arctic took a considerable length of time to develop such an adaptation, and that the development took place in stages. For the purposes of this chapter the following five components together comprise the complete adaptation and their identification in the archaeological record will form the basis of this analysis: inhabiting coastal areas; hunting smaller sea mammals in the open water from the shore or from the floe edge; hunting larger sea mammals in the open water, especially from boats; hunting seals at their breathing holes; and living out on the sea ice. Each of these components potentially has an archaeological signature as discussed in the following paragraphs.

25.4.1.1 Inhabiting Coastal Areas

Evidence of people inhabiting coastal areas would include the identification of sites adjacent to contemporaneous coastlines or the identification of faunal remains (animal bones) from marine species in sites. But for the earliest cultures in this region these kinds of evidence are unlikely to be available. If there ever were any coastal sites from the Paleo-Arctic tradition (when sea levels would still have been much lower than they are today due to the effects of the last glaciation), they would now be deeply submerged. All the Paleo-Arctic sites that have been found and excavated would have been located far inland from contemporary coastlines and so represent terrestrial, non-coastal adaptations. Organic preservation at these sites is poor so there is no possibility of finding most of the kinds of diagnostic artefacts discussed below. Most interpretations of these sites emphasize their continuity with earlier cultural manifestations of the Upper Paleolithic, implicitly or explicitly suggesting that they represent a continuation of the terrestrial big-game hunting adaptation that has been well documented for those earlier cultural manifestations. As such, it is likely that the Paleo-Arctic peoples did not have anything approaching the kind of sophisticated adaptation to a frozen coastal environment that their successors would eventually develop. Similarly, sites of the Northern Archaic tradition are mostly found in interior locations that would have been boreal forest kinds of environments. There is a fairly widespread consensus that Northern Archaic peoples did not inhabit frozen coast environments.

The earliest good evidence for the habitation of a frozen coast environment comes from the Arctic Small Tool tradition. The earliest sites from this tradition are found from coastal west and northwest Alaska, as well as in the interior. The houses at the interior sites are substantial semi-subterranean structures that are interpreted as winter dwellings. These contrast with the light tent rings of coastal sites which, based on the limited organic artefactual and faunal data that have been recovered, are interpreted as having been occupied only during the spring or summer. Thus, it seems possible that the traditional use of the coast was not year-round, but only seasonal. However, even the earliest Arctic Small Tool tradition sites in the Canadian Arctic and Greenland tend all to be located on or near the coast—there is no evidence for inland winter occupation sites. This may be due to the fact that, in contrast to the situation in western and northern Alaska, in the High Arctic the inland areas are not game-rich. Thus, there was no incentive to retreat from the coast in the winter. However, a winter focus on terrestrial species is still evident in the earliest Arctic Small Tool tradition sites in the High Arctic and current interpretations suggest that these populations survived through the winter largely on stored muskox meat. Later Arctic Small Tool tradition sites, and at least some sites from all subsequent cultures throughout the entire North American Arctic, are found in coastal regions although in some cases it remains difficult to determine the precise season of occupation of particular sites, or the degree to which marine resources were exploited while people lived there.

25.4.1.2 Hunting Sea Mammals

The primary benefit of inhabiting coastal areas would have been to take advantage of marine resources, especially sea mammals. Apart from their bones, archaeologically visible evidence for the hunting of sea mammals from the shore or floe edge comes primarily from the initial appearance of harpoon parts, especially harpoon heads. Unlike harpoons, stabbing or throwing weapons such as spears or lances are often ineffective when hunting sea mammals because one thrust will rarely kill an animal outright. A wounded animal will usually dive and swim away immediately making it almost impossible for the hunter to deliver a second, killing blow or to follow the animal to recover its carcass if eventually it dies from the wound. The invention of the harpoon solved that dilemma. With skill or luck a single throw or thrust with a harpoon might kill an animal but the harpoon design really comes into play when the animal is not instantly killed. The harpoon head is designed to detach from the shaft and become firmly embedded in the flesh of the animal where via the harpoon line it acts as a kind of anchor. To prevent the harpoon head from pulling out of the wound, some are equipped with backward-pointing barbs while others are equipped with basal spurs that cause them to toggle—rotate ninety degrees within the wound. By holding on to the other end of the harpoon line or attaching it to something else the hunter is able to prevent the wounded prey from escaping while he gets close enough to kill it. The initial appearance of harpoon parts in the archaeological record, especially toggling harpoon heads, would seem to be good evidence of open-water hunting of sea mammals, at least from shore or from the floe edge, or as they basked at their breathing holes in the late spring and early summer (see below for hunting from boats). This kind of open-water hunting could also take place during the winter at leads in the ice or at polynyas (Schledermann 1980; Henshaw 2003).

Both non-toggling and toggling harpoon heads are known from archaeological sites of the Arctic Small Tool tradition in the Canadian High Arctic and Greenland and the faunal remains from some sites include sea mammals. Organic preservation at Arctic Small Tool sites in Alaska is generally poor so harpoon parts do not survive but undoubtedly they were in use there too. Harpoon heads of a wide variety of styles, along with other harpoon parts, are extremely common finds from all subsequent cultures inhabiting the frozen coast regions of Arctic North America and Greenland (Fig. 25.3).

25.4.1.3 Hunting Sea Mammals from Boats

The hunting of sea mammals from boats, especially larger sea mammals such as whales, requires additional technology (that is, knowledge, skills, and associated material culture) beyond harpoons. There are of course the boats themselves, diagnostic parts of which may show up archaeologically. However, the simple transference of shore or floe-edge based harpooning to a boat presents a problem. When hunting from a kayak or umiak it would be dangerous for the hunter to hold on to the harpoon line himself or attach it to his vessel since the sudden pull on the line of the harpooned and panicking animal might overturn the boat or damage it. Instead, the hunter attaches the harpoon line to a drag float made from inflated sealskin, much like a big balloon. As soon as the animal is harpooned the hunter tosses overboard the drag float that is now tethered to the animal. The direction of movement of the drag float on the water, or its reappearance after being dragged underwater by the diving animal, tells the hunter where to paddle his boat in order to harpoon the animal again or stab it with a lance when it resurfaces to breathe. Even if the hunter is unsuccessful in wounding the animal again when it resurfaces, he tries to force it to dive again before it has time to recover its breath. By doing this repeatedly the animal eventually becomes so exhausted and winded that it can no longer dive. The hunter can now approach and deliver a killing wound with the lance. The drag float attached to the harpoon head further ensures that the carcass can be recovered even if it sinks. Drag floats themselves almost never survive in identifiable form in the archaeological record. However, because the floats needed to be easily inflatable, a small bone or ivory nozzle was sewn into the sealskin bag that formed the drag float. These drag float inflator nozzles are very distinctive and their appearance in the archaeological record is thought to mark the advent of open-water hunting of larger sea mammals from boats (Fig. 25.3g).

A very few finds from Greenland and elsewhere do make it clear that the peoples of the Arctic Small Tool tradition possessed kayak-like boats (Grønnow 1996). However, it is not clear how important boat-based

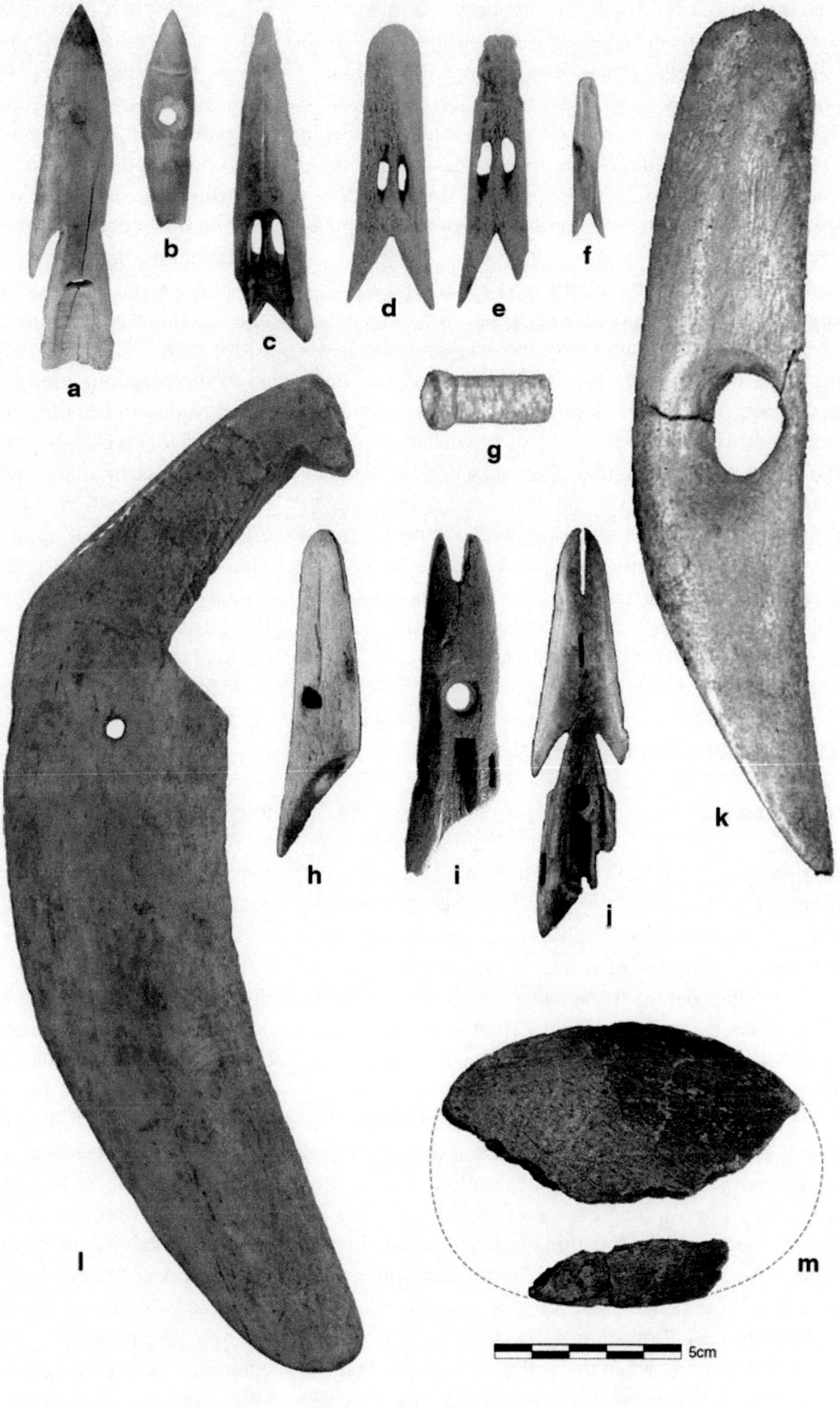

Fig.25.3 Arctic material culture. **a**, **b** Arctic Small Tool tradition harpoon heads. **c**–**f** Dorset culture harpoon heads. **g** Thule tradition drag float inflator nozzle. **h**–**j** Thule tradition harpoon heads. **k** Thule tradition whaling harpoon head. **l** Thule tradition snow knife. **m** fragments of an Arctic Small Tool tradition lamp. (**a**, **b** and **m** from Igloolik; **c**–**h**, **j**–**l** from Devon Island; **i** from Baffin Island)

hunting was in their economy. In at least one region it has been observed that Arctic Small Tool tradition sites are concentrated in areas where sea ice would have formed relatively early in the autumn and broken up relatively late in the summer, which contrasts to the site distribution of later cultures that are known to have relied heavily on hunting from boats. This may indicate that pedestrian hunting of seals at the floe edge or basking on the ice was more important than hunting sea mammals from boats, and that during the season when the ocean was free of ice the people lived inland, hunting terrestrial species or fishing. Similarly, there is no evidence that the Dorset people made extensive use of skin boats and in fact there is reason to believe that hunting on the sea ice was even more important for them than for their Arctic Small Tool tradition predecessors (see below).

It is in the cultures of the Thule tradition that we first see the appearance of drag float inflator nozzles, showing that the technology necessary for the hunting from boats of very large sea mammals had been developed. The appearance of extremely large harpoon heads and of numerous bones of bowhead whales in their sites corroborate this inference (Fig. 25.3k). The ability to hunt these huge sea mammals reliably led to a distinctive way of life: in the winter family groups occupied large semi-subterranean houses in coastal locations and sea mammal meat and blubber obtained during the summer and autumn contributed greatly to their winter food and fuel requirements.

25.4.1.4 Breathing Hole Sealing

In comparison with the open-water harpooning of even the largest species of sea mammal, the hunting of ringed seals at their breathing holes requires yet more complex technology. In areas where the breathing holes are completely covered in snow and are thus invisible, a situation common in the channels within the Canadian Arctic Archipelago, dogs were used to locate them by scent by Historic Inuit. Even in regions where the breathing holes are visible, such as in areas where the ice is swept bare of snow by the wind, or along a newly frozen lead, if the breathing hole is disturbed at all the seal will recognize that fact and avoid it. The temporary insertion and rotation of a breathing-hole probe—a long curved bone pin—into the air hole at the top of the breathing hole allows the hunter to ascertain the precise location and dimensions of the breathing hole beneath the ice without otherwise disturbing it. This allows him to know exactly where to thrust his harpoon when the seal arrives. However, the air hole itself is very small so the hunter is unable to see when a seal is rising in order to know when to thrust. The solution, especially with snow-covered breathing holes, is to use some sort of indicator which informs the hunter when the seal is rising in the hole. Ethnographically, indicators could take the form of a sinew device to which a piece of down was attached, which fluttered as air was expelled from the breathing hole by the rising seal, or a long and thin bone pin whose bottom end rested on the thin layer of ice that formed on the water's surface within the breathing hole while its top end protruded from the hole. When the seal rose in the breathing hole the ice would shift and then break, causing the pin to move up and down. Both types of indicators alerted the hunter to the arrival of a seal, at which he would plunge his harpoon down into the center of the breathing hole in order to harpoon it. Unfortunately, both types of indicator are unlikely to survive in a form that would be recognizable archaeologically. In the absence of unambiguously diagnostic artefactual evidence for breathing hole sealing, archaeologists are often forced to draw upon more inferential arguments. The time of year that a seal was killed can be determined from the analysis of thin-sections of its teeth. If mid-winter-killed seals are found at an archaeological site located at a great distance from areas where there was likely open water at that time of year, such as the floe edge, then the likelihood of open water hunting techniques being used at that location would have been low and breathing-hole sealing can therefore be inferred.

As noted above, in Alaska the coastal sites of the Arctic Small Tool tradition peoples are inferred to have been occupied during the spring or summer only, so on that basis they probably did not practice breathing-hole sealing. Early sites of the Arctic Small Tool tradition in the High Arctic that are inferred to have been occupied during the winter are located near the coast but the faunal remains suggest that stored food, including but not restricted to muskox, formed the basis of the mid-winter diet. There is somewhat more reason to infer that the Dorset culture practiced breathing-hole sealing, based largely on evidence that suggests that using the sea ice environment was an extremely important activity for them (see below). However, the Dorset appear to have lacked one important part of the technology that was an important feature of breathing hole sealing as practiced in historic times in areas where the ocean freezes completely and the breathing holes become snow-covered: dogs. If breathing hole sealing in such regions was important for the Dorset, then they must have had some other means of locating the breathing holes beneath the snow.

It is assumed that the Thule people in the Canadian Arctic and Greenland practiced breathing hole sealing in mid-winter as a supplement to the stored food they had accumulated during the open water season, as some sites located well away from the floe edge contain some winter-killed seals. However, opinion varies on when and how the Thule developed the technique. Some archaeologists have speculated that they learned it from encounters with the Dorset but, even setting aside the chronological problems with this scenario, it seems improbable. The one aspect of hunting ringed seals at their breathing holes with which the Thule arriving from Alaska might have been unfamiliar—the locating of breathing holes beneath snow-covered expanses of ice—was something that their descendants would do with dogs. Since the Dorset didn't have dogs, it is difficult to imagine the Thule learning the technique from them. Instead, it is likely that the Thule simply extended a hunting technique with which they had already become familiar back in Alaska.

25.4.1.5 Living on the Sea Ice

The most extreme component of an adaptation to the Arctic coastal environment in the winter is the ability to actually live out on the sea ice for extended periods of time. As documented ethnographically, in the central part of the Canadian Arctic archipelago this adaptation was largely predicated on the ability to hunt ringed seals at their breathing holes since there was no floe edge nearby and no other reliable food resources were available there at that time of year. In addition to the technology required for breathing hole sealing, two additional items of technology would seem to be necessary for a successful adaptation to living on the sea ice environment: snow houses (igloos), and lamps. Snow houses are both thermally and logistically very efficient. Snow provides considerably more insulative effectiveness than even a double-walled tent and the ability to quickly create a new snow house when needed means that heavy tents do not have to be transported, lightening the load on the sled. Snow houses and campsites out on the sea ice obviously do not survive archaeologically but the one diagnostic implement necessary to construct snow houses does: the snow knife. Snow knives (Fig. 25.3l) were used to cut the blocks of snow to construct snow houses, and their appearance in the archaeological record shows that snow houses were being used. Snow probes—long straight bone probes designed to ascertain the consistency of snow for construction purposes—are another item of material culture associated with snow houses. Lamps are perhaps the most important item of technology that is necessary for living in snow houses. Semi-subterranean houses and tents can be heated by open fires using driftwood or dwarf willow twigs. However, such wood is obviously not available out on the sea ice and it is too difficult to control the heat generated from an open wood fire with enough precision to make use of such a fire in a snow house without melting the roof. Lamps designed to burn oil derived from sea mammal fat both use a resource that was available out on the sea ice—blubber from seals—and allow very fine control of the amount of heat that is generated. Although lamps can be created from a number of materials, stone, especially softer soapstone (steatite), was preferred. Archaeologically, lamps often survive only as fragments and sometimes these cannot be differentiated reliably from the fragments of stone cooking vessels, but the appearance of identifiable lamps in the archaeological record at least indicates that one of the items of technology necessary for living in snow houses on the sea ice was present (Fig. 25.3m).

No snow knives have been identified from Arctic Small Tool tradition sites. Lamps are not found in early Arctic Small Tool tradition sites anywhere in the Arctic, and are only infrequent finds on later sites. One conclusion that can be drawn from these facts is that snow houses either were not used or were used very rarely. Stone lamps do become more common in the Dorset culture along with the very first snow knives, consistent with the at least occasional use of snow houses. However, the Dorset also constructed substantial semi-subterranean winter houses heated by open box hearths so some substantial portion of the winter was presumably spent onshore in these coastal sites. Other evidence for the importance of the sea ice environment to the Dorset comes from the appearance of small sled shoes. The almost complete absence of dog remains from Dorset sites and the small size of the sled shoes suggest to most researchers that the sleds themselves were very small and hand-drawn. Ice-creepers—essentially crampons—also appear with Dorset, suggesting that being able to walk on the ice was important in the Dorset adaptation. Taken together, this evidence suggests that the Dorset people lived and hunted on the sea ice to a greater extent than earlier populations, or at least possessed a more technologically sophisticated adaptation to the sea ice environment.

Despite their demonstrated ability to accumulate large quantities of meat and blubber prior to the winter by hunting large sea mammals on the open ocean, the Thule people appear to have made extensive use of the sea ice environment as well. Evidence for this comes from very common occurrence in Thule archaeological assemblages of snow knives, lamps, sled shoes from large sleds, and dog harness parts. Items that are either snow probes or breathing hole probes are also common finds although since the only significant difference between them is the curvature of the breathing hole probes, in fragmentary archaeological specimens it is difficult to distinguish them reliably. Seal bones from winter and spring-killed seals are also evidence that the Thule made extensive use of the sea ice environment despite living in their large winter houses on the coast. This clearly set the stage for some of their Inuit descendants in the central part of the Canadian Arctic adopting a settlement pattern that involved spending almost the entire winter out on the sea ice and relying almost exclusively during those months on breathing-hole sealing.

25.5 Landscape and Ideology

As noted in the ethnographic summary presented above, the land–sea dichotomy figured prominently in Inuit ideology in historic times. Hints of the ideological importance of the sea ice environment and its resources are found in some of the cultural traditions for which we have adequate artefact preservation. For example, the Dorset culture is well known for exquisite small carvings of animals, both sea mammals and land mammals (Taylor and Swinton 1967; McGhee 1985; LeMoine et al. 1998) and it is possible that some Dorset structures are metaphorical representations of the sea ice environment (Park 2003). However, the best evidence for the prehistoric existence of the ideology of the land-sea dichotomy comes from the Thule culture in the Canadian Arctic. In the course of studying the artefact assemblage from a Thule site, McGhee (1977) discovered that all of the material culture associated with hunting land mammals, primarily arrowheads, was manufactured from a terrestrial product (antler), whereas all of the material culture associated with hunting sea mammals, especially harpoon heads, was made of walrus ivory or whale bone—in other words, materials from marine mammals. There being no obvious functional advantage explaining why those materials should be preferred for those functions, he concluded that what he was observing was an extrapolation of the rules found among the historic Inuit governing the separation of things of the land from things of the sea. His finding was especially intriguing because no similar injunction about raw material source for hunting implements existed in historic times. McGhee's study has not been duplicated or expanded upon largely because the findings at the site he studied appear to be unique. At other Thule sites harpoon heads are often made of antler and arrowheads can be made of sea mammal bone or ivory, as they sometimes were in historic times. This in itself is intriguing, however, because it demonstrates that variation existed within cultural traditions with respect to ideas concerning the landscape. Future research is likely to increase our understanding of how the Arctic landscape was conceptualized by the various prehistoric cultures that lived there.

25.6 Change in the Arctic Coastal Landscape

In much of the preceding discussions the Arctic coastal landscape has been treated implicitly as unchanging over the course of human occupancy of the Arctic, but of course it has changed in important ways over the millennia. Changes in climate and in isostasy are especially relevant in understanding the history of human use of the landscape.

Changes in the climate have undoubtedly had profound impacts on the ways societies used the Arctic landscape, although many of the linkages that have been proposed between specific instances of climate change and specific instances of culture change are best characterized as speculative rather than well established. Nonetheless, many of the cultural changes summarized above may have been driven by or made possible by climate changes (McGhee 1969–70; Schledermann 1976; Jacobs 1979; Mason and Gerlach 1995; Freskild 1996; Jacobs et al. 1997; Ferguson et al. 1998; Murray et al. 2003;). Our knowledge of the history of climate

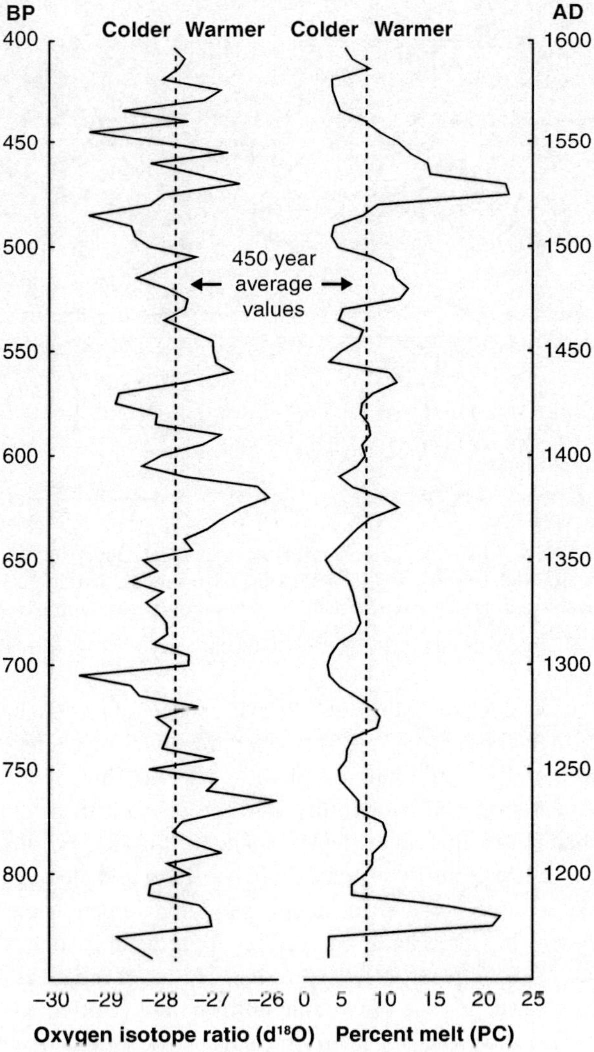

Fig. 25.4 Comparison of oxygen isotope (reflecting mean annual temperature) and percent melt (reflecting summer warmth) values obtained from the Devon Island ice cap for the period 850–400 BP. (After Alt et al. 1985)

change in the Arctic comes primarily from ice cores taken from ice caps in Greenland and in the Canadian Arctic. Figure 25.2 presents the general climate change trends that have been documented for the time humans have lived in the North American Arctic contrasted with the cultural sequence; Fig. 25.4 presents much more detailed information about the climate in the Canadian High Arctic during the period 850–400 BP. Two aspects of this figure are worth commenting on. First, the most common measure of ancient temperature derived from ice cores, oxygen isotope ratios, provide information primarily about temperatures during the seasons when the precipitation was accumulating; another measure, percent melt, provides more information about summer temperatures. As Fig. 25.4 illustrates, these two measures did not always change in unison, meaning that the climate could differ over time in very complex ways. The figure also demonstrates that there could be significant differences in annual and summer temperatures over very short time spans. Unfortunately, the chronological resolution that is possible from ice core data is far more accurate than is possible with archaeological dating techniques, meaning that it is practically impossible to correlate specific archaeological sites with any of the specific climate fluctuations that show up in this figure.

With those caveats, it is possible to identify ways in which climate changes must have affected human use of the Arctic landscape. The most significant effect would have been on the extent and annual duration of the sea ice. Generally, climate warming would have resulted in earlier break-up of the ice in the summer and later freeze-up in the autumn. Such changes could have major impacts on human groups relying on sea mammals. A longer open-water season would increase the duration of time each summer that migratory species such as bowhead or beluga whales were available to be hunted from boats. It has been speculated that the warming trend that culminated approximately 1000 years ago extended the range and availability of bowhead whales within the Canadian Arctic archipelago, and that their increased availability encouraged or facilitated the Thule migration from Alaska. However, a longer open-water season would also reduce the time available for hunting sea mammals from or through the sea ice. It has been speculated that the Dorset culture, for which the sea ice environment seems to have been especially important and for which there is little evidence of boat use, was best adapted to a colder climate with only a short open-water season (Maxwell 1985). The general warming trend that took place in the centuries preceding the disappearance of Dorset may have put the culture's adaptation under severe stress. A longer open-water season could also, if it involved a significantly earlier break-up of the sea ice, reduce the year-round abundance of one of the most important non-migratory species (ringed seals) at least on a regional basis. Ringed seal pups are born on the sea ice in March and if the ice breaks up before they are weaned they are unlikely to survive. More than a century ago Boas (1888) identified a correlation between the distribution of human populations and the shapes of coastlines due to this same factor—in areas of convoluted coastline the sea ice tended to survive later into the summer and a higher proportion of ringed seal pups were successfully weaned, leading to larger numbers of seals overall and attracting human populations to those areas.

Isostatic adjustment is another factor that has changed the Arctic coastal landscape dramatically over the course of human occupation in the North American Arctic. Beach ridges are a feature of many Arctic coasts that is relevant both for its prehistoric inhabitants and for contemporary archaeological research. In much of the Canadian Arctic, which was covered by the massive Laurentide Ice Sheet during the last glaciation, isostatic adjustments since deglaciation have gradually raised many former beaches well above today's coastline. Because coastal inhabitants reliant on sea mammals normally camped very near the water edge during open water seasons, a useful horizontal stratigraphy was created in many locations. As one walks inland and uphill from today's coastline one encounters increasingly old archaeological sites (although in some regions such as east coast of Baffin Island the postglacial isostatic adjustments have been more complex and old sites are now submerged). Such isostatically created beach ridges are not found in unglaciated Alaska but in a few parts of that region the effects of tides and storms have over millennia resulted in the regular addition of new accretional beach ridges. In those locations the same kind of horizontal stratigraphy has been created, with older archaeological sites being situated further inland than more recent sites.

In addition to providing archaeologists with a useful means of assessing the relative ages of archaeological sites, the process of isostatic uplift would have changed the coastal landscape gradually but relatively dramatically. Figure 25.5 illustrates the island of Igloolik as it is today and as it would have been during its intensive occupation by peoples of the Arctic Small Tool tradition when the land was approximately 30 m lower relative to sea level than it is today. The fact that the island was much smaller at that time is relevant, but of perhaps even more relevance for humans living there would be the changes in the sea environment. The bathymetry of the waters adjacent to the island would have changed just as dramatically as the land and this would have had important effects on the summer and winter distribution of sea mammals. Some important

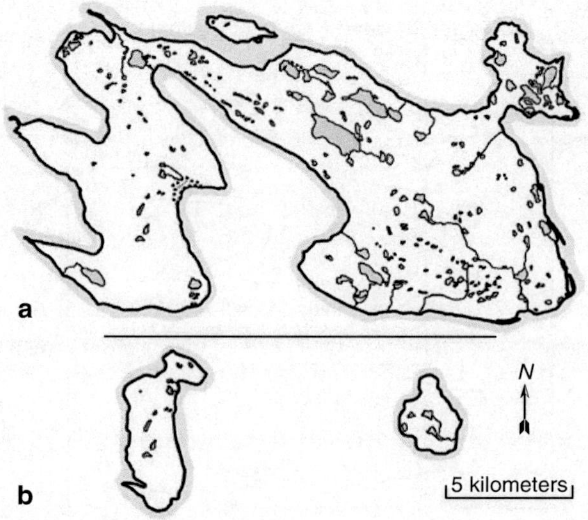

Fig. 25.5 Changing landscape in the Canadian Arctic due to postglacial isostatic uplift. **a** Igloolik Island today. **b** Igloolik Island during the Arctic Small Tool tradition period approximately 3800 BP. (After Dredge 1991)

species such as walrus and bearded seals obtain much of their food from sea-bottom resources such as clams, so they frequent relatively shallow waters. Thus, their distribution and availability can be expected to have changed as the water depths changed. Further, ocean currents are partly responsible for opening and closing leads in the ice, and for keeping polynyas open. This means that leads and polynyas occur predictably in the same locations from year to year. However, changes in water depth due to isostatic uplift will have affected the location and speed of currents, meaning that the distribution of leads and polynyas, and thus of sea mammals and other important sea ice resources, will likely have been quite different in the past.

25.7 Conclusion

Although historic Inuit ideology often drew a profound distinction between the land and the sea, in fact a successful adaptation to frozen coastal regions required that resources of both be drawn upon. Through increasingly complex technology, the ancestors of the Inuit developed ways to reliably exploit the resources of this daunting environment. It appears that successful adaptations to the extreme sea ice environment of the central and western part of the Canadian Arctic Archipelago were developed at least twice, first by the peo-

Glossary

Glossary

Abbasid caliphate third caliphate of the Islamic Empire. It reached its greatest extent in the middle of the ninth century AD.

acidification a progressive change towards a low pH especially in soil.

aeolianite a sandstone (usually calcareous) formed by wind transported sediment.

Agatha's veil during eruptions of Mt. Etna, the people of Catania have exposed St. Agatha's veil for public veneration to ask for protection.

agent orange code name for a herbicide and defoliant used by the U.S. military in its Herbicidal Warfare program during the Vietnam War.

agora meeting place, Greece; especially a market place.

agrobleme "agricultural scar"—physical effects of agriculture on a landscape.

aguada perennially moist sink or depression, and one often altered for water storage around ancient Maya sites.

Alfisols a soil order of the US Department of Agriculture (USDA) soil taxonomy that has relatively high fertility and subsoils enriched in clay. Alfisols occur all over the world and often form under forest vegetation.

alluvial fan a cone shaped accumulation of sands and gravels that occurs at the junction between a mountain or hill front and the facing plain.

alluvial soils a soil type found on floodplains with variable drainage conditions classified as Tropaquents in Sri Lanka and Fluvisols (WRB: World Refernce Base for soil resources), Fluvents (USDA: United States Department of Agriculture soil taxonomy classification).

altar an elevated place or structure for religious ceremonies.

Alþing ancient parliament of Iceland.

alveolar pyorrhoea a destructive disease of the supporting structures of the teeth.

Andosols Term used in the WRB Classification for soils that form on volcanic regions. Equivalent to Andisols in the US Department of Agriculture classification of soils, Soil Taxonomy.

ante quem the date (time) before which an archaeological artefact or event must have been deposited or occurred.

I. P. Martini, W. Chesworth (eds.), *Landscapes and Societies*,
DOI 10.1007/978-90-481-9413-1, © Springer Science+Business Media B.V. 2010

Antigori facies undecorated ceramic of the Bronze Age Nuraghic culture of Sardinia (1800–600 BC).

anthrobleme "human scar"—physical effects of all human activities on a landscape.

Anthropocene originally the roughly 300 years since the Industrial Revolution. Used here to mean the last 10,000 years of the Holocene, the period during which agriculture came to prominence.

Anthropogenic climate change long-term changes in climate driven by human activity, associated with increases in concentrations of greenhouse gases in the Earth's atmosphere which result in increased absorption of outgoing long wave radiation and a warming of the coupled land-atmosphere-ocean system.

aquifuge impermeable body of rock or of sediment.

aquifer underground layer of water-bearing permeable rock or unconsolidated materials.

Aramaean ethnic groups, some of whom were mobile pastoralists, that occupied extensive areas of Syro-Mesopotamia and who spoke languages of a west Semitic dialect. They were most prominent at the end of the second millennium and the early first millennium BC.

Archaeobotany the study of plant remains recovered from archaeological sites.

Archimedes' or Archimedean screw a screw within a hollow pipe used to lift water. Used to drain land in the Netherlands, powered by windmills. The development of powerful pumps during the Industrial Revolution rendered the technique obsolete.

Ard a primitive, light plow, without a mould board to turn the furrow.

arenaceous sandy.

arrowhead pointed striking tip of an arrow.

artefact an object produced or shaped by human craft, especially a tool, a weapon, or an ornament of archaeological or historical interest.

***Artemisia* steppe** a vast semiarid grass-covered plain dominated by herbs and shrubs of the genus *Artemisa.*

Asclepion A sanctuary dedicated to the Greek god of healing, Asclepios.

Austronesian a group of languages derived from Taiwan and spoken in many Pacific islands and Madagascar.

avulsion abrupt change in the course of a stream.

bajos low-lying depressions or sinks of the Maya Lowlands that range greatly in size from less than a hectare to thousands of hectares.

baroni landowners in Sicily, the lowest order of nobility in the feudal system in Europe; baron in English.

base flow the portion of streamflow that comes from groundwater and not runoff.

bastion projecting part of a fortification.

beach ridge wave-deposited ridge running parallel to a shoreline.

Beringia the name given to the land bridge that joined Siberia to Alaska during glaciations.

bethma practice a cultivation practice that temporarily redistributes plots of land among shareholders (paddy landowners) in part of the command area of a tank (reservoir) during drought periods, Sri Lanka.

betils conical stones near Giant's tombs of the Nuraghic culture of Sardinia.

bifacial tool a two-sided stone tool, manufactured through a process of lithic reduction, that displays flake scars on both sides.

biome climatically and geographically defined area of ecologically similar conditions reflected in communities of plants and animals.

biostasis the ability of an organism to tolerate environmental changes without having to actively adapt to them. A steady state in the biosphere.

biotope an area of uniform environmental conditions providing a living place for a specific assemblage of plants and animals.

biscotto hard bread, literally "twice cooked".

bisokotuwa A valve pit developed by Sri Lankans during the first century AD using the principle of the valve pit to discharge water from large reservoirs, Sri Lanka.

Black Death one of the deadliest a pandemics in human history of the fourteenth century, thought to be an outbreak of bubonic plague.

blubber the thick layer of subcutaneous fat characteristic of sea mammals. This was the primary source of fuel in Arctic regions that, when burned in lamps, allowed humans to live out on the sea ice in snow houses.

bog acid, *Sphagnum*-dominated wetland.

Bond events North Atlantic climate fluctuations occurring approximately every 1470 years throughout the Holocene.

bone tuberculosis contagious, wasting disease caused by any of several mycobacteria.

Book of Fire detailed (typically day-to-day) account on the progress of a volcanic eruption based on visual observations by a particular individual.

bottini tunnels that also form the underground aqueduct in Siena, Italy.

breathing-hole sealing a technique used by Inuit to hunt ringed seals through the holes they maintain throughout the winter in the sea ice. A vital technique for adaptation to the sea ice environment.

broad-leaved tress with relatively broad leafs rather than needle-like or scale-like leaves.

Bronze Age a period of human culture between the Stone Age and the Iron Age, characterized by weapons and implements made of bronze. In the Near East it lasted from about 3200 to 1200 BC.

Buxton line marks the modern boundary of the geographical distribution of malaria in the southwest Pacific.

castellari castle-like communities, Italy.

cenote vertical collapse sinkholes that have open, perennial water at a range of depths in the coastal plains and karst plains of the Maya Lowlands.

centuria a Roman unit of land area assigned to military veterans.

cerrado the form of savanna found in Brazil, the largest savannah region in South America and biologically the richest savannah in the world.

Chalcolithic (of Copper Age Chalcolithic) Copper Age beginning with copper mining in the Balkans about 7000 years BP and being succeeded by the Bronze Age.

chalklands the part of southern England underlain by chalk beds of Cretaceous age.

chena shifting cultivation, Sri Lanka.

chultuns underground Maya cisterns.

city-state a sovereign state consisting of an independent city and its surrounding territory.

civilization conventionally a complex human society with a centralized authority responsible for the organization of daily life and an advanced administration. It includes division of labor, cities supported by rural populations, and a public infrastructure. Civilization is 'structure', it may include several cultures.

Climatic Optimum the Holocene Climate Optimum was a warm period during roughly the interval 9000–5000 years BP.

comb facies cultural aspect characterized by decorated impressed pottery realized through the use of toothed instrument ("comb"), Sardinia during Nuragic age.

compaction the action of compressing soil particles together by loading the surface, thereby diminishing pore space and adversely affecting hydrodynamic properties.

comuni local authority areas.

contadi territories belonging to a family, or city.

crampons footwear to provide traction on snow and ice.

cuesta a ridge or hill characterized by a steep incline on one side and a gentle slope on the other.

cultivars a variety of a plant that has been created or selected intentionally and maintained through cultivation.

culture the distinctive ideas, customs, social behavior, products, or way of life of a particular society, people, or period.

cuneiform texts early form of wedge-shaped writing, frequently on clay tablets, in use in the Near East from the end of the fourth millennium BC until the first century AD.

cutbanks an erosional feature of streams - river cliff.

Daco-Roman Civilization a term used by Romanian historians to refer to a civilization formed by the combined Dacian traditions and Roman culture.

dandan-batake stepped fields, Japan.

Darwinian competition individuals or groups achieve advantage over others as the result of biological superiority.

deciduous plant shedding foliage at the end of the growing season.

default conditions non-negotiable principles that must be met in order for a system to exist in the real world.

deflation basins a topographic depression formed by wind erosion.

dendroclimatology the science of determining past climates from the width of tree-rings.

degradation (of soil) changes in a soil that diminish its ecological functioning in the biosphere.

desertification the transformation of arable or habitable land to desert by natural (change in climate) and/or anthropic (destructive land use) means.

dolii large ceramic vases, Sardinia of the Nuragic age.

drag float an inflated skin bag attached to a harpoon line, which permitted the hunting of large sea mammals from boats.

DRE (~20 km^3) = dense rock equivalent a volcanologic calculation used to estimate volcanic eruption volume.

drg or edeyen Arabic words for dune field: a large, relatively flat area of desert covered with wind-swept sand in form of dunes.

drought a long period of abnormally low rainfall.

drylands areas of low rainfall.

dustbowl (related to soil erosion) a region subject to drought, usually where agriculture has removed the cover of vegetation, leaving the soil prone to erosion by wind.

ecoedaphic relating to soil conditions, especially as it affects living organisms.

ecological footprint the total area used by a human being or a society to provide resources and to receive wastes.

ecotonal a transitional zone between two communities containing the characteristic species of each.

ecumene inhabited areas of the world.

El Niño a warm current that flows along the coasts of Ecuador and Peru when trade winds weaken and warm surface water flows from the western to the eastern tropical Pacific, overlying the normally cold waters of the Peru current.

emishi peoples of the northeast, Japan.

enclosure surrounding a piece of land with a fence, hedge or boundary; especially the act of converting common land into private property.

enclosure (in England and Whales) used to refer to the process that ended the ancient system of arable farming, mowing meadows, o rgraing lierstock in open field or owned by others.

endorheic closed basin with internal drainage.

Eneolithic transitional period between the Neolithic Period and the Bronze Age, during which the earliest metallic (copper); synonym of Chalcolithic.

entablocure horizontal superstructure supported by columns and composed of architrave, frieze, and cornice.

entropy a thermodynamic property of a system, commonly taken to be a measure of disorder, which increases in spontaneous processes.

erosion the process of loosening and transporting materials of the surface of the earth by the agency of wind, water, freeze and thaw or organisms including humans.

estuary part of the wide lower course of a river where its current is met by and mixes with seawater.

European Enlightenment an intellectual movement that was predominant in the Western world during the eighteenth century during which tradition and authority in western European society were questioned by philosophers devoted to the idea that human reason should be the basis of improving the lot of humanity.

evapotranspiration the loss of water from soil by evaporation and by transpiration from plants growing in the soil.

evergreen having foliage that persists and remains green throughout the year.

ex nihilo out of nothing (Latin).

exedra semicircular area in front of tombs delimited by upright stone slabs or walls.

eyden or drg Arabic worlds for dune field: a large, relatively flat area of desert covered with wind-swept sand in form of dunes.

fabric a cloth produced by knitting, weaving, or felting of fibers.

fast wheel a sort of fly wheel made of wood for making pottery during the Nuragic age in Sardinia, Italy.

Fertile Crescent a crescent-shaped region of the Middle East with one wing stretching along the eastern shores of the Mediterranean to Anatolia, the other reaching to the Persian Gulf. Encloses the valleys of the Tigris and Euphrates.

fire hearths places where a fire was confined for domestic or other purposes.

fissure eruption volcanic eruption where lava flows through fissures in the earth's crust rather than central vents.

flint a very hard, fine-grained quartz that produces sparks when struck.

flood lava eruption enormous emission of lava amounting to millions of tonnes per hour.

floodplain a plain bordering a river and subject to flooding.

floodwater farming the practice of planting crops in areas that are flooded every year in the rainy season.

foggara a traditional systems of water catchments made up of horizontal underground shafts that drain water and convey it by gravity. North African usage, equivalent to qanat of Middle East.

foredeep a long, narrow depression that borders an orogenic belt, or an island arc, on the convex side.

foredune first wind-blown dune ridge bordering the sea.

frame of gold coastal and near coastal area of Sicily with well developed agriculture.

gabelloti agents who rented small plots to peasant farmers in Sicily.

garrigue scrubland made-up of evergreen species of Mediterranean affinity; equivalent ot marquis in France.

gasgommana an upstream tree belt of small tanks grown naturally, Sri Lanka.

gathering foraging, the action of collecting food from wild plant, such as berries, roots, and grains.

ghelta semi-permanent water pool typical of desert regions.

ghost acreage the amount of land needed to support a crop, over and above the land on which the crop is actually growing. A mean of determining what extent of land is needed to meet the growing human needs.

Giant's tombs typical collective tomb of the Nuragic period; it consists of a long burial chamber preceded in front by a semicircular ceremonial area (exedrae).

gley a soil or part of a soil that has been reduced by the exclusion of oxygen caused by waterlogging.

godawala an upstream water hole constructed to trap sediment entering into the tank from its catchment, Sri Lanka.

gomito elbow (Italian).

graben a usually elongated depression between geologic normal faults.

green revolution the significant increase in agricultural productivity, particularly of rice and wheat, brought about by the introduction of high-yielding varieties of crops, the intensive use of fertilizers and pesticides, and the development of a supportive infrastructure.

groyne a protective structure of stone or concrete; extends from shore into the water to prevent a beach from being washed away by waves and longshore currents.

gualcherie fulling mills to make woollen cloth, Italy.

Harmattan dry wind from the Sahara.

harpoon a spearlike weapon with a barbed head used in hunting fish.

heathlands open ground, often described as waste land or wilderness, covered by species of *Erica* and similar acid-tolerant plants.

heiya plain, Japan.

hemp the tough, coarse fiber of the cannabis plant, its seeds can be eaten.

highland elevated land.

Histosol a soil order of the US Department of Agriculture (USDA) soil taxonomy that is composed dominantly of organic matter and form in wetlands. Common names for these soils are peats and mucks. The WBA Classification uses the same term.

hoe a tool for digging soil with a flat blade attached approximately at a right angle to a long handle.

Holocene Climatic Optimum the period lasting from approximately 9000–5000 year BP during which conditions were somewhat warmer than at present in some mid-latitude regions and considerably more humid in many sub-tropical regions.

Holocene geological epoch which began approximately 11,700 years ago (10,000 ^{14}C years ago).

hum a residual limestone hill in the middle of a karstic depression.

husbandry The act or practice of cultivating crops and breeding and raising livestock; agriculture.

hydraulic civilization generally applied to the civilizations that developed in major river valleys of the Old World (for example the Nile, Tigris and Euphrates, Indus and Huang He rivers). Less commonly applied for the period 500–1200 BC in Sri Lanka, where civilization developed based on an intrinsic network of water storage, conveyance and utilization.

hydraulic concrete cement that hardens under water.

hyperarid extremely arid area, with an aridity index of less than 0.05.

hyperostosis is an excessive growth of bone.

hyperostosis porotic (porotic hyperostosis) a disease that causes bone tissue to appear spongy and become soft.

hypohaline lower salinity than normal sea water.

ice-creeper shoe fitting to avoid slippage on ice.

igloos commonly used to mean domical Inuit houses built of blocks of compact snow. In Inuktitut the word refers to any kind of house.

illuviation deposition in an underlying soil layer of compounds leached out of an overlying layer.

imbricated to be arranged with regular overlapping edges like roof shingles.

insolation a measure of the energy of solar radiation received on a given surface area in a given time.

intramontane basins geological/topographic lowland basin between mountain/hill ranges.

IOM (Indian Ocean Monsoon) a wind from the southwest or south that brings heavy rainfall to southern Asia in the summer.

Iron Gates identifies the Danube's defile sector in Romania designating the passage between the Haţeg Depression and the Bistra Corridor.

irrigation artificially supplying water to land, usually by means of channels or streams, in order to support a crop.

intensification (of agriculture) the action of increasing agricultural productivity from a given plot of land.

isostasy general equilibrium of the forces tending to elevate or depress the Earth's crust.

isostatic adjustment concomitant adjustment in the relative elevation of land and ocean as load on the lithosphere changes: for example as very extensive glaciers accumulate or decay.

iswetiya* or *potawetiya an upstream soil ridge constructed at either side of the tank bund to prevent entering eroded soil from upper land slopes, Sri Lanka.

ITCZ Intertropical Convergence Zone near the equator where the northeast trade winds and southeast trade winds converge; solar heating in the region forces air to rise through convection resulting in intense precipitation.

jökulhlaup (glacier burst) is a glacial meltwater outburst flood; in Iceland they are commonly triggered by volcanic activity beneath a glacier.

jori gridded field system, Japan.

karahana a water distribution device fixed across a canal made up of log with two weir shape cuts, Sri Lanka.

karst landscapes formed primarily by water dissolution of bedrock, commonly limestone.

kattakaduwa a reserved land below the tank bund to safeguard the tank bund from breaching, Sri Lanka.

kayak a type of small, usually one-person covered skin boat utilized by many Arctic cultures.

kekulama a cultivation practice that advances the cultivation time using early seasonal rains whenever they feel that tanks would not get enough water to cultivate the command area, Sri Lanka.

kijiya wood-workers, Japan.

Kladéos right bank tributary of the Alpheios River, Greece.

kobiki lumberjacks, Japan.

koku (1 koku = 180 l), Japan.

kulinas the dry zone nobility who possessed irrigation expertise, Sri Lanka.

land (in ecology) the complex of landcape components such as soil, water, flora and fauna, that make up the terrestrial, as opposed to marine, biosphere.

landfast ice in Arctic waters, a strip of ice that can be many kilometers wide, extending out from the coast to the floe edge.

Landnámabók Book of Settlement, compiled in the twelfth Century, is the oldest document containing information on historical eruptions (Iceland).

Lapita cultural complex a prehistoric archaeological culture whose bearers colonized Melanesia and Polynesia.

latifundia an extensive agricultural estate owned by wealthy landowners. Roman in origin, the usage extended to other places such as in Latin America by the Spanish conquests.

lava aa (aa = means hard to walk on) lava field with jagged surface (Hawaiian).

lava pahoeho*e* lava field with smooth, lobate and undulating surface (Hawaiian).

lava molten rock that reaches the surface and is emitted through a volcano. The surface expression of magma.

lead in Arctic waters, a linear stretch of open water within an area of otherwise continuous sea ice.

Levant the eastern part of the Mediterranean including adjoining countries and islands.

liminal status temporary status of workers.

lithic industry the part of an archaeological assemblage of artefacts, manufactured of stone.

Little Ice Age a period of cooling dating from about 550 to 1850 AD.

loess a pale yellow, usually calcareous clayey-silt deposit mostly transported and deposited by wind.

Low Humic Gley soils a poorly drained soil type found in the dry zone of Sri Lanka classified as Tropaqualfs in Sri Lanka.

mace the symbol of royal authority, originally used as a battle weapon carried by the royal bodyguard.

macehead the stone or metal top of a mace.

mafia a loose, secret association of criminal groups that share a common organizational structure and code of conduct. Thought to have originated in Sicily.

maha season the major cultivation season in Sri Lanka prevails from September to February.

Malthusian concerning the ideas of Malthus—that population increase always has the power to outstrip any increase in the food supply.

Malthusian disaster a return to subsistence-level conditions as a result of agricultural or other economic production being eventually outstripped by growth in population.

marling artificial addition of a carbonate rich material to soil.

matagi hunters, Japan.

Medieval Warm Period period from about 1000 to 1400 AD in which global temperatures were a few degrees warmer than those of today.

megaron temples religious edifice characterized by a rectangular floor plan and the extension of the side walls beyond the façade.

Melanesianization a concept to explain various differences observed between the Austronesian cultures of firstly, Vanuatu and New Caledonia, and secondly, Polynesia.

mesa a broad, flat-topped elevation with clifflike sides.

Mesopotamia literally "The Land Between the Rivers", referring to the Tigris and Euphrates. Used here to denote the greater Mesopotamian region encompassing most of Iraq, northeastern Syria, parts of southeastern Turkey and southwestern Iran.

metope entablature of building.

midden a mound or deposit containing shells, animal bones, and other refuse that indicates the site of a human settlement.

Mistral wind from the NW, Western Mediterranean area.

moernering getting salt by burning peat previoiusly inundated by salt water in the Netherlands.

Mollisol a soil order of the US Department of Agriculture (USDA) soil taxonomy that has thick topsoils and is nutrient rich.

monsoon domain geographic area where the rainfall regime regulated by a seasonal prevailing wind which lasts for several months.

mortar a vessel in which substances are crushed or ground with a pestle.

NAWQA National Water-Quality Assessment Program, USA.

nematode unsegmented worms with elongated rounded body pointed at both ends. Ubiquitous in soil.

Neolithic literally the New Stone Age, the period that started about 10,000 BC and when a revolution in food production took place involving the cultivation of soil.

Neolithization related to the Neolithic Revolution that marks the transition from hunting and gathering communities and bands, to settled agriculture.

Nunamiut people of the land, referring to natives of the Arctic.

nuraghe conical, several-storied, dry-stone, truncated tower.

nympheion a monumental fountain, Greek.

oasis a fertile or green spot in a desert or wasteland, made so by the presence of water.

observatory a building designed and equipped for making observations of astronomical or other natural phenomena.

Oldest Dryas the coldest climatic period that occurred after the maximum of the last major Pleistocene glaciation.

ophiolites assemblage of metamorphic and igneous rocks, found in orogenic belts and interpreted as remnants of pre-orogenic oceanic crust.

orti fields, Italy.

OSL (optically stimulated luminescence) a method of dating quartz-grain bearing sand.

palaestra in ancient Greece rectangular enclosure attached to a gymnasium where athletes competed in various sports before an audience.

Paleolithic literally the Old Stone Age, the earliest period of the use of stone tools.

palynofacies an assemblage of palynomorphs in a portion of a sediment, representing local environmental.

Palynology the study of pollen and pollen-like fossils that serve as proxy evidence for past environmental changes.

pantropic distributed throughout the tropics.

Parallel Climate Model The Parallel Climate Model (PCM) a joint effort to develop a US Department of Energy (DOE) sponsored parallel climate model between Los Alamos National Laboratory (LANL), the Naval Postgraduate School (NPG), the US Army Corps of Engineers' Cold Regions Research and Engineering Lab (CRREL) and the National Center for Atmospheric Research (NCAR).

pastoralism the branch of agriculture concerned with the raising of livestock. Also the type of nomadic, non-industrial society that this implies.

Pax Romana the Roman peace; the long period of stability under the Roman Empire.

pedogenesis the processes by which soil orignates.

pedogenetic adjectival form of the previous term of pedogenesis.

perahana an upstream meadow formed as undergrowth of *gasgommana*, Sri Lanka.

permafrost subsurface deposits that remain frozen year-round for multiple years, beneath a thin surficial active layer that melts each summer.

pescaie lowlying overflow dams used to deepen the river flow and deviate some of the water to mills, Italian.

pesticide a biocide used to kill pests.

pestle a club-shaped, hand-held tool for grinding or mashing substances in a mortar.

petroglyphs engravings on a natural rock face.

Phoceans Ancient Greek navigators and colonizers from the island of Phocea, now Foça.

piedmontese adjectival form of piedmont (mountain foot region). originally in Northern Italy (regione piemontese), hence the adjective piedmontese.

placer alluvial deposit that contains particles of some valuable mineral, for example gold, cassiterite, or diamond.

planation flattish plain resulting from erosion; also the process by which such a plain forms.

plassen lakes left after peat-cutting, Netherlands.

playas a dry or ephemeral lakebed usually characterized by salt deposits.

Pleistocene the first epoch of the Quaternary period, lasting about 2 millions to about 11,700 years ago. It was characterized by extensive glaciations in the northern hemisphere and the evolutionary development of man.

polder tract of land protected from inundation, in which the water level can be regulated.

polje a large flat plain in karst territory with areas usually 5–400 km^2 (areas of coalescent karst sinkhole or dolines).

polity an organized society having a specific form of government.

polynya a year-round or seasonal area of open water within sea ice, often resulting from strong ocean currents or winds.

ponor swallowing karstic hole, Greece.

Portus Pisanus medieval principal sea-port of Pisa, Italy.

psl relative present sea level.

pyroclastic ash very fine rock fragments of volcanic origin.

qanat underground aqueduct in which water is gathered from the ground water table by means of a gently sloping tunnel ventilated by vertical access shafts. Developed in the first millennium BC, and particularly common in Iran and Oman.

rapid climate change change in climate and climate-sensitive natural systems of sufficient rate and magnitude to be represented by decade to century scale anomalies in proxy records, noticeable over timescales of the order of a human lifetime, and potentially problematic for the functioning of extant ecological and social systems.

Reddish Brown Earths a well drained soil type found in the dry zone of Sri Lanka classified as Rhodustalf.

regione piedmontese mountain foot (piedmont) region of Northern Italy.

regolith a layer of loose, heterogeneous lithic material covering solid rock.

Rendolls a soil of the US Department of Agriculture (USDA) soil taxonomy that is as type of Mollisol. Thin soils, usually under forest covers and usually over limestone sequences.

rickets is a softening of bones in children, predominantly caused by vitamin D deficiency, and potentially leading to fractures and deformity.

riparian relating to or located on the banks of a river or stream.

rock varnish dark coating found on exposed rock surfaces in arid environments, formed under slightly wetter environmental conditions.

sabkha coastal plain with a salt crust.

sacred spring religious edifice constructed around a ground level spring.

sacred well construction around a subterranean water-source composed of an atrium, a stairway and an underground chamber often covered with an overhang.

Sahul continent combined land mass of Australia, New Guinea, and Tasmania at times of low sea level during the Ice Ages.

salinization the deposition of salts, sodium chloride and carbonates predominantly, in the soil, by evaporation. Tends to be exacerbated by irrigation in arid and semi-arid regions.

sand sea extensive assemblage of sand dunes of several types.

sapropel a mud rich in organic matter formed at the bottom of a body of water.

savannah or savanna a flat, commonly sparsely treed grassland of tropical or subtropical regions.

scagliose scaly, Italy.

Scirocco wind from the south in the Western Mediterranean area.

sedentarization term applied to the transition from nomadic to permanent, year-round settlement.

semiarid climatic conditions characterized by relatively low annual rainfall of 250–500 mm and having scrubby vegetation with grasses.

semi-subterranean houses pit dwellings with low raised walls, surmounted by raftered roofs characteristic of some Arctic cultures.

seriation in archaeology is a relative dating method in which assemblages or artifacts from numerous sites, in the same culture, are placed in chronological order.

Shamal northwesterly wind, often strong, that blows in the Arabian Gulf region.

shaman a member of certain tribal societies who acts as a medium between the visible world and an invisible spirit world and who practices magic or sorcery for purposes of healing, divination, and control over natural events or people.

shrubland a habitat type dominated by woody shrubs.

silica karst processes and landforms of great similarity to limestone karst, found on many of the world's quartz sandstones.

slag the vitreous mass left as a residue by the smelting of metallic ore.

slash and burn practice consisting in cutting and burning of woodlands to create fields for agriculture or pasture for livestock (swidden).

sled shoes strip along the bottom of a sled runner.

snow house a kind of dwelling constructed entirely out of snow blocks and used by many Arctic cultures. Commonly known as igloos, although in Inuktitut that word refers any kind of houses.

soapstone (steatite) a soft metamorphic rock composed mostly of the mineral talc.

society organizational form in which individuals of a species live together.

Society, human a group of people living together for their mutual benefit, with a distinct culture.

soil a deposits composed of loose mineral particles and and/or organic matter, mostly layered (horizonated), generally plant bearing that covers most of the Earth's land surface.

soil micromorphology (or micropedology) study of soil and sediments (including archaeological sediments) in thin section using a petrographic microscope.

solares kitchen gardens.

solum the surface (A horizons) and subsoil (B horizons) layers that have been subjected to the same soil forming conditions.

spherulite (fecal) microscopic mineral features found in animal dung and formed of microcrystalline calcium carbonate.

splash dam temporary wooden dam used to raise the water level in streams to float logs downstream to sawmills.

stellingen area artificially inundated to no more than a few cedimenters (to impede movements of troups but avoid floating of boats), for defensive purposes, Netherlands.

steppe a vast semiarid grass-covered plain (prairie in North America, pampa in South America).

Sub-Carpathians peri-Carpathian physiographic unit consisting of an alternation of hills and depressions marking the transition between the Carpathian Mountains and their surrounding plains.

subsistence practices human activities such as hunting, gathering or agriculture that provide the necessities for living.

Sundaland a biogeographical region of Southeastern Asia that comprises Malaysia, Sumatra, Java, and Borneo, which were joined at low sea level during the Ice Ages.

sustainability an ecological state that can be maintained for a defined length of time.

swidden an area cleared for temporary cultivation by cutting and burning the vegetation.

tafoni small, shallow cavities with overhanging roof formed by weathering processes.

tank cascade system a connected series of tanks organized within small catchments of the dry zone landscape, storing, conveying and utilizing water from an ephemeral rivulet, Sri Lanka.

Tareumiut people of the sea.

tell prominent mounded archaeological site consisting of superimposed levels of occupation, usually rendered more prominent by an encircling fortification wall.

tent ring the rocks that were used to weigh down the edges of a skin tent. Tent rings are all that remain archaeologically of many campsites in the Arctic.

tephra fragmented volcanic rock and lava regardless of size that are emitted by a volcanic eruption.

terp artificial mound for surviving inundations in areas not protected by dikes, Netherlands.

territorial empire territorially expansive polities in which a ruling power effectively controls, dominates and sometimes unifies a number of smaller and often weaker subordinate societies and their territories.

tethering stone stone slab, either roughly smoothed on their central part or finely pecked to obtain a groove to which a long rope was tied used to fasten an animal so that it can range or feed only within certain limits.

thalassemia an inherited form of anemia caused by faulty synthesis of haemoglobin.

tholos construction with a corbelled roof or "false cupola" obtained by the laying of successive stone courses. Circular domed building.

tidal marsh coastal zone inundated at particularly high tides.

Tifinagh a Berber alphabetic script.

till (agricultural) to prepare land for rising rock.

tis-bambe a fertile land strip found around the settlement area and the resting place of buffaloes, Sri Lanka.

tor a loose pile of rock formed by weathering of an originally coherent outcrop along bedding and/or joint planes and, usually found on or near the summit of a hill.

totemic image an animal, a plant, or a natural object serving among certain tribal or traditional peoples as the emblem of a clan or family and sometimes revered as its founder, ancestor, or guardian.

transfer fault a vertical or sub-vertical geological fault that, via primarily lateral movements, allows the juxtaposition of two geological terrains with different characteristics.

transhumance the seasonal transfer of livestock from one grazing ground to another, as from lowlands to highlands, with the changing of seasons. Hence the adjective transhumant: migrating between regions with differing seasonal climates.

tundra a largely treeless biome in high latitude regions characterized by the presence of permafrost.

turquoise a blue to blue-green mineral of aluminium and copper, prized as a gemstone in its polished blue form.

umiak a type of large, multi-person open skin boat utilized by some later Arctic cultures.

umi-no-sachi sea treasures, Japan.

une ridge, Japan.

upland land or an area of land of high elevation in relation to a local datum.

Vedda a tribe living in Sri Lanka that evolved in the middle Stone Age dependant on griculture, gathering and hunting, Sri Lanka.

VEI volcanic explosivity index to provide a relative measure of the explosiveness of volcanic eruptions. It has a maximum value of 8.

Vertisol a soil order of the US Department of Agriculture (USDA) soil taxonomy that is composed of deep, expanding and contracting clays. The term is also used in the WRB Classification.

Via Francigena major pilgrimage route through northern Italy from France (and other European localities) to Rome during medieval times.

vitric compound resembling, or having the nature of glass.

volcanic system active volcano-tectonic system up to 100 km-long and 20 km-wide, with a lifetime of 0.5–1.5 million years and featuring a fissure swarm or a central volcano or both.

volcanic zone a discrete 15–50 km wide belts of active faulting and volcanism.

wadi Arabic term traditionally referring to a valley, commonly a dry riverbed.

Wallace line the zoogeographic boundary between Asiatic species to the west and north, and a mixture of Asiatic and Australian species to the east. It approximately follows the edge of the Asian continental shelf.

waterlogging (of soils) saturating soil, that is filling the pore-space, with water.

Watt engine a stationary steam engine utilized for pumping water in the Netherlands in the late eighteenth and early nineteenth centuries.

welded soil buried soil overlapped by an upper soil.

yala season the minor cultivation season in Sri Lanka prevails from March to August.

yama-no-sachi mountain treasures, Japan.

Younger Dryas a phase of reduced temperatures and glacial advance between approximately 11,000 and 10,000 BP.

zigurrat an Assyrian or Babylonian temple-tower with the form of a pyramid in which each successive storey is smaller than the one below so as to leave a terrace all round.

Index

I. P. Martini, W. Chesworth (eds.), *Landscapes and Societies*,
DOI 10.1007/978-90-481-9413-1, © Springer Science+Business Media B.V. 2010

N

O

Color Plate Section

Plate 1 **a** Burial tumulus near Tifariti, Western Sahara, one of over 400 funerary monuments recorded in an area of some 9 km^2 (Brooks, Chap. 4). **b** Pillar of Holocene lake sediment in the central Sahara of south-western Libya, an outcrop of a much more extensive series of deposits, around which are numerous Holocene stone tools (Brooks, Chap. 4)

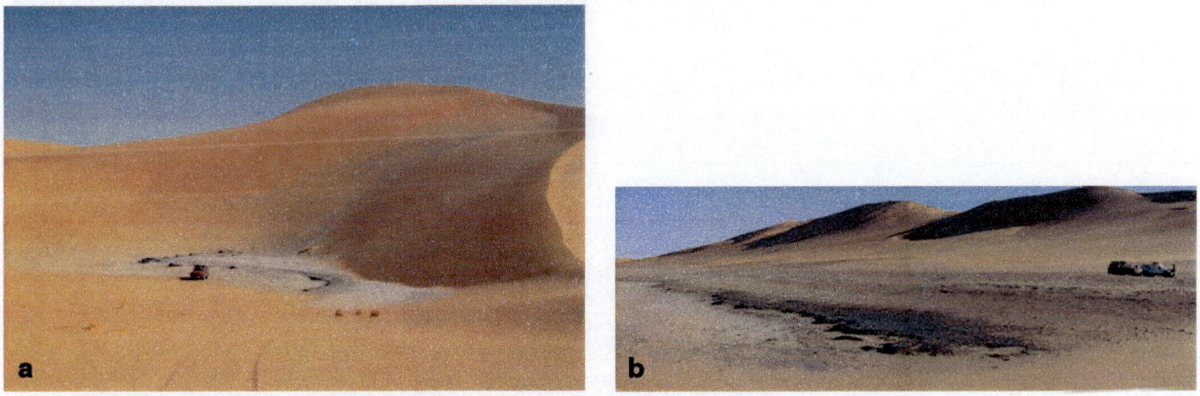

Plate 2 **a** A former interdune lake basin in the Erg Uan Kasa, Libia. Notice the black organic matter-rich sand representing the shore facies and the carbonatic mud corresponding to deep water facies (Cremachi and Zerboni, Chap. 5). **b** Organic sand deposit at the base of the dune, corresponding to shore facies, in the Erg Uan Kasa (Cremachi and Zerboni, Chap. 5)

I. P. Martini, W. Chesworth (eds.), *Landscapes and Societies*,
DOI 10.1007/978-90-481-9413-1, © Springer Science+Business Media B.V. 2010

Plate 3 **a** The oasis of Ghat (SW Fezzan, Libya), remains of the mid-late Holocene Wadi Tanezzuft paleo-oasis. **b** The origin of the Wadi Teshuinat in the central part of the Tadrart Acacus massif. Notice the characteristic shape of the slopes, the stony pavement (hamada) covering the terraces, and the sinuous pattern of the valley (Cremachi and Zerboni, Chap. 5)

Plate 4 **a** The Wadi Afar cave (southern Tadrart Acacus); its anthropogenic fill dates to the early Holocene. **b** The rock paintings at In Taharin (Wadi Tashuinat, central Tadrart Acacus). This gallery probably date to the early Holocene, and represents both the *Ammotragus* and the *Orix* (the animal on the right) (Cremachi and Zerboni, Chap. 5)

Plate 5 **a** Libyan Desert east of Gilf Kebir – dune filled wadi. **b** Libyan Desert – tool site on gravel terrace (Brookfield, Chap. 6)

Plate 6 A large dune section in the United Arab Emirates of the type which has been dated by OSL (Goudie and Parker, Chap. 7)

Plate 7 **a** View northwards across part of the WF4 field system from the hill south of WF5720, mining area of Wadi Faynan, SW Jordan. In this part of the system, water runs from the slope in the foreground and was guided by the wall network to where it was needed. Excess water drained to the *left*, eventually entering the Wadi Faynan braidplain, which is visible beyond the field system. **b** A Classical-period mine-shaft excavated by the Bochum Mining Museum in the Wadi Khaled. The site is blanketed by highly toxic minespoil, so there is no vegetation, ~1500 years after the end of copper extraction (Hunt and el-Rishi, Chap. 8)

Plate 8 **a** A paleochannel of the River Jaghjagh in NE Syria being examined by Katleen Deckers. **b** The Roman Dam at Harbaqa in central Syria (Wilkinson, Chap. 9)

Plate 9 **a** Temple of Zeus at Olympia, Greece, with columns destroyed by an earthquake at the end of fifth century AD (Fouache and Pavlopoulos, Chap. 10). **b** Single tower tholos nuraghe (tower) of the Crabia Nuraghe at Paulilatino, Sardinia (Depalmas and Melis, Chap. 11)

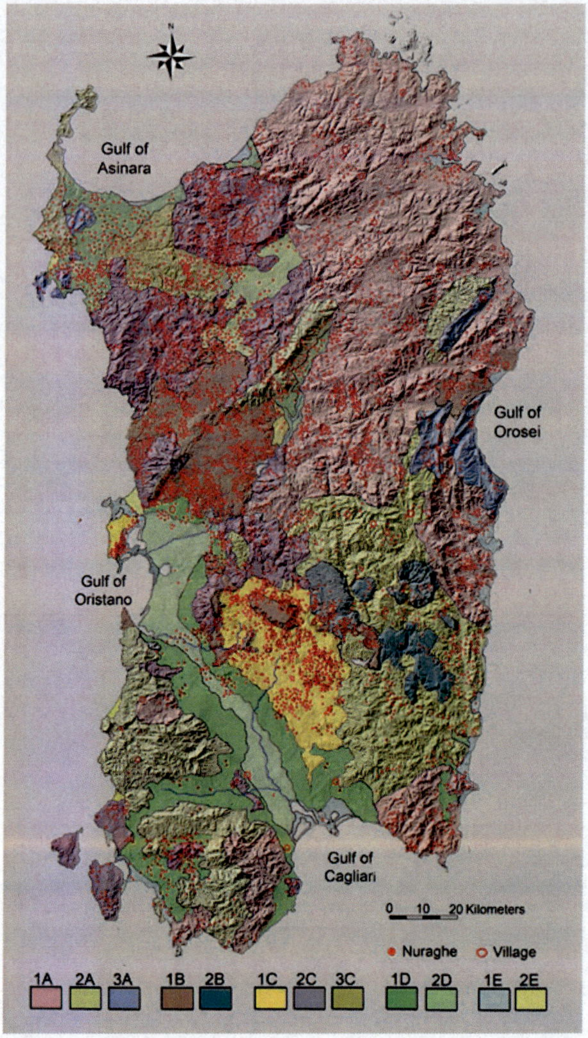

Plate 10 Geomorphologic units map of Sardinia, Italy, and distribution of nuraghe and Bronze Age villages. **Mountain Landscape:** 1A peneplain remains, ridges, escarpments, mountains on predominantly granite Paleozoic rocks; 2A peneplain remains, smooth summits, ridges, escarpments, mountains on predominantly metamorphic Paleozoic rocks; 3A ridges, escarpments, mountains on predominantly limestone Mesozoic rocks. **Plateau landscape:** 1B basalt plateaus (Plio-Pleistocene); 2B limestone plateaus (Mesozoic). **Hilly landscape:** 1C complex hills on predominantly marine sedimentary Cenozoic rocks; 2C complex hills on predominantly volcanic Cenozoic rocks; 3C predominantly tabular hills on marine sedimentary Cenozoic rocks. **Plain Landscape:** 1D dissected alluvial fans, piedmont plain (Pleistocene), 2D alluvial plain (Pleistocene–Holocene). **Coastal Landscape:** 1E Coastal plain (Pleistocene–Holocene), 2E Coastal dune (Pleistocene–Holocene). Sardinia, Italy, with superimposed the distribution of nuraghe and villages of the Bronze Age (Depalmas and Melis, Chap. 11)

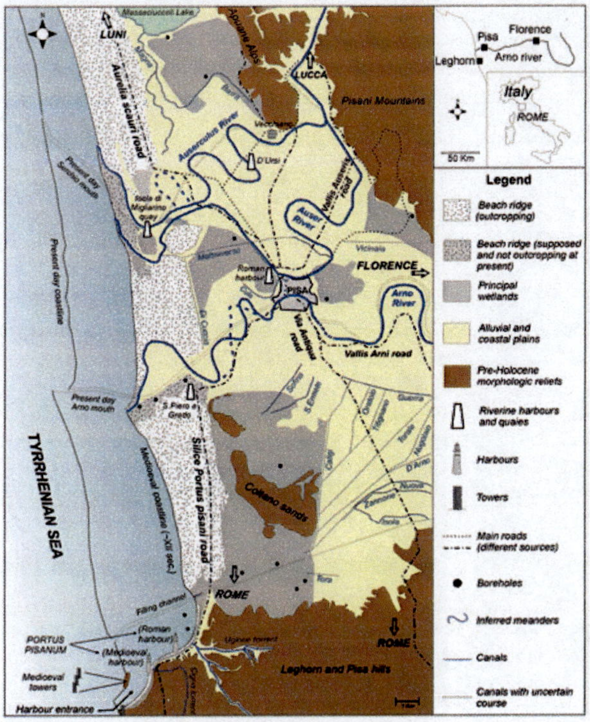

Plate 11 Map of the Pisa plain, Italy, for the tenth to twelfth centuries schematically showing the various elements of the landscape of those times (compiled by Sarti, after Carratori et al. 1994) (Martini et al., Chap. 13)

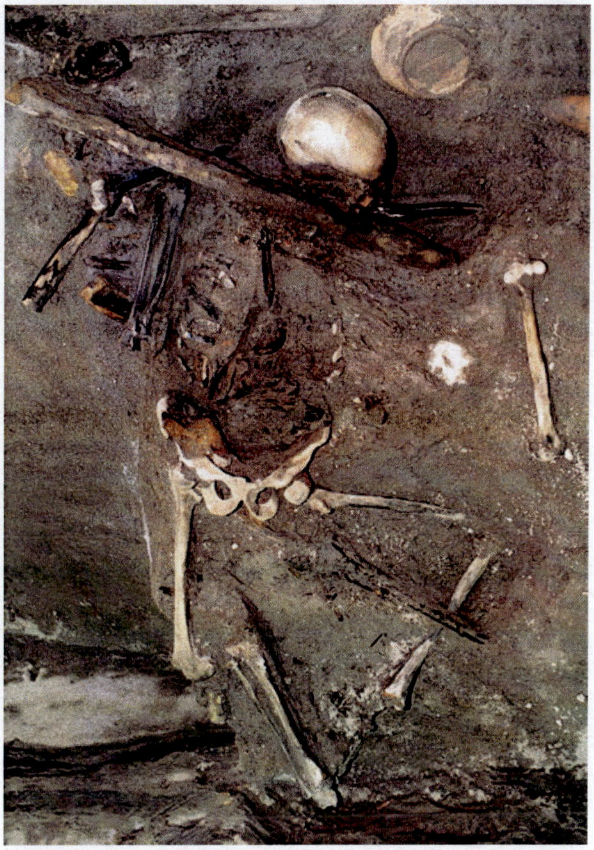

Plate 12 Human skeleton found below the ship B in the Pisa-S. Rossore site of the recently discovered Roman harbor, originally buried by the cargo of this vessel. During the shipwreck this man, evidently onboard, was swept away and rapidly covered by the cargo material removed from the ship by a violent flood-flow. The scene brutally freezes the timeless, fragile, relationship between the man and its environment, which, although deeply shaped and transformed by civilizations, is still regulated by natural, sometimes catastrophic, processes (Benvenuti et al., Chap. 12)

Plate 13 **a** The so called 'chain map' of the medieval–Renaissance city of Florence (1470s) and its surroundings, also showing some of the human activities along the Arno River. **b** Map of ancient Pisa (after Betelli, 1616/1629) showing navigation on the Arno River and moat along left city-wall, possible trace of a ancient river. Recently discovered Etruscan-Roman harbor is located near left part of this map. **c** Details of the fresco (1338–39) by Ambrogio Lorenzetti in the Palazzo Pubblico of Siena showing activities in the countryside and transport of produces to the city of Siena during 'good government' (Martini et al., Chap. 13)

Plate 14 **a** Roman temple of Pozzuoli, Italy: the columns perforated by lithodomi recond variations in sea level (Morhange and Marriner, Chap. 14). **b** Catania and the south eastern flank of Etna, Italy (from Sir William Hamilton, 1776, Campi Flegrei) (Chester et al., Chap. 15)

Plate 15 **a** Intermontane depressions of the Eastern Carpathians, Romania .Terracing of the piedmonts and slopes in the Giurgeu Depression at the foot of the Curmăturii Mountains. Piatra Singuratică (1587 m) in the foreground is a large limestone outlier. **b** Forested Prahova Valley at the boundary between the Southern Carpathians and the Curvature Carpathians (the Baiului Mts in the background). It has been a route of commercial exchanges between Braşov and Wallachia since early times. The town of Sinaia (in the forefront) is currently a major centre of cultural tourism and winter sporting activities (Cioacă and Dinu, Chap. 16)

Plate 16 **a** An octagonal mill (*left*) and a stockmill, with a modern pumping station in the *middle*, The Netherlands. **b** Maps of Amsterdam and territory, mid 1700s (I Cóvens and C Mortiner) (Jungerius, Chap. 17)

Plate 17 **a** Laki volcanic-cone row indicated with white arrow on east side of Laki Mt., Iceland. **b** Laki lava flow and Heidi farm (Thordarson, Chap. 18)

Plate 18 **a** Loess hills in the eastern Gansu-Qinghai region, China. Dadiwan culture site is located on the second terrace south of Qingshui River. The loess hills are about 300 m higher than the river valley. **b** The stilt-style buildings are reconstructed on Hemudu site after excavation. The site is located on the first terrace of Yuyao River, 3~4 m above sea level, China (Mo et al., Chap. 19)

Plate 19 **a** AsukaTanada, Japan: Terraced upper river valley with village on adjoining hillside; the 'tanada' are watered from the diverted mountian stream. Unterraced mountain in the background (authors's photo, 1979). **b** TsukuriyamaTomb, Japan: Tsukuriyama (meaning "artificial mountain"), in picture center, standing in a sea of paddy fields, is a fifth century keyhole-shaped mounded tomb. *Upper left* is a natural wooded hill. In the *lower* foregound is another natural hill that has been converted to dry fields. Note the traditional villages nested at the foot of the hills (courtesy of Richard K. Beardsley, Okayama 1954) (Barnes, Chap. 20)

Plate 20 **a** Village tank and upstream catchment, Sri Lanka. **b** Ancient sluice of Kalawewa, Sri Lanka (Dharmasena, Chap. 21)

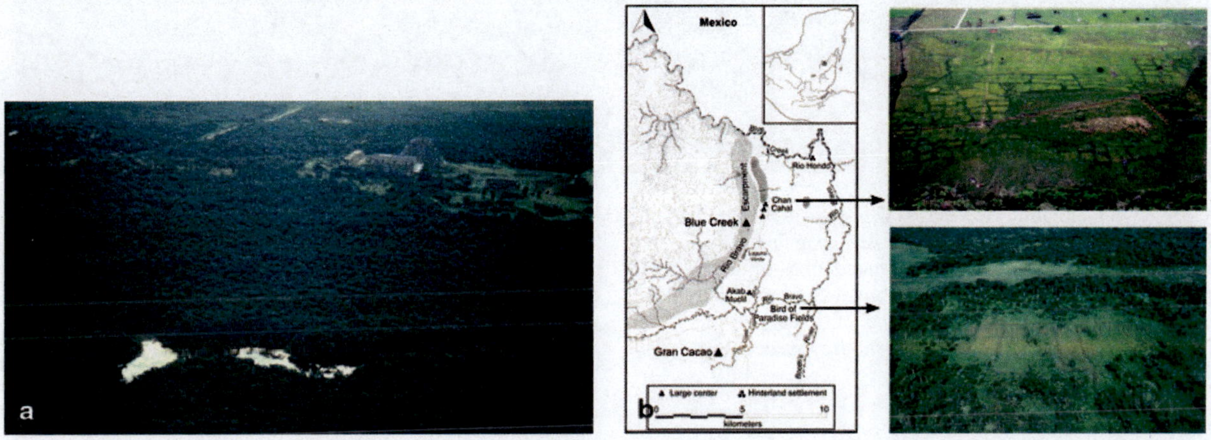

Plate 21 **a** Air photo of the site centre of the ancient Maya city of Uxmal, Yucatan. Uxmal was surrounded by a system of reservoirs, like Chanchen Reservoir in the foreground, which likely helped it withstand droughts which may have felled many neighboring towns. **b** Wetland field systems at Chan Cahal and the Birds of Paradise Fields, Belize. The Chan Cahal rectilinear canals have a range of shapes and sizes indicating piecemeal evolution over the Late Preclassic through Classic, whereas the Birds of Paradise Fields have distinctly rectilinear and evenly spaced canals indicating a preplanned system in Late to Terminal Classic (after Beach et al., 2009) (Dunning and Beach, Chap. 23)

Plate 22 **a** Panoramic view across the banks of the South Platte River in Fort Morgan, Colorado, USA. The main channel is out of view to the right. All of the vegetated areas in these photographs were historically part of the broad, unvegetated, braided channel (Wohl, Chap. 24). **b** Remains of a 700-year-old Thule tradition semi-subterranean winter house located on Devon Island, prior to excavation. The roof framework was constructed from the bowhead whale bones visible in the interior of the house, NW Territories, Canada (Park, Chap. 25). **c** Remains of an Arctic Small Tool tradition tent ring approximately 3500 years old, located on the island of Igloolik, prior to excavation, NW Territories, Canada (Park, Chap. 25)